GLACIAL AND
QUATERNARY
GEOLOGY

GLACIAL AND QUATERNARY GEOLOGY

RICHARD FOSTER FLINT

Henry Barnard Davis Professor of Geology,
Emeritus, Yale University

John Wiley and Sons, Inc.
New York · London · Sydney · Toronto

Frontispiece:
Mt. Robson, British Columbia, August, 1964.
(Austin Post, U. S. Geol. Survey)

Library of Congress Catalogue Card Number: 74-141198

ISBN 0-471-26435-0

Printed in the United States of America

10 9 8 7 6 5 4 3 2 1

To the Friends of the Pleistocene,
the greatest Quaternary nonorganization,
this book is dedicated with loyalty and affection.

Preface

Most general works on the Quaternary Period have included a rather full treatment of glacial geology, apparently because glacial sediments and the classic terrestrial glacial stratigraphy have played a large part in the study of the Quaternary.[1] Pollen stratigraphy, sea-floor stratigraphy, and isotopic geochemistry including radiometric dating are only a few of the significant studies that have contributed increasingly to research on the Quaternary since 1950. The present volume, which contains about 80 percent new material, attempts to take account of studies in these newer fields, integrating them into the classical discipline. Yet despite the obvious strains that result, the work retains the conjunction of two entities: systematic treatment of "glacial geology" involving process, and stratigraphic, environmental, and historical discussion of the Quaternary. This has been done for the practical reasons that many academic courses, organized on a similar basis, need such a book, and that a general reference work on matters relevant to the Quaternary can usefully include "glacial geology." The fact that physical features and processes are emphasized more than biologic ones implies no judgment on priorities; it merely reflects my professional interests.

Like its two predecessors (Flint, 1947; 1957) this book is a compromise between an encyclopedic treatment and a generalized summary. In its field, embracing areas of several sciences, the literature is huge. The annual increment of relevant papers greatly exceeds that of twenty years ago. In view of the rapid increase of knowledge no single book could even approach full and detailed coverage, and this one must be selective. Its aim is a discussion of features, relationships, and problems that is fairly comprehensive, accompanied by an earnest effort to cite the titles

---

[1] The use of this term in preference to *late Cenozoic* is defended in chap 1.

in the literature that will give the reader, regardless of the level of his interest, readiest access to greater detail.

Because of the two terms in the book's title—*glacial geology* and the *Quaternary*—it could be argued that the book is a two-headed monster and that each head deserves its own separate book. The argument is valid, and separate books are precisely what the future holds in store. Defense of a single work, albeit two headed, is based partly on the long tradition established by James Geikie (1874; 1894) and W. B. Wright (1914; 1937) and continued by me. But a single work can be defended more practically on the ground that knowledge of glacial geology is essential to an appreciation of Quaternary events, because glaciation created distinctive sediments and other features on nearly one-third of the world's land area. Furthermore, the vast glaciated region includes densely populated areas in which the fact of former glaciation affects agricultural soil and the foundation conditions for a wide variety of man-made structures.

In the study of the Quaternary half a century ago, glacial geology absorbed the lion's share of attention. Since then the study has grown mightily and its interdisciplinary aspects have steadily increased. Today conferences and symposia on Quaternary problems typically include representatives of several sciences. We realize that the climatic events of late-Cenozoic time lie at the root of natural activities we observe today, such as fluctuation of sealevel and of certain lakes, distinctive movements of the Earth's crust, and a wide variety of ecologic changes. We realize also that not all the features of today's landscape and ecology are the products of activities that operated at today's intensities. Some of those features are relict from former times when intensities were greater or smaller than they are now.

These fluctuations and relict features are directly or indirectly attributable to changes of climate, which in recent Cenozoic time seems to have varied more frequently and perhaps also in greater amplitude than in the preceding time. The variations repeatedly affected environments, altering the balance of forces that act on them and thus altering, at times profoundly, the distribution or the forms of living things.

Quaternary environments, therefore, deserve intensive study both as a part of geologic and biologic history and for what they tell us about today's environments. In addition their study broadens the base of the principle of uniformity of process, first set forth by Hutton at the end of the 18th century, by approximating the amplitudes within which temperatures and precipitation values have fluctuated within geologically recent time. The recognition that climate does fluctuate has resulted mainly from the study of relict features (including both glacial features and the records of distribution of organisms). Thus far at least, such recognition is only secondarily a contribution from meteorologic analysis and study of the instrumental record.

I sincerely thank these persons who critically read manuscript, supplied data or illustrations, or gave other kinds of help: Björn Andersen, J. T. Andrews, A. L. Bloom, R. B. Colton, H. B. S. Cooke, H. C. Coulter, Willi Dansgaard, G. H. Denton, W. O. Field, Margaret C. H. Flint, R. P. Goldthwait, Vance Haynes, Gunnar Hoppe, John Imbrie, Vojen Ložek, Franz Mayr, B. C. McDonald, G. F. Mitchell, N.–A. Mörner, J. G. Ogden III, J. H. Ostrom, S. C. Porter, John Rodgers, Barry Saltzman, H. D. Scammell, J. P. Schafer, K. K. Turekian, A. L. Washburn, Ann Weippert. I especially thank J. H. Hartshorn, who read the entire manuscript, with very beneficial results. Other contributors of tables, illustrations, or other information are acknowledged in text or captions. I alone am responsible for any errors in such material as well as in the rest of the book.

All stated measurements are metric and all temperatures are Celsius. The uncertainty symbol (~) appearing in the text should be read *approximately*. It is employed to conserve space, where a round number is adequate for the discussion.

Richard Foster Flint

New Haven, Connecticut, December 15, 1970

Contents

GLACIAL AND QUATERNARY GEOLOGY

CHAPTER 1

Overall View of Late-Cenozoic Climate and Glaciation

In vast areas of middle and high-middle north latitudes we see about us the effects of former extensive glaciers. Curiosity about the characteristics and origins of these effects sparked the development of the science of glaciation, which we call *glacial geology*. The first part of this book is a description of the modifications by glaciers of the surface of the Earth. The second part gives a historical analysis of the climatic events of which glaciation was but one manifestation. These events occurred throughout late-Cenozoic time, but the clearest part of their record lies in sediments of the Quaternary age. Hence the book is appropriately titled *Glacial and Quaternary Geology*. Even though labeled geology, it deals also with biology, especially where the record of fossil pollen comes to prominence, and touches upon other disciplines as well.

The relative freshness and continuity of the Quaternary strata afford an unmatched opportunity for their study, necessitating minimal interpolation across erosional gaps. In addition, radiometric dating, by C^{14} at the near end and by K/Ar through much of the rest of the time range, make possible a beginning at time calibration of these strata. As a result of these favorable factors, the Quaternary strata constitute a bridge between the sediments being deposited today and much older strata—an intermediate step between the present and the more distant past—enabling us to interpret earlier strata with greater confidence. In this respect the value to geologists of studying the Quaternary extends well beyond its own boundaries.

We shall attempt a sketch of the Quaternary of various parts of the world, emphasizing in general the changes of climate for which the Qua-

1

ternary is famous, and in particular the glaciations that directly and indirectly affected large parts of the Earth's surface.

CHRONOLOGY OF GLACIATION

Perhaps the most stirring impression produced by recent great advances in the study of the Quaternary Period is that the Quaternary itself is losing its classical identity. It was long supposed that at the near end of geologic history major glaciation was confined to Quaternary time, distinguishing that period in terms of climate from its immediate predecessors. Now, however, it is evident that extensive glaciers had formed in high latitudes of both polar hemispheres before the end of Miocene time. They seem to have persisted, with fluctuation, through the Pliocene into the Quaternary. Hence it is no longer possible to equate the Quaternary alone with glacial ages or to define it on a basis of temperature. We now think in terms of late-Cenozoic cold climates and glaciers.

This lengthened concept corresponds better with geologic history than did the long-held earlier belief. Evidence is accumulating that the Miocene Epoch witnessed widespread uplift of lands and localized orogeny as well, and that such movement continued through the Pliocene into later time. Temperatures on newly elevated lands would have been lowered, and on highlands in the highest latitudes glaciers would have formed. Glaciers and the snow associated with them would have increased the albedo of land surfaces in their vicinity, favoring the spread of snow and ice.

The inception of high-latitude glaciers in late-Cenozoic time, therefore, is expectable rather than mysterious. What we do not yet fully understand is what, precisely, caused the spreading of great ice sheets into middle latitudes of the northern hemisphere. It was that spreading that led, early in the 19th century, to the concept of former glacial ages, at a time when the vast ice sheet in Antarctica still awaited discovery. Spreading alternated with shrinking, although not in a pattern that is rhythmic in an obvious way. The times of spreading and shrinking are labeled glacial ages and interglacial ages. The mechanism by which the mid-latitude ice sheets expanded and shrank repeatedly is not yet known. The durations of the changes seem to be variable and seem to be measured individually in terms of several tens of thousands rather than in hundreds of thousands of years.

The development of radiometric dating has enabled us to appraise the chronology of glaciation in a more realistic way than was possible earlier. We now see that the "four classical glaciations" of the older literature (ac-

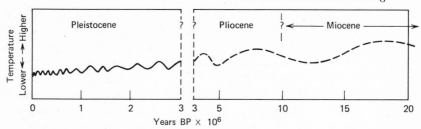

Figure 1-1 Curve suggesting general trend of secular temperature since earlier Miocene time. Fluctuation is superposed on a general downward trend. Lack of detail toward right may reflect merely lack of data. Scale changes at 3×10^6 years BP. (Generalized from figs 16-2, 16-3, 16-6, 16-7.)

tually there were more) represent only a small and very late fraction of the time involved in the late-Cenozoic glacial events. Those events began at least 10 million years ago, whereas the classical events seem to have embraced little more than one-tenth of that span. At present, then, we discern two broad phases within late-Cenozoic glacial history. One was a long-enduring phase in which glaciers both large and small existed but seem to have been confined mainly or wholly to highlands and their surroundings, within high latitudes. The other was a (thus far) much shorter phase in which broad ice sheets repeatedly spread over middle latitudes in the northern hemisphere. We are able to sketch these phases only in barest outline (fig 1-1). It may well be that the distinction between them is artificial and that the one will later prove to grade smoothly into the other.

These time relationships mean that the Quaternary, or the Pleistocene part of the Quaternary, loses the peculiar identification with glacial conditions it formerly enjoyed, and with that loss the long controversy about where to place the Pliocene/Pleistocene stratigraphic boundary diminishes in importance. Despite this fact the word *Quaternary* appears commonly in the names of scientific organizations and in the titles of monographs, in implied identification with the study of conspicuous climatic change. Hence we employ the word in the title of the present book even though in a strict sense *late Cenozoic* would be more appropriate.

EVIDENCE OF CLIMATIC CHANGE

The formation, change in mass, and disappearance of glaciers are responses to change of climate. We are sure of this because fluctuation of

climate or of a climatic parameter, measured independently even in a region where there are no glaciers, is broadly synchronous with fluctuation
of glaciers. Also the timing of the glacial variations throughout the last
20,000 years or so indicates fairly clearly that most glaciers, regardless of
geographic location, respond to climatic change in the same general way
more or less contemporaneously. Although there are local exceptions,
some of which are easily explained on a basis of altitude, form, and other
very local factors, this synchronous behavior leads to the belief that the
changes are of worldwide rather than local origin.

The basic parameter of climate is temperature. Although glaciers do
vary with changes in precipitation, most of the major changes recorded
by glaciers are related to temperature variations. The relationship is clear
despite the fact that temperature changes in the climatic record also involve precipitation change as well. Evidence of former change of climate
is many fold. Highly important among the kinds of evidence is the indication that during glacial ages the snowline everywhere stood lower than
it does today and that although not uniform from point to point, the
change was systematic. The amplitude of change was not much less in
equatorial regions than in northern middle latitudes. This argues that
climatic change was worldwide and was not primarily the result of proximity to ice sheets.

The statistics of distinctive "cold" and "warm" microfaunas recovered
from long cores that represent succesive layers of sediment from the deep-
sea floor suggest that during glacial ages even tropical ocean waters were
chilled. Pluvial lakes that in the latest glacial age occupied hundreds if
not thousands of basins in dry regions of middle latitudes similarly indicate lowered temperatures in addition to probably increased rainfall.
Other sorts of evidence of lowered temperatures include extension of
areas underlain by permafrost and by frost-action effects such as geifluc-
tion. Finally, in the biosphere are seen worldwide changes caused by glacial-age climatic stress. The changes consist mainly of wholesale displacements and mixing of populations, most clearly indicated in relict
marine-planktonic organisms as well as pollen and spores of land plants
where microscopic size and large numbers of individuals make for studies
of broad scope.

The glacial/interglacial climatic changes seem to have occurred within
moderate amplitudes. The ranges varied from place to place as seasonal
changes do today. Temperature ranges were greater in continental than
in maritime climate. Thus extreme glacial/interglacial ranges over lands
at low altitudes in maritime situations, may have been as much as 6° to
as little as 3° in mean-annual terms, increasing to 12° or more in continental situations. Today's temperatures lie much nearer to interglacial
values than to glacial values.

THE GLACIERS

During glacial ages the aggregate area occupied by glaciers was nearly three times as great as that occupied by glaciers today. The aggregate volume of glacier ice likewise is estimated at nearly three times that of existing glaciers. If the Antarctic ice is excluded, glacial-age glaciers occupied an area about 13 times greater than do today's glaciers. Thickness varied enormously from one glacier to another, but we estimate average thicknesses of around 2km for the large ice sheets, with maximum thicknesses perhaps amounting to 3km or more. The aggregate volume of glacier ice in the northern hemisphere was roughly twice that in the southern. This difference arose from the fact that Antarctic ice could not spread far beyond the Antarctic Continent; it broke up in the ocean and floated away. In North America and Eurasia, in contrast, ice sheets had far more ample space for spreading over land areas. The Scandinavian ice was able to spread southward nearly to lat 48°; the Laurentide (North American) ice, because of moister climate and greater snowfall, spread to lat 37°.

It has sometimes been argued that the accumulation of the huge volume of ice that existed in glacial ages would have demanded much greater rates of precipitation than those measured today. However, although precipitation may have been greater, calculation shows that the argument is not valid. It appears entirely possible for the ice sheets to have been built by precipitation at today's rates, provided only that temperatures were moderately lower.

The positions and extent of the former ice sheets are known because the recognizable effects of thick ice spreading over the land have been observed in detail and carefully mapped. Such effects include scouring and rupture of bedrock, and deposition of nonsorted rock particles to form broad layers of glacial drift spread over the surface. Direction of flow of the ice masses is consistently shown by a variety of linear features. The aggregate effect of thick ice sheets spreading over the lands was not great. With local exceptions, thicknesses both of bedrock eroded and of drift deposited were not large.

Layers of glacial sediment and nonglacial sediment in some places containing fossils are widespread on the lands. Where two or more glacial layers occur superposed, they indicate repeated glaciation. In many areas ancient soils are developed in sediments of both kinds. They record former land surfaces that were long exposed to weathering, and yield indications of the climates that prevailed while the soils were forming.

A noteworthy fact about the distribution of local glacial-age glaciers on highlands throughout the world is that although, of course, they were larger than today's glaciers on the same highlands, the asymmetry of their distribution (greater extent on the wetter side of the highland) is similar to that of today's glaciers. This similarity indicates that the sources of moisture and the wind systems were the same in glacial times as they are today, and so the major relief features have remained little changed from then to now.

The ice sheets constituted an extra load on the crust so great that the crust subsided beneath their weight, by an amount (supposed to be about one-third of the thickness of the overlying ice) sufficient to maintain isostatic equilibrium. The displaced subcrustal material surely flowed outward but the areas of its residence during the existence of the ice sheets have not been identified clearly by geologic observation. The ease of the displacement is reflected in the rapid response of the crust to deglaciation. Thinning and contraction of the ice sheets caused gradual uplift of the underlying crust. Where lakes or seawater lay adjacent to the wasting ice, deformation of their strandlines, time calibrated by the C^{14} dates of shells and driftwood, can be reconstructed at least for the later part of the process. It is seen that the deformation, most rapid at first, gradually relaxes. Movement caused by deglaciation that began some 15,000 years ago is still in progress.

FLUCTUATION OF SEALEVEL

As the growth of glaciers affected the exchange, within the hydrologic cycle, between liquid and solid water substance, the level of the world ocean stood below its present position during glacial ages and somewhat above it during interglacial times. Because numerous factors hinder determination of the amplitude of fluctuation of sealevel through late-Cenozoic time, our data at present include more estimates than measurements. General estimates of the altitude of glacial-age sealevel fall between values less than −100m and about −140m. That of interglacial sealevel may have reached about +20m, but at present we must view all estimates with skepticism. The lowering of sealevel converted very large areas of the continental shelves into land, more than offsetting the areas temporarily buried beneath large glaciers and affording new habitats for plants and animals.

Before the Antarctic ice came into existence (that is, at some time before the late Miocene) we suppose that few glaciers, none of them of large size, existed. The world ocean then must have stood even higher than it did during interglacial intervals. From the estimated volume of glacier ice existing today we calculate a theoretical height of +65m. This value has only limited meaning, however, because crustal movements of various kinds have so changed the continental coasts that we have little idea of the position of any Miocene sealevel relative to the coasts. A more significant calculation from the volume of existing glaciers is the future effect of further deglaciation, should it occur. For example, if we neglect the Antarctic Ice Sheet and assume the Greenland Ice Sheet were to melt completely, world sealevel would rise some 6m. The result in terms of human affairs would be very great. We cannot say the submergence would be catastrophic because it would be very slow, but the economic losses that would result would be colossal.

Conversely, glacial-age lowering of sealevel of the order of 100m would have converted enormous areas of the continental shelves into land. The change would have enlarged the habitats of land plants and animals at the expense of shelf-zone marine life. Where the emergence of land was sufficiently great to connect two substantial land masses, creating bridges such as the Bering land bridge, the bridge centering in the Sunda Shelf, and the bridge at the Strait of Dover, the resulting movements of terrestrial organisms created an impressive fossil record.

SEA-FLOOR SEDIMENT CORES

Important aspects of late-Cenozoic history are illuminated by sediments recovered through coring of the deep-sea floor. Some cores from high-latitude regions contain layers of "grit" and pebbles dropped on the ocean floor during melting of floating ice. Such layers define large areas of floating ice shelves, and icebergs derived from shelves and from glaciers themselves. More significant in the long run is the information derived from fine-grained layers of nonglacial sediment in all latitudes. Critical here are the tests of planktonic micro-organisms which are manipulated in various ways, counted, and treated statistically. The proportions, at each level in a core, of taxa of "warm" or "cold" affinity yield temperature data, as do also the results of oxygen-isotope measurements on the calcareous tests themselves. Individual layers in a series of cores are capable of being correlated, thus contributing importantly to worldwide stratigraphic sequences.

STREAMS, WINDS, AND SOILS

What effect did glacial-age climates have on streams and stream systems? In the regions of former ice sheets, blocking by ice or glacial drift obliterated pre-existing valleys and created new ones, some of them major. In many valleys, including some that traversed nonglaciated terrain, outwash sediments created fills that in places attained great thickness. Most such fills are represented today by pairs of terraces. In regions not covered by ice sheets drainage patterns were generally unaltered, but the regimens of streams were affected to varying degrees by fluctuation of climate. The results included cuts, fills, and changes, in places conspicuous, in the geometry of channels. Many such changes are complex and the details of their relationship to climatic parameters remain obscure.

Changes of these kinds raise the question of the overall changes in the hydrologic cycle that occurred when climates were altered from nonglacial to glacial. Evidence on this question is very meager and, as is often true when data are scanty, apparent conflicts appear. We do not know whether precipitation values increased or decreased overall, although there is reason to believe that increase in one region may have been contemporaneous with decrease in another.

Winds are a different matter. We suppose that temperature gradients between lower and higher latitudes were steeper in glacial ages than they are today. If they were, wind velocities must have been greater on the average, but criteria for inference of former velocities are lacking. On the other hand, scattered data of various kinds do suggest that within specified regions average wind directions changed. Most obviously the middle-latitude belts of westerly winds seem to have shifted toward the equator. During later Pleistocene time a considerable number of dune fields were created, but in most of them the rather sparse studies that have been reported have not developed unequivocal evidence as to the time or times when the sand accumulated. Thus in some instances we are ignorant as to whether the dunes are of glacial or of interglacial origin.

With regard to relict soils, however, knowledge is a little better. The stratigraphic positions of major relict soils in glaciated regions indicate that such soils are interglacial. Indeed we know of no thick, mature relict soils that postdate the latest interglacial age. Glacial-age soils are rather numerous, but are of quite different character.

STRATIGRAPHY

Charts showing stratigraphic sequences and correlation are condensed reports of progress that reflect the state of knowledge of sequences of events. The wider the region it represents, the less accurate a chart is likely to be. Despite their inaccuracies, correlation charts are of great value. Because they compress and condense information, they permit identification of similarities in sequences from widely separated localities or regions. Many such charts are included in this book. They should be read for what they are, with the realization that in another year better ones will become possible. Those which can be time calibrated carry at least some dates, but as the dates vary in quality even the calibration is subject to change.

In setting up regional sequences we have followed published opinions of authorities on the regions. In general, however, we have avoided interregional and intercontinental correlation because such correlation is still very uncertain ground. We can be reasonably confident that the sequences representing the latest (Wisconsin, Weichsel) glacial age are broadly correlative worldwide because these sequences are developed in greatest detail and are comparatively well supported by C^{14} dates. There is still no certainty, however, as to the wide correlation of earlier glacial and interglacial units. Even the number of major glacial events is unknown.

In this connection we can only deplore the widespread practice of correlating strata and events anywhere in the world with the sequence (Günz, Mindel, Riss, Würm) developed for the Alps soon after the beginning of the present century. Because that sequence is poorly known in most respects, its use as a general reference is misleading, and confuses rather than aids the progress of stratigraphic knowledge.

PLANTS AND ANIMALS

The fossil organisms of the late Cenozoic constitute an extremely important source of information on the history of the last 10 million years or more. Both animals (especially mammals) and plants (chiefly megafossils) are significant guides to correlation, and both (chiefly microfossil marine organisms and pollen) likewise provide important data on former climate. Such data are generally compatible with data derived from physical features and help to define the ranges within which temperature and rainfall have fluctuated in one region or another.

Furthermore, the fossil record indicates that at least within Quaternary time, if not earlier, climatic fluctuation was more rapid than the general progress of evolution. Many taxa seem to have survived repeated temperature fluctuation without much obvious adaptation to cold environments until several hundred thousand years ago. In contrast, changes in the ranges of particular populations, under the influence of changing temperature or rainfall, are detected in sediments belonging to much older parts of late-Cenozoic time.

Extinction of faunas and floras occurred through the late Cenozoic at rates compatible with earlier times. An exception consists of massive extinctions of mammals in North America and other parts of the world, between about 20,000 and 6000 BP. Although the destruction of certain specialized mammals may have been climatically induced, the weight of evidence suggests that human activities directly and indirectly influenced the process.

Hominids are apparently of late-Cenozoic origin. Their evolution seems to have been influenced by climatic change, the result, in part, of orogenic movements and perhaps also of the factors that brought on the glaciations. The making of artifacts has been in progress during at least the last 2.5 million years and the species *Homo sapiens* has been in existence through at least the last 0.4 million years. Much of the early history of man, as represented by the genus *Australopithecus*, appears to have taken place in Africa, from which peoples migrated into Europe and Asia and there developed. The peopling of the Americas, however, seems not to have occurred until some time in the second half of the Wisconsin Glacial Age.

BASIC CAUSES OF CHANGING CLIMATE

The causes of climatic fluctuation and the causes of glacial ages, although related, are not necessarily the same. Fluctuation of climate has occurred without the creation of extensive ice sheets. Although many factors that may have influenced climate importantly are known, the basic factors that brought about the glacial ages are largely matters of hypothesis and opinion. In this book two factors are preferred: (1) worldwide creation of highlands and (2) assumed solar variation. Highlands, a *sine qua non* for extensive glaciation, were created and elevated in Miocene, Pliocene, and Quaternary time, making possible the formation of glaciers. The extension and fluctuation of glaciers once formed could have been caused by fluctuation of radiant energy emitted by the Sun. This second factor, as we have said, is an assumption; other possibilities are discussed in chap 30.

Some have ·asserted that curves showing fluctuation of temperature with time (e.g., fig 16-12) contain rhythmic elements. If they do, the possibility of extrapolation appears, with consequent prediction of future climatic events.

The foregoing account, essentially a highly selective summary of the substance of this book, is so condensed that to some readers it may appear breathless. Despite its short length, at least it should make it easier for a reader to place each chapter, as he comes to it, within a broader framework.

CHAPTER 2

Development of Concepts

THE GLACIAL THEORY[1]

In 1839 Sir Charles Lyell named the Pleistocene ("most recent") strata on the basis of their content of fossil mollusks. At that very time the Glacial Theory was being announced in Switzerland; it embodied the idea that glaciers were formerly more extensive on a huge scale than they are at present. This theory of a former ice age, or glacial age, was so dramatic that it focused on climate a widespread attention not accorded to fossil mollusks. Later it appeared that the Pleistocene strata and the glacial deposits represented, at least in part, the same bracket of geologic time. This resulted in the dichotomy in the definition of Pleistocene and Quaternary which we noted in chap 1.

The Glacial Theory was based principally on the occurrence of erratic boulders, consisting of rocks unlike the bedrock on which they lie, and secondarily on deposits of till and on other glacial features. Erratics and till had long been known, but the prevalent theories of their origin had been based on the idea of transport by water. The widespread occurrence of till across northern Europe led first to the notion of a universal flood, which many identified with the biblical Deluge. An example is the excellent description of till written in 1603 by a Welsh country gentleman (John, 1964). Early in the 19th century there arose the concept of a myriad of icebergs floating down from the far north, dropping foreign boulders and other rock material through the universal waters on to the bottom.[2] It was through this concept that the glacial deposits came to be

---

[1] Historical summaries in Klebelsberg, 1948–1949, p. 7–13; North, 1943; McCallien, 1941; Charlesworth, 1957, p. 614–633.

[2] Gillispie (1951) made an illuminating study of this view. See an American variant in Flint in H. E. Wright and Frey (1965a, p. 3).

called *drift,* because they were thought to have drifted, in bergs, upon the waters. That term, in wide use today, has survived its abandoned context by more than 150 years.

The iceberg concept may have been stimulated by the very large sizes of some of the boulders; certainly it was promoted by an interest in polar exploration then rapidly expanding. Whalers and others had long recognized the rock fragments in icebergs as having been brought from land, and in the early 1830's Charles Darwin, noting erratic boulders at a locality in Tierra del Fuego scores of miles from their probable source, ascribed them to "transport by ice while the country was under water."

Erratics of granite, overlying limestone in the Jura Mountains, played a particularly important part in this history. The celebrated traveler Saussure described them in 1779, and apparently was the first to use the term *erratic,* but he subscribed to the idea, then popular, that they had been water transported, although they lay nearly 100km from their source in bedrock of the Alps. Sixteen years later James Hutton (1795, v. 2, p. 218) interpreted such boulders as having been carried by former glaciers larger than those of his day, and this idea was reiterated by John Playfair (1802, sec. 348). Hutton may have been the first to visualize a former expansion of glaciers, although his concept was limited to the region of the Alps.

Despite Hutton's opinion, Buckland, Lyell, and Darwin in England, von Buch and von Humboldt in Germany, and even Agassiz (of whom more later) in Switzerland leaned to the view that the erratics being noticed in more and more parts of Europe had been transported by or in water. Among these believers was the poet Goethe, who in 1830 had written of "an epoch of great cold," during which floating ice had transported the boulders (Cameron, 1965), and who therefore may have been the first to express the idea of an age of cold climate, although not apparently that of an age of glaciers.

We can speculate that long before Hutton's time Swiss peasants, who lived their lives in valleys and on mountainsides freshly abandoned by Alpine glaciers, had drawn the fundamental inference from the evidence that lay everywhere about them. The scored and polished bosses of bedrock, projecting through the turf of Alpine meadows, were unmistakable. The profusion of smoothed and faceted boulders spread wide through the fields and in the forests, and the ridgelike, abandoned end moraines on the valley floors were identical with those at the margins of the glaciers themselves and must have been interpreted correctly by many who could see both boulders and glaciers in a single view, but who would not have written down their thoughts.

In 1815, however, one such peasant, a man named J. P. Perraudin, communicated this inference orally to a Swiss civil engineer named Ignaz Venetz-Sitten, who, impressed, read a paper in 1821 before the Society of Natural History (the Helvetic Society) at Luzern. In it he argued that the glaciers of the Alps had at some former time been expanded on a large scale. In 1824, Jens Esmark, in Norway, reached a similar conclusion concerning the glaciers in the mountains of Norway. He thought the ice was more than 1km thick. In 1832 a German professor called Bernhardi (1832), knowing of the views of both Esmark and Venetz, inferred from the distribution of erratics and till that glacier ice from the far north had once extended across Europe as far south as Germany. The idea of glaciation on a large scale was beginning to take hold. Apparently it was Bernhardi who first conceived of a "polar ice cap" forming and extending into Europe. Although a South Polar ice sheet was discovered later, the notion of one in the north was shown to be erroneous. The term, however, still appears in popular speech, always with vague definition.

Like the generations of people before him, Venetz seems not at first to have projected his imagination beyond the Alps. However, in 1829, after longer reflection and probably a good deal of discussion with his scientific friends, he stated the belief that not only the Alps but the plains north of them, and the whole of northern Europe as well, had once been glaciated. A colleague in the Helvetic Society, Jean de Charpentier, though at first incredulous, was stimulated to further observation, and in 1834 he read a paper strongly supporting Venetz's views and strengthening their proof.

Louis Agassiz, a young zoologist and a member of the Helvetic Society, doubted so sweeping an inference, and in the summer of 1836 he visited the Diablerets Glaciers and the moraines of the Rhône Valley with de Charpentier, primarily in order to convince his friend of his error. But after some weeks of study and discussion in the neighborhood of Bex, it was Agassiz who realized that Venetz and de Charpentier were right. Karl Schimper, the botanist, spent much time with Agassiz, both in the field and in discussion of the meaning of widespread glaciation. As Schimper wrote little on the subject, the magnitude of his contribution to the concept of a glacial age may never be known.

The next year, 1837, Agassiz addressed the Helvetic Society, picturing "a great ice period" caused by climatic changes and marked by a vast sheet of ice extending from the North Pole to the Alps and to central Asia, all *before* the Alpine region had been lifted up to mountainous heights. The substance of the address was published three years later (Agassiz, 1840), a year before the amplified views of de Charpentier (1841) were issued in book form. De Charpentier resented this, believing that Agassiz should have deferred publication until after the appearance of his own book.[3] Not until 10 years later did Agassiz (1847) recognize that the former glaciers of northern Europe were quite separate from the former Alpine glaciers, and that the Alpine ice postdated rather than antedated the making of the Alps themselves.

Agassiz's exposition of the glacial-age concept was stated so strikingly that it commanded the instant attention of a far wider group of readers than earlier contributions had done. The greatest proponent of the submergence concept had been William Buckland, professor of geology at Oxford. Interested by the new discoveries in Switzerland, he visited Agassiz in 1838 and soon realized that the British features were similar to those in the Alps. Convinced that his own earlier ideas would no longer stand and that the Swiss were on the right track, Buckland persuaded Agassiz to visit the United Kingdom in 1840, and there accompanied him into the field. Though warmly supported by Buckland and in part by Lyell (who still preferred the iceberg theory for most of Britain), the conclusion reached by Agassiz that Britain had been glaciated was at first ridiculed by many British geologists. The concept of marine origin of the drift lingered (B. Hansen, 1970) in many minds until it was firmly dispelled by two classic papers, one by T. F. Jamieson (1862) and the other by Archibald Geikie (1863). These papers mark a turning point in the development of the study of glacial geology. During the decades of controversy real study had stagnated. Investigation now began anew and an epoch of discovery began.

Whatever unpublished thoughts about the origin of the drift may have been entertained in America before 1839, the numerous published ideas, dating from as early as 1793, were strongly influenced by religious bias or by British ideas of submergence. American recognition of the theory of Agassiz began as early as 1839, with a published statement by Timothy Conrad (1839, p. 241). Conrad's was probably the first American acceptance of the Glacial Theory; it was stimulated by Agassiz's 1837 paper alone, for it antedates Agassiz's widely known book.

Two years later, when the results of Buckland's and Agassiz's 1840 trip through northern Britain reached America, the Glacial Theory was taken up by Edward Hitchcock (1841) in a "First Anniversary Address" before the newly formed Association of American Geologists. The State Geologist of Massachusetts strongly supported Agassiz, though later, curiously, he recanted in part. Agassiz himself arrived in the United States in 1846 to become a professor at Harvard. His presence there undoubtedly hastened wide acceptance of the Glacial Theory, but with many, belief in the iceberg theory lasted as long as it did in Britain. By 1852 Danish sci-

[3] The story of Agassiz, de Charpentier, Schimper, and others involved in the birth of the Glacial Theory, is well told by Carozzi (1966).

entific exploration in Greenland had established the existence of a huge ice sheet, furnishing a convincing analogy with the former ice sheet in northern Europe, postulated on geologic evidence. Slowly the glacial view progressed against resistance; one of the best American statements in advocacy of it was made by J. D. Dana (1863, p. 541). The last scientific opposition to it in North America died in 1899 with J. W. Dawson; in England Sir Henry Howorth published a 1000-page argument in opposition as late as 1905.

Who, then, were the founders of the ideas we accept today? We must accord a First to Hutton and to his exponent Playfair for the idea that erratics have been carried by former glaciers. Bernhardi seems to have been the first to recognize former widespread glaciation of lowland Europe. Venetz, de Charpentier, and Agassiz (and perhaps Schimper as well), forming a sort of chain reaction, developed the theory of even wider glaciation during a "great ice period," taken from Goethe's "epoch of great cold." However, it was Karl Schimper who, after discussions in the field with Agassiz, coined the term *Eiszeit* (ice age).

In a sense Bernhardi deserves more credit for his contribution to the concept of an ice age than do any of the three Swiss. They had living glaciers before them and could make direct comparison between areas now occupied by ice and those evidently abandoned by ice. Bernhardi's evidence was less fresh, less distinct, less abundant, and above all, geographically removed from direct comparison with the work of living glaciers. In bridging the gap between Germany and the Arctic with glacier ice, Bernhardi displayed a lively imagination. It is remarkable that no one in Germany paid serious attention to the idea (despite Agassiz's conversions in Britain and America) until 1875, when Otto Torell, a Swedish geologist, brought to Berlin overwhelming evidence that Bernhardi had been right, and that to him rightfully belonged the distinction of having been first to recognize an ice age.

NONGLACIAL EVIDENCE OF COLD CLIMATE

Quite apart from glacial deposits, facts that imply a former cold climate were accumulating even before the Glacial Theory received broad attention. The widespread occurrence of tusks and bones of the extinct woolly mammoth (*Mammuthus primigenius*), chiefly in Siberia, was well known in Europe since the 18th century. The tusks, a source of ivory, were a recognized article of commerce and the climatic implication of the fossils was understood. Indeed Bernhardi appealed to mammoth fossils as supporting his concept of polar ice in Europe. Analogously, one of the facts that led Agassiz to include Asia as well as Europe in his "ice period" was the occurrence of mammoths and other mammals in frozen sediment in northern Siberia, presumably killed by intense cold.

Before 1834 various people including Lyell, William Smith (the English "father of stratigraphy"), and Agassiz himself had remarked the presence in Quaternary sediments of fossil shells that implied cold climate.

Conrad, whom we identified earlier as perhaps the first American to accept the glacial theory, described bedrock shattered from the surface down to depths of several feet, yet without great displacement of the fragments. "I think it impossible to account for this . . . ," he wrote, "except by the agency of intense cold, freezing the water which filled the fissures, and thus forcing the rocks into tabular fragments, and disturbing their position by the lateral and upward pressure." (Conrad, 1839, p. 243.) Possibly this passage constitutes the earliest interpretation of frost wedging during a glacial age, although of course Conrad did not visualize a glacial age as such.

It was R. A. C. Godwin-Austen (1851, p. 130) who first perceived that the mantle of colluvium covering hillslopes in much of southern England is the result of cold-climate processes, including frost wedging, no longer active in that region. He thought, however, that the cold climate resulted from regional uplift, which had also caused the building of glaciers in Britain. Ramsay and Geikie (1878) identified relict sliderock at Gibraltar, which they attributed to frost wedging under a glacial-age climate. However, wide, systematic study of relict cold-climate features did not begin until 1910, when the International Geological Congress put on an excursion to Spitsbergen. There for the first time a number of scientists were enabled to see polar activities such as frost wedging and solifluction in actual operation. The sight opened the eyes of many scientists, who began to observe relict features in temperate latitudes.

The deduction that lowered temperature would result in lowering the snowline was made at least as early as 1868 (Whittlesey, 1868, p. 93). Realization that the altitudes and sizes of glaciers vary systematically with variations in temperature and precipitation was first clearly stated by Penck (1882, p. 437–440), who drew the correct conclusion that fluctuation of glaciers was a phenomenon of secular change of climate. This statement perhaps marks the beginning of the shift of emphasis from glaciers present and past to the more fundamental phenomenon of change of climate.

GLACIAL/INTERGLACIAL STRATIGRAPHY

Bernhardi, Agassiz, and other pioneers of the glacial theory had not thought of the ice age otherwise than as a single event, probably because the erratic boulders and till on which the idea was principally based had always been thought of as a contemporaneous group of deposits. But the concept of a single ice age was not to last long. The publication of Agassiz's book in 1840 and other papers it engendered stirred up field study. As early as 1847 Collomb (1847) reported two distinct layers of drift in the Vosges Mountains, and inferred two separate glaciations. Shortly thereafter, A. C. Ramsay (1852) in Wales, Chambers (1853) in Scotland, and Morlot (1856, p. 39) in Switzerland inferred two glaciations on similar evidence. The strata between two layers of till were called *interglacial* because of their intertill positions.

Although he did not clearly and specifically identify two layers of till, Godwin-Austen (1851) recognized a rather complex Pleistocene sequence in southern and central England, which he represented in a correlation table. He identified, beneath till and above the Crag strata, a layer containing fossil elephants and other large mammals, as well as pine and fir trees. Although he correlated the layer incorrectly with the submerged floor of the North Sea, from which fishing boats had brought up fossil mammals, he inferred correctly that the North Sea floor had formerly been emerged and that England had then been connected with the European mainland. He described the "pleistocene" of Lyell as the period "marked by the first appearance of a northern marine fauna in sea-beds in our latitudes" and held that it was both long and characterized by "more contrasting conditions in physical geography than those of any other period."

Before 1882 three successive layers of glacial drift had been identified in the northern Alps (Penck, 1882, table 2) and later there was recognized the four-layer sequence that came to be widely referred to. In America the four major drift layers still recognized today had been identified before the end of the 19th century.

The identification of successive major glaciations was closely interlocked with the recognition of layers of nonglacial sediments occurring between layers of glacial drift. Interglacial sediments containing fossils seem to have been first recognized by Heer (1858) in Switzerland. Although at first these sediments, which included peat with a fossil flora, were thought to be preglacial, later discovery of till beneath them showed them to be truly interglacial. After study of the fossil flora Heer demonstrated, a few years later, that the climate it implied was similar to or slightly cooler than today's. Although this was not the earliest use of the ecologic method of extracting climatic information from fossils, it has been followed routinely ever since.

Possibly the earliest recognition of a soil developed in glacial drift and buried beneath younger sediments was that of Worthen (1873) in Illinois. In the same year fossil plants occurring between two glacial layers were identified as interglacial by Orton in Ohio and by Winchell in Minnesota, apparently the earliest such finds in America.

LIMITS OF GLACIATION

The concept of an ice cover over Europe, first put forward by Bernhardi, raised the question: Over *how much* of Europe? The wide notice given to Agassiz's book in 1840 resulted in widespread search in many countries. In the United States the systematic tracing and mapping of the outer limit of glacial drift (Flint in H. E. Wright and Frey, eds., 1965a, p. 5) began soon after 1860. By 1874 a primitive but reasonable compilation of the limit from New Jersey to Kansas had been published. A more detailed map published in 1878 extended the limit westward into Montana and eastward along the continental shelf as far as Newfoundland. The accompanying text mentioned centers of outflow of the glacier ice in Greenland, Labrador, and the Rocky Mountains.

The general limits of glaciation in Europe and in some other parts of the world had been established well before the end of the 19th century, as can be seen on a map published by Penck (1882, fol. p. 483). On both continents refinement of glacial maps has continued up to the present day.

Study and mapping of end moraines built by continental ice sheets seems to have begun first in central United States in 1871, when G. K. Gilbert described and interpreted the origin of what are now called the Wabash and Fort Wayne Moraines in western Ohio. But it was not until around the turn of the century, beginning with the many classical reports by Leverett, that American moraines were mapped in great numbers. Extensive detailed mapping of continental moraines in Europe came even later.

PLUVIAL CLIMATES

Hardly had the fact of widespread glaciation become generally established when the effect of glacial climates on saline lakes in arid regions was perceived. Writing in Edinburgh, T. F. Jamieson (1863, p. 258) said: "Now this heat and dryness [of the arid regions] being much lessened during the glacial period, there must have resulted a much smaller evaporation, which would no longer balance the inflow. These lakes would therefore swell and rise in level." He mentioned specifically some of the great saline lakes of Asia: the Caspian, Aral, Balkhash, and Lop-Nor lakes.

Very soon afterward, and apparently independently, J. D. Whitney (1865, p. 452) in California reached a similar conclusion.

Meanwhile in Paris, Louis Lartet (1865, p. 798), the archeologist, published the results of studies he had recently completed in the region of the Dead Sea, another large salt lake. He had found deposits, evidently made by the lake, high above its present shores and inferred that the lake had expanded during the ice age, which he believed was recorded by glacial features on Mt. Lebanon, west of the Dead Sea. The deposits are those of what we now call Ancient Lake Lisan. Lartet's inference, which Jamieson's deduction anticipated by two years, was later confirmed and established. Today it is widely evident that glacial ages were marked in many dry regions by the expansion of lakes which shrank during interglacial times. Except for repeated studies of the Aral-Caspian system, examination of this problem in the Old World has lagged. The most detailed and reliable evidence now available comes from western North America; yet even in that region a great deal remains to be done.

Though Jamieson deduced that arid-basin lakes should have expanded during a glacial age, and though Lartet established that both glaciation and lake expansion had in fact occurred in the Dead Sea region, proof of the simultaneous occurrence of both events did not appear until much later. In 1889 I. C. Russell (1889, p. 369; also Putnam, 1950) showed, through the relation of the shoreline of the expanded former Mono Lake in eastern California to moraines of a Sierra Nevada glacier, that lake and glacier were apparently contemporaneous. Also, in his classic study of ancient Lake Bonneville in Utah, Gilbert (1890, p. 318) demonstrated the probability of a similar relationship between lake sediments and glacial sediments on the flank of the Wasatch Mountains. The close relation in time between the ancient Searles Lake in southeastern California and the last great glaciation was later established through radiocarbon dating (Flint and Gale, 1958).

It is unlikely that a similar relationship will soon be demonstrated for most lakes, but evidence of former expansion (in some cases two or more expansions) is so well-nigh universal that few doubt that lakes in the middle-latitude dry regions of all continents expanded and shrank nearly synchronously with the onset and waning of the successive glacial ages, or that both lakes and glaciers were the result of a single underlying cause, climatic fluctuation.

We are confident that when the lakes were expanded the now-dry regions were characterized by increased rainfall, as well as by decreased temperature. The lakes are therefore generally known as *pluvial* lakes and the periods of their expansion have been called *pluvial ages*. The term pluvial seems to have been first applied to an expanded lake by Edward Hull (1885, p. 182) in a report on the geology of a region that includes the Dead Sea. But the term itself originated in a different concept. Alfred Tylor (1868, p. 105) applied the term to coarse alluvium occurring in England and France as massive valley fills, now generally dissected. In earlier papers he stated his belief that the transport of these sediments would have required more than 750cm of rainfall annually; hence he wrote of a "Pluvial period." Others had the mistaken notion that a general increase of glaciers demands a general increase of precipitation. Today, however, we realize that although pluvial climates prevailed in certain regions that are now dry, they do not necessarily imply overall increase in precipitation. They can be explained by changes in the distribution of precipitation resulting from changes in the circulation pattern of the atmosphere.

FLUCTUATION OF SEALEVEL

While the glacial theory was in its infancy Charles Maclaren (1841, p. 60; see also Tylor, 1868), perceiving the role of glaciers in the hydrologic cycle, correctly deduced that if ice had been widespread over the lands, the sealevel must then have been drawn down to a position much lower than that of today. He knew that the amount of water substance at the Earth's surface is nearly constant; when glaciers and lakes are formed on the lands, water in the sea must diminish by a corresponding amount. When the glaciers melt and the lakes dry up, the water is returned to the sea. On assumptions as to the extent and thickness of former ice sheets, Maclaren calculated that the sealevel might have been lowered in glacial ages by as much as 350 to 400 feet.

On steep rugged coasts the resulting changes in land areas were small. But on the broad continental shelves comparatively small reductions in sealevel resulted in the addition of large areas to the adjacent lands, in corresponding local changes in climate, and in forced migrations of animals and plants.

Not until later was it recognized that in at least some interglacial ages the sea must have stood higher than it is today. Whatever its amplitude, the fluctuation of sealevel, synchronously with the coming and going of the glacial ages, is a relationship now universally acknowledged. There is reason to believe that along parts of the Atlantic Coastal Plain of the United States the shore has migrated east-west from a glacial minimum to an interglacial maximum sealevel, through a distance of more than 250km.

Subsidence of the Crust Beneath Ice Sheets

As long ago as the 17th century it was realized that the coasts of Sweden and Finland were emerging from the Baltic Sea. About 1740 this movement was attributed by Celsius and by Linné to subsidence of sealevel caused by reduction in the quantity of terrestrial water. But the corollary, that the subsidence then must be worldwide, was gradually seen not to meet the facts. The obvious alternative, that emergence was being caused by crustal uplift, was perceived by Runeberg in 1765, and thereafter Playfair, von Buch, Lyell, and Berzelius entertained the concept. They supposed the cause might be related to progressive secular cooling accompanied by shrinkage of the crust.

That the emergence is the result of regional crustal warping is established by many facts, but the explanation of the probable cause had to await the announcement of the theory of glaciation in the 1830's. Impressed by the recognition of similar marine sediments lying above sealevel along the Scottish coast, and aware of the theory of isostasy, which had been proposed in 1855, Jamieson (1865, p. 178) perceived the true cause. So weak and flexible was the material beneath the upper crust, he believed, that the weight of accumulated glacier ice had caused the crust to subside; as the ice melted away, the crust slowly rose and regained its former position. This view, according to which the northern Baltic region has not yet recovered its preglacial altitude and is still rising, has been universally adopted. Among its early proponents were Whittlesey (1868) and Shaler (1874, p. 338), who applied it in explanation of observed emergence of the coast of Maine. During his classic study of pluvial Lake Bonneville, Gilbert (1890, p. 362–383) tested the principle by measuring strandlines. Finding that they are warped in domelike form, he inferred isostatic recovery following removal of the very considerable weight of water by evaporation as the lake dried up.

Scandinavian and Finnish scientists have measured the attitudes of deformed postglacial standlines of the Baltic Sea and have unraveled their stratigraphic relations. Americans have made similar studies, by which they established the history and deformation of the Great Lakes region as well as of seacoasts. All these studies have shown that the form of the upwarping, both past and present, is domelike, and that the centers of the warped areas lie in the central regions of the former ice sheets.

As the great glaciers thickened and spread, the crust subsided and as the ice melted, the deformed crust rose to assume its former shape. Such subsidence and elevation on a majestic scale must have taken place with the waxing and waning of each glacial age, synchronously with the fall and rise of the surface of the sea.

LATE-CENOZOIC MOUNTAINS

No one who examines the present-day distribution of glaciers can fail to realize that glaciers are related to highlands. Without high and extensive mountains, some of them situated in the paths of moist winds, extensive glaciation cannot occur. Late-Cenozoic time seems to have been characterized by more extensive mountains and higher continental altitudes than those which existed earlier in the Cenozoic Era. Much of the uplift and mountain-making movement responsible for the present highlands are of Miocene and later date, and uplift with amplitude amounting in places to thousands of meters has occurred within post-Miocene time. The Cordilleran mountain system in both North and South America, the Alps-Caucasus-Central Asian system, and many others have attained the greater part of their present height during the late Cenozoic. But more significantly for general glaciation the mountains of Scandinavia, Greenland, and northeastern Canada, which are believed to have contributed importantly to the building of some of the largest ice sheets, are thought to have reached their present heights in late-Cenozoic time.

Glaciers existed on lofty mountains in high-middle and high latitudes, including the Antarctic Continent, at least as early as late-Miocene time. Thereafter, further uplifts and increased albedo in areas of glaciers may have set the stage for the large ice sheets in middle latitudes.

ANIMALS AND PLANTS

Long before the Glacial Theory appeared, fossil animals and plants of Quaternary age were being collected and studied, and in both England and France geologists were recognizing marine strata in which they discerned differences in the kinds and proportions of marine invertebrates the strata contained. Looking back over these and subsequent studies, we can perceive two basic generalizations, each of which emerged only gradually. These pertain to evolution and to repeated shifts of population.

That evolution continued during Quaternary time is well established.

Of course it continued, as it has always done since life on the Earth began. A better way to put the matter would be to say that Quaternary time was long enough so that the effects of evolution within it can be clearly recognized. Many extinctions occurred and many new species appeared, although the number of new species seems to have been smaller than the number of extinctions. Attempts have been made to delimit stratigraphic subdivisions of the Quaternary in the classical way, on a basis of its fossils, chiefly land mammals; the results are set forth in chap 14.

At first, shifts of populations of animals and plants were not recognized as such. Long before the Glacial Theory was formulated, fossils of cold-climate animals had been found in middle latitudes. They were left unexplained or were explained in various odd ways. But soon after the Glacial Theory appeared the finds were attributed to cold glacial-age climates. Later warm-climate fossil organisms were recognized in middle latitudes, and were gradually seen to be related to interglacial ages in which, at times, temperatures were higher than they are at present. The early discoveries of both "cold" and "warm" fossils were chiefly those of land mammals. Marine invertebrates and plants entered the field later.

The bulk of our organic information on former climates has been drawn from plants. An early milestone in their study is represented by the pioneer work of Oswald Heer (1865), who attempted to reconstruct the ecology of Switzerland at times in the Quaternary when the climate differed from that of today. This was perhaps the first work of its kind in any country.

Some of the shifts of populations of animals and plants imply the former presence of land bridges that do not exist today. Examples are the former bridges between Siberia and Alaska and between France and England. The creation of such bridges resulted in part from fluctuation of the level of the sea. That their presence was required by the distribution of terrestrial animals was recognized by the middle of the 19th century.

To many people the most fascinating of the Quaternary mammals is man. Whether viewed as the genus *Homo* or under a variety of generic names, man is regarded by many as an organism pertaining exclusively or primarily to the Quaternary Period. Although flint artifacts were found in association with extinct fossil mammals in southern England in 1825 and in Belgium in 1833, and although the association was recognized by the discoverers, the implication that man must have existed long before the Creation specified in the Book of Genesis was opposed by widespread doctrine, and was not accepted by the scientific world until after the publication of *The Origin of Species* by Charles Darwin (1859).

NONGLACIAL STRATIGRAPHY

Over substantial parts of the two-thirds of the Earth's land area that was not glaciated, and beneath most of the sea floor, are strata of Quaternary age and nonglacial character. They were deposited by marine, fluvial, lacustrine, and eolian agencies and in some areas they include volcanic rocks. Most such strata do not differ in origin from the similar strata of pre-Quaternary age, and in themselves do not reflect the peculiar conditions of the Quaternary. Some, however, occur in belts of latitude where they would not have been deposited had Quaternary climatic changes not occurred. Once their origin is recognized, their significance for us lies mainly in their secular distribution.

With most such sediments we are not concerned here. Were it possible to describe, by regions, all the known Quaternary sediments, the result would be a sort of encyclopedia without sharp focus. Our concern is with the repeated fluctuation of climate that characterized the Quaternary, and especially with the widespread glaciation caused by it. Hence we shall deal only with those aspects of nonglacial stratigraphy which bear significantly on Quaternary climates. Among these are loess, peat and pollen, and sediments beneath the floor of the deep sea.

Although much of the world's Quaternary loess overlies areas not glaciated, much of it would not have been formed had glaciation not occurred. Although its thickness has not been compiled for wide regions, the loess shown on existing maps exceeds, in area, $1.8 \times 10^6 km^2$ in Europe and $1.5 \times 10^6 km^2$ in North America. Most of it was derived secondarily from glacial sediments.

Loess in Europe was recognized early in the 19th century, and was first ascribed to great floods (possibly the same floods visualized as having deposited the glacial drift) or to submergence beneath the sea. That idea was abandoned when it was perceived that loess contained neither coarse sediments nor marine fossils. A theory introduced by Lyell in 1834 and popular for four decades (its advocates included de Charpentier, Collomb, Heer, and Agassiz) ascribed loess to large lakes or flooded rivers. The eolian hypothesis was proposed in 1872 and brilliantly argued by Richthofen (1882—a later reference widely available) from his observations in China, his main support coming from the nonstratified character of the sediment and its extraordinary range of altitude.

Once the origin of loess was understood, research in both North America and Europe began to turn up two, three, and more layers of loess in stratigraphic superposition, separated by sediments of other origins or simply by soils developed in the loess itself. Indeed, probably a large proportion of the buried Quaternary soils now known are developed in loess. The stratigraphy of loess strata has acquired great importance, especially in semiarid regions.

Another sediment that is stratigraphically important consists of bog peat, most of which embraces the time since the last glaciation but some of which carries a far longer record. The basis of its study was laid by the Norwegian botanist Axel Blytt (1876), who recognized that the stratigraphy of bogs and the ecology of the fossil plants in them contain the story of the history of vegetation and hence of climate. He developed the now-familiar sequence of strata in Scandinavian bogs which, modified by Rutger Sernander, is widely known as the Blytt-Sernander sequence (table 24-C).

The sequence was greatly enriched by Lennart von Post (1916), who introduced the technique of statistical manipulation of the relative quantities of pollen in the bog strata. On the basis of such statistics Knud Jessen subdivided the very late Quaternary in Denmark into nine pollen zones (table 24-C), whose boundaries are based on the immigration, culmination, and decline of plant species.

The collecting and stratigraphic interpretation of fossil pollen was extended from bog peat to lake sediments, and more recently to alluvium and even to sediments beneath the deep-sea floor.

The physical study of Quaternary alluvium in various parts of the world dates mainly from the 20th century. It has developed that at least some streams have undergone fluctuation of regimen, which is reflected in a record of alternate trenching of valleys and filling or partial filling of valleys with alluvium. It is believed that in some valleys the changes were brought about by changes of climate, but alternative possibilities make reliable interpretation difficult, so that the study of alluvium and related stream terraces progresses slowly.

Research on the Quaternary sediments underlying the deep-sea floor began in the 1930's, as soon as the development of coring devices had progressed far enough so that sediment cores adequate for study could be raised. The stratigraphic potential of the long cores that are now possible would be hard to exaggerate. Such cores, together with those taken from thick lacustrine sediments, open up the possibility of continuous stratigraphic sections through the entire Quaternary System. Apart from this, alternating "cold" and "warm" microfaunas are now well recognized in marine cores, implying new possibilities of stratigraphic correlation. Also, coupled with radiometric dating, such cores offer possibilities of measuring rates of sedimentation at times during the Quaternary, with the corollary that overall rates of erosion of the lands could also be determined.

INTERRELATIONS OF QUATERNARY EVENTS

Having traced the beginnings of some of the chief concepts in Quaternary research, we can see that both the concepts and the data on which they are based are closely interrelated, constituting a consistent though still very incomplete picture. If world events during the later part of the Cenozoic Era could somehow be shown in generalized form on motion picture film, and the action enormously speeded up, the synchronous operation of the great movements involved could be quickly perceived. The scene would begin with gradual and piecemeal upheaval of lands and the raising of high mountain chains. As the temperature became lower and again higher the most conspicuous events would be the waxing and waning of glaciers. Accompanying this development would be the subsidence and later recovery of the crust beneath them, the slow fall and rise of the surface of the sea, and the growth and shrinkage of pluvial lakes. Moving in harmony with these changes in inanimate things would be the gradual, irregular movements of animals and plants driven from their habitats by temperature changes, inundations, desiccations, or incursions of glaciers.

The data extracted from the study of Quaternary sediments and other features define a climatic problem of great importance. If we can solve the problem of the alternating glacial and interglacial climates we will have taken an important step forward in understanding the climatic changes recorded in older sedimentary rocks. Also we will better understand the meaning of the climatic changes that have occurred during the last few thousand years.

One of the main objectives of the study of the Quaternary is historical. It is the reconstruction of the climatic events and their sequels that have affected the world during the latest part of geologic time. The reconstruction follows the classic method of geologists: inference from the evidence of former events, and comparison with events now taking place.

Our discussion, then, begins with glaciation and continues with closely related features made by streams and lakes, by wind, and by the sea. It considers the sequence of Quaternary strata and the methods used to establish a true calendar of related events. It attempts to describe significant Quaternary features in various parts of the world. Finally, it attempts to evaluate what is known and what has been speculated about the causes of the climatic oscillations that have been so conspicuous a feature of the Quaternary Period.

CHAPTER 3

Glaciers of Today[1]

Our interest in existing glaciers lies in the fact that they provide numerous analogies with the vanished glaciers of the Pleistocene. Study of the movement and the regimens of today's glaciers has greatly enhanced our understanding of the distribution of former glaciers in area and altitude, and of the erosional features and deposited sediment that constitute their principal record. Such study constitutes an important segment of the science of *glaciology*, the scientific study of snow and ice. Glaciology is concerned mainly with process, in contrast with glacial geology which, although relating in part to process, relates also to history. For extended treatment of existing glaciers the reader is referred to a textbook of glaciology such as Lliboutry (1964–1965) or to a condensed summary such as Mellor (1964) or Sharp (1960a). Our book is concerned only with those features of glaciers which bear directly on the distribution, fluctuation, erosional features, and deposits of glaciers in the past.

A glacier is not easy to define precisely. A serviceable, though not exclusive definition, which we shall use for lack of a better, is: a body of ice and firn consisting of recrystallized snow and refrozen meltwater, lying wholly or mostly on land and showing evidence of present or former flow.

Glaciers, as well as other bodies of ice and snow, represent the solid-state portion of the hydrosphere, which exchanges with the liquid and vapor portions. Glaciers develop where and when accumulated snow exceeds ablation and persists from year to year. Once created, a glacier per-

---

[1] Properties of ice: Kingery, ed., 1963; physics of glaciers: Paterson, 1969; glaciers and climate: Meier in H. E. Wright and Frey, eds., 1965a, p. 795–805. Antarctic Ice Sheet: Hatherton, ed., 1965, p. 199–278; Greenland Ice Sheet: Fristrup, 1966; popular presentation: Dyson, 1962; excellent descriptions and photographs of Alaskan glaciers: Tarr, 1909.

sists for a time even where, because of change in climate, these conditions are reversed (fig 3-10).

A glacier is an active and very sensitive body. We can view it as an open system, with input of snow which is then converted to ice, transfer of ice downward and laterally by flow, and output of water, water vapor, or ice. The system is controlled chiefly by climate through a chain of processes (fig 3-1). We shall begin with the forms of glaciers, next mention features common to all glaciers, then explain the economy of a glacier and discuss its relation to climate, next treat the diagenesis of snow and the movement of glaciers, and finally discuss some special features and describe the two existing ice sheets as a basis of comparison with the vanished ice sheets in the geologic record.

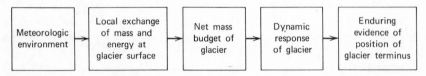

Figure 3-1   Chain of processes by which the position of a glacier terminus is related to climate. (After M. F. Meier in H. E. Wright and Frey, eds., 1965a, p. 795.)

CLASSIFICATION ACCORDING TO FORM

Glaciers have been classified on a basis of temperature, and in other ways (cf. Ahlmann, 1948, p. 59) but the most generally useful basis is form. Adopting it, we can group glaciers into three broad categories in order of increasing size, cirque glaciers, valley glaciers, and ice sheets; and two intermediate types, piedmont glaciers and mountain ice sheets.

A *cirque glacier* (fig 3-3) is a small glacier that is wholly or largely confined to a cirque. Many hundreds of cirque glaciers exist in the conterminous United States and hundreds more are found in the Alps and in Norway.

A glacier that flows down a valley bounded by exposed rock is a *valley glacier* (fig 3-2). Generally its width is small in proportion to its length. Like streams of water, such glaciers can be short or long, wide or narrow, single or with branching tributaries. The lengths of some are measured in hundreds of meters, of others in scores of kilometers. One of the longest valley glaciers yet measured is the west branch of Hubbard Glacier in Yukon Territory and Alaska, 112km overall. Common to lofty highlands

Figure 3-2 South Cascade Glacier, a small valley glacier in the North Cascade Range, Washington, in September, 1956, showing the annual snowline separating area of net accumulation from area of net ablation. (R. C. Hubley, courtesy of M. F. Meier.)

in many parts of the world, these glaciers have also been called *mountain* and *Alpine glaciers.* Some originate in cirques or groups of cirques; others (*outlet glaciers,* fig 3-3) originate in ice sheets and ice caps. But all, as they flow down the usually steep slopes, mold themselves to the shape of the underlying ground and thus generally follow pre-existing valleys. The Beardmore Glacier, an outlet glacier of the Antarctic Ice Sheet, is about 220km long and 40km wide.

In contrast, an *ice sheet* is an enormous, cakelike mass that buries all but the very highest parts of the underlying ground and flows outward from one or more central areas. Although others have existed in the past, only two glaciers reach the ice-sheet status today, the vast Antarctic Ice Sheet with an area of more than 12,500,000km² and the smaller Greenland Ice Sheet, 1,726,000km² in extent. Within them, in places, are *ice*

Figure 3-3 View near Pangnirtung Pass, Baffin Island, showing, in the distance, part of the Penny Ice Cap and, in the foreground, several of its outlet glaciers. End moraines fringe the termini, surface moraines form stripes on the glacier surfaces, and meltwater streams with braided channels are depositing sediment in the right foreground. Cirques, some with small cirque glaciers, are conspicuous in the central area. (Courtesy Canadian Armed Forces.)

streams, narrow zones within an ice sheet along which flow is much faster than in adjacent broader zones, in many cases because buried valleys are present at depth. An *ice cap* resembles an ice sheet but is much smaller. Penny Ice Cap (fig 3-3) on Baffin Island, Vatnajökull in Iceland, and others on Spitsbergen and Nordaustlandet are large ice caps. Smaller ones overlie plateaulike uplands in the Norwegian mountains and in other highland regions. Where their margins encounter well-defined valleys, both of today's ice sheets and many ice caps give origin to outlet glaciers.

Intermediate between valley glaciers and ice caps are *piedmont glaciers,* which occupy broad lowlands at the bases of steep mountain slopes. Each is the spread-out, expanded, terminal part of a valley glacier descending the highland, or the coalesced combination of two, three, or many parallel valley glaciers. The large Malaspina and Bering Glaciers on the coast of Alaska at lat 60° and the Frederikshaab Glacier at lat 62°30' on the west coast of Greenland are examples.

Another intermediate type is the *mountain ice sheet*. As defined originally (Kerr, 1934, p. 19) the term referred to vanished, former glaciers, but it is applicable to glaciers of today. A mountain ice sheet is a glacier that overlies or originates in two or more mountain masses, burying most interfluves, but with many high crests and peaks exposed; outflow takes place in two or more directions. Examples are the glacier complexes on Ellesmere Island, and in the St. Elias Mountains, Alaska, Yukon, and British Columbia, around lat 60°.

All these types of glaciers constitute a gradational sequence, and a genetic one as well. Major ice sheets seem to have originated through the coalescence of many valley glaciers that flowed down from mountains, spread out on lower lands as piedmont glaciers, and gradually thickened until the combined mass of ice submerged most or all of the mountain summits. For this reason a more detailed classification of glaciers according to their form is not necessary, and perhaps not even desirable. Every kind of glacier existing today has been identified among the glaciers of former times.

CONVERSION OF SNOW TO ICE

Regardless of their form, glaciers consist mainly of snow transformed into ice. Ice is both a mineral and a rock, which under the temperatures at the Earth's surface is exceptionally unstable. Wherever water freezes, in streams, lakes, the sea, the atmosphere, and the ground, the resulting kinds of ice have distinctive characteristics. The most distinctive and in many ways the most remarkable kind is *glacier ice*,[2] which is not only a rock but also a metamorphic rock. It can be defined as firm ice made by recrystallization of fallen snow and refreezing of meltwater, and having undergone deformation.

In pits and borings made in the higher parts of glaciers the snow at the surface is observed to grade downward through compact, very granular snow into firm ice. Field and laboratory studies have shown that the gradation is genetic; it occurs in the following general manner (fig 3-4). Newly fallen dry snow has low specific gravity (in some cases no more than 0.05) and high porosity (as high as 95%). These properties, together

---

[2] This term parallels the terms sea ice, lake ice, and river ice. Some authors write *glacial* ice, although they do not also write marine ice and lacustrine ice. *Glacier ice*, the form used by Sharp (1960a), Mellor (1964), Gow *in* Hatherton, ed., 1965, p. 223, and German authors (*Gletschereis*, not *Glazialeis*), seems clearly preferable.

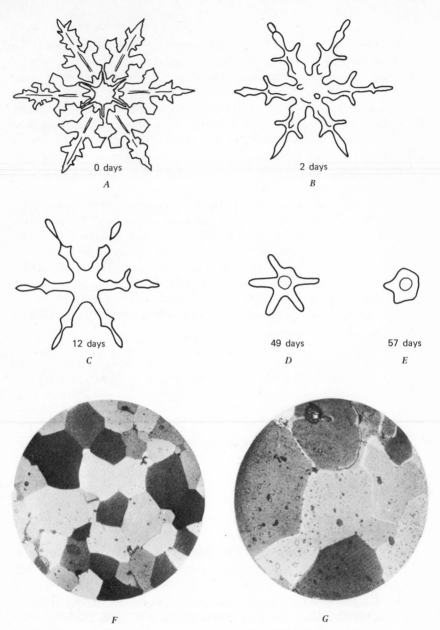

Figure 3-4 Conversion of snow into ice. A-E. Changes in snowflake after 2, 12, 49, and 57 days respectively. E is essentially granular ice (x8). F. Ice made artificially from snow soaked in water; analogous to ice made from firn. G. Ice after shear deformation during six days; analogous to glacier ice. Increase in grain size as compared with F is evident. (x8.) (Bader et al., 1939.)

with the irregular shapes of the flakes give snow an enormous internal surface area. Snow exchanges moisture readily with the adjacent air, at low temperatures by sublimation, and at higher temperatures, near the melting point, by evaporation. As the activity of molecular exchange between the solid phase and the vapor phase is greater at the projecting points of a snowflake than at its flat or concave surfaces, the snowflakes gradually change shape, and clusters of them are converted into nearly spherical grains. The snow settles and becomes more compact. If temperatures are high enough the growing granules begin to melt at points of mutual contact, because there pressure is slightly increased; the meltwater, augmented in some glaciers by meltwater percolating downward from surface snow, moves downward and refreezes, still preserving the crystalline form of each granule. Large granules initially grow at the expense of adjacent smaller ones; hence all tend to become about equal in size as they increase in average diameter. Granules that are considerably less than a millimeter in diameter when first converted from snowflakes, may grow to 1mm or more during a single season.

The process of growth and change is one of recrystallization, and the result is compact granular snow. When more than one year old, such snow is known as *firn* (Ger. "of last year") or *névé*. At temperatures near the melting point a body of snow can develop into granular snow within a matter of weeks, as can be seen in an old snowdrift at winter's end, although at very low temperatures, as in polar regions, the conversion may require many years. Porosity of granular snow is about 50%; specific gravity is around 0.5; grain size varies from one to several millimeters.

The next stage is the conversion of firn into ice. As compaction continues, air is forced out from the diminishing intergranular spaces. When permeability to air becomes zero the firn, by definition, becomes ice. This change occurs when specific gravity has increased to about 8.3, normally after burial, by further accumulation above it to depths of 30m or more. The texture continues to be granular, and each granule is still a crystal. However, average grain size has increased further, ranging from several millimeters to more than 1cm. All that remain of the once extensive pore space are bubbles of trapped air. The specific gravity of glacier ice varies, but approximates 0.9.

The time required for the conversion of firn into ice varies with rate of accumulation of snow and with temperature. It is estimated to range from (rarely) one year or (more commonly) two or three decades up to 300 years or even more, depending mainly on temperature and on rate of accumulation of snow. The depth at which conversion to ice becomes complete also varies greatly; in the Antarctic Ice Sheet it reaches large values, more than 150m in some places.

The final stage is slow plastic deformation of the ice under the stress resulting from its own weight. Such deformation begins very early, because even a thin layer of snow can flow down a sloping surface. However, deformation takes on much greater importance in thick ice.

The changes that produce the sequence snow→firn→ice are analogous to the changes that convert sediments into sedimentary rocks. Snow and firn are sediments; undeformed ice produced from them can be thought of as a sedimentary rock, if we neglect the recrystallization of its individual grains. Deformation by flow converts the ice into glacier ice, a metamorphic rock. This is why the successive annual layers of firn, which are distinct in many places immediately beneath the area of accumulation, are not seen in the area of ablation because they have been destroyed and replaced by foliated structure.

COMMON FEATURES

Each valley glacier has a head, lateral margins, and a terminus; an ice sheet or ice cap has a central area rather than a head, and a terminus or margin; a piedmont glacier has a feeder or feeders, and a terminus or margin.

Many valley and outlet glaciers, like those shown in fig 3-3, are systems consisting of main stream and tributaries. Although tributary ice streams that have joined the main glacier become physical parts of it, they do not mix with it; each maintains its identity as the combined glacier flows down its valley (fig 3-5).

The long profile of a valley glacier approximates the smoothed-out long profile of its bed. As the bed is usually a stream valley, most such profiles are concave-up, but where the bed is convex, convexities occur, some of them so steep that the ice breaks up and falls in chunks to the foot of the slope. Near the terminus the profile steepens, as shown in fig 3-6.

The termini of glaciers that end on land conform to the surface beneath them but in plan are convex downstream,[3] owing to the tendency of melting to reduce the surface area of the ice as much as possible, and to the thrusting of plates of ice upward and outward over underlying ice, in directions radial to the central axis of the glacier.

Termini that end in water are subject to calving (mechanical breaking off) of pieces of ice with a wide range of sizes. In plan, calving termini

---

[3] The useful terms *downstream* and *upstream*, with reference to direction of flow in glaciers, were introduced by Gilbert (1906a, p. 303).

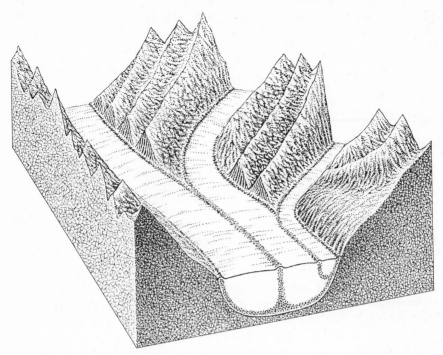

Figure 3-5   Idealized complex valley glacier, showing that tributary bodies maintain their identities after joining the main glacier.

tend to be concave downstream, as mentioned in a later section. In uncommon glaciers on very steep slopes, calving occurs on land.

The upper surfaces of glaciers are cut in places by *crevasses* (fig 3-11), which vary in width up to 15m or more, in measured depth down to about 35m, and in length from tens to many hundreds of meters. A glacier, like the crust of the Earth, can be said to possess a thin, rigid, outer shell (a zone of fracture, in which tensional stresses cause rupture), and an underlying zone of flow, in which material responds to stresses by flowing. In places where the slope of its bed steepens, and in other, more complex situations, the flowing ice extends or spreads, setting up tensile stress that opens up cracks in the surface ice. Where the slope becomes less steep, the ice becomes subject to compression, and the crevasses in its surface zone tend to close. However, the changed stress conditions may be met by closure of some crevasses only, leaving others open as they move downstream to the terminus. In the common case, crevasses are approximately transverse to direction of flow, but they can trend in various directions. Crevasses are complex features and their mechanics form the subject of special study (Meier, 1958).

Crevasses become significant in glacial geology when they travel downstream into the zone of net ablation. There they promote ablation because of the greatly increased surface area they create, promote separation of parts of a glacier in the terminal zone, and invite localized deposition of rock waste in them, in the form of ice-contact sediment, both stratified and nonstratified.

ACCUMULATION, ABLATION, AND THE ANNUAL SNOWLINE

A conspicuous feature of a glacier seen at the end of the summer melting season is the annual snowline (fig 3-2). It separates the snow-covered upstream part of the glacier (the area of net accumulation) from the downstream part, which at that season is bare ice (and is called the area of net ablation). The relation between the two areas is shown in fig 3-6. During the winter season snowfall, wind transport, and other processes generally cover the entire glacier with snow. During the ensuing summer,

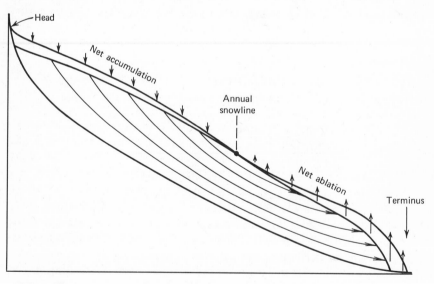

Figure 3-6 Long section of ideal valley glacier, showing area of accumulation and area of ablation, separated by the annual snowline. Long arrows suggest streamlines of flow. (Adapted from H. F. Reid.)

ablation (consisting chiefly of melting and evaporation) reduces the snow cover over the whole glacier, and on the downstream part commonly removes not only the previous winter's snow but some of the underlying ice as well. The area thus bared increases upstream until the end of the ablation season. The altitude of the annual snowline can then be measured and compared with that of other years.

The annual snowline has been termed the *firn limit* by some glaciologists. We note this term because it appears commonly in the literature, and although there is in fact a small difference between the two parameters, for general purposes we can take them as synonymous.

THE NET MASS BUDGET

Through instrumental measurements on a glacier it is possible to calculate gains and losses of mass from one year to the next.[4] The two basic values measured are *accumulation,* the solid water substance added to the glacier, and *ablation,* the substance lost from the glacier. Because the specific gravity of snow and ice varies, both values can be expressed in terms of height, in centimeters, of a column of liquid equivalent per unit area. The difference between the two values is the *net mass budget.* When in any year this parameter equals zero, the open system represented by the glacier is in a steady-state condition. However, because weather and climate are changing continually, persistence of a steady state is rare from year to year, although in some glaciers values averaged over a period of years approximate it.

The budget year begins and ends with the end of the summer ablation season, because only at that time can measurements made in two successive years be compared.

What are the actual accumulation values measured on glaciers? Obviously they vary with climate. On dry central parts of the Antarctic Ice Sheet, accumulation is less than 5cm, although at the coast it is many times greater. On the Vatnajökull, a large glacier in Iceland, in a very moist climate, mean accumulation exceeds 300cm. We can generalize by saying that in a maritime climate accumulation is large whereas in a continental climate it is small.

Ablation results from the application of solar heat to an ice or snow surface by two processes: (1) direct radiation, and (2) conduction, convection, and condensation. Latent heat also enters into ablation, but in such small proportion that for general purposes it can be neglected.

[4] For such measurements see articles and references in the *Journal of Glaciology.*

Direct radiation accounts for more than half the heat applied to ablation on most glaciers and for more than 80% on some. It reaches its largest values on the glaciers of high mountains and on the high, central areas of the two ice sheets. This happens despite the very high reflectivity of snow-covered surfaces, which can turn back more than 80% of incoming radiation.

In contrast *conduction, convection,* and *condensation,* which apply sensible heat to the surfaces of glaciers, reach their largest values (which, however, rarely amount to half the total heat applied) in warm, moist climates, notably those of maritime type, and are most effective at medium and low altitudes. Warm, moist air overrides a glacier and loses heat by conduction to the ice, with resulting ablation. If chilling of the air in contact with the ice causes atmospheric moisture to condense, the liberated heat of condensation causes further ablation. Warm rain originating in this way at times accounts for rapid ablation of the maritime glaciers of Iceland. Atmospheric turbulence renews the supply of warm air to the surface, prolonging the process of condensation, which will proceed even in dry air if the air temperature is high enough. For example, even though relative humidity is no more than 51%, condensation can occur at temperatures as low as 10°.

Regardless of relative proportion among the processes that cause it, ablation occurs at rates that are related closely to temperature. However, the greater the proportion of sensible heat applied, the more rapid the ablation, because rate of heat exchange is greater.

In temperate maritime climates glaciers—even those at high altitudes —are ablated by melting, with evaporation and sublimation accounting for an insignificant proportion of the total. In contrast, outlet glaciers of the Antarctic Ice Sheet produce relatively little meltwater at their termini; their ablation consists principally of calving, deflation, evaporation, and sublimation. The difference is one of temperature of the surface air.

Glaciers seem to be generally more sensitive to variations in ablation than to variations in accumulation. Hence temperature rather than precipitation seems to be the climatic factor primarily responsible for the fluctuations observed in glaciers. In theory, annual ablation should be greatest when summer-temperature and rainfall values are large and when winter snowfall is small. However, precipitation is thought to be the chief factor in the fluctuation of at least some glaciers (Lamb et al., 1962; Osmaston, 1965, sec. 6.141).

MOVEMENT

Ice is so weak a crystalline solid that it flows easily through the action of gravity on its mass. There are at least two kinds of glacier movement. One is sliding of the body of ice along the ground; the other consists of plastic flow, or creep, analogous to the creep of metals. The creep of ice is very complex, but a conspicuous feature of it consists of minute slipping along basal planes within the lattices of the crystals of ice.

Stress and Strain

The motion of a glacier is best visualized if we relate it to the mechanical behavior of solids in laboratory experiments. An external force acting on an unconfined solid sets up in it an internal *stress* (force per unit area) of three kinds (fig 3-7A). In the directions of maximum shearing stresses the particles of the solid tend to slip past each other. As a result the solid changes shape or volume or both. The change is expressed in a ratio termed *strain:*

$$\text{Strain} = \frac{a - b}{a}$$

where *a* is the original shape or volume.
   b is the changed shape or volume.

In a flowing glacier strain occurs continuously, and its rate governs the velocity of the glacier.

Stress in a glacier can be calculated by considering the forces acting on a cube of ice at the base of a glacier (fig 3-7B). The weight of the overlying column of ice (thickness of glacier × density of ice × acceleration of gravity) acts vertically downward. Its component parallel with the slope, W_t (weight × sine of slope angle α) is the shearing stress on the cube, expressed as

$$\tau = \rho\, g\, h\, \sin\, \alpha$$

where τ is the shearing stress at the base of the glacier (kg/cm^2).
   ρ is the density of the ice (g/cm^3).   (0.9)
   g is the acceleration of gravity (cm/sec/sec).   (980)
   h is the thickness ("height") of the glacier, measured perpendicular to the upper surface (cm).
   α is the slope of the upper surface ($°$).

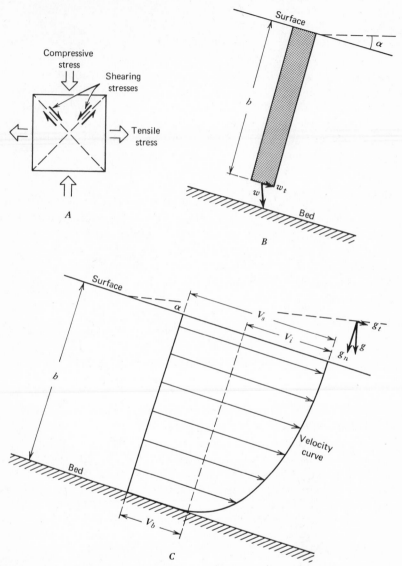

Figure 3-7 Motion of a glacier. *A.* Geometric relationships of stresses in a solid being deformed by external forces. Only two dimensions are shown. *B.* Schematic section showing the forces, acting on a 1 cm cube of ice at the base of a glacier, caused by the weight of the overlying column of ice (shaded). *C.* Long section of a glacier having thickness *h* and angle of slope α. Velocity curve, for any depth, has two components, V_b and V_i.

If we apply this equation to a glacier with thickness 200m (20,000cm) and surface slope 2° (an angle whose sine equals 0.035), we need only substitute numbers for the terms on the right-hand side of the equation in order to find the value of τ:

$$\tau = \rho \quad g \quad h \quad \sin \alpha$$
(Substituting) $\tau = 0.9 \quad 980 \quad 20,000 \quad 0.035 = 0.62 \text{kg/cm}^2$

This equation tells us that shearing stress increases directly with thickness of the glacier.

Flow Law

Laboratory measurements on ice samples show that rate of strain, which, as we have noted, governs velocity due to internal motion, is related to shearing stress according to the *Glen flow law*,[5] which, though empirical and only approximate, seems to agree with observation of glaciers:

$$\dot{\gamma} = (\tau/B)^n$$

where $\dot{\gamma}$ is the rate of strain.

B and n are constants.

When the time base for the rate of strain is one year, the value of B is about 2 and that of n is about 3, for ice near its melting temperature. In other words, $\dot{\gamma} = (\tau/2)^3$.

Using this relationship, we can calculate the rate of strain on our cube of ice, subject to a shearing stress, τ, equal to 0.62 kg/cm²:

$$\dot{\gamma} = (0.62/2)^3$$

or about 0.03 per year.

The velocity, V_s, of the surface part of a glacier consists of two components (fig 3-7 C). These are velocity, V_b, of sliding on the bed, and velocity, V_i, of internal motions. The flow law allows us to calculate only V_i. It is equal to the sum of the strain rates of *all* the 1cm cubes in the column overlying the original cube. As we go upward from cube to cube, the strain rate becomes progressively less. However, each cube moves forward bodily by an amount equal to the sum of the strain rates of all the cubes below it. Therefore, the velocity due to internal motion, V_i, increases upward. Without entering into the mathematical manipulations involved, we can express this relationship in a statement of surface velocity (V_i in cm/sec), as follows:

[5] See Nye, 1959, and references therein.

$$V_i = \frac{1}{32}(\rho g)^3 \sin^3 \alpha \, h^4$$

Because gravity and ice density are essentially constant, we can consider the value $(\frac{1}{32})$ $(\rho g)^3$ as a constant, K. For most glaciers this value works out to be: $K = 0.68 \times 10^{-18}$. Hence it is evident that the surface velocity of a glacier, caused by its internal motion, is proportional to the third power of the sine of the slope angle and to the fourth power of ice thickness.

Using this relationship, and assuming $V_b = O$, we can calculate the surface velocity of the glacier on which we determined basal shearing stress. Substituting the values given above, we get

$$0.68 \times 10^{-18} \times 0.035^3 \times 20,000^4$$

This comes out to 1.2×10^{-4} cm/sec, or 38m/yr.

This rapid increase in surface velocity of a glacier with increase in thickness or slope agrees with laboratory experiments in which increasing pressure is applied to ordinary ice. The results suggest that ice undergoes little deformation until shear stress reaches a (rather low) critical value; thereafter, rate of flow increases rapidly.

Let us consider the relationship between velocity of internal motion and the flow law in the substantial part of a glacier that lies within the area of net accumulation shown in fig 3-6. If the thickness of the glacier were doubled by the addition of a very large amount of snow to its surface, the velocity equation tells us that V_i would increase 16 times. A glacier flowing so rapidly would soon be extended or "stretched" in the accumulation area unless a large amount of snow were continually added to it. It would become thinner, V_i would decrease exponentially, and finally h and V_i would come into dynamic equilibrium, somewhat as the hydraulic factors in a stream come into a new equilibrium at the end of a flood. The thicknesses of the Greenland and Antarctic Ice Sheets, which are rather comparable, may be explained by the relationship between V_i and h. In these glaciers ice spreads out so rapidly that the existing rate of snowfall cannot build them up to greater thickness. Indeed, if the influence of irregularities of the ground underlying them is neglected, the thicknesses and slopes of large ice sheets should be rather similar, despite differences in accumulation.

The flow law also makes clear why the surface slope of a glacier increases greatly near the terminus (fig 3-6). The increased slope is necessary to maintain the value of τ in compensation for diminishing thickness caused by ablation. The flow law pertains to a glacier with uniform gradient. Actual gradients are nonuniform. Where gradient steepens, the glacier flows faster and "stretches" or extends, and thins. Where gradient becomes less steep, the glacier tends to be compressed, and thickens. The theory of *compressive and extending flow* was elaborated by Nye (1952).

Basal Sliding

The flow law assumes that all the motion of a glacier consists of creep. But basal sliding occurs also. The mechanics of the process is not yet understood, partly because the base of a glacier is far less accessible to study than is the upper surface. Apparently the sliding process consists partly of regelation (pressure melting of ice followed by refreezing), in which, therefore, transport of water is involved. In some glaciers, at least at some times, basal sliding may possibly account for most of the motion that occurs. This is discussed in a following section.

THERMAL RELATIONS

Many glaciers are at the pressure-melting temperature throughout their thicknesses, at least in summer, although in winter their uppermost parts lose heat to the atmosphere and so attain negative temperatures. If summer heat received at the upper surface is sufficient to melt snow, the meltwater percolates into the glacier until it refreezes. In glaciers in very cold situations, negative temperatures extend far downward, in some glaciers right to the base.

Geothermal heat passes at a slow rate from the ground into the base of every glacier. Such heat warms the ice or, if the ice is already at the pressure-melting point, causes melting in a very thin zone at the base of the glacier; as a result, water can accumulate there. In 1968 water was encountered in the base of the Antarctic Ice Sheet, in a drill hole more than 2.1km deep.

Ice at the pressure-melting point flows more rapidly than ice at a negative temperature. There is a possible significance here for glacial geology in that glaciers of former times probably differed thermally from one region to another. Such differences may be reflected in the kind and intensity of glacial erosion of the underlying rocky floors.

SURGES

Although glaciers respond sensitively to climate, many of them also experience sudden, spectacular movements that are induced not by climate but by dynamic conditions within the glacier itself (e.g., Post, 1960; 1966–1967; 1969; Meier and Post, 1969; Bayrock, 1967). These movements, which have been called *surges,* seem to follow phases of inactivity extending over many decades. During an inactive phase the downstream part of a cirque, valley, or outlet glacier or an ice cap thins and may even become stagnant, reflecting a negative economy in that part. Then suddenly a surge begins. A middle segment farther upstream becomes thinner, while the terminal part thickens and becomes bulbous. Active ice overrides the ice downstream. Crevasses form in very large numbers. Evidently there has been a massive transfer of ice from the middle to the terminal part, yet without any net change in the mass of the glacier as a whole. In some surges the transfer involves distances of >10km and changes in thickness of at least 60m; the time periods of most surges are 2 to 3 years. Rates of movement are known to have exceeded 6km/yr. These values imply that the velocity of a surging glacier is many times greater than its internal velocity v_i. Presumably most such motion is attributable to v_b.

It appears that while the terminal part stagnates, the middle part constitutes a reservoir that gradually thickens as it accumulates ice from upstream. Stresses are gradually built up to a critical point, at which they are released. Instability results and the surge occurs, thinning the ice at the source. With relaxation of stresses, there begins a new phase of diminished activity in the terminal part of the glacier and renewed thickening upstream. This phase lasts two or three decades, or longer. The cause of this apparently cyclic movement is unknown. No measurement or observation links it to climatic change. It has been thought that the solution may lie in basal melting, by which basal ice that was frozen to the subglacial floor is suddenly "uncoupled," permitting a substantial segment of the glacier to slide bodily on its bed, only to freeze again. This idea, however, is purely speculative.

Surges are not difficult to recognize, even from air photographs, and hundreds of glaciers are known to have surged—most of them repeatedly —within the recent past. The prevalence of surges in recent time prescribes caution in the interpretation and correlation of ancient glacial advances recognized from moraines and other geologic features, in terms of minor changes of climate. We cannot say even that a substantial part of an ice sheet, existing or former, might not be capable of surging—at least until we understand the cause.

GLACIERS AND CLIMATE

We have said that glaciers respond sensitively to variations of climate. When the net mass budget shows a surplus, the economy of a glacier is

positive; when the budget shows a deficit, the economy is negative. Positive or negative economy can be identified regardless of the absolute
quantity of water substance that is passing through the open system. In
the moist climate of Iceland, the Hoffell outlet of the Vatnajökull shows
accumulation values of around 400cm at the annual snowline, whereas in
the dry climate of East Greenland, the Fröya Glacier shows only about
30cm. Yet both glaciers, in common with many others in the North Atlantic region, experienced negative economies in the 1930's (Ahlmann,
1948, p. 67). That decade formed part of a much longer period during
which temperatures increased and precipitation decreased on a secular
scale (Mitchell in Furness, ed., 1961, p. 248); in other words, climate underwent a change. Probably the general change in these glaciers was a result of the change in climate.

Because determination of economy is laborious and time consuming,
another, less accurate means of roughly determining the state of a glacier
is employed. This involves measurement of the areal position of the terminus or of the vertical position of the upper surface in the area of ablation from year to year (fig 3-8). In a few glaciers such measurement has
been correlated with determinations of economy. The correlation shows
that, for these glaciers, negative economy is accompanied by shrinking.

Between about 1850 and about 1950, when, as noted above, tempera-

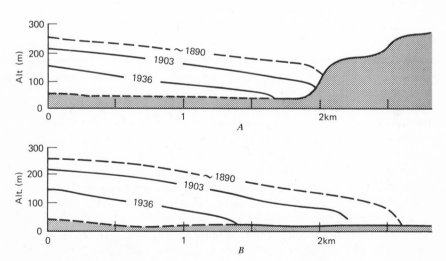

Figure 3-8 Partial cross-profiles (A) and long profiles (B) within area of ablation on the Hoffell outlet of the Vatna Glacier, Iceland, in 1890, 1903, and
1936, showing progressive thinning. (After Ahlmann and Thorarinsson, 1937, p.
191.)

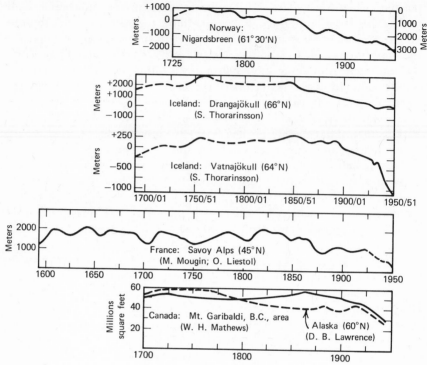

Figure 3-9 Curves assembled by Ahlmann (1953, p. 15), showing fluctuations of the terminal positions of selected glaciers in the northern hemisphere through the period A.D. ~1700 to ~1950. Details are given in the source.

tures generally rose, pronounced shrinkage of glaciers occurred in most parts of the world (fig 3-9). Most of this change occurred after 1900. That there is a causal relation between the climatic variation and the shrinkage of glaciers is fairly obvious. Yet, whether shrinkage is measured in terms of retreat of their termini, lowering of their surfaces, or negative economy, no two glaciers shrink at the same rate, and a few actually have advanced. A well-known example is the Taku Glacier near Juneau, Alaska, whose terminus, during the period 1900–1952, advanced about 5.5km because of peculiarities in its form and geographic relations (Heusser et al., 1954). Another example is the Franz Josef Glacier in New Zealand. Most individual variations are reasonably explained as the consequences of variations in thickness, relative size of area of accumulation, altitude relations, and other factors.

By analogy we can expect that whereas the former glaciers of a particular region should have shown general similarity of fluctuation, there have been many individual departures from the general pattern, owing to differences in the time of response to changes in climatic factors. Nye (p. 151 in Kingery, ed., 1963) attempted mathematical treatment of the differences in response; to be useful, however, such treatment must be based on substantial quantitative data on the glaciers involved.

One of the prime objectives of research in glacial geology is to determine the time relations of conspicuous advances of glaciers in the past. This involves both calibration with isotopic dates and correlation with climatic data derived mainly from fossils. But as yet, degree of resolution in both regards is poor; hence the time relations of major glacial events are known only in a very general way.

It will be useful to follow the practice of engineers in reference to streams of water, and refer to the system or activity of the glacier as a whole, based on its meteorology, economy, rate and possible type of flow, and fluctuation, as the *regimen* of the glacier. The term, applied to glaciers as well as to streams, is not quantitatively precise; it is broadly descriptive. Among the glaciers mentioned in the foregoing discussion, the Vatnajökull could be said to have a notably active regimen, the Fröya Glacier in East Greenland a much less active one. We shall use the term *regimen* repeatedly throughout the book.

THE TERMINAL ZONE

The profile of the terminus of a glacier with a strongly positive economy is a smooth, steep curve, so that the ice is comparatively thick even a short distance upstream from its margin. However, termini of glaciers with a negative economy or with an inactive regimen can present a different appearance.

The margin of the Greenland Ice Sheet near Thule in northwestern Greenland has been the subject of detailed study. According to B. C. Bishop (1957), when this part of the ice sheet is in a steady-state condition, a narrow zone at the margin will be so thin (65 to 80m thick) that rate of creep within the zone will be negligible. The zone acts as a stagnant barrier that impedes the flow of the thicker ice upstream; that ice, flowing compressively, rides up over the barrier, forming steeply dipping shears (fig 5-14). With a negative economy (ablation>accumulation) the marginal ice melts more rapidly and the glacier terminus retreats, and with it the stagnant barrier and the shears at its upstream limit. Rock debris brought up from the subglacial floor by the shears reaches the ice surface, covers it at least in part, and is let down onto the ground as melting proceeds. We shall deal more fully with such deposits in chap 5;

Figure 3-10 Wood Glacier, Geikie Inlet, Glacier Bay district, Alaska in 1929, a separated residual mass of ice with an area of about 5km ^{2}. Photographs made in 1941 show a little ice still remaining; in 1950 it was evident that the ice had disappeared several years earlier. Foliation in the ice is visible. For further details on this district see Field, 1947. (U.S. Navy.)

here we are concerned with the presence of a stagnant glacier margin (see also Sharp, 1951, p. 108–110). Such a margin can become separated from the main body of the glacier through retardation of melting by a covering of rock debris (fig 5-14) or by isolation in a local basin (fig 3-10).

In districts with strong relief and steep slopes, thinning accompanying negative economy can cause separation regardless of whether terminal ice continues to flow or becomes locally stagnant. The ground appears progressively above the surface of an ice sheet or ice cap, beginning with the highest points which constitute *nunataks* ("islands" in the ice, fig 3-14). These islands of rock enlarge and coalesce, and the result is local separation of bodies of ice.

Meltwater

Most of the meltwater (J. R. Davis and Nicholas, 1968) commonly seen in glaciers forms in the zone of net ablation. During the summer season streams of meltwater may be visible on the surface or along the sides of a tonguelike glacier, or emerging from the terminus and flowing away. The abundance of meltwater at any time depends on temperature, quantity of drift on the glacier surface, width of the zone of net ablation, and aspect of the terminus. Meltwater at the base of or within the glacier is rarely visible, but by inference can be very abundant. Meltwater erodes ice by melting, and aids in causing separation of terminal ice. It is the chief agent by which sediment of glacial origin is carried outward beyond a glacier, and it is responsible for most of the sediment deposited at the termini of some glaciers. In warm, dry regions whose drainage originates in mountains with glaciers, meltwater sustains stream discharge during dry seasons.

Termini in Water

Many glaciers terminate in the sea or in lakes (fig 3-11). Because the specific gravity of glacier ice is less than that of water, the terminal part of such a glacier may float, but only if the water is deep enough to submerge 75 to 90% of the thickness of the ice, depending on the abundance of rock debris in the glacier and on other circumstances. Many calving termini stand 30 to 60m above the waterline, and some reach greater heights. In very cold climates the glacier proper may be fringed by an ice shelf like those in the Antarctic, mentioned in another part of this chapter. Geophysical theory of such termini is discussed by Carey and Ahmad (1961).

Ice at the terminus is melted by conduction of heat from the water, and calving is caused by the buoyant rise of masses of ice broken off from the submerged part of the terminal cliff (commonly along crevasses) and by gravitative sliding of ice bodies broken off above the waterline. The presence of near-vertical crevasses hastens calving and also influences the sizes of the icebergs created.

The profile of a calving terminus is rarely continuous from top to base. There is usually an offset at or below the waterline, creating (1) a submerged "foot" of ice that may extend far outward from the visible ice cliff, or (2) overhang of the cliff. The offset depends on the thermal gradient and convection in the water, the effectiveness of waves on the visible cliff, the number and depth of crevasses, and perhaps other factors. This matter was discussed by Tarr (1897; 1909, p. 31), who described calving termini and other features of Alaskan glaciers with a wealth of detail.

It seems likely that where fractures in a terminus are closely spaced (fig 3-11), calving can occur about as rapidly as the glacier, flowing into deepening water, is buoyed up by flotation. If this is the case, the termini

Figure 3-11 Hubbard Glacier, Disenchantment Bay, Yakutat Bay district, Alaska in August, 1965. Its terminal cliff is more than 5km long and 100m high. Small icebergs are visible. An intricate pattern of crevasses is visible throughout the area of net ablation, and a conspicuous dark band of rocky debris marks the line of contact between the main glacier and a tributary. See maps and description in Tarr (1909, p. 42–46.) (Austin Post, U.S. Geol. Survey.)

of fractured, calving glaciers should generally be aground. Actual measurements, however, are difficult to make and are very few.

The termini of glaciers that end in water and calve icebergs are irregular or nearly straight, and many are concave downstream. The cause of the concave form has not been closely investigated; it may be related to water depth and to distribution of fractures in the glacier, among other factors. In glaciers otherwise roughly comparable, it has been observed repeatedly that recession of the termini of those which end in water is much more rapid than in those which end on land. For example, in the Glacier Bay District, Alaska, through the 56-year period 1907–1963, the Casement Glacier, ending on land, retreated ~6.5km, while the Muir Glacier, ending in water, retreated ~14.5km. Tarr (1909, p. 37) thought the difference lay mainly in the clifflike termini of calving glaciers, which yield to ablation in large units and which prevent accumulations of rock debris on the ice surface, accumulations that on land protect the terminal area from ablation.

Icebergs that calve into the sea float with the currents; those which move toward lower latitudes gradually melt in the warmer water. Some large, tabular icebergs in Antarctic waters are huge; apparently these originated in ice shelves. Geologic interest in floating termini and icebergs lies mainly in the rock debris they transport. The amount of such material transported depends mainly on aggregate volume of floating ice. As such ice melts, its rock content drops to the bottom, where it mingles with the other sediments of the sea or lake floor. Although the average amount of sediment per iceberg thus transported and deposited is small, the total volume of sediment rafted out and dropped onto wide lake- and sea-floor areas is enormous.

GEOLOGIC EFFECTS

Glaciers perform conspicuous geologic work in eroding the floors beneath them, transporting the eroded products, and depositing them elsewhere. When, in a warming climate, a glacier disappears, all the rocky debris it was carrying is deposited on the ground in various characteristic accumulations, some of which are visible in fig 3-3. But because our concern is with former glaciers we shall discuss the geologic effects of glaciers in subsequent chapters, making comparisons, where appropriate, with observations on existing glaciers. This procedure should preclude unnecessary duplication.

GREENLAND ICE SHEET[6]

General Description

We shall need to examine the principal features of the two existing ice sheets because of the useful analogies they offer for a study of the former ice sheets of North America and Europe.

The Greenland Ice Sheet,[7] with an area of 1,726,400km^2, occupies

---

[6] Detailed recent data are published in the reports of the Expéditions Polaires Françaises, issued in parts since 1948. Useful summaries include Bauer (1954) and Bader (1961).

[7] Also called the *Inlandsis,* a proper name of this ice body only, sometimes applied incorrectly to the Antarctic Ice Sheet and to former ice sheets elsewhere.

about 80% of the area of Greenland, all except a narrow coastal belt (fig 3-12). It has the shape of an elongate inverted dish, with a central dome reaching a maximum altitude of nearly 3.3km and a lower, southern dome; both domes lie east of the median axis of the ice sheet. Although nearly all of the body consists of ice, the upper surface is a layer of snow, not more than a few meters thick, the base of which is continually being converted to ice. The surface of the snow layer is added to each year, so that a pit dug into the snow reveals annual strata, each marked by a thin, more compact zone created by the effect of summer melting.

The slopes of the broad interior region are gentle (between 1:100 and 1:2000), but near the margin slopes increase to as much as 1:5. The margin is very irregular because of the presence of mountains along the east and west coasts; the inland parts of these mountains are buried beneath the ice sheet. Those on the east coast are high, with alpine summits reaching an extreme altitude of 3.5km. Ice discharges through deep valleys transecting the coastal mountains as outlet glaciers as much as 10km in individual width; many of these reach the sea. The western mountains are lower. They, too, are transected by outlet glaciers and, in addition, by ice streams (Bader, 1961, p. 10), some of which are marked off by transverse crevasses and by an elongate depression in the ice surface. They are seen to extend inland (80km in places) at least as far as the firn limit, beyond which they are covered with snow. One of them seems to follow the line of a broad, shallow, subglacial valley. Such ice streams can be compared with currents within the body of the sea and are evidence of differential streaming flow within the ice sheet.

Possibly the fact that the highest part of the ice sheet lies east of the center of Greenland results from outflow that is easier toward the west than toward the east, where mountainous obstructions are much higher. A steeper gradient on the eastern slope of the ice may be required to maintain flow over and through the barrier.

In places near its margin the ice sheet does not slope smoothly outward but is characterized by low domes as much as several kilometers in individual diameter. Some of the domes expose bedrock at one or more places in their sides, and although many do not, probably most are localized by bedrock masses high enough above the general surface of the ice sheet to cause greater accumulation than on the surrounding ice. Beyond the margin of the ice sheet are local ice caps and other glaciers, some of them 60km or more in diameter, connected with the main ice in one sector or entirely detached.

The domes and ice caps form a genetic sequence, implying that during shrinkage of the ice sheet, mentioned in the following section, positive regimens persist at least for a time in high areas, complicating the general pattern of ice-covered areas and locally reversing the directions of

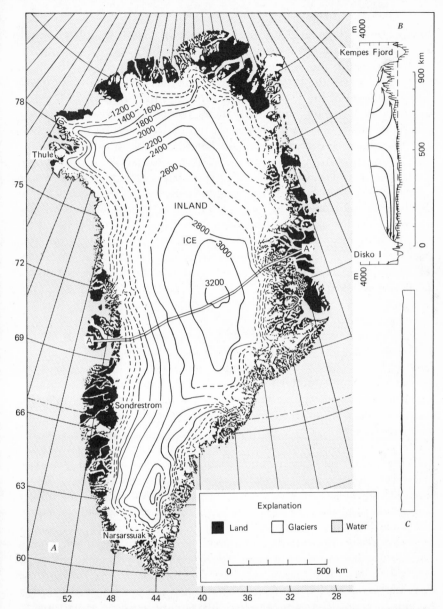

Figure 3-12 *A.* Sketch map of Greenland showing outline of ice sheet (the many other glaciers and all nunataks are omitted) and contours on ice sheet. (Map data compiled from Albert Bauer; S. J. Mock and W. F. Weeks; and Bader, 1961.) *B.* Cross section of Greenland Ice Sheet along line AA′ (map), showing subglacial floor determined from seismic and gravimetric measurements. (Data from Albert Bauer.) Dotted lines suggest hypothetical streamlines of flow like those in figure 3-13. *C.* Same cross section without vertical exaggeration.

53

flow of ice. Geologic evidence indicates that analogous complications affected the shrinking great ice sheets of former times.

Thanks to the pioneer seismic work of the Wegener expeditions of 1929–1931 and to the much more detailed 1948–1953 seismic and gravimetric traverses of the Expéditions Polaires Françaises, the British North Greenland Expedition, and U.S. agencies as well as radar traverses of the U.S. Army, we now have a fair idea of the position and form of the bedrock floor beneath the ice sheet, and therefore of the thickness of the ice (fig 3-12*B*). Evidently the ice sheet occupies a lowland, which in places lies below present sealevel and which is probably in part a result of subsidence of the crust beneath the weight of the ice. The lowest point on the floor as yet measured lies at −400m; the maximum measured thickness of ice is 3.4 km, at a locality within the central dome. The far-northern part of the ice sheet is thinner, with a mean thickness, along the traverse measured, of only 1.5km, although 3.3km was found at one place. The volume of the ice sheet as a whole has been estimated to be 2,600,000km³±5% of ice.

Regimen [8]

The energy of the atmospheric circulation comes from temperature differences between warm and cold source regions of the Earth. Heat transfer occurs through interactions between polar and tropical air masses, which create storms and cause precipitation of moisture. Belts in which temperature contrasts are great are therefore stormy; such a belt includes southern Greenland.

Except for Tibet, the Greenland Ice Sheet forms the most extensive high plateau in the northern hemisphere. At its crest minimum temperatures are extremely low. This plateau is a great wedge projecting into a relatively warm and ice-free sea, across which most cyclonic air masses move northeastward from the North American continent and the northwestern Atlantic Ocean. Cyclonic activity is most marked in winter, and its result is the precipitation of much moisture on Greenland. Weak storms generally move north along the western or eastern coast and often finally become occluded; during their progress they bring snow to the marginal areas of the ice sheet. The stronger storms, whose circulation extends to great heights, pass right across the ice sheet, bringing snow to its central area as well.

Accumulation of snow (Bader, 1961, p. 6, 7), much of it by drifting, is

[8] Meteorology: Matthes and Belmont, 1950; Hare in Malone, ed., 1951, p. 961–963; general data: Bauer, 1954; many publications of the Expéditions Polaires Françaises; Hamilton et al., 1956, p. 236; Diamond, 1960.

greatest along the southeast and southwest coasts, decreasing both north-ward and upward toward the crest of the ice sheet; mean annual accumu-lation ranges from more than 84cm near the south coast to as little as 15 cm on the far-northern part of the crest. Along parts of both the western and eastern flanks of the ice sheet, accumulation increases inland to a maximum and then decreases toward the crest. This reflects a similar in-crease and decrease of precipitation, and is an "orographic effect" caused by condensation of moisture from relatively warm air forced, by the high steep margin of the ice sheet, to rise abruptly. The northern slopes of the ice sheet receive their moisture from eastward-moving cyclonic storms and from rising air masses that move out of the Arctic Ocean basin.

The climate over the northern part of the ice sheet is cold, −20° to −30° mean annual values, but dry. In contrast, the southern periphery of the glacier has a maritime climate, and summer air temperatures near the coast are sometimes high. The cold dry climate of the interior is partly a result of altitude and therefore of the presence of the ice sheet, without which the climate of Greenland would be warmer and, in the north at least, moister.

In addition to the frequent passage of cyclonic storms along and across it, the ice sheet is characterized by outward-flowing air. Such air, cooled by contact with the ice, flows radially down the surface of the glacier under the stress of gravity, like a sheet of water. The layer, only 200 to 300 meters thick, creates *drainage winds* (*katabatic winds*) which, when channeled into deep valleys and fiords, locally possess great force. When such winds reach sealevel their energy diminishes rapidly, although they are still felt at distances of 50 to 100km beyond the ice sheet.

The altitude of the snowline varies from less than 600m to more than 1500m. In places, therefore, the width of the zone of ablation is as great as 150km, although its area is only about 15% of that of the whole ice sheet. Summer melting soaks the surface snow with water even above the snowline, throughout more than two-thirds of the area of the ice sheet. Only in a high-central and northern region does surface snow remain dry throughout the year. In this region, even in summer, persistent winds drift large quantities of snow, creating low, dunelike ridges on the low-re-lief surface.

Existing data are inadequate to support a firm conclusion as to the economy of the ice sheet as a whole. Although estimates have been at-tempted, it is not yet known whether the economy is positive or negative. However, like most glaciers elsewhere, those Greenland outlet glaciers which have been measured have been shrinking for several decades, and other parts of the ice sheet have shrunk within the recent geologic past.

The state of knowledge of movement in the ice sheet is inadequate to permit a synthesis. Probably the general pattern of movement resembles that sketched roughly in fig 3-12*B*. Beneath the broad interior area, through a depth of 200m or more, movement probably consists mainly of subsidence, perhaps 20 or 30cm/yr, with only minor lateral displacement. Farther toward the margin, flow is outward (fig 3-13) and at least in some sectors is differential, giving rise to ice streams. The ice sheet has an active regimen in its southern part where snowfall is greatest, and a slower regimen at its northern extremity where snowfall is least.

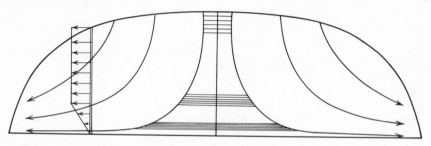

Figure 3-13   Cross section of an idealized ice sheet overlying a horizontal surface. Snow accumulates in annual layers, which become thinner, by plastic deformation, with increasing depth. Ice particles follow paths suggested by the long arrows. The small arrows define a velocity curve somewhat like that in fig 3-7C. (After Dansgaard et al., 1969.)

Many outlet glaciers discharge icebergs into the sea; those of the west coast are the most productive. Icebergs from Greenland and other North Atlantic coastal sources have been known to drift as far south as 35°N lat before being destroyed by ablation.

Among the data on the Greenland Ice Sheet, those most useful for comparison with the former great ice sheets are the following.

1. The ice sheet owes its snow supply to mainly warm, moist air masses and enjoys a maritime climate in the southern coastal areas where those masses reach Greenland.

2. Climate over the northern interior is cold and dry, in part because of high altitude and therefore of the presence of the ice sheet itself.

3. The reflectivity of the ice-sheet surface is very great. This helps maintain the ice body, once formed.

4. Vigor of the regimen varies with accumulation; the less well-nourished northern part has a sluggish regimen.

5. Differential streaming movement within the body of the ice sheet, independent of outlet glaciers, occurs.

6. The ice sheet occupies a basin that may be, at least in part, the result of crustal subsidence under the load of the ice.

7. During the shrinkage of the ice sheet, domes and ice caps persist over highlands, entailing local reversals of directions of flow.

In addition to these data, we may call attention again to the probability that the ice sheet originated through the coalescence of small glaciers that formed in the mountain areas and passed through the intermediate stage of piedmont glaciers.

ANTARCTIC ICE SHEET [9]

Form and Dimensions

The Antarctic Continent (fig 3-14) nearly approximates the area lying inside the Antarctic Circle. The area of the continent is around $14 \times 10^6 km^2$, more than 1.5 times that of conterminous United States. About 90% of its area, $12.53 \times 10^6 km^2$, is occupied by the ice sheet (not including ice shelves) and about $50,000 km^2$ by other glaciers. Mean altitude of the continent, with its ice sheet, is about 2km; hence the continent is the most extensive high plateau in the southern hemisphere. The large area, polar latitude, and high altitude insure a cold climate, and a dry one as well, and the low temperatures and low humidity are enhanced by the surrounding sea ice which, especially in winter, increases the area from which incoming radiation is strongly reflected and also the distance from the Southern Ocean, the source of heat brought to the continent by conduction. As an important heat sink and as a source of cold seawater that flows toward low latitudes, the ice sheet significantly affects the climate of the rest of the world.

Figure 3-14 includes a representative profile across the ice sheet. Its central and eastern parts are far less dependent on the topography of the floor than is the profile of the Greenland Ice Sheet. Yet this and other Antarctic profiles (Bentley et al., 1964, pl. 2) do reflect the form of the floor to some extent. Seismic traverses in which both surface and floor profiles were determined suggest strongly that the greater the slope of the floor, the less the thickness of ice over it. This relationship has been treated mathematically by Nye (1959, p. 502).

We can think of the continent as subdivided by the meridian 0°–180° into two parts. East Antarctica includes the greater part of the ice sheet, whose smooth, domelike surface reaches an extreme altitude of 4km in a region well east of the Pole. In West Antarctica the form of the ice sheet

[9] Bentley et al., 1964; Mellor, 1961; Hatherton, ed., 1965; Priestley et al., eds., 1964; Mercer, 1967.

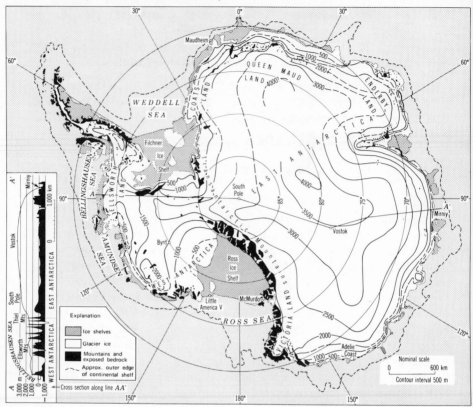

Figure 3-14 Sketch map of Antarctic Continent. (Base, cross section, and some data from Bentley et al., 1964; other data from Mellor, 1961.)

is complicated by great irregularities in the coast and in the relief of the subglacial floor. In many places mountains project as nunataks above the ice sheet (fig 3-15). In both halves of the continent, the ice flows outward from various ice divides, under control of the bedrock topography, to the coast, forming numerous ice streams and outlet glaciers. One of the latter, the Beardmore Glacier, is possibly the largest outlet glacier in the world.

The inset in fig 3-14, a typical cross section, shows the striking difference between West Antarctica, with its numerous nunataks, lower bedrock floor, and lesser volume of ice, and East Antarctica. Indeed, much of the bedrock floor of West Antarctica lies well below the existing sealevel, and even if the Antarctic ice melted away, after all adjustments much of the region probably would be submerged. In this section maximum ice thickness is between 3 and 4km; the thickest ice thus far measured

Figure 3-15 Antarctic Ice Sheet at lat 85°40', long 140°00'; the mountains are part of the Queen Maud Range. View northeast, showing several nunataks. The directional features are probably the work of wind rather than of flow of the ice sheet. (U.S. Navy for U.S. Geol. Survey, February 1963.)

(by seismic techniques) approaches 4.3km, of which about 2.5km lies below sealevel.

Ice Shelves. Along about one-third of the Antarctic coast, glacier ice flowing out into bays merges into ice shelves, of which the Ross and Filchner Ice Shelves are the largest. The shelves as a group cover about $1.4 \times 10^6 km^2$; the area of the Ross Ice Shelf alone nearly equals the area of Texas. They are essentially floating sheets of ice, spreading under their own weight and flowing seaward, and are perennially attached to the land along one or more sides but with a free face toward the ocean. From their thick landward parts they thin seaward, with upper surfaces nearly level or undulating, to less than 250m, and terminate in cliffs that may be 30m high above the waterline. They break off to form tabular icebergs, some of which are enormous; at least two have been reported with estimated areas of $> 30{,}000 km^2$.

Four factors control an ice shelf. (1) Low, polar temperatures. (2) Nourishment from snowfall, by freezing of seawater to its base, and by flow of the adjacent glacier into it. (3) Presence of a bay, which protects the shelf against breaking loose and drifting seaward. (4) Anchorage of the shelf to shoals beneath it; this promotes stability. The thickness of a shelf at any time represents approximate equilibrium among these factors.

Ice shelves erode bedrock along their sides and in the shoals beneath them. They transport seaward small quantities of rock waste, which eventually drops down onto the sea floor. Along much of the coast no bedrock is exposed, and broad, clifflike termini calve glacier ice rather than shelf ice into the sea.

Even during the maxima of former glacial ages, temperatures were probably not low enough to permit the creation of ice shelves in middle latitudes. If shelves existed then, they are likely to have fringed parts of the west coast of Greenland and islands of Arctic North America.

Regimen

As mentioned above, the polar position, large area, and high altitude of the Antarctic Ice Sheet and the wide belt of sea ice that surrounds it guarantee a cold, dry climate. Mean annual temperatures range from $-5°$ to $-15°$ along the coast to about $-55°$ in parts of the interior, where temperatures as low as $-88°$ have been recorded. Mean annual precipitation is as small as that on most of the world's deserts.

Like its counterpart in Greenland, the ice sheet is characterized by persistent, usually strong katabatic winds, effective up to a height of as much as 1km; these are the chief agents of the frequent blizzards encountered on the ice. The Antarctic convergence, the mean position of the unstable zone of conflict between South Polar and tropical air masses, forms a subcircular line enclosing the Antarctic Continent between about $47°$ and $61°S$ lat. Always unstable, it bulges far southward at times, with cyclonic storms moving in from the Southern Ocean especially over the lower parts of West Antarctica. They bring heat and precipitation, but rarely penetrate over the high region. As a result annual precipitation reaches maximum values of more than 50cm at places on the west coast, but everywhere diminishes rapidly from coast toward interior where, over the polar plateau, it is less than 5cm. The computed annual average for the continent is between 14 and $15g/cm^2$. Because ablation values are small (chiefly calving at the margin, deflation, and some summer melting in the coastal zone), accumulation on the ice sheet is nearly the same as precipitation. Almost the entire ice sheet lies above the annual snowline.

Despite the low temperatures of surface ice, a borehole completed in 1969 in the central part of the ice sheet, penetrated to the base of the ice at a depth of 2164m, and there encountered water under hydrostatic pressure. If this condition is not purely local, it suggests the likelihood that basal melting might uncouple the ice sheet from its bed and thus permit surging.

Because accumulation is far greater near the coast than in the interior, it can be deduced that most of the ice reaching the sea has flowed, not from the geographic center of the ice sheet but from areas of outflow in submarginal regions. This concept implies that in the far interior the ice sheet is maintained largely by internal spreading, and that its regimen is extremely inactive, whereas the peripheral parts of the glacier, receiving more nourishment, have more active regimens. One could add, as was first suggested by David (1914, p. 567), that in East Antarctica the ice sheet probably was built up from local glaciers that accumulated on high mountains, especially those near the coast, and that in time enlarged, coalesced, and buried the mountains, shifting the loci of greatest accumulation and outflow. In contrast, in West Antarctica the evolution (Bentley and Ostenso, 1961, p. 895) of growing glaciers in mountainous areas would have gone through a phase of ice shelves in the intermountain areas that lay below sealevel, before coalescence and final buildup of the West Antarctic part of the ice sheet could occur. We can suppose that similar development occurred, at least in part, in other countries where two mountainous areas were separated by seawater. Possible examples are Greenland vs Ellesmere and Baffin Islands, and Arctic Coastal Europe vs Arctic islands.

Rates of flow in the peripheral belt, where the ice is not constrained by topographic controls, are as great as 130m/yr, although in well-defined ice streams rates may exceed 1km/yr.

The data, particularly ablation values, are too scanty to permit accurate determination of net budget for the ice sheet as a whole. Positions of glacier termini in the McMurdo Sound region have shown little change through a 50-year period. It is suspected that the regimen at present is positive. Data on rates of accumulation and local rates of flow, bolstered by assumptions, have led to the speculation that to create an ice sheet similar to the existing one, under today's climate, some 15,000 to 50,000 years would be needed (Hollin, 1962, p. 175).

Changes in Volume

Information is scanty regarding recent history; what there is has been assembled by Mercer (1962). It has been claimed that there is geologic evidence of glacial erosion and deposition on nunataks near the periphery, which implies the ice surface there was formerly higher, perhaps by hundreds of meters, and suggests that volume of ice was formerly greater.

What does control changes in volume of the ice sheet, however those changes may be distributed spatially? The answer still lies in the field of speculation. The classical concept assumed that control was directly by the long-term changes of climate that are thought to have controlled ice sheets in temperate latitudes.

A second concept views control through rise and fall of sealevel. Hollin (1962, p. 189, 186) set up a model, based on an analysis by Albrecht Penck, in which the radius of the ice sheet is determined by the line along which the terminus, moving into the sea, becomes buoyant, loses contact with the ground beneath it, and so begins to calve rapidly. Hence the radius lengthens or shortens with fall or rise of sealevel. According to this model, conspicuous change in thickness of the Antarctic ice, caused by change of climate, occurs mainly near the periphery, with little change in the central region. This relationship is easier to visualize when we remember that as much 90% of all ablation consists of calving. If sealevel should fall by 100m, the ice sheet could enlarge across the gently sloping continental shelf (fig 3-14), with an aggregate area of nearly four million square kilometers, and thereby thicken its peripheral part.

A third concept appeals to surges, thought of as occurring from time to time in one part or another of the ice sheet. Surges on so large a scale, it can be supposed, would result in conspicuous changes in thickness of ice, which, however, probably would not affect all parts of the ice sheet at the same time. According to the surge concept the changes would not be related basically to fluctuation of climate.

Whatever the factors that have controlled its fluctuation in the recent geologic past, the Antarctic Ice Sheet and the smaller ancestral glaciers that are believed to have formed it by their coalescence have been in existence for a long time. The K/Ar dates of igneous rocks that overlie, and are therefore younger than, glaciated surfaces or sediments of glacial origin show that glaciers in Antarctica existed at least 10 million years before the present. The ancestral Antarctic glaciers, it appears, may be the earliest to have formed since the Mesozoic Era.

CHAPTER 4

Spatial Distribution and Volume

of Glaciers

Quantitative comparison of the world's glaciers today with those of earlier Quaternary times yields data which, although partly speculative, are useful in various ways. Such data guide inference as to the climates of earlier times and yield a first approximation to the amplitude of fluctuation of sealevel. In this chapter, therefore, we shall compare existing glaciers with those of glacial ages as to vertical distribution, areal distribution, and volume, and shall try to frame significant generalizations from the comparisons.

Their areal distribution shows that glaciers are favored by low temperature and abundant snowfall. Because their distribution is intimately related to the snowline, we shall examine the snowline before turning to other aspects of the localization of glaciers.

THE SNOWLINE

The Regional Snowline Today

As we noted in chap 3, glaciers are linked together genetically by the annual snowline, the altitude of which is determined by the relative amounts of accumulation and ablation. Because ablation depends chiefly on mean temperature during the ablation season, mean summer temperature is closely related to accumulation at the annual snowline (e.g., Ahlmann, 1948, p. 48). Indeed, measurements on representative glaciers have established that the greater the accumulation the higher the mean summer temperature can be, with glaciers still maintaining a steady state.

The annual snowline varies from one glacier to the next. Furthermore, in many areas the lower limit of patches of perennial snow is traceable along the surface between glaciers. This limit, called the *orographic snowline* by Ratzel in 1886, because local variations of its altitude are controlled mainly by local topography and by orientation, is an important parameter in mountain areas of limited size. When viewed on a regional or secular scale, the local irregularities of the orographic snowline integrate into a band whose width varies from near zero to as much as 200m or even more. This band is widely called the *regional snowline,* a term introduced by Matthes in 1940.

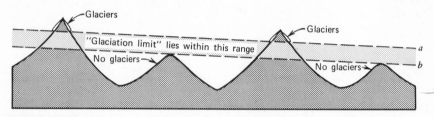

Figure 4-1 Method of approximating the regional snowline by plotting (*a*) summit altitudes of the lowest peaks bearing small glaciers and (*b*) summit altitudes of lower peaks with slopes gentle enough to retain snow, yet with no glaciers. The zone between (*a*) and (*b*) has been termed the "glaciation limit." The regional snowline must lie within or slightly above it. Distortion caused by local topography is reduced by plotting, in (*a*), only the lowest peaks that have glaciers on a single flank, that facing the sun. In the northern hemisphere this would be the southern side.

A method of approximating the regional snowline, eliminating many local irregularities, is illustrated in fig 4-1. First used by Partsch in 1882, it has been more recently refined, as discussed in Østrem (1964, p. 327). The altitude of the "glaciation limit," shown in the figure, is minimum for the regional snowline. Although the term is ambiguous, we use it here for lack of a more distinctive one. The method is best adapted to highlands that carried a large number of small, local glaciers but that were never completely covered with glacier ice. The results are generally shown by contour lines (fig 4-2).

When local irregularities, caused mainly by topography, are smoothed out in a broader view, the regional snowline is seen to vary systematically with temperature and precipitation (fig 4-3). Still another parameter, the climatic snowline, has been defined as the lower limit of perennial snow on fully exposed flat surfaces. This, however, is a theoretical concept,

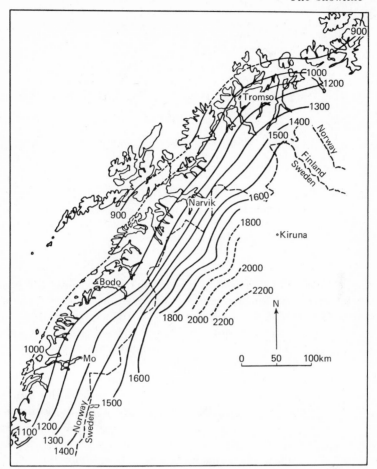

Figure 4-2 Contemporary "glaciation limit" in northern Norway and Sweden, shown by contours (Østrem, 1964, p. 334).

whereas the annual and orographic snowlines are measurable. Hence we can confine our attention to the latter, more useful parameters. The regional snowline, as a mean, has great value for wide-ranging comparisons. We shall frequently refer to it simply as "the snowline," in contexts that show that the regional snowline is meant.

A large number of measurements combine to show that the altitude of the regional snowline coincides with or lies slightly below the summer (or July) isotherm of 0° (e.g., L. B. Leopold, 1951, fig 1). This makes it possible to approximate, by free-air temperature readings from balloons,

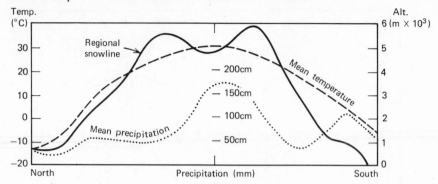

Figure 4-3 Generalized curves showing fluctuation of temperature, precipitation, and regional snowline with latitude. In the equatorial zone the snowline falls, despite increased temperature, apparently because of increased precipitation. (After Paschinger, 1912, pl. 4.)

the altitude of today's snowline in places where it lies above the highest land. More important for our purpose, this relationship gives us a means whereby a local temperature pertaining to a glacial age can be roughly determined. Such a temperature cannot be determined exactly, because altitude of the snowline is influenced by precipitation as well as by temperature. High precipitation depresses the snowline; low precipitation lifts it. The broad effect of precipitation on the snowline, seen in fig 4-1, is evident also in a characteristic rise of the snowline from the windward coast of a land mass toward the interior (fig 4-4). In northwestern United States, for example, the snowline rises eastward from around 1800m in the Olympic Mountains, Washington, near 48°N lat, to around 3000m in Glacier Park, Montana, 800km inland at about the same latitude. In Asia, similarly, the snowline rises from 3800m in the Pamir to 6400m in western Tibet. Even on a single mountain range the snowline rises from the windward flank toward the center of the range, mainly because on most mountains precipitation decreases in that direction (figs 4-2, 4-5, and 4-6). Seen in three dimensions, then, the snowline is really the trace of a surface having complex undulations determined by temperature, precipitation, and the form of the land. The surface is only approximate, because even though we neglect the effect of precipitation, the altitude of the regional snowline is rarely determinable to within less than about 100m. Because our present interest is in former positions of the snowline, let us turn to the glaciers and snowlines of glacial ages.

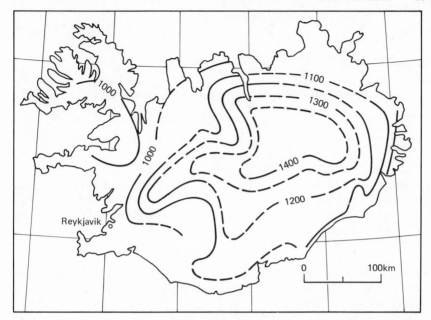

Figure 4-4 Sketch map showing generalized altitude (in meters) of regional snowline in Iceland. (After Ahlmann.)

The Snowline in a Glacial Age

Various methods for fixing points on the snowline of a glacial age have been devised,[1] but in only two of them are the uncertainties small enough to make possible reasonably good results. These two consist of plotting altitudes of the floors of cirques, and plotting a "glaciation limit."

Cirque-Floor Surfaces. The factors that control the excavation of cirques are given in chap 6. The dominant process is freeze-and-thaw, which is most frequent at an average temperature of $0°$. Hence cirques tend to cluster near (actually below) the $0°$ summer isotherm. As we have seen, that isotherm approximates the orographic snowline, which normally lies slightly above the floors of active cirques. Plots of the altitudes of the floors of abandoned cirques throughout a region (fig 4-5) makes it possible to draw contour lines and thereby construct a theoretical cirque-floor surface (fig 4-6). Although regionally consistent, such a surface lies below the lowest points on a former orographic snowline, not a regional

[1] The methods are set forth in Osmaston, 1965, chap 5; less fully in Dainelli and Marinelli, 1928, p. 86–98.

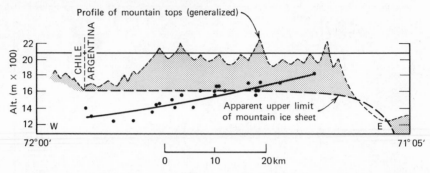

Figure 4-5 Relation between cirques and configuration of former mountain ice sheet in Argentine Andes between 41°00' and 41°20'. Small disks are cirque floors, which become higher from west to east. (Based on Flint and Fidalgo, 1964, fig 2.) (Compare fig 6-7.)

snowline, because cirques occupy those points on a highland which are best protected and where ablation values are therefore smallest. The regional snowline lies higher.

A further restriction on the use of cirque floors to approximate a snowline is that in many highlands cirque glaciers form early in a glaciation, and increase in size, becoming valley glaciers. During the early, cirque-glacier phase the snowline approximates the cirque floors but during the valley-glacier phase it is lower. The cirque-floor measurement, therefore, is valid only in highlands where former glaciers never grew beyond the cirque-glacier condition.

For maximum consistency of results it is advisable to consider only the lowest cirques, and only cirques having similar aspect (that is, facing in the same direction). The data in fig 4-7 show the variations commonly encountered. Even then the results may reflect only a generalized "glacial age" because of the possible reoccupation and deepening of some cirques during two or more glaciations.

"Glaciation Limit." The method of approximating today's regional snowline by constructing a "glaciation limit" (fig 4-1) has been mentioned. This method can be applied to a glacial-age snowline by plotting peaks that are marked by abandoned cirques instead of peaks marked by small existing glaciers. This method gives more consistent results, and approximates the regional snowline more closely, than does the cirque-floor method which, we have noted, relates to the orographic snowline. In the Pacific Mountain System in North America the glacial-age "glaciation limit" lies about 450m higher than the cirque-floor surface pertaining to the same age.

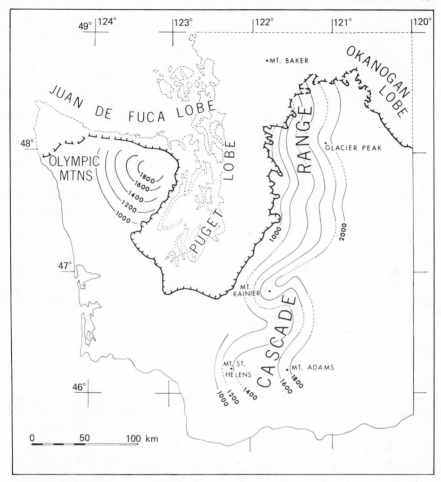

Figure 4-6 Isolines connecting floors of lowest north-facing inactive cirques in parts of the state of Washington. Hatched line represents southern limit of large mountain ice sheet during Vashon glaciation. (S.C. Porter, 1964.)

Other Reconstructions. Other methods of reconstructing former snowlines are reviewed by Paschinger (1912), but there are large inherent errors in some of them. A method long used in Europe consists of plotting the altitude of the outermost terminal moraine of a particular glaciation, and that of the top of the headwall of the cirque in which the related glacier originated. The orographic snowline is assumed to have stood midway between the two altitudes.

Examples of reconstructed glacial-age regional snowlines include Wahr-

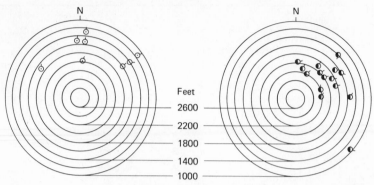

Figure 4-7   Polar diagrams showing unsymmetrical aspect of inactive cirques (circles) in two individual mountains in the Lake District, northwestern England. Concentric circles represent altitudes of cirque floors. (Temple, 1965.)

haftig and Birman in H. E. Wright and Frey, eds., 1965a, p. 304, 309 ("glaciation limit") and Richmond in idem, p. 225 (moraine/headwall method).

All these methods of reconstructing a former snowline involve cirques, directly or indirectly. The probable reoccupation of cirques by glaciers of later date, with renewed invasion, makes difficult the dating of reconstructed snowlines. Other possible sources of error include isostatic uplift of cirques formed late in a glaciation, carrying them to altitudes higher than those at which they were formed, and tectonic uplift of cirques in young, rising mountain belts. This matter is discussed at length by Machatschek (1944).

A method of fixing a former snowline which avoids the difficulties inherent in relying on cirques was developed by Porter (in course of publication). He plotted the altitudes of what are essentially steady-state annual snowlines for a group of existing glaciers and constructed an area/altitude histogram for each. A single, integrated snowline curve and histogram for the group were then fitted to an area/altitude histogram constructed from detailed mapping of a former glacier in the same district. This makes possible the reading of snowline altitude for the former glacier, if it is assumed that the range of altitudes of steady-state annual snowlines now and in the past were comparable. Apart from this assumption, the method is limited to a homogeneous district, beyond which new data on existing glaciers must be developed. It has not been tested against the "glaciation limit" method described above, but should be subject to a smaller margin of uncertainty.

Comparison of Contemporary and Former Snowlines

Various local or regional comparisons of the contemporary snowline with that of a glacial age, reconstructed by one of the methods outlined above, have been made. The most comprehensive was that of Klute (1921), whose altitude/latitude curves, touching four continents in both polar hemispheres, were excellent in their time but are now out of date. Nevertheless the substantial body of newer data has not altered two main relationships shown by Klute's curves, and by the curves in fig 4-3. These are that the former snowline is everywhere lower than today's snowline, and that the two surfaces are substantially parallel. In detail, however, there are systematic departures from parallelism in that the vertical distance between the two surfaces is greater in middle than in low- or very high latitudes, and greater in moist than in dry areas. On moist highlands in middle latitudes, the inferred glacial-age snowline lay below that of today by about 1.1km in New Zealand (R. W. Willett, 1950, p. 28), 1.2km or more in the Alps (Penck and Brückner, 1909, p. 1145), and 1.4 to 1.5km in the Sikhota-Alin Range north of Vladivostok (Chemekov, 1961a, p. 86). In the Brooks Range in northern Alaska the comparable value is at least 0.65km (Porter, 1966, p. 93). In the southern hemisphere as in the northern, the departures of glacial-age values from today's values decrease with increasing latitude.

In these main relationships we see additional evidence that during glacial ages topography, wind belts, and precipitation were rather similar to what they are today.

Temperatures Derived from Snowline Fluctuation [2]

The downward movement of the snowline in glacial ages clearly indicates lowering of temperatures or, more specifically, summer temperatures. Ever since Penck (e.g., Penck and Brückner, 1909) calculated the movement of the snowline in the Alps, attempts have been made to derive from such movement the amount of corresponding change of temperature. Common practice has been to apply a standard lapse rate (generally $0.6°/100m$, although both lower and higher values have been used) to the altitude difference between a calculated glacial-age snowline and today's snowline. Results for a number of areas are shown in table 4-A. They are, however, artificial for at least two reasons. (1) Altitude of the snowline depends not only on temperature but on precipitation as well. (2) Although apparently nearly uniform in some regions, the lapse rate

[2] General references: Morawetz, 1955; Penck, 1938a; Weischet, 1954; H. E. Wright, 1961a, p. 966.

Table 4-A

Representative Glacial-Age Temperatures Expressed as Negative Departures from Today's Temperatures. Most Values Calculated Merely by Applying Lapse Rate to Differences in Altitudes of Orographic or Regional Snowlines.[a]

Area	Source	Neg. departure of mean temperature (°C)	
		Annual	Summer
A. World average	Penck, 1928	4.0	
	Klute, 1930	4.0 to 5.0	
	Penck, 1932	5.0 to 6.0	
	Klute & Wilhelm, 1934	7.0	
Conterminous western United States	Louis, 1926	5.5	11.0
	Klute, 1928	2.0	3.9
Southwestern United States	Antevs, 1935	4.4	5.6
New Mexico	Leopold, 1951	6.6	9.1
	Harris & Findley, 1963		5.8
Northwestern Texas	Wendorf, 1961		8.4 to 11.2
	Reeves, 1965	5.2	10.1
B. Andes, equatorial	Meyer, 1904	3.0 to 4.0	
lat. 23° to 29°	Rohmeder, 1942	3.0 to 5.0	
lat. 33° to 35°	Polanski, 1957	3.6 to 4.2	
East Africa	Hume & Craig, 1911	5.0 to 7.0	
Alps	(Snowline data of		
	Penck & Brückner, 1909	7.2 to 8.4	

[a] Group *A* from Reeves (1965a, p. 188), who cites the listed sources in full. Group *B* added by R. F. Flint, with sources cited in the present volume.

can vary from one area to another, especially in the vicinity of mountains. More sophisticated study is needed in order to obtain more realistic values than those shown in table 4-A. Perhaps the best that can be said for those in the table is that their extreme range, in the annual rather than in the seasonal category, is around 6° and that their mean is 5°, so that they are at least fairly consistent internally. However, Penck (1938a) thought that in moist regions generally, glacial-age temperatures were as much as 8° lower than today's.

Apart from the rigidity of an arbitrary value for the lapse rate, such determinations suffer from ambiguity as to whether annual or summer values are being derived. As we have noted, summer temperature is more critical than mean annual temperature as the chief influence in fixing snowline altitude; decrease of summer temperature, with little or no change in annual values, could lower the snowline on a highland.

Despite their artificial character, such values derived from snowline measurements have been widely relied on as a basis for calculating glacial-age temperatures. Many other kinds of data have been employed also; some of these are set forth in chap 16.

GLACIERS TODAY AND IN A GLACIAL AGE

Existing Glaciers

The existing Antarctic Ice Sheet centered near the South Pole, and the Greenland Ice Sheet centered near lat 72°N, are shown in figs 3-13 and 3-12, respectively. Other glaciers, extending from the polar regions to the equator, are far smaller; most of them would be no more than dots on a large-scale map of the world. Some of them are shown in fig 4-8. Their number is unknown but runs into the tens of thousands. Although many, perhaps most, of them are not even mapped, inventories have been begun in many countries [3] and a few maps, by countries, have been published.[4]

A general inventory for the world, based on data of greatly varying accuracy, appears in table 4-B, which shows that the aggregate area of the world's glaciers today is nearly $15 \times 10^6 km^2$. Hence glaciers cover about 10% of the existing land surface ($\sim 150 \times 10^6 km^2$) of the world.

Distribution During Glacial Ages

The distribution of glaciers during a glacial age is shown, for selected regions,[5] on the sketch maps in figs 4-9, 18-1, 18-5, 23-1, 24-1, 26-4, and 25-7. The data are compiled from many sources, both cartographic and textual, presenting geologic evidence of former glaciation. Many small areas are lost because of the scales of the maps, and the details of large areas are generalized. However, there are still so many sectors on which adequate information is lacking that world maps on a larger scale would exaggerate errors in those sectors while it refined the data in others.

[3] A preliminary guide to inventories, with references, appears in Annals of the I.G.Y., 1967, v. 41, p. 14.

[4] Representative maps include those of Canada (Prest et al., 1968) and Scandinavia (Østrem, 1964, p. 314). A brief discussion, without maps, appears in Klebelsberg, 1948–1949, p. 408–420. Meier (1960, p. 425) inventories the glaciers of conterminous United States.

[5] World coverage is given in Antevs, 1929 which, despite being out of date in some areas, is still serviceable.

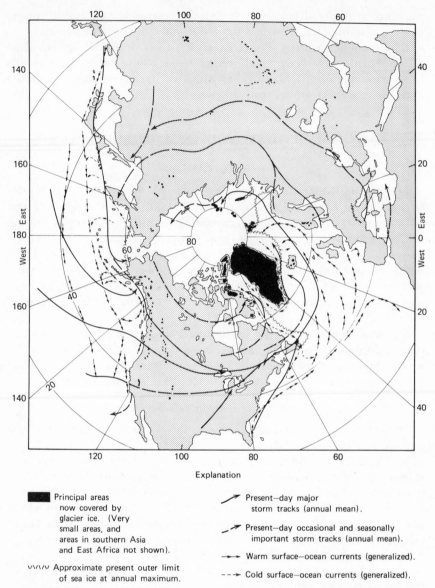

Explanation

■ Principal areas now covered by glacier ice. (Very small areas, and areas in southern Asia and East Africa not shown).

↗ Present–day major storm tracks (annual mean).

↗ Present–day occasional and seasonally important storm tracks (annual mean).

⌇⌇⌇⌇ Approximate present outer limit of sea ice at annual maximum.

→ Warm surface–ocean currents (generalized).

⇢ Cold surface–ocean currents (generalized).

Figure 4-8 Part of the northern hemisphere today, showing existing large glaciers, sea ice, selected ocean currents, and storm tracks. (Data from various sources.)

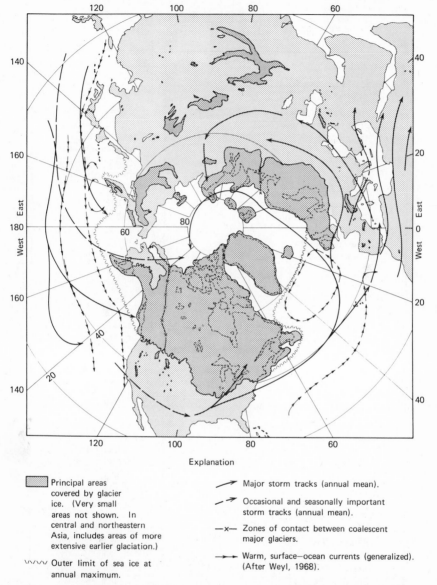

Explanation

▨ Principal areas
covered by glacier
ice. (Very small
areas not shown. In
central and northeastern
Asia, includes areas of more
extensive earlier glaciation.)

〜〜〜 Outer limit of sea ice at
annual maximum.

⟶ Major storm tracks (annual mean).

⟶ Occasional and seasonally important
storm tracks (annual mean).

—x— Zones of contact between coalescent
major glaciers.

⟶ Warm, surface–ocean currents (generalized).
(After Weyl, 1968).

Figure 4-9 Part of northern hemisphere during a glacial age, showing large glaciers, sea ice, and storm tracks. (Data from various sources.)

Table 4-B
World Distribution of Existing Glaciers

Region	Area (km^2) (approx.)	Authority[a]
North Polar Region		
Greenland Ice Sheet	1,726,400	Ba
Other Greenland glaciers	76,200	Ba
Baffin Island	41,260	A
Bylot Island	4,921	A
Queen Elizabeth Islands	106,988	A
Iceland	12,173	A
Jan Mayen Island	117	A
Spitsbergen and Nordaustlandet	58,016	S
Zemlya Franza Yosifa	13,735	GK
Novaya Zemlya	24,300	GK
Severnaya Zemlya	17,500	GK
Ostrov Viktoriya	6	GK
Subtotal	2,081,616	
North America (continental)		
Alaska	51,476	A
Yukon Territory and District of Mackenzie	12,060	Bo
Alberta and British Columbia	12,820	Bo
United States (conterminous)	513	M
Mexico	11	L
Subtotal	76,880	
South America Subtotal	26,500	Ll
Europe (continental)		
Scandinavia	3,810	Sc, Ø
Alps	3,600	K
Pyrenees	33	
Caucasus	1,805	GK
Ural	28	GK
Subtotal	9,276	
Asia		
Byrranga Mts.	50	GK
Verkhoyansk Mts.	23	GK
Koryak Mts.	650	GK
Cherskiy Mts.	162	GK
Kamchatka Mts.	866	GK
Suntar Khayata Mts.	206	GK
Kodar Mts.	15	GK
East Sayan Mts.	32	GK
Altai Mts.	646	GK

(continued)

Dzhungar Mts.		956	GK
Tyan Shan		6,190	GK
Alay & Pamir Mts.		9,375	GK
Mts. in Turkey & Iran		50	GK
K'un Lun chains, incl. Nan Shan		16,700	W
Inner Tibet, incl. Aghil Mountains		9,100	W
Hindu Kush and Hindu Raj		6,200	W
Karakoram and Ghujerab—Khunjerab Ranges		16,000	W
Ladakh, Deosai, Zaskar, and Rupshu Mountains		1,700	W
Trans—Himalaya		4,000	W
Himalaya		33,200	W
Mountains SW of Salween River		7,500	W
Mountains S of Eastern K'un Lun		1,400	W
	Subtotal	115,021	
Africa	Subtotal	12	Os
Pacific Region			
New Guinea		15	
New Zealand		1,000	T
	Subtotal	1,015	
South Polar Region			
Antarctic Ice Sheet (excluding ice shelves)		12,535,000	BS
Other glaciers in Antarctic Continent		50,000	
Sub—Antarctic islands		3,000	H
	Subtotal	12,588,000	
	Total	14,898,320	

[a]A: Data on file at World Data Center A,
       Am. Geogr. Soc.;
Ba: Bauer, 1954;
Bo: H. S. Bostock (unpubl.);
BS: Bardin and Suyetova, 1967;
GK: Grosval'd and Kotlyakov, 1969;
H: Hess, 1933;
K: Klebelsberg, 1948;
L: Lorenzo, 1959;
Ll: Lliboutry, 1965, v. 2,
       p. 495—504;
M: Meier, 1960;
$\emptyset$: Østrem, 1959—60;
Os: Osmaston, 1965, table 6—11;
S: Sharp, 1956;
Sc: Schytt, 1959;
T: Thorarinsson, 1940;
W: von Wissmann, 1956.

Table 4-C

Glaciated Areas of the World (maximum Pleistocene extent regardless of age) [a]

Glaciated Region	Area (km^2)	Authority
North America		
Greenland Ice Sheet	2,295,300	g
Laurentide Ice Sheet (incl. Ellesmere-Baffin Glacier Complex)	13,386,964	p
Alaska		
Brooks Range	230,220	·p[10]
Seward Peninsula	11,872	p[5]
Mountains east of Fairbanks	500	e
Area fronting Gulf of Alaska	790,470	p[10]
Subtotal	1,033,062	
Cordilleran Canada	1,441,030	p[3]
Cordilleran Ice Sheet in northwestern U. S.	136,035	p
Separate areas in conterminous U. S., Hawaii, Mexico, & Costa Rica	220,000	e
Subtotal North America	18,512,391	
Europe		
Scandinavian Ice Sheet, west of Ural Mts. and including all British ice	6,666,708	p
The Faeroes	13,000	e
Iceland and Jan Mayen Island	142,000	e
Spitsbergen and Nordaustlandet	298,662	p, x
Zemlya Franza Yosifa	48,775	p
Alps, Pyrenees, and other separate continental areas	39,000	e
Subtotal	7,208,145	
Asia		
Ural–Siberian ice sheets (minimum estimate)	2,707,000	S
Mountains in northeastern USSR	707,000	S
Mountains in southern USSR:		
Caucasus	16,000	S
Central Asia and Kazakhstan	176,000	S
Altai and Sayan	70,000	S
Region of L. Baykal	68,000	
Southern part of Far Eastern Territory	18,000	S
Himalaya and other mountains outside USSR	189,000	
Subtotal	3,951,000	

Glaciated Region		Area (km²)	Authority
South America			
Coalescent glaciers in Chile and Argentina		700,000	e
Separate areas		170,000	e
	Subtotal	870,000	
Africa	Subtotal	1,900	Os
Australasia			
New Zealand, Australia,			e
New Guinea	Subtotal	30,000	
Antarctica			
Antarctic Continent (all ice bodies)		13,800,000	BS
Sub—Antarctic islands		10,000	e
	Subtotal	13,810,000	
	Total	44,383,436	

a : Areas that include coasts are considered arbitrarily to extend to the −200m isobath.
e : Estimate.
g : Present area of Greenland plus 5% for submerged area within 200m isobath.
p : Planimetry by R.F.Flint.
p^3: Planimetry, less 3% for nunataks.
p^5: Planimetry, less 5% for nunataks.
p^{10}: Planimetry, less 10% for nunataks.
x : If these glaciers were coalescent with the Scandinavian Ice Sheet, about 8% would be
       added to the area of the latter.
BS: Bardin and Suyetova, 1967, p. 93, item 2b.
Os: Osmaston, 1965, tab. 6—11.
S : Suyetova, 1961, p. 54.

Except for central North America and much of Europe, the extent of the areas glaciated during the last glacial age is not clearly distinguished from their extent (commonly greater) during one or more earlier glacial ages. Accordingly, only the maximum extent of former glaciers, regardless of date, is shown on most of the sketch maps mentioned. Such maps are therefore artificial in that they neglect time differences. They are nevertheless useful pending the appearance of better data.

Table 4-C lists areas occupied by glaciers when at their greatest Pleistocene extent, determined by planimeter from maps of glaciated areas on various scales. The table is not uniformly reliable. The least reliable values are those from Asia, where information is particularly scanty. Although the areas do not represent the extent of ice at any single time, values are believed to be not far from those attained during the Illinoian maximum in North America and during the Saale maximum in Europe.

For the purpose of the table a "glacial age" is defined as a condition in which ice is assumed to have covered all the territory recognized as having been glaciated at any time or times within the Pleistocene. This definition is employed because stratigraphic correlation of the various layers of glacial drift is not yet refined sufficiently to permit realistic subdivision into glacial stages on a worldwide scale. Therefore, the areal values quoted are probably somewhat greater than the actual values for any single glacial age. Despite these uncertainties the confidence limit is probably no greater than 20% for the aggregate value shown in the table, namely $\sim 44 \times 10^6 \mathrm{km}^2$.

We noted earlier that glaciers now occupy about 10% of the world's land area. In a glacial age about 29% of the same land area was ice covered. The former increase of glaciers to nearly three times their present area was brought about chiefly through the creation and growth of ice (principally as the Laurentide and Scandinavian Ice Sheets, which together occupied an area of nearly $20 \times 10^6 \mathrm{km}^2$) in high-middle-latitude regions that today possess only a little ice. In contrast, the glacier-covered area in Greenland was only about 25%, and in the Antarctic around 10%, more than the corresponding areas of today's inventory. The discrepancy results partly from temperature and precipitation values that were small, but mainly from the effect of calving into the deep water surrounding both lands, which limited the radius of any ice sheet to little more than that of the land it covered.

If we exclude the Antarctic Ice Sheet, whose area has not changed greatly, we find that the area of all other ice was greater in a glacial age than it is today by a factor of ~ 13.

Comparison of the areal positions of glaciers now and in a glacial age leads to an important generalization. Nearly all the highlands that carry glaciers today carried larger glaciers formerly, and the pattern of the former glaciers shows asymmetry like that of the existing glaciers. These relationships suggest that the configuration of the lands and that of the atmospheric circulation pattern were then much the same as those of today.

Factors in the Distribution of Glaciers

The differences in distribution of present and former ice bodies lead us to consider the factors that localize large glaciers. Contrary to a still-persisting popular belief dating from the early 19th century, glaciers are not primarily "polar." They are best developed on highlands under maritime climates. That the Antarctic Ice Sheet is centered at a pole is mainly the result of the presence of a high continent there. If the continent were centered, not at the South Pole but, say, some 40° farther north, it would still be extensively covered with ice. There is no "north polar ice cap" because the north polar area is a deep ocean incapable of supporting land ice. In the northern hemisphere the largest glaciers surround this ocean, principally in areas of maritime climate and moderate to high altitude.

From table 4-B it is apparent that, in terms of area, about 85% of the world's existing land ice resides in the Antarctic and a little less than 9% in the Greenland Ice Sheet. All the rest of the world's glaciers together constitute about 6%, by area, of the total. It follows that most of the ice —that in the Antarctic and in such north-polar areas as the northern part of Greenland—lies in those regions where reception of solar radiation is minimal, where atmospheric pressure is predominantly high, and where cyclonic disturbances are less effective in causing precipitation than is the case in the middle latitudes. The high albedo values of these vast glaciers, a measure of their ability to reflect solar radiation, is a further conservative influence acting to perpetuate the ice.

In the middle and low latitudes where such conditions do not obtain, glaciers should be more unstable, and the record of their past fluctuation suggests that this may be the case. For example, during the expansion of glaciers to their last (Wisconsin) maximum in the Great Lakes sector of North America alone, an area of more than 200,000km² was added to the Laurentide Ice Sheet within a period of 6000 to 7000 years, and within the following period of equal length an area of more than 500,000km² was lost by ablation within the same sector. We cannot be wholly sure that such changes of area, even in a large ice sheet, were controlled by climate until we can confidently eliminate the possibility that surging, unrelated to climate, was involved. However, pending clarification of the role of surges, we can still consider it likely that a very slight change in, for example, the mean temperature of the ablation season, could push a glacier in a steady-state condition over a threshold value, and trigger rapid expansion or shrinkage.

If surges are set aside for the moment, factors other than change of climate can influence the area of a glacier. An ice sheet or piedmont glacier whose advancing margin encounters a topographic barrier may respond by thickening accompanied by little further spreading. A glacier may encounter lakes or seas that check its advance by causing calving. Or a glacier may spread into a warmer, drier region, where ablation comes into equilibrium with ice supply and further spreading is checked.

Apart from the general control of ice by latitude, there is also a great difference between the two polar hemispheres as to ice cover. During glacial ages nearly all the ice, over and above that in the Antarctic and

Greenland, was built up in the middle and high latitudes of the northern hemisphere; ice in corresponding latitudes of the southern hemisphere was negligible. It seems rather likely that if the rapid growth of science in the 19th century had centered in the middle latitudes of the southern hemisphere instead of in the northern, the Glacial Theory would have developed later than it actually did.

The principal factors in this difference between the hemispheres were (1) the situations of the continents with respect to the belts of planetary winds, and (2) the situations of highlands on the continents. North America is ideally placed to receive and maintain an extensive cover of ice. Cold highlands in the northeast, with continuous transport of tropical-oceanic moisture via the Gulf of Mexico clockwise toward the northeast, afford nearly optimum conditions for glaciers, once they have formed on highlands, to expand southward and southwestward. Europe has the Scandinavian mountains on its western (windward) flank, with continuous transport of moist Atlantic air from the southwest. Scandinavian ice calved into deep water on the Atlantic side, but on the continental side flowed far eastward, aided by a new, temporary highland. This consisted of the southern margin of the ice sheet itself, which induced precipitation all along the southern front of the ice body. This European system was a good one for building ice, but was less effective than the system enjoyed by the North American Ice Sheet, whose area exceeded that of the Scandinavian Ice Sheet by a factor of 2. The difference is discussed further in chap 23.

In contrast, in narrow, middle-latitude South America plenty of moist oceanic air came in from the Pacific, but the Andes Cordillera is so high that it creates a virtual desert on its eastern side. On the west, glacier ice calved into the Pacific, but toward the east it made little headway under the prevailing dry climate with high summer temperatures.

As for Africa, that continent simply lacks high land in sufficient area to generate much ice. Glaciers occupied some high mountains, but they were hardly more than pinpoints. In addition the proportion of ocean to land in the southern hemisphere is great enough to raise temperatures above those in comparable latitudes of the northern hemisphere. This constitutes an additional factor that tends to limit glaciers.

It is worth repeating that on virtually all highlands the areas of the extended ice-age glaciers show asymmetry having the same orientation as the asymmetry of the smaller glaciers of today. From this pattern we infer that planetary wind belts, although they shifted position somewhat, had the same orientations relative to the continents in glacial ages as those they have in the contemporary world.

The Problem of Volumes

Hitherto we have been dealing with areas, which in some regions at least have been measured. But for a quantitative evaluation in terms of the hydrologic cycle we must deal with volumes, based not only on areas but also on average thicknesses of glaciers. Data on thickness of the existing Antarctic and Greenland Ice Sheets permit preliminary calculation of their volumes. On the other hand the vanished Laurentide and Scandinavian Ice Sheets present a more difficult problem, because their thicknesses cannot be measured directly. Probably thickness values could be deduced from models carefully constructed on a basis of analogy with existing ice sheets, but pending the appearance of such data we shall have to be content with estimates, recognizing that they are speculative, crude, and subject to error that is possibly large.

Starting with existing glaciers, we have in table 4-D a world water inventory, in which the aggregate volume of today's glaciers is estimated at $24 \times 10^6 km^3$, about 1.7% of the world's water. A tentative volume inventory of land ice today and in a glacial age is contained in table 4-E. On the assumption that the volumes given in it are real rather than speculative, the volume of land ice during a glacial age would have been $\sim 78 \times 10^6 km^3$, about three times that existing at present, the same propor-

Table 4-D
World Water Inventory (contemporary)

Residence	Volume $(km^3 \times 10^6)$	Authority	Percent of Total (approx.)
World ocean	1,350	MS	97.6
Rivers, lakes, and ground water	8.6	N	0.6
Glaciers (water equivalent)	24.	BS	1.7
Atmosphere	.013	N	trace
Total:	1,382.613		99.9 +

BS: Bardin and Suyetova, 1967.
MS: Menard and Smith, 1966, J. Geophys. Res., v. 71, p. 4315.
N: Nace, 1967, Envir. Sci. and Technol., v. 1, p. 58.

Table 4-E

Estimated Volume of Land Ice Today (T) and in a Glacial Age (G) (speculative and subject to substantial refinement)

1	2	3	4	5	6	7
Glacier	Time	Area (10^6 km^2)	Thickness (km)	Volume Ice	(10^6 km^3) Water[a]	Sealevel equivalent[b] (m)
Antarctic ice	T	12.53 [c]	1.88 [BS]	23.45 [BS]	21.50	59
	G	13.81 [BS]	e	26.00 [e]	23.84	66
Greenland ice	T	1.73 [c]	1.52 [B]	2.60 [B]	2.38	6
	G	2.30 [d]	1.52	3.50	4.01	11
Laurentide Ice Sheet	G	13.39 [d]	2.20	29.46	27.01	74
Cordilleran Ice Sheet	T	(negl.) [d]				
	G	2.37	1.50	3.55	3.25	9
Scandinavian Ice Sheet	T	(negl.)				
	G	6.66 [d]	2.00	13.32	12.21	34
All other ice	T	0.64		0.20	0.18	0.5
	G	5.20		1.14	1.04	3
Total	T	14.90 [c]		26.25	24.06	65
	G	43.73 [d]		76.97	71.36	197
Difference (approx.) between T and G				50.72	47.30	132

[a] Calculated as 91.7% of ice volume (Lawson, 1940).

[b] Based on area of world ocean = 362 × 10^6 km^2 (Menard and Smith, 1966, p. 4315). Not adjusted for isostatic changes.

[c] From table 4-B

[d] From table 4-C

[e] No thickness quoted; volume calculated as today's volume plus 10% (Hollin, 1962, p. 192).

[f] Exclusive of sub-Antarctic islands.

[g] Includes sub-Antarctic islands.

[B] Bauer, 1955, p. 67.

[BS] Bardin and Suyetova, 1967.

tion that is calculated for the relative areas given in tables 4-B and 4-C. The exchange of so large a quantity of water substance between the ocean and land areas obviously exerts an important influence on the level of the ocean. The significance of the exchange for sealevel is discussed in chap 12.

CHAPTER 5

Glacial Erosion and Transport[1]

GLACIATION

The process of *glaciation* is the covering of any part of the Earth's solid surface with glacier ice, or of its water surfaces with floating glacier ice or shelf ice.[2] Although we frequently speak of glaciation as a physical process, we also employ the term in a stratigraphic or chronologic sense. In this sense glaciation is a specific event, in which a particular area is overspread by ice, and is later uncovered by deglaciation. Thus we speak of the Illinoian glaciation, during which, in the Illinoian Age, glaciers formed, spread to maximum area within their valleys or other terrain, and subsequently shrank or completely disappeared.

Erosional alteration may consist of anything from inscribing minor scratches to profound excavating of valleys or denuding wide surfaces (fig 6-5); depositional alteration may range from the lodgment of a single particle of foreign rock to the building of a thick and extensive mantle of glacial sediments. By *deglaciation* is meant the process of uncovering of any area by waning glacier ice or shelf ice.

How is the former glaciation of a given area established? The features that led early scientists to the knowledge that glaciers in the Alps had formerly extended far beyond their 19th-century limits was the presence,

[1] General references are cited under the various subjects treated in this chapter.

[2] C. S. Wright and Priestley (1922, p. 134) distinguished between *glaciation* ("the erosive action exercised by Land-Ice upon the land over which it flows") and *"glacieri-zation"* ("the inundation of land by ice"). Though followed by some geologists, particularly in Great Britain, this distinction is of questionable value. *Glacier covered* is a clearer and simpler term than *glacierized* for describing areas now covered with glacier ice. As for areas formerly so covered, they must be glaciated in the sense used in the text above; otherwise the fact of former overspreading by ice could not be established. Therefore a term such as *glacierization* seems unnecessary.

far beyond the glaciers themselves, of features recognized as the work of glaciers. These features were of three kinds: (1) bosses of grooved, scratched, and polished bedrock, (2) boulders and other fragments of rock foreign to the localities of occurrence, and (3) deposits consisting of rock fragments of all sizes, from boulder size down to that of clay particles, without sorting or stratification. Although other kinds of evidence of glacial activity have been recognized since, and although more detailed knowledge has qualified to some extent the interpretation of all these features, they are still the chief basis of determining the extent of former glaciers throughout the world. In the present chapter we shall examine (1), and shall discuss (2) and (3) later. We shall begin with small erosional features and proceed to larger ones, then discuss the transport of rock waste in glaciers, and, finally, comment on quantities and rates.

ACTIVITIES AT THE BASE OF A GLACIER

From studies made hitherto, we can suppose that basal ice at its pressure-melting point is metastable and tens to melt as it receives heat resulting from friction and from geothermal heat flow. With changes in local rate of movement, topography of the subglacial floor, and number, size, and shape of rock particles in the ice, we can imagine that local stresses would change continually, and so would localized melting, refreezing, and intensity of attrition between particles and floor and between the particles themselves. But very little measurement of such relationships has been done. We do not know to what extent particles on the bed are moved (1) by scraping and dragging, (2) by being enveloped by ice during its creeping motion, or (3) by being frozen into ice created by refreezing of basal water that has flowed from other areas in the base of the glacier. Nor do we know to what extent erosion is concentrated near the margin of a glacier rather than occurring in all parts of the base of the glacier at the same time. The possibility is strong that many rock particles, instead of being quarried from bedrock by the glacier itself, are loosened by frost wedging immediately in front of the glacier, under a rigorous climate, and thus are ready for transport when encountered by the moving ice.

If erosion of rock material by glaciers is to be understood, it should be studied from two viewpoints: observation and measurement of activities at the bases of existing glaciers, and inference from erosional features exposed to view by deglaciation. Clearly the research first mentioned is by far the more difficult. Tunnels have been driven into various glaciers to expose their bases and permit the direct observation of dynamic processes. Such research, however, is still in its infancy and as yet has attained results which, although valuable, are limited. At present we must rely mainly on the exposed record of former glaciers, supported by as much theory as seems helpful.

SMALL-SCALE FEATURES OF GLACIAL ABRASION

Bedrock in wide areas of glaciated regions has been abraded by rock particles included in the base of glacier ice, quite apart from the problem of how the particles got into the ice. In most such areas the bedrock has been partly blanketed by glacial and other deposits, which conceal the ice-abraded surfaces, but usually enough are exposed to permit determination of the extent of glaciation. However, as mentioned hereafter, mechanical wear of rock on rock is caused also by natural processes other than glaciation. Some of the resulting abraded surfaces are so much alike that the process responsible is not always identifiable. Hence other kinds of evidence must be sought before glaciation can be established.

Striations, Polish, and Small Grooves [3]

The most common and conspicuous unit of glacial abrasion (fig 5-1) is the striation (stria or scratch). Chamberlin (1888, p. 216) described striations succinctly as "fine-cut lines on the surface of the bedrock which were inscribed by the overriding ice." The description is not quite accurate in that it is improbable that ice itself can make striations. This is because its scratch hardness at $0°$ is about 2 and at $-44°$, about 4—less than that of most rocks. Chamberlin's implication was that scratches are made by rock particles held in the base of a glacier. The prevailing angularity of the fine particles carried by glaciers (fig 7-11) helps to explain the sharp definition of most striations.

Probably the finer scratches were made by particles of sand and silt; with decreasing diameter the scratches grade into a general polish, whose smoothness and brilliance are limited only by the fineness of the abrading grains of silt and by the textural and mineralogic properties of the abraded rock. Most polished surfaces also bear larger scratches, because rarely do all the fragments in the base of the glacier lie within a single narrow range of grain size. Very fine striations generally appear only on fine-grained and mineralogically soft rocks such as limestones and shales. On harder and coarser grained rocks such as granites and coarse-grained volcanics only the coarser striations, probably made by pebbles, are

[3] General references: Chamberlin, 1888; Carney, 1909a; Edelman, 1949.

Figure 5-1   Glacial striations and polish on Ventersdorp Diabase, with overlying Dwyka Tillite (Permian), Harrisdale, South Africa. (R. F. Flint.)

found. However, in some places veins and other bodies of quartz in such rocks retain very fine striations, some of which are visible only under a hand lens, because of the resistance of quartz to weathering.

With increasing diameter, striations grade into grooves, giant furrows observed commonly only in soft rocks such as carbonate rocks and claystones. The grooves appear to have been made by the fortuitous enlargement of single striations as angular pieces of rock carried in the base of the ice gouged them deeper. Ice appears to have molded itself to fit the slight depressions thus made, further localizing abrasion. Individual grooves reach depths of 1 to 2m and lengths of 50 to 100m. Many have overhanging sidewalls, and on the undersides of some of the overhanging walls are fine striations parallel with the groove. A famous locality for grooves is Kelleys Island, Ohio, where a broad, flat-lying surface of limestone subjected to vigorous glaciation afforded conditions ideal for striation and grooving. Giant grooves reaching depths of 30m and lengths of 1.5km, but without overhanging sidewalls, occur in the Mackenzie River valley in northwestern Canada. These suggest gradation between small-scale features and the much larger ones described elsewhere in this chapter.

Striations and grooves are not confined to bedrock. They have been observed on compact loess and, on the Plains in Alberta (Westgate, 1968b), they occur on the surfaces of contact between sand and overlying till, with the till squeezed into the grooves to form molds. Orientations of the features are compatible with those of other directional parameters in the vicinity.

The relation between striations and the processes that together constitute the movement of glaciers is little known. Striations could be caused by bodily sliding of a glacier, but some diverge around minor obstacles, implying that in these, at least, flow played a part. In either case a single rock particle imbedded in the base of a glacier might not produce a scratch if pressure melting caused it to be retracted upward into the ice. But the presence of many particles of various sizes, in mutual contact, commonly visible in basal ice in a glacier terminus, should act against retraction and make possible individual striations of great length.

Striations and Direction of Movement

In themselves, striations are not criteria of glaciation because they can be made by agencies other than glaciers. A nonglacial agent common in high latitudes is floating ice in rivers, lakes, and the sea (I. C. Russell, 1890, p. 117–120; Dawson, 1893, p. 105–110; A. L. Washburn, 1947, p. 47–48; Clayton et al., 1965; Schulz, 1967). Indeed, this fact was recognized in both Europe and America even before the glacial theory was thought of. At that time striations were attributed to the action of floating Arctic ice during a universal submergence of the lands. In certain districts in Arctic Canada, much of which was submerged at the end of the last glaciation, it is difficult to distinguish between glacier-ice scratches and floating-ice scratches on bedrock. In general, floating-ice striations on a single exposure of bedrock are shorter and parallelism among individuals is far less than in glacial striations. Also there is less consistency of orientation from one exposure to another within a given district.

Apart from the striationlike features of slickensides and wind-cut flutings, true striations have been made also by snow avalanche, by rock avalanche, by debris flow, and most remarkable of all, by flows of fluidized tephra down the slope of a volcanic cone. Because striations can be made by almost any flowing or floating heavy mass, reasonable evidence of glacial origin must be established before a striation is interpreted as glacial.

In every glacial striation it is possible to determine *orientation*, but only in a few instances is it possible to determine *sense* from evidence furnished by the striation itself. Features diagnostic of sense are rat-tail ridges streaming from protuberances in the bedrock (fig 5-2).

Figure 5-2 Glaciated surface with whaleback form, cut in conglomeratic sandstone. View looking south, parallel with striations and grooves. Long "rat tails" stream southward from pebbles of quartz, where the finer-grained matrix was protected by the hard pebbles. Although not themselves striations, the rat tails are diagnostic of southward sense of all the linear features in the view. Broad gouges parallel with hammer handle were made by teeth of power shovel. Scattered stones represent a lag concentrate, by recent rain, from overlying till. East Haven, Connecticut. (R. F. Flint.)

The asymmetry of some striations, which are blunt at one end and taper toward the other, does not surely indicate direction of flow. Unsymmetric striations on slopes that face upstream [4] lie with their blunt ends in the downstream direction, apparently as a result of gouging, whereas most of those on lee slopes are blunt at the upstream end, apparently as a result of quarrying (H. C. Lewis, 1885, p. 557). Again, striations having two or more orientations occur within a single limited area and even crossing each other on the same exposure of bedrock.

All this might lead one to conclude that striations, as individual features, are of very limited value as direction indicators. Nevertheless detailed studies applied to large groups of striations and employing simple statistical treatment have proved successful, especially where combined

[4] Upstream-facing slopes are also referred to as *stoss* slopes (Ger. *Stoss* impact or thrust).

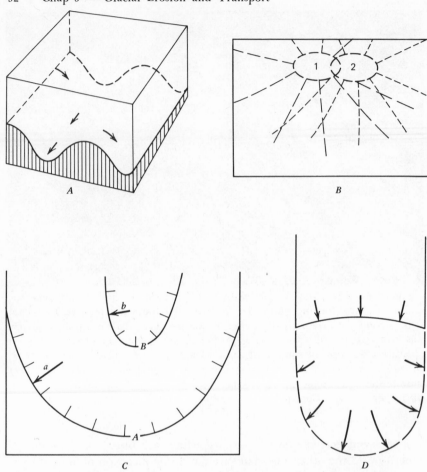

Figure 5-3 Possible ways by which divergent striations within a single area can be made. *A*. Simultaneous divergent flow controlled by topography. *B*. Shift in center of outflow through time. *C* and *D*. Change in direction of local flow during deglaciation on land (*C*), and with change to water environment (*D*).

with the measurement of other directional indicators. Some of the chief results of such studies are given below (fig 5-3).

1. Even very small topographic differences can cause local divergent flow in a thin glacier. In a thick glacier overlying a dissected highland, basal ice can flow along deep valleys while ice higher up in the glacier, unimpeded by topography of the bedrock, can flow across and over interfluves.

2. Direction of flow along a radius of an ice cap or ice sheet can change through time, owing to shift in the nearest center of outflow of ice.

3. Local direction of flow near the margin of a glacier lobe can change, through time, with change in thickness and radius of curvature of the lobe margin.

4. Flow near a glacier terminus can change from divergent to convergent directions as the environment changes from land to water and the terminus changes from convex to concave.

Figure 5-4 illustrates results obtained from measurement of 1300 striations and other directional indicators in an area of 1200km². Three different movements are indicated, all apparently related to a single glaciation, the latest one. Schulz (1967, p. 135) compiled striation orientations for three glacial ages in a part of central Europe, although there could be doubt, in some cases, as to which age is represented. Probably most of the striations recorded on glacial maps relate to the latest glaciation of the relevant district, simply because the probability is good that a striation made during one glaciation will be erased during the next. Another reason is that unless protected by burial beneath glacial or other sediments, glaciated surfaces of bedrock become so weathered, at least under a humid climate, that striations are soon lost. For example, in south-central Connecticut (ann. precipitation >1150mm) the majority of natural exposures of basalt, diabase, and coarse arkosic sandstone which possess a generally glaciated form show weathering but no striations. Yet the same rock types, where artificially stripped of overlying sediment, display fresh striations and other marks of glacial abrasion. The surface in fig 5-2 was striated at least 14,000 years ago, buried, and re-exposed recently. The striated surface in fig 5-1 has been protected beneath a thick overburden through more than 200 million years.

Striated surfaces are seen most clearly after a rain. A few minutes' work with a whiskbroom followed by a bucketful of water can reveal fine scratches invisible under dry, dusty conditions and prepares the surface for a photograph.

Marked local variations in distribution of striations occur. At Kelleys Island, Ohio, stripping operations removed till overlying limestone bedrock and exposed a smoothly polished area, about 250m long, between two rough areas bearing no trace of glacial erosion. Beyond one of the rough areas is a larger surface showing deep grooves and striations. All three areas lie in nearly the same plane. The basal ice is thought to have become so heavily loaded with rock fragments that it stagnated in places, the ice above flowing over it and eroding the bedrock beyond (Carney, 1909b, p. 640).

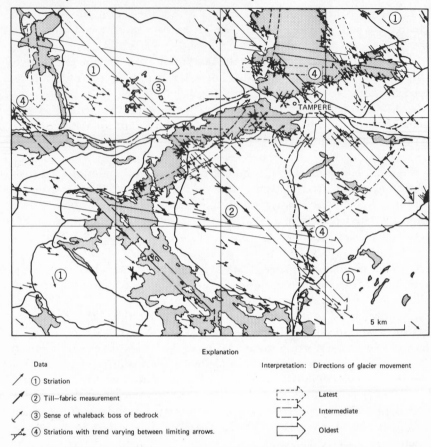

Explanation

Data   Interpretation: Directions of glacier movement

/ ① Striation

/ ② Till—fabric measurement   Latest

/ ③ Sense of whaleback boss of bedrock   Intermediate

/ ④ Striations with trend varying between limiting arrows.   Oldest

Figure 5-4   Map of the Tampere area, southwestern Finland, showing three successive directions of glacier movement (not separated in time by deglaciation), inferred from the features listed in text. (Virkkala, 1960.)

Plotted on a continental scale, striations are distributed very irregularly. The spotty distribution shown on the Glacial Map of the United States East of the Rocky Mountains (Flint et al., 1959) and on the Glacial Map of Canada (Prest et al., 1968) is the result of a number of highly variable factors such as: (1) lithology, (2) destruction by postglacial erosion, chiefly weathering, (3) concealment beneath a blanket of drift, (4) debris content of the glacier ice, and (5) variable observation and mapping. Countless numbers of striations exposed to view on the resistant rocks of the Canadian Shield have never been recorded or even observed by geologists.

Summarizing the foregoing discussion, we conclude that (1) not all natural striations on bedrock are of glacial origin, (2) few striations indicate sense of glacier motion, (3) strong divergences in the trends of striations are produced by lobation of glacier ice and by topographic irregularities, and (4) probably most of the striations that have been mapped were made near glacier margins during deglaciation and therefore do not indicate the directions of flow that characterized glacial expansion. All this adds up to the opinion that striations must be interpreted with a great deal of judgment. A few examples mean little, but statistical analysis of many hundreds of striations, accurately mapped, yields at least a general idea of the movement of the ice concerned.

Crescentic Marks

In some places the striations and grooves on surfaces of hard, brittle bedrock such as quartzite are accompanied by crescentic marks (Gilbert, 1906a; S. E. Harris, 1943). Such marks are also observed on large clasts, possibly derived from former boulder pavements. Four distinct kinds have been recognized, three of which are shown in fig 5-5. All are alike in

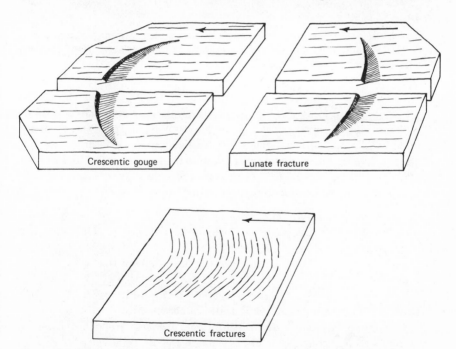

Figure 5-5 Crescentic gouge, lunate fracture, and crescentic fractures shown diagrammatically. Arrows indicate direction of movement of glacier.

Figure 5-6 Striations, crescentic gouges (center and right) and crescentic fractures (center and left) on surface of granite, Mount Desert Island, Maine.
Former glacier moved away from the observer. (F. D. Larsen.)

that they are commonly lunate, are oriented at right angles to the direction of movement, mostly range in length from 10 to 12cm and in depth
from 10 to 25mm, and are generally arranged in nested series (fig 5-6)
varying in length up to 30cm, with component marks spaced at intervals
of 2 to 4cm. Long axes of the series commonly parallel striations on the
same surface. However, the four kinds differ in these ways:

1. *Crescentic gouges* are concave upstream and consist of two fractures,
from between which the rock has been removed. Gouges 1m long have
been recorded.

2. *Lunate fractures* resemble crescentic gouges except that they are concave downstream.

3. *Crescentic fractures* are concave downstream and consist of a single
fracture, without removal of rock. Individuals 40cm long are known.

4. *Chattermarks* occur only within large grooves, are concave downstream or are not concave at all, are made by the removal of a chip of
rock, and have no fracture that extends deeper than the scar left by removal of the chip.

It has been asserted that in the three kinds of marks showing fracture, the principal fracture dips in the downstream direction. However, many crescentic marks dip upstream (Flint, 1955b, p. 59; Dreimanis, 1953). In summary, neither the direction of concavity of a mark nor the dip of its fracture seems to be reliable evidence of direction of movement.

Evidently the marks are made by sizable rock fragments held in the base of a glacier and moved across the surface of bedrock or of a boulder pavement. Experiments made by moving steel spheres under pressure, across glass, suggest that the direction of concavity of a crescentic fracture may be determined by whether the engraving rock fragment is sliding or rolling. A sliding sphere produced crescentic fractures as defined above; a rolling sphere produced fractures concave upstream. Possibly therefore all known crescentic fractures were made by sliding tools.

LARGE ICE-FLOW FORMS

Whaleback Forms; Stoss-and-Lee Forms

In figs 5-1 and 5-2 the abraded bedrock surfaces are not plane but curved. Indeed, glaciated surfaces are plane most commonly where stratification, sheeting, or joints created the planes prior to glaciation. Glaciers, by abrasion of bedrock, apparently tend to create curved surfaces. Such surfaces (Agassiz, 1840, atlas, pl. 6, 8, 9, 13, 15, 17) are difficult to describe. In their simplest form they are bosses and hillocks of smooth whaleback form having lengths greater than widths and merging laterally into smooth intervening troughs. In a group of individuals there is little uniformity of length, height, angle of side slope, or ground-plan pattern (fig 5-7). The greatest approach to uniformity lies in orientation of long axes, which tend to be subparallel in the direction of flow of the glacier. Heights range from less than 1m to tens of meters. Clearly lithology and rock structure play a significant part in the dimensions of these features, but the mechanics of moving glacier ice must play a dominant role in the creation of whaleback forms which are evidently of streamline character, as mentioned under *drumlins* in a later section. As yet, little attention has been given to analyzing the forces involved and no definitive study can be cited. However, tunnels have been driven through various glaciers to encounter bedrock floors beneath, and systems of measurement have been undertaken.

Families of these whaleback forms were recognized in the Alps by Saussure (1786–1796, v. 4, § 1061), who described the rippled, glistening effect produced by a whole series of them (not the individual bosses) as "roches moutonnées," in fancied resemblance to contemporary wigs

Figure 5-7   Whaleback forms in granitic bedrock, Mono Pass, Sierra Nevada, California. (F. E. Matthes, U.S. Geol. Survey.)

slicked down with mutton tallow. As the term has been widely misapplied and mistranslated, it is not used in this book.

Some whaleback forms and also hills of larger size and apparently preglacial origin possess pronounced asymmetry, with steep sides facing in the downstream direction of former glaciation. Many, though not all, of these forms are glacially abraded on their upstream sides. The downstream sides may form cliffs or series of cliffs, coincident with joints or other structures in the bedrock; they are rarely abraded. Such cliffs are the work of mechanical erosion, termed *glacial quarrying* (or glacial plucking) by analogy with the tearing off and lifting out of blocks in a stone quarry. The quarried product, in the form of cobbles and boulders, some of which may reach several meters in diameter, still lies in the lee of some of the cliffs.

The asymmetry of these features, and the systematic distribution of abrasion and quarrying on them, affords evidence of direction of glacier movement. It indicates which of the two directions recorded by the striations was taken by the ice. The persistently asymmetric arrangement of bedrock bosses and small hills in a strongly glaciated district, each hill having a comparatively gentle abraded slope on the upstream (stoss) side and a somewhat steeper and rougher quarried slope on the lee side, has been termed *stoss-and-lee topography* (fig 5-8).

Because of their pronounced asymmetry, stoss-and-lee hills are more reliable guides to the direction of glacier motion (at least within 10° to 15°

Figure 5-8 Stoss-and-lee form, showing both abrasion and quarrying, in coarse-grained volcanic rock, Aquarius Plateau, Utah. Erosion was the work of an ice cap of Wisconsin Age, which at this place flowed from left to right. (R. F. Flint.)

of arc) than are most striations, and also withstand postglacial erosion through a longer time.

Such topography is ideally developed in the Scottish Highlands in Ross Shire, through a distance of nearly 50km from The Minch on the west to the head of the Moray Firth on the east. On both flanks of the Highlands the steep sides of the rock bosses face outward, but in the high central part they are abraded on all sides and exhibit no pronounced asymmetry. This suggests that the ice sheet centered on the Highlands shifted its center of outflow from time to time, though within moderate limits.

The mechanism of glacial quarrying is still unknown. One theory proposes "mechanized quarrying," consisting mainly of frictional drag and adhesion as ice passes over bedrock. Another appeals to frost wedging in openings in bedrock beneath (Carol, 1947) or immediately in front of the glacier (Yardley, 1951). Blocks loosened in this manner are then included in the moving glacier and carried away. Still another theory advocates release of pressure immediately downstream from a bedrock protuberance in the subglacial floor, thereby actuating the release of stresses inherent in the rock with consequent cracking, and removal of the debris. (W. V. Lewis, 1954).

Whatever the mechanism of quarrying, it appears that whereas the orientation of abrasional features is determined primarily by direction of movement of the glacier, that of quarried features is determined primarily by the distribution of susceptible rocks. In both cases a secondary factor is local topography.

Streamline Molded Forms

Some glaciated areas having low or moderate relief are characterized by groups of ridgelike and groovelike forms that impart a fluted pattern to the surface. Although their sizes and shapes vary greatly, these forms belong to a single family in that all are products of the streamline flow of glaciers, which molded the subglacial floor through erosion, deposition, and various combinations of both processes.

This family of features, which we can label *streamline molded forms* or simply *streamline forms,* includes alternating negative and positive elements. The negative elements constitute a gradational series ranging from simple striations through grooves and flutings of increasing size. The positive elements constitute another series, including the features known as *drumlins.* They merge into hills, at least partly shaped by ice, having a variety of irregular forms. Both ridges and grooves range in composition from 100% bedrock to 100% glacial drift. Those consisting of bedrock are entirely the work of erosion. Those made largely of drift may be the work of accumulation and erosion combined.

The family as a whole, therefore, is difficult to define. The one characteristic its members share in common is streamline form, free of discontinuities. Such forms offer minimum resistance to fluids or fluidlike substances flowing past them, as illustrated by the fuselages and wings of airplanes, the hulls of ships, and the bodies of racing automobiles, and are an expectable product of erosion and deposition by actively flowing glacier ice. The resemblance between drumlins and airfoils is elaborated by Chorley (1959).

Drumlins. The classical members of the family are drumlins. The word is Irish (Gaelic *druim,* the ridge of a hill), and was first applied to streamline hills by M. H. Close (1867). It is now common to many languages. Drumlins are widely distributed in both North America and Europe and, though conspicuous forms constitute rather distinct fields, less ideally shaped drumlins are present in intervening districts where they have escaped attention. Among the conspicuous fields in North America are those in central-western New York (about 10,000 drumlins), east-central Wisconsin (about 5000), south-central New England (about 3000, many of which consist of rock), and southwestern Nova Scotia (2300). Large groups of remarkably long narrow forms occur in various parts of

Figure 5-9 "Ideal" drumlin molded in glacial drift, near Madison, Wisconsin. View over part of the drumlin field shown in fig 5-10. Effective glacier flow was toward the southwest. (C. C. Bradley.)

the Great Plains in Canada and northern United States (e.g., Lemke, 1958). Some are chiefly bedrock; others are chiefly drift. Possibly such forms outnumber "conventional" drumlins (fig 18-7).

The "ideal" drumlin approximates half an ellipsoid in form, like the inverted bowl of a spoon, with the long axis paralleling the direction of glacier flow and the transverse axis situated upstream from the midpoint of the long axis. Most "ideal" drumlins (fig 5-9) fall within these dimensions: length 1 to 2km, width 400 to 600m, and height 5 to 50m. Within a single field the individuals tend to be rather similar. B. Reed et al. (1962) dealt with the geometry of form, orientation, and spacing of selected drumlins; Aronow (1959) discussed some of the many local controls involved.

"Ideal" drumlins constitute a standard from which individual units depart widely. Proportions range from nearly circular to a length/width ratio of 50:1. Single isolated forms are rare; usually drumlins occur in groups of scores, hundreds, or even thousands, each group constituting a distinct field. In some places adjacent drumlins are separated from each other only indistinctly. They form "double or triple ridges united at the steeper end with the tails only distinct; doublets en echelon, the tail of one rising from the flank of the other as an inclined terrace or shelf; small drumlins plastered on the side of larger ones, giving a grooving effect to the flanks of the latter; two-tiered drumlins . . ." (Hollingworth, 1931, p. 326). Despite these variations, however, long axes still parallel the direction of flow. This is strikingly evident where radial flow occurred in the terminal part of a glacier lobe (fig 5-10).

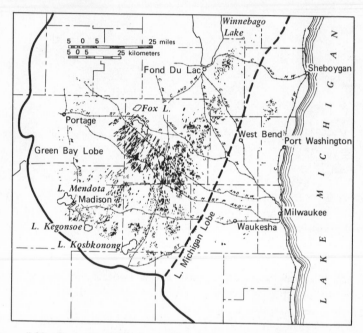

Figure 5-10 Drumlin field, Green Bay lobe of Wisconsin drift, southeastern Wisconsin. Radial flow within the former glacier lobe is indicated by diverging axial trends of drumlins. The east-west trend of the small group west of Milwaukee reflects westward flow in the adjacent Lake Michigan glacier lobe. (W. C. Alden.)

Drumlins have been locally reported to be more numerous and more fully formed on regional slopes facing upstream than on those which face downstream, but whether this is true generally is not known.

Drumlins whose composition is known, at least in part, consist predominantly of clay-rich till, packed closely so that it is distinctly tough. However, in a few drumlin fields the till is sandy. Other drumlins consist partly of sorted and even stratified drift, overlain by till around most or

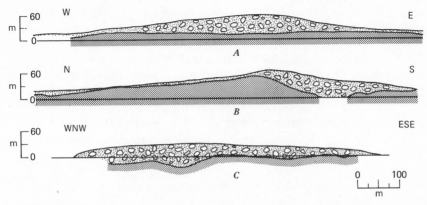

Figure 5-11 Sections through three drumlins showing bedrock surface beneath till. Horizontal and vertical scales are identical. The irregularities in the surfaces of the drumlins are typical. *A.* Oblique section through Mt. Ida, Newton, Massachusetts. The drumlin trends northwest-southeast. Bedrock surface reconstructed from borings and a tunnel. (Redrawn from I. B. Crosby.) *B.* Transverse section through Parker Hill, Roxbury, Massachusetts. Bedrock surface reconstructed from borings and a tunnel. (Redrawn from I. B. Crosby.) *C.* Long section through Governors Island, Boston Harbor, Massachusetts. Steepening of the surface profile at the two ends is the result of erosion by waves. Bedrock surface reconstructed from seismic soundings. This drumlin was later destroyed during construction of an airport. (Data from F. W. Lee.)

all of the surface (e.g., Slater, 1929). Many drumlins have cores of bedrock, commonly at or near their upstream ends (fig 5-11); at least 25% of ~200 drumlins in the Boston, Massachusetts, district have been estimated to have rock cores (B. Reed et al., 1962, p. 205). Others consist almost entirely of bedrock discontinuously veneered with till. Related features are the *crag-and-tail* units (fig 5-12), each consisting of a hill of bedrock with a long sloping tail of drift streaming behind it. Although not

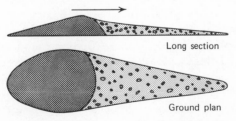

Figure 5-12 A common form of crag-and-tail. Lengths range from a few meters to 2km or more.

classed as a drumlin by most writers, crag-and-tail would be difficult to distinguish from a drumlin with a large rock core at its upstream end, if the core were veneered with till.

If we attempted to arrange drumlins and drumlinlike features in a logical sequence based on composition we should have something like this:

BEDROCK: whaleback bedrock forms including stoss-and-lee forms → drumlins of bedrock with veneers of till → crag-and-tail → drumlins with bedrock cores → drumlins with cores of stratified drift → drumlins consisting of till only: DRIFT

There seems to be complete gradation, independent of outward form and within a single field, from rock to drift. This suggests that any one group was molded contemporaneously under a single set of controls, in which both erosion and deposition could operate across some sort of threshold condition, to add or subtract material locally and thus achieve a streamline interface between flowing ice and the ground beneath.

Smalley (1966; 1968) suggested a model in which flowing ice, basal load, and underlying ground are considered as separate units. The basal load is a thin, mobile layer of saturated clay and larger particles. The clay-water system is both dilatant and thixotropic. When expanded it flows readily with the glacier, but when an obstacle is encountered and behind it flow stress declines below a critical value, the dilated material tends to pack tightly and stabilize. Meanwhile ice and its thixotropic load continue to flow around the obstacle where local stresses have increased. In this way local packing can cause cumulative local accretion, while flow continues nearby. The form of the ground becomes streamlined and local stress differences become minimal. When flow ceases at the close of glaciation, dispersed clay then in the basal load is soon washed away, leaving the tightly packed clay as a streamlined hill.

One might add that where the basal load is deficient in clay, flowing ice should be capable of grinding bedrock hills to streamlined form through direct abrasion with silt, sand, and coarser particles. Close comparison of bedrock lithology, till lithology, and composition of streamline hills in selected areas should afford tests of these ideas.

Fluted Surfaces. In North Dakota, Montana, and adjacent parts of Canada, fields of straight parallel grooves with intervening ridges form fluted surfaces. These features (fig 5-13) range up to 20km in individual length, up to 100m in width, and up to 25m in height, although average dimensions are much less, some grooves being no more than 2m deep. Fluted surfaces, like drumlins, occur in fields and are developed in bedrock, in till, and in stratified drift. Some of those in till are separated by

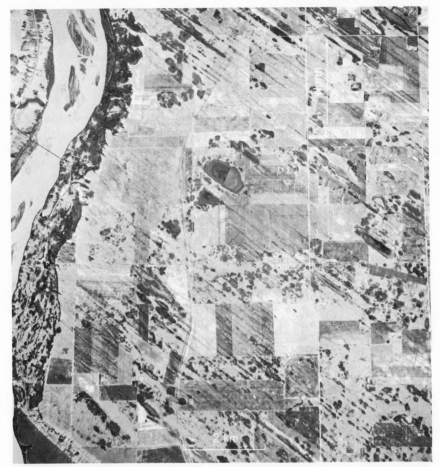

Figure 5-13   Fluted surface of till sheet, Battleford, Saskatchewan. As relief is estimated at less than 2m, and slopes of ridges are very gentle, it might be difficult to identify these features from a position on the ground. The flutings are believed to represent very late, localized flow toward the southeast. (Canadian Armed Forces.)

ridges that have massive knobs of bedrock at their upstream ends and that are essentially very elongate drumlins. Those in bedrock are developed best in soft or mechanically weak lithology and are essentially independent of bedrock structures.

These forms parallel the local direction of glacier flow as determined from other evidence. They constitute a series which in one direction grades into drumlins. Indeed some drumlins themselves have grooves and flutings on their sideslopes.

Two general inferences can be drawn from streamline molded forms.

1. The presence of such forms establishes the existence of an actively flowing glacier at the time of formation, in contrast with ice-contact drift, which indicates comparatively inactive ice.

2. The long axes of streamline forms are a more reliable indicator of the general direction of movement of a former glacier than are striations, because they are less influenced by local topography (fig 18-7).

TRANSPORT

Data from Existing Glaciers

Rock particles in glaciers constitute glacial drift in transport. They are spoken of individually as *inclusions* and collectively (especially in quantitative statements) as the *load* of the glacier. The load is concentrated mainly at the contact between glacier and bedrock (fig 5-14) as is seen where the basal parts of glaciers are exposed at their termini or along their lateral margins. Such concentration reflects the chief source of the rock material, the ice-rock interface where abrasion and quarrying are in progress, and is analogous to the bed load of a river. Drift tends to accumulate also on the upper surfaces of some glaciers (though obviously only in the area of net ablation) and within the bodies of glaciers close to their termini. With these exceptions most glaciers consist of nearly pure ice, free of drift.

A part of the drift on the upper surfaces of valley and piedmont glaciers consists of rock fragments avalanched from steep valley sides; it is recognizable through its lithology and because most of the fragments are not modified by glaciation. Another part consists, in favorable places, of windblown silt, and even fine-grained tephra that blanket some glaciers in the vicinity of volcanoes. But the bulk of the surface drift originates as does the drift within the glacier: near the terminus it is carried upward along the oblique shears shown in fig 5-14, is released from the ice through ablation, and accumulates on the surface of the glacier.

Distinct, elongate accumulations of drift on the surfaces of glaciers constitute one class of moraines.[5] On a simple valley glacier the moraines occur along its lateral margins, at the outcrops of steeply dipping, drift-rich zones within the ice. These are *lateral moraines*. On a compound

---

[5] Moraines are discussed more fully in chap 7; for surface moraines see V. Okko, 1955, p. 27–35.

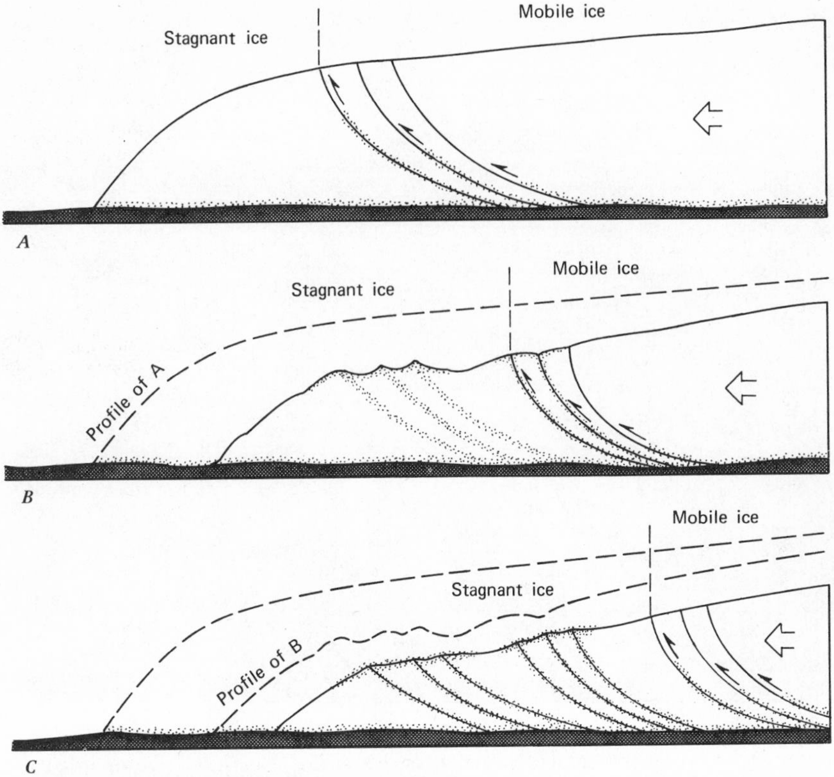

Figure 5-14 Radial section of ice sheet at a locality in northwest Greenland at three successive times. A narrow terminal zone is so thin that it stagnates, impeding the flowing ice upstream, which shears forward over it. Rock particles are brought to the surface, where they spread out. As deglaciation proceeds, superglacial drift slides, is flushed down, or is let down onto the ground. Not to scale. (After B. C. Bishop, 1957.)

valley glacier, that is, one fed by two or more tributaries, junction of two lateral moraines creates a *medial moraine* downstream from the junction (fig 5-15). Tributary units and their moraines maintain their identities as they flow in close contact down the main valley. Near the terminus, ice shearing upward and forward brings drift to the surface where it can spread out (fig 5-14). Both kinds of superglacial moraines have a ridge-like form, but very commonly this form consists of a mantle of drift concealing a ridge of ice. The belt of drift protects the ice beneath it from rapid ablation which, however, lowers the adjacent surface of clean ice. As the ridge steepens, the drift slides away, exposing the ice beneath. In

Figure 5-15 Barnard Glacier, Alaska, showing relation of lateral and medial moraines to tributary glaciers. Ablation moraine, possibly accompanied by stream deposits, is developing along sides of main glacier. The high peak is Mt. Natazhat (4084m) on the Alaska-Yukon boundary, 40km distant. (Bradford Washburn.)

this way, little by little, the drift cover spreads and the moraine widens. Some ridges of this kind exceed 40m in height.

A glacier having several medial moraines is likely to be transporting a volume of drift sufficient to cover the surface in the terminal zone almost completely. Here, ridging of the ice through the protective action of surface debris becomes complex, and the resulting lateral sliding of rock fragments aids their spreading over the surface. Areas of ice not coated with debris melt down to form basins that fill with meltwater and accumulate sediment, only to be destroyed. Individual medial moraines lose their identity and merge into an irregular superglacial veneer of rock fragments, referred to as an *ablation moraine*. (For discussion, see V. Okko, 1955, p. 52–57.) The component assemblage of rock fragments constituting the various moraines described above is *superglacial drift* or *ablation drift*. We use the latter term for drift of this origin that remains after the glacier ice has melted away from beneath it.

Superglacial drift occurs on both valley glaciers and ice sheets. Probably most of it accumulates in the manner shown in fig 5-14. Possibly some of it was quarried from hills that project into a glacier without being high enough to break the surface of the ice. Quarrying at such places would add to the drift within the ice, and the drift would reach the surface ultimately by ablation.

Although there is some identity of material, there is little or no identity of form between superglacial moraines and the moraines deposited on the ground and described in chap 17. In most glaciers not only is a large part of the drift deposited on the ground from the base of the ice rather than from its top, but continuous ablation and sliding repeatedly rework superglacial drift. The forms of the final moraines are determined by movement of the ice and by the positions of the glacier margin at the times of final deposition (e.g., R. P. Goldthwait, 1951).

The appearance of glacier ice covered with ablation drift is one of chaotic ridges, knolls, and depressions, well described by Tarr (1909, p. 85–88), Mannerfelt (1945, p. 12–16), Sharp (1949b), and Ward (1952). This relief, however, is mainly in the ice surface itself and not in the drift. After the drift has been let down onto the ground the relief is reduced or eliminated, particularly where the deposit consists chiefly of till. Where the material was reworked into stratified drift in the presence of the last wasting remnants of the glacier ice, the surface form is likely to be one of knolls and depressions.

Repeated slumping and sliding of drift on slopes of melting ice affords so much opportunity for washing by meltwater that much superglacial drift consists only of the coarser rock fragments originally present in the drift, the finer elements having been flushed away. In a similar manner stratified drift may be deposited on thin ice and subsequently let down onto the underlying surface. The ice must be very thin so that the process of collapse does not destroy the stratification.

So complete is the coating of ablation drift over parts of the margin of the Malaspina piedmont glacier in coastal Alaska that an extensive forest of alder and spruce has grown up on it. Elsewhere slump reveals the presence of glacier ice two or three meters beneath the surface. An extreme form of coarse ablation drift may be represented by some rock glaciers (chap 10).

Data from Deposited Drift

Significant inferences drawn from deposited drift are discussed in chaps 7 and 8, but some observations pertinent to an understanding of glacial transport are mentioned here.

One observation is that glaciers carry only a minor proportion of their load, as coarse fragments, far beyond the places where it is picked up; there is a generally close relationship between the composition of coarse fragments in the drift and that of the local bedrock. This relationship is compatible with the concept of glacier ice loaded with drift only in the basal zone. The great bulk of the ice has little opportunity to acquire a load. In loaded basal ice attrition is rapid; most particles are crushed and are worn down to small sizes within a short distance of travel.

A small proportion of the rock matter picked up does, nevertheless, travel long distances. Stones and boulders from Scandinavia and Finland were carried in the Scandinavian Ice Sheet through hundreds of kilometers to points in Britain, Germany, and Poland, and (1250km) to Russia. Stones from Ontario were carried by the Laurentide Ice Sheet as much as 1000km to positions in Missouri. Most such stones and boulders consist of durable rock types containing hard, resistant minerals, and with few joints or other surfaces of weakness. They may have survived long-distance travel in the base of the ice at the expense of considerable loss of size by attrition, or may have traveled in englacial positions where there were few other rock fragments to abrade them.

A second observation is that some rock fragments, surviving long enough to be carried great distances, are lifted through considerable heights above their inferred places of origin in the bedrock (table 5-A). Such lifts, of course, are determinable only where the ice was flowing against the slope of the land.

Presumably the vertical component of transport results from oblique upward shear in the zone of net ablation. In some of the examples cited in the table the critical stones are confined to the upstream-facing slopes of mountains. Such distribution suggests that the stones were dragged upslope while embedded in basal ice rather than having been emplaced from positions higher within the glacier.

QUANTITATIVE ASPECTS

Abrasion and Quarrying Compared

Abrasion, the scour of rock by rock, is evident not only in the striated, polished, and fluted surfaces common to most glaciated regions but also in the large quantities of fine sand and silt, "rock flour," visible in the streams that drain living glaciers and in the deposits of former glacial streams. Obviously the grinding of bedrock into fine particles is an important element in glacial erosion. Quarrying, however, is evident in the steplike profiles of the floors of some glaciated valleys and of the lee slopes of glaciated hills and bosses, as well as in the coarse clasts that

Table 5-A
Examples of Vertical Movement of Drift in Glacial Transport

District of Occurrence	Approximate Minimum Transport Distance (km)	Approximate Minimum Vertical Lift (m)	Authority
Mount Katahdin, Maine	18	1,000	A. S. Packard, 1867, Boston Soc. Nat. H., Mem. 1, p. 239.
Adirondack Mountains, New York	100	900	(Secondary source)
Allegheny Plateau, central New York	160	500	(Not published)
Killington Peak, Green Mountains, Vermont	80?	900?	Ed. Hungerford, 1868, Am. J. Sc., 45, p. 4.
Presidential Range, New Hampshire		1,200	R. P. Goldthwait, 1940, GSAB 52, p. 2016.
Kittatinny Mountain, New Jersey	5 or 100	400 or 160	G. F. Wright, 1911, Ice age in N. Am., 5th ed., p 247–249.
Shamattawa River district, Manitoba	240	150	J. B. Tyrrell, 1913, Ont. Bur. Mines, v. 22, p. 199.
East base Rocky Mountains, Alberta	1,300	1,300	G. M. Dawson, 1896, GSAB 7, p. 37–38.
Porcupine Hills, Alberta	800	1,060	A. Stalker, 1960, Geol. Surv. Can. Mem 306, p. 71.
Mackenzie and Franklin Ranges, Northwest Territories	600	1,200	M. Y. Williams, 1923, Can. G. S. Sum. Rp, 1922, p. B71.
Wicklow Mountains, Eire	15	400	(Secondary source)
Moel Tryfaen, Caernarvonshire, Wales	10	400	(Secondary source)
Northern Germany	1,000	400; probably 900	V. Milthers, 1950, Geol. För. Stockholm, Förh, 72, p. 257; P. Woldstedt, 1952, Peterm. Mitt. p. 269.

make up a small but conspicuous element in the deposits of glaciers. Therefore, quarrying, like abrasion, is a large element in glacial erosion. Which has contributed the larger share to the sum total of rock eroded?

Generally, and with local exceptions, far more rock has been removed by quarrying than by abrasion. This is shown by the long profiles of steep glaciated valleys, which demonstrate greatest excavation where joints and other surfaces of weakness are most closely spaced (fig 6–1), and by the greater amount of erosion evident on quarried lee slopes than on abraded stoss slopes of hills in areas eroded by ice sheets. A quantitative study of granite hills in eastern Massachusetts (Jahns, 1943) established the relation between glacial erosion and preglacial sheet structure roughly paralleling the ground (fig 5-16). Although on the stoss sides of hills such structure is essentially parallel with the present surface, on their lee sides the sheeting is inclined more gently than the present surface. Clearly the stoss slopes were mainly abraded while the lee slopes were mainly quarried. An empirical curve showing thickness of sheets relative to depth makes it possible to determine the approximate amount of glacial erosion at each point. Abrasion took more off the sides of hills than off their stoss slopes, but quarrying took several times more rock from lee slopes than abrasion took from stoss slopes. Erosion was greatest where joints cutting the sheet structure are most closely spaced.

This group of determinations is more exact than is usually possible, but the results are similar to those obtained elsewhere by less refined means (e.g., Putnam, 1949, p. 1288). They show that, in general, quarrying is quantitatively more effective than abrasion.

The conclusion is in accord with theory. It takes much less work per unit volume to quarry out most kinds of rock than to wear rock down by abrasion. This is particularly true of hard, well-jointed rocks and is less true of weak, poorly consolidated rocks. Clay-rich till and claystone are

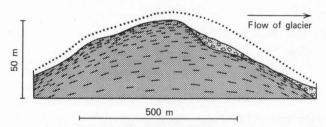

Figure 5-16 Section through glaciated hill of granite, showing relation of existing surface to sheet structure, and inferred preglacial surface (dotted line). The ice sheet quarried the lee slope more than it abraded the stoss slope, and deposited drift locally on the lee side. (Jahns, 1943.)

well consolidated in relation to glacial erosion. Where free of joints they resist quarrying effectively, and where glacially eroded they show chiefly the effects of abrasion. As a rule, glacial erosion is expressed by quarrying so long as joints and similar surfaces are closely spaced and somewhat open and cut a surface that is subject to tensional stress as the ice moves across it.

However, it does not follow that where ice-abraded surfaces are widespread, abrasion exceeded quarrying. Such surfaces show only that abrasion was the latest significant process to affect them; actually this process may have merely polished off a surface that previously had been reduced mainly by quarrying. Probably this occurred over large areas as is indicated hereafter.

With regard to the aid afforded by sheet structure in glacial erosion, sheeting might develop contemporaneously with glaciation of large, deep mountain valleys, aiding erosion through parallelism of the sheets with valley floor and sides (W. V. Lewis, 1954).

A part of the erosion seen in mountainous terrain may have been accomplished not by ice but by subglacial streams, many of which may have been subject to confined flow. This idea has been put forward from time to time, but it has not yet been shown that more than a small proportion of the overall erosion is the work of subglacial water. Holtedahl (1967, p. 195) cited evidence that such erosion had been "major" in a particular Norwegian valley.

In cold, moist climates the upheaval, by frost wedging of joint-bounded blocks from flat-lying surfaces of bedrock has been observed. Such activity, occurring immediately before glacial invasion, might possibly have produced the abundant boulders in the youngest drift sheets south of the Canadian Shield, independently of glacial quarrying. If the Shield had been worn nearly smooth by abrasion during earlier glacial ages, frost wedging might explain the relatively large proportion of boulders and cobbles in the youngest drifts.

Relation of Rate to Depth of Erosion

Rates of erosion by two large glaciers in Iceland, calculated from suspended load in proglacial streams and cited by V. Okko (1955, p. 32), are respectively 6.4cm and 55cm/100 yr. At these rates, the glaciers would have reduced the surface beneath them by 6.4m and 55m during a glacial age lasting 10,000 years.

The rate of glacial erosion depends on (1) thickness of the glacier, (2) rate of movement, (3) abundance, shape, and hardness of the rock particles with which the base of the glacier is armed, and (4) erodibility of the ground beneath the glacier. Over broad areas erodibility of the ground can be expected to vary with depth and therefore with time. After a glacier has removed the weathered regolith and has worked down through open-jointed bedrock into fresh bedrock with fewer and tighter joints, rate of quarrying should decrease. The decrease should reduce the armament of drift in the base of the ice, thereby reducing the slower process of abrasion. As evidenced by the Canadian Shield, northwestern Finland, and other strongly glaciated areas of resistant rocks, the surface that has undergone deep quarrying should have been worn down to a series of irregular bosses having abraded surfaces, especially on their upstream sides.

The actual depth at which joints become tight enough to inhibit glacial erosion must vary with varying lithology, topography, and preglacial climate. The available direct evidence, some of it detailed hereafter, suggests that in a region of low relief this depth may be as little as 10m. The occurrence, in mountain valleys, of glacially eroded rock basins with closures measured in hundreds of meters indicates that under favorable circumstances depth of erosion can be great. In some mountain valleys this depth certainly exceeded 600m. Some of the factors that make this possible are: (1) steep preglacial gradients, assuring high velocity to the flowing ice and making quarrying easier by minimizing the support that joint-controlled blocks of rock in place receive on their downstream sides; (2) concentration, within narrow valleys, of ice discharge from wide sectors of the snowshed, thus favoring rapid flow; and (3) frost wedging of rocks above the snowfields along range crests, and valley widening by glacial erosion, providing increments of tools for abrasion even after quarrying beneath the glacier has been much reduced. Daly (1912, p. 581) noted that abrasion was predominant over quarrrying in the deep glacial erosion of the low-gradient major valleys of the Rocky Mountains near the 49th parallel. If we grant that the abrasion was anything more than a final polish on an excavation made chiefly by quarrying, we may suppose that, after all the quarrying that was possible had been accomplished, the ice completed the excavation by the slow process of grinding down the surface.

Depth of Erosion: Examples

The glacial deposits on the plains of Germany, Poland, and Russia contain "immense quantities" of rock of Scandinavian origin. A. M. Hansen (1894, p. 123) calculated that the volume of this drift is so great that were it removed it could be used to fill up the basin of the Baltic Sea and the basins of all the lakes in Scandinavia, and that enough would be left over to add a layer 25m thick to the surface of the entire Scandinavian peninsula. Although based on a possibly exaggerated estimate of drift volume, this calculation implies considerable erosion in the Scandinavian mountains, especially as a large part of the rock glacially eroded from the mountains was transported, not southeast to Germany and Russia, but westward into the Atlantic Ocean, where it is concealed from view.

The depth of apparent glacial erosion of the Sognefjord, Norway, amounts to 1.2km measured below present sealevel plus 1.2km measured above it. Total apparent depth is therefore 2.4km. However, it is not proved that a part of this depth is not attributable to stream erosion occurring at times when the valley was not occupied by glacier ice.

Kerr (1936, p. 681) believed that glacial scour had deepened the large valleys of northern British Columbia and southern Alaska by at least 600m. It is thought that valleys in the Scottish Highlands have been locally deepened by a nearly equal amount. Reid (1892, p. 51) concluded that the average load of glacial rock flour carried by the streams draining Muir Glacier in southern coastal Alaska corresponds to an annual loss of 1.9cm of rock from the entire area beneath the glacier. At this rate of erosion 600m of rock could be removed in about 30,000 years.

These values are large, but both the Scandinavian mountains and the mountains of coastal Alaska afford optimum conditions for glacial erosion. Both are high and steep, and both have maritime climates which during a glacial age would provide abundant snowfall. Together with the similar mountains of Chile and the South Island of New Zealand these regions should be expected to exhibit profound glacial erosion.

The Canadian Shield is an example of the other extreme. Local evidence of slight depth of glacial erosion has been reported from many different districts. Unequivocal evidence is furnished by the Flin Flon district, 600km southwest of Hudson Bay. Here marine sedimentary rocks of Ordovician age overlie Precambrian rocks with irregular unconformity involving relief of about 15m. Before the Pleistocene, the Ordovician cover had been stripped back, exposing an extensive area of the irregular sub-Ordovician surface. The whole region was then glaciated. But the stripped and glaciated surface has almost the same appearance and relief as the surface still unconformably covered by Ordovician strata. Evidently glaciation failed to modify more than the details of the relief. An intricate pre-Ordovician drainage pattern closely adjusted to weak belts in the Precambrian rocks remains unaltered by the glaciation save for the excavation of shallow rock basins in some of the larger valleys (Ambrose, 1964, p. 821).

Indeed, the detailed adjustment of drainage to lithology, long antedating the glaciation and yet not destroyed by that event, is a feature that characterizes wide areas of the Canadian Shield. Widespread also in the Shield and in the resistant rocks of New England region is glacial removal of the weathered regolith. Out of thousands of exposures of bedrock examined in New Hampshire only 46 include any chemically decomposed rock (J. W. Goldthwait and Kruger, 1938).

Little is known as to the thickness of the altered regolith that in preglacial time overlay areas such as the Canadian Shield, New England, and Scandinavia. Bedrock altered to slight depth is reported from various localities in all these regions. Altered bedrock as much as 100m thick is recorded at Boston, Massachusetts (Kaye, 1967), and at localities in northern Finland (Donner, 1965, p. 203). If the alteration at such places represents normal weathering and not a hydrothermal origin, substantial glacial erosion is suggested, but data are insufficient to serve as a basis for estimating depth of glacial erosion. Pertinent to this question are the values for thickness of glacial drift cited in table 7-A.

Whatever its aggregate depth, glacial erosion probably occurred chiefly beneath the marginal parts rather than the central parts of the ice sheets. Glacial erosion was also probably promoted by the cumulative results of weathering during interglacial ages.

Generalizing, we can say that in mountains with abundant snowfall glacial erosion has been deep, whereas in regions of slight relief overspread by ice sheets glacial erosion has been comparatively slight. Locally, however, topography plays an important part in diminishing or increasing erosion. Beneath an ice sheet a rock floor with a relief of many hundreds of meters and steep slopes as well interferes with and reduces the radial flow of the ice mass. This was apparently true of the part of the Cordilleran Glacier Complex that invaded southern British Columbia and Washington, where, despite the presence of ice 1km or more in thickness, very little erosion occurred (Daly, 1912, p. 581; Johnston, 1926). Yet in this same region a large valley glacier, fed by abundant snowfall in the crest of the Cascade Range, excavated a rock basin (Lake Chelan) with a closure of 315m. Even individual hills have a protective effect, as shown by preglacially weathered regolith in the lee of hills whose tops are scraped clean.

A conspicuous example of slight glacial erosion is the wide region extending from the crest of the Shickshock Mountains, the backbone of the Gaspe Peninsula, southward across Chaleur Bay through western New Brunswick. Here glacial erosion was so slight that the terrain has been considered by more than one geologist to have escaped glaciation. Two views as to the slight erosion have been expressed. One holds that the latest glaciation of this part of America failed to cover this high terrain, which therefore has undergone great modification since it was glaciated at some earlier time. The other refers to the position of the terrain in the lee of a formidable barrier, the Shickshock Mountains, in places reaching altitudes of more than 1200m, lying at right angles to the regional direction of flow of the Laurentide Ice Sheet. As this glacier flowed southeastward from the highlands of Labrador and southeastern Quebec into the capacious trough of the lower St. Lawrence, basal ice was diverted northeastward down the trough, scouring it vigorously, while the upper ice was barely able to clear the barrier. In consequence rate of flow downstream from the barrier was effectively diminished and erosion inhibited (Flint et al., 1942).

At the opposite extreme is the valley trending parallel with the local radius of the ice sheet that enters it. The thicker ice over the valley than over the adjacent uplands makes for increased erosion of the valley floor, and the lack of topographic obstruction enhances still further the capacity of the ice to excavate the valley. The result is valley deepening, commonly coupled with excavation of rock basins in favorable segments of the valley floor. The floors of tributaries, excavated little or not at all, are likely to be left hanging above the deepened floor of the main valley.

Classic examples are the troughs of the Finger Lakes in central New York State. These valleys were first cut by preglacial streams flowing north. They were widened, greatly deepened, and straightened by subglacial erosion when the Laurentide Ice Sheet repeatedly filled and overran them from the north. Dams of drift at their northern ends created the present lakes. The valleys of both Seneca and Cayuga Lakes have tributaries that hang above the main valley floors by as much as 120m. The steep-sided trough of Seneca Lake is more than 300m deep *below the lake surface;* its bedrock floor has not been reached by deep borings. In this region conditions for ready glacial erosion were nearly optimum: the valley floors are excavated in shales that dip gently south, so that they continually presented a hackly surface to the base of the ice.

Despite deep erosion of the valleys, the tops of the adjacent uplands in some places still retain preglacially weathered regolith, showing that the ice at this altitude (450 to 600m above the excavated valley floors) was either too thin or too poor in basal load to accomplish much erosion.

Another example of exceptionally great glacial erosion controlled by topography is the Highlands segment of the Hudson River valley between Newburgh and Peekskill, New York (Berkey, 1911, p. 81). Here the river follows a preglacial gorge through a massive highland, 420 to 500m high, which formed a barrier transverse to the direction of flow of the Laurentide Ice Sheet. Throughout the gorge the bedrock floor of the Hudson River valley is a rock basin at altitude −232m to −288m; tributary valleys hang above it. South of the gorge, floor depth decreases to −212m at Tarrytown, −83m at the Holland Tunnel opposite Lower Manhattan, and to −61m near the Verrazano Bridge (Newman et al., 1969, fig 1). Upstream from the gorge, floor depth decreases likewise. The rock basin, like those near the mouths of many fiords, is attributed to increased glacial erosion caused by the more rapid flow required for the ice from upstream to pass through the gorge and the relatively high land south of it.

The valley of the North Branch of the Susquehanna River between Pittston and Wilkes-Barre, Pennsylvania, consists of a series of rock basins with closures of as much as 68m. The valley of the Connecticut River in Connecticut and Massachusetts contains a basin more than 60km long, excavated in a north-south belt of relatively erodible bedrock. Near Hartford, midway between its eastern and western sides, the basin floor lies at altitude −60m. Farther south, at Middletown, the bedrock floor rises to ~ −35m. Both here and in the Susquehanna valley the basins are believed to have resulted from glacial erosion.

Great Slave Lake (fig 18-7), in Canada's Northwest Territories, illustrates differential glacial erosion induced by topography. The narrow eastern part of the basin, occupying a glacially deepened preglacial valley in resistant rocks, has a floor more than 600m below the lake surface; the broad western part, excavated in terrain consisting of more erodible rocks with smaller relief, is less than 150m deep.

Indeed, many of the lakes in northern North America and Europe occupy rock basins excavated by ice sheets or valley glaciers in places where lithology, topography, or both combined, made erosion especially easy. Probably the Great Lakes of North America owe their basins chiefly to local glacial excavation of the floors of broad preglacial valleys, although other factors have operated locally (Shepard, 1937; Thwaites, 1947). The general distribution of basins in North America and Europe was mapped by Shepard (1937); most of these are believed to be of glacial origin, although not all are the work of glacial erosion (e.g., Woldstedt, 1952).

It has been asserted that the region covered by the Laurentide Ice Sheet consists of three concentric zones: an outer zone of predominant deposition (mainly the Great Lakes and Plains regions), an intermediate zone of predominant erosion (mainly the greater part of the Canadian Shield), and a central zone of little erosion (mainly the region east of Hudson Bay). It is implied that these zones are related to varying intensity of erosion along a radius of the ice sheet. However, lithology appears to have played a part in the pattern.

Unquestionably a broad marginal belt extending westward and northward from the eastern Great Lakes does have thicker drift than the region north and east of it. But this belt coincides generally with outcrops of comparatively erodible Paleozoic and Mesozoic strata. Eastward from the Great Lakes, though still in the marginal belt, the drift becomes thinner and less continuous, and the change coincides with a change to much less erodible rocks. The intermediate zone of erosion coincides generally with the very strong rocks of the Canadian Shield. The only area within this zone where relatively thick drift has been reported is that immediately southwest of Hudson and James Bays, where the rocks are weak Paleozoic strata. These facts suggest that the determining factors are lithologic rather than related to position with respect to the ice sheet. Finally, the observations on the basis of which the supposed central zone of little or no erosion is inferred—surface rocks broken up by frost wedging rather than conspicuously eroded by glacier ice—may indicate merely postglacial destruction of glaciated surfaces by frost wedging rather than the absence of glacial erosion.

The complex history of the Laurentide Ice Sheet is opposed to any simple view of the distribution of erosion. At one time or another the marginal part of the ice sheet swept over nearly the whole region covered by the ice when at its maximum. Furthermore, there appear to have been various centers of outflow, which shifted their positions from time to time and which should have given rise to complex patterns of erosion and deposition. However, as lithology and topography appear to be the dominant factors in controlling erosion and deposition, such patterns would be indistinct if recognizable at all.

The zonal concept of the distribution of glacial erosion seems to fit the Cordilleran region somewhat better than the Laurentide region. In British Columbia both the Coast Ranges and the Rocky Mountains bear the marks of much deeper glacial erosion than do the lower mountains of the "Interior Plateau" region that separates them. This results partly from preglacial topography, in which narrow valleys favored deep erosion in the high mountains, whereas the valleys of the plateau are more open. It results also from much longer occupation of the mountains than of the plateau by glacier ice. The glaciers formed in the mountains and widely invaded the intervening country only near the glacial maxima. The invading glaciers consisted of thick piedmont ice masses which at times coalesced into an ice sheet. During the occupation of the plateau by widespread ice, movement was slow, and over large areas erosion was slight. Indeed, the fact of glaciation in the centrally placed Cariboo and Cassiar districts was questioned for a time, though detailed study later established it beyond doubt. Other areas within this interrange plateau, although glaciated, evince only moderate alteration of the terrain by ice (e.g., N. F. G. Davis and Mathews, 1944; W. A. Johnston, 1926).

Relative Rates of Glacial and Subaerial Erosion

Both before and for some years after the end of the 19th century a lively controversy among geologists centered on the question of whether glacial erosion is more or less effective than subaerial erosion within a given time. Good summaries of the argument include those by Bonney (1910), Carney (1909a), and Garwood (1932). As has occurred time and again in similar controversies the matter was complicated unnecessarily, and general agreement was delayed, by the fact that not all those concerned were considering the same aspect of the problem. Some drew inferences from localized zones of specially effective erosion while others based their arguments on areas of little erosion or dominant deposition. The outcome gradually solidified into a general opinion that overall, glacial erosion removes more rock material from a region than subaerial erosion would remove in the same time interval. This opinion, widely held today, is not qualified as to climate or predominant type of glacier, and whereas it is very likely true, it lacks a good quantitative basis.

Two obvious means of measurement are available: comparison of the loads carried by normal streams with those carried by glacial-meltwater streams in similar terrain, and relation of the total volume of glacial drift in a given region to the area from which it was eroded. Considering stream loads first, we take the finding of Judson and Ritter (1964) that at present rates, erosion by weathering, mass-wasting, and streams is removing an average of ~6cm of rock material per thousand years. To compare with this value we have only spot sampling of individual meltwater streams, most of them in the Alps, Norway, and Iceland. An average of 13 of these, stated in terms of lowering of the watershed area, amounts to 70cm/1000yr, larger by a factor of more than 10. However, these meltwater measurements are not comparable because they are not continuous, because many of them consider only suspended load and neglect bed load, and because they relate chiefly to mountainous topography drained by valley glaciers. A carefully planned sampling system would probably yield results of value, but at present the examination of stream loads has not answered the question.

The second approach, via measurement of the volume of glacial drift, is less accurate for various reasons, of which three predominate. (1) Estimates must replace measurements in large part. Thickness of drift in a glaciated region is estimated from the records of borings, but the drift carried seaward and now submerged is very difficult to estimate. (2) Glaciers did not remove rock matter from one side of a fence and deposit it on the other. Substantial volumes of drift were derived from bedrock that now lies beneath a drift cover; this makes it difficult to calculate the area of the region from which any large body of glacial sediment was derived. (3) The drift in extensive glaciated regions is of more than one age; hence there are difficulties in calculating the period of time during which erosion occurred. In view of these uncertainties there is little profit at present in trying to calculate rates of glacial erosion from volume of drift.

In summary, it is generally supposed that valley glaciers erode mountain terrain more rapidly than subaerial erosion would the same terrain. The supposition, though possibly true, is not based on measurement. As to the comparison between ice sheet and subaerial erosion on a terrain of low relief, no quantitative comparison has been attempted.

ICE-THRUST FEATURES

Hitherto we have been dealing mainly with the erosion of consolidated bedrock by glaciers and have hardly considered what happens to unconsolidated sediment as it is glaciated. Surface material, if unconsolidated, can be pushed or scraped away or can be enveloped by the basal ice. In many places negative evidence of any glacial removal of unconsolidated sediment has been described. At the time of glaciation, such material may have been frozen and therefore consolidated, and so it behaved for the time as a resistant rock. Erosion of a frozen mass must occur as abrasion, for unless the temperature were extremely low, permitting shrinkage to create frost cracks, there would be no joints to permit quarrying. Another possibility is that regolith was protected by a covering of snow, which acted as a buffer between it and the moving glacier.

In other places unconsolidated sediments of both glacial and nonglacial origin have been deformed to create well-defined structures as well as accompanying land forms. (General references include E. Hansen et al., 1961; Kupsch, 1962; Lamerson and Dellwig, 1957; Lessig and Rice, 1962; Mackay, 1959; Mathews and Mackay, 1960; Slater, 1927; Viete, 1960.) In some of them glacial erosion as well as glacial deformation has been involved. Most such structures seem to have been made at or near glacial termini. We can appropriately class such structures and land forms as *ice-thrust features*. A partial list of them follows.

1. Ridges, generally curvilinear and in places complexly arcuate; some of them occur in nested series (fig 5-17). In broad outline they resemble end moraines, from which they have been carefully distinguished, although in Germany such features are known as *Stauchmoränen* (Woldstedt, 1954–1965, v. 1, p. 105–110).

2. Structures seen in vertical exposures (fig 5-18). Compared with most tectonic structures they are shallow (mostly <100m and possibly not >200m in amplitude) and die out downward. Some of them coincide with (1). Among those reported are:

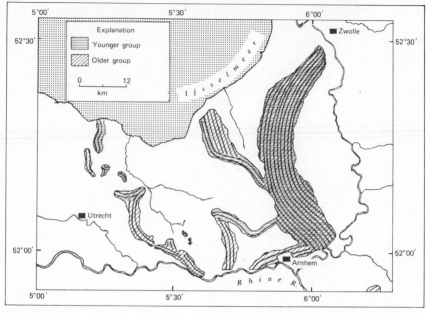

Figure 5-17 Anticlinal ridges in the Netherlands, formed by thrust of the Ijssel lobe of the Scandinavian Ice Sheet at the maximum of the Saale glaciation. The deformed beds are chiefly middle Pleistocene alluvial sand. Ridges attain maximum heights of nearly 100m. (After Netherlands Geological Survey, A. J. Pannekoek, and G. C. Maarleveld.)

    a. Folds, some overturned and many isoclinal, mostly no more than a few meters in amplitude and rarely with axial-plane foliation in fine-grained sediments (fig 5-19).

    b. Faults, chiefly imbricate thrusts, the blocks or plates rarely more than 100m thick though some are more than 200m thick. Such blocks grade into detached, (ice-?) transported masses, some with internal deformation. The much-studied exposures in the cliffs near Cromer in East Anglia expose masses of Cretaceous chalk, as much as 540m long, that have been lifted 20m from their preglacial positions, contorted, and inclosed in till (fig 7-4).

    c. Fracture and brecciation.

    d. Attenuated strata drawn out plastically, mainly by drag at base of glacier.

    e. Material forced downward plastically between blocks of more competent substance, forming what are essentially clastic dikes.

   3. Erosion along ice/ground interfaces, locally decapitating structures.

Figure 5-18 Folds, thrust-faults, and breccia caused by glacier ice overriding lacustrine sediments. Below the contorted zone is lake sediment with ice-rafted stones; above it is nondeformed till of the thrust sheet. Movement was approximately toward the observer. Near Glens Falls, New York. (E. Hansen et al., 1961.)

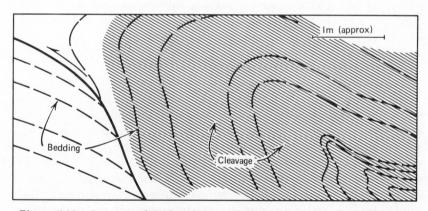

Figure 5-19 Overturned fold and thrust fault in lacustrine silt and fine sand, created by a glacier flowing compressively against a massive end moraine. Axial-plane foliation has developed in the silty layers. Outlet of L. Huechulafquén, Neuquén, Argentina. (Flint and Fidalgo, 1964, pl. 1, loc. 23.)

Most such structures are compatible in form and attitude with the position and direction of movement of ice determined from independent criteria, although local variations occur (E. Hansen et al., 1961). Most occur in areas in which the ice was moving compressively against an opposed slope.

The mechanical processes invoked to explain the observed structures include:

1. Frictional drag along the interface between glacier and floor.

2. Pushing (bulldozing) by the glacier terminus, in conjunction with (1). To judge from the heights of pushed-up mounds of debris, the process occurs only on a relatively small scale.

3. Envelopment of underlying material by flowing ice and confinement within the ice, making possible deformation within large masses of such material.

4. Frictional drag by compressively flowing ice on a layer of frozen ground aided by high neutral stresses in pore water trapped and sealed beneath the frozen ground. This reduces shearing strength, slices of the frozen body become detached, and are moved forward over a nonfrozen floor. Thus, in effect, the base of the glacier becomes the base of the moving part of the frozen material beneath the glacier ice (Mackay, 1959).

In this connection we note that Slater (1927) viewed intensely convolute structures in drift as inherited from structures within drift-loaded moving ice, preserved with little change through the process of ablation. He termed such structures "glacial pseudomorphs," and in a sense that is what they are, inasmuch as we consider the drift a part of the glacier. However, they are better described simply as ice-thrust structures.

ICE-PRESSED FORMS

Apart from ice-thrust features, which result from dynamic pressure, forms and structures resulting from flow and also compaction of sediment under the static pressure of glacier ice have been recognized. This origin was proposed around 1900 for certain eskers in Denmark, which contain at the center of the base a low ridge of clay-rich till paralleling the trend of the esker. The ridge was believed to have been squeezed up into the esker tunnel before deposition of the esker sediment proper began (fig 5-20). On Long Island, New York, broad low arches of large dimensions involve beds of clay, whose compaction under the weight of the edge of the ice sheet, and consequent flow, may have been chiefly responsible for the flexures.

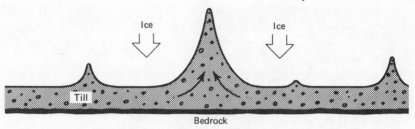

Figure 5-20   Model showing the mechanism of "ice pressing." (After Hoppe, 1952.)

The concept first outlined in Denmark was applied in detail by Hoppe (1952) and Schytt (1959) to features in Sweden and later to features in Alberta by Henderson (1959b, p. 48–56) and by Stalker (1960a), who called them *ice-pressed forms*. The fabric of the till is compatible with lateral and upward flow, as are some of the details of the many forms produced. The latter, most of which imply the presence of stagnant glacier ice, are logically described in chap 8 in connection with moraines. Here we mention two ice-pressed forms believed to have been made under ice that was still active. One consists of parallel flutings (figs 5-21, 8-5) in till rich in clay or silt, described first by Dyson (1952) and later with more detail by Schytt (1959), beyond the termini of both existing and former glaciers. Ranging in height from a few centimeters to 40cm and in exposed length up to more than 300m, they are believed by Schytt to be water-saturated till at the pressure-melting point, statically pressed up into cavities in the glacier base. The cavities form downstream from fixed boulders, and till is continually fed into a cavity as it lengthens with motion of the glacier and freezes to the base of the glacier. Downstream, where the ice is thinner, negative winter temperature reaches the base of the glacier and the ridged till freezes to the material beneath it. The glacier then slides along the parallel ridges. According to this theory of origin, the presence of flutings indicates basal ice at the pressure-melting point. Pressure is equivalent to about 92 metric tons per square metter per 100m thickness of pure ice.

Stalker (1960a, p. 27) invoked ice pressing as a possible origin, at least in part, of very long, low drumlins. Whatever its other limitations, ice pressing seems to be confined mainly to drift rich in clay and to areas of low relief.

Some of the many kinds of clastic dikes, in which upward flow of clay-rich sediment has occurred, may likewise have resulted from differential static pressure of overlying, possibly stagnant ice.

Figure 5-21   Fluted Dwyka Tillite, Harrisdale, C. P., South Africa. Notebook is 22 cm long. (R. F. Flint.)

Compaction. Another effect of the static pressure of ice is the compaction of sediment beneath it (Viete, 1957; Harrison, 1958; Legget, 1962, p. 119, 127; Rominger and Rutledge, 1952, p. 164–166). Sediment thus compacted is literally another "ice-pressed feature" but terminologically it is not so considered. Responding to the load, interstitial water is forced out and the sediment becomes more closely packed. When a sample of such sediment is subjected to a rigorous test (a preconsolidation test), the resulting data have been employed as a basis for determining the maximum effective stress to which the material had been subjected previously. Systematic testing of a thick body of clay showed that preconsolidation pressures increase with increasing depth below the original surface of the clay, despite the fact that erosion has removed part of the clay overlying the sampled points (Crawford and Eden, 1966). This substantiated the theory that consolidation is irreversible. Hence it has been thought possible to translate such tests on glacial drift into terms of thickness of former overlying glacier ice or a former cover of sediment. Such calculation, however, involves assumptions; furthermore, both freezing and desiccation likewise can cause significant compaction in some sediments. Therefore unusual compaction in a sediment does not necessarily imply that it formerly carried a substantial overburden.

CHAPTER 6

Glacially Sculptured Terrain[1]

Chapter 5 describes the basic features of glacially eroded surfaces of bedrock. The present chapter views glacial sculpture on a larger scale, examining valleys and groups of valleys and the inferences that can be drawn from them. For example, a glaciated valley can be fashioned either by a valley glacier or by an ice sheet. Other features, such as cirques, are fashioned by cirque and valley glaciers. Specific groups of features are made by complex systems of valley glaciers and related small glaciers occupying mountainous terrain, part of which stands above the upper surfaces of the glaciers. Other groups are made by glaciers that cover all or most of the terrain. The differences are therefore significant in determining the character and extent of former glaciers.

GLACIATED VALLEYS

Valleys bearing the stamp of glacial modification occur in many parts of the world, and in some regions exist by the hundreds. They were recognized in the Alps by Venetz and by Charpentier early in the 19th century. However, probably none of them are primarily the work of glaciers; rather they are pre-existing valleys, mostly stream valleys, occupied for a time and remodeled with various degrees of thoroughness by glaciers. At first this concept was not recognized as a general principle. But the gradual identification of valleys glacially eroded in their upstream parts, yet having in their downstream parts the characteristics of normal stream valleys free of any glacial features, led to the establishment of the principle. Furthermore we find little evidence that glaciers do cut or have cut

<hr />

[1] General references include Blache, 1952; Cotton, 1942, p. 147–299; Klebelsberg, 1948–1949, p. 349–371; Embleton and King, 1968; Sissons, 1967, p. 35–63.

valleys into areas of low relief; instead they spread out and erode the surface with broad uniformity, provided the rocks are of uniform erodibility. Finally, theory also leads to a similar conclusion, that glaciers do not create valleys.

Long Profile. The simplest way to analyze the erosive effects of glaciers on valleys is to compare glaciated valleys with normal stream valleys, noting the significant differences. In general the glaciated valley has a steplike long profile, steeper in the headward part than that of the stream valley. Between the steps, or *riegels*, and farther down the valley where pronounced steps are rare or absent, there may be rock basins.

The common occurrence of rock basins in glaciated country attracted attention long before riegels came under close observation, and A. C. Ramsay (1862, p. 188) may have been the first to call attention to it. The significance of rock basins as records of glaciation emerged through the elimination of other origins, such as solution and crustal movement, in specific cases. Such basins are more obvious and probably more abundant than are riegels.

For many years the development and maintenance of the steep faces of the riegels were believed to result from frost wedging, caused by refreezing of meltwater as it percolates down through a transverse crevasse. But as few crevasses are known to extend to the rock floor beneath the glacier, this cannot be the main factor in developing riegels.

A satisfactory explanation of some riegels is based on their coincidence with poorly jointed parts of the bedrock. It attributes the steps to the resistance to erosion of massive, poorly jointed rock compared with rapid quarrying of well-jointed rock immediately down the valley (fig 6-1). This

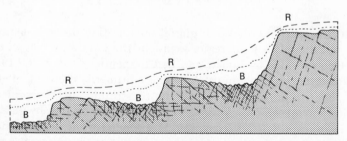

Figure 6-1 Long section of part of a glaciated valley, showing riegels (R) and basins (B). Surfaces smoothed by abrasion alternate with slopes steepened by quarrying, imparting to the valley a steplike profile. In this valley the steps are shown to be controlled by the unequal distribution of joints and other surfaces of weakness throughout the bedrock. The profile before glaciation is suggested by the dashed line. The dotted line suggests a profile made early in the process of glacial erosion. (After Matthes, 1930, p. 96.)

appears to be the origin of many if not of most riegels, but some may be formed in other ways (reviewed in Cotton, 1942, p. 256–271). Among these is the steepening of slope caused by increased erosion where large tributaries join a main glacier.

In some valleys riegels occur at places where the valley narrows. This relationship might be explained by the theory of extending and compressive flow (chap 3), with erosion of bottom and sides greater at the steeper places, where flow is extending, than in the gentler places, characterized by compressive flow.

Some basins, too, are related to erodibility, for they are developed in transverse belts of rock exceptionally well provided with planes of weakness. Other basins record excavation accompanying the narrowing of the cross-sectional area of a valley, where a glacier flowed through specially resistant rock. Still other basins may have developed immediately below the points of entrance of tributary glaciers, where ice discharge was increased substantially.

Many glaciated valleys, among them a large number of Norwegian fiords, have in their outer or downstream parts long, gently sloping rock basins. The closures of some of the basins exceed 900m. In Norway the basins and the rock thresholds at their downstream ends occur at so nearly the same segment of each fiord that a common explanation of their origin seems probable. Brøgger attributed the basins to the greater amount of glacial erosion localized at the base of the glacier-filled valley vertically beneath the firn limit (A. M. Hansen, 1894, p. 124). Some basins at or near the mouths of mountain valleys are more simply explained as the result of decreased rate of erosion with decreased velocity at the point where a narrow valley glacier merges into a broad, thick piedmont glacier. Even more simply, some basins may mark the places where flotation in the sea ended the erosive capacity of valley glaciers.

Some thresholds, especially those known only as a result of submarine soundings in fiords, may consist wholly or partly of moraines and other glacial deposits rather than of bedrock.

It is worth noting that both steps and basins occur in the floors of stream valleys that have never been glaciated, although they are far less common in such valleys than in glaciated troughs. At any rate, steps and basins are not in themselves criteria of glacial action.

Cross Profile. Glacial alteration of a stream valley involves both deepening and widening. In some valleys the volume of rock excavated by deepening exceeds that excavated by widening, but in many valleys widening has been the more important process (Matthes, 1930; Crosby, 1928; Cotton, 1942, p. 296–299). In most valleys these changes result in a cross profile of pronounced U-shape, approaching that of a parabola. To be sure, the cross profiles of some stream valleys take the form of a U, but the slopes are generally mantled with regolith, whereas in the glaciated U valley the slopes expose much bare bedrock.

Most glaciated U valleys are made by alteration of youthful V valleys originally cut by streams and mass-wasting. Before glaciation, channel cutting occurs only on the valley floor where the stream is flowing; the sideslopes are eroded chiefly by mass-wasting. When the valley is filled with flowing ice, rapid glacial scour replaces the slower process of mass-wasting along the entire ice-covered perimeter of the valley cross section. The result is a trough, essentially a channel analogous to the channel of the former stream.

Glacial erosion tends to remodel a valley toward a semicircular cross section, in part because that shape has the least area in proportion to the volume of ice flowing through the valley, so that drag is at a minimum.

Some glaciated valleys, particularly some of those in the Alps, have composite cross profiles, that is, profiles consisting of a narrow U set within a higher, broader U, the shoulder between the two being unrelated to lithology. Several explanations of this feature are summarized by Cotton (1942, p. 284-296). Of them, two appear probable. Penck and Brückner (1909, p. 288, 305, 376, 617, 837) believed these valleys to be preglacial stream-cut two-story valleys later modified by glaciation without obliteration of the preglacial two-story character. Visser (1938, p. 140) explained them as preglacial stream valleys later converted by glaciation into U valleys, still later subjected to interglacial rejuvenation by stream erosion, and finally slightly modified by renewed glaciation. Very likely both explanations are applicable.

Hanging Tributary Valleys.[2] The deepening or widening of a main valley by glacial erosion, at a rate more rapid than that at which the tributary valleys are cut (whether or not the tributaries are filled with glacier ice), leaves the tributary, at its junction with the main, higher than the main, that is, hanging above it. Upon deglaciation, the stream in the hanging tributary falls or cascades down to the main valley floor. The vertical drop may amount to 200m or more. Even where all tributaries carry glaciers, the normal relationship of their floors to the floor of the main valley should be a hanging one, even though the upper surfaces of tributary and main are accordant at the point of junction, because ice thickness presumably is an important, perhaps the dominant, factor in the rate of glacial deepening of a valley floor.

The mouths of some hanging tributaries project somewhat beyond the sidewall of the main valley; the hanging floor therefore constitutes the

---

[2] Good general discussion in Cotton, 1942, p. 216-234.

top of a sort of pedestal in bas-relief that bulges from the sidewall and narrows the main valley. The pedestal, termed a *bastion*, is believed to result from reduced flow and reduced erosion by the main glacier as it encounters the laterally directed stress of the emerging tributary glacier.

That deepening and not merely widening of the main valley is responsible for some hanging tributaries is said to be shown by the fact that long profiles of the tributaries, projected into the main, fall well above the floor of the main at its center line. If it can be shown that this lack of accordance is the result of glacial erosion rather than preglacial stream erosion, the case is proved. However, in view of the abundant evidence of great deepening of valleys by glaciers, the matter is academic.

The reservation mentioned carries with it the implication that hanging tributaries are not criteria of glaciation. They have been observed in non-glaciated highlands (Matthes, 1930), where they can result from any process, such as crustal warping, that rejuvenates the stream in the main valley while affecting the tributaries less or not at all. Hanging tributaries, even though glacial, do not constitute evidence of erosion by valley glaciers as compared with ice sheets. If in an extensive ice sheet the ice over a main valley is thicker and less impeded by topographic barriers than the ice over tributaries, the tributaries can be left hanging. Examples include hanging tributaries of the Finger Lakes valleys in central New York State.

Fiords. (See Hubbard, 1934; Peacock, 1935; Cotton, 1942, p. 273–283; Holtedahl, 1967.) Many of the strongly glaciated valleys of high-standing coasts underlain by resistant rocks in high latitudes [3] are partly submerged, and constitute fiords. A fiord is a segment of a glaciated trough partly filled by an arm of the sea. It differs from other strongly glaciated valleys only in the fact of submergence. The floors of many though not all fiords contain basins, some of which are both deep and long.[4] As a result many fiords are shallower at their mouths than at some distance inland. Most such basins are believed to be eroded in bedrock, and some are known to be so (e.g., Holtedahl, 1967, p. 190). Others may have been created by the building of massive end moraines that formed their seaward ends. Still others may be rock basins with moraines built upon their seaward rims (e.g., Holtedahl, 1967, p. 194).

In the past, extreme views on the origin of fiords have been held. Fiords have been regarded as tectonic, controlled by faults, fractures, and joints, and later somewhat modified by streams and glaciers. Fiords have also been thought of as almost wholly the product of glacial erosion, and

[3] Norway, Scotland, Iceland, Greenland, Labrador, many Arctic islands, Alaska, British Columbia, southern Chile, and New Zealand.

[4] See representative long profiles in Holtedahl, 1967.

by others as deep stream valleys only lightly glaciated. Indeed every fiord is the result of both stream erosion and glacial erosion, but the relative importance of each varies from region to region. It is no more possible to generalize about the origin of fiords than about the origin of other types of valleys, beyond the statement that all fiords were stream valleys before glaciation and all have been partly submerged. Some of these valleys were conspicuously controlled by rock structures; others are independent of them. Some were only moderately glaciated whereas others were profoundly widened, deepened, and basined by glacial erosion.

The glaciation of some valleys took place while their floors stood above sealevel, submergence occurring later. Other valleys received their glacial deepening largely below sealevel. A glacier flowing down a valley and advancing into tidewater along a partly submerged mountainous coast excludes the sea and continues to exert stress on its floor until flotation occurs. As the density of glacier ice is about 0.9, a valley glacier 1000m thick would continue to erode its floor even when submerged to a depth of nearly 900m. As Crary (1966) showed, even after the glacier tongue begins to float, it can continue to rasp the mounds and hilltops of the valley floor until it has lost all contact with the floor. Both increase in thickness of ice and local uplift of the crust would bring increased floor area into contact with the base of the floating ice and permit renewed erosion.

Probably the deep basins in the floors of fiords were excavated well below sealevel, because it is unlikely that net postglacial submergence of most fiord regions has greatly exceeded 100m. Yet table 6-A shows that fiord depths are much greater than 100m.

Table 6-A
Greatest Known Depths of Fiords (meters). (Values are depths of water; depths to bedrock may be greater)

In British Columbia	780	(Finlayson Channel)[a]
In Alaska	878	(outer part of Chatham Strait)[a]
In Europe	1308	(Sogne Fjord, Norway)[b]
In South America	1288	(Messier Channel, Chile)[a]
In World	1933	(Skelton Inlet, Antarctica)[c]

[a] Peacock, 1935, p. 669.   [b] Holtedahl, 1967, p. 192.

[c] Crary, 1966, p. 926.

CIRQUES [5]

Form and Dimensions. The heads of many glaciated valleys other than valleys formerly occupied by outlet glaciers are shaped like a theater,[6] or half a bowl. Such valley heads vary in form: their floors are narrow to broad, their sideslopes moderately steep to nearly vertical. The feature is widely known as a *cirque*.[7] It occurs not only at the head of a valley but also independently as a major indentation in an otherwise smooth slope. Broadly, then, a cirque can be defined as a deep, steep-sided recess roughly semicircular in plan, cut into a slope by erosion beneath and around a bank of firn or a glacier. It may reach a diameter of much more than 1km, and its floor may contain a rock basin, commonly holding a lake. As Willard Johnson said, such a cirque has a down-at-the-heel appearance. Indeed, a cirque can be of any size or proportion, depending on such factors as preglacial form of the valley head, regimen and duration of the glacier and, above all, composition and structure of the rock from which it is cut. Cirque diameters range from 20 or 30m up to tremendous values. The Walcott cirque in Victoria Land, Antarctica, is said to be 16km wide, with a headwall 3km high.

There appears to be a crudely systematic relationship between the length of a glacier and the volume of the cirque or cirques at its head. Single cirques form the heads of small glaciers only. Long valley glaciers head, not in single cirques, but in many small tributaries each with its own cirque, or in a single broad upland basin formed by the coalescence of several adjacent cirques.

Origin. Before we discuss the altitudes and orientations of groups of cirques it may be helpful to review the origin of these features. At least two groups of processes are involved in cirque making: (1) frost wedging accompanied by mass-wasting, and (2) glacial erosion including transport. The first is dominant early in the growth of the cirque; the second comes later and plays a part, dominant or auxiliary, in those cirques that come to be occupied by glaciers. The cirque, therefore, is in large part the product of mechanical weathering and mass-wasting, but it is localized at topographically favorable places near the lower limit of perennial snow.

The essential factor in localizing a cirque is the presence of a banklike

[5] Tricart and Cailleux, 1953, p. 177–202; Cotton, 1942, p. 169–188; W. V. Lewis, 1949b.

[6] Frequently miscalled amphitheater, which is like two halves of a bowl put back together, or a whole bowl.

[7] British *corrie* or *cwm;* German *Kar;* Norwegian *botn;* Swedish *nisch.*

patch of firn [8] occupying any slight pre-existing depression or niche. The essential activity is melting by day and freezing by night during the summer ablation season. Chapter 10 mentions some of the factors that control frost wedging. Meltwater penetrates the underlying mantle and crevices in the bedrock beneath the whole area of the firn, freezes at night, expands, and wedges out rock particles of sizes that depend on local lithology. These are moved downslope by creep, gelifluction, and small rills and trickles of water. The effect is to deepen the depression near its margins, flatten its floor, steepen its sidewalls, and reshape its ground plan (fig 6-2). A bank of firn lying on a flat surface tends to develop a circular form, the form having the smallest periphery per unit of area. On a steep slope, however, only the upslope half of the circle can be eroded effectively. Hence the ground plan, regardless of its original form, gradually approaches the semicircular.

This combination of freeze-and-thaw with mass-wasting has been called *nivation* (Matthes, 1900, p. 183), and the cirques are termed by some authors *nivation cirques.* Movement within firn banks is believed to consist mainly of settling and compaction, with creep becoming conspicuous only when the firn grows thick. However, the transport and striation of stones observed in firn banks is evidence of effective creep (Costin et al., 1964), and thus we should think of nivation and glacial erosion in a growing cirque as overlapping with time, with no distinct point at which one process ends and the other begins.

The rock waste, especially the fine-grained material produced by frost wedging is moved downslope out of the nivation cirque by mass-wasting, which is made more effective in many areas by the fact that beneath a thin surface layer of thawed material the ground beyond the firn is frozen. In some nivation cirques coarse fragments slide down the firn bank and accumulate at its toe as a ridge 1 to 6m high and as much as 300m long. First described and interpreted by Daly (1912, p. 593), these ridges were later named by Bryan *protalus ramparts* (fig 6-2). They occur in many cirques that no longer contain firn, and hence imply a former climate more nearly glacial than that of today.

Nivation, then, consists of weathering and mass-wasting localized by banks of firn that play the passive role of reservoirs of water substance repeatedly changing phase from solid to liquid and vice versa. But to the extent that creep occurs in the firn the bank becomes a glacier and glacial erosion, including transport, contributes to the further evolution of the cirque. Rock fragments frost wedged from the cirque walls are incor-

---

[8] Firn banks are also called *névés* and "snowbanks," but as they consist mainly of firn, we use this appropriate term.

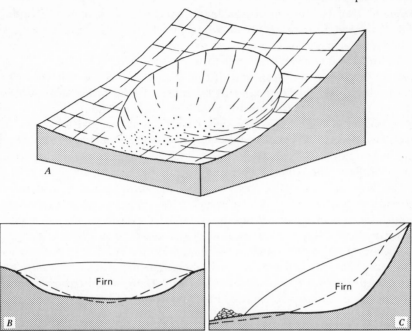

Figure 6-2 Incipient cirque seen in perspective (*A*), cross profile (*B*), and long profile (*C*). Broken lines represent profiles before cirque erosion began. The ridge at toe of firn bank is a protalus rampart.

porated into the glacier and moved down its length, and an end moraine, differing from a protalus rampart in the wear of its constituent fragments and possibly also in its fabric, may be built at the terminus.

Because it is doubtful that nivation alone can create a closed depression, probably we must attribute rock basins in cirque floors to quarrying and abrasion by moving ice. The competence of glacial erosion versus nivation in the development of cirque floors is not known, although probably retreat of walls is far more rapid than deepening of floors. Where cirques are closely spaced they tend to intersect, so that their floors constitute a shelflike platform along the side of a valley or near the crest of a highland.

Rates at which nivation cirques are enlarged are not known, although they are believed to be rapid; they depend on several factors. Rapid erosion was indicated in the Wasatch Plateau in central Utah in the exceptional year 1952, which experienced the greatest snowfall on record up to that time. Numbers of large banks of firn and snow persisted through the summer. During that one season banks overlying weak materials undermined the sod and created incipient cirques 15 to 500m in diameter and nearly 1m in maximum depth.

Altitude and Aspect; Relation to Snowline. When cirques are examined in families or groups, their altitudes and aspects (i.e., directions toward which they face) are seen to be arranged in a rather orderly way. Cirques occur at all altitudes from high mountain summits down to sealevel and even below it, where they have been submerged by the rising postglacial sea. The relation between the altitudes of cirque floors and the orographic snowline is significant because it is a basis for estimating the altitude of the climatic snowline during glacial ages.

Cirques form at or close to the annual snowline because they demand physical conditions that occur only near that limit. For a persistent bank of firn or a very small glacier to come into existence, the orographic snowline must lie below the summit of the highland in which the firn or ice occurs. Essentially coincident with the annual snowline on a glacier, the orographic snowline must lie higher than the glacier terminus but must coincide with the toe of a bank of firn. Hence the orographic snowline lies at or slightly above the cirque floor.

This being true, the altitudes of the floors of small independent cirques no longer containing firn should approximate the orographic snowline that existed when the cirques were made. The average altitude of the floors of a related group of cirques selected in this way constitutes a theoretical cirque-floor surface, which should be a general approximation of the climatic snowline of the time of cirque making (fig 4-6; Louis, 1944, p. 477).

In many mountain masses the altitudes of cirque floors rise systematically in a particular direction. The rise is related to systematic change in the values of climatic parameters such as precipitation, temperature, and wind directions. In fig 4-5 an eastward rise of floors of inactive cirques, across the Argentine Andes is shown in cross section. In fig 4-6 a similar relationship is shown by form lines on maps. In fig 4-7 altitudes of moraines related to cirques are represented by stereographic plots. In each of these examples rise of cirque floors in the inland direction is coupled with rise of the orographic snowline. This in turn is related primarily to decreasing snowfall, as moist westerly winds interacted with highlands standing across their path, and to increasing temperature eastward, as the climate became less maritime in that direction.

As regards the aspect of cirques, extensive observation has established that northeast aspect predominates in the northern hemisphere and southeast aspect in the southern. Three factors that control aspect, listed in order of decreasing probable importance, are (1) protection from insolation, especially in summer, (2) distribution of snowfall, and (3) wind-drifting of snow, effective only where the snow is cold and dry. Snow drifts across a highland and drops into pockets on the lee side. A fourth factor, very conspicuous in some areas, consists of lithology and structure of the local rocks. In the Scottish Highlands and in southern Norway quantitative studies of minor structures in the bedrock showed that in those districts cirque orientation is closely controlled by rock structure. Aspect is conveniently represented on polar diagrams (fig 4-7; also R. P. Goldthwait, 1970). Such diagrams have also been used to show (Seddon, 1957) that end moraines of former cirque glaciers in Wales are lowest where the aspect of related cirques is northeast, and become higher with variation of aspect in both directions.

Cirques now inactive (that is, free from summer firn) appear to be grouped, as to altitude, in two different ways. In the first, cirques occur in vertical sequences; in the other, they occur singly. The former arrangement consists of "flights" of two to five individuals, their floors separated from each other by vertical distances of as much as 200m. For example in the Boundary Range, Alaska-Yukon, the members of a five-fold sequence of this kind have been found to occur at comparable altitudes in a number of valleys (M. M. Miller in Raasch, ed., 1961, p. 833), suggesting that each records a phase in a general fluctuation of the snowline. In the second arrangement only a single family of cirques has been discerned, but where this occurs, most of the cirques are believed to have been active at some time during the last major glaciation.

Possible History. We can rationalize the two configurations by considering what would happen during a glacial age, bearing in mind that the floor of a cirque approximates the position of the orographic snowline at the time the cirque was active. As the snowline first falls and later rises, both configurations could develop, depending mainly on rate of change of temperature and on erodibility of the local rocks. With slow change of climate and/or erodible rocks we might expect cirques to form at various altitudes as the snowline shifts. With more rapid change or more obdurate rocks, possibly only the longest enduring position of the snowline would be reflected by the making of cirques. Such a relationship would be favored by the self-extension of a cirque once it is formed. As it enlarges, the cirque becomes increasingly protective, increasing the probability that thereafter snow will accumulate in it rather than in some adjacent area. If the cirque were a low member of a series, it might engulf higher ones as it enlarged. This might explain, at least partially, the common occurrence of only a single cirque in the head of a mountain valley.

We could extend this concept to a succession of two or more glacial ages. It seems unlikely that a cirque could outlast a long interglacial age without substantial modification by weathering, whether mechanical or chemical. Weathering might even reduce the sharp crests of some highlands to positions near the floors of existing cirques before the advent of the next glacial age. Cirques that were not obscured or thus destroyed would be likely sites for the accumulation of firn and ice under a later cold climate, causing old cirques to persist from one glaciation to the next, being enlarged and renewed during each episode of cold climate. Indeed many cirques outside the areas of existing and former ice sheets appear to represent a single orographic snowline at or near its minimum position—that is, a snowline related to a glacial maximum. Possibly the cirques that constitute the second or "single" configuration are a composite produce of two or more glacial ages and reflect a composite snowline rather than that of a single glacial age. Notwithstanding this, it is possible in places to distinguish among cirques dating from two different glacial ages (Dort, 1962, p. 902).

In some districts the record of cirques is more complex. The relations shown in fig 4-5 suggest that in one mountain region, at least, the cirques pertain to an early phase of glaciation. Those in the western part of the district shown must have been engulfed rather rapidly and their further growth halted by the mountain ice sheet that grew up through coalescence of local glaciers. An analogous although not identical sequence of events was visualized by Dort (1957, p. 540) for the Rocky Mountains in northern Idaho.

Again, in highlands in West Greenland and in the Presidential Range in New Hampshire cirques of moderate size were strongly eroded by an ice sheet that overran the entire highland. This implies that the cirques formed or re-formed early in the last glacial age, before the main ice sheet arrived at and covered the locality. Such occurrences are an expectable result of descent of the snowline, followed later by its rise, during the complete course of a glacial age. Possibly their comparative rarity results in part from dearth of critical observations, but it suggests also that the waxing phase of the last glacial age, at least, was short enough to permit only moderate cirque development before the ice sheet arrived and put an end to nivation.

In other highlands cirques have developed wholly or partly since the highlands were deglaciated by an ice sheet that formerly covered them. Such cirques are more numerous and better developed than those of the first kind, and in their case we are sure that their growth or renewed growth has involved periods of only a few thousand years.

Evidently the study of cirques has several facets. Deeper and more extensive studies of cirques are needed before we shall understand the full significance of these features.

ALPINE SCULPTURE [9]

Evolution of Land Forms. A highland affected by a glacial climate is sculptured in a characteristic and broadly predictable way. Sculpture results from a combination of frost wedging and rapid mass-wasting at the heads of glaciers and above their surfaces, and glacial erosion of the valleys. Two predominant land forms result: cirques and U troughs. Combinations of these forms, repeated again and again throughout the highland, constitute the characteristic features of alpine sculpture, so called because it has been studied longer and in more detail in the Alps than in any other major highland.

The broadly predictable sequence of forms, every stage of which has been observed in glaciated highlands, is shown in fig 6-3. In fig 6-3*A* is seen a mountainous highland before the advent of a glacial climate, in fig 6-3*B* and *C* the climate is becoming increasingly glacial, and in *D* it has become nonglacial once more. In *B* nivation is in progress around perennial firn banks, some of which have enlarged to form cirque glaciers and short valley glaciers. Firn and ice are shown as being more abundant on slopes facing toward the right, where incidence of solar heat is least. Meltwater from ice and firn carries rock waste down the steep tributary valleys and deposits it in the less steeply sloping main valley.

Expansion of the glaciers results in their coalescence, forming a large trunk valley glacier in the principal valley (fig 6-3*C*). Glacier-occupied valleys are widened and deepened, tributary valleys are left hanging above them, and spurs between tributaries are blunted and beveled as the larger glaciers grind past them. As the trunk valleys are enlarged they tend to become straighter.

While the valleys are being enlarged the crests of the mountains are sharpened by frost wedging. Continued growth of cirques on opposite sides of a crest eventually reduces the crest to a knife-edged form (an *arête,* so named by climbers in the Alps) kept sharp by frost wedging. As a result the mountain range develops a sharp main ridge with sharp lateral spurs.

Where two cirques enlarging toward each other cut through the ridge that separates them, a sharp-edged gap (a *col* in alpine climbers' terminology) with a smoothly curved profile results. Many alpine passes have this origin. Some of them have lost their sharp crests as a result of glacial

[9] General references include W. M. Davis, 1906; Hobbs, 1910; Ahlmann, 1919; Cotton, 1942, p. 189–205; Tricart, 1963.

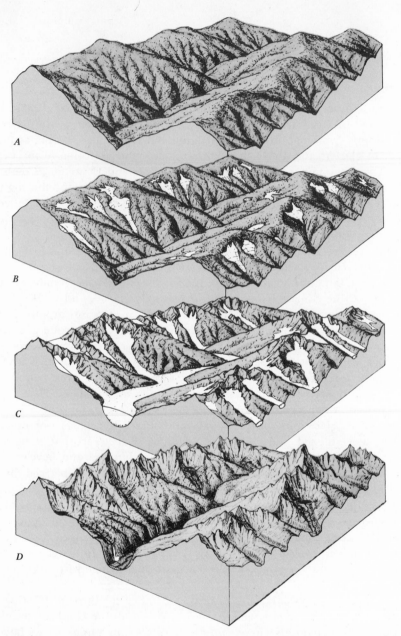

Figure 6-3 Alpine sculpture. *A*. A mountain highland before glaciation. *B*. Growth of firn banks and small glaciers. *C*. Development of a network of valley glaciers. *D*. The same area after deglaciation, showing glaciated troughs, rock basins, faceted spurs, hanging tributary valleys, cirques, arêtes, cols, and horns.

scour at times when glaciers expanded and flowed through them, converting them into smoothed U troughs.

Three or more cirques being eroded inward against a single high part of the mountain crest can sculpture the high part into a pyramid (called a *horn* by climbers) with several facets, each facet being the headwall of one of the cirques. These pyramidal peaks standing above jagged crests are the dominating feature of mountain ranges sculptured by alpine glaciers. All these features appear in fig 6-3D, which shows the mountains immediately after the glaciers have disappeared.

Alpine sculpture is characteristic of mountainous areas occupied now or formerly by a network of valley glaciers that never grew thick enough to bury or nearly to bury the ridges and higher peaks. It also characterizes mountains formerly buried beneath glacier ice but later occupied for so long by a network of valley glaciers that rapid frost wedging of the higher ridges sharpened them and re-created the alpine forms. Alpine sculpture is therefore not in itself evidence that the mountains were never buried beneath glacier ice.

Significance of Glaciated Valley Form. None of the features of a glaciated valley is in itself diagnostic of glaciation. Each individual feature is found in regions never glaciated. The U-shaped cross profile is seen in many mature valleys sculptured only by streams and mass-wasting. The floors of river valleys are commonly marked with basins. Hanging tributaries are created through deepening of a main valley by streams alone. Valley heads with perfect cirque shapes are numerous in tributaries to the Grand Canyon of the Colorado River. But the combination of these features has never been reported from a valley not glaciated, and as each feature, if not glacial, is the product of a different cause, the occurrence of all of them in a single nonglaciated valley is improbable. The weight of evidence, rather than the presence of a single diagnostic feature, justifies the inference of glaciation.

SCULPTURE BY MOUNTAIN ICE SHEETS [10]

The morphology of some highlands shows the effects of glacial abrasion superposed on alpine forms, even on the highest crests. From this relationship it is inferred that a system of valley glaciers like that shown in fig 6-3 enlarged to form an ice sheet, burying the summits, except for scattered nunataks marking the positions of the highest peaks, and subjecting the whole terrain to glacial erosion. This is believed to be what

[10] Kerr, 1936; N. F. G. Davis and Mathews, 1944; Ström, 1945; Flint and Fidalgo, 1964.

has occurred in the part of Antarctica shown in fig 3-15. This kind of glacier is a mountain ice sheet, described in chap 3. Such glaciers existed formerly over extensive districts in British Columbia and Alaska, over parts of the southern Andes, and possibly over some of the mountainous parts of northeastern Siberia. It is shown in fig 6-4, together with the resulting morphology exposed by deglaciation. Horns and arêtes are ground down to make domelike peaks and rounded ridges, and cols are smoothed, deepened, and broadened, in some places linking in a continuous trough two valleys whose heads were formerly separated from each other by a wall-like col. The intensity of sculpture of all these features varies from place to place beneath the mountain ice sheet according to the distribution of snowfall on its upper surface and according to the form of the rock surface beneath it. If, as in the interior of British Co-

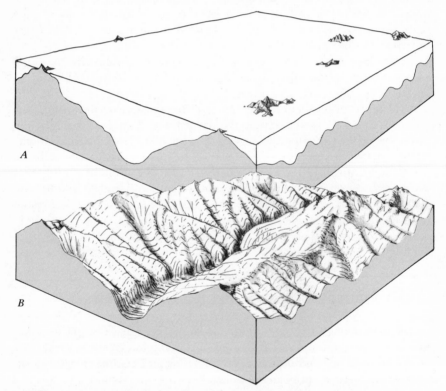

Figure 6-4 Sculpture by a mountain ice sheet. *A.* The area of fig 6-3 *D,* buried beneath a mountain ice sheet. *B.* The same area after rapid deglaciation, showing smoothing and rounding of the formerly buried surface, contrasted with the reduced but frost-sharpened former nunataks.

lumbia, the ice sheet fills a basinlike region having only a few narrow outlets, the intensity of erosion will be less than on a surface that does not confine the tendency of the ice sheet to spread.

This brief description of an ideal sequence is not meant to imply that every alpine district submerged beneath an ice sheet becomes altered to the extent shown in the illustration. Short duration of burial, little movement within the ice sheet, or resistant lithology might result in little modification of the details of the buried terrain, which when deglaciated would show few of the features described above. A related example summarized by Hole (1943, p. 513) is the problem posed by sandstone buttes standing within a glaciated district in central Wisconsin and themselves lacking evidence of having been glaciated. Were they not modified by the ice sheet or were they modified and later "recragged" following deglaciation? Some areas of volcanic rock, marked by buttes and pinnacles and also by patches of drift and erratic stones, present a similar problem. In some such areas the pinnacles are best developed near the limit of glaciation and disappear in the former upstream direction. This suggests that these features antedate the glaciation and were left unmodified only in the zone where the former ice was thin and perhaps also short lived.

SCULPTURE BY PIEDMONT GLACIERS AND LOWLAND ICE SHEETS

The sculpture wrought by a large piedmont glacier differs little from that accomplished by an ice sheet on a lowland. We need not mention individual features, which are described in chaps 5, 7, 8, and 9. But we can note in broader view that a belt of glaciated lowland terrain representing the outer or marginal part of a former ice sheet is commonly constructional, reflecting the predominant deposition of drift. An inner belt, upstream from the margin, commonly contains less drift and reflects predominant glacial erosion. The inner belt may be blurred and indistinct, owing to displacement of the ice-sheet margin during growth and subsequent shrinkage, for most glaciated lowland areas are the product of repeated growth and shrinkage accompanied by repeated overlapping of one belt upon the other. Figure 6-5 illustrates dominant erosion; fig 5-9 shows dominant deposition of drift.

The outer limit of the outer belt likewise can be blurred. The absence of evidence of glacial erosion and (or) deposition within a restricted area does not in itself establish that the area escaped glaciation. In some areas the concept of glaciation is compelled by sparse, scattered erratics or, in their absence, by the presence of erratics on adjacent higher areas. Such

Figure 6-5 Granite terrain deeply scoured by glacial erosion. Rae Isthmus, Melville Peninsula, Northwest Territories, Canada. Erosion has removed the regolith and has exposed the structural features of the denuded bedrock. (Canadian Armed Forces.)

circumstances dictate appeal to postglacial erosion or, alternatively, to special conditions within the former glacier, such as the exceptional deposition of all basal load immediately upstream, mentioned in chap 7.

UPPER LIMIT OF FORMER GLACIATION

The upper limit of former glaciation in highlands cut by large valleys presents analogous problems. Probably the simplest situation is that of readvance of a glacier, down a valley marked by forest or other vegetation, at some time within the last few hundred years. The vegetation is trimmed off neatly through the entire area of contact between ice and ground; the upper limit of this area is a *trimline*. Measurement of the growth rings of trees toppled by the glacier makes it possible to date the re-advance (Lawrence, 1950).

0   1   2 km

Figure 6-6 Trimline and end moraine (dotted line) made during readvance, probably within the last 300 years, of a glacier in the Colombian Andes. Bare drift and recently exposed bedrock (quartzite) are pale colored, showing high reflectivity. Trimline and end moraine of a readvance many thousands of years earlier (dashed line) represent rock material that has been slightly altered, that carries sparse vegetation, and that is less reflective. (After Gonzalez et al., 1965.)

In another situation the re-advancing glacier erodes vegetation (in some cases consisting of nothing more than lichens) and some earlier drift and/or bedrock, changing the appearance of the surface. Because the change involves changes in reflectivity of the surface material, it generally shows up clearly in air photographs (fig 6-6).

In many mountain areas formerly affected by valley glaciers or a mountain ice sheet it is possible to fix approximately, though rarely precisely, the upper limit of glaciation. Where glaciation was vigorous, probably because rate of movement of ice was rapid, this limit is approximated by the highest record of glacial abrasion of bedrock (fig 6-7). Above it, bedrock surfaces generally show the effects of frost wedging. Because in most cases it can be determined only to within about 100 to 200m, the limit is not a line but a zone which, however, is consistent through considerable distances, sloping in the downstream direction of the former glacier. In any case the altitude of the limit represents only a minimum value for the upper surface of the glacier.

Figure 6-7 Apparent upper limit of glaciation (dotted line) separates strong glacial erosion of bedrock, below, from well-marked frost-wedging, above. Cerro Los Angeles, Lago Huechulafquén, Argentina. (R. F. Flint; see Flint and Fidalgo, 1964, p. 341.) (Compare fig 4-5.)

Where glaciation was weak, abrasion of bedrock may not be strongly evident, and the upper limit of glaciation may be recorded best by the altitudes of the highest erratics identified on mountainsides. The results generally have a larger inherent error than do those obtained from the erosional record but, if enough points are determined, they show a consistent rise upstream. Contours on the upper surface of the Cordilleran ice body in northern Washington and Idaho (Richmond et al. in H. E. Wright and Frey, eds., 1965a, p. 232) were constructed in part on a basis of such data.

CHAPTER 7

Glacial Drift: Sediments

ORIGIN AND MEANING OF THE TERM DRIFT

Long before the Glacial Theory was announced, the earth and stones spread over large parts of the surface of Britain were recognized as deposits brought from elsewhere. They were thought of as having been "drifted" into their positions by water or by floating ice. In 1839 the term *drift* was proposed for such sediments; its meaning was essentially stratigraphic. Although rooted in a concept that was in error (except for some of the drift on the deep-sea floor, which is a genuine iceberg deposit) this name was so well established by the time the true origin of the sediment became widely recognized that it persisted in the standard glacial vocabulary. As used today, the term *glacial drift* [1] embraces all rock material in transport by glacier ice, all deposits made by glacier ice, and all deposits predominantly of glacial origin made in the sea or in bodies of glacial meltwater, whether rafted in icebergs or transported in the water itself. It includes till, stratified drift, and scattered clasts that lack an enclosing matrix.

Drift, thus defined, is widespread. It covers large parts but by no means all of the glaciated regions of the world, and extends beyond them along stream valleys and on the floors of lakes and the sea, wherever water could carry drift away from the margins of the glaciers themselves.

THE SEQUENCE: TILL→STRATIFIED DRIFT

Geologists early subdivided drift into two supposedly distinct kinds: *till* or nonstratified drift, and *stratified drift*. Thus they recognized a dis-

---

[1] The short form, *drift*, is used in this book.

tinction that is fundamental although not universal. Till was soon seen to be a direct glacial deposit, for its content of a wide range of grain sizes and the common lack of obvious arrangement of its component particles show that the selective activity of water had played a minimum part in its deposition. It is commonly defined as nonstratified sediment carried or deposited by a glacier; hence the term is both sedimentologic and genetic. The frequently used term "glacial till" is redundant.

In contrast stratified drift, as its name implies, is sorted according to size and weight of its component fragments, and thereby indicates that a fluid medium far less viscous than glacier ice, in other words water or air, was responsible for depositing it.[2]

Till, like drift, is a term that long antedates the Glacial Theory. It is a Scottish word, used by generations of Scots countryfolk to describe "a kind of coarse, obdurate land," the soil developed on the stony clay that covers much of northern Britain. The earliest detailed areal glacial studies published in Britain (e.g., A. Geikie, 1863) were Scottish. Hence the Scots term came into wide use (perhaps its first use in the geologic literature was by James Geikie, 1874), rather than the English term *boulder clay,* put forward at a time when *boulder* referred to a clast of any size rather than to a large clast (H. Miller *fil.,* 1884).

Tillite, a name introduced into the literature by Penck (1906, p. 608) is indurated till.

No sharp dividing line separates till from stratified drift; one grades into the other. The gradation is easy to understand. Throughout the zone of ablation meltwater is present at least seasonally on, in, and under a glacier. In some places such water flushes away some of the finer sediment fractions during or even before deposition by the glacier of the residual coarse sediments. In others, pockets of stratified sediment accumulate in glacial pools and are incorporated in masses of nonstratified sediment. Yet the whole is considered as till.

Hence we regard till as drift dominantly nonsorted according to grain size. It is the non-size-sorted end member of a series whose opposite end member is well-size-sorted stratified drift. Ideally till is formed without the cooperation of water, but actually size sorting is present to an indefinite degree in deposits to which the term is applied. Some till is nonsorted in the field sense, yet is at least partly sorted in the laboratory sense, because its grain-size distribution is bimodal or polymodal.

---

[2] Actually, *washed drift, sorted drift,* or the ancient *modified drift* would be a better term than stratified drift because it is more comprehensive. A great deal of drift has had the fine components washed out of it by water without having been actually stratified. But it is hardly likely that any attempt to supersede the well-established term would be successful.

GENERAL LITHOLOGY AND THICKNESS
OF DRIFT BODIES

Whether stratified or not, drift consists predominantly of rock material that was not decomposed before deposition. Minerals such as hornblendes, micas, and plagioclase feldspars, notably susceptible to chemical decay, are conspicuous in drift derived from rocks that contain those minerals. Most of the clasts, regardless of size, are mechanically broken or abraded. These characteristics indicate that the related glaciers were eroding fresh rock rather than decomposed regolith. Chemical freshness characterizes the earlier glacial deposits as well as the later ones. The inference is plain that weathered regolith did not contribute importantly to drift; the ice soon worked down through the zone of weathering into firmer rock.

Because sheet structures in granites become more widely spaced with increasing depth, the sizes of glacially quarried monoliths increase with depth below the preglacial surface from which they were derived (Jahns, 1943). Hence till in which granitic clasts are of pebble and cobble sizes suggests erosion of a surficial zone, whereas large boulders suggest erosion to greater depth.

Although the thickness of drift is so variable and is dependent on so many factors that local thickness values have no great significance, average values can be meaningful. The largest values in table 7-A occur in buried valleys and represent ancient valley fills. Other areas of thick drift occur in massive end moraines; examples of thick drift in both valleys and end moraines are seen in fig 7-1. Some valleys are so deeply buried beneath drift that their presence is not indicated by the existing topography (for examples see Norris and White, 1961). They are discovered through study of the logs of wells and other borings and through seismic and electric-resistivity depth measurements and other geophysical techniques. Interest in buried valley fills centers mainly in potential water supply. In many districts thick fills of gravel and sand, whether or not covered by till, constitute important aquifers. Other valleys, including several in eastern South Dakota (Flint, 1955b, p. 140, 147–152), were only partly filled with drift, or were re-excavated after having been filled (for examples in Ohio see Stout et al., 1943, p. 51 ff). Such valleys are reflected in the modern topography.

Volume of deposited drift is determined by load, velocity of flow, and time. With high load and velocity values, the time can be short. For example the Sefström Glacier in Spitsbergen built a pile of till 30m thick in less than 10 years.

Table 7-A
Thickness of Drift in Parts of North America and Fennoscandia

Area	Thickness (m)	Estimate (E) Measurement (M)	Authority
1 Spokane Valley, Idaho—Washington	180 to 400	M	A.L.Anderson,unpublished
2 Southwestern Alberta: range	0 to 75	M	J.Westgate,1968,Alta.Res.Coun.
3 av.	15	E	B.22, fig. 7.
4 Central & southern Alberta	~7	E	A.Stalker,1960,GSC B 57,p.3.
5 Great Lakes region	12	E	T.Quirke,1925,Ill. Acad. Sci.Tr.,v.18,p.394.
6 Iowa	45 to 60	E	Kay & Apfel,1929,Ia. G.S.v. 34, p.181, 256.
7 Southeastern Wisconsin	14	E	W.Alden,1918,USGS P. P.106, p.151.
8 Illinois: av.	35	E	F.Leverett,1899,USGS Mon. 38, p.542—549
9 range	0 to ~180	M	K.Piskin et al.,1967,Ill. G.S.Circ.416,p.17.
10 buried valleys	up to ~ 180	M	do.
Central Ohio (~115,000 km^2)			K.Ver Steeg,1933,Sci.,v. 78,p.459.
11 entire area, av.	29	M	
12 buried valleys, av.	60	M	
13 do., max.	231	M	
14 Kirkland Lake, Ontario: gen.range	0 to>75	M	Lee,1967,G.S.Canada Map11—67
15 one point	>223		
16 Western New York: buried valley	>283	M	E.J.Muller, unpubl.
17 Central New York: av.	18	M	Coates,1966,Sci.,v.152,p.1619.
18 Seneca Valley	>327	M	R.Tarr,1904,J.Geol v.12,p.71.
19 New Hampshire: av.	10	M	R.Goldthwait,1949,Artes. wells in N.H.,Concord,p.11.
20 buried valley	121	M	
21 East-central Massachusetts	3 to 5	E	W.Alden,1925,USGS B 760, p.39.
22 Central Quebec—Labrador	<2 to 3	E	E.Henderson,1959,GSC B 50,p.14.
23 Denmark: total range	2 to 28	M	K.Milthers,1959,D.G.U.ser.1,
common range	20 to 40	E	no.21—A,P.91. p.92.
24 Sweden: total range	0 to 194	M	G.Lundquist,1959,S.G.V. ser
av.	5 to 15	E	Ba. no. 17, p.8—9
25 Finland: av.	2 to 3	E	E.Hyyppä,1960,IGC, 21st Excursion C35,p.6.

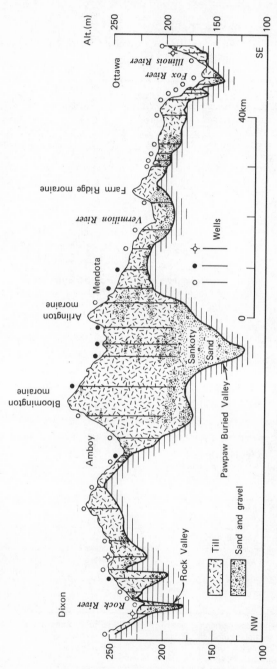

Figure 7-1 Cross section showing thickness and character of drift through a part of northwestern Illinois, reconstructed from logs of wells. Thickest drift lies in buried valleys and in end moraines. (Piskin and Bergstrom, 1967.)

151

Table 7-B
Characteristics of Tills and Tillites Compared with Those of Other Diamictons and Diamictites.[a]

Common Characteristics / Sediment	Clasts			
	Composition	Provenance	Shape	Surface Markings
Subaerial — Till: Drift deformed and injected by ice-pressing	Any composition possible. Clast size related to crushing strength of rock types	Mainly local; minor fraction exotic	Some with facets separated by rounded edges. A few 'flat-irons'	Striations and polish on many
Subaerial — Sliderock, Rock-avalanche debris, Debris-flow sediments, Solifluction sediments	Any composition possible	Local	Predominantly angular	Striations exceptionally on soft-rock clasts
Subaerial — Mudflow sediments, including "flowtill", Flash-flood alluvium, Fan alluvium		Confined to drainage basin	Predominantly worn	
Subaqueous — Slumped debris ("Fluxoturbidites," etc.)		Inherited	Inherited	Inherited
Subaqueous — Glacial-marine drift, including deposits from shelf ice, bergs, and sea ice; sea-floor till; subaqueous sediments bulldozed by tidewater glaciers and floating ice; in part modified by slump and sliding		Resembles that of related till	Some with facets separated by rounded edges	Some with striations
Subaerial or subaqueous — Volcanic breccia	Volcanic	Local	Angular; not faceted	
Subaerial or subaqueous — Fault breccia	Any composition possible	Local	Angular; slickensided clasts faceted	Slickensides may resemble glacial striations and/or polish, but may include internal as well as external ones
Subaerial or subaqueous — Collapse breccia	Any composition possible; commonly carbonate rocks	Local	Angular; some may have solution surfaces	Solution markings

152

Stratification	Fossil Content (exclusive of secondary fossils)	Color	Thickness Extent and Shape of Body	Stratigraphic Relationships
None, apart from thin lenses and transported bodies of stratified sediment. May possess distinctive fabric	None, apart from rare, broken, transported individuals	Inherited, mostly light colored	Broad blankets, tongues; may be discontinuous. Thickness rarely exceeds 100 m	Underlying floor commonly polished, with striations, grooves and crescentic marks
Graded layers in some volcanic—mudflow sediments			Tongues, blankets; may be thick	On, or near bases of, slopes
Graded layers in some volcanic—mudflow sediments			Commonly tongues; relatively thin	On, or near bases of, slopes
Graded layers in some volcanic—mudflow sediments			Tongues and fan forms; may be thin	
Distorted or completely destroyed	If present, may include mixture from several depth zones	Mainly dark colored	Local masses; thickness variable up to 500 m	May be interstratified with graded layers and fine-grained marine or lacustrine sediments
Non-stratified to stratified, with or without graded layers and/or distortion	May be present, broken and unbroken	Color of matrix governed mainly by depth conditions of normal marine sediment. Color of glacial fraction inherited	Broad blankets. Thickness may exceed 300 m	Commonly associated with till in landward direction
May be crudely stratified; more commonly not stratified	Charred logs in some volcanic-mudflow sediments		Local, irregular bodies and extensive layers	Local bodies are related to volcanic vents
			Body narrow, thickness rarely greater than 15 m	May be associated with slickensided surfaces resembling glaciated floors
			Local, irregular bodies, mostly thin	May overlie solution remnants of soluble rocks

a After R. F. Flint *in* Nairn, ed., 1961, p. 148. Assembled with collaboration of J. E. Sanders.

In some areas drift is so thin and discontinuous that it fails to mask the irregularities of the underlying bedrock; in others it has its own topographic form. End moraines are distinct from ground moraine, from outwash, and from sediments built in glacial lakes. Yet although ground moraine and end moraines ordinarily consist mainly of till, some of them are built partly of stratified drift, whereas some masses of outwash include till near their upstream ends. These various forms are discussed in detail hereafter. We are concerned now only to point out that there is no simple and general relationship between composition and surface form, and for that matter, origin.

TILL [3]

Till as a Diamicton. Although commonly defined as nonstratified sediment (or rock) carried or deposited by a glacier, till and tillite have been frequently confused with nonstratified sediment (or rock) having quite different origins. The terms diamicton and diamictite were proposed (Flint et al., 1961a, 1961b) for nonsorted terrigenous sediments and rocks containing a wide range of particle sizes, regardless of genesis. Table 7-B compares some of the characteristics of till with those of other diamictons, many of which have nothing to do with glaciation. Incorrect identification of a diamicton therefore could lead to incorrect inference as to the history of climate. We shall therefore describe till more fully and then attempt to summarize its diagnostic characteristics.

Texture; Grain Size. Till is possibly more variable than any other sediment known by a single name. Its outstanding characteristic is that it is not obviously sorted [4] (figs 7-2, 7-3, 7-4). It may consist principally of clay particles, or principally of large boulders, or of any combination of these and intermediate sizes. Grain-size analysis of a sample of till is accomplished by disaggregation followed by passing the particles through a series of sieves of diminishing mesh. For analysis of the clay-size particles more complex technology must be used. The particles stopped by each sieve are then weighed and calculated as percentages, by weight, of the sample as a whole. The result can be expressed in a ternary diagram (fig 7-5A) whose three apices represent 100%, respectively, of the sand, silt,

---

[3] A general plan of the study of till, with a large bibliography, is given by Lüttig, 1964a. Criteria for identifying tills are listed by Harland et al., 1966, p. 240.

[4] The term "well sorted," widely used by geologists, implies that some grain sizes are not present in the sediment concerned. An engineer would describe the same sediment as "poorly sorted," because for him perfect sorting means the presence of the full range of grain sizes.

Figure 7-2   Massive clay-rich till, Fishell's Brook, Newfoundland. (Sanborn Partridge.)

and clay sizes, all larger sizes usually being neglected. More commonly it is expressed in a cumulative curve, fig 7-5*B*, in which grain size is plotted against percentages that are smaller than stated diameters.

The curves for most tills are bimodal and even multimodal. One cannot confidently explain such asymmetry merely by saying that some sizes were flushed away before deposition. Possibly the asymmetry is inherent in the process of crushing of rock particles during glacial transport; analogs have been observed in experiments with ball mills (cf. H. Lee, 1963, p. 16).

Figure 7-3   Fissile silt-rich till, Fair Haven, Connecticut. (R. F. Flint.)

Figure 7-4 Complex till exposed in wavecut cliff, East Runton, Norfolk, England. It contains a mass of Cretaceous chalk with overlying Weybourne Crag (another sedimentary unit), 54m long and 7 to 10m thick; its width is concealed. The chalk, technically a boulder, was torn from outcrops seaward of the cliff and lifted 18m vertically during glacial transport. Though deformed it still shows stratification, indicated by parallel rows of flint concretions. (Hallam Ashley.)

The slope of the curve of grain-size distribution for a particular till is likely to be distinctive, making it possible to distinguish, in a single district, between two tills that are similar in appearance (Murray, 1953). An empirical coefficient, derived from size distribution determined from bulk samples of till, is the standard means used in Denmark for identifying and correlating till sheets of various ages; a similar basis has been employed in Ohio (Shepps, 1953) for correlation of tills. The reason why this is possible seems to be that an ice sheet transports drift having a characteristic assortment of grain sizes, which it has mixed, through its travel, into a rather uniform mechanical composition.

From such determinations, likewise, rates of decrease of grain size (resulting from crushing, attrition, and dilution) of a particular lithology with increasing distance downstream from its source can be determined (Gravenor, 1951, tab 1; C. D. Holmes, 1952; Gillberg, 1965). The results are fairly systematic. According to Dreimanis and Vagners (1970) these rules govern the distribution of particles in till. (1) In a till that is monomineralic or that consists of minerals having similar physical properties, the distribution of every component is bimodal, with one mode in the coarse (pebble-cobble-boulder) fraction and the other in the fine (sand-silt-clay) fraction. If the parent rocks consist of minerals with a variety of physical properties, the fine fraction is polymodal. (2) Near the source of a till the coarse fraction is large, but decreases downstream and may even disappear. (3) In the fine fraction the modes are typical of the minerals

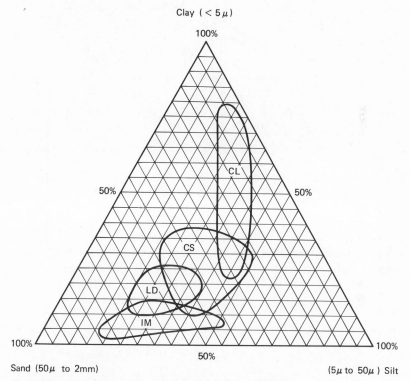

Figure 7-5A Ternary diagram expressing grain-size distribution in four types of till in Ontario, based on analyses of ~400 samples (Aleksis Dreimanis).

IM = Tills consisting of particles of igneous and metamorphic rock. (Tills consisting of reworked glaciofluvial sediments are similar.)

LD = Tills consisting of particles of limestone and dolostone and containing as much as 60% particles of igneous, metamorphic, and clastic rocks.

CS = Tills consisting mainly of claystone and siltstone.

CL = Tills consisting chiefly of clay and silt derived from lacustrine sediments.

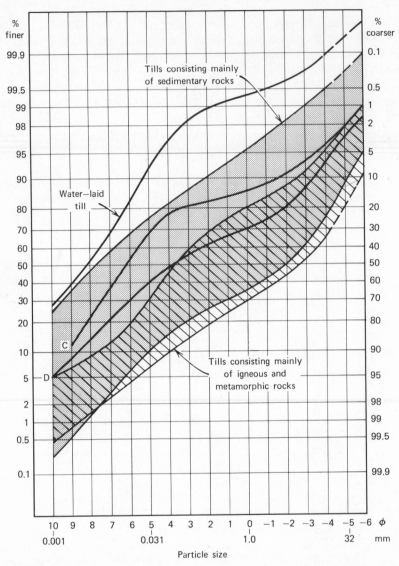

Figure 7-5 B Cumulative curves showing grain size, for particles having diameters smaller than 32mm, in several kinds of till from Ontario (Aleksis Dreimanis). The two areal patterns represent ranges of grain size for tills derived from two distinct classes of rocks. The width of each patterned area at each vertical line indicates the range of percentage occurrence of each grain size in a number of samples analyzed.

D = Till rich in dolostone.

C = Till rich in claystone.

present. Thus a silt-size mode characterizes carbonate minerals, and a sand-size mode characterizes feldspars and heavy minerals. It appears, however, that the distribution of very far-traveled clasts (say from >200km distant) is nonuniform and apparently unrelated to that of clasts from nearby sources. This suggests that the distant travelers may have been carried in positions above the base of the ice, where they escaped crushing and attrition (Flint, 1955b, p. 61).

Degree of compaction in till, and in some other glacial sediments as well, varies considerably. A blow with a pick may make only a dent in one till but may cause another till to crumble. Among the causes of greater-than-normal compactness, are: (1) Grain-size distribution such that pore space is minimum. Most tills containing less than 10% clay-size particles are loose and friable; those with more than 10% clay or more than about 40% clay and silt combined tend to be very compact. (2) Physical settling, progressing with time. (3) Cementation. (4) Pressure of overlying glacier ice. (5) Dewatering. Forcing of water out of a deposited sediment can result in rearrangement of clay particles around the larger grains so as to reduce pore space (i.e., increase compactness) permanently. Exceptionally compact till is often called "hardpan" by well drillers and farmers.

Structure. Although many tills exhibit no obvious structure, some tills rich in silt and clay possess fissility (fig 7-3), which at first sight looks like stratification consisting of laminae a few millimeters in thickness. However, it has the lenticular, flaky quality of light pastry. In places the laminae bend around large clasts. Fissility is not stratification but structure, the cause of which is not certainly known. It has been thought to be induced by the accretion of successive layers of drift from the base of the glacier as it moved forward. Similar fissility has been considered a soil structure, formed long after deglaciation. Probably fissility of both origins exists.

Virkkala (1952) supported the accretion concept for till exposed in eastern Finland. He noted not only fissility but distinct layers as much as 1m thick, separated by very thin layers of till notably rich in clay (fig 7-6). In his view these layers were made by the successive stagnation of thin, overloaded, basal layers of ice in the terminal zone of the glacier. The base of the moving glacier shifted upward in jumps, leaving beneath it drift with interstitial ice. Another mechanism that might explain fissility and thicker layering in till is the accumulation of drift along gently dipping shear planes near a glacier terminus, followed by deposition on the subglacial floor as the ice melts.

Much till, especially till rich in clay and silt, is cut by joints. Some joints are developed weakly whereas others (fig 7-7) are as distinct as those

Figure 7-6 Layered till exposed in a drumlin, Nurmijärvi, Finland. (S. Kilpi, from K. Virkkala.)

in bedrock. Many if not most joints in till are the result of shrinkage caused by desiccation. Some, however, appear to be original structures, because they occur at considerable depths and appear to be related to the fabric of the till in which they occur. In till as in bedrock, joints facilitate quarrying, as is shown by the occurrence of joint-bounded blocks of till incorporated in till of younger age.

Fabric. (General references: C. D. Holmes, 1941; West and Donner, 1956; Glen et al., 1957; H. E. Wright, 1957; 1962, p. 89–94; Kauranne, 1960; Penny and Catt, 1967.) The fabric of till—that is, the arrangement of its component rock particles—is not necessarily confused and chaotic. In many places it is organized. In some exposures organization is apparent on inspection, whereas in others it appears only through measurement of the positions of many stones. It was first noted by Hugh Miller Sr. (1850) that many stones in till lie with their longest dimension paralleling the direction of ice flow inferred from striations. Richter (1932) showed that such fabric can be regionally consistent and can thereby aid the reconstruction of former glaciations of large areas.

C. D. Holmes (1941), analyzing till fabric, found that in some cases stones lie at an angle (commonly 90°) to the direction of glacial movement. He concluded that "parallel" stones had probably slid along the

Figure 7-7 Joints in Kansan till, New Cambria, Macon County, Missouri. Oxidation (dark stain) has penetrated the till inward from the joints which, in turn, have been filled with calcium carbonate (white) deposited by capillary water percolating into them from the till. (S. N. Davis.)

subglacial floor and then lodged in upgrowing till, whereas "transverse" stones had rolled along the floor just before lodging in the till. On the other hand Glen et al. (1957) found experimentally that the "parallel" clasts become oriented rapidly, whereas a transverse pattern develops only after a longer time. Galloway (1956) measured the fabric in englacial drift and in an end moraine of an active glacier. The results included both parallel and transverse fabrics. Andrews and King (1968) encountered and warned against local inconsistencies in measured results.

Fabric studies are useful, because from them the direction of movement of a glacier at a time of till deposition should be determinable with possibly greater accuracy than from streamline forms, striations, and other features, and even indeed where such features are absent. However, results from a single station can be misleading, differing from the average determined from measurements made throughout a district. Also, as noted by Holmes, shape and roundness of a clast are significant in determining its orientation in till. Extensive fabric data should lead to improved understanding of the mechanism by which till is deposited at the base of the ice.

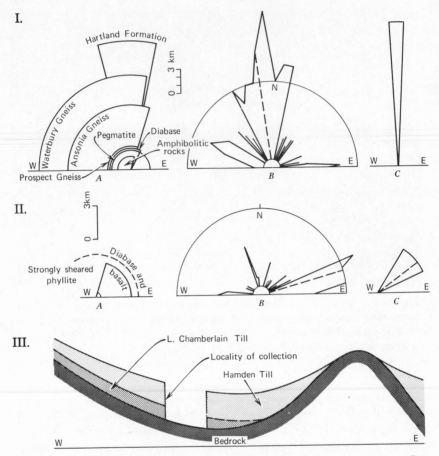

Figure 7-8 Directional data on two successive glaciations in southern Connecticut, expressed in azimuths of three parameters (Flint, 1961).

I. Lake Chamberlain glaciation.

A. Probable directions of transport of 103 stones from Lake Chamberlain Till, represented by radial lines connecting locality of collection with lateral limits of nearest outcrop, upstream, of each bedrock type.

B. Till fabric, based on 87 stones.

C. Directions of striations on bedrock underlying the till at locality of collection.

II. Hamden glaciation.

A. Probable directions of transport of 15 significant stones from Hamden Till.

B. Till fabric, based on 41 stones.

C. Direction of striations on bedrock underlying the till within the district.

III. Idealized stratigraphic relations.

Presumably sliding, shearing, and even flow in till as a body (long ago suggested by Thwaites) could be factors in the creation of a fabric. Measurements of the inclination as well as azimuth orientation of long axes could be useful in determining direction of movement, but although it appears reasonable that long axes should be inclined upstream in an imbrication pattern, data are inadequate to establish the relationship.

Most of the measurements reported in the literature deal with macrofabric, in which pebbles and cobbles are used; techniques are described by Krumbein (1939), C. D. Holmes (1941), and Schmoll and Bennett (1961). Others are concerned with microfabric, with technique described by Ostry and Deane (1963). Results are expressed graphically as rose diagrams or contour diagrams, the latter desirable when inclinations of long axes are plotted. Measurement of macroscopic fabric in drill cores of known orientation has been done successfully.

Fabric measurements have aided in determining that specific drumlins were molded by the glacier that deposited the till rather than having been shaped by erosion by a later glacier (H. E. Wright, 1962, p. 89). Likewise they have aided in differentiating two or more tills in a single sequence (Penny and Catt, 1967; Flint, 1961b) (fig 7-8). Similarly they have helped differentiate between glacially created fabric in till and fabric reoriented by postglacial mass-wasting of the surficial portion of till (Rudberg, 1958) (fig 7-9). Figure 7-10 illustrates the way in which re-

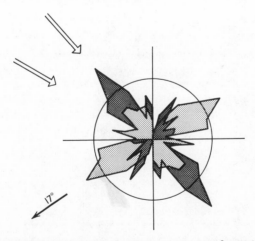

Figure 7-9 Fabric diagram from stones in till on a 17° hillslope in northern Sweden. Open arrows indicate sense of striations; thin black arrow shows direction of hillslope. Dark tone shows fabric of till at 40 to 60cm depth; lighter tone at 0 to 20cm depth. Apparently the upper part of the till has moved down the slope and so has become converted to colluvium; its stones have been reoriented to a new fabric. (Rudberg, 1958.)

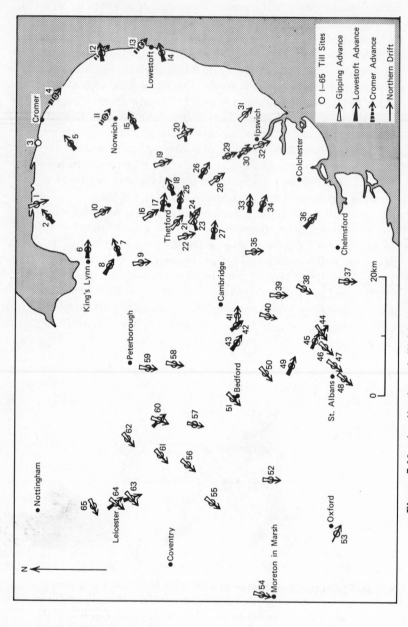

Figure 7-10 Application of till-fabric measurements to several glaciations on a regional scale. Fabric diagrams for 65 sites in central and eastern England are summarized by arrows, which give a consistent picture of two glacial invasions that spread over the region with somewhat different patterns of flow. (West and Donner, 1956.)

164

gional till-fabric analysis can aid in the reconstruction of entire segments of ancient ice sheets.

A concept repeatedly put forward is that the fabric of some tills is the little-altered fabric of the basal drift in former glacier ice. According to this view ice in the base of a glacier becomes so fully charged with drift that it loses its plasticity and can no longer flow. The ice overlying it may or may not continue to flow, overriding the stagnant drift-laden ice underneath. Ultimately ablation of the basal ice leaves drift content essentially in the same arrangement as when it was incorporated in the glacier, an arrangement that might be called a "fossil glacier fabric." The concept dates at least as far back as detailed studies by Slater (1927 and other papers) in Denmark, and was re-examined and strongly favored by Harrison (1957).

Shapes of Clasts. (References: Wentworth, 1936; C. D. Holmes, 1960.) The shapes of the clasts in till vary widely. The majority, especially those consisting of hard rock, possess shapes inherited from the clasts torn out or picked up by the glacier. Many are bounded by joint and stratification surfaces, others by surfaces of irregular fracture, indicating they were reduced by crushing rather than by abrasion (fig 7-11). Still others are rounded, having been worn by streams of meltwater before being picked up by the ice and incorporated in till, or during ice melting. The prevalence of inherited form corresponds with common observation of existing glaciers: some clasts travel for miles on or in a glacier without the slightest modification in their shapes. Clasts can be modified only by crushing or by grinding, and these processes cannot operate on a clast unless it comes into contact, in a moving glacier, with other clasts or with bedrock. We conclude that clasts are altered most readily at the base of the glacier.

Some clasts are cut by fractures radiating from a single point or area. Such fractures are not inherited from bedrock but were made after the stone had been shaped. They must result from crushing, probably between larger stones and under high stress. The progressive abrasion of pebbles of soft rock through minimum transport of 19km is illustrated in fig 7-12. Expectably, crushing of clasts is more common in glacial transit than in streams of water, where attrition is the chief erosional process.

In most tills a small proportion of clasts bear facets apparently unrelated to structures inherited from bedrock. Such facets were made by grinding as the clasts were carried forward in the sole of the glacier in contact with bedrock beneath. Some facets, especially in soft carbonate rocks, are very flat; others, especially in harder-rock types, are somewhat curved. Commonly the edges and corners between adjacent facets are rounded. Facets, and striations as well, are much more common on soft-

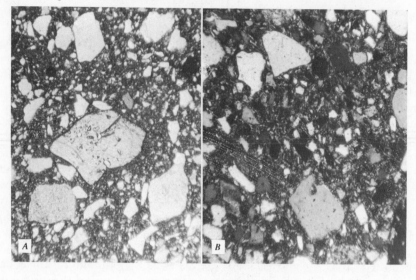

5mm

Figure 7-11 Photomicrographs of till and tillite, showing pronounced angularity of particles and lack of stratification. Most of the particles consist of quartz and feldspars.

A. Dwyka Tillite (Permian), South Africa.

B. Lake Chamberlain Till (late Pleistocene), Bethany, Connecticut.

rock clasts than on clasts derived from harder rocks. Wentworth (1936) found that in 300 cobbles from till in Wisconsin,~10% of the carbonate rocks and~1% of other rocks bore striations. Alimen (1961) found that of 1200 clasts of various lithologies from till in the Pyrenees, 14% bore striations.

Many glacial geologists agree that a pebble or cobble with certain shape characteristics is diagnostic of glacial transport. In its ideal form (fig 7-13) such a clast can be described as roughly pentagonal or triangular in plan. The "base," [5] broad at one end and narrow at the other, is subflat, generally slightly convex in both the transverse and longitudinal directions. At the narrow end the longitudinal convexity increases strongly, like the sole of a well-worn shoe. The sides and "top" are smooth, either rounded or faceted. The clast may be striated on any part

[5] Placed between quotation marks because C. D. Holmes (1960, p. 1657) argued that during its abrasion, this facet was uppermost.

0   3 cm

Figure 7-12 Progressive abrasion of pebbles during transport in the base of an ice sheet. Pebbles of limestone collected from till south of Syracuse, New York, at three points (*A.*~5km. *B.*~13km. *C.*~19km) downstream from the outcrop of the limestone as bedrock. (C. D. Holmes.)

of its surface, most commonly on the "base," and in many cases parallel with the long axis of the clast.

Even though it is agreed that such a clast—even one occurring in an ancient sedimentary rock—is diagnostic of glaciation, opinions differ as to whether the shape described is an end product of erosion during glacial transport or merely a conspicuous intermediate form created during the shaping process. Analyzing the shapes of 626 cobbles collected from till in a single region, Wentworth (1936) found that pentagonal, flatiron-like forms predominated. Alimen (1961, p. 7) found that of 900 clasts from till in the Pyrenees, roughly 50% possessed four to six sides, with pentagons predominating. On the other hand, Holmes (1960), analyzing clasts from till in central New York, concluded that with continued wear, these tend to become rounded, and that although pentagons may be conspicuous in any till, they represent intermediate forms; the final form will

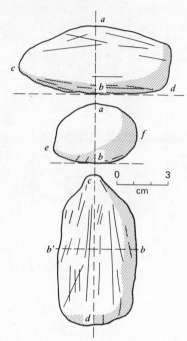

Figure 7-13 Outlines of one form of ideal glaciated clast, viewed in each of three dimensions. Identical points are indicated by similar letters. Drawn from actual specimens. Striations are represented by fine lines. See text for description.

be rounded. Here the matter rests at present. The problem is evident and should be capable of solution.

Comparison of clasts in outwash with those in till of the same age indicates that transport in a proglacial stream modifies glacially shaped stones rather quickly (e.g., Alimen, 1961). Striations disappear within a short distance, particularly on boulders, and the process of rounding of faceted stones begins. Outwash sediments rarely yield clasts having striations or well-developed facets.

Striations and Other Surface Textures of Clasts. Striations can occur on any part of the surface of a stone but are more common on facets than on edges. They occur commonly with a random or scatter pattern, criss-crossing each other (fig 7-8), but are found also with subparallel striations on one facet (fig 7-9). The fact that only a minor percentage of the stones in till of varied lithology bear striations has already been stated.

Compact, fine-grained, far-traveled stones are striated and faceted more readily than friable, coarse-grained stones close to their sources. Some claystones are so soft that their striations disappear during laboratory washing and sieving. Brittle rocks such as quartz, quartzite, and chert tend to be fractured rather than abraded during transport.

The younger Hugh Miller (1884) long ago observed that depth of striation is nearly proportional to the diameter of the striated clast. This results, as C. D. Holmes (1960, p. 1656) pointed out, from the relative instability of small clasts, which change position before deep striations can occur.

Very minute surface textures, visible only with an electron microscope at magnification of ×5000 or more, have been recognized as characteristic of sand grains transported by a glacier. Eight such textures were described by Krinsley and Donahue (1968). They include microstriations, characteristic conchoidal-fracture patterns, steplike fractures, and other peculiarities, groups of which are said to differentiate glacial from littoral and eolian transport. Such textures, like the ideal glaciated stone, make possible the recognition of former glacial transport whether or not the clasts were collected from glacial drift.

It might be added that the making of both striations and facets on clasts depends on distance of transport, abundance of load, and physical properties such as hardness, brittleness, cleavage and texture, and grain size.

Lag Concentrates; Striated Boulder Pavements. Exposures of till show concentrations of pebbles, cobbles, and boulders at the surface (fig 7-14). The sizes, shapes, and lithology of the stones being the same as those in the till beneath, it is inferred that their concentration resulted from sheet erosion of the till on gentle slopes, either immediately after deglaciation and before vegetation covered the till, or, in steppe climates, during postglacial dry times when the vegetation cover was weakened. A widespread pebble concentrate, including many ventifacts, on the "Iowan" (Kansan?) Till in Iowa has been attributed chiefly to deflation. Either sheet erosion or deflation could remove the fines, leaving the coarse fraction unmoved as a lag concentrate, which could then develop until it covered the surface and inhibited further erosion.

Some lag concentrates consisting mainly of boulders and cobbles have a sharply marked upper surface that bevels the stones and in some cases removes the entire upper half of a boulder. The beveled stones are striated in parallel fashion, all the striations being oriented in one direction. The beveled surface is overlain by till, in many instances distinct as to lithology from the beveled till beneath the concentrate. A lag concentrate of this kind is a *striated boulder pavement* (fig 7-14). The feature was first recognized in 1828, before the Glacial Theory was established.

Figure 7-14 Two kinds of lag concentrate. *A*. Pebbles and cobbles overlying Wisconsin till near Vermillion, South Dakota, probably concentrated by runoff from rain. (R. F. Flint, U.S. Geol. Survey.) *B*. Boulder pavement overlying clay-rich till and overlain by sand-rich till. Near Greenway, Manitoba. (J. A. Elson, Geol. Survey of Canada.) A third kind of lag concentrate appears in fig 7-26.

Some striated boulder pavements appear to have resulted from reglaciation of terrain that had been subject for a time to sheet erosion or deflation, and that may have been frozen during the second glaciation. Others may have resulted from rejuvenation of a dwindling glacier, with renewed subglacial erosion succeeding an episode of deposition. At any rate the presence of a pavement is not in itself proof of an interval of deglaciation.

Genetic Classification. The classification of till began far back in the 19th century. Goodchild (1875, p. 75, 94, 96) recognized a difference between compact till and different till overlying it. Torell (1877) laid the groundwork for a twofold classification now in wide use. He distinguished between two kinds of till occurring one above the other in some exposures. The "Upper Till," as he called it, was richer in coarse clasts and poorer in fines than the till beneath it. The distinction was described more fully by Stone (1880, p. 433), and was detailed by Upham (1891). Inherent in the distinction, from the beginning, was the concept that the "lower," "basal," or "subglacial" unit had been deposited from the base of the flowing glacier, whereas the "upper" or "englacial" unit was essentially ablation drift, now recognized as formerly superglacial and to some extent englacial in position, and later let down to the ground during ablation. A good comparative description of the two types, with photographs, was published by Gillberg (1955, p. 497–510).

With evolution of observation and discussion, the two units are now generally termed *lodgment till* and *ablation till* (figs 7-15, 7-16). Lodgment till is thought of as deposited from the base of a glacier. Pressure melting of the flowing ice frees drift particles and allows them to be plastered, one by one or in aggregates as layers, under pressure, on to the subglacial floor and there *lodged* (in Chamberlin's words) in the accumulating drift.[6] No size sorting is involved, but stones tend to lodge with their long axes paralleling the direction of flow. Crushing and abrasion of particles is intense, and the till is compact and may acquire fissile structure as it is built up. The deposition of lodgment till could occur anywhere beneath the zone of ablation, and probably also upstream from the annual snowline. Given time enough, such till could be built up to any thickness.

There seems to be no reason why lodgment till cannot be deposited during both expansion and shrinkage of a glacier, provided the terminal zone is actively flowing during shrinkage. Radiocarbon dates of wood from logs imbedded in lodgment till of Late-Wisconsin age in Ohio indicate deposition during expansion rather than shrinkage, which occurred some thousands of years later (R. P. Goldthwait, 1958, p. 216).

Ablation till is deposited from drift in transport upon or within the terminal area of a shrinking glacier. As the ice melts inward from terminus, top, and base, this drift slides, flows, is dumped, or subsides onto the ground. The resulting till is therefore loose, noncompact, and nonfissile, and its clasts are less strongly abraded than those in lodgment till. During the process of settling, fines are washed away selectively, and all parti-

[6] However, Harrison (1957) argued that the fabric of such till in one district has been inherited with little change from the glacier itself.

Figure 7-15 Ablation till about 1m thick overlying lodgment till, Willimantic, Connecticut. (R. F. Flint.)

cles are reoriented by settling as their matrix of ice melts. Unlike lodgment till, such till consists only of the load existing at the time of ablation, and hence is likely to be thin. However, ablation till as much as 40m thick has been reported (Clayton and Freers, eds., 1967, p. 38). Conceivably such thicknesses are made possible by local sliding or flowing of superglacial drift. That debris flow and mudflow do occur during ablation is indicated by the presence of "flowtill," mentioned hereafter.

Although the separate identity of the two kinds is widely recognized, specific occurrences of till regarded by some as ablation till may be no more than a surficial zone of lodgment till reworked by strong frost action (chap 10). Another possible source of confusion lies in failure to distinguish between ablation till and a layer of lodgment till deposited during a later glaciation. Criteria for discrimination, in which lithologic provenance plays an important part, were set forth by Howard (1965).

A third genetic variety of till is implied by the process of pushing (bulldozing by the advancing terminus) recognized by Chamberlin (1894, p. 528). This process does heap up or steepen small moraines of limited height, but has not yet been identified with a particular variety of till. A fourth variety consists of till in ice-pressed features, on which unfortunately the literature is still scanty.

Relation to Topography. We have noted already that records of borings indicate generally thicker drift (including both till and stratified

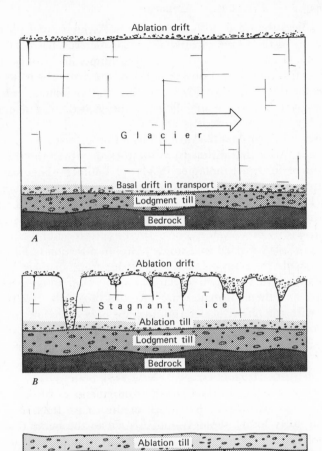

Figure 7-16 Origin of lodgment till and ablation till. *A.* Basal drift in transport lodges over bedrock to form lodgment till. *B.* Later, thin, nearly stagnant ice near the glacier margin wastes away beneath a cover of superglacial drift, from which water has removed the fines. Ablation till is deposited from basal zone of ice. *C.* Postglacial condition. Ablation drift forms thin layer of ablation till over lodgment till.

drift) in valleys than over hills and uplands. Asymmetry of till thickness with relation to the slopes of hills of pre-till material has been observed also. In central New York State, data from numerous borings show that the average thickness of till on lee slopes of bedrock hills is nearly eight times greater than on stoss sides. Whether the asymmetry represents a sort of crag-and-tail, or is analogous to the growth of a drumlin that began with a bedrock nucleus is not known.

Lithology and Provenance. Most till contains at least some particles derived from bedrock that differs from the rock on which the till lies. The recognition of foreign lithologies in glacial boulders dates back to the 18th century. Later, tracing of foreign stones in the Alps to their sources in bedrock (by Arnold Guyot between 1842 and 1848) was a significant factor in confirmation of the Glacial Theory. The source of a transported rock particle is its *provenance*. The earliest application of the provenance of a suite of clasts collected in a single place, to direction of flow of an ice sheet, was made by Hampus von Post (1856) from an esker in Sweden. Later, quantitative treatment of large numbers of units developed, in pebble sizes and then in smaller sizes. Results of counts were treated statistically, with the object of determining not only direction and distance of travel but also patterns of deposition by former ice sheets. Significant modern studies of pebble lithology of drift include: V. Milthers, 1909 and K. Milthers, 1942 (Denmark); MacClintock, 1933 (Illinois); G. Lundqvist, 1935 (Sweden); C. D. Holmes, 1952 (New York State).

The average distance of travel of the components of till is not known. It was long ago established that much of the coarse fraction consists of material of fairly local lithology, and this led to the belief that till as a whole is of predominantly local origin. As an example Alden (1918, p. 220–222) found that of 8000 pebbles from 60 localities in southeastern Wisconsin, 87% were derived from local bedrock, implying an average distance of transport of no more than 64km. However, it has been argued (Shepps, 1953) that the comparative homogeneity of a till throughout a considerable area implies thorough mixing and hence long distance of travel, at least of the fine fraction. Provenance of fines is difficult to establish because these components consist mainly of crushed quartz and feldspar, which are common to many rocks. Hence, although the argument is reasonable, it has not yet been fully established. A method of reducing uncertainty as to direction of glacier movement, by matching indicators to the widths of their outcrop areas upstream, is shown in fig 7-8.

Study of the composition of tills underlying wide regions leads to an understanding of the varying directions of movement of an ice sheet during successive glaciations, without the necessity of pinpointing localized sources of drift. An example is a study of tills of Wisconsin age in Minnesota (Arneman and Wright, 1959) in which glacial invasions from the northeast were discriminated from invasions from the northwest, on a basis of both lithology of coarse particles and mineralogy of fines. A somewhat similar study (Willman et al., 1963) of tills of several ages in Illinois discriminated, on a basis of mineralogy of the fine-sediment fractions, among glaciations that entered the state from three major directions.

Erratics; Indicators. Whether contained in drift or lying free on the surface, a clast that differs in lithology from the bedrock underlying it is an *erratic.* The name derives from an adjective used by Saussure to describe material of foreign origin overlying the local bedrock. *Terrain erratique* was his term for such material. In 1830 De La Beche described as "erratic blocks" boulders lying on the surface at points far distant from their source rocks. The shortened term *erratic* is applied to a particle of any size. Although used mainly for particles transported by glaciers, it is not confined to glacial transport; erratics are identified in shore deposits made by grounded sea ice and in other sediments of unknown origin. Products of mudflow and debris flow, as well as fanglomerate, for example, might include erratics. The origin of a sediment containing erratics would be sought on a basis of criteria other than content of erratics.

Although the locality of provenance of most erratics is unknown, that of many can be placed within a specific geologic province. An erratic whose point or area of origin is known by direct comparison with outcropping bedrock is an *indicator* (V. Milthers, 1909, p. 1). Indicators were first recognized in the western Alps by Saussure (1786–1796, v. 1, p. 201), a sharp-eyed observer whose classic description of them has never been improved upon.

Although many indicators have traveled as far as 500km, none are known to have traveled more than about 1200km. None of those identified in North America between the Atlantic and the Rocky Mountains were derived from any point farther away than the outer fringe of the Canadian Shield (fig 7-17). A line drawn from the locality of an indicator back to the locality of its origin broadly parallels streamline forms and other directional features between the two points. Generally indicators occur in considerable numbers, trailing downstream from their source; their positions are mapped. These families of indicators form linear bands, but much more commonly are fanlike in form, spreading out in the downstream direction (fig 7-18). Although these features were called "boulder trains" in the 19th century when a "boulder" was a clast of any size, we term them *indicator trains* and conveniently call the common fanlike trains *indicator fans.* Examples of linear trains are the Snake Butte train in Montana (Knechtel, 1942) and a much larger train in Al-

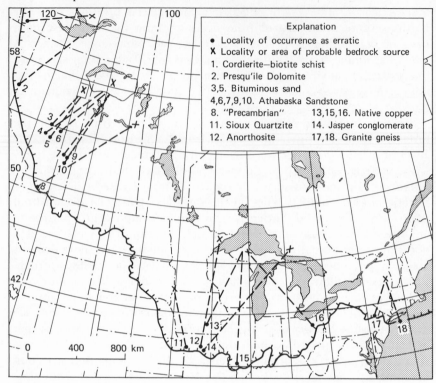

Figure 7-17 Apparent paths of selected far-traveled indicators in North America. True paths are commonly curves (figs 7-18, 7-19). Sources of data: E. Antevs, 1928, Am. Geog. Soc. Res. Ser. 17, p. 65; S. Davis, 1955, unpubl.; Gravenor and Bayrock, 1955, GSAB, v. 66, p. 1326; R. Rutherford, 1941, Roy. Soc. Can. Tr., v. 35, p. 119; C. Slawson, 1933, J. Geol., v. 41, p. 546; C. Whittlesey, 1866, Smiths. Misc. Colls., no. 197, p. 11.

berta (Stalker, 1956). The angular width of an indicator fan can result from one or both of two factors: (1) rate of spreading of an ice sheet per unit of distance traveled beyond the source, and (2) change in direction of flow with time. Probably an indicator fan develops from an initial linear form. The few recorded linear trains may have been made at times so late in a glaciation that no shift in direction of movement occurred subsequently.

Not shown on fig 7-18 are small but wide indicator fans in eastern Finland, consisting of copper ore. Detailed plotting of such ore in glacial drift led to the discovery of copper sources in bedrock hidden beneath till, and to the opening of successful mines (V. Okko and Peltola, 1958).

Indeed the success of such discoveries stimulated the development of computer programs for analysis of indicator fans (Häkli and Kerola, 1966).

The overlapping flow lines plotted in fig 7-18 cannot have been contemporaneous. They must be paths followed at various times, with changes in thickness of the ice sheet; possibly two or more glaciations were involved. More indicator fans have been identified in New England (fig 7-19) than in central North America because the bedrock is much

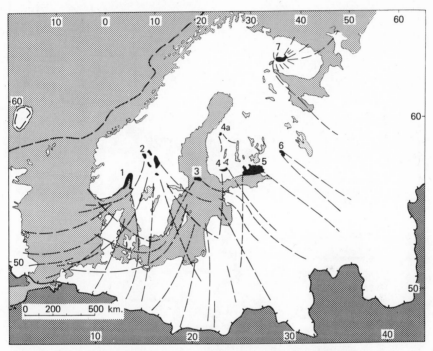

Figure 7-18 Dispersal of indicators by the Scandinavian Ice Sheet. Only the lateral limits of the indicator fans and a few representative inferred paths within them are shown. (After Atlas of Finland, 1910, and M. Saksela.)

Key to Lithology

1. "Oslo district bedrocks."
2. Dala porphyries.
3. Åland Islands Rapakivi granites and quartz porphyries.
4. Satakunta olivine diabase; Hameenlinna uralite porphyry.

4a. Lappajärvi karnaite.
5. Viipuri Rapakivi granites.
6. Lake Ladoga Rapakivi granites.
7. Umptek and Lujavr-Urt nephelite syenite.

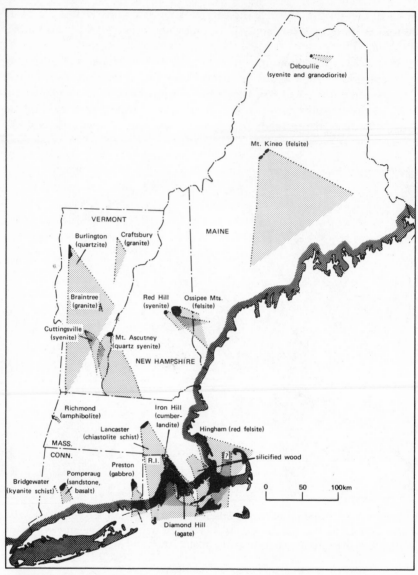

Figure 7-19 Dispersal of indicators in New England. The Richmond feature consists of two distinct, nearly linear trains. The other features are fans, shown on the map by pairs of discontinuous lines representing their apparent lateral margins. Data from published sources, mainly Flint and others (1958), with extensive modifications by J. P. Schafer, and from these unpublished sources: (Deboullie): G. M. Boone; (Pomperaug, Preston): Fred Pessl, Jr.; (Richmond): G. W. Holmes.

more varied and the region is only thinly covered with drift.

Since 1863 at least 82 diamonds of excellent quality have been found in glacial drift at the localities shown in fig 7-20. The diamonds, mostly of pebble size, are erratics but not indicators, as their source is not known, although it is confidently believed to lie in the Precambrian rocks of the Canadian Shield. The broken lines on the map represent possible paths of travel inferred from independent data; they were originally plotted in an effort to locate the diamond source.

In that effort, several attempts resulted in failure, but in 1964 a large construction project coincidentally exposed kimberlite, a type of bedrock known to contain diamonds in South Africa. Thereafter, renewed exploration resulted in the discovery of kimberlite at localities near Montreal, near Kirkland Lake, and west of Port Arthur. Because these three places span an east-west distance of nearly 1000km, the possibility appears that the diamonds found in the drift were transported from several sources in the bedrock.

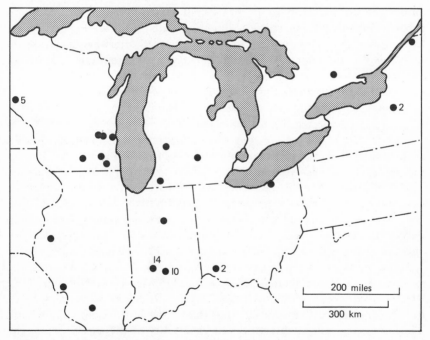

Figure 7-20 Localities (disks) of finds of diamonds in drift in the Great Lakes Region. Numbers are numbers of finds; unnumbered disks represent single finds. (After Gunn, 1968.)

Indicators also demonstrate a vertical component of motion in an ice sheet. A distinctive granite rock occurs uniquely in Ailsa Craig, an island 330m high, in the Firth of Clyde. This rock type is found as erratics in the drift of Wales and of eastern Ireland, as much as 500km to the south and as much as 75m vertically above the highest altitude of the source (W. B. Wright, 1937, p. 67–68) (table 5-A).

Another special type of indicator consists of microfossils sieved from samples and traced to source areas in bedrock. Detailed study (Veenstra, 1963) of till of Saale age in the Netherlands made use of microfossils and also degree of rounding and frosting of quartz grains. Fossil pollen grains, also, have been isolated from till and used as indicators of a regional rather than a point source. Heavy-mineral particles, separated from the quartz and feldspar, are commonly useful as indicators (e.g., Dreimanis et al., 1957), but their use involves uncertainties (e.g., Bayrock, 1962). Finally, where till contains no single distinctive indicator, the proportions of various minerals or lithologies can be significant (e.g., R. C. Anderson, 1957).

The curves in fig 7-5 show that in its type area, the Valders Till contains more silt- and clay-size particles than does pre-Valders till. The fine particles include red lacustrine sediments, eroded from the Lake Superior basin, that did not accumulate in that basin until after the glaciation responsible for the older till. The red particles are therefore indicators.

As yet, however, the statistical significance of indicators is not well known. We can speculate that the proportion of a given kind of rock in a till should be controlled by at least four factors.

1. *Area of outcrop of the source rock upstream.* For example, Eskola (1933) found that in a collection of 961 stones from till near Riga, Latvia, the percentage of each rock type was similar to the percentage of the area of Finland underlain by that rock type. The similarities do not imply that the Latvian stones were concentrated from the whole area of Finland; actually they could have come only from southwestern Finland (fig 7-18). All the rock types occur in that part, and the Scandinavian Ice Sheet appears to have quarried stones from all of them.

A similar relationship between till in Indiana and bedrock along the assumed path of the related ice was found by Harrison (1960, p. 443).

2. *Erodibility.* As shown in table 7-A, the estimated thicknesses of till in New England are substantially less than those in the Central Lowland. The difference reflects differences in erodibility of the source rocks; the igneous and metamorphic rocks of New England yield less readily to mechanical erosion than do the predominant shales and limestones of the country farther west. The greater relief and steeper slopes of New Eng-

land may be a factor also. Again, the predominant particle size into which the rock breaks plays a part. New England granites were glacially quarried along joints to create boulders; boulders in transit were crushed and disaggregated into their component grains; this is the origin of much of the sand-size fraction in New England till.

Indeed, all conspicuously jointed hard rocks yield till composed of boulders and large stones in a scanty matrix of pronounced sandiness. Sandstones yield stones in an abundant matrix rich in sand. Limestones and dolostones yield boulders, cobbles, and pebbles in an abundant matrix of clay and silt that consists largely of minute calcareous fragments. Shales yield till that is chiefly clay. The stones present are mainly rocks other than shale and are derived from elsewhere.

3. *Durability in transport.* The greatest discrepancies between the proportions of rock types in till and the kinds of bedrock from which they were derived are in nondurable rocks such as sandstones and schists, which evidently broke up or wore out in transit to a greater extent than did their durable neighbors. Indeed most far-traveled stones in till are of durable types. A process of natural selection takes place in transit, resulting in survival of the toughest. It is not uncommon to find tough indicators hundreds of kilometers from their sources (fig 7-17).

4. *Distance of transport.* This factor is obviously related to durability, but as pointed out earlier, it is not yet known with certainty how far most of the sediment constituting till deposited by ice sheets has traveled. However, in the collection of pebbles from Latvia, already mentioned, a considerable number of indicators were from bedrock in Finland, and had therefore traveled at least 320km. The bulk of the material, however, was derived from the more erodible rocks of Estonia, Latvia, and the Baltic Sea floor.

The fines, more difficult to evaluate as to source, could be of more distant origin, ground up to small sizes by mechanical wear en route. Analysis of till from eight localities in southern Ontario, a region of sedimentary rocks south of the Precambrian crystalline rocks of the Canadian Shield, showed more material of Precambrian origin in the sand/silt fraction than in the pebble fraction. Hence at least parts of the fine-grained fraction had traveled a minimum of 50 to 200km according to locality (Gravenor, 1951).

It is not surprising that till, particularly the coarse fraction, tends to take on the complexion of the bedrock from which it is chiefly derived. Changes in the character of the underlying bedrock constitute the chief reason why till varies in composition from one area to another (e.g., Flint, 1955b, p. 137–139; Låg, 1948, p. 202, figs 1, 8).

In some districts till is predominantly local at its base, with the proportion of foreign elements increasing upward. This suggests the possibility that the glacier first stirred up the local regolith, mixed it with drift, and redeposited much of it nearby; the redeposited layer protected the ground from further erosion and permitted deposition of far-traveled drift. Such a sequence should be related only to the peripheral part of a former glacier. An alternative explanation was suggested by Dort (1965b).

RECOGNITION OF GLACIAL ORIGIN ⋅ OF A DIAMICT

Having outlined the characteristics of till, we digress at this point to consider the features by which an unknown diamict can be identified as of glacial origin (Flint in Nairn, ed., 1961), p. 142).

1. Abraded surfaces of bedrock, commonly with striations and (or) grooves, underlying the diamict.
2. Whaleback (streamline) form of the upper surface of the diamict or of that of the underlying bedrock.
3. Diamicton or diamictite with:
    a. Wide range of grain size.
    b. Lack of sorting.
    c. Coarse particles not rounded; some faceted; some pentagonal, conspicuous basal facet.
    d. A few particles striated.
    e. Characteristic microtextures on surfaces of sand grains.
    f. Presence of a characteristic fabric.
    g. Lithology generally variable; clasts predominantly fresh.
    h. Thickness and lateral extent variable.
    i. Upper surface possessing constructional features.
    j. Boulder pavements with upper surface uniformly abraded.

If characteristics (3a), (b), (c), (d), and (e) were present in an unknown diamict, and if (1) underlay it, the probability that the sediment is till or tillite would be very high. If (3i) were likewise present, the probability would approach certainty.

FREE BOULDERS

In some areas erratic boulders (as well as smaller particles) overlie till or bedrock, free of any matrix of till. Their distribution is very irregular. In some places they are free because a former enclosing matrix of fine-grained sediment has been removed by wave erosion, sheet erosion, or deflation. In others they were deposited originally in a free condition, either because the glacier contained little drift or because most of the drift in the base of the ice was cleated out by a steep topographic obstacle, as mud is removed from the sole of a shoe by a scraper. An extensive field of free boulders in South Dakota is thought to be the result of such cleating (Flint, 1955b, p. 85–87).

Some free erratic boulders are very large,[7] but these are not numerous, probably because the chances of fracture in transport are great. One of those earliest described (Charpentier, 1841, p. 125), a limestone block near Les Devens in Switzerland, measures 18.5 × 16 × 15m. The Madison erratic near Conway, New Hampshire, is a chunk of granite measuring 27 × 12 × 11.5m and weighing about 4700 metric tons. The Okotoks erratic (Stalker, 1956), 50km south of Calgary, Alberta, consists of quartzite derived from at least 80km to the west. It consists of two blocks that together measure 50 × 16.5 × 7.5m and have a calculated weight of 18,280 metric tons. More remarkable, though slightly smaller, is a tabular erratic of limestone lying free on till in Warren County, Ohio. Its area is more than 1850m², yet its thickness averages only 1.5m. Its weight approximates 13,700 metric tons, and it was transported by ice at least 7.2km. How this was accomplished without breaking the monolith is possibly explained by mechanism 4, listed for ice-thrust features in chap 5 (p. 124). A similar but larger erratic mass of chalk overlying till near Malmö, Sweden, is nearly 5km long, 300m wide, and 30 to 60m thick. Although extensively fractured, the chalk is coherent.

Some large free boulders lie on hills or bosses in such unstable positions that they are easily dislodged. Most such *perched* boulders owe their positions to accident. At every point on the ground the motion of a wasting and thinning ice sheet decreases gradually, until sooner or later it reaches zero. At some points motion will cease at a time when a boulder is perched there. Melting will then free it in the perched position. No doubt many more almost-perched boulders, with even greater instability, slid or rolled away when released by melting, and reached stable positions. Perched boulders, like ablation till, imply deposition from thin, motionless ice.

STRATIFIED DRIFT

Stratified drift (or more appropriately *washed drift*) falls into two classes, based on characteristics that indicate environment of deposition. One class, built upon or immediately adjacent to glacier ice, is *ice-contact*

[7] A catalog of large free boulders in New Hampshire appears in C. H. Hitchcock, 1878, p. 263–275.

Figure 7-21 Ice-contact stratified drift, showing wide range and abrupt changes of grain size, clasts with little rounding, distortion of layers (R), and "flowtill" (dark layer, center). Nonstratified layer at top is windblown sand and silt. Suffield, Connecticut. (F. D. Larsen.)

stratified drift, or more simply, *ice-contact features,* a term that emphasizes surface form as well as internal character. The other class consists of *proglacial sediments* which, although of glacial origin, are built beyond the glacier itself, in streams, in lakes, and in the sea. The two classes grade into each other at the terminus or margin of the glacier.

Ice-Contact Stratified Drift

The characteristics of ice-contact stratified drift consist of distinctive surface form (described in chap 8) and internal features. Internally, three general characteristics distinguish it from outwash: (1) extreme range and abrupt changes in grain size, (2) included bodies of till, "flowtill" (Hartshorn, 1958; see also Boulton, 1968) resulting from superglacial mudflow and debris flow, and marked deformation. In addition, individual clasts tend to be less well abraded than in outwash.

Fig 7-21 illustrates these characteristics, each of which reflects the immediate proximity of glacier ice, generally stagnant, during deposition. Whether accumulation takes place upon, against, or underneath the wasting terminal zone of the glacier, it is likely to be sporadic and irregular, with no intervening distance to smooth out diurnal and seasonal differences in rate of melting and release of sediment. The same site may suc-

Figure 7-22 Ice-contact laminated silt on Columbia River near Bossburg, Washington. Successive bodies of silt, each several meters thick and each deposited in a pool, have collapsed, mostly with accompanying contortion, as supporting ice melted away. (R. F. Flint.)

cessively see a swift stream, a quiet pool, a debris flow, a small avalanche of boulders, and overriding by ice, folding or faulting the layers of sediment or smearing till upon them. Ice is likely to melt out from beneath the accumulating sediment, or from a supporting position beside it, causing sagging, collapse, slump, and debris flow. In such a place anything can happen, and it often does.

A common kind of structure displayed by ice-contact stratified drift is caused by removal of the support of layered sediments by melting of the ice upon or against which they were built. The resulting structures show dips too steep to be primary (fig 7-22); vertical dips are not uncommon. In fine sediments such structures may be accompanied by intense local contortion with recumbent folds and thrusts caused by gravity sliding. Normal faults, mostly with small displacement, occur in sediments both coarse and fine.

Outwash Sediments [8]

General Description. The term *outwash sediments*, or simply *outwash*, is used for stratified drift built by streams ("washed out") beyond the gla-

[8] See a detailed study of the sedimentary geometry of an outwash body 115km long (Jewtuchowicz, 1953).

Figure 7-23 Outwash sediments of Wisconsin age, showing cut-and-fill stratification exposed in long section. Mississinewa River Valley, Wabash County,
Indiana. (W. D. Thornbury, Indiana Geol. Survey.)

cier. Such sediment is *glaciofluvial,* but so are ice-contact sediments deposited by streams. Outwash consists mainly of sand and gravel—bed-
load material that comes to rest on the beds of streams of meltwater as
they flow away from a glacier. The silt-and-clay fraction, carried as suspended load, is generally exported from the region and so is little represented in outwash which, however, shows decreasing grain size in the
downstream direction. Outwash is well sorted, and is commonly stratified
in thin courses of foreset layers, none of which has great continuity because each is partly cut away by younger courses (fig 7-23). Variations in
grain size are numerous both horizontally and vertically, but are far less
marked than they are in ice-contact sediments, and they decrease downstream.

Most streams that build outwash originate, not at the glacier terminus,
but within or upon the glacier itself. The head of the drainage basin may
lie as far up the glacier as the firn limit or even beyond, in extreme cases
many kilometers from the terminus. The drainage system takes the form
of many small streams upon, beneath, and along the margins of the glacier; stream loads are acquired upstream from the glacier terminus. In
some valleys with steep gradients meltwater can sluice away the entire
load picked up in the terminal zone. But ordinarily the gradient is insufficient, the coarser part of the load is dropped, and aggradation occurs;

hence the upstream parts of some outwash bodies are built up to thicknesses exceeding 100m.

Aggradation occurs chiefly because some streams flowing through the glacier have steep gradients and efficient channels, whereas beyond the glacier, slopes are less steep and the proglacial channels are broad, shallow, and inefficient, and are also more numerous because the water, unconfined by walls of ice, can spread out in a pattern of braided channels. However, on inactive, low-gradient glaciers such as the Malaspina Glacier, many channels are inefficient, and aggradation can begin far upstream from the terminus.

Average grain size of outwash diminishes downstream, whereas roundness of particles increases. Rounding of boulders and cobbles takes place even within the terminal zone. The facets and striations occurring on some of the glaciated stones are quickly worn away, for they are rare in outwash.

In the downstream direction, outwash is diluted by an ever-increasing proportion of nonglacial alluvium derived through tributary streams that did not originate in glaciers. Great areas of such alluvium accumulated in eastern Nebraska, and the Minford Silt in southern Ohio (chap 9) is mainly of this origin. Such material has been termed *inwash*. In a long stream, the outwash is eventually lost in this alluvium. But in rare streams like the Mississippi River, which at the maximum of the Wisconsin Glacial Age drained a sector of the Laurentide Ice Sheet more than 2700km in length, the massive body of Wisconsin outwash can be traced down to the delta at the river's mouth, an airline distance of 800km from the mouth of the Ohio, a principal outwash-carrying tributary, and 1100km from the position of the nearest glacier terminus. At this distance the outwash sediment consists chiefly of fine sand and silt.

However, erratic clasts of much larger size occur in outwash. For example, erratic cobbles and boulders are imbedded in Mississippi Valley outwash sand at localities more than 300km downstream from the position of the nearest contemporary glacier ice. Such erratics are believed to have been rafted downstream on blocks of ice, either glacier ice or, more often, river ice carried away during the spring thaw and breakup.[9] No thick blocks of ice could travel far down the shallow water of a typical proglacial stream that is actively building outwash. However, as McGee once pointed out, a rafted boulder $2 \times 2 \times 1.2$m, found in the Chesapeake Bay region, could have been floated on a slab of ice $10 \times 12 \times 0.8$m, the greatest thickness observed in river ice there today.

Surface Form. The simplest form of a mass of outwash is that of a sin-

[9] For a description of this process in operation see A. L. Washburn, 1947, p. 83.

gle fan banked against the terminus of the glacier and with its apex at the point of emergence of a glacial stream from the ice. Small fans are present on the slopes of some large end moraines. Other, larger fans head in pronounced gaps in the moraine. More commonly the outwash mass consists of a row of coalescent fans whose individual apices are still recognizable after outwash deposition has ceased. Massive outwash bodies on Long Island and Cape Cod have this form. Where both meltwater and drift are very abundant, outwash is built up so that it extends gradually headward over the terminus and lateral margins of the glacier, merging with ablation drift in the terminal zone and with kame terraces along the sides of the ice tongue. There is thus a complete gradation from small distinct outwash fans to a great mass of sediment that may bury not only any end moraine that may be present but also the terminal zone of the glacier itself, through many kilometers upstream from the terminus.

Many outwash bodies take the form of valley trains (fig 7-24). We can define a valley train as a long, narrow body of outwash with or without a zone of closely related ice-contact stratified drift in its headward part. If the ice-contact sediment extends from side to side of the valley it is considered part of the valley train. If not, it is classed as kame terraces (chap 8). The boundary between the ice-contact part and the outwash part of a valley train is placed at the down-valley limit of evidence of residual ice during sedimentation.

Both valley trains and fan forms of outwash have braided stream patterns while active deposition is in progress. The braided pattern characterizes all streams, glacial or nonglacial, whose bed loads reach the limit of carrying capacity of the streams. Gradients of 0.5 to 1% in the headward parts of valley trains are common, and exceptional gradients as steep as 7% have been recorded. Figure 22-1 shows representative profiles.

Although valley trains commonly lead directly away from the glacier terminus, some follow the terminus through long distances, depending on the local slopes (Flint, 1955b, pl. 1).

The filling of a valley with outwash takes place more rapidly near the center than along the sides. Accordingly an active valley train reaches its greatest height along its middle line and slopes slightly toward its lateral margin as well as down valley. Furthermore, outwash is built up in a main valley faster than most large tributaries, not fed by meltwater, can aggrade their valleys. In consequence outwash is built into the mouths of tributaries in fanlike or deltalike forms. Many tributary streams with small discharge are dammed and ponded, some of them deeply, by the fills at their mouths, forming lakes that may have a greater aggregate area than that of the outwash itself (Thornbury, 1950, p. 10).

Figure 7-24 Valley train with braided channel pattern. View upstream to the terminal zone of the glacier, blanketed with ablation drift. The principal streams emerging from the ice are lateral. Fedtchenko Glacier, Pamir Mountains, Tadzhik S.S.R. (W. Rickmer Rickmers.)

In some large valleys the gradual accumulation of outwash buries local drainage divides. It also diverts the proglacial stream into adjacent valleys having less outwash and hence lower surfaces. In this way the proglacial Mississippi River was diverted permanently into the valley of the adjacent Ohio River.

The upstream parts of some outwash masses are pitted with *kettles*—basins created by melting of buried blocks of ice after sedimentation had ceased at the site of the kettle. Very small kettles in outwash may be attributable to pieces of ice floated away from the glacier by the proglacial stream. Such streams, however, are shallow, and as any block of ice when afloat is largely submerged, no very large depression could be made by ice floated into place. Most of the kettles in outwash, including all the large ones, result from gradual upstream migration of the head of the outwash over a thin irregular terminal zone of the glacier, which subsequently melts out (chap 8). Apart from trenching and terracing by major streams, most outwash is long enduring because, being very permeable, it resists erosion by local runoff.

Most outwash masses are terraced. In many places terracing occurs as a result of shrinkage of the glacier at the head of the outwash mass. The shrinkage creates a settling basin, generally filled with water, in which sediment brought from the glacier is deposited. The overflow, passing over the abandoned outwash mass, is underloaded, so that instead of adding to the outwash, it trenches it. A valley train trenched in this way is left as a pair of terraces along the valley sides. A nonglacial stream reoccupying a valley filled with outwash, and carrying a load brought from farther up the valley, cuts downward more slowly. It carves the outwash mass into nonpaired stream terraces, with decreasing gradient from the highest and earliest to the lowest and latest. Gradual excavation of thick outwash fills exposes buried bedrock spurs upon which the stream may become superposed, with resulting rapids and falls.

Repeated glaciation of a valley results in the building of successive outwash fills, separately identifiable in exposed sections and in surface form. Examples are valley trains in the northern Alps (Penck and Brückner, 1909) and in the Susquehanna valley (Peltier, 1949).

Time of Deposition. The bulk of the outwash visible today was built during the shrinkage of glaciers of the last glacial age. Whatever outwash was built within the glaciated region during the expansion of those same glaciers was overridden and destroyed or buried before the glaciers reached their maxima. Such overridden outwash has been identified in many districts. Beyond the glaciated region the outwash may consist of basal sediment laid down as *advance* outwash, overlain in turn by *recessional* outwash built up during glacial shrinkage.

However, there is reason to believe that throughout much of the glaciated region less outwash was built during expansion than during shrinkage. Lower regional temperatures and high steep termini of the actively expanding glaciers would have resulted in a zone of ablation of minimum width, with correspondingly small meltwater discharges capable of picking up and reworking the drift contained in the ice. Yet the higher temperatures that later caused shrinkage, the gradual thinning of the glaciers as shrinkage proceeded, and reduced rate of movement of the ice must have conspired to cause much larger discharges of meltwater. Within the belt of conspicuous drift that forms the marginal part of the region covered by the Laurentide Ice Sheet and within the corresponding belt abandoned by the Scandinavian Ice Sheet, the inner, younger part seems to contain more outwash and other stratified drift than does the outer, older part. This fact appears to support the deduction.

Relation to Till Lithology and to Climate. We have noted in this chapter that the lithologic composition of till reflects the composition of the bedrock. Studies in Sweden (Gillberg, 1968) show that the same is true of stratified drift, especially the ice-contact type. The similarity is ascribed to the derivation of stratified drift from till, and to evidence that deposition took place rapidly and without later repetition, affording little opportunity for spreading and mixing.

The fact that some glaciated districts have very little outwash despite an abundance of till has been attributed to preponderance of evaporation over melting in the ablation of glacier ice. This explanation is improbable because the measured values of evaporation in existing glaciers are small, and because channels cut by meltwater streams are present, usually in abundance, in every instance. Evidently meltwater was not lacking. Furthermore, outwash is meager only in districts in which the till consists predominantly of silt or clay. These fractions of the drift could be carried in suspension in the meltwater streams and probably were exported from the district. The sand grains, pebbles, and larger particles, constituting bed load, were so scanty that the resulting outwash is thin. Where these conditions exist in South Dakota, typical outwash bodies average no more than a couple of meters in thickness.

However, other factors that influence abundance of outwash are time and glacier regimen. Apart from till composition, bulk of outwash should be proportional to the time during which a glacier terminus occupied a particular valley, and also to the rate at which drift was delivered to the terminal zone to be released by ablation. Both factors are related to climate. On the northeastern (landward) side of the Coastal Mountains of Alaska, end moraines are bulkier than those on the seaward side of the same mountains, where precipitation is much more abundant. A number of factors are involved, but one may be the greater volume of meltwater on the seaward side. This water could have washed away so much drift from the terminal zones of the glaciers that less drift was available for building end moraines.

Compaction. Although, as has been stated, an active valley train is a little higher at its center than at its sides, the surfaces of many thick outwash fills long abandoned by meltwater streams slope gently from sides to center, reversing their presumed original slopes. Where a thick fill is built into and up a tributary valley, its surface may slope back toward the main mass of outwash, despite the fact that its stratification shows that formerly it must have sloped in the opposite direction. Such reversals of slope are believed to be the result of differential compaction of sediments of the fill. The porosity of such sediments when first deposited is 25 to 50%, but this is gradually reduced throughout the entire mass as settling takes place. If the fill is thick and if the bedrock floor beneath it slopes in some direction (as from the side toward the center of a valley), net compaction will be greatest where the floor is lowest. In this way an initial slope can be reversed through distances up to several kilometers, and the resulting surface may slope backward at a rate as great as 0.5%.

Sediments Built in Standing Water

Lake Basins. The distribution and internal character of stratified drift show that much of it was deposited in lakes held in temporary basins, of which one side consisted of glacier ice. The majority of such basins were formed by the margin of a glacier advancing over ice-free ground that sloped down toward the glacier. This situation occurred commonly in mountain valleys [10] and on broad regional slopes. Perhaps the grandest example on record is that of the glacial Great Lakes. On a front more than 2000km long from the upper St. Lawrence to Saskatchewan, great lakes fringed the shrinking Laurentide Ice Sheet. Throughout much of this distance the waters at their northern shores lapped against glacier ice from which bergs broke off and floated southward. Fed chiefly by melt-water, the lakes overflowed through a varied succession of spillways across the divide that separated valleys draining toward the ice from those draining directly southward or eastward to the sea. The sequence of lakes and spillways is described in chap 21.

Another type of basin, smaller and less common, was created by a dam of drift or other deposits across a valley fed with meltwater by a glacier upstream. A conspicuous example is the basin of Lake Hitchcock, which stood in the Connecticut Valley in Massachusetts and Connecticut during the Late-Wisconsin deglaciation. The lake was at least 80km long and possibly much longer; it endured through hundreds if not thousands of years. The dam was a thick mass of stratified drift at Rocky Hill, Connecticut. The outlet was a spillway over a bedrock threshold west of the dam, so that the lake was analogous to a modern reservoir held in by a high dike of earth, with a lower concrete spillway to one side. Outflow from the lake poured over the ready-made spillway, detouring part of the drift-filled Connecticut Valley, until the slow processes of local erosion succeeded in breaching the outwash fill. The outlet then shifted to this breach and the lake was drained. While the lake was in existence, however, thick deposits of sand, silt, and clay accumulated in it. When the Connecticut River, replacing the lake, eroded them, they were left as terraces.

Similar relations characterize the basins of Okanagan and Skaha Lakes occupying the Okanagan Valley in southern British Columbia. The drift dam and the abandoned bedrock spillway near the town of Oliver, as well as extensive deposits of laminated silt made in the glacial lake itself, at least 120km long, are clearly evident.

[10] See the classic paper by Jamieson (1863) on Glen Roy in Scotland; also discussions of former glacial lakes Missoula (Alden, 1953, p. 154–165) and Coeur d'Alene (Dort, 1960) in Montana.

Figure 7-25   Outwash delta exposed in radial section, showing foreset, bottomset, and topset beds. Salmon Brook, Glastonbury, Connecticut. (W. E. Reifsnyder.)

Still another sort of basin is created by a valley glacier flowing out of a tributary valley into a main valley which it blocks completely. The main valley is then filled with meltwater coming from a second glacier farther upstream. These relations were common in the North American Cordillera during the Wisconsin Age.

A large number of glacial lakes, usually small and inconspicuous, are created where a glacier in a main valley forms a dam across the mouth of a tributary valley and ponds the water in it. Nearly every glaciated region of considerable relief shows examples of this kind of lake, and 53 existing lakes of this kind have been inventoried (K. Stone, 1963) in one segment of coastal Alaska.

Glacial-Lake Sediments. The sediments of glacial lakes include five principal kinds: till, delta sediments, bottom sediments, rafted erratics, and littoral sediments.

Till occurs among the deposits of the former Lakes Agassiz and Dakota (chap 21). It appears to differ little from subaerial till, but is identified as lacustrine by interbedding with lake sediments.

Some deltas built into lakes by meltwater streams consist of coarse, bed-load sediments; their foreset layers are steep (fig 7-25). They include normal deltas, built into glacial lakes by streams of nonglacial origin, and ice-contact deltas built at a glacier terminus and in places connected with eskers (Trotter, 1929, p. 573). Massive examples occur in the Salpausselkä ridges in Finland, described in chap 8.

Sediments of lake floors consist of the finer products of glacial erosion, chiefly silt and clay, whereas most deltas consist mainly of gravel and sand. A small delta can be made very quickly, perhaps in a single season, but in the same period only a thin deposit of fine sediment can be spread over the floor of an extensive lake. Many glacial lakes are drained before more than a thin deposit of sediment has accumulated on their floors. Others, particularly small lakes, have been completely filled with sediment. In these the deposits become coarser near the top, indicating gradual shoaling and correspondingly increased capacity of lake currents to sweep suspended sediment toward the outlet.

Fine-grained bottom sediments of some glacial lakes are rhythmically laminated, having alternating coarser and finer laminae. Some lamina couplets of this kind are believed to represent annual cycles of deposition and are used as a basis for time calculation (chap 15).

The bottom sediments of many glacial lakes include erratic clasts, both singly and in clusters. Apparently these were rafted on floating ice, either on bergs broken off from the glacier or lake ice broken up and set free during spring thaws. In some sections of lake-floor deposits erratics diminish from the base upward, suggesting that floating ice was common at first but later disappeared as the glacier shrank. Deposits made in the earlier glacial Great Lakes include nests of erratic boulders and stones, each nest apparently marking the site of an overturned or grounded berg that left its load on the spot (von Engeln, 1918). Preliminary measurements (Schmoll, 1961) of rafted stones in lake-floor sediments suggest a significant orientation. When developed further, measurements of this kind may serve to distinguish such sediments from till.

Till-like sediments lying upon or between lake-floor sediments may have been deposited from grounded bergs in which, during melting, there was no opportunity for sorting of the contained drift. Abundant berg deposits, however, imply deep water, because glaciers ending in shallow water can discharge only small pieces of ice.

The shores of former large glacial lakes such as the glacial Great Lakes are marked by shore features and littoral sediments. On slopes steeper than about 1.5°, wave-cut cliffs (in unconsolidated sediments but not in bedrock) occur. On gentler slopes beach ridges built of sand and gravel are characteristic; some may be longshore bars, but probably the majority are subaerial at least in part, and include storm beaches.

As such features are directly or indirectly the result of wave work, they are best developed in wide, enduring lakes. Around the shores of many former small or narrow glacial lakes shore features are not present, as a result of (1) feeble waves and currents, (2) short life of lake, (3) fluctuation of water level by changes in the outlet or by crustal warping, (4) susceptibility of faint shore features to destruction by mass-wasting, or (5) presence of bare firm bedrock or glacier ice along parts of the shores.

Massive beach ridges are likely to be associated with rising water levels and advance of shorelines upon the land, because of continuous reworking and concentration of beach sediment. Small ridges are more com-

Figure 7-26 Lag concentrate of boulders at the shoreline of a former glacial lake. Hyde County, South Dakota. (R. F. Flint.)

monly associated with receding shorelines, when sediment is left behind mainly as beaches resulting from individual storms.

The shores of some former glacial lakes are marked in places by low, steep ridges caused mainly by the shove of ice pans, driven by the wind during storms, against littoral sand and gravel (Peterson, 1965). These are distinguishable from beach ridges by their discontinuity, more irregular form, and less well-sorted character.

Other shore features consist of linear concentrates of boulders, left as a lag along former shorelines, by wave erosion of till from which the smaller size grades were removed (fig 7-26).

Glacial-Marine Sediments.[11] *Nomenclature.* Glacial drift (in the original meaning of the term) is widespread on and beneath the sea floor. First recognized in the late 19th and very early 20th centuries, in samples dredged up from the sea floor around the Antarctic Continent and called by Philippi *glazialmarine Ablagerungen* (glacial-marine deposits), such sediments have since been identified both at and near continental coasts and beneath the open ocean far from land. Observation of glaciers, icebergs, and sea ice today indicates that large quantities of rock detritus are deposited beneath floating glaciers and ice shelves, by the icebergs calved from these, and by sea ice of nonglacial origin. Such transport and deposition of detritus must have been far greater during glacial ages, and geologic discovery is yearly confirming the truth of this statement.

It is not easy to apply names to such sediments without ambiguity. Glacial sediments grade into those of ice shelves and icebergs, and indeed into those of sea ice as well. The sediments deposited on the sea floor by glaciers, shelves, and bergs are certainly glacial drift. But sea ice, being frozen seawater, is not likely to acquire and transport much glacial drift. Most of its detritus is likely to be alluvium washed onto it at river mouths by floods resulting from spring thaw in high latitudes. Yet, when

[11] See discussion in chap 27.

sampled on or beneath the sea floor, far from its source, such detritus may be difficult to distinguish from sediment of glacial origin.

Furthermore, we can expect gradation in the character of sediment, from glacier termini at a coast seaward into the marine realm. At the coast the sediment should be almost entirely glacial, but should gradually acquire normal marine characteristics in the seaward direction.

In this discussion the whole sequence is considered to be glacial-marine sediments in the sense of Philippi, W. H. Bradley et al. (1942, p. 4), Menard (1953, p. 1282), and later authors. Such sediments, likewise, can be considered glacial drift, except for much of the sediment dropped from sea ice. Many authors have referred to the deposits made from floating glacier termini, ice shelves, and icebergs as glacial-marine drift, or marine drift. The term is a good one, but we exclude from this category till deposited on land and subsequently submerged in the sea; such till is likely to be widespread in glaciated areas of the continental shelves. It is drift but not marine drift, which differs from it as noted hereafter. The area of uncertainty between terrestrial drift and marine drift is much like that between till and stratified drift. It seems a useful distinction to restrict the term till to the terrestrial diamictons to which it was applied originally, and not to apply it to glacial-marine diamictons.

Characteristics. Glacial-marine diamictons resemble till in possessing some clasts that are faceted and striated with no secondary rounding. They differ from till in two main respects. (1) They include, in places, lentils of sorted and stratified sediment, chiefly sand, some of which grade laterally into the diamicton facies and which probably were reworked on the sea floor. Indeed the diamicton itself may be recognizably bedded. Such stratified material, it is true, does occur in some tills, especially in tills deposited in glacial lakes, but it is less abundant than in glacial-marine diamictons. (2) They contain, in places, marine fossils such as bivalves in growth position (fig 7-27), and barnacles and worm tubes attached to clasts. In addition such sediments may have larger void ratios and lesser bulk densities than have comparable tills (Easterbrook, 1964). Finally they should contain a larger proportion of fines than does till derived from the same province, because in the marine realm export of fines by proglacial streams is not possible.

Some marine diamictons have been found interbedded with and overlain by turbidites, implying the activity of slump and turbidity currents during and after peaks of glacial activity.

Descriptions and discussions of glacial-marine sediments exposed on land include D. Miller, 1953, p. 22–35 (old drift >1100m thick); J. E. Armstrong and Brown, 1954 (young drift >135m thick); Easterbrook, 1963; sediments raised on oceanic cores are discussed in W. H. Bradley

Figure 7-27   Articulated pelecypod in glacial-marine diamicton of Late-Wisconsin age near Bellingham, Washington. (Easterbrook, 1963.)

et al., 1942; Menard, 1953; Holtedahl, 1959. Sedimentation from ice shelves is discussed in Carey and Ahmad, 1961 (theory); Drewes et al., 1961, p. 663–665 (morphology, including "pseudo end moraine"). Glacial-marine sediments deposited in local basins, such as the St. Lawrence Valley in Late-Wisconsin time, are described in Karrow, 1961, p. 101). The distribution and stratigraphy of glacial-marine sediments are discussed in chap 27.

Erratics that could resemble sparse glacial-marine drift are rafted and deposited along coasts in cold climates by sea ice, which also abrades bedrock in places, creating features that at first sight suggest glacial abrasion. Activity of sea ice, and its products, are conspicuous along the shores of the St. Lawrence River estuary in Quebec (e.g., Dionne, 1968).

CHAPTER 8

Morphology of Glacial Drift

CLASSIFICATION AND TERMINOLOGY

Having examined glacial drift as a sediment, we are ready to examine drift in terms of its morphology. We have noted that because intimate mixtures of till and stratified drift as well as gradations between one and the other are common, clear-cut distinctions are difficult to make. Such distinctions are even more difficult in discriminating among morphologic types of drift, again because of mixtures and gradations. This fact compels caution in the use of technical terms, because when these are not based on clear definitions they are likely to be more confusing than helpful. The few basic terms that are widely used are rooted in the early history of research in glacial geology. They were applied when understanding of glacial activities was comparatively primitive, and as understanding grew, definitions, whether explicit or implicit, had to be revised, although they were kept as close as possible to the original meanings. We still have a long way to go before we can achieve a terminology that is widely acceptable. Already various new terms have been introduced with various degrees of success. A new term should be proposed only after much consideration and then only if accompanied by a carefully worded definition, because it is difficult to extirpate a term once it has entered the literature. Terms applied to drift morphology should restrict variations as little as possible because there are likely to be many potential variations. True, the variations and gradations among rock types are treated terminologically through the use of arbitrary quantitative boundaries. But it is very difficult to quantify the diameters, slopes, and shapes of groups of hills and wrap up the result in a practical field classification. Until a workable classification has been achieved we shall have to content ourselves with the use of descriptions wherever possible, and make as few alterations as we can in the use and meanings of classical terms. These precepts form the basis of the following discussion.

MORAINE [1]

Moraine is an ancient French word long used by peasants in the French Alps for the ridges and embankments of earth and stones around the margins of the glaciers in that region. It appeared in the literature as early as 1777 and was taken up and used by Saussure, and later by Venetz and Charpentier, and was given wide currency by Agassiz. Its early use carried a topographic implication. The recognition later of a wide variety of forms of drift fashioned by large ice sheets made it necessary to extend the original quite limited meaning of the word. Accordingly we now think of moraine as an accumulation of drift deposited chiefly by direct glacial action, and possessing initial constructional form independent of the floor beneath it. This definition is preferable to that implied in literature of Scandinavia and some other parts of Europe, in which *moraine* is a synonym of *till*. It is preferable because it is both unambiguous and closer to the original meaning of the word.

Ground Moraine

Moraine is subdivided into two chief kinds, ground moraine and end moraine. Ground moraine has always connoted the concept of accumulation beneath the glacier. The early term *moraine profonde* was translated into the German *Grundmoräne* and so into the English *ground moraine*. We now define ground moraine as moraine having low relief devoid of transverse linear elements (Flint, 1955b, p. 111); this definition distinguishes it clearly from end moraine. The term is used without an accompanying article.

Ground moraine is at the surface throughout wide areas south and west of the Great Lakes and in the Great Plains region of west-central Canada, as well as in many parts of northern Germany, Poland, and western Russia. It forms undulating plains marked by gently sloping swells, sags, and depressions (closed or not) with apparently random pattern and local relief generally less than about 6m. The chief material composing ground moraine is till, mainly lodgment till but in many areas ablation till as well. In places such moraine is broken by bedrock hills that project through it and, as in New England, by areas of till so thin that they lack

---

[1] The basic reference for historical data is the voluminous work by Böhm (1901). Physical relationships and current terminology are discussed in Flint, 1955b, p. 111–117. Good descriptions of recent moraines appear in V. Okko, 1955, p. 27–69.

topographic expression other than that of the immediately underlying bedrock surface. Thin drift of this character is not properly moraine; it is merely part of a drift sheet, a stratigraphic but not a morphologic unit.

Areas of till with grooved or fluted surfaces (figs 5-13 and 5-21) and groups of drumlins (fig 5-9) are not ground moraine; terms such as "fluted ground moraine" and "drumlinized ground moraine" are undesirable because if they were accepted, ground moraine would be merely a synonym of drift sheet instead of retaining the character implied by its distinctive topography.

End Moraine [2]

Varieties. The term *end moraine* (*moraine terminale* of Agassiz, 1840, p. 97) has always connoted the concept of accumulation at the glacier margin. We define it as a ridgelike accumulation of drift built along any part of the margin of an active glacier. Its topography is primarily constructional. The word is often prefixed by an article (for example *the Bloomington moraine*), indicating that a particular ridge is meant. It is used also without the article, to refer to a part of an extensive end moraine or a complex of two or more ridges. A well-defined complex of many ridges, having great linear extent, is an *end-moraine system*.

A *terminal moraine* is an end moraine built along the downstream or terminal margin of a glacier lobe occupying a valley.[3] A *lateral moraine* is an end moraine built along the lateral margin of any glacier lobe occupying a valley. Ideally, therefore, a lateral moraine grades into a terminal moraine (fig 8-1). A special variety of end moraine is an *interlobate (end) moraine* (fig 5-10) built along the line of junction of two adjacent glacier lobes.

The essential thing about end moraine is its close relationship to the margin of the glacier. Many end moraines are no more than a marginal thickening of a drift sheet, which may or may not be characterized upstream by ground moraine. Although normally it forms part of a drift sheet, which is a stratigraphic unit, an end moraine is not a stratigraphic unit but a morphologic unit (Richmond et al., 1959, p. 671).

Form. The form of an end moraine is the initial, ice-built form modified by erosion during and after building. The initial form results from (1) amount and vertical distribution of drift in the glacier, (2) rate of movement, (3) rate of ablation, and (4) amount of meltwater. The greater

[2] General references include V. Okko (1955, p. 57–69) and Flint (1955b, p. 111–112).
[3] This term and its relationship to lateral and end moraine date from 1863. Subsequently it was defined in several ways, some of them vague. The usage of Chamberlin (1883, p. 302), in which the term was restricted to "the termination of important glacial advances" is especially undesirable because it is ambiguous.

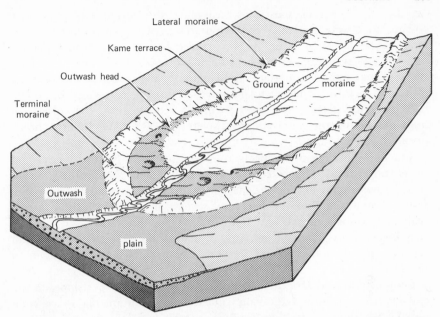

Figure 8-1 Simplified sketch of a valley lobe of drift, showing the terminal and lateral elements of an end moraine, an area of ground moraine, and a body of outwash. At first the latter was fed by streams originating along an extensive segment of the end moraine. Later, after the margin of the glacier had retreated a short distance, outwash was deposited inside the end moraine, over the ice. After further retreat, the proximal part of the outwash collapsed, forming an outwash head. Not to scale.

the load, movement, and ablation of the glacier, the more massive the end moraine. The greater the amount of meltwater, as in climates characterized by warm rains, the greater the proportion of stratified drift to till in the moraine. The rapidity with which the ice-built form is destroyed by erosion depends on its slope, on the permeability of the drift, and on the climate.

The topography of end moraines varies from smooth, gently undulatory surfaces, common in moraines built of clay-rich till, to sharply irregular surfaces marked by knolls, hummocks, and closed depressions, common in moraines in which coarse-grained till and stratified gravel are conspicuous. Some of the gravel knolls are fans built out from the front of the ice by meltwater, and then destroyed by collapse of their upstream parts when the ice melted.

The ridged form of an end moraine is usually discontinuous. It is broken by gaps (fig 8-1), most of which mark the positions of meltwater streams that were generally contemporaneous with the building of the moraine. Some of the gaps, however, appear to mark sectors in which no moraine was ever built, either because of scanty load in the ice or because the terminal part of the glacier had become stagnant at the time. The ground-plan form of most end moraines is curved, reflecting the lobate margin of the glacier that built them. In some regions the curves are multiple, giving the moraines the appearance of festoons, clearly shown on many glacial maps.[4]

The end moraines of ice sheets, common in central North America and in central Europe, rarely exceed 30 to 50m in height, though the two sub-parallel bases of each moraine may be many kilometers apart. Some moraines have a single crest; others have several, indicating that they are in reality successive end moraines built in such close proximity that together they form a unit. The end moraines of many valley glaciers in mountain regions are much higher and steeper, reaching 150 and even 300m in height.

Composition and Origin. The greater height and steeper slopes of the end moraines of some large valley glaciers in mountain regions as compared with those of ice sheets result in part from the steep gradients, rapid flow, and rapid erosion associated with valley glaciers. A greater load of drift scoured from the valley floor and avalanched from the valley sides is spread through a thicker zone at the base of the glacier. Thus an initially high morainic ridge is likely to result. The ridge is added to rapidly because rapid flow permits the basal ice to carry its load high on to the growing moraine before ablation brings about deposition. Sluicing away of fine sediments is more effective in steep mountain valleys than on lowlands; hence the residual mass is likely to be coarse, permeable, and resistant to erosion.

The clay-rich till of many lowland regions, however, forms lower moraines partly because it is built up chiefly by lodgment beneath the ice and is therefore low to begin with. Also, because of its high content of clay and silt, it yields readily to solifluction and creep, which further reduce the initially gentle slopes.

End moraines consist of till or of various proportions of till and stratified drift. Much of the stratified drift was deposited (1) in ephemeral ponds and puddles between glacier and moraine or in small basins in the surface of the moraine itself, and (2) as fans and stream-channel fillings,

---

[4] Maps of large areas which represent end moraines well include central and eastern United States: Flint et al., 1959; Canada: Prest et al., 1968; Ohio: R. P. Goldthwait et al., 1961; Northwest Pennsylvania: Shepps et al., 1959; eastern South Dakota: Flint, 1955b, pl. 1.

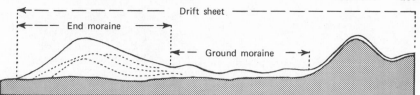

Figure 8-2   Ideal radial section showing relations of an end moraine, ground moraine, and drift sheet. Dotted profiles suggest three earlier phases of upbuilding of the end moraine. (A more complex sequence of end moraines is shown in Lüttig, 1964b, pl. 7.)

mainly on the distal slope of the moraine. Through fluctuation of the position of the glacier terminus, such material can be overridden and deformed, and thus be incorporated into a complex mixture of till and stratified drift.

Regardless of composition, the concentration of drift in narrow end moraines indicates that the glacier terminus occupied these positions longer than neighboring positions. Thus at times of moraine building the position of the terminus represented equilibrium between rate of advance and rate of ablation. An exception is the *push moraine,* as Chamberlin (1894, p. 525) called it, a narrow, steep-sided ridge not known to exceed 9m in height, made by bulldozing by an advancing terminus.[5] Push moraines are small, apparently because a more massive heap of drift will cause the ice to fail and override it, thereafter depositing drift mainly by lodgment.

Although no quantitative study has been published, it has been reported that end moraines generally contain a greater concentration of boulders than does till of the same age in the upstream direction. Possibly the difference results from selective flushing away of finer particles at the glacier terminus, while drift of all diameters continues to be delivered there.

Rapid advance or retreat of the glacier terminus spreads drift over a wide zone and in normal circumstances would create ground moraine. Slow advance or retreat across a particular zone builds up thick drift in that zone. The resulting end moraine should act as a cleatlike impediment to the moving terminus and should induce further accretions of drift upon itself, particularly on its upstream side (fig 8-2). Repeated advances of a terminus are therefore stopped by the barrier; only when shrinkage becomes rapid does growth of the moraine cease. Probably this

[5] Bayrock (1967, p. 16, pl. 10) described in detail a push moraine being built by a surging glacier.

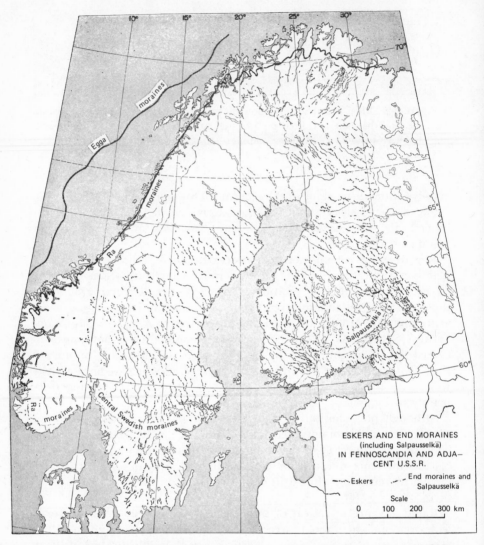

Figure 8-3 Map showing eskers in Fennoscandia and adjacent area of USSR. End moraines and Salpausselkä built at the Fennoscandian maximum are shown also. (Compiled from published sources and from unpublished data from B. G. Andersen.)

is the manner in which massive end-moraine systems, such as the Valparaiso system beyond the southern end of Lake Michigan, were built.

Radiocarbon dates of end moraines built during the last 10,000 years or so suggest that many, possibly most, end moraines result, not from pauses during general shrinkage of a glacier, but from culminations of re-expansions during general shrinkage.

Composite End Moraines. Some end moraines are composite, having been built at one time and then added to by later overriding, during the same or a later major glaciation. The addition commonly consists of a layer of till that conceals the older drift beneath and creates the false impression that the later glacier constructed the moraine. This is seen to have been far from the truth when the unconformity separating the superficial drift from the main mass of the moraine is exposed. Examples are the Fort Wayne, Middle, and Defiance moraines in the Ann Arbor district, Michigan (Leverett, 1908) and the Kent and Ashtabula moraines in Ohio (G. W. White, 1962). According to Totten (1969) some end moraines in north-central Ohio are overlain by as many as three layers of younger till. The moraines themselves probably date from an Early-Wisconsin glaciation. Others are the Harbor Hill moraine on Long Island (W. L. S. Fleming, 1935) where outwash heads seem to have been covered with till during later readvance.

Salpausselkä. In this connection we can note the controversial ridges in southern Finland, which are, in places, >2km wide and >100m high, and are known as the Salpausselkä. Their pattern (fig 8-3) is that of end moraines, and they have been so considered (e.g., Sauramo, 1931, p. 301). According to Virkkala (1963, p. 68), however, the Salpausselkä consist almost exclusively of ice-contact stratified drift, and ice-contact morphology characterizes both their proximal and distal sides in some sectors. In his view they were formed under a wide variety of local conditions, in fracture zones parallel with but not at the ice-sheet terminus and therefore, although they delineate a zone of ablation, they are not end moraines.

The outer margins of drift sheets are not necessarily marked by end moraines. As a matter of fact the outer limit of the Wisconsin drift sheet in North America between the Atlantic Ocean and the Rocky Mountains is marked by end moraine throughout not much more than half its total extent. This drift is so recent that the moraines cannot have been destroyed by later erosion; they simply never were built. In a few places an end moraine was not built because meltwater was so abundant and flowed so rapidly that the drift was sluiced away as fast as it was delivered at the glacier terminus. This was notably true of glaciers occupying mountain valleys with steep gradients. In most places, however, the glacier did not maintain a steady-state position long enough to build an end moraine.

"Washboard Moraines" and Related Features

In many districts in northern North America and northern Europe small ridges built of till occur in series. One meter to 5m high, a few meters to >2km in individual length, and spaced 30 to 200m apart, they occur singly or in groups of as many as 30 units. Variously termed "washboard," "annual," "cyclic," "winter," "cross-valley," and "De Geer" moraines, they have been discussed since about 1936 (e.g., Lawrence and Elson, 1953, p. 95; Gravenor and Kupsch, 1959, p. 54; Strömberg, 1965, p. 81; Westgate, 1968a, p. 24). Although these features have not been fully explained, it seems likely that genetically they are of two or more kinds. Some appear to be small end moraines. Others are not end moraines, but features built at the bases of thrust planes or of cracks in a glacier back of its terminus. In this respect—even in convexity in plan—they resemble one view of the origin of the Salpausselkä. Some of those in Sweden seem to have been formed back of the margin of an ice sheet that calved into a lake, whereas ridges in Alberta and Saskatchewan are related to an ice sheet that did not terminate in water. Gravenor and Kupsch (1959, p. 54) considered the ridges in that region to belong among the ice-disintegration features described hereafter. Clearly these ridges need further sorting out, and until their various origins are more fully understood the names mentioned above should be avoided or used with proper restrictions.

Ice-Cored End Moraines

In Scandinavian mountains as well as in other parts of the world small end moraines consisting of drift concealing a core of ice have been found in substantial numbers. The basic reference on these *ice-cored moraines* is Østrem, 1964. In most of them the core consists of former banks of firn, buried during advance of a glacier over an older moraine. In some it consists of glacier ice possibly covered with drift and so preserved.

Although ice-cored moraines are by definition end moraines, most of them would cease to be ridges, and therefore end moraines, upon complete ablation of their ice cores. Yet they are not ephemeral features. In a cold, wet climate they can persist through a long time. Radiocarbon dates of windblown organic particles embedded in some of the cores of firn-bank origin show that they are many thousands of years old (Østrem, 1965, p. 4, 34). It is unlikely, however, that ice buried beneath drift in low-altitude regions of middle latitudes could have persisted through so long a time.

ICE-DISINTEGRATION FEATURES

Separation of Ice Masses Near the Glacier Terminus

In some districts there occur suites of distinctive related features that differ from end moraine. They are clearly related to former thin, virtually stagnant ice in the terminal zone of a glacier, commonly wide and overlying terrain of low relief. To these features as a group the term *ice-disintegration features* (Gravenor and Kupsch, 1959) has been applied. The term derives from the disintegration of a thin, stagnant, marginal part of a glacier by separation of ice masses along structural and other surfaces that transect the thin ice. The terminal part of the glacier separates into many blocks or "pieces," throughout a zone that may be 10km in width or even much more. The result is that drift within, upon, or beneath the ice, instead of being carried forward by movement of the glacier to the terminus and there deposited, accumulates by various processes in the channels and other openings between and beneath the separating ice blocks, as well as widely over the surface of the disintegrating ice. In such a situation not only is the locus of deposition of drift not concentrated at the glacier terminus, but the deposited drift assumes distinctive forms most of which are very different from those of end moraines. The forms tend to reflect the patterns of the structures along which the ice separates into blocks, and so they are uncommonly angular and in some areas even display repetitive geometric patterns.

Study of ice-disintegration features has not reached the point at which all the individual features can be defined and assigned a logical place in a system. We shall therefore treat them in terms of two groups, a classical group of features consisting mainly of stratified drift and a more recently identified group consisting mainly of till, noting however that the two groups are apparently interrelated in a gradational fashion. We can add that at least some ice-disintegration features (though by no means all of them) have been observed forming today in broad areas of stagnant or nearly stagnant ice, for example the Malaspina piedmont glacier on the southeast coast of Alaska.

Features Built Mainly of Ice-Contact Stratified Drift

Characteristics. The first or classical group of ice-disintegration features represents mainly the result of deposition of stratified drift, by running water and in places in standing water, in openings in glacier ice and in spaces between ice and slopes of bedrock or sediment from which the ice is melting away. Although this group of features consists primarily of stratified drift (notably ice-contact stratified drift), in some places ablation till and "flowtill" are secondary constituents. These features generally possess some of the internal characteristics of ice-contact stratified drift, described in chap 7. Likewise they possess ice-contact morphology, at least in some part of their extent, if such morphology has not been altered by mass-wasting or other processes of erosion. Ice-contact morphology consists of surfaces that are either constructional or the result of deformation through the melting of supporting ice.

The typical constructional surface is the sideslope or face of a mass of sediment that was deposited against a steep supporting wall of glacier ice. The ice melted away and the adjacent part of the sedimentary mass slumped down. Although slump destroys the actual surface of contact with the former ice in detail, the larger features of the surface of contact are commonly preserved. Thus indentations remain where great protuberances in the ice once stood, and projections, mostly cuspate, remain where the ice was marked by reentrants. Such features as these characterize the faces of kame terraces, described later in this chapter, and of isolated masses of sediment accumulated in openings surrounded by thin wasting ice. They were long ago aptly termed *ice-contacts* (Woodworth, 1899). Recently formed ice contacts are well described by Tarr (1909, p. 87, 96, 98, 110).

Where permeability is great and where rates of infiltration are rapid, ice contacts tend to persist. Under other conditions ice-contact morphology is modified by sheet erosion, solifluction, and development of gullies. Ice-contact origin then must be inferred from internal character and from position and areal relations, rather than from details of morphology.

A second kind of constructional surface consists of the outer part of a body of stratified drift that was built up beneath an overhanging wall of ice. Depressions in such a surface may be the direct molds of downward projections of the ice. Caution is necessary in interpreting a steep hillslope marked by knolls and basins as an ice contact. Slump and debris flow, particularly on river bluffs and stream terraces, create features that superficially resemble ice contacts.

In contrast with constructional surfaces are collapsed surfaces created by uneven subsidence of drift as ice melts out from beneath it. Such surfaces range from single basins sited above former buried blocks of ice, to broad irregular surfaces of unsystematic form. Broadly speaking, collapsed surfaces develop where stratified drift was comparatively thick and where underlying ice was comparatively thin. In contrast, ablation till, primarily a collapsed deposit, is drift (mostly nonstratified) that was thin in relation to the ice upon which it formed.

Features of ice-contact stratified drift constitute an essentially gradational series, classified on a basis of morphology. Although described vari-

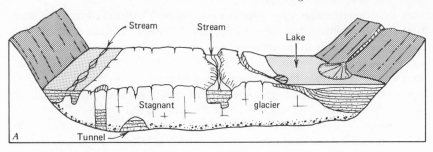

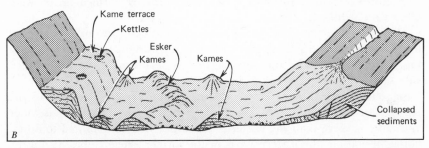

Figure 8-4 Origin of bodies of ice-contact stratified drift. *A.* Stagnant glacier ice affords temporary retaining walls for bodies of sediment built by streams and in lakes. *B.* As ice melts, bodies of sediment are let down and in the process are deformed.

eties are many, we list here seven principal kinds: kame terraces, outwash heads, kames, kettles, collapsed sediments, and eskers. All these features are seen forming near the termini of modern glaciers, and their genesis has been well described (e.g., R. P. Goldthwait, 1959).

Kame Terraces. The origin of terraces with ice-contact faces was first recognized (Jamieson, 1874) in valleys in the Scottish Highlands, although a description of the actual deposition of stream sediments along the sides of an existing glacier did not appear until later (I. C. Russell, 1893, p. 236). Salisbury et al. (1893, p. 156) applied to such terraces the name *kame terrace.* This feature is now widely recognized as an accumulation of stratified drift laid down chiefly by streams between a glacier and the side of a valley and left as a constructional terrace after disappearance of the glacier (fig 8-4). The significant distinction between kame terraces and ordinary stream terraces is that they are not the remnants of former valley fills but were never much more extensive than they are today.

Commonly narrow, a kame terrace slopes downvalley. If very narrow it may be discontinuous where spurs and knobs of bedrock interfered with deposition or where the depositing stream flowed across ice for a short stretch. The sediments composing it were derived chiefly from the drift in the adjacent glacier (mostly well upstream from the site of deposition) and secondarily from tributary streams and from colluvium. The face of the terrace initially is of ice-contact form, although later this is likely to be destroyed as the terrace is cut by the postglacial stream. The top of the terrace is likely to be pitted by kettles. These are more numerous near the face of the terrace, the part built most recently, than near its inner or landward edge where buried ice is more likely to have melted out before stream deposition ceased, so that the early-formed kettles were filled with sediment.

Many kame terraces occur in pairs along the two sides of a single valley. Not uncommonly these grade downvalley into outwash, thus extending the long profiles of the outwash bodies headward along both valley sides.

In some valleys oriented north-south, as in New England and in the Appalachian Plateau, the eastern member of a kame-terrace pair is wider and more massive than the western. The difference is attributed to differential insolation, the eastern parts of the former glacier lobes having received solar radiation during the afternoon when the western parts lay in the shadow of the western sides of the valleys.

At some localities (e.g., Daly, 1912, p. 590) kame terraces occur in series, one below the other, marking the successive positions of ice-margin streams as the former glacier thinned. In places a terrace of this kind ends downstream in a saddle or notch in a bedrock spur. In such places the irregular bedrock surface acted as a series of baselevels for the streams that built the terraces.

Most kame terraces are short; even the longest hardly exceed 10km. In theory a terrace could extend headward along a glacier to the annual snowline, but little farther, because upstream beyond that point meltwater rapidly diminishes. In actual fact probably no terrace extended even that far.

Outwash Heads. In some valleys an outwash valley train ends headward in an ice contact, either constructional or collapsed, with or without a pair of kame terraces extending the valley train upstream. Such a form is an *outwash head* (fig 8-1). Built against or upon the thin terminal part of the glacier, the sediments here were left as an ice contact or else gradually collapsed to form a very irregular slope inclined upvalley. Very likely end moraines lie buried beneath some outwash heads.

Kames. (Woodworth, 1899; Fairchild, 1896; C. D. Holmes, 1947.) *Kame,* a word of uncertain origin, describes a moundlike hill of ice-con-

Figure 8-5 Features near terminus of Woodworth Glacier, Tasnuna Valley, Alaska, in 1938. Ice-contact stratified drift forms include a fan with feeding esker, and kettles containing ponds. Fluted drift (fig 5-21), exposed by retreat of glacier terminus, is overlapped by outwash, entering from right. Much of the drift at upper left is underlain by stagnant ice. (Bradford Washburn.)

tact stratified drift, of any size. Kames originate in at least two principal ways. Some are bodies of sediment deposited in crevasses and other openings in or on the surface of stagnant or nearly stagnant ice which later melted away, leaving the accumulated sediment in the form of isolated or semi-isolated mounds. Kames of this kind are common at and just beyond the faces of kame terraces. Had the terraces grown wider the kames would have been incorporated in them. Related forms having the shape of short ridges are *crevasse fillings* (e.g., Flint, 1928); long, sinuous ridges are the *eskers* described hereafter. Another type of kame consists of deltas (called by some *kame deltas*) and fans (fig 8-5) built outward from ice, or inward against ice, which later melted, collapsing and isolating the mass of sediment to form an irregular mound.

As a group, kames grade into kame terraces, collapsed masses, ablation drift, and some eskers, and also form integral parts of some end moraines.

The majority of kames are intimately associated with kettles and with them record stagnant, disintegrating ice.

Kettles. (Fuller, 1914, p. 38–44.) A *kettle* is a basin in drift, created by the ablation of a former mass of glacier ice that was wholly or partly buried in the drift. Few kettles exceed 2km in greatest diameter, although some large ones in Minnesota attain diameters of 13km. Whereas most kettles are less than 8m deep, some exceed 45m. They may have any shape in plan, but most tend to approach circularity, the shape any mass of ice separated from a glacier will tend to assume as ablation progresses. Kettles occur singly, as groups (especially linear groups of coalescent basins), or in such profusion that the body of drift in which they occur appears as a maze of mounds and basins sometimes described as "kame-and-kettle topography."

Kettles occur in stratified drift and less commonly in till. Those in till are most irregular; it is particularly difficult to differentiate these from basins having other origins.

At least three common kinds of kettle are recognized (fig 8-6). The largest and deepest kettles result from the ablation of thick bodies of ice that projected above the accumulating stratified drift. They have steep sideslopes caused by sliding and creeping of the sediment as its ice support was removed. Thinner, buried ice masses create shallow kettles formed by collapse rather than by sliding (fig 8-7).

In a broad sense kettles are the counterparts of kames. Both are the product of deposition in contact with disintegrating ice.

Kettles are peculiar to the terminal zone of a glacier where thinning is actively in progress. Many occur at the proximal bases of end moraines where thin glacier termini stagnated and became detached and covered

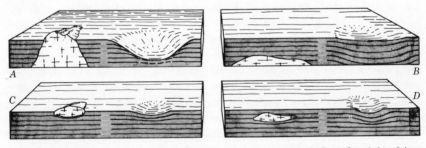

Figure 8-6 Three common kinds of kettles and their inferred origin. Lines on faces of blocks suggest stratification. *A*. Projecting ice mass and kettle formed from it. *B*. Buried ice mass and kettle formed from it. *C* and *D*. Floated-in or dropped-in ice masses and kettles formed from them. (In part after Fuller, 1914.)

Figure 8-7 Part of a small kettle in sand, formed by collapse and later filled with glaciofluvial gravel, in a kame terrace, Boxborough, Massachusetts. Handle of tool is 30cm long. (J. P. Schafer.)

with till, possibly by overriding ice from upstream. In some areas a peripheral belt of glacier ice many kilometers in width becomes stagnant and separates into isolated masses chiefly through meltwater ablation. Stratified drift deposited upon and between such masses creates extensive complexes of kettles.

How long a detached and buried mass of glacier ice can endure before ablation destroys it is not known. A sedimentary cover impermeable to water should delay ablation. Seventy layers of silt, believed to be annual, were laid down in a Swedish lake overlying buried ice and later collapsed; hence apparently the buried ice endured at least 70 years (De Geer, 1940, p. 47). Probably, however, beneath a cover of sand or gravel, the usual sort of sediment in which kettles occur, buried ice would not last as long. According to J. H. Hartshorn (unpublished) buried ice may have remained under outwash of the Malaspina Glacier for at least 30 years and possibly 45 years.

Although it was once thought that ice found buried beneath sedimentary layers in Arctic regions now free of glaciers was "fossil" glacier ice inherited from a colder period, such ice is now believed to be ground ice formed in place, probably in comparatively recent time.

Basins in drift that have origins different from the one described above are not kettles. The most common are basins of nondeposition, usually irregularities in the surfaces of moraines. Others include basins created by the blocking of pre-existing valleys by drift (Flint, 1955b, p. 67), basins caused by differential thawing of perennially frozen ground (Wallace, 1948), and sinks and deflation basins in bodies of silt (e.g., Flint, 1955b, p. 73).

Collapsed Sediments. Bodies of stratified sediment, deposited upon ice, are let down onto the ground as the ice melts. Variations in thickness of ice and of sediment combine to give the collapsed sediment steep dips (fig 7-21), folds, faults, and an unsystematic undulatory upper surface, resembling that of ground moraine, characterized by low mounds and shallow basins with gentle sideslopes. Stratification parallels the upper surface, at least in a general way.

Collapse is one of the processes involved in the making of ice contacts and is clearly evident in the abandoned heads of bodies of outwash, but collapsed masses also occupy other areas, some of which reach many square kilometers in individual extent.

Eskers. (Giles, 1918; Chadwick, 1928; J. E. Armstrong and Tipper, 1948, p. 293–299; Henderson, 1959a, p. 27–34; Bergdahl, 1953; Eriksson, 1960.) A remarkable form of ice-contact stratified drift is the *esker,* a long narrow ice-contact ridge commonly sinuous and composed chiefly of stratified drift (figs 8-5, 8-8). Some authors (e.g., Hartshorn, 1958, p. 477) group eskers and crevasse fillings under the general name *ice-channel fillings.* The term esker is derived from a Gaelic word, first used in a technical sense in 1867 by Close in Ireland. The word *ås* (pl. *åsar*) is widely used in Scandinavia for these features.

Eskers range in height from 2 or 3m to >200m, in breadth from a few meters to as much as 3km, and in length from less than 100m up to more than 500km if gaps, which occur in every long esker, are included. Sides are generally steep, approaching the angle of rest of the esker sediments. Crests are smooth or broadly hummocky. Kettles pit the broader parts of some esker tops.

Eskers are sinuous through nearly the same range as normal streams. That is, some are only slightly sinuous whereas others have great curves resembling meanders. Many eskers, including most long ones, have tributaries, forming with them patterns like those of streams. A few appear to have distributaries. These features are seen in regions where eskers are both numerous and large, such as Fennoscandia (fig 8-3), the Canadian Shield (Prest et al., 1968), and Maine (Flint et al., 1959).

Most eskers occur in regions of rather low relief, and in broad outline their trends parallel the direction of flow of the latest of the former glaciers (fig 8-3; see also Prest et al., 1968). Where conspicuous valleys are present, some eskers follow valley floors (well shown in J. D. Ives, 1960a,

Figure 8-8 Esker in the barren region east of Great Bear Lake, Northwest Territories, Canada. Segment shown is about 6km long. (Canada Dept. Mines and Resources.)

fig 3), but not where valley trends diverge greatly from the direction of former glacier flow. Esker gradients are gentle but not necessarily continuous. That is, an esker may climb up a valley, pass over a low divide, and descend the far slope. A single esker may have a range in altitude amounting to nearly 250m. Crossings of divides occur at saddles or gaps, and at such a crossing the esker is likely to be discontinuous.

Some eskers are compound, forming braided or reticulated systems. Some connect downstream with fans or deltas built at the glacier terminus (fig 8-5), but many end without apparent relation to other features. Sand and gravel are the chief constituents of most eskers, although both silt and boulders are present in places. Consequently eskers are a source of material for road surfacing, and many pits expose their internal character. In most places the sediments are cross-stratified as though deposited by streams. The general direction of foreset layers and consistent decrease in grain size indicate the direction in which the stream flowed. Cross-stratification, however, is much less regular than in outwash; lenses and pockets of till and only slightly washed drift are present. Particles coarser than sand are generally waterworn.

Regardless of the stratification in the central part of an esker section, the material at the sides tends to lie nearly parallel with the steep side-slopes. This structure results in part from sliding, probably during the melting of retaining walls of ice. In some eskers it is attributable also to original deposition at angles oblique to the stratification of the central part of the esker. Cross-section exposures of such eskers have a false appearance of possessing anticlinal structure.

The sediments of which an esker is built are closely similar to the lithology of the till or bedrock in the vicinity. The similarity suggests that they have a common origin in the drift carried in the ice. Detailed study (Hellaakoski, 1931; see also V. Okko, 1945; Lee, 1965) of an esker north of Åbo in southwestern Finland indicated that its constituents are of local origin. Through a distance of 29km this esker overlies a single kind of bedrock covered with till of somewhat variable composition. The pebbles in the esker had clearly been derived from the till and on the average had traveled less than 5km from their former positions in the till. Comparable results were obtained from study of an esker in the Kennebec valley in Maine (Trefethen and Trefethen, 1945). Pebbles in another Finnish esker had traveled farther than pebbles in nearby till (Virkkala, 1958).

Few eskers have been observed during the process of construction. They occur on the Malaspina (piedmont) Glacier in Alaska, and nascent eskers have been observed in Norway (W. V. Lewis, 1949a), Spitsbergen (Szupryczyński, 1963) and Wyoming (Meier, 1951).

As Hummel (1874) was apparently the first to point out, most eskers are the deposits of glacial streams confined by walls of ice and left as ridges when the ice disappeared. Because the eskers in many districts parallel striations and other records of the latest direction of glacier flow, and because they are closely related to deltas, kames, and outwash, they are late deposits, built in the terminal zone of the glacier.

Eskers are formed in several distinct ways, of which two seem to be more common than the others. The most common mode of origin appears to have been in tunnels (less commonly in open canals) at the base of the glacier, during so late a phase of deglaciation that the ice was thin and stagnant. It is unlikely that tunnels could easily form or, once formed, stay open unless the ice that enclosed them was nearly motionless. If stagnant, the ice must also have been thin. Water derived chiefly from surface melting worked its way downward through crevasses and other openings and at the base of the ice enlarged systems of openings to form tunnels. The lowest possible channelways were sought, which is why eskers generally occupy valleys. Passing through these openings, chiefly by confined flow, the water emerged in ponded bodies (a glacial lake or the sea) at the glacier terminus. There is little in the eskers to indicate whether a long individual was formed at the same time throughout its length, or whether its downstream part was built first and was gradually added to in the upstream direction.

Perkins (Leavitt and Perkins, 1935, p. 71; also S. A. Andersen, 1931) observed that the long eskers in Maine have increasing continuity, with fewer gaps, as they are traced upstream, and that in the same direction the proportion of erratic stones in them increases. He suggested that this might mean a stagnant terminal zone narrow at first but increasingly wide as thinning progressed. It has been suggested (Carey and Ahmad, 1961, p. 881) that the basal tunnels in which such eskers formed originated by piping, the tunnels growing headward from their exits.

A second way in which eskers originate, first suggested by Shaler (1884), was later detailed almost simultaneously by De Geer (1897) in Sweden and by Hershey (1897) in North America. De Geer showed that certain eskers in Sweden consist of short segments, each segment beginning upstream with coarse gravel and grading downstream into fine sediments. The coarse upstream part is narrow, but downstream the esker broadens into a distinct delta. From these facts he inferred that each segment represented the deposit made during one year, chiefly in the summer ablation season. The narrow part of the esker was made in a short subglacial tunnel leading to the terminus, beyond which the stream was free to spread beyond the confining walls of the tunnel and build a delta. Small eskers in northern England (Trotter, 1929, p. 573) and in Quebec (Norman, 1938) have this character, broadening into deltalike mounds at intervals of 100 to 200m. The examples cited offer clear evidence that the ice terminus constituted one shore of an extensive glacial lake. All the important conditions favor this mode of origin: (1) the well-defined cliff-like terminus characteristic of a glacier margin in standing water, (2) the rapid wastage that occurs under these conditions, and (3) the preserving effect of submergence of the esker segments as fast as they are abandoned. Probably all clearly segmented eskers, and perhaps some others, were formed in this way, though they constitute a minority of all the eskers on record. On the assumption that each segment was made in a single year this type of esker affords an interesting possibility of determining the average rate of deglaciation during the period measured by the number of segments in the esker.

Two other hypotheses are that eskers form in superglacial stream valleys (Holst, 1876–1877; Tanner, 1937; Sproule, 1939) and in englacial tunnels (Alden, 1924, p. 54; V. Okko, 1945) respectively. Both hypotheses state that as the glacier thins the deposited sediments are gradually let down onto the ground beneath. One merit claimed for these hypotheses is that they explain the tendency of some eskers to climb over divides, without the necessity of supposing that the streams were controlled hydrostatically. Against these views are two principal facts.

1. Most large eskers do not trend indiscriminately across country, as they should do if superposed from upon or within the ice. They are highly selective, following valleys through long distances and crossing divides at conspicuous low points. This could happen only if they were built on the ground, under the guidance of the local topography. Indeed the englacial hypothesis is an attempt at a compromise by keeping the ice tunnel close enough to the ground to be influenced by the terrain.

2. The process of superposition involves collapse on a large scale. It seems likely that surface form and internal stratification could be preserved through this process only under special conditions. However, collapse is minimized in the concept that superglacial streams nearly free of sediment can cut down through nearly clean ice into the basal drift-rich zone, where they pick up a load and redeposit some of it to form eskers.

Superglacial stream-channel sediments occur on the very thin terminal parts of some glaciers, and although repeated topographic inversion happens as differential ablation proceeds, small eskers of superglacial origin might persist.

Near Strö, in Denmark, a smooth-crested esker with no discontinuities is cut through by a deep gap. The gap is obviously stream made, but it was not made by a modern stream because the floor of the gap lies above the surrounding country. The esker, built in a subglacial tunnel, was gradually exposed by thinning of the enclosing ice, but as exposure began, a superglacial stream was superposed across the esker, gradually cutting the gap.[6] The stream was diverted or ceased to flow before the ice at the sides of the esker had entirely disappeared.

In summary it appears likely that many eskers, particularly long ones, are subglacial tunnel deposits, that some were built headward in successive segments each marked by a delta where the esker stream entered a glacial lake, and that others have been let down, from superglacial and possibly englacial positions, through very thin ice.

The trends of eskers make it possible to reconstruct former glacier margins in some districts where end moraines are lacking (e.g., Brenner, 1944, fig 3).

Features in Which Till Predominates

In distinguishing between disintegration features built of stratified drift and those built mainly of till we are not drawing a sharp line. Many such features are gradational, and the terms used for them actually overlap to some extent. The state of understanding of the genesis of these features is not so great as to permit the clear-cut definition of a set of mutually

[6] S. A. Andersen, unpublished.

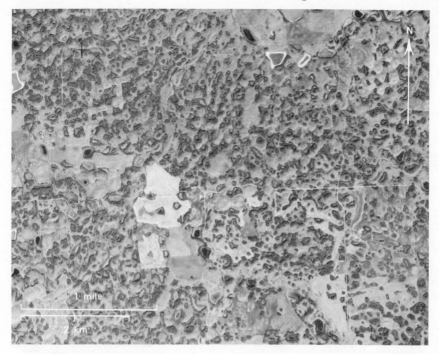

Figure 8-9 Vertical air photo showing hummocky ablation drift on the Canadian Plains, 125km SE of Edmonton, Alberta. Absence of lineation is evident. (Alberta Department of Lands and Forests, courtesy C. P. Gravenor.)

exclusive forms. The terms used here are therefore provisional, adhering as closely as possible to those proposed in the literature, yet avoiding use of the word *moraine* for features not built at the margin of a glacier, because that is the classical implication of the word.[7] Useful references on the description and genesis of this group of features include Gravenor and Kupsch (1959), Clayton and Freers, eds. (1967, p. 25–46), Stalker (1960a), Hartshorn (1958), Kaye (1960), and Parizek, p. 44 in Schumm and Bradley, eds., 1969. We subdivide the features into two broad classes, *hummocky ablation drift* and *disintegration ridges*.

Hummocky Ablation Drift. This first class consists of an apparently random assemblage of hummocks, ridges, basins, and small plateaus, without pronounced parallelism of these elements and without significant form or orientation as an assemblage (fig 8-9). Slope angles vary, but

[7] However, where hummocky ablation drift, described hereafter, constitutes an elongate ridge parallel with end moraines in the same region (as in an example in Rhode Island discussed by Kaye, 1960), it could reasonably be considered end moraine.

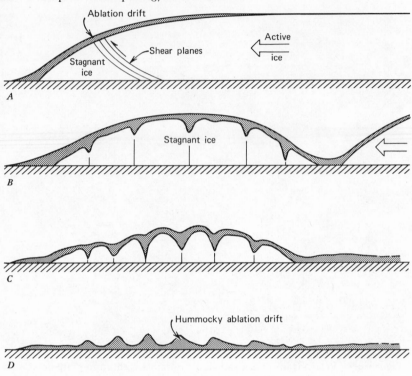

Figure 8-10 Four schematic stages in the creation of hummocky ablation drift by mass-wasting, with filling from above. (Figure 5-20 illustrates how a similar complex might be formed by ice pressing.)

many slopes are steep. Such assemblages are considered to have originated in one or both of two ways. In the first way, ablation drift, which earlier possibly constituted ablation moraine on stagnant ice, was slowly let down as the ice melted, accompanied by repeated slump, debris flow, solifluction, and other processes of mass-wasting, and reached the ground after repeated redeposition en route (fig 8-10). In the second way, ice pressing, described in chap 5, emplaced wet, clay-rich material by transfer both laterally and upward from beneath disintegrating ice. Such activity might occur sequentially or nearly simultaneously over an ever-widening marginal zone of an ice sheet.

Disintegration Ridges. The second class of features is orderly rather than chaotic. The simplest units are long, straight, or curved ridges of till or stratified drift, 1m to 10m high and ranging in length to >10km. In some areas the ridges form intersecting systems consisting of two or even three sets (fig 8-11). They are apparently the fillings of crevasses and other openings in disintegrating ice, made from above by mass-wasting,

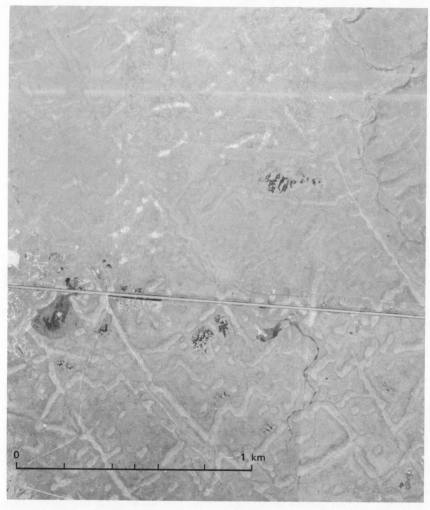

0   1 km

Figure 8-11 Vertical view showing disintegration ridges, consisting mainly of till and about 3m high, forming a reticulated pattern near Nashua, Montana. Probably they reflect the pattern of crevasses in a former glacier. (U. S. Geol. Survey.)

from below by ice pressing, or by both. Other disintegration ridges are circular, and for this reason have been called ice-contact rings and even "doughnuts." A possible origin for them is suggested in fig 8-12; it is less easy to see how ridges of this form could be made by ice pressing. They are described in detail and analyzed by Parizek (p. 49 in Schumm and Bradley, eds., 1969).

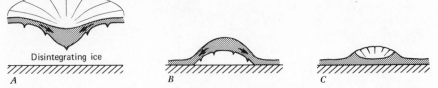

Figure 8-12   Three schematic stages in the creation of a circular disintegration ridge.

In some districts disintegration features occur intermingled with the forms of ice-contact stratified drift described earlier. Among these, collapsed sediments are particularly common.

Summary

As a large group, ice-disintegration features result from a single general condition: the disintegration, by separation, of a marginal belt of ice. This condition occurs where ice is thin and where rate of flow is not high. Possibly this is why such features are common on the northern Great Plains of North America. Till is a common constituent of the features in that region, whereas in the Appalachian region and New England these features consist predominantly of stratified drift. Whether the difference reflects mainly differences in regional lithology, slope, or seasonal volume of meltwater is not known.

SIGNIFICANCE OF THE MORPHOLOGY AND INTERNAL CHARACTER OF DRIFT

Some inferences can be drawn from drift and other geologic evidence as to the regimens of former glaciers at particular times and places. As regimens are in part a direct reflection of climate, such inferences have value for reconstructing former climates, but the limitations of the evidence must be recognized.

Lodgment Till and Stratified Drift. Lodgment of till is favored both by rapid movement in a glacier and by low temperature, which reduces ablation. This combination of conditions is met with during rapid expansion of a glacier, regardless of its type or location. Probably at such times a maximum proportion of the drift in transport is deposited as lodgment till, while a minor proportion is deposited from meltwater. Also, part of the stratified drift laid down in front of the expanding glacier is reworked and redeposited as till, further reducing the proportion of stratified sediment later exposed to view. Such reworking is reflected in tills that contain a large number of rounded stones.

In places during general deglaciation, till was being lodged on the ground by actively flowing ice in a district characterized by streamline molded forms (Hoppe, 1951). The till fabric shows that the long axes of stones parallel the long axes of the topographic features, uniformly to the top of the till; no younger deposit is present.

Deposition of stratified drift, however, is favored by high temperature and slow movement within the glacier. Accelerated ablation during the shrinkage of a glacier should lead to thinning and reduced rates of flow. Abundant meltwater should spread through a zone of ablation of increased width. In consequence, much drift that in a glacier with positive economy would be lodged on the subglacial floor as till would be reworked into ice-disintegration features including a substantial proportion of ice-contact stratified drift.

However, scarcity of stratified drift does not necessarily imply scarcity of meltwater; it may reflect nothing more than scarcity of grain sizes coarser than silt in meltwater transport.

End Moraine and "Attenuated" Drift Borders. Most if not all end moraines imply an actively flowing glacier in a steady-state condition. This is implicit both in their ridgelike form and in the presence of lodgment till within them.

However, the absence of end moraine at the margin of a drift sheet does not necessarily indicate that the glacier was not actively flowing at the time the drift was laid down. It may be that the drift in transport was too scanty to build a moraine, or that retreat of the glacier terminus began as soon as the ice had spread to its outer limit, or that meltwater had sufficient discharge and gradient to carry away all the drift released from the ice.

Drift borders, the outer limits of drift sheets, unmarked by end moraine, have long been recognized and described as "attenuated" drift borders. In both North America and Europe such borders appear to be at least as extensive as those characterized by end moraine. In old drift sheets and in mountain valleys with steep gradients, attenuated borders may be in part the result of erosional destruction of formerly existing end moraines. But in large part they appear to result from lack of a steady-state condition in the glacier, leading to the diffusion of deposition over a wide belt of territory rather than concentration in a narrow one.

Lobate and Arcuate Drift Borders. Lobation of a drift border, whether or not the border is marked by an end moraine, has significance for the regimen of the glacier. The thicker the ice, the smaller the influence of given subglacial relief in making the terminus lobate. This is why the outer limit of the Wisconsin drift sheet is less conspicuously lobate than the drift borders created later in Wisconsin time (Flint et al., 1959). The glacier grew progressively thinner, and its margin became correspondingly more lobate as relief features assumed increasing relative importance in the direction and rate of flow of the glacier.

The concave-outward form of the terminus of a calving glacier is reflected in striations and other direction indicators that converge, instead of diverging, in the downstream direction. An example (Sauramo, 1929, p. 13; fig 5) in the Tampere district, Finland, shows that convergence is toward the axis of a major valley, where the water of the Baltic Ice Lake, into which the ice sheet calved, would have been deepest. The concave-outward patterns of minor end moraines constitute another example (Hoppe, 1948, fig 9, pl. 1).

If former depth of water is known from strandlines, thickness of ice at the terminus of a calving former glacier can be approximated. Values of 50 to 100m have been calculated for the ice-sheet terminus in Finland, during the retreat following the readvance to the Fennoscandian moraines and the Salpausselkä (Sauramo, 1923, p. 147).

Positions of former calving termini can be fixed approximately by the iceward terminations of strandlines marking the water body into which calving occurred. Examples are seen in the proximal endings of strandlines of the earlier Baltic water bodies (Baltic Ice Lake, Yoldia Sea, Ancylus Lake) and of some of the ancient Great Lakes. Determinations are not exact because strandlines tend to become obscure in the proximal direction, possibly as a result of damping of wave action by the presence of abundant floating ice. Further evidence of calving termini exists in the coarse glacial-marine deposits described in chap. 7.

Regional Thinning of Ice Sheets. A survey of wide areas reveals two contrasting styles in sediments and morphology developed during deglaciation. A sequence of deposits consisting chiefly of lodgment till with end moraine records the work of actively flowing and therefore comparatively thick ice. In contrast, a sequence consisting of ablation till and ice-disintegration features is produced by ice that is thin and either flowing very slowly or entirely stagnant. In some regions ice-disintegration features occur in broad belts generally parallel with drift borders. Within such belts local sequential relations establish the former existence of stagnant terminal ice through widths of as much as 20 to 30km. However, as it is difficult to discriminate between features made contemporaneously and those made successively, especially where they are gradational, the maximum width attained by belts of thin ice is not known. Brief descriptions of such belts, unaccompanied by numbers, are fairly numerous (e.g.,

R. P. Goldthwait, 1959, p. 203; Henderson, 1959b, p. 26; Penttilä, 1963, p. 59–65).

Marked relief of the surface beneath the ice should hasten the process of ablation and should extend the zone of thinning, both by imposing topographic obstacles to flow and by reflecting heat onto the ice from nunataks and other exposed rock masses. It is difficult to assess the relative influence of these factors, but both may have been great. The close association of ice-contact stratified drift with fine-textured topography and considerable relief in regions such as New England, the Appalachian Plateau, Norway, and the Welsh and Scottish highlands suggests that the influence of these factors is substantial.

The kinds of drift built by the latest great ice sheet in North America have been so inadequately mapped that firm inferences as to their broad significance are not yet justified. However, there appears to be a zonal arrangement of drift in central North America and in the Baltic region of Europe. In each of these regions, through an irregular outer belt 300 to 500km wide, the drift normally consists chiefly of clay-rich till in the form of ground moraine with many concentrically arranged end moraines. Inside this belt lies a wide zone in which, although till and end moraines are still present, ice-disintegration features are prominent. Inward toward the central part of the glaciated region, end moraines become fewer and great systems of eskers (fig 8-3) occur.

From this zonation we can draw the speculative inference that during the early part of the last deglaciation of North America and Europe the two ice sheets were large and vigorous, albeit with a net deficit most of the time. Hence thinning was not conspicuous, and drift was actively lodged on the ground to form till and end moraines.

Gradually, however, climatic change resulted in increased shrinkage, especially by thinning. After the building of the Lake Border moraines in North America and the Brandenburg moraine in Europe, the net deficit became acute, [8] and ice-disintegration sediments replaced till in more and more areas. As thinning slowed rates of flow, re-expansions of the glaciers still occurred, as indicated by the Port Huron, Valders, and other drift borders. Thereafter shrinkage reduced the ice to such small area and thickness that nourishment was drastically curtailed, widespread thinning occurred, re-expansions gradually ceased, and finally eskers formed throughout the residual ice. In New England massive eskers occur nearer to the outer limit of the glaciated region, possibly because of relatively great relief, fine-textured topography, and steep slopes in that region, in

[8] It is not implied that the moraines named are time correlatives.

addition to the regional regimen conditions sketched in chap 18.

If more detailed examination of the drift should justify this specula-tion, the scattered occurrence of ice-contact stratified drift in the subpe-ripheral belt might be regarded as the forerunner of the great esker sys-tems farther toward the interior. It is only fair to say that the relative dearth of till and end moraines in the great central areas (Finland–north-ern Sweden and the Canadian Shield) must be in part the result of lack of materials with which to build such features, the rock types in both re-gions being unfavorable for the production of bulky drift. However, as this condition applies equally well to ice-contact features, the general speculation still seems to be justified.

CHAPTER 9

Drainage; Eolian Features

GLACIAL DRAINAGE

Meltwater generated by glaciers is a principal means by which water substance is returned to the ocean after residing on land in the solid state. Channels, large and small, cut by meltwater streams, and sediments deposited by such streams are an important part of the record of present and former glaciers. Of the world's rivers, the Mississippi and the Volga have been among the greatest drainageways for meltwater during glacial ages. In discussing glacial drainage we begin with meltwater in contact with the glacier itself and then follow it outward, away from the glaciated area.

Drainage in Contact with a Glacier [1]

Meltwater can form anywhere on the surface of a glacier downstream from the snowline, as well as temporarily upstream from it. Flowing away as intermittent streams, it forms drainage systems that include superglacial, englacial, subglacial, marginal, and proglacial segments. Any of these segments can include lakes as well as streams. After the glacier has disappeared, superglacial and englacial drainage leave slight traces or none at all, but in the other segments streams leave erosional features or sediments from which the drainage can be at least partly reconstructed. A classification of glacial drainage channels was proposed by Derbyshire (1962).

Any turbulent stream, whether glacial or not, that drives sand or silt against firm bedrock smooths the rock or fashions its surface into an irregular network of small coalescent cup-shaped depressions separated by narrow cusplike salients. Surfaces of this kind appear in many places be-

[1] General references: Von Engeln, 1911; Tarr and Martin, 1914.

neath thin stratified drift in natural or artificial exposures, and testify to abrasion by the stream that deposited the drift. Although stream-abraded surfaces have been mistaken for glacially abraded bedrock, they differ in that they lack striations and are smoother in the concave than in the convex areas, whereas with glacial abrasion the reverse is commonly true.

Potholes. A common effect of glacial drainage is groups of potholes, some of which are >10m deep, drilled into bedrock, accompanied by more widespread water-worn rock surfaces. Many such groups occupy the sides and tops of hills, where they could not have formed in normal streams. When they were cut, ice must have formed part of the local topography. According to Brøgger and Reusch (1874) and later Gilbert (1906b) the potholes formed at the bases of waterfalls localized by crevasses. Alexander (1932) and Streiff-Becker (1951), among others, thought that origin improbable, in that most potholes require jetstreams oriented in a low-oblique rather than a vertical direction, and that most crevasses would not remain in one place long enough to permit the drilling of bedrock. Experimental work has shown that the conditions necessary for pothole drilling are met with in subglacial, glacier-margin, and even englacial streams, and that crevasses are unnecessary, although a pothole would be possible at the base of an inclined crevasse. Valleyside potholes and abraded bedrock in a Norwegian fiord are described and illustrated by Holtedahl (1967, p. 200).

Subglacial Channels. Channels of various sizes have been presumed to have been cut by streams flowing in subglacial courses, generally with confined flow. The diagnostic feature of those cut in bedrock is an undulatory long profile, which seems to require confined flow (Derbyshire, 1962, p. 1118). Those cut in drift not only possess undulatory profiles, but also, although they postdate the drift, do not follow the general slope of the surface; hence their courses are supposed to have been influenced by surrounding ice that had ceased to move. These are the *tunnel valleys* of Denmark (Madsen et al., 1928, p. 152) and the *Rinnentäler* (Woldstedt, 1954–1965, v. 1, p. 137) of the North German Plain. If they have been explained correctly they are analogous to eskers, but differ from eskers in representing underloaded, eroding subglacial streams.

Glacier-Margin Streams [2]

Hillside Channels. The erosional effects of glacier-margin streams in glaciated terrain are striking. They would be less well understood had they not been seen forming along the margins of existing glaciers. The following description (Tarr, 1908, p. 101) is illuminating.

[2] General references: Von Engeln, 1911; Henderson, 1959a, p. 40–46; Penttilä, 1963, p. 29–39.

"Along the margin of every glacier that reached well down into the zone of melting, in the Yakutat Bay region, there is a marginal channel, with ice for one wall and the mountain side for the other. These marginal channels are noted for their lack of continuity. In some places, after following the margin of the ice for awhile, and trimming the mountain slope, they disappear beneath the ice; or they may flow on a gravel bed which rests on buried ice; or they may leave the ice margin entirely and cut a gorge across a rock spur, perhaps returning to the ice margin lower down, or even going off to the sea by an independent course. Where engaged in gorge cutting these streams work with great rapidity, for the volume is great, the sediment load heavy, and therefore, with sufficient grade to prevent deposit, they are active agents of erosion.

When such channels are found in process of formation their characteristics are easily observed; but above these, marking former high levels of the ice, are others now abandoned. These are far less easy to understand except in the light of those below, now forming along the existing ice margin. There may be a short section of gorge, contouring the hillslope and open to the air at both ends; or the gorge may extend, on its lower end, across a divide into another valley, and in its lower course appear to be a normal stream gorge though its upper course is most unnatural without taking into account the former presence of the ice. There may be steep cliffs, evidently trimmed by stream erosion, whose formation is difficult to understand unless one can restore the old ice wall that once formed the other bank of a stream valley. Sometimes these trimmed cliffs end in a gorge like the above. Sometimes they die out without any apparent cause; but this is easily understood when one pictures the stream disappearing under the ice, or flowing on it. At times there is a perfect stream bed at the base of the trimmed cliff; but at other times there is only a narrow terrace, often of very irregular form. Then it requires much imagination to restore a stream valley here where we must supply not only one wall but even the stream bed. The vanished glacier readily accounts for it, however; and if one doubts the explanation, he has but to go a few hundred feet downward to the place where a stream is actually flowing on ice veneered with gravel, and walled by a moraine-covered ice bank on one side and a trimmed cliff on the other. Here it is all clear, for all the elements are there; but in the channel above three of the four elements of the ancient valley are gone—one wall, the stream bed, and the stream itself."

Hillside channels in an area of 400km^2 in Finnish Lapland were described and mapped in great detail by Penttilä (1963, p. 29–39; app. 1). In that area the channels have these dimensions:

Figure 9-1   Glacier-margin stream channels, cut into till and bedrock, Stor-närfjället, western Sweden (lat 61°13′, long 13°03′). Channel depths, ~5m; width of view in central part of photograph, ~1km. (Gösta Lundqvist.)

Length: a few hundred meters to >1km.
Width: a few meters.
Depth: <1m to a few meters.
Spacing: In places no more than 15 to 20m apart.
Abundance: As many as 70 on a single hillslope.
Gradient: 1 to 4%.

Common characteristics of the channels are (1) they are typically oblique to the hillslope and (2) open at both ends; (3) their downhill sides are lower than uphill sides; (4) they contain little or no sediment. Channels possessing these characteristics must have been cut at times when ice pressed firmly against the hillsides (figs 9-1, 9-2). Some of the closely and evenly spaced channels are believed to be annual, reflecting yearly lowering of the upper surface of the ice. In Lapland this idea is supported in part by agreement of channel slopes with measurements of other features recording the upper surface of the ice. In some other areas, however, hillside channels are fewer and less continuous, and after deglaciation are difficult to reconstruct.

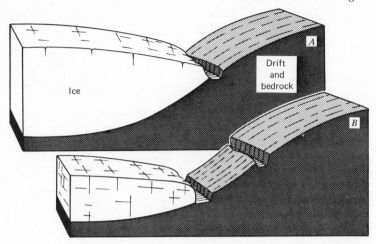

Figure 9-2 Idealized sketch showing development, during deglaciation, of two glacier-margin hillside channels.

In an early paper Kendall (1902) explained an extensive system of discontinuous channels on steep hillsides in northern England as representing valleys of which one side was bedrock whereas the opposite side was ice. Discontinuities were explained as valley segments, both sides of which consisted of ice. Implicit in this explanation is the concept that the ice was sufficiently impermeable to retain flowing water against the hillside. It was argued later that in places, waning glaciers in hilly terrain contained openings of various kinds to lead marginal streams into and beneath the ice. This is undoubtedly true, but the subparallel, closely spaced channels met with in Fennoscandia and the Canadian Shield indicate that in those areas, and in a particular phase of deglaciation, the ice was impermeable or nearly so. Possibly relatively low summer temperatures played a part.

It seems likely also that the angle of slope of an interface between rock and ice is critical. With a nearly vertical interface a glacier-margin stream should be able to cut through the ice to the base of the slope with maximum ease, whereas with a flat-lying interface, as on a bedrock spur, a similar stream would more likely incise its channel into the bedrock.

Rates of cutting by marginal streams vary widely, but are likely to be high because of steep gradients, ample loads, and frequent, short-lived flood discharges. A small stream from Casement Glacier, Glacier Bay, Alaska, was superposed across resistant bedrock in 1935. When seen in 1941 it had cut a gorge 15m wide and more than 8m deep.[3] This average

[3] W. O. Field, unpublished. For similar occurrences see Von Engeln, 1911.

rate of deepening, a little more than 1m/yr, is compatible with the concept of annual hillside channels in Lapland, if only the shallower units are considered. However, because rates of erosion are variable, time estimates based on gorge cutting alone are likely to be unreliable.

The fact that most hillside channels postdate the last passage of ice over the localities of their occurrence, and the lack of recognizable channels that have been glaciated, imply that glacier-margin drainage performs more erosion during deglaciation than during the expansion of glaciers.

Large Glacier-Margin Rivers. Abandoned segments of large river valleys, some of them filled with drift, show that at times an ice sheet filled part of a valley, ponded the drainage, and forced the water to spill over and cut a new course detouring the ice sheet. No two cases are exactly alike because of variations in the relation of the ice margin to the general slope of the land and to other pre-existing valleys.

A conspicuous example is the temporary diversion of the Columbia River through the Grand Coulee in the State of Washington. During a pre-Late Wisconsin glaciation of northern Washington, the Okanogan Lobe of the Cordilleran Glacier Complex blocked the east-west canyon of the Columbia River and more than filled its depth of about 450m. The lake created upstream from the dam rose until it spilled over a local threshold into a pre-existing valley, enlarging it to form the present Grand Coulee. The great discharge that passed through the coulee formed several conspicuous falls (Bretz, 1932) and, following existing low-level routes, detoured the glacier and flowed back into the Columbia at several points. The farthest of these was nearly 150km from the point of detour. Similar features, collectively known as channeled scabland, were widely created east of the Grand Coulee by overflow of an ice-margin lake upstream (Bretz, 1969).

Farther east the middle and upper segments of the Missouri River underwent great changes because of blocking by an ice sheet (fig 9-4). The changes are most clearly evident in South Dakota (Flint, 1955b, p. 139–156; pl. 7) where the river possesses three anomalous features: (1) it flows at right angles to the regional slope, (2) its trenchlike valley is narrower and less well developed than are the valleys of its chief tributaries, and (3) it receives no important tributaries from the east. The major tributaries entering the Missouri from the west are continued eastward by valleys now wholly or partly filled with drift and alluvium. At least two former major divides, now mostly covered with drift, cross the Missouri at right angles; one of them is the former continental divide that separated Arctic drainage from drainage to the Gulf of Mexico. The existing drainage, therefore, has been completely reoriented (fig 9-3).

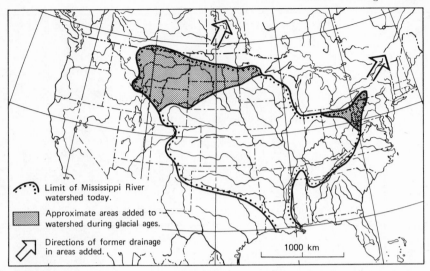

Figure 9-3 Effect of glaciation on the area of the Mississippi River watershed. Compare fig 9-4.

One of the ice sheets before the last one, which invaded South Dakota from the northeast, reached its maximum extent along the site of the present Missouri River (fig 21-1) and blocked the former east-flowing streams. The steepest remaining component of the former eastward slope of the land became a slope toward the southeast. Apparently each ponded stream rose and spilled over the lowest part of the interfluve that separated it from the major valley southeast of it. Most of the low points occurred between the heads of opposed minor tributaries. The overflow rapidly trenched the erodible shale bedrock. In effect the water flowed up one former tributary and down another opposed tributary. When the glacier margin retreated it failed to uncover the former east-draining valleys until after the temporary trenches across the interfluves had become so deep that the diverted waters were unable to return to their former routes. The floors of the ancient valleys, filled with drift, were left standing somewhat above the profile of the new river, the present Missouri. Although not dated with certainty, the diversion probably occurred during one of the later glaciations. The evidence consists of fossil vertebrates and geomorphic relations (Warren, 1952).

Along the upstream segment of the Missouri in eastern Montana, similar abandoned and partly buried valleys lead northward to the Milk River, and other abandoned valley segments suggest that before glaciation the Missouri, joined by the Yellowstone and the Little Missouri,

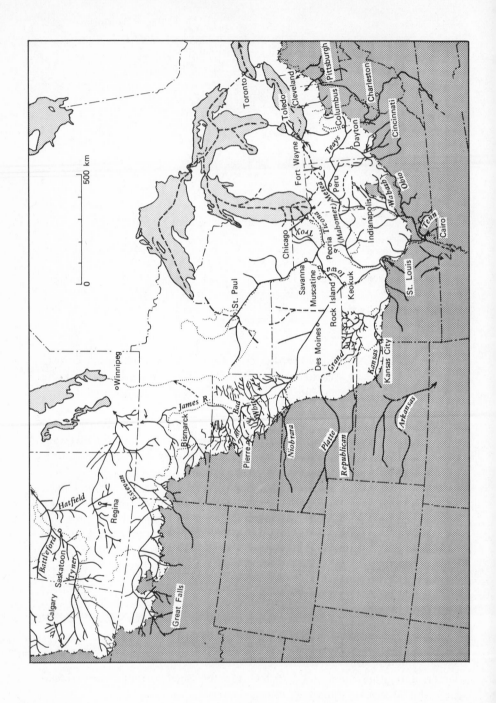

flowed north into the Souris-Assiniboine system, perhaps discharging via the Lake Winnipeg depression, the Nelson, and the Hudson Bay region. The advent of an ice sheet from the northeast blocked all this drainage and detoured the Missouri along the glacier margin. That this kind of event happened more than once in western North Dakota is indicated by an abandoned channel system constituting a temporary diversion valley, parallel with the Missouri trench but southwest of it, having an aggregate length of more than 240km (Colton et al., 1963). Throughout much of this distance the diversion valley approximates the margin of the glaciated region. Channel cutting was easy because the bedrock is erodible, with shale predominating.

Farther south, at Kansas City, the Missouri River emerges from the glaciated region (figs 9-4, 21-1). Here its trench abruptly widens and turns sharply eastward. At the same point the Kansas River enters from the west. These relations are believed to mean that the Missouri River below Kansas City is the preglacial continuation of the Kansas River, and that the segment above Kansas City came into existence much later, as a glacier-margin stream (Heim and Howe, 1963).

A less spectacular but more complex example of diversion of a large stream by an ice sheet is that of the Mississippi River in Iowa and Illinois. Here the detour segments have been in part filled with drift during subsequent glaciations and still later in part reexcavated. From the drift fillings and the vertical relations of the bedrock floors of the various valley segments, the general sequence of events has been provisionally reconstructed (Leverett, 1921; 1942). Earlier routes of the Mississippi and some of its tributaries are shown in fig 9-4. In the Kansan or Nebraskan Age the ice sheet forced two detours, one of which (between Rock Island and Muscatine) is followed by the river today. In the Illinoian Age the ice sheet, reaching the Mississippi River from the east, forced a temporary

---

Figure 9-4 Major known stream valleys that antedate at least one glaciation in central United States and adjacent Canada. (Probably some are preglacial.) Temporary glacier-margin channels are not included. Not all the valleys shown were active contemporaneously. Most of the valleys shown, where not reoccupied by modern rivers, are partly or wholly buried beneath drift. Dotted lines show selected major existing rivers; ticked line represents outer limit of glaciation. Names of valleys (slant letters) are based on local maps. (SOURCES: *Alberta:* Westgate, 1968a, p. 20; *Saskatchewan:* Kupsch, 1964, p. 282; E. A. Christiansen, 1965, Sask. Res. Coun. Map 3; *Montana:* Colton et al., 1961; *North Dakota:* Colton et al., 1963; *South Dakota:* Flint, 1955b, pl. 7; *Missouri:* G. E. Heim and Howe, 1963; *Illinois:* Piskin and Bergstrom, 1967; *Indiana:* Wayne, 1952; *Ohio:* Stout et al., 1943, p. 51; *Kentucky:* Leverett, 1929; *General:* Horberg, 1950.)

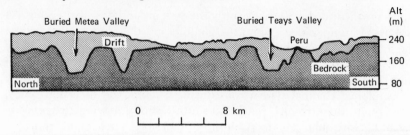

Figure 9-5 Buried Teays Valley and its tributary, Metea Valley (defined by subsurface exploration), at and near Peru in north-central Indiana. (Thornbury and Deane, 1955.)

detour from Savanna, Illinois, to Keokuk, Iowa. In the Wisconsin Age a shorter diversion occurred in the same sector (Horberg, 1950).

In the drainage of the Pennsylvania–Ohio–Indiana region successive ice sheets made even more extensive changes. The logs of thousands of borings as well as subsurface geophysical exploration have aided in the recognition of a system of valleys that have been filled or partly filled with drift related to one of the earlier glaciations. From this the main outlines of the "preglacial" drainage have been reconstructed (fig 9-4). The Ohio River then was shorter, and what is now the upper Ohio flowed north, possibly to a river leading northeastward through the Lake Erie basin. Larger than either of these streams was the Teays [4] River, a major prolongation of the Kanawha River. The Teays flowed northwest across Ohio, Indiana, and Illinois, occupying a wide, deep trench comparable in size with the trench of the Ohio River today (fig 9-5).

With the advent of the ice sheet during a pre-Illinoian (probably the Kansan) Glacial Age, the downstream part of the Teays drainage basin in Ohio was ponded. Valleys were filled with laminated silt (the Minford Silt), in places more than 60m thick, and drainage was extensively diverted so that some valleys remain buried beneath silt. Within the region covered by glacier ice at the time of ponding, the Teays valley system was filled with drift, locally more than 120m thick. Later erosion reexcavated and deepened these valleys and instituted some others, giving origin to the buried "Deep Stage" valley system, the members of which seem to have originated at various post-Teays times, some as late as Illinoian. The "Deep Stage" valleys are buried beneath Illinoian and (or) Wisconsin drift.

[4] Pronounced *Tāz;* named for an abandoned segment of its former valley in western West Virginia.

Among them, then, the ice sheets of three or more glacial ages made profound alterations in the drainage pattern of North America, diverting important Arctic and Atlantic drainage toward the Gulf of Mexico and enlarging the existing Mississippi River watershed so that it drained a sector of glacier terminus more than 2700km in length.

In Europe also a major glacier-margin river was created, altering the pre-existing drainage. Prior to the Saale glaciation the Rhine River, joined by its tributary the Maas (Meuse), flowed northward across the floor of the North Sea and emptied into the ocean at a point east of northern Scotland.[5] In Saale time the Scandinavian Ice Sheet, encroaching southward, blocked this drainage and all the drainage of northern Germany as well, including the Weichsel, Oder, and Elbe river systems. In the Netherlands the blocked waters turned westward along the glacier margin and followed what later became the extreme southern part of the North Sea and the English Channel (fig 9-6). This "Channel River" entered the sea, which then stood much lower than now, between Brittany and Cornwall. Later, as the ice sheet shrank, rise of sealevel submerged all but the eastern segment of this route, part of which is still occupied by the Rhine. The abandoned valleys in the southern Netherlands, with depths of around 10m, are still visible in places at the surface. Toward the north they become deeper and are buried completely.

Possibly the Saale glaciation merely re-created a "Channel River" that had existed earlier, during the Elster glaciation. Saale events have so obscured the evidence that the earlier history is not definitely known, but as the ice sheet blocked the North Sea basin in Elster time, the southern part of that basin should have been ponded, the water escaping over a structural divide between Dover and Calais, the lowest point available, and cutting the valley which, when submerged, became the Strait of Dover.

In southern England also, glacier-margin rivers are recorded by large abandoned valleys. The Thames appears to have been "pushed out of its course" by two successive glaciations. Its pre-Lowestoft course was through the Vale of St Albans, a capacious abandoned valley floored with outwash. In Lowestoft time the Scandinavian Ice Sheet flowed southward over the valley and established the Thames along a glacier-margin route, the Finchley Vale, that parallels the older route farther south. In Gipping time the ice sheet overran the Finchley Vale and forced the establishment of a second glacier-margin route, essentially the one followed by the Thames today (fig 9-7).

Stream-Trench Systems. The route of the "Channel River" across

<hr />

[5] Based on sea-floor topography. See maps in Gregory, 1931 and R. G. Lewis, 1935.

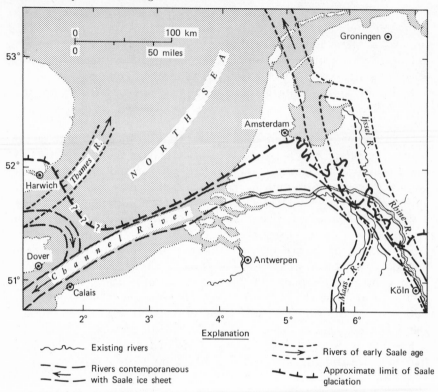

Figure 9-6 Former drainage of the Rhine, Maas (Meuse), and Thames Rivers. After the Holstein Sea had been lowered far below present sealevel, Rhine, Maas, and Thames flowed northward during early Saale time. With the arrival of the Saale ice sheet the Rhine and Maas took a new, westward course, forming the Channel River. Presumably the Thames joined them. (Data mainly from J. I. S. Zonneveld.)

northern Germany and Poland is characterized not by a single continuous trench but by a system of subparallel interconnecting trenches, partly filled with outwash and other sediments, that form a complex network. As the Scandinavian Ice Sheet shrank, its retreating margin continually uncovered new and lower ground and thus opened new and lower routes for the glacier-margin drainage that was escaping westward. Routes were successively abandoned in favor of parallel routes farther north. The diversions took place partly along the ice margin and partly along *Rinnentäler,* believed to have been excavated by subglacial streams. The resulting complex of interconnecting trenches (fig 9-16) is striking; German geologists call the trenches *Urstromtäler* (Woldstedt, 1956a). Four east-west units can be discerned. Of these, three are of Weichsel age; the outermost one is older.

In the southwestern part of the Soviet Union, the area glaciated by the Dnepr and Don lobes of the Scandinavian Ice Sheet during the Saale Glacial Age includes a similar but less extensive system of stream trenches, most of which were cut along lines determined by temporary positions of the glacier margin (Spreitzer, 1941, p. 15).

Another system of drainage trenches, many of them marginal, is cut into the plains of southern Alberta. These, like the European systems mentioned, are in a region of low relief and erodible rocks, two factors which permitted the rapid cutting of deep trenches.

The chief value of ice-marginal streamways is that they afford a means, in addition to end moraines and outwash heads, of determining the successive positions occupied by glacier termini during deglaciation.

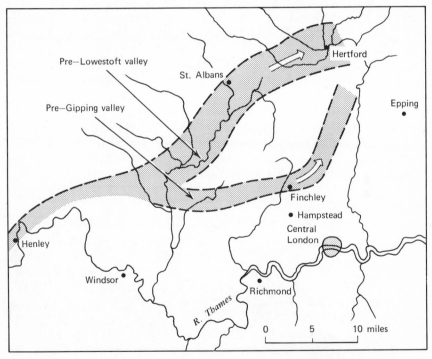

Figure 9-7 London district, showing two successive glacier-margin positions of the River Thames. (After Wooldridge, 1960, Proc. Geologists Assoc., v. 71, p. 122.)

Glacier-Margin Lakes

Stratigraphic Position. In chap 7 temporary lakes created by the block-ing effect of glacier ice were discussed, and several possible configurations were mentioned. Today, lakes formed by the blocking of normal drain-age by glaciers are numerous, but most of them are small, occupying mountain valleys. This results from the fact that most existing glaciers occupy high land that drains away from the ice. However, during former times when ice sheets were extensive, many lakes, some of them enor-mous, were formed by ice dams.

The stratigraphic positions of most extensive bodies of lake sediment show that the lakes came into existence during deglaciation rather than during glaciation. We do not doubt that water was ponded during gla-cial advances, but lake deposits then would have been vulnerable to over-riding, destruction, and burial. Furthermore, the volume of meltwater was greatest during deglaciation. Similarly the great majority of glacier-margin lakes (widely termed *glacial lakes*) in the geological record belong to the Wisconsin glaciation and its correlatives. This is mainly a function of degree of preservation of the record, greatest for the youngest glacia-tion.

Nevertheless, in favorable places we find buried or dissected remnants of lake sediments pertaining to earlier glacial ages. Prominent among such remnants is the extensive Minford Silt, mentioned in the preceding section. Analogous with it is the high-level Calcutta Silt in southwestern Ohio, the record of an early Pleistocene lake created apparently by gla-cial blocking of the former, north-draining Old Monongahela stream sys-tem (fig 9-4).

The Glacial Maps of the United States and Canada show a fringe of glacial lakes along the margin of the Wisconsin Laurentide Ice Sheet when at its maximum, and a truly spectacular array of big lakes along various later positions of the ice margin, reflecting increased production of meltwater during deglaciation. The length of Lake McConnell (B. G. Craig, 1965a, p. 13), in northwestern Canada, was more than 1100km, and that of the slightly younger Baltic Ice Lake, which fringed the south-eastern margin of the Scandinavian Ice Sheet, was more than 1500km.

Outlets; Spillways. Most of the larger, former glacier-margin lakes pos-sessed outlets through which overflowing water escaped from the lake basin. A term widely used for such an outlet is *spillway,* the engineers' term for a passage for superfluous water in a reservoir. Another term seen in the literature is "overflow channel." The natural spillway of a glacial lake is apt to be marked by erosion, inasmuch as the water entering it, having passed through a settling basin, is relatively free of sediment.

Hence the outflowing stream is able to channel erodible material, and in very resistant rock, at least to create recognizably abraded surfaces. Rarely is substantial alluvium present, other than lag concentrates ("pavements") of coarse material.

The position of a spillway is determined by the configuration of the surface (including that of the blockading glacier). Hence overflow can occur along the margin of the ice, under some conditions over the ice itself, or more commonly over the lowest point on a former divide at some other place along the lake shore. The dimensions of a spillway generally are proportioned to the size and duration of the related lake, the gradient offered to the escaping water, and the erodibility of the local materials.

The spillways of some small glacier-margin lakes can closely resemble parts of discontinuous glacier-margin stream channels, and are generally recognized by the presence upstream of lake sediments at appropriate altitude.

Proglacial Streams

Many of the streams that flowed away from the former ice sheets carried loads of sediment greater than the nonglacial loads of the same streams today. This is shown by bulky fills of outwash deposited by meltwater; modern streams are removing the fills. Large rivers such as the Columbia and the Susquehanna, whose downstream courses extend beyond the glaciated country, as well as streams occupying territory uncovered by the glaciers, were profoundly altered by the temporary deposition of voluminous sediment.

Of course, the discharges of the proglacial streams cannot be measured. The prevailing opinion is that proglacial discharges were greater than those of today, at least during deglaciation, mainly because runoff from the ablation of ice was added to that from "normal" precipitation on each watershed. During late phases of deglaciation, when ice sheets had become thin through ablation and the snowline had risen substantially, the area of net ablation on an ice sheet may have attained widths of hundreds of kilometers. At such times the discharges of proglacial streams, at least temporarily, must have been great. Seasonal variations of discharge, too, must have been exaggerated. Very likely the great thickness of many outwash bodies and the relatively coarse grain size of their constituent rock particles are chiefly the product of times of maximum discharge, and give few clues as to mean-discharge values.

Some authors have employed the term "sluiceway" in reference to a proglacial stream, apparently without specific definition. Because the term, from its context, appears to mean nothing more than the valley of a former major proglacial stream, it seems unnecessary and is not used here.

Proglacial Mississippi River Drainage System.[6] The largest proglacial drainage system in North America was the Mississippi–Missouri–Ohio River system, which drew its discharge from a sector of the ice sheet 2700km in length. In it, proglacial regimens differed from present conditions. Instead of being deep and confined in most places to a single channel, each stream consisted of a braided network of shallow channels. This is inferred from the stratification and the channel pattern of the fill built up by each river within its trench (Fisk, 1944). The fill consists mainly of sand and gravel. Even as far downstream as Louisiana, about 25% of the outwash consists of gravel. Today, in contrast, the Mississippi River in Louisiana is transporting sand and silt.

In the Mississippi trench the fill built during the Late Wisconsin glaciation is the principal one whose remnants are exposed to view. In northern Illinois the upper surface of the Wisconsin fill stands 30m above the present river, which in turn flows on sediments that in places reach more than 30m above the bedrock floor of the trench. As far downstream as northern Arkansas, outwash is 50m thick and its top stands 18m above the stream. Still farther south the Pleistocene fill, presumably outwash but certainly not all pertaining to a single glacial age, reaches thicknesses greater than 100m. In the delta, an area that is slowly subsiding, thicknesses are even greater.

Local Consequences of Outwash Fills. Three special features of thick valley trains are noteworthy: (1) outwash-margin lakes, (2) "spillovers," and (3) superposition of stream channels.

1. When the rate of aggradation of a valley train exceeds the capacity of a tributary stream for erosion, the mouth of the tributary is blocked with outwash and a lake may form in the tributary valley. A number of such lakes formed along the margins of valley trains in the Ohio, Wabash (Thornbury, 1950), Mississippi, and Missouri River Valleys, and are shown on the Glacial Map of the United States. This type of occurrence was demonstrated first by E. W. Shaw (1911a; 1911b) along the Allegheny River and along the Mississippi in southern Illinois on a basis of stratigraphic relations and provenance of sediments. Postglacial dissection of the valley train exposes outwash interfingered with lacustrine sedi-

---

[6] Details of outwash in various parts of the Mississippi drainage system appear in several publications. A few representative titles are Ohio River: Leverett and Taylor, 1915; Allegheny River: E. W. Shaw, 1911a; Wabash River: Thornbury, 1950; Mississippi River: Leverett, 1899; Fisk, 1944; Wisconsin River: MacClintock, 1922; Missouri River in South Dakota: Flint, 1955b; Missouri River in Montana: Alden, 1932.

ments. In wide outwash-margin lakes, beaches and other shore features develop.

2. Upbuilding of a valley train also enables the proglacial stream to spill over, across low bedrock interfluves, into adjacent valleys. The proglacial Mississippi in Missouri and Arkansas spilled repeatedly across the long Crowleys Ridge, the interfluve separating the former Mississippi River Valley from the valley of the Ohio River, adjacent on the east (Matthes, 1931). Today the Mississippi occupies the latter valley, but the "spillover" segments, incised into bedrock during degradation, remain.

3. Other channels are incised into bedrock by superposition during the postglacial dissection of a valley train. Some have been abandoned whereas others are still occupied by the postglacial streams. Such channels can be confused with hillside channels and other glacier-margin features. Nevertheless the distinction is important, because a channel cut by superposition from outwash may lie in a district that was never glaciated.

Proglacial Volga River. The greatest of the European proglacial streams (as distinct from ice-marginal streams) was the Volga. During the Saale Glacial Age it drained a sector of the ice margin nearly 1500km wide, reaching from the Ural Mountains to the region of Volgagrad. During the Weichsel Glacial Age the sector drained by the Volga was almost as wide, but the Volga system was much longer than in the earlier time because the ice sheet did not reach as far southeast. During glacial ages the Volga was considerably shorter than now, because the greatly increased Caspian "Sea" then submerged the lower course of the river through a distance that at one time, at least, approached 1000km. It was as though the Mississippi Valley had been submerged as far north as St. Louis.

Summary. From the examples cited it is apparent that meltwater streams greatly extended the indirect influence of glaciation over a large territory that was not itself glaciated. The outwash fills of these streams constitute a means of correlation of great potential value between the drift sheets on the one hand and the fluctuating sealevel on the other. Hitherto the fills have been given less study than they deserve, but the results of future study will go far to establish relationships between glaciation and events in the extraglacial regions.

EOLIAN FEATURES

Probably most of the widespread nonconsolidated wind-blown sediments at or near the surface on the lands are of Pleistocene age. The physical character of those sediments and their stratigraphic and areal distribution show that many of them are related, directly or indirectly, to glaciers. The relationship is indicated further by the distribution of ventifacts and other features of wind abrasion. A study of eolian features makes possible valid inferences as to climatic conditions that prevailed in some regions during various times in the Pleistocene. In the following discussion the sediments are subdivided into sand and loess.

Wind-Blown Sand

Some of the eolian sand occurs along coasts and is clearly related to marine beaches at various altitudes and of various dates. Such sand, which includes both quartz sand and calcareous sand, is discussed in chap 12, in connection with fluctuation of sealevel. The remaining sand is continental and consists chiefly of quartz. Such sand was blown from three principal source areas: (1) outwash plains and valley trains, (2) sandy alluvium in dry regions, and (3) the beaches of glacier-margin lakes, outwash-margin lakes, and pluvial lakes. It is evident from this list that most such sand was moved by the wind during glacial ages, although some was deposited during interglacial times.

A minor proportion of the continental sand occurs in featureless sheets little more than 1m thick, spread over rather wide areas. Such *coversand,* as it has been called by geologists in the Netherlands, is poorly stratified subparallel with the ground. Its subparallel laminae are probably the deposits made during individual storms or storm seasons. It has been suggested, in accordance with observation during modern winters, that sand grains and cold, dry snow particles move together during winter storms, the sand grains later "collapsing" as the snow melted. Whatever its origin, coversand is so thin that it is rarely mapped, and enters but little into the literature.

Sand Dunes.[7] Most continental sand, however, takes the form of thicker bodies possessing constructional topography, and consisting of groups of dunes. Major groups of dunes are shown on a map of Pleistocene eolian deposits in the United States (Thorp et al., 1952). Most of those in the glaciated regions are built wholly or partly of reworked glacial drift. Some have ceased to accumulate and have become covered with vegetation. Others are in process of construction. But all are in the immediate proximity of the glacial deposits that are their source of supply.

Most of the dunes occur in groups, clusters, or "fields," the majority at the margins of conspicuous masses of outwash. In northern Germany, for example, many large dune groups fringe the Urstromtäler. This suggests that the sand consists largely of redeposited outwash. The suggestion is confirmed in some areas by comparison of the size distribution and min-

---

[7] For a critical review of Pleistocene dunes in Europe, see I. Högbom, 1923.

eralogy of the constituent rock particles. Other groups have formed in the lee of valley fills of coarse alluvium; some of those on the North American Great Plains south of the glaciated region are examples. Still others, including most of those in the Basin-and-Range region farther west, stand in the lee of areas where pluvial lakes existed.

On a basis of form, five principal kinds of dunes are recognized (fig 9-8).

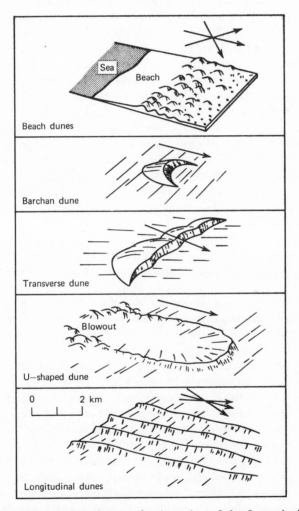

Figure 9-8 Characteristic form and orientation of the five principal types of dunes. Arrows show wind directions. Steep, straight slopes are slip faces. (After Longwell, Flint, and Sanders, 1969, Physical geology, p. 307.)

1. *Beach dunes.* Groups of hummocks of various sizes bordering beaches. Inland part of dune area is generally covered with vegetation. Built by variable winds with abundant sand supply.

2. *Barchan dune.* Crescent-shaped dune with horns pointing downwind. Occurs on hard, flat floors in deserts; built by constant wind with limited sand supply. Height 1m to >30m. Bare; migrates readily.

3. *Transverse dune.* A dune forming a wavelike ridge transverse to wind direction. Occurs in areas with abundant sand and little vegetation. In places grades into barchans. Generally bare; migrates readily.

A variant is the crescent-shaped *lunette* dune, which wraps around the lee shore of a small, subcircular lake that is seasonally or perennially dry. Lunettes consist either of quartz sand or of sand-size pellets that are aggregates of clay particles with silt grains, derived from the exposed lake floor. They have been recognized in coastal Texas, North Africa, and southern Australia. Those in Australia are discussed in detail by Bowler (1970), and are mentioned in chap 25 of the present book.

4. *U-shaped dune.* A dune of U shape, opening upwind. Some individuals form by piling of sand along leeward and lateral margins of a growing blowout in older dunes. A common type that forms in semiarid and moist areas; usually partly clothed with vegetation while forming.

5. *Longitudinal or linear dune.* A long, straight, ridgelike dune parallel with wind direction. Occurs in deserts with scanty sand supply and strong winds varying within one general direction (chap 25).

Because all five kinds of dune occur in an active state today, dune orientation relative to effective wind direction is known. Also, in these active, modern dunes the steep foreset laminae invariably dip downwind (fig 9-9), although in some dunes, particularly U-shaped dunes, some laminae dip gently upwind. From these data we can determine former wind directions in dunes no longer active, and in sand that through erosion has lost its constructional form.

Attempts to reconstruct factors of the climate that prevailed during the building of dunes no longer active have had two main objectives: first, former wind directions, and second, former atmospheric humidity. Both parameters can help elucidate former patterns of atmospheric circulation.

Wind directions are inferred from: (1) positions of dune fields with respect to obvious source areas, (2) orientation of dune forms, and (3) directions of dip of steep foreset laminae. In both Europe and North America, contrary to some early beliefs, in part based on erroneous interpretation of transverse dunes as longitudinal dunes, all three lines of evidence indicate that the same planetary circulation prevailed, at and after the last glacial maximum, that prevails today (Büdel, 1949). That is, in the mid-

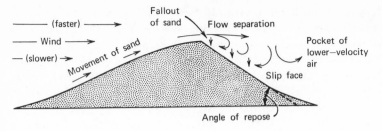

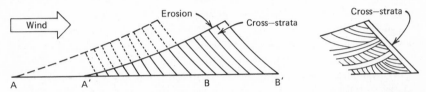

Figure 9-9 If wind had uniform direction and velocity, it would build (*left*) simple, parallel cross-strata while dune migrates from **AB** to **A′B′**. But variations in direction and velocity are common and result in cross-strata like those at right.

dle latitudes in both regions the predominant winds were westerlies. This generalization refers to the weather pattern as a whole and does not preclude local and temporary differences such as the former greater frequency of NE winds, in and south of the glaciated regions.

Some evidence of minor change, however, between glacial and post-glacial conditions has been found. In the Great Plains region in northwest Texas, groups of inactive dunes related to small pluvial lakes indicate former wind direction by their occurrence relative to sand source. Groups formed during the Late Wisconsin deglaciation imply effective NW winds, whereas those dating from within the last 5000 years imply SW winds (Reeves, 1965b).

The largest single area of dunes in North America is the Nebraska Sand Hills in northern Nebraska. Eolian sand there covers an aggregate area of > 35,000km² to an average thickness of about 8m. The dunes constitute at least three groups, of different though unknown late-Pleistocene ages, consisting of at least three morphologic kinds (H. T. U. Smith, 1965; E. C. Reed in C. B. Schultz and Frye, eds., 1968, p. 25; H. T. U. Smith in idem, p. 29). The oldest group indicates N winds; the others NW like the effective winds of today.

In the reconstruction of wind direction, the effective wind (the wind that built a dune) must not be confused with the prevailing wind of the district. For example, north of Spokane, Washington, a large group of dunes, still active, caps the scarps of a terraced outwash body. Since the dunes lie south of the souce of the sand, they must have been transported by winds with a northerly component; yet the winds that prevail today are SW. The SW winds are perhaps stronger as well as more frequent than all other winds, but not having access to a source of sand, they have not built dunes. The less frequent northerly winds, sweeping across an area of abundant sand, succeeded whereas the prevailing winds failed.

Theoretically wind force could be inferred only from longitudinal dunes, which imply strong winds. Such dunes occur in both eastern Europe and southwestern North America, but are active today only in very dry parts of North America. An example of inactive longitudinal dunes is the group of *paha* lying upon and beyond the "Iowan" drift sheet in northeastern Iowa and northwestern Illinois (Thorp et al., 1952).

Paha are elongate ridges, some of them complex, as much as 15m high and as much as 16km long. Their long axes trend NW–SE. Sections through them reveal small hills of the underlying drift buried at or near their NW ends. Paha appear to be longitudinal dunes whose constituents were transported and deposited, not as individual particles of silt and clay, but as aggregates having the diameters of sand grains, as inferred from lack of sorting of the particles (Scholtes, 1955). Such eolian transport of clay aggregates was described from Texas as long ago as 1909.

Deposition was localized in part by small steep-sided knobs and hillocks in the drift sheet. As longitudinal dunes imply strong winds and a surface nearly free of vegetation, the paha are believed to have been built during a deglaciation of Wisconsin age, when the surface immediately beyond the ice was still nearly bare. This was the time when the lag concentrate with ventifacts developed on the "Iowan" (Kansan?) drift sheet (chap 7).

The orientation of the paha clearly implies effective NW winds, strong enough to build longitudinal dunes. But in the same region strong NW winds occur in winter today, and it is not certain that if the drift-sheet source of sediment were stripped of its cover of vegetation, similar dune building would not be resumed.

Are dunes in moist regions an indication that the climate was formerly dry? The affirmative opinion was once widespread; it embraced also the notion that absence of vegetation in the source area is also a requirement for the accumulation of dunes. The dual concept was refuted by Cooper (1935, p. 108), who demonstrated that though both factors favor dune building, neither is necessary for it. Coastal dunes are forming today in very moist regions such as the coasts of Oregon and Washington, and southern Alaska, including areas with a well-developed forest cover. Cooper believed that although dune building can be promoted by dry climate and absence of vegetation, it can be brought about also by subsidence of the water table as a result of stream intrenchment. In fact, the extensive dune system north of Minneapolis, Minnesota, was developed following intrenchment of the Mississippi River, into a body of outwash. Because such intrenchment is a normal event in the history of nearly all outwash bodies when the supply of glacial sediment diminishes, and because a very large number of dune areas are located on outwash, intrenchment is probably an important cause of dune building.

We can say, then, that extensive dune fields not associated with sand-rich glacial, lacustrine, or marine sediments probably imply either (1) the appearance of a rich source of sand not present earlier, or (2) subsidence of the water table through desiccation. Many of the extensive dune fields of the Great Plains are believed to have resulted from the first cause, when recurrent Pleistocene fills of sand and gravel were deposited in the valleys of streams draining the Rocky Mountains (Flint, 1955b, p. 155–160). In many areas dunes are believed to have resulted from the second cause, during the Hypsithermal and at other times of climatic drought. A notable example is the occurrence in northern Arizona of longitudinal dunes now fixed by forest vegetation in districts receiving 300 to 380m of precipitation annually. Dunes of this kind are being built in that region today only in districts receiving less than 250mm of precipitation. It is concluded that the vegetation-covered dunes indicate a climate drier than that which now prevails (Hack, 1941, p. 262).

In summary, neither lack of vegetation nor arid climate is required for the accumulation of the common U-shaped dune; a sufficiently dry sandy surface is all that is necessary. Most of the ancient dunes recorded in the literature of the Pleistocene were built in the lee of outwash bodies and other rich sources of sand. Some accumulated during outwash aggradation, probably during deglaciation; others did not form until postglacial dissection of the outwash had begun. Later both source and dunes became fully clothed with vegetation, and dune building came to an end. Other dunes are unrelated directly to glacial deposits but are a result of the repeated climatic changes that affected nonglaciated regions during the Pleistocene.

Various attempts have been made to date dunes relative to glacial-age climates. C^{14} dates have been obtained in a few cases. Stratigraphic tracing of the Nebraska Sand Hills into the Peorian Loess of the glacial sequence in central United States has been attempted. Age attributions based on limonitization of the sand have been suggested. The Algodones field of coastal dunes, on the eastern flank of the Salton basin in southeastern California, has been treated on a speculative volumetric basis. Sand volume in the dunes, estimated at ~10.8 × 10^9 m³, would have required at least 160,000 years to accumulate, if unchanged climate and other conditions are assumed (McCoy et al., 1967). Despite such efforts, we lack a clear general picture of the relation of Pleistocene dunes to glacial ages. Many groups, clearly linked with outwash bodies and pluvial lakes, must be glacial but other groups, including many coastal dunes, are probably interglacial. The latest phase of building of longitudinal dunes near the SE margin of the huge dune field in the interior of Australia is dated indirectly at 17,000 to 15,000 years BP (Bowler, 1970).

Deflation Basins

Both wind-blown sand and loess imply the effective deflation of outwash and other masses of sediment containing much sand and silt. Outside the glaciated region large volumes of silt and clay were deflated from nonglacial sedimentary bodies and even from the surfaces of claystone bedrock. The evidence consists of basins, as much as 1km long and 5m deep, in many semiarid regions. In the Great Plains of west-central United States such basins occur by the tens of thousands (e.g., Judson, 1950), and during the droughts of 1930–1940 were observed to form and to deepen. Some basins occur in parallel rows independent of structural control; it has been suggested that they were excavated in the floors of troughs between former longitudinal dunes (Flint, 1955b, p. 157). Similar deflation basins occur in central Asia, Australia, and Patagonia; in southern Africa, where they are known as *pans,* they are numerous, reaching diameters of many kilometers and depths of 60m. They occur typically in clay, silt, or fine sand or in rocks that yield such sediments as they are weathered.

Little is known about the dates of pan excavation, although there is little doubt that, in semiarid grassland desiccation, impairment of the cover of vegetation is a prerequisite to deflation. Hence we can regard deflation basins as one of the many records of climatic fluctuation in regions now semiarid.

Ventifacts

Ventifacts (stones abraded by wind-driven rock particles) occur in some localities not now characterized by sand blasting. This implies an incomplete cover of vegetation to expose sand and silt, and winds of considerable force. From facets, flutings, and grooves on some ventifacts, direction of the effective wind can be inferred. From ventifacts in Wyoming four episodes of wind abrasion, presumably related to glacial maxima, were inferred (Sharp, 1949a), but the wind directions did not differ from those that prevail today. Studies on modern beaches show that wind abrasion

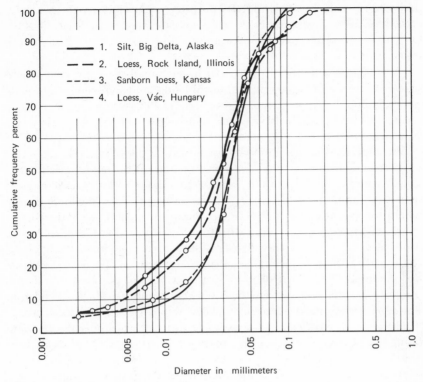

Figure 9-10 Curves showing grain-size distribution in four samples of loess. (1 to 3: T. L. Péwé; 4: Márton Pecsi.)

of hard-rock stones is effective within a period of 100 years, but that well-developed facets require more time.

Exposed bedrock likewise bears flutings that indicate the direction of temporary effective winds. Such flutings in various parts of eastern Massachusetts and Rhode Island imply NE winds during deglaciation, compared with W winds today (J. H. Hartshorn, unpublished).

Loess [8]

Character. One of the most remarkable Pleistocene sediments, around which an extensive literature has accumulated, is *loess,* a sediment, commonly nonstratified and nonconsolidated, composed dominantly of silt-size particles (fig 9-10), ordinarily with accessory clay and sand, and de-

[8] General references: C. B. Schultz and Frye, eds., 1968; Stuntz and Free, 1911; Grahmann, 1932; Scheidig, 1934; Condra et al., 1950, p. 33–46; Poser, 1951; Krinitzsky and Turnbull, 1967.

posited primarily by the wind. Its porosity is extremely high, generally exceeding 50%. Loess underlies an aggregate area of at least 1.6×10^6 km^2 in North America and an area of more than 1.8×10^6km^2 in Europe.

Loess consists mainly of quartz, with very small, variable proportions of clay minerals, feldspars, micas, hornblende, and pyroxene. Carbonate minerals may be present, ranging as high as 40% by weight of the sediment. This composition is so elastic that it tells us little about the rocks in which the minerals originated.

Although loess is "commonly nonstratified" it does have well-laminated facies. Some of these may represent sedimentation in small, perhaps temporary, ponds, but others are so situated topographically as to suggest that the lamination is eolian, reflecting variations in wind force during deposition. In any body of loess such occurrences lie close to its source.

Loess is not easily distinguished from water-laid silt. The resemblance is reflected in the widespread use of such terms as "loesslike silt." According to Millette and Higbee (1958), sediments of the two origins are not separable on a basis of grain-size distribution, porosity, or mineral composition alone. However, these workers found, within the sand fraction, greater angularity of grains in loess than in water-laid sediment.

The color of most exposed loess ranges through various hues of yellow, orange, and brown, due to the presence of ferric oxides; where loess is unoxidized or reduced it may be grayish. Loess, especially where its dominant grain sizes lie near the finer end of the grain-size range of loess, has crude vertical cleavage which in some exposures is platy. This results from tension cracks and causes the material to form cliffs (fig 9-11).

Loess occurs as layers or sheets which, like sheets of till, may overlie surfaces having steep slopes and a local relief of 100m or more. Such occurrence has led to the expression that loess lies "draped over the hills." Individual loess sheets underlie areas of tens of thousands of square kilometers. Near its source a sheet of loess may exceed 50m in thickness; indeed, a thickness of 180m has been reported from one locality in China. In the downwind direction a loess sheet thins down past the limit of mappability (say 40cm). This limit may be reached within a few kilometers from the source, but in some regions it is not reached within hundreds of kilometers.

Where loess blankets a steep slope it is moved downward by mass-wasting and sheet erosion and accumulates at the base as colluvium. Loess in place can be distinguished from loess in secondary transport only after close examination, and on slopes the true thickness of a loess sheet is almost impossible to measure with accuracy.

Grain size as well as thickness diminishes with increasing distance from

Peorian Loess
(Late Wisconsin)

Loveland Loess
(Late Illinoian)

Kansan Till

Figure 9-11 Type exposure of Loveland Loess, with Peorian Loess and Kansan Till at Loveland, Iowa. (A. L. Lugn.)

the source. Detailed studies [9] show that within limits the relation of texture and thickness to distance from source can be expressed mathematically. However, grain size varies locally with the topography on which the loess was deposited.

The fossil fauna of loess is terrestrial; it consists mainly of land snails and far less commonly, small and large land mammals. Fossil snail shells are rarely found in thin loess, where leaching of carbonates has destroyed them; in thick unleached sections, however, shells may be very abundant (cf. Brunnacker, 1957, fig 1).

In some places fossil snail shells occur in loess with a frequency approaching $175,000/m^3$. The ecology of the fossil snails in the region west of the Missouri River in central United States implies a somewhat cooler and moister climate than that which now exists. In Illinois and in Iowa, loess of Wisconsin age contains fossil mollusks that record a progressive change upward from a cool climate to one like that of today (Baker, 1936; Kay and Graham, 1943, p. 167, 190–191).

Sources. The chief sources of fine sediment for wind transport, exposed at the surface without the protection of a continuous cover of vegetation, are desert basins, active bodies of outwash alluvium, and areas of till recently uncovered by deglaciation. The distribution of loess is so closely

[9] G. D. Smith, 1942; Swineford and Frye, 1951; Simonson and Hutton, 1954; Lugn in C. B. Schultz and Frye, eds., 1968, p. 139. In the Smith reference changes in thickness, texture, and carbonate content furnish a basis for locating the source.

Figure 9-12   Silt being deflated from outwash valley train, Delta River, central Alaska. View looking north (downstream), summer 1948. The valley train is 2 to 3km wide. (T. L. Péwé, U.S. Geol. Survey.)

related to the distribution of such sources that there can be no doubt the relationship is genetic. Granted a source, the optimum conditions for its deflation are merely a dry continental climate with moderate to strong winds. In continental-desert regions these conditions exist; in central Asia, the largest area having such a climate, deflation is spectacular. Glaciers create such conditions, temporarily at least, in regions where they would not otherwise exist (fig 9-12). Because of this, central North America and eastern Europe, where deflation today is normally inconsiderable, received huge volumes of loess during glacial ages. The chief contribution of the glaciers was the temporary creation of large areas of bare outwash and secondarily of bare till. Also, higher wind velocities presumably accompanied the glaciers, but this was probably a secondary factor; given the bare areas there is little reason to doubt that today's winds could deflate them and build loess.

We have, then, loess of nonglacial origin and loess of glacial origin, though in many regions loess was built from both sources. According to Grahmann, loess derived from deserts can be distinguished from loess derived from outwash by means of the range of its grain size. Loess derived from deserts has a much wider range of sizes, including a conspicuous quantity of very fine grains. This comparatively poor sorting reflects the fact that the sediment has been sorted only once, during its transport by the wind. Loess derived from outwash, in contrast, has but a narrow range of grain size, the coarser and finer particles having been screened out. This more thorough screening reflects a double sorting, first by melt-water streams and then by wind.

That till as well as outwash contributed to the loess in Iowa is shown by the mineral content of the loess, which resembles that of the corresponding size-grade fraction of the till in the same region (Kay and Graham, 1943, p. 183).

Wind blowing across outwash sets sand grains into saltatory motion. The jumping grains bombard the surface and kick silt particles into the airstream. If the sand is sufficiently abundant, dunes may accumulate at the leeward margin of the outwash. The silt particles are carried beyond and gradually settle out over a wide area. Once they have settled, the cohesion inherent in their small diameters stabilizes them, and together with the local cover of vegetation, prevents their being moved again by the wind, unless bombarded by sand grains. Probably this explains why loess possesses conspicuous constructional topography, if at all, only where it is coarse grained and near its source; elsewhere its surface is featureless.

Silt particles in transport are lifted to altitudes approaching 3km, probably in outbursts of cold polar air, and are deposited as the wedge of air spreads out and loses energy. Settling is by dry sedimentation, but if the cold wedge is overrun by warm maritime air, silt particles may form raindrop nuclei and reach the ground in rain.

In analogy with conditions in central Alaska today, we can suggest that loess accumulated chiefly during the summer, the season of ablation of glaciers, when sedimentation of outwash was at a maximum, and when the ground surface was neither frozen nor blanketed with snow. According to Hjulström et al. (1955), most loess is sedimented during storms of peak wind velocity.

The great thickness (30m) of loess of Wisconsin age along the valley of the Missouri River in Kansas, decreasing to 6m or less through no more than 10km from the river is attributed to the presence close to the river of a forested belt whose trees reduced wind velocities and so trapped airborne silt (Leonard and Frye, 1954, p. 403; G. D. Smith, 1942, p. 163). This inference is confirmed by the fossil mollusks in the loess: near the river they imply a forest ecology; beyond they imply grassland.

In moister regions decrease in thickness downwind from the source (e.g. Schönhals, 1953) is less abrupt, presumably because ecological differences were less pronounced. In fig 9-13 we see this, and also that, assuming westerly winds like those of today, loess of Wisconsin age in Illinois is related with remarkable fidelity to two major bodies of outwash along the Mississippi and Illinois Rivers. But because its relation to the border of the Wisconsin drift is minor, no large proportion of this loess could

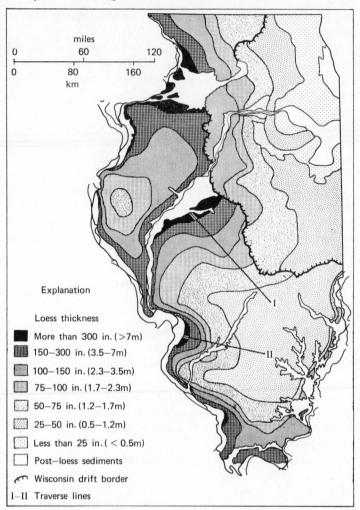

miles
0   60   120
0   80   160
km

Explanation

Loess thickness

More than 300 in. (>7m)

150–300 in. (3.5–7m)

100–150 in. (2.3–3.5m)

75–100 in. (1.7–2.3m)

50–75 in. (1.2–1.7m)

25–50 in. (0.5–1.2m)

Less than 25 in. (< 0.5m)

Post—loess sediments

Wisconsin drift border

I–II Traverse lines

Figure 9-13 Thickness of loess of Wisconsin age in Illinois, showing relation of thickness to outwash bodies (white) and to border of Wisconsin drift. (After G. D. Smith, 1942.) Prevailing wind today is westerly.

have been derived from the contemporaneous till. We conclude that the till rapidly acquired a protective covering of vegetation, whereas the outwash source was continuously renewed by the proglacial streams. Only through such renewal can the large volume of major loess sheets be explained. Renewal was long-continued; radiocarbon dates suggest that the loess related to outwash of Late Wisconsin age in Illinois was deposited during a period of nearly 15,000 years, although the bulk of it may have accumulated during less than 10,000. Measurements of degree of leaching of carbonates in this loess, likewise, indicate that rate of accumulation was slow. The progressive change of climate implied by an upward change in the snail fauna further indicates slow accumulation. Variations in rate of accumulation based on variations in degree of leaching have been inferred in each of two loess bodies in Kansas.

Because loess is confined almost entirely to Pleistocene strata, it has sometimes been thought to record a unique (and in some opinions a mysterious) set of climatic conditions peculiar to an "ice age." But with more detailed study loess has lost whatever mystery may have attached to it. To explain loess of glacial origin no such condition as a "loess climate" is needed. Such loess occurs in periglacial regions, not by virtue of former special climatic conditions peculiar to such regions, but mainly because within them continuously renewed outwash sediments afforded rich source areas for wind-blown silt.

True, the Pleistocene, with its high land barriers, its desert basins on their lee sides, and its extensive areas of rapidly accumulating outwash, afforded opportunity for the deflation of silt and clay on a scale far greater than that which existed with the lowlands and maritime climates of much of geologic time. On a reduced scale, however, winds probably have made extensive deposits of silt throughout geologic history. Pre-Pleistocene rocks believed to be consolidated loess have been reported. But in general the older Pleistocene and pre-Pleistocene loess sheets have been removed by erosion or have lost their identity through mixing with other kinds of sediment. The Pleistocene Epoch provided conditions ideal for loess making, but it was not the unique workshop for eolian activity that it was once thought to be.

With this general statement we turn to the loess of the glaciated regions, for it is these with which we are chiefly concerned, although desert winds have given rise to large volumes of loess. The glaciated regions in which loess has been most thoroughly studied are central North America and central Europe.

Loess in North America. The map, Pleistocene Eolian Deposits of the United States, summarizes existing knowledge of loess distribution. Two minor areas of loess, each of $100,000km^2$, occupy eastern Washington and northeastern Oregon, and southern Idaho. Both areas are characterized by steppe climate and extensive basins in which Pleistocene sediments, primarily nonglacial, were deposited by major streams. Another area of loess occupied central Alaska, where maximum thicknesses of 60m are measured. Sources consist chiefly of outwash.

By far the largest area of loess occupies the central region of North

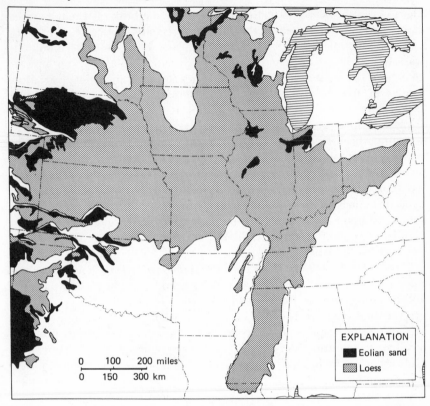

Figure 9-14 Distribution of windblown sand and loess of Wisconsin age in part of central North America. (Adapted from Thorp et al., 1952.)

America (fig 9-14). It extends from the Rocky Mountains across the Great Plains and the Central Lowland into western Pennsylvania, and east of the Mississippi River trench constitutes a broad belt that reaches south to the delta. Extending far south and west of the glaciated region, loess in this vast area blankets the exposed pre-Wisconsin drift sheets and overlaps the older part of the exposed Wisconsin drift as well. Most of the loess shown on the map is of Wisconsin age.

Probably the broad areas of loess on the Great Plains are derived from various sources. The loess in Nebraska has been attributed to sources in outwash along the Platte River, and also to Pleistocene and pre-Pleistocene alluvium, some of it partly consolidated (Condra et al., 1950, p. 42; Lugn, p. 139 in C. B. Schultz and Frye, eds., 1968). That in Kansas is shown to have come mainly from the Platte valley and secondarily from the Missouri Valley (Swineford and Frye, 1951).

If the Platte River source consisted chiefly of outwash, the volume of the related loess, estimated at more than 235,000km^3 in Kansas alone, is surprisingly great to have been derived from glaciers in the Rocky Mountains.

Most of the remainder of the loess in the central region is certainly of glacial origin, and there the quantitative adequacy of its chief source, outwash, is beyond question. Although some of this loess was derived from bare till exposed briefly during deglaciation, the great bulk of it is so clearly related to stream valleys containing outwash that it obviously represents the fine fraction of the outwash reworked by the wind. All this loess implies dominant W, NW, and SW winds like those of today. However, belts of loess along the western sides of the Mississippi and Missouri valleys, becoming thinner and finer textured toward the west, imply the existence of occasional "minority" winds with easterly components.

Thicknesses of 20 to 35m are encountered frequently along the eastern bluffs of these valleys, and thicknesses of more than 8m are common within distances of 40 to 50km eastward from the bluffs.

Even as far west as eastern Indiana and Ohio loess, although extensive, is generally less than 1m thick. Farther east it is rare and always thin, as in the Boston district, the Connecticut Valley, and along the Delaware River in Pennsylvania and New Jersey. All are adjacent to outwash sources.

At least three factors explain the scarcity of loess in eastern United States:

1. The drift in the Missouri River-Mississippi River region is rich in silt and clay, the chief components of loess, whereas the drift in the Appalachian region and on the Atlantic slope is richer in sand and poorer in silt and clay.

2. The Missouri River-Mississippi River region has stronger winds, less relief, and wider valleys than the Appalachian region and New England; these conditions favor deflation.

3. From the Missouri River region to the Atlantic coast the climate changes from steppe to moist-continental. The greater moisture, forest cover, and lower wind velocities of the moist-continental climate operate to reduce surficial drying of the outwash in preparation for deflation.

The stratigraphic positions of the loess sheets in central United States are summarized in table 9-A, which shows that loess occurs in every glacial stage except the Nebraskan. The thickest and most extensive unit is the Peorian Loess, of Wisconsin age. Very likely the pre-Wisconsin loess sheets, including a Nebraskan loess, were originally nearly as thick and

Table 9-A

Generalized Stratigraphy of Loess Sheets in Central United States.

State / Stage	Nebraska	Iowa	Illinois	Indiana
Wisconsin	Bignell Loess Peorian Loess Gilman CanyonLoess	Wisconsin loess	Peoria Loess Farmdale Silt Roxana Silt	Peoria Loess Member silt bodies
Illinoian	Loveland Loess Beaver Creek Loess Grafton Loess	Loveland Loess	Loveland Silt Petersburg Silt	unnamed loess
Kansan	Sappa Loess	unnamed loess	unnamed loess	Cagle Silt
Nebraskan				

extensive as the Wisconsin loess is now. No doubt much pre-Wisconsin loess has been removed by erosion, buried and lost to view, or decomposed and incorporated into zones of weathering. Indeed the stratigraphic relations of loess sheets confirm the inference that loess is glacial and not interglacial.

In some districts a layer of loess splits at the outer limit of a sheet of till, so as to lie both above and beneath the till (fig 9-15). This geometry shows that deflation of outwash began before the related glacier reached its greatest extent. In some exposures the loess beneath the till contact demonstrates plowing by the advancing ice.

Although abundant at the stratigraphic position of the Late-Wisconsin maximum, loess decreases stratigraphically upward. In the Plains region the widespread but rather thin Bignell loess implies that with its steppe climate that region continued to afford opportunity for deflation of silt even after active deposition of outwash had ceased. In the Great Lakes region, however, the newly created lakes acted as traps for silt formerly washed out along valley floors. Furthermore, as the annual floods of silt ceased, valley trains became more stable and were converted into grassy meadows not subject to deflation. In the region north of the Lakes the

Figure 9-15   Idealized section of the order of 100km in length, showing loess sheet "split" by till layer. (Not to scale.)

rocks, unlike those of the Plains and the Central Lowland, are productive of less silt and, furthermore, had been largely swept bare of loose surficial material by the ice sheet. Finally, the climate is moist and many of the streams consist of chains of lakes in which the stream sediments settle, safe from seizure by the wind.

The loess in central North America, then, is explained consistently as sediment deposited by strengthened winds, during glacial maxima, mainly from sources in outwash.

In some regions the clay and carbonate minerals in loess of glacial origin are indicators of source. In central United States, for example, three sources have been identified: a northwestern source (montmorillonite and calcite), a northern source (vermiculite and dolomite), and a northeastern source (illite) (Glass et al. in Bergstrom, ed., 1968, p. 35). The relations among outwash contributions made by the Missouri, Mississippi, and Ohio Rivers are interestingly demonstrated by the mineral compositions of suites of samples.

Loess in Europe.[10] *Distribution.* The distribution of loess in Europe is shown on the generalized map, fig 9-16. Basically the distribution is similar to that in North America, because loess is most abundant in the east, which has a plains and prairie terrain and a steppe climate, and is least abundant in the west, where the terrain is dominantly hilly and the climate maritime. As in North America, loess is related to extensive outwash valley trains along principal rivers, to the border of the glaciated region, and to dry climates. The vertical range of the loess approaches 600m, so far up the flanks of the highlands has it lodged.

In European Russia, as in Iowa and Nebraska, loess is continuous over wide areas and is thick: sustained thicknesses of 10 to 15m continuously over many thousands of square kilometers are found. Thickest in the region west of the Dnepr, it gradually thins eastward. Farther west it breaks up into smaller discontinuous areas related principally to outwash masses and to highlands against which it lodged. Westward, too, thickness diminishes, commonly reaching only 2 to 3m but rising to 5 to 10m along major streams and approaching 30m along the east flank of the Rhine valley, in its finest development in western Europe.

The outwash origin of loess in Europe was first recognized in 1889, and most geologists regard this origin as valid for the bulk of the European loess. The close association of loess with outwash is clearly apparent in Germany. In France this association is seen along the Rhône and Garonne Rivers, which carried outwash from glaciers in the Alps and Pyre-

[10] Kriger, 1965; Grahmann, 1932; Woldstedt, 1954–1965, v. 1, p. 168–186; Poser, 1951; Büdel, 1953; H. E. Wright, 1961a, p. 946, 961; Ložek in Schultz and Frye, eds., 1968, p. 67; various authors in idem, p. 185–369.

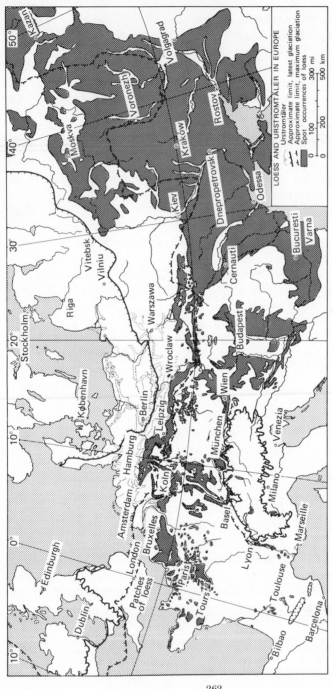

Figure 9-16 Distribution of loess and *Urstromtäler* in Europe. (Loess after Grahmann, 1932; in Britain from Tilley, 1964, fig. 1.) *Urstromtäler* from unpublished map by K. Keilhack. For a detailed map of loess in France, see Alimen, p. 70 in INQUA, 1969a.

nees respectively. In northern France the relationship to outwash is much less evident. Probably the "Channel River" was partly responsible; the relationship of loess to the present coasts of Brittany and Normandy suggests this. It has been held that even where there is no outwash, spring floods would have been more sudden and abundant during glacial ages than under present European climates, and that this should have resulted, in major valleys, in the spreading out of fine sediment for seizure by the wind, particularly in the spring. Such sources may have supplied much of the loess in France.

In western Germany the relation of loess to the Rhine, with its great valley train of Alpine outwash, is like that of loess in western Illinois, Kentucky, and Tennessee to the Mississippi River. Loess is thicker and more widespread east of the Rhine than west of it, and is thicker on west-facing slopes than on slopes facing east. The inference is drawn that westerly winds (the cyclonic planetary winds like those of today) were the chief agents of transport. In contrast, in north-central Germany the loess tends to be thicker on northeast-facing slopes and is thought to have been transported by NE winds from outwash in and around the *Urstromtäler*.

In south-central Europe the chief source of loess was outwash in the valley of the Danube, derived from glaciers in the Alps and in the Carpathians. From central Hungary, where Danube outwash was spread out wide, silt was blown onto the slopes of hills and mountains in Czechoslovakia and was carried across Hungary into northern Yugoslavia. From the lower Danube, silt was spread over eastern Rumania, Bulgaria, and the southern part of the Ukraine. (Penck, 1936; Milojević, 1950, p. 18–23; Różycki, 1967a, who draws attention to widespread deposition of loess in low ridges parallel with the direction of the effective wind.)

A richer source of the widespread loess in southwestern Russia consisted of great volumes of outwash deposited along the valleys of the Bug, Dnepr, Don, and Volga Rivers. As with the Mississippi and the Missouri, loess is thicker (in places 50m) in the vicinities of the Russian rivers than far from them.

Scandinavia and Finland, like central and eastern Canada, are areas of little loess. As in North America, this fact is explained partly by a succession of large water bodies that followed the shrinking ice sheet in the Baltic region, partly by moist climate, and partly by bedrock that does not yield abundant silt.

In the Mediterranean region there is little loess because of a conspicuous scarcity of outwash. The chief accumulations are in northern Italy (derived from outwash in the Po valley) and in southern France (derived from outwash spread out by the Rhône). In the British Isles loess, known in some localities as "brickearth," is not present in volume, is best developed near the valley of the Thames, and is believed to date from a glaciation earlier than the last one (Tilley, 1964). Conceivably the prevailing maritime climate may have maintained carpets of vegetation over considerable areas of valley trains even while outwash was actively accumulating, thereby reducing the area of dry silt exposed to the wind.

In summary, the areal distribution and the variations in thickness of European loess indicate that it was derived from widely and irregularly distributed sources, and that it reached its present distribution mainly through the agency of westerly winds.

Stratigraphy. Stratigraphically also the positions of European loess sheets (tables 24-A and 24-B) are much like those in North America. In Germany loess has been identified on till of Elster, Saale, and Weichsel age. In northeastern France, loess sheets correlated similarly are recognized. In the northern and western Alps region loess occurs on outwash of Mindel, Riss, and Würm age. The ancient Günz deposits are so thoroughly eroded and so poorly exposed that Günz loess, like Nebraskan loess in America, could have been widely present formerly and yet be rarely apparent today. Along the Danube in lower Austria, the oldest of several loess sheets is of Mindel age. An indurated loess in the Rhône valley region in southeastern France contains a Villafranchian vertebrate fauna, and is therefore no younger than earliest Pleistocene. Unfortunately it is not exposed in contact with any glacial deposit.

The Würm drift in the Alps (except in southeastern France) and the Weichsel drift in northern Germany have little associated loess, although loess in Belgium and northeastern France, related to the valley train of the "Channel River," is believed to be of Weichsel age. The rapid decrease of loess deposition following the Weichsel maximum is related to the great bodies of water that succeeded each other in the Baltic region, forming traps for the sediment that would otherwise have been dropped on outwash plains. Meanwhile the *Urstromtäler* and other former sources, deprived of annual spring floods bearing heavy loads of sediment, became covered with vegetation and hence ceased to furnish silt to the winds. This development closely parallels the events of the Wisconsin Age in North America.

In European Russia loess overlies drift sheets of Saale and Weichsel age. It is thicker and more widely distributed than farther west, reflecting the influence of a steppe climate.

Loess in Other Parts of the World. In northern Asia the extent of loess is still poorly known. Although from fig 9-17 no close relation between former ice sheets and loess is apparent, this may result in part from inadequate information. In central Asia the mapped areas of loess coincide

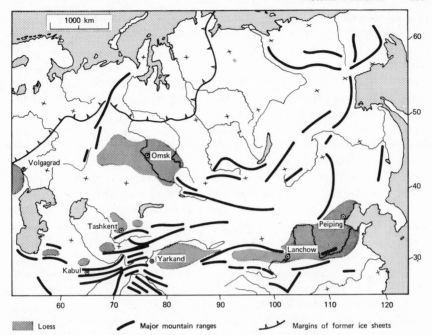

Figure 9-17 Sketch map showing principal areas of loess in central and northern Asia. Distribution of former mountain glaciers is not shown, but most major highlands carried glaciers, Loess areas from Kriger, 1965, who does not show many small areas reported in the literature. Discrepancies in loess boundaries between this map and fig 9-16 are not resolved.

with tectonic basins in arid and semiarid terrain. There is little doubt that in such environments the source of loess consists of sediment washed down by streams from adjacent mountains. To what extent such sediment was proglacial is not known; some of it definitely was, but a substantial part of it may have been nonglacial alluvium. Still another likely source consists of relict sediments of pluvial lakes. The great area of loess in eastern China is thought to be mainly nonglacial, and indeed the aggregate area of glacier ice in the region was very small.

According to Liu et al. (1958) there are two distinct layers of loess. The older, locally 120m thick, overlies the early Pleistocene Sanmen strata and is considered Middle Pleistocene. It includes soil zones and is therefore multiple. Unconformable upon it is Malan loess, as much as 40m thick, which decreases in grain size from NW to SE.

In South America the plains region of Argentina and Uruguay, especially between lats 30° and 40°, has thick, widespread loess that forms a broad belt around and south of the Rio de la Plata. Thicknesses are said to reach 10 to 30m over wide areas and to exceed 100m in places. But because loess has not been studied in great detail eolian silt has not everywhere been clearly distinguished from water-laid silt. Furthermore, although more than one sheet of loess has been identified, stratigraphic relations are not yet well understood. Probably a small proportion of the loess represents reworked outwash; most of it, however, is the product of deflation of nonglacial alluvium from arid and semiarid country to the west, in the rain shadow of the Andes.

On the extensive plains on the South Island of New Zealand, loess, locally as much as 18m thick, occurs in stratigraphic section and as surface mantles on terraces of outwash and alluvium. It is clearly related to glaciation. At least three loess sheets have been identified, two of them postdating the last interglacial of the New Zealand time sequence and one antedating it (Young, 1964). Thinner loess on the west coast is also related to outwash.

In both Australia and Africa, loess is rare. It may be significant that although both continents possess extensive arid regions, neither was more than slightly glaciated in Pleistocene time.

Summary. Our knowledge of the composition, distribution, and stratigraphy of loess leads to the conclusion that near the maximum of each glacial age windblown silt accumulated at the periphery of each of the great ice sheets wherever bedrock, climate, and distribution of drift were favorable. In general the favorable regions were beyond the southern and western sectors of the Laurentide Ice Sheet and the southern and eastern sectors of the Scandinavian Ice Sheet. The development of loess around the Siberian Ice Sheet is still little known. A large amount of loess occurs in and to leeward of desert basins, particularly in Asia, but its source is mainly nonglacial alluvium.

It is doubtful that loess possesses consistent climatic significance. Its accumulation today under steppe conditions and its comparative scarcity in moist climates suggest that dry conditions are optimum. Probably, however, the continual deposition of fine alluvium such as characterized major proglacial rivers, coupled with strong winds, would have resulted in the transport of great quantities of silt by wind regardless of climate. Snail faunas in North American loess suggest climates not unlike those of the present day; the rodent fauna of European loess suggests both arctic tundra and steppe conditions. As we have noted, there is evidence of gradual change of climate during the accumulation of a single body of loess. In view of these facts it does not seem possible to associate any particular climate with the times of great accumulation of loess.

CHAPTER 10

Frost-Action Effects[1]

This chapter deals with the effects of freeze-and-thaw action of moisture in bedrock and regolith. The natural processes in which freezing and thawing of moisture in the ground play a dominant part include frost wedging, frost cracking, frost heaving, and some kinds of mass-wasting, notably gelifluction. The effects, which involve sediments, structures, and land forms, include frost-wedged debris, gelifluction sediments, frost cracks, frost-churned ground, patterned ground, rock glaciers, and other features as well. Before examining these individually let us outline the spatial distribution of frost action and its effects under the climates of today.

Frost action is controlled primarily by climate (low temperature and adequate moisture) and secondarily by the character of the rock material at the surface. The zone within which the processes are effective includes the polar regions and, like the snowline, rises toward the equatorial belt, where it occurs only at high altitudes. Beneath large areas in middle latitudes the ground freezes seasonally to shallow depths. Beneath extensive parts of high-latitude lands the ground remains perennially frozen, in some places to depths of hundreds of meters. More than 4% of the Earth's land area is underlain today by permafrost (perennially frozen ground); this frozen area is nearly four-tenths as great as that now covered by glaciers and perennial snow. In summer the frozen ground generally thaws to shallow depths of a meter or so, creating an annual cycle of freeze and thaw. Cycles of seasonal, daily, and other lengths occur in various latitudes, with the result that frost action performs much geologic work and creates effects, some of which are conspicuous.

[1] General references include Bertil Högbom, 1914; Troll, 1944, 1958; Büdel, 1953; A. L. Washburn, 1968, p. 45–105; Permafrost International Conference, 1966; H. E. Wright, 1961a, p. 940–947; Tricart, 1963, 1970; Tricart and Cailleux, 1967; Péwé, ed., 1969; H. E. Wright and Osburn, eds., 1967.

Frost wedging depends less on the frequency with which temperature passes through O°C than on amplitude and rate of the change (Battle in W. V. Lewis, ed., 1960, p. 83). Freezing must occur from the surface inward, to seal openings and create closed spaces. Under specific circumstances, pressures as great as $70kg/cm^2$ can be generated.

Beginning in the mid-19th century it was recognized that features characteristic of much colder climates occur in a "fossil" or relict condition in areas where they are not forming under today's climates. It was soon realized that such features reflect the lower temperatures of the glacial ages. Subsequent research in this field has been directed toward specifying the limits of temperature, moisture, and other climatic parameters within which the various processes of frost action operate. With reliable results one should then be able to approximate the glacial-age climates in the areas where relict features occur. However, so many uncertainties are involved that this research has proceeded slowly. One of the common hazards is the danger of confusion of relict features that imply cold climate with those that do not, because resemblances between the two groups can be close. Therefore, interpretation of former climate must be made with extreme caution.

This group of features and the climates under which they formed came to be thought of as "periglacial," a term introduced by Lozinski in 1909 to designate the climate and related geologic features peripheral to the ice sheets of the Pleistocene. The "periglacial" concept came to be identified, through loose usage, with frost-action effects and also with loess, without regard to the relation of these features to ice sheets.

However, frost-action effects may have little in common with eolian sediments derived from outwash, and important sectors of former ice sheets seem not to have been fringed with extensive features created by frost action. For example, at glacial maxima the oceanic climate of coastal New England differed from the more continental climate of the Dakotas, and the climate of southern European Russia differed markedly from that of coastal Norway. Hence the term "periglacial" in this sense is ambiguous, at least until it can be defined precisely. The word has come to imply rigorous, cold-climate conditions regardless of proximity to glaciers, but it does not necessarily imply that conditions fronting all glaciers are of this character. By some European students the term *frost climate* has been employed to describe climates in which temperatures fluctuate so as to freeze and thaw ground moisture repeatedly, and we shall use the term in this meaning. Such a climate may differ from one in which, merely, air temperatures at the surface traverse the freezing point.

Despite the difficulties involved in the study of frost-action effects, at least a beginning is being made at the task of defining parameters of glacial-age climate in territory that was not covered with ice. In the discussion that follows we shall be concerned, not with analysis of the processes of frost action but with the relict features that have resulted from them and with the possibilities they offer for reconstructing former climates. Treatment of the topic has been helpfully influenced by publications and an unpublished syllabus by A. L. Washburn, long a student of frost action.

Permafrost [2]

Permafrost is neither a process nor an effect; it is a frozen condition in regolith or rock. As we have just noted, this condition is believed to exist today as continuous belts (exclusive of discontinuous patches) beneath about 4% of the world's land area. Permafrost is concentrated mainly in high northern latitudes; there is little in high southern latitudes because that region consists mostly of glacier ice and sea water. Permafrost thicknesses as great as 1000m have been measured, although average thickness is far less.

Figure 10-1 shows, for the northern hemisphere, the present southern boundaries of continuous and discontinuous permafrost, broadly paralleled by the 0° annual isotherm in surface air. In the zone of continuous permafrost ground temperatures generally range between −5° and −12°, rising to −1° or higher in the discontinuous zone where mean air temperatures are higher.

Permafrost develops where and when mean air temperatures are negative; the temperature of the surface portion of the permafrost itself fluctuates seasonally, so that a surface layer (the *active layer*) thaws and refreezes to depths ranging from <1m to >3m. At the depth of no seasonal change, ground temperature generally approximates mean annual air temperature; hence much permafrost is believed to be essentially in equilibrium with existing climates. However, minor fluctuations occur, as indicated by slight shrinkage of permafrost areas within the past 100 years, at least in USSR territory.

Little is known about the time of inception of permafrost. Radiocarbon dating of peat in permafrost in southern Alaska (D. R. Nichols, 1966) suggests at least partial thawing at the time of warmer climates a few thousand years ago. Still earlier fluctuation is implied by the occurrence of extinct Pleistocene mammals in permafrost, in both Siberia and Alaska.

The fossil record (chap 28) implies that in early Cenozoic time there could have been no permafrost in the Arctic region, but we have no di-

[2] General reference: S. R. Stearns, 1966.

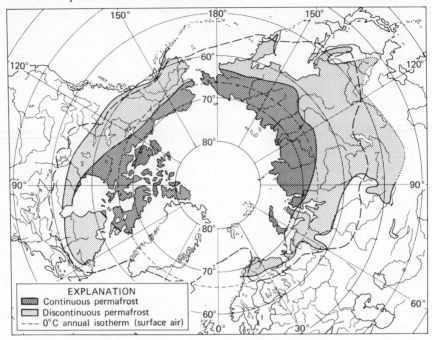

Figure 10-1   Present extent of permafrost in the northern hemisphere. (After S. R. Stearns, 1966, fig. 1; Canada Geol. Survey Map 1246A, 1967.)

rect evidence as to whether widespread permafrost antedated the invasion of high-latitude lowlands by Pleistocene ice sheets. A comparison of fig 10-1 with fig 4-8 shows that the areas of former ice sheets bear little relation to distribution of permafrost. This casts doubt on the suggestion that the ice sheets, acting as thermal insulators, inhibited the development of permafrost beneath them. It seems more likely that the glaciers spread outward over Arctic areas already underlain by permafrost. However, some of the permafrost in polar regions probably has developed since the retreat of ice-age glaciers.

Pollen studies indicate that belts of frozen ground, which in some sectors extended beyond the peripheries of major ice sheets when the latter were at their maxima, followed the shrinking glaciers by thawing at the outer limits of the frozen areas. During interglacial ages probably the areas of discontinuous permafrost that exist today decreased substantially or disappeared. Whether at those times the existing areas of continuous permafrost either became discontinuous or disappeared altogether is not known.

In the Arctic region today, thawing of perennially frozen ground in flat areas results in the development of pits that may grow in diameter and become lake basins. Such thaw basins impart to the surface a peculiar pattern that characterizes many Arctic lowlands. After the permafrost has thawed, however, the basins do not persist as relicts. Being temporary features that exist only by virtue of partial thawing, they disappear by the time thawing has become complete. Indeed, once permafrost has thawed, it leaves, in itself, no direct trace of its former presence. Yet former permafrost can be identified by the presence of features that must have been associated with it, such as ice-wedge casts, pingos, and some rock glaciers.

SEDIMENTS CREATED BY FROST ACTION

Frost-Wedged Debris

Angular fresh debris, mostly coarse grained and derived from local bedrock, covers a very large aggregate area in polar and alpine regions. This material is thought to result mainly from *frost wedging* (also referred to as frost shattering and frost splitting), the prying apart of rock by ice as it forms during the freezing of water, usually along joints, stratification planes, and interfaces between grains. Both volume increase and the directional growth of ice crystals contribute to the forces exerted. Much of this patchy mantle of debris postdates the latest glaciation, because in places it is clearly derived from rocks having glaciated surfaces. On nearly flat surfaces the waste accumulates as mantling sheets; at the bases of slopes it forms taluses. Optimum conditions for frost wedging today seem to be a moist climate and frequently recurring freeze and thaw. However, the process occurs, although probably more slowly, in dry climates as well.

It seems obvious, then, that in any region the inception of a "frost climate," one in which the freeze-thaw cycle is conspicuous, should accelerate frost wedging. An example of a country that during glacial ages was both cold and moist is Scotland. There the depth to which bedrock is shattered reaches 10m in some places. Because such a depth exceeds that reached by annual or daily freeze-thaw cycles at present and probably also in glacial ages, it has been suggested that rupture occurred in permafrost, and that the frost wedging was accomplished by water drawn upward from beneath the permafrost (Galloway, 1961).

Taluses, Block Fields, Protalus Ramparts. Frost wedging was formerly significant even in climates that are dry today and that, even in glacial ages, were drier and probably less cold than Scotland. In western United

States, for example, most taluses in the south are inactive today but increase in size and abundance toward the north and toward higher altitudes. The inference is drawn that former colder, moister climates reflected in the pluvial lakes in that region promoted frost wedging and the production of sliderock, activities that are now generally decadent. Indeed the flash floods that now characterize much of the region have dissected cliffs in the Colorado Plateau country to such an extent that remnants of cliff face, protected by mantles of sliderock, now stand out from the present cliffs as relics of a time when taluses were more active.

Another kind of relict feature, differing in appearance from taluses but having a somewhat similar origin, consists of detached blocks of bedrock lying in groups and swaths on the slopes of highlands in middle latitudes. They are angular in form, are commonly bounded by joint and other structural surfaces, and exceptionally may reach 10m or more in diameter. Commonly they form a continuous armor or mantle over the surface; this fact has given rise to the name *block fields* for accumulations of them.

In most middle-latitude occurrences the boulders that constitute block fields are now stable; in one locality or another they are being weathered chemically, or covered with lichens, or overgrown by trees. They are regarded as relics of a time or times when the climate was colder, and when the boulders, together with smaller particles, were frost wedged from bedrock farther upslope and moved downslope, probably by frost heaving. Relict block fields have been reported from the Appalachian Mountains (cf. Potter and Moss, 1968) as far south as North Carolina and from highlands in middle Europe and in eastern Australia. In some instances block fields have been confused with surface sheets consisting of crudely rounded, weathered boulders. As such features could have resulted from chemical weathering accompanied by creep, they are not necessarily relicts indicating a former frost climate. The distinction, together with a detailed quantitative study of block fields in Tasmania, was discussed critically by Caine (1968).

The climatic changes implied by block fields and some taluses have not been evaluated quantitatively, but the fact of change is evident. Some taluses record present activity separated from an earlier longer episode of activity by a period of stability.

Another form in which frost-wedged rubble accumulates is that of ridges along the lower edges of snowbanks that nestle at the bases of cliffs. The detached rock fragments skid down the snowbank and are thus localized at its base as *protalus ramparts*. These ridges, as much as 10m high, are commonly seen in cirques. They resemble small end moraines, from which they can be distinguished by lack of glacial abrasion of any of the component fragments and by differences in fabric. Destruction of the snowbank by ablation under a warmer climate ends the growth of the rampart, though the altered climate may still be such as to permit frost wedging to continue.

Frost wedging is implicit also in the occurrence of gelifluction sediments that include coarse fragments of mechanically disrupted bedrock. These are discussed in a subsequent section.

In northwestern Canada frost wedging has lifted blocks as much as 8m in diameter from flat exposures of crystalline rocks, along foliation planes and joints through vertical distances of more than 4m (Yardley, 1951). Apparently the process occurs today. It can be assumed to have been active also before the ice sheet reached its maximum extent during the latest glaciation. If this was the case, it may help explain the abundant presence, west of the Great Lakes region, of large boulders in Wisconsin till, in contrast with pre-Wisconsin tills which have conspicuously fewer large boulders. The implication is that the interval immediately preceding the Late-Wisconsin glaciation was characterized by a frost climate in parts of Canada, whereas at least some earlier nonglacial times were not. The mantles of frost-wedged blocks mentioned in chap 18 also might date from the same time.

Rock Glaciers. Still another form in which frost-wedged rubble is found is that of the *rock glacier,* a lobe-shaped accumulation of angular boulder rubble originating at the toe of a cliff. The form (fig 10-2) is remarkably similar to that of a glacier. Active rock glaciers occur near the snowline in mountainous highlands and are most abundant at cliffs, many of them cirque cliffs, that face away from the Sun. Some of the largest rock glaciers are more than 3km wide, although most are much smaller, and average around 50m thick. Margins, 5m to more than 30m high, are abruptly steepened to approach the angle of rest of the rubble, in places as much as 40°. Inside the margins there are clearly marked concentric lobes or wrinkles, as well as crevasses and pits, all suggesting flow. Upstream from the margin the surface is hummocky, sloping between 5° and 20°. The component rubble commonly consists of joint-faced boulders as much as 8m in diameter, derived from the cliff. The boulder rubble grades downward into a frozen mixture of boulders and particles of other sizes, including silt. An active rock glacier flows. Rates between 50 and 70cm/yr have been obtained by measurement in Alaska and more than 150cm/yr in the Alps. According to the most detailed publication available (Wahrhaftig and Cox, 1959), which deals with active rock glaciers in Alaska, motion is the result of flow of interstitial ice and low temperature is therefore essential to it. During flow, boulders are fed to the rock glacier by frost wedging in the cliff at its head. Additional

Figure 10-2 Rock glacier, now inactive, at head of Silver Basin, Silverton, Colorado. (Whitman Cross, U.S. Geol. Survey.)

data supporting this view are set forth by Blagbrough and Farkas (1968). Another theory, advanced as early as 1910, views some rock glaciers as being the residual superglacial loads of small true glaciers that have melted away; Outcalt and Benedict (1965) detail the data supporting this view.

Some rock glaciers, including many of those in middle latitudes, are clearly inactive and therefore relict, on the evidence of weathering, growth of lichens, and stratigraphy. Some, for example, overlie end moraines dating from late in the last glacial age, and in turn are overlapped by taluses consisting of fresh sliderock. Most of the relict forms are considerably thinner than the active ones, probably reflecting compaction during loss of interstitial ice. It is not known whether these inactive rock glaciers originated as did the active forms studied in Alaska. Some may have been small true glaciers with extraordinarily coarse superglacial loads, some may have moved in part by landsliding or by other processes of mass-wasting, and various combinations of movement probably have occurred. Indeed, one active rock glacier in Alaska grades upstream through a body of ice-cored ablation drift 3km long into a small active cirque glacier (Foster and Holmes, 1965). Until rock glaciers have been investigated more fully we can say only that many relict rock glaciers indicate a frost climate.

Rubble in Caverns. Former frost wedging within caverns is recorded by layers of rubble derived from the cavern roofs, now stratified in the floor accumulations of wide-mouthed caverns in which today only fine sediment is accumulating. Such occurrences are reported from many caverns, especially those in which detailed archeologic excavations have been made. A carefully studied example controlled by isotopic dates is Haua Fteah, a cavern on the coast of Libya near lat 32°. Studies by MacBurney and Hey (Hey, 1963) showed that the strata beneath the cavern floor include two layers containing angular rubble derived from the roof, separated by layers without rubble. The C^{14} dates of both layers, the very different types of artifacts the layers contain, and in one layer the character of the fossil fauna as well, indicate that each layer corresponds approximately with one of the glaciations that occurred in Weichsel time (table 24-B). Furthermore, the local valleys contain two generations of angular sliderock in the form of relict taluses, correlated by means of the two assemblages of artifacts with the two layers of rubble in the cavern. Sliderock does not appear to be forming today. Therefore, the evidence is strong that in this district, despite its low-middle latitude, frost wedging was active during at least two Weichsel glaciations. Relict taluses occur likewise at Gibraltar.

Solifluction Sediments [3]

The sheetlike or lobelike movement of saturated regolith down slopes, mainly by fluid flow, was called *solifluction* by J. G. Andersson (1906, p. 95) in a classical study of relict features in the Falkland Islands. Solifluction is widely active today in high-latitude regions of frozen ground including permafrost, where rates of movement of a few centimeters per year have been measured by annual surveys of stakes and painted boulders in selected localities (Rapp, 1960, p. 182; A. L. Washburn, 1947, p. 92; 1967). In continuous permafrost, with summer thawing limited to shallow depth, thaw water cannot infiltrate the permafrost beneath it and so remains in the thawed layer, saturating it and permitting rapid flow of that layer downslope. The process can occur on slopes as gentle as 2°. At rates so rapid, slopes could become mantled with flowed·debris within a geologically very short time.

Solifluction sediments (also called *flow earth* and *Head*) are characterized by lack of sorting and by lithologic provenance confined to the slope on which the deposit occurs (fig. 10-3). Grain sizes ordinarily include a large proportion of fines, but the coarse fraction may include very large boulders, which may have moved mainly by frost creep. In some sediments the platy constituents are imbricated. Like till the material is a

[3] General references include Troll, 1944; A. L. Washburn, 1947, p. 88–96; Cotton and Te Punga, 1955, p. 1003–1012; Büdel, 1959.)

Figure 10-3 Sediment created by gelifluction active today. Schuchertdal, northeast Greenland. The base of the flowed mass is not visible; the stake is 61 cm long. (A. L. Washburn.)

diamicton, but it differs from most till not only in the local provenance of the particles, but also in more angular shapes of the coarse particles, looser, less compact texture, and fabric in which long axes are oriented predominantly downslope (Galloway, 1961).

Individual solifluction sheets are generally lobate and range in thickness from less than 1 m on most slopes to several meters on flats or in basins, where two or more sheets can pile up to a total thickness of several meters. In some lobes cobbles and boulders can be segregated and concentrated at the downstream margins of the lobes. As many as five episodes of solifluction have been established from the shatigraphy in a part of Alaska (Wahrhaftig, 1949).

Solifluction sediments other than those derived from the erosion of glacial drift or other unconsolidated material imply the continuous production of a supply of waste by frost wedging. No cliff is demanded, however, because the supply could have been derived from buried bedrock.

Relict solifluction sediments in the form of a discontinuous mantle are widespread on hillslopes in nonglaciated districts of western and central Europe and have been described from New Zealand. In North America they are locally developed in the Appalachian region, within a zone >100km wide beyond the outer limit of glaciation (Denny, 1951) and in places extending back into the glaciated region. In parts of Europe the corresponding zone is as much as 300km wide. Not only are solifluction sediments frequently exposed in section, but they also impart to the hill-slopes smooth, flowing profiles that commonly are lacking outside the zone of solifluction. The smooth profiles tend to be preserved best in areas where the solifluction sediments are permeable, minimizing surface runoff.

In middle-latitude regions solifluction sediments as a group have been thought of as relicts of former frost climates. However, solifluction sediments have been identified in wet, tropical highlands, and it is evident that solifluction can occur without any floor of frozen ground beneath it provided only that the regolith is saturated. Consequently such sediments in themselves are not diagnostic either of permafrost or of ground that freezes only seasonally. The term *gelifluction* has been applied to solifluction that moves over a frozen floor; thus the term excludes "tropical" solifluction.

Whereas relict solifluction sediments in middle latitudes do not in themselves prove a former frost climate, they are regarded as relicts of former colder climate because of their regional distribution and their association with features such as frost cracks that do indicate frozen ground. Although the ground could have consisted of permafrost in some areas, the observed features could have been the result of seasonal freeze-thaw cycles alone. Because gelifluction is most active today in treeless areas, it seems likely that the regions of intense gelifluction in glacial ages were treeless when the process was most active. The analogy is supported by the occurrence of dominantly nonarboreal pollen grains in organic sediments closely associated with gelifluction features.

In northwestern and central Europe two distinct layers of gelifluction sediments are recognized in many places; in northern France and in Poland several layers have been identified, the oldest related to early glaciations (Breuil, 1939; Jahn, 1960).

STRUCTURES RESULTING FROM FROST ACTION

In addition to creating the sediments described, frost action also creates characteristic structures, including frost cracks, pingos, frost-stirred ground, and patterned ground. We shall outline the characteristics of each and indicate the climatic inferences, if any, that can be drawn from them.

Frost Cracks

Thermal contraction of frozen ground at substantially subfreezing temperatures can result in *frost cracks*. These are irregular, nearly vertical fissures, single or multiple, of variable depth. When filled with ice or sediments they can attain widths of a few centimeters to a meter or so. In some places cracks unite to form polygonal patterns on the ground surface. The cracks form most readily when temperatures drop suddenly to low values, causing contraction of frozen ground to variable depths. Such cracks occur in middle as well as high latitudes, and so are not characteristic of any single zone. Some of them close up without leaving any durable record of their existence.

However, many of the frost cracks observed in vertical exposures in permafrost are filled with ice, to form downward-pointing wedges as much as 5m wide at the top and as much as 10m long. Frost accumulates in the crack which, reopening at times of low temperature, acquires more frost, and so the ice wedge is gradually built through a period that may amount to hundreds or even thousands of years. With climatic warming ice wedges melt from the top downward. Ice wedges in central Alaska penetrate alluvial silt having an age of $>$30,000 C^{14} years at the position of the point of one of the wedges. The melted tops of the wedges are uniformly overlain by a different body of silt, the base of which dates approximately 4000 C^{14} years. Thus the period of relative warmth when the melting occurred ended before 4000 C^{14} years ago, and is therefore believed to correlate with Hypsithermal time.

Because the growth of an ice wedge implies its existence without thawing for many years, ice wedges indicate the presence of permafrost.

In a number of lowland areas in middle latitudes, vertical sections have exposed wedge-shaped bodies of silt or sand that transect the stratification of enclosing sediment and that resemble ice wedges in form and dimensions (fig. 10-4). In nonstratified material such as till, bodies of this kind are far more difficult to recognize. Such features are described representatively from localities in Wisconsin (Black, 1965), Sweden (Johnsson, 1959) and central Poland (Dylik, 1966), three regions characterized by very low temperatures during glacial ages. They are present also in joints in bedrock, in areas such as southern New England, penetrating to depths as great as 8 to 10m. Many of these wedges of silt or sand are believed to be the casts of former ice wedges. When a wedge of ice melts, sediment from above can trickle into the cavity and create a pseudomorph of the former wedge. A true *ice-wedge cast,* then, like an ice wedge itself, implies former permafrost, the thickness of which was at least as great as the length of the cast. According to Péwé (1966) the mean annual tempera-

Figure 10-4 Ice-wedge cast consisting of fine sand transecting stratified silt and fine sand. Exposed length of wedge is about 5m. Growth of the wedge has resulted in small-scale normal faulting. Near Lublin, Poland. (A. L. Washburn.)

ture of regions in which ice wedges are actively growing today is around
$-6°$ to $-8°$, and therefore true ice-wedge casts imply comparable temperature values at former times. If this is the case, glacial-age mean temperatures were lower than today's means by as much as $15°$ to $18°$ in
western and central Europe, and by $10°$ to $15°$ in parts of central North
America.

Ice-wedge casts occur in more than one stratigraphic position. In western and central Europe wedges of both Late and Early Weichsel age are
identified. The map, fig 10-5, shows that ice-wedge casts (undifferentiated
as to age) occur in a wide belt south of and overlapping on the outer
limit of ice-sheet glaciation. Ice-wedge casts of two ages are known likewise near Calgary, Alberta, and in central Alaska.

Closely resembling ice-wedge casts in both form and dimensions are
sand wedges, which likewise form in frost cracks and which have been observed in the arid climate of McMurdo Sound, Antarctica (Péwé, 1959).
Instead of filling with ice, the cracks fill with clean sand that falls into
them from the surface. The wedge-shaped fillings of sand lack structure
of any kind, and like ice-wedge casts, may form polygonal patterns on the
surface. Their implication for former temperature is the same as that of
ice-wedge casts.

However, not all wedgelike bodies of sediment are either true ice-
wedge casts or sand wedges. Structures that in vertical section resemble
such features are created by the penetration and decay of large tree roots,
by leaching as a part of the general process of weathering, and probably
in other ways (Johnsson, 1959). These possibilities must be eliminated before firm climatic inferences can be drawn from wedges.

Pingos

Associated with existing permafrost in Arctic North America and Eurasia are large domelike mounds, some of which reach 30m in height, with
cores of ice or frozen sediments. Known as *pingos* (an Eskimo word),
these features, which consist of both structures and land forms, are also
called hydrolaccoliths. Like true laccoliths they develop through injec-
tion and deformation. In pingos the injected material consists of ground
water, confined either by freezing of ground around it or by artesian conditions. The injected water freezes, creating an ice-cored mound. When
the ice melts, the land form partly or entirely disappears but the deformed structures remain as relics.

Relict pingos from middle-latitude Europe have been reported (for a
brief review of the literature, with references, see Maarleveld, 1965). Such
relict forms imply permafrost, but they are so few and in some cases so
equivocal that no reconstruction of glacial-age climate has yet been based
on them.

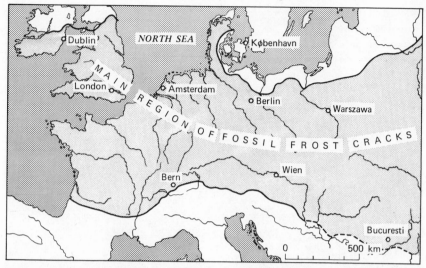

Figure 10-5   Sketch map of western and central Europe, showing (light-shaded area) the main region in which ice-wedge casts occur. Northern limit of region is drawn at the southern limit of glacier ice at the Lammermuir, D-line, and Pomeranian maxima (figs 23-4, 23-6). (After Dylik and Maarleveld, 1967.)

Cold-Climate Patterned Ground [4]

Patterned ground is the general term applied to a group of features, including aggregates of polygons, circles, stripes, and steps, that form distinct patterns on the surface, particularly though not exclusively in cold regions. Some patterns are created by the systematic segregation, or sorting, of coarse particles from fine in the regolith. Large sorted forms are closely related to frost climates, as is demonstrated by the distribution of those which are active now. Other patterns are nonsorted, and some of these, likewise, are confined to frost climates.

Although the origin of many patterned features is still obscure, it is believed that the main processes of origin include frost cracking, frost heaving, changes in water content, and hydrostatic pressure in unfrozen material confined by frozen ground.

Some large, sorted forms occur as relicts on highlands in southeastern Canada, New England, and the central Appalachian region (figs 10-6 and 10-7), and on many highlands in Europe and other parts of the world. In these regions minimum altitude at which the forms occur increases toward the equator, well below though roughly parallel with the snowline,

[4] General references: Maarleveld, 1965; J. Lundqvist, 1962; A. L. Washburn, 1956.

Figure 10-6 Relict sorted polygons consisting of blocks of sandstone, summit of Negro Mountain (alt. 969m) near Meyersdale, Pennsylvania. Diameter of the polygon in left center is 8.5m. (G.M. Clark, 1968.)

and above though roughly parallel with the upper limit of trees. Such relict large, sorted forms indicate former frozen ground, and in most places probably indicate permafrost. More precise indications are likely to emerge from more thorough study of the various kinds of patterned ground.

Frost-Stirred Ground

Frost-stirred ground consists of regolith, the components of which have been stirred by frost action. It implies the presence of moisture and is characterized by the distortion and mixing of strata or soil horizons or by the mixing of ventifacts, which formerly lay on the surface, down through the regolith. In places the stirring action probably reflects the same processes that are responsible for patterned ground. If this is correct, then frost-stirred ground is an indication of frozen ground. A special variety, consisting of interpenetrating tongues of frost-stirred regolith and forming what are known as *involutions* (fig. 10-8), may be related to permafrost. Johnsson (1962) critically discussed these features. The stratigraphic relationships of involutions in northern Illinois suggest a time of formation close to that of a glacial maximum dating from less than 20,-000 years ago.

Figure 10-7 Localities of occurrence of cold-climate patterned ground (small disks) in the Appalachian region south of the glaciated region. Altitudes of occurrence increase irregularly with decreasing latitude, but it has not been established that all the occurrences are of the same age. (G. M. Clark, 1968.)

Figure 10-8 Frost-stirred ground. Involuted contact between eolian silty sand (above) and glaciofluvial gravel (below). The digging tool is 50cm long. Near Wickford, Rhode Island. (J. P. Schafer.)

However, features similar to those of frost-stirred ground can be created by differential loading (McKee and Goldberg, 1969, pls.), ice-thrust features, mass-wasting processes, solution, growth and decay of tree roots, and burrowing organisms. Therefore, great care is necessary in establishing frost action as the cause of movement if climatic inferences are to be drawn. Relict frost-stirred ground occurs widely in central and western Europe, but has been reported less frequently from middle-latitude North America. If the difference is real, it may indicate that during glacial ages frost climate in North America was restricted to a rather narrow belt peripheral to the ice sheet, because the country was more open to warm air masses flowing in from the south than was the case in Europe.

LAND FORMS SHAPED BY FROST ACTION

Frost action takes a hand not only in creating distinctive sediments but also in shaping land forms. The array of sculptured forms—cirques, horns, and arêtes—that characterize mountains of the Alpine type are in large part the work of frost wedging as an accompaniment to sculpture by glaciers (chap 6). Mountain summits that reach up into a frost climate but that do not reach the firn limit are lightly but still recognizably affected by frost wedging. In niches where banks of snow and firn can accumulate seasonally, nivation operates effectively, creating incipient cirques.

Another sort of feature, occurring at much lower altitudes, that has been cited as related to frozen ground, consists of streamless valleys. Systems of dry valleys in districts where drainage at present is wholly subsurface exist in Germany and Britain, and are developed best in the South Downs in southeastern England (summary in Bull, 1940), where the surface bedrock consists of chalk. The valley floors are free of insoluble residues, but the chalk beneath valley sideslopes is locally shattered as though by frost wedging. Some valley systems extend below sealevel, suggesting that they date from times when sealevel was lower than now.

It was inferred as early as 1887, by Clement Reid, that such valleys are relict, having been cut by streams during glacial ages when the ground was frozen (although not necessarily perennially), inhibiting infiltration, and when large discharges, probably early in the season of thaw, swept the growing valleys clear of frost-wedged debris. At the end of a glacial age, thawing caused water tables to subside, permitted ready infiltration into the underlying chalk, and put an end to sustained surface runoff.

Alternative explanations of the dry valleys have not been investigated very thoroughly, and we are not in a position to evaluate the significance of the valleys. All we can say at present is that some of them may have developed in former frozen ground.

SUMMARY

The potential afforded by relict frost-action effects for approximating former temperatures is good. However, it is still largely potential, because the origins of many such features are not yet understood well enough to permit firm inferences as to temperature. At present it seems possible to infer only two environments: permafrost and frost climate, the latter being much the less well defined of the two. We have already noted that ice-wedge casts, pingos, and probably some rock glaciers and large sorted polygons indicate permafrost. Because areas of continuous permafrost are mostly treeless, general inferences as to former plant cover should be possible if areas of former permafrost can be delineated. Sediments resulting from frost wedging and gelifluction also indicate a frost climate, but not necessarily permafrost.

From the available data derived from relict frost-action effects, it is clear that during glacial ages the frost-climate zone expanded into lower latitudes and descended highlands to lower altitudes than those it occupies today. This change resembled in a general way the descent of the climatic snowline that occurred at the same times. Despite the rather small number of stations for which data on relict frost-action effects exist, their rather orderly relation to latitude and altitude leaves little doubt as to the general nature of the climatic changes.

Nevertheless, in assessing the significance of relict features in middle latitudes we must bear in mind that glacial-age climates in Europe and in southern North America were not quite the same as existing Arctic climates having comparable isotherms, because "the Sun was higher." The winters were shorter and the summers longer and warmer. Temperature fluctuation probably was more frequent, giving rise to a more rapid rate of frost action than occurs in the Arctic today.

CHAPTER 11

Weathering and Soils; Caverns; Stream Regimens

WEATHERING AND SOILS

In this chapter we are concerned with processes of weathering, the activities of streams, and dissolution and deposition in caverns. We deal mainly with the way in which change of climate may be reflected in such activities and with the stratigraphic aspects of their effects. For example, soils and zones of weathering develop in glacial drift and in loess and other deposits related to drift. In many places such zones are buried beneath younger drift, indicating that nonglacial conditions intervened between two glaciations. Not only do these zones establish the fact of multiple glaciation, but also they afford at least some basis for inferring conditions of topography, drainage, and ecology under which they formed. Furthermore, zones of weathering and soil zones constitute horizon markers useful in stratigraphic correlation and are broadly helpful in the discrimination of late-Pleistocene sediments from earlier ones.

In discussing soils we must clarify an important matter of terminology, because geologists use the term *soil* differently from the way engineers use it. To pedologists and most geologists *soil* means that part of the regolith that can support rooted plants. To the engineer it means the entire regolith and therefore includes all nonconsolidated sediments. We shall follow the geologist's definition of soil, without prejudice and for the practical reason that the majority of readers of this book are likely to have become accustomed to the geologist's usage.

Weathering Profiles

Zone of weathering is a geologic term. It refers to the zone, broadly parallel with the surface, within which exposed bedrock or regolith be-

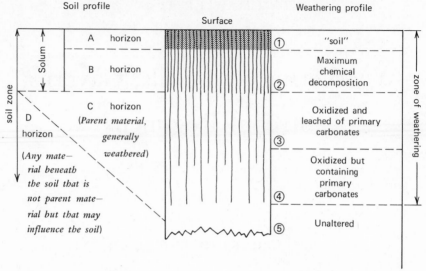

Figure 11-1 Ideal section of glacial drift in which primary carbonates are present, and in a region of moist climate. It illustrates differences between the weathering-profile concept and the soil-profile concept. The circled numbers are those of the horizons recognized by Leighton and MacClintock (1962). (Adapted from various sources.)

comes weathered, especially by chemical alteration. As weathering proceeds downward from the surface, progressive changes occur in bedrock or regolith, and in fairly homogeneous material these result in the development of successive subzones, generally parallel with the surface, each altered to a different degree. The sequence of subzones as a group constitutes a *weathering profile*. The thickness and character of the subzones depend on such factors as mineral composition, texture, structure, position of water table, climate, and topography. A common kind of weathering profile developed in calcareous drift and under a moist climate is shown diagrammatically in fig 11-1, center and right. In this as in many profiles, oxidation of iron minerals outpaces dissolution of carbonates; in others the oxidized and leached subzone is very thin or is lacking altogether. Thicknesses are extremely variable from <1m to >30m, but in many sections of glacial drift, zones of weathering may be 2 to 5m thick.

Beginning at the bottom of the profile and going upward, we can trace the succession of changes that have occurred. *Oxidation* is the pioneer reaction, taking place first and extending deepest. Ferrous iron compounds are altered to oxides whose brown, yellow, and red tones impart a distinctive appearance to the lower subzones. *Leaching* of the primary carbonates in the drift, by percolating water charged with carbon dioxide, follows close behind the oxidation process. In time leaching dissolves not only the finely divided carbonates (mostly pulverized limestone and dolomite) in the till matrix but even boulder-size clasts of carbonate rocks as well. *Complete decomposition* of the drift follows slowly as percolating water attacks and alters not only the rather unstable micas, ferromagnesian minerals, and feldspars, but also in time the stubborn, far less soluble silicates. The residual alteration products are mostly hydrous aluminum silicates collectively termed clay minerals. These are deposited little by little at increasing depth, with decay of the primary minerals above. In a till or other sediment containing pebbles and larger particles, weathering usually alters each pebble from the surface inward. At an early stage a pebble may acquire a *weathering rind* in which brown iron oxides are conspicuous. Ultimately the pebble becomes completely decomposed, loses its separate identity, and furnishes clay minerals to the accumulating residuum.

In some compact tills and other jointed materials weathering proceeds along joint interfaces and gradually penetrates into the joint blocks. An early phase of this process is seen in fig 7-7.

Soil Profiles

In the sense in which the term is used by pedologists and geologists, *soil* is difficult to define precisely. By some it is defined as a natural body of minerals and organic matter, subdivided into horizons and differing from the material underlying it as to morphology and physical, chemical, and biologic properties. By others it is defined as the material at the Earth's surface that can support the growth of rooted land plants. However he may define soil, the pedologist is concerned with what lies immediately below the surface. He recognizes that the factors that influence soil development are climate, topography, organisms, original (parent) material, and time.

The soil lies within the zone of weathering and is a part of it. It may coincide exactly with the zone of weathering or may constitute only an upper fraction of it. It is subdivided into *soil horizons,* which usually are related to each other in a gradational way. The sequence of horizons together constitutes the *soil profile,* which is analogous to but not necessarily coincident with the weathering profile of the geologist's terminology. The distinction between the two profiles, somewhat artificial to be sure, is the result of the same sequence of features having been studied with different emphasis by two different groups of scientists. Let us repeat that the development of soil is the result of weathering.

Fig 11-1 illustrates the relation between the two profiles. The soil profile consists basically of four horizons, designated A, B, C, and D, only the first two of which are included in the soil proper or *solum*. The A horizon is uppermost; from it clay minerals have been removed and carried downward, and in it, commonly, organic matter has accumulated, imparting to it a black, gray, or brown color. The B horizon, next underlying it, is that in which descending iron and aluminum minerals, including clay minerals, have accumulated. Beneath this is the C horizon, a part of the soil profile but not a part of the soil proper. The C horizon is the parent material from which the soil has developed; therefore it lacks a lower boundary. To the pedologist, parent material means that in which there has been no movement of clay minerals and in which original characteristics are still recognizable. Therefore it may be somewhat oxidized or leached or both, or apparently unaltered. The D horizon, not present in some profiles, is not parent material but may influence the soil by affecting its drainage or some other process. Where the C is not present, the D may underlie the soil directly. The A, B, and C horizons are commonly subdivided, and are labeled with subscripts. More details, unnecessary for our purpose, are given in a standard manual (U. S. Bur. Plant Industry, 1951).

Each of the two concepts shown in fig 11-1 has its own usefulness. The differences between them are partly differences in the terms used and partly differences of emphasis, for the soil-profile concept introduces far greater detail into the upper parts of the weathering profile than is shown in the figure. However, both concepts refer to *a single natural profile of weathering*. Actually, geologists quite commonly refer to a profile of weathering, where encountered in a stratigraphic sequence, as "a soil," even though erosion may have removed the upper horizons that make it a soil by definition. Possibly this only means that the geologic and pedologic concepts are in process of merging. At any rate we shall freely use the term *soil* to denote a profile of weathering wherever it is convenient.

Development of a Soil Profile. A body of fresh rock material newly exposed at the surface (for example, till uncovered by deglaciation) is attacked by processes of weathering and develops a soil profile, which approaches maturity after the passage of sufficient time. Attempts to establish recognizable phases in the development of a soil profile generally recognize three gradational phases.

1. Weakly developed. (An A horizon overlying a C horizon, the top part of which is slightly oxidized, denoting the inception of a B.)

2. Moderately developed. (Between the A and the C is a B horizon in which clay can be recognized; the oxidized part of the C is thicker; in dry climates secondary carbonate is commonly abundant.)

3. Strongly developed. (The B is thicker, richer in clay, and commonly more reddish; clay is likely to be conspicuous. Oxidation and, in dry climates, accumulation of carbonates are represented by thicker zones.)

Although objective, these differences are qualitative rather than quantitative, and their proper recognition requires an experienced observer.

Topography and Time in Soil Making. Topography influences weathering in several ways; chief among these is the direct effect of slope. On a steep slope erosion may remove the products of weathering as rapidly as they are created. Depending on relative rates of surface erosion and alteration, a profile may be poorly developed, truncated, or completely lacking. In contrast, a flat surface accompanied by poor subsurface drainage results in a soil profile with a B horizon extraordinarily rich in secondary clay.

Time exerts an influence on weathering that has great interest for stratigraphy. In some profiles individual horizons are thin and indistinct, whereas in others they are thick, well developed, and strongly contrasted. The former are said to be immature, the latter mature; the difference is in part attributable to differences in the time during which the profile developed. The validity of time as a factor is established in central North America by conspicuous differences between weathering profiles developed on Wisconsin drift and profiles developed on Illinoian and older drift. The latter are generally considerably more mature, having thicker horizons and greater development within them. Similar differences exist in Europe, between soils in young drifts and those in much older drifts.

Even though a soil of Wisconsin age can usually be discriminated from one of greater age in the same region, it has not yet been found possible to isolate the time factor from the other factors that influence weathering. Hence attempts to discriminate among weakly developed soils of Wisconsin age on the basis of their character alone have met with only partial success.

It has been possible to estimate the maximum time required for the development of some postglacial soils in glacial drift, in areas where deglaciation has been dated by C^{14} (Dimbleby, 1965; Frye et al., 1968, p. 16) or by otherwise bracketing a soil between two relevant C^{14} dates. Thus the buried Farmdale Soil in Illinois can be said to have formed between 22,000 and 28,000 years BP, although it need not have required 6000 years to develop. Indirect evidence suggests that some Chernozem soils have developed to maturity within only 100 to 200 years, that a thin Podzolic soil in New England developed in about 1500 years, and that many Podzolic soils in moist-temperate environments seem to have matured in 2000 to 4000 years. In contrast, very thick mature soils rich in clay or iron or both, represent far longer periods, possibly running into tens of thousands of years. Yet on hard, glaciated bedrock postglacial soils have not developed because in the last 10,000 years or so, weathering in such material has been negligible.

Although truly quantitative data are scarce, it appears likely that during weathering, rates of change are greatest rather early in the process and decrease with time.

Classification of Soils

The greatest differences among soils now forming are related to climate and vegetation. These differences have led to a broad classification [1] of the group of soils called *zonal soils,* which owe their characteristics mainly to these two factors. This is well illustrated in a southwest-northeast section across North America (fig 11-2). As the climate changes from cold and wet to warm and dry, vegetation, weathering characteristics, and soils change facies correspondingly, much as an extensive sedimentary deposit may change facies from one region to another. In the northeast, with abundant precipitation, soluble cations such as Ca^{2+} are carried downward out of the weathered zone; but in the dry southwest, water in that zone tends to evaporate, precipitating Ca^{2+} as a carbonate in the B horizon of the soil, which is therefore alkaline. Again, in cool, moist regions humus is present in the A horizon, making the soil acid. Under the acid conditions, feldspars and ferromagnesian minerals are decomposed, and resulting secondary clay and iron oxides accumulate conspicuously in the B horizon.

These differences of weathering process between cool moist and warm dry climates form the basis of a subdivision of the zonal soils of North America into two groups, pedocals and pedalfers. Pedocals are soils with calcium-rich upper horizons, shown on the left side of fig 11-2. Pedalfers are soils in which secondary clay and iron oxides are conspicuous. Like the latitudinal change of facies seen in fig 11-2, a change of soil facies occurs on many mountain ranges. In mountains in western United States we may find a Desert soil in dry climate at the mountain foot, grading upward through Chestnut, Chernozem, and Prairie soils shown in the figure, to a Podzol at moist higher altitudes, and, in the high alpine zone, to a Tundra soil. A change of this kind is shown in detail, for a single mountain mass, by Richmond (1962). The change may occur through a range of altitude of ~1000m, accompanied by increase of precipitation from ~250mm to ~800mm. On a much smaller scale, facies changes are

[1] See U. S. Department of Agriculture, 1938, p. 970–1001, with a large soil map of the United States; Soil Survey Staff, 1960; Kubiëna, 1953, in which the plates in color are particularly helpful for identification of European soils.

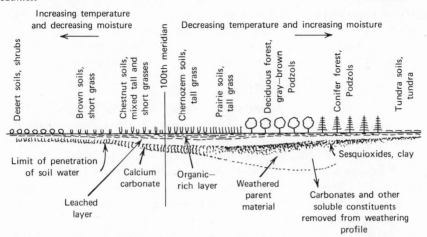

Figure 11-2   Relation of weathering profiles and zonal soils to climate and vegetation through North America, from deserts in southwestern United States to tundra in arctic Canada. (After C. B. Hunt, 1967, Physiography of the United States: C. W. H. Freeman and Co., p. 93.)

observed from top to base of a single hillslope through a distance of 100m or less, due partly to differences in rate of erosion.

Among the major groups of zonal soils are these, grading from pedalfers on the right to pedocals on the left in fig 11-2.

A. Pedalfers

1. *"Tundra soils,"* immature, stony soils resulting from much mechanical weathering under a very cold climate and greatly reduced chemical activity.

2. *Podzols,* a varied and extensive group having a gray bleached zone in the A horizon; they occur beneath cool-temperate forests in regions of moderate rainfall. Brown, yellow, and red soils of podzolic character, containing secondary iron oxides, appear with increasingly warm climate in southeastern and southern United States.

3. *Prairie soils* are related to Podzols, but occur beneath tall grass under cool-temperate climates.

B. Pedocals

4. *Chernozems* have a black or dark gray A horizon rich in organic matter, and a B horizon containing secondary carbonates derived from above. They occur beneath tall grass in the less humid, treeless parts of central United States.

5. *Chestnut* and *Brown soils* are like Chernozems but are brown rather than black, with less organic matter and much carbonate. They occur under semiarid climates, with a vegetation of short grass and shrubs.

6. *Desert soils* are brown, gray, or reddish. They are thin, with much secondary carbonate at shallow depth, and occur under arid climates.

In the pedocals the secondary carbonates may take the form of nodules, small or large, or of massive, continuous layers. Such material is called *caliche* (North America), *calcrete* (southern Africa and Australia), *croûte calcaire* (French lands), and *kunkur* (India and Pakistan).

In Europe a similar succession of zonal soils ranges from Podzols in the west into Chernozems and even Chestnut soils in the far southeast, whereas in the Mediterranean region, as in southern United States, there are red soils as described hereafter.

The section shown in fig 11-2 grades from warm, dry climate through temperate climate into cold, moist climate, and does not pass through the warm, moist region of the southeastern quarter of the United States. Figure 11-3 is a more complete representation of the changes in soils that accompany the transition from one climatic zone to another. It shows that with increasing warmth soils take on yellowish and reddish colors, owing to the accumulation of secondary iron oxides created by the breaking down of iron-bearing minerals. In their most intense development, such soils are called *Lateritic soils*. They consist of a layer of hematite with or without clay minerals, overlain in some areas by quartz sand, and overlying gradationally a zone of pronounced kaolinization of the parent bedrock. It is thought that ground water dissolves all the bases, all the iron, and in some instances even silica. The entire zone of weathering may exceed 30m in thickness; an extreme thickness of 250m has been reported. The iron is thought to be reprecipitated at the water table as hematite, while kaolinization proceeds at greater depth. Any quartz in the parent rock is not dissolved, but accumulates as a lag concentrate, usually of sand-size particles, above the layer of secondary iron. Less mature Lateritic soils have less pronounced characteristics, but all are red or reddish owing to the presence of hematite.

Lateritic weathering (called by some "red weathering") is in progress today in the wet tropics. Hence high temperature and high rainfall, at least seasonally, have been rather widely believed to be controlling conditions. However, the process is not well understood, and given substantial time, it could perhaps be accomplished under a wider range of climates than has been generally believed. Furthermore, Lateritic soils apparently grade into the Red-yellow Podzolic soils shown in fig 11-3, making it dif-

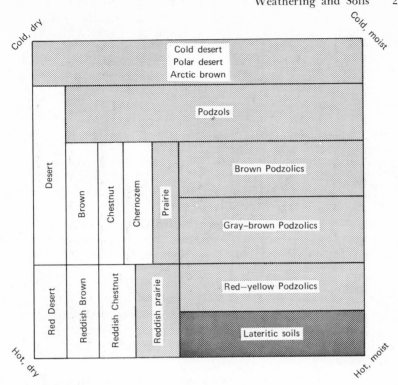

Figure 11-3 Idealized relationships among zonal soils. Large, shaded area = pedalfers; white area = pedocals. Although related more closely to pedalfers than to pedocals, Lateritic soils are in a class by themselves. (Compiled from various sources.)

ficult to relate Lateritic soils to specific minimum values for the temperature and precipitation necessary to bring about lateritic weathering.

Apart from Lateritic soils, which require special conditions of climate for their development, other special kinds of soils result from local environmental factors. An example that depends primarily on local bedrock is the group called *Rendzina soils*. These have gray or black humic surface layers overlying calcareous material derived from limestone or chalk. When mature they develop secondary clay tinged with red iron oxide, resulting from a climate that is moist, at least seasonally. Rendzina soils occur beneath calcareous prairies near the Gulf of Mexico and on calcareous rocks in coastal California. In southern and central Europe they occur in many areas of limestone and chalk.

Another kind of soil, or rather B horizon of a soil, that depends on local environment is *gley,* a sticky, gray or mottled, clay-rich horizon developed under the continuously wet conditions of swamps and other poorly drained areas. In this case drainage rather than type of bedrock is the controlling factor.

Ancient Soils [2]

Relict Soils and Buried Soils. In order to clarify the significance of soils for study of the Quaternary, we need to define three terms for the present discussion.

Relict soil. A soil that is not adjusted to its present climatic or topographic environment and has undergone internal change. It is relict from an earlier different climate or topography.

Buried soil. A soil that has been buried beneath sediment or rock, whether or not subsequently re-exposed. It may or may not be relict. Some soils, formerly buried, have been exhumed.

Ancient soil. A soil that is not forming today. It includes relict soils and some buried soils.

We here avoid the term "fossil soil," and also the term *paleosol,* which has been used in all three senses and has been defined in one place (J. V. Howell et al., 1960) as a buried soil. The term paleosol does not appear in table 11-A.

Ancient soils are widely present in middle latitudes. In very high and very low latitudes they are much less evident, possibly because response to Quaternary fluctuation of climate in those environments has been less sensitive than in middle latitudes, where not uncommonly differences between ancient and modern soils are considerable. Under such circumstances ancient soils constitute a basis for interpretation of the former environments in which the soils developed, and also are a useful element in stratigraphic correlation.

Interpretation of Environment. The environmental factors that can be inferred from ancient soils, at least in an elementary way, are topography, plant cover, and climate. The top of a buried soil is a surface of unconformity, and if the soil can be traced through closely spaced exposures the form of the surface at the time of soil development can be inferred. Where a soil is confined to interfluves and is lacking on the sides of intervening valleys, we are led to suggest that soil-making processes and valley-shaping processes (mainly mass-wasting) constituted a system in a steady state, but that now valley shaping predominates. The change

---

[2] General references: Morrison and Wright, eds., 1967; Simonson, 1954; Hole, ed., 1965; Richmond, 1962; Birkeland, p. 469 in Morrison and Wright, eds., 1967.

might result either from climatic change, affecting rates of weathering and mass-wasting, or from some dynamic event such as crustal uplift, affecting the stream in the valley.

Again, the changed conditions may be independent of topographic changes. A soil of one zonal type superposed upon or partly obliterating a soil of another zonal type suggests a general climatic change. When a soil is thus altered by the development of a younger one it is said to be *degraded*. An example is a former Chernozem partly converted to a Podzolic, but with remnants of the secondary carbonate that characterizes the Chernozem still visible. A change of this kind suggests a change of climate from semiarid to humid, with invasion of grassland by forest, such as might occur at the beginning of a glacial age. Another example is the podzolization of red and yellow soils on the fringes of the subtropics, suggesting the advent of cooler climate at the onset of a glacial age. In southern England, patches of relict red soils of early Pleistocene or even Pliocene date have been degraded to soils related to Podzols, probably under cooler climates of glacial ages.

A complex Desert soil in west Texas reveals a sequence of events comprising: (1) accumulation of secondary carbonate, (2) solution of the top of the carbonate, and (3) redeposition of carbonate, implying climatic change within the last few thousand years (K. Bryan and Albritton, 1943). Somewhat similar changes are implied by ancient soils exposed in the Mexican plateau (K. Bryan, 1948). Climatic change, possibly correlative with this, is indicated by a partly podzolized Prairie soil in central Iowa, implying two invasions of prairie by forest, unaccompanied by significant change in the local topography. A long Quaternary history is inferred from the sequence and character of soils exposed in the Canberra district in eastern Australia, with alternation of climate from more humid and stable to more arid and stormy (van Dijk, 1959). In areas near the boundaries between climatic belts, even slight changes in climate can result in visible changes in the soil.

On a broader scale, red soils are forming in the moist parts of Burma today, whereas the soils in the dry parts of the country are yellowish. By analogy, alternating red and yellow soil zones in Burma are interpreted as recording significant climatic fluctuations (De Terra and Movius, 1943).

Stratigraphy. Many though not all relict Pleistocene soils have been buried; they appear today in natural and artificial exposures and in the logs of borings. The sediment that covers them commonly consists of loess, till, outwash, or alluvium, and, in the Mississippi River delta, marine sediments. In many parts of western United States, buried soils occur in alluvium, where they afford evidence of alternate filling and cutting

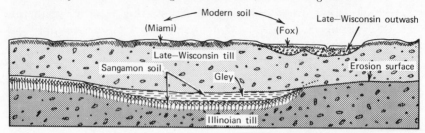

Figure 11-4 Generalized section showing glacial sediments and soils in the Richmond, Indiana, district. (After Thorp in Hole, 1965, p. 3.) Compare fig 11-6.

by streams. Wherever found, buried soils form part of the stratigraphic sequence.

The stratigraphy of ancient soils has done much to clarify the succession of Quaternary events in the glaciated regions. A typical occurrence is illustrated in fig 11-4. At the present surface is the modern soil, changing facies with changing local environment; the Miami facies, developed in Wisconsin till, differs from the Fox facies, developed in well-drained gravel and sand. Underlying the Wisconsin till is interglacial soil of Sangamon age, developed in Illinoian till. In places interglacial erosion has removed this soil, which is thicker and more mature than the modern soil. It is thoroughly weathered and consequently is rich in clay. In the B horizon almost all clasts in the till (other than highly stable quartzites and quartzes) have been destroyed. The till, too, changes facies. In the shallow, wet former basin shown in the center of the section the B horizon is a thick gley, whose layering and other characteristics suggest that fine soil particles were slowly washed down the surrounding slope into the basin and were incorporated into the maturing soil there. With or without the gley facies, interglacial soils of this kind, developed on drift, were once widely termed "gumbotil" [3] (Kay and Pearce, 1920) before they were placed (Simonson, 1954) within the framework of soil terminology. Many of them lost their friable A horizons, and also more substantial thicknesses, by erosion before burial beneath younger drift. Others were never covered by younger drift, and if not eroded, these are still at the surface, generally in a degraded state (fig 11-5). The proportion of gley facies visible today is quite large, because wet basins were better protected against erosion than uplands were, and because many gleys are relatively thick and hence durable.

[3] The word is an adaptation of the popular word *gumbo*—a highly descriptive name, understandable to anyone who has tried to drive a car on a nonsurfaced road in "gumbotil" country on a rainy spring day.

Although the most extensive soils of this kind are of Sangamon age, similar soils are developed in Kansan till and in Nebraskan till; at rare localities two interglacial soils occur. In all three of the recognized inter-glacial soils developed in gravel, loess, or other material in which drain-age is rapid, the B horizon is rich in brown or red iron oxides, constitut-ing the *ferretto zone* of early geologists. Indeed both secondary clay and iron oxides are present in abundance only in the interglacial soils; in soils dating from the latest glacial age such materials are only weakly de-veloped. To what extent the difference is determined by elapsed time (several tens of thousands of years at least), and to what extent by slightly higher temperature is not known.

An example of soil stratigraphy in Europe is that set forth by Brun-nacker (1964a, 1964b), in two districts in Bavaria where a sequence of alternating cold- and warm-climate soils goes back into the early Pleisto-cene.

Not shown in fig 11-4 are the rather weak, thin soils that characterize parts of the stratigraphy of the Wisconsin Glacial Age. The thinner and weaker an apparent soil, the more care is needed in distinguishing it from layers that resemble it superficially. For example, a dark, humic layer in alluvium does not necessarily imply a soil; it may be transported sediment. Again, a sheetlike accumulation of carbonates or iron oxides is not necessarily a B horizon; it may have been localized by an abrupt

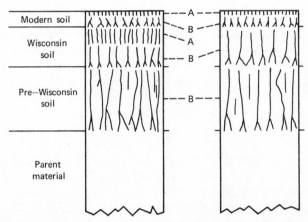

Figure 11-5 Diagrammatic sections illustrating the complicated stratigraphy that results from repeated superposition and degradation of soils. A modern soil has developed in the A horizon (left), and elsewhere (right) in the B horizon of a Wisconsin soil, which in turn had developed in the B horizon of a still-older soil. (After Hunt, 1954, p. 130.)

change in texture or by a former water table, without being a true soil horizon. Hence it is essential at the outset that clear evidence of profile development be established. Furthermore, the profile need not be complete; indeed, in most buried soils the A horizon is lacking, because it is especially erodible.

Even though soils are not sediments in the classical stratigraphic sense, they have a place in formal stratigraphy as soil-stratigraphic units (Richmond et al., 1959, p. 669), useful for lithologic correlation. Three sorts of hazards create difficulties in correlation: soil-like layers that are not soils, superposed and degraded soils, and facies changes. Despite these hazards many a soil has proved to be a good stratigraphic datum; the sequence of loess sheets in east-central Europe has been clarified with the aid, in places, of more than ten distinct soils, most or all of which represent nonglacial times of relative stability when loess was not accumulating.

Depth of Leaching. Two sorts of techniques have been applied to the differentiation of soils formed in glacial drift since the last interglacial. The first involves determination of depth of leaching of carbonates. It is reasoned that in calcareous drift or loess not subsequently buried, thickness of the zone leached of carbonates (that is, the combined thicknesses of subzones 1, 2, and 3 in fig 11-1) should reflect the time elapsed since deposition of the sediment. Therefore the average of a large number of measurements of this zone should make it possible to compare, as to age, two sedimentary bodies otherwise similar. However, such measurements in till have yielded aberrant frequency-distribution curves (Flint, 1949), apparently because of variation in the amount of primary carbonates present. However, reasonably good results have been obtained in comparing Wisconsin drift with Illinoian drift as to depth of leaching.

Weathering of Clasts. Just as leaching ultimately destroys entire boulders of carbonate rock, so the combined forces of chemical weathering ultimately destroy large clasts that form part of a body of drift. The process is slow and its progress, when carefully measured, affords a basis for differentiating two or more otherwise similar drift sheets that occur within a given climatic province. Granitic rocks, whose constituent plagioclase, amphiboles, and biotites decompose fairly readily, loosening the framework of the rock and freeing the quartz grains, are especially well adapted to such measurement.

In studying tills on the east slope of the Sierra Nevada in California, Blackwelder (1931, p. 877) found that the ratio of weathered granitic clasts to fresh clasts is related systematically to stratigraphic position. He distinguished among three tills in this way. The method was further developed in the Rocky Mountains in Colorado by Nelson (1954), and in the Argentine Andes by Flint and Fidalgo (1964). The latter workers counted samples taken from the zone of weathering at localities that met rigid specifications and found that the percentages of weathered granitic clasts, in three drift bodies identified by independent criteria, were 93, 78, and 55, from oldest to youngest. At all localities the drift measured is at the surface; hence the numbers reflect uninterrupted weathering throughout the time since deposition. Measurement of a deeply buried drift would reflect only the weathering that occurred before burial.

The technique makes possible the identification of a drift body (subject to the underlying assumptions) where other criteria are not available. It is most useful where applied to drift of the last glacial stage, and its value diminishes with increasing age of the drift. This is true because the differences in state of weathering between two drifts decrease as time progresses.

Another parameter of weathering of clasts consists of *weathering rinds*. These consist of thin "shells" of conspicuous oxidation that form at the surface of a clast and extend inward. By breaking the clast into halves one can measure the thicknesses of rinds, which commonly range between <1mm and 1cm or more. Rinds are most conspicuous in massive, dark-colored crystalline rocks. Measurement of the rinds of a good-sized family of clasts of uniform composition and texture, across a series of end moraines in a mountain valley, followed by statistical treatment, is a promising way of distinguishing among drift bodies that differ in age by a few thousand years or more.

CAVERN REGIMENS

A cavern, like a lake, is ephemeral in geologic terms; as crustal uplift occurs erosion eventually reaches the cavern and destroys it. The cavern fillings at Chou Kou Tien and in the Transvaal, which have been the sites of important archeologic excavations, merely record the former presence of caverns well back in Pleistocene time. From such fillings, as well as from the fillings of younger caverns that still exist, it is possible to extract much information about former climate. The archeologic and paleontologic record yielded by sediments in caverns forms a part of standard regional Pleistocene stratigraphy, with which we are not concerned here. We are concerned, however, with the stratigraphic record of deposits peculiar to caverns, with fluctuation in rates of dissolution and reprecipitation of calcium carbonate, with sediments washed in from surrounding slopes and preserved, and above all with analysis of chemical precipitates. Because much of the record is related to the general subject of weathering, it can be discussed appropriately here. The record to date suggests that caverns will repay a great deal more study than they have yet received.

Fluctuation in Rates of Dissolution and Precipitation. An early analysis of cavern stratigraphy in terms of its climatic implications was made at Sandia Cave in north-central New Mexico by K. Bryan (1941), who attributed the precipitation of calcite in the form of dripstone and flowstone, together with iron oxides, to a relatively moist climate thought to have been contemporaneous with the Late Wisconsin Glacial Sub-age. In other caverns layers of breccia occurring between thick layers of fine sediment have been explained as evidence of episodes of frost wedging of roof rocks in a glacial-age climate; fillings of eolian sand have been thought to indicate intervals of warm dry climate. Some such inferences receive support from the fossil pollen and fossil vertebrates occurring in the cavern sediments. Three other cavern sequences in western United States, with climatic implications, were reviewed by Hunt (1953).

A cavern in a limestone hilltop near Lusaka, Zambia, has a watershed <800m² in area; hence few factors other than intensity of rainfall are likely to have played a part in the changes of regimen that have affected the locality (Flint, 1959a, p. 357). The cavern was excavated by dissolution and then partly refilled with calcite. Later it was at least partly unroofed and filled with breccia and redeposited sediment apparently produced by red weathering and washed in. Still later it was further unroofed, and calcite formed again. This calcite, and earlier soluble deposits, were locally dissolved, after which the resulting cavities were partly refilled with washed-in material free of evidence of red weathering. Artifacts indicate that this sequence, which seems to represent fluctuation of climate, extends well back into the Pleistocene.

Climatic inference has been drawn from spongy travertine, which incrusts the walls of many caverns in central Europe. Such travertine is forming today in caverns in the limestone of the Carpathians, at altitudes of ~1200m and under precipitation of >1500mm. However, in caverns at much lower altitudes (~750m) with precipitation ~900mm, the travertine is not now forming but is relict, occurring at a stratigraphic position corresponding to the lower part of the Atlantic unit (table 24-C) and dating from some 7000 years BP. Ložek (1963) inferred that early in Atlantic time precipitation was higher at the cavern localities than it is today. Possibly a change of temperature, too, may have played a part.

Variations in rate of deposition of travertine unrelated to caverns have occurred likewise. A scarp of Precambrian dolostone in southern Botswana, in Africa, is characterized locally by remarkable accumulations of travertine. The material was precipitated from springs in small valleys, as the water evaporated or as its CO_2 was absorbed by aquatic plants. The travertine filled and blocked the valleys, diverting the small streams. Likewise it forms great carapaces as much as 30m thick, built outward and downward, with initial dips, over cliffs, as great as 75°. At some places at cliff edges the deposits are outright dripstones. Peabody (1954; see also Flint, 1959a, p. 358) identified four travertine units separated by erosional unconformity. Today spring discharges are small and little travertine is being deposited; the climate is semiarid, with evaporation ~250cm annually. This might suggest the carbonate deposits date from a wetter climate; yet fossil plants in the travertine represent the semiarid assemblage that flourishes today. Possibly therefore the travertine reflects a climate more strongly seasonal than today's, with deposition confined to the waning segments of relatively rainy periods. Hossfield (1951) drew a similar inference from relict travertine in New Guinea.

The oldest travertine unit contains a cavern, in the breccia of which *Australopithecus africanus* and other extinct mammals have been found. As the cavern antedates valley cutting, which in turn antedates the next oldest of the four travertines, several alternations of climate extending back into early Pleistocene time are indicated.

The Cave of Hearths, at Makapansgat in the Transvaal, has yielded an analogous record, elucidated by quite different means. At that place the country rock is dolostone. The surrounding slopes are mantled with angular grains of quartz sand, concentrated as a residue by dissolution of the dolostone. Regional study shows that today, where precipitation is $>$ ~1000mm, dissolution of the dolostone is rapid and the released quartz grains are washed downslope quickly, some of them entering caverns. These grains retain their angularity. With precipitation values of between 500 and 1000mm, dissolution is slower, giving time for the quartz grains to become less angular. But with $<$ ~400mm of rainfall, the quartz grains en route downslope acquire coatings of $CaCO_3$, which protect the grains against erosion. When angularity of quartz grains in cavern fillings is measured at close vertical intervals, the resulting plot shows a rather systematic fluctuation of inferred climate with time as measured by artifacts (Brain in W. W. Bishop and Clark, 1967, p. 298). Some might discern in the plot changes equivalent to middle and late Weichsel pluvial maxima.

Geothermometry in Speleothems

A first approximation to determining paleotemperatures in caverns in western United States is based on the crystal chemistry of the $CaCO_3$ in stalactites and other speleothems (G. W. Moore, 1956). In the northern part of the region the $CaCO_3$ in speleothems consists of calcite, in the southern part it consists of aragonite, and in an intermediate belt it includes both. The three belts seem to be related systematically to climatic isotherms. Calcium carbonate crystallizes as aragonite at high temperature and as calcite at lower temperature; the critical temperature is thought to be ~17°, and the control may be the influence of temperature on the effective ionic radii of calcium ion and carbonate ion. The difficulty is that probably there are determinants other than temperature; if so, the method is not accurately calibrated as to temperature. However, it offers promise for thermometry of the time span represented by speleothems.

Another approach is the application of the $0^{18}/0^{16}$-ratio technique used in calcium carbonate in deep-sea cores (chap 27) to the thermometry of speleothems. A first approximation (Hendy and Wilson, 1968) is based on measurement of samples taken at intervals along the radii of stalactites from caverns in the North Island of New Zealand. It was established first that the calcite was deposited in equilibrium with the solution that passed over it. The measured samples were C^{14} dated, the dates having had an arbitrary 5000 years subtracted from them to allow for the presence of ancient, dead carbonate. The resulting arbitrary time scale fits the most recent segments of paleotemperature curves for the past 30,000 to 40,000 years, derived from other parts of the world by means of $0^{18}/0^{16}$ ratios and by pollen analysis.

STREAM REGIMENS

Climatic Influence on Stream Regimen in Dry Regions

Indicators of Climatic Change. In and close to a glaciated region the influence of changing Quaternary climate on the regimens of streams is masked by the deposition of outwash sediments and by other direct and indirect effects of glaciation. Farther from extensively glaciated territory, however, changing climate likewise has influenced streams, and strong attempts are being made to identify and organize the data there. Significant studies have been made in regions such as southwestern United States, northern Africa, and Australia, where glaciation has been minimal and where, because today's climates are mainly semiarid and arid, the response of streams to moderate change of climate is more sensitive than it is in moist regions and is therefore more evident in the features present in stream valleys.

In dry regions many stream valleys contain remnants of dissected fills of alluvium, some of which form terraces. The stratigraphic relations of the fills imply repeated changes in the hydraulic regimens of the streams, which have experienced alternating phases of net erosion and net deposition, cutting and filling. Some valleys, furthermore, contain ancient alluvium that is of very different grain size from that which is being deposited today. Change of grain size in itself has often been taken to be evidence of climatic change. However, cutting, filling, and change of grain size in alluvium do not in themselves demand a climatic cause. They can result from quite other events, for example the creation of a local dam by avalanching or by the building of a fan, from tectonic movement, from the incursion of a glacier into the watershed, or from change in the level of a lake or sea to which a stream is tributary. Most such causes are local. They must be carefully eliminated individually, but if they are disposed of, there remain wide areas in which all the streams, regardless of orientation, display obvious histories of similar alternating cutting and filling; radiocarbon dating has aided substantially in local establishment of this relationship. The explanation is believed to lie in repeated, possibly minor changes of climate.

The linkage through which climatic change is felt by such a stream consists of change in the ratio of load to discharge. By altering either of these factors, a changing climate is thought to be capable of inducing net trenching or net deposition in a stream system that was previously in a steady-state condition.

A succession of cuts and fills is likely to result in the formation of paired stream terraces, but the presence and character of any such terraces will depend on the heights and sequence of fills and the amount and character of the intervening erosion. Therefore, although the sequence may include terraces, basically it consists of alluvial stratigraphy involving repeated erosional unconformity (fig 11-6). It involves not only ordinary alluvium but also fan alluvium, in which evidence of alternating growth and trenching of fans has been discerned (Lustig, 1965). A difficulty inherent in the evaluation of an alluvial fill is that sediments accumulated during a long period of aggradation must be distinguished from those deposited during a single exceptional flood.

Once nonclimatic causes have been eliminated, the aim of study of changed stream regimens is to establish former changes of climate from cuts and fills and, of course, to correlate contemporaneous sequences of events from one district to another. It is difficult to draw inferences as to climate from alluvial stratigraphy because of variable factors such as amount of precipitation, distribution of precipitation throughout the year, mean and seasonal temperature, and the like. The response of a stream, in terms of discharge and load, to a change in one or more of the climatic variables will be affected by area and range of altitude of watershed, local topographic texture, slope angles, character of vegetation cover, and other circumstances. Hence a change in only one climatic fac-

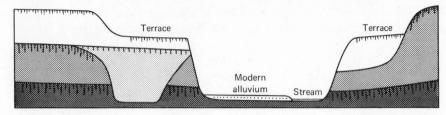

Figure 11-6 Idealized section of a small stream valley in a dry region, showing alluvial stratigraphy. Four bodies of alluvium older than the modern alluvium, three interfaces of unconformity, and three ancient soils older than the modern soil are present, as are a filled and buried channel and a pair of stream terraces cut from the youngest body of alluvium. In such situations the geometry can be developed from exposures in the sides of the valley and its tributaries. The fills and cuts shown are not necessarily all that were created; they are those which are exposed at the present surface.

tor might lead to different responses in two different streams, and even in two different segments of a single long stream.

It has been argued that in semiarid country like much of that in southwestern United States, vegetation is a key factor. According to this theory impoverishment or strengthening of plant cover alters the ratio of infiltration to runoff and so promotes or inhibits flood peaks, sediment loads, and erosion on slopes. That changes in total precipitation or in distribution of precipitation through the year would affect these factors can hardly be doubted, but in what direction and by how much are questions on which opinions differ.

Table 11-A represents an attempt to evaluate the alluvial stratigraphy in an extensive part of southwestern United States. It illustrates by implication some of the methods used in correlation, including C^{14} dating.

Attempts to Identify Former Climates. It seems probable, through elimination of other causes, that in some dry regions repeated climatic fluctuation has been the basic factor in synchronous cutting and filling by streams since the end of the last glacial age. Even if this is accepted, there remains the major difficulty of fixing the kind and quantity of change in climate implied by cutting or filling a particular area. This difficulty was attacked first by deduction from theory; the results have varied. At least three different opinions have been expressed. First, moister climates result in stream trenching; drier climates result in filling. This opinion has been applied to streams in Nevada, Nebraska, Arkansas, and Virginia. Second, moister climates result in filling; drier climates in trenching. This view has been expressed as to streams in Texas, New Mexico, Arizona, Wyoming, and western China. Third, dry climates result in trenching in upstream valley segments, and filling in segments farther down-stream. During transition to a moister climate, upstream segments become filled while downstream segments begin to be trenched, a process that continues through the period of moist climate. This opinion has been applied to southwestern United States generally.

Possibly all three opinions are valid for particular situations, provided a large enough range of streams and climates is considered. Indeed there is support for at least two of them in local evidence. As long as this possi-bility is admitted, an obvious recourse is the laborious one of detailed field examination of each case, in the hope that independent evidence of climatic change can be found, so related to evidence of regimen change that the two changes can be demonstrated to be contemporaneous and therefore probably in cause-and-effect relationship. Such evidence, in re-gions now dry, includes: (1) deposits recording ponds of standing water, (2) soils, suggestive of moister climate, developed in fills, (3) fossils, espe-cially pollen, with ecologic implications, and (4) archeologic data imply-ing ecology. In moister regions eolian sediments, especially sand, in-terbedded with alluvium, imply impairment of vegetation cover and hence probably drier conditions; but as we have noted earlier, extreme aridity cannot be inferred from eolian sand alone, exclusive of other evi-dence. Because all such information is difficult to find and to evaluate, deduction from theory has not yet succeeded in fixing the major relation-ship between climate and stream regimens.

A different approach is represented by attempts to determine relative rates of erosion today, in terms of sediment yield, in various climatic provinces in which today's rainfall and runoff data are known. This amounts to the close comparison of meteorologic data with data from gaging stations and from siltation of reservoirs. The objective is the rough determination, by analogy, of former rainfall values on a given watershed, from geologic evidence of change from filling to cutting, or vice versa, on that watershed. Although valid in theory, this method is subject to several errors. The largest error lies in the assumption that modern stream loads represent natural erosion unaffected by human ac-tivity. It is estimated (Meade, 1969) that mechanical loads of streams in eastern United States have been increased, through human influences, by a factor of 4 to 5. With quantitative correction for human influence on present-day sediment loads, the calculation of former rainfall values may become possible.

The Nile River. A detailed study (Giegengack, 1968) of the alluvial ge-ology of the Nile in Egyptian Nubia revealed a sequence of alluvial bod-ies antedating the modern alluvium. The latest of these bodies implies an episode of bed-load aggradation by the Nile and its local tributaries.

Table 11-A

Correlation of Alluvial Bodies and Significant Soils in Southwestern United States, Based on Archeology, Fauna, Physical Stratigraphy, and C^{14} Dates [a]

Age (C^{14} years BP; scale not linear.) — markers: 100, 1,000, 4,000, 7,000, 11,500

DEPOSITION	HOPI COUNTRY, ARIZONA (Hack, 1942)	BIG BEND REGION, TEXAS (Albritton & Bryan, 1939)	MEXICAN SPRINGS WASH, NEW MEX. (Leopold & Snyder, 1951)	EASTERN WYOMING (Leopold & Miller, 1954)	KASSLER AREA, COLORADO (Scott, 1963)	NEBRASKA (Schultz et al, 1951)	LA SAL MOUNTAINS, UTAH (Richmond, 1962)	DOMEBO SITE, OKLAHOMA (Albritton, in press)
ε	Naha Fm.	Kokernot Fm.	Upper Nakaibito		Post–PineyCreek alluvium		Gold Basin Fm. / soil	Q3
δ	Tsegi Fm.	Calamity Fm.	Lower Nakaibito Fm.	Kaycee Fm.	Piney Creek alluvium		Gold Basin Fm.	Q2
soil				soil	soil		soil	
γ	(erosion; dunes)	(erosion; dunes)	(erosion; soil)		Pre–Piney Creek alluvium	Soil Y		Upper Domebo Fm.
β					Broadway alluvium	Bignell loess and T2 terrace	Beaver Basin Fm.	Lower Domebo Fm.
soil			soil	soil	soil	Soil X	soil	
α	Jeddito Fm.	Neville Fm.	Gamerco Fm.	Ucross Fm.	Younger loess	loess	Beaver Basin Fm.	Bedrock

Age (C^{14} years BP; scale not linear.) — markers: 100, 1,000, 4,000, 7,000, 11,500

DEPOSITION	TULE SPRINGS, NEVADA (Haynes, 1965c)	LEHNER (Haynes, 1965a)	NACO (Haynes, 1965a)	BLACKWATER NO. 1, NEW MEXICO (Haynes and Agogino, 1965)	LINDENMEIER SITE, COLORADO (Haynes and Agogino, 1960)	SISTER'S HILL SITE (Haynes and Grey, 1965)	UNION PACIFIC MAMMOTH SITE (Haynes, 1965b)	BREWSTER SITE (Haynes, 1965b)	HELL GAP SITE (Haynes, 1965b)
ε	G	H	H	H			E		H
δ	F	G₃ / G₂	G	G	G	G	D	G	G₂
soil	Soil 5	soil	soil	soil	soil	soil	soil	soil	soil
γ		G₁	F	F	F	F			G₁
β	E₂	F₄ / F₁₋₃	E / D	E / C–D	E / D	E / B–D	D / C	E / B–D	F / E
soil	Soil 4	soil	soil	erosion	soil	soil	soil		soil
α	E₁	E	C	B	C	A	A	A	B–D

(NACO, LEHNER: ARIZONA. SISTER'S HILL, UNION PACIFIC MAMMOTH, BREWSTER, HELL GAP: WYOMING.)

[a] Triangles indicate stratigraphic positions of dated samples. Letter designations of units are field identifications and apply only to the locality specified. Correlation with the generalized scheme (Greek letters) is shown in column at left. Episodes of erosion and soil formation commonly occurred at the end of each depositional period. Because soils are not sediments, the upper boundary of each unit is placed at top of the soil. Soils are normally truncated by immediately subsequent erosion. (After Haynes, p. 600 in Morrison and Wright, eds., 1968.)

References: Albritton, C. C., 1966, Mus. of the Great Plains, Lawton, Okla., Contribs., no. 1, p. 11–13. Albritton, C. C., and Bryan, K., 1939, G.S.A.B., v. 50,

This suggests substantially greater rainfall and discharge than those prevailing today. Radiocarbon dates place the episode within the time of the late Weichsel glaciation in Europe. These data support the hypothesis that the latest glaciation in Europe was accompanied, in the now-dry region south of the Mediterranean, by enhanced rainfall as well as by lower temperatures—in other words, by a pluvial climate.

The Riverine Plain Problem. The elucidation of former climates from alluvial stratigraphy and morphology constitutes an active problem in the Riverine Plain north and west of the Great Dividing Range in southern New South Wales and northeastern Victoria, Australia. With an area of ~65,000km², the Plain is a gently sloping, semiarid lowland drained by west-flowing streams, chief of which is the Murray River. Its surface sediments consist of alluvium, predominantly sand and fines derived from the highlands adjacent on the east and built out fanwise from points of exit from the highlands.

In addition to the few existing streams there are many ancient stream courses. Some are represented at the present surface by shallow channels with bordering natural levees, others by buried channels as much as 15 to 20m deep, filled with complex layers of alluvium. Some ancient stream courses die out westward; others diverge from and rejoin one another. Some ancient patterns are braided, others meander, and there are marked differences in meander wavelength among the various streams, as well as differences in the grain size of their alluvium. There is evidence that some ancient streams did not function concurrently with others.

In the literature on the Riverine Plain are papers such as those by Butler (1958), Pels (1966), Bowler and Harford (1966), Schumm (1968), and Bowler (1970), which together summarize many of its puzzling features. Radiocarbon dates suggest that significant hydrologic changes have taken place within the last 3000 to 4000 years and that part of the alluvial sequence antedates the Late Wisconsin glaciation in North America.

The opinion is general that the major Riverine features are related to climatic changes, but it is uncertain whether ancient streams were active under climates moister or drier than the semiarid climate of today. It is not known, either, whether the ancient streams were contemporaneous

1423–1474; Hack, J. T., 1942, Harvard U., Peabody Mus. Pap., v. 35; Haynes, C. V., 1965a, A.A.A.S., 41st Mtg., Flagstaff; 1965b, Soc. Am. Archaeol., 30th Mtg., Urbana; 1965c, Univ. of Arizona, Ph.D. dissert; Haynes and Agogino, G. A.., 1960, Denver Mus. Nat. H. Proc., no 9; 1965, G.S.A. Mtg., Rocky Mtns Sec., Ft. Collins, Colo.; Haynes, C. V., and Grey, D. C., 1965, Plains Anthropologist, v. 10, no. 29; Haynes, C. V., p. 591 in Morrison and Wright, eds., Richmond, G. M., 1965, in Wright, H. E. and Frey, eds., p. 217; Schultz, C. B., et al., 1951, Nebraska State Mus. B. 3, no. 6; Scott, G. R., 1963, U.S.G.S.P.P. 421-A.

with glacial or nonglacial maxima, although some systematic relation is considered to be probable.

Frost Climate and Stream Regimen

Changes of regimen have occurred in streams that drain areas formerly affected by a frost climate, and the cause seems fairly evident. A clear example is that of streams in the eastern Alps (Büdel, 1944). The streams drain areas that were not glaciated; they contain bulky fills of alluvium which, being graded to outwash in larger valleys, date from glacial times. The alluvium is thought to have been furnished by frost wedging and solifluction, and to have been deposited by floods of springtime thaw water, despite a probably reduced overall precipitation as compared with that of today. Alluvial fills occurring generally in central Europe are correlated, by analogy, with glacial ages, times of trenching with interglacial ages.

Another special case consists of streams draining nonglaciated terrain in areas where loess was deposited abundantly during glacial ages. Such streams may have acquired massive loads of silt, through erosion of loess-covered slopes, with resulting accumulations of silty alluvium in their valleys. Under these special circumstances slope erosion may have been abnormally great regardless of the vegetation cover.

Stream-Channel Geometry

The geometry of stream channels, likewise, throws light on former climates. Many streams possess braided channel patterns that have been superposed on former meandering patterns. Others meander within valleys that are themselves meandering, but the ratio of wavelength of the former (valley) meanders to existing meanders is large, being in many valleys ~10:1. Still other streams meander on floodplains beneath which are buried meandering channels, again with large wavelength ratio. This ratio defines the modern streams as underfit, a relationship long ago described by W. M. Davis and attributed by him to loss of discharge through stream capture. Indeed meander wavelength is related systematically to channel width and hence to discharge, and therefore the concept of former greater discharge—at least peak discharge—is demanded. However, a history of capture is demonstrated in very few underfit streams.

The problem has been dealt with intensively by Dury (1964–1965), who computed discharge ratios from wavelength ratios and obtained values, when adjusted for change in slope and channel form, commonly ~20:1 and in some cases >50:1. Underfit streams occur through a wide range of latitude (fig 11-7), but in those situated close to former ice sheets, discharge ratios are higher than in those in lower latitudes. Where stratigraphic correlation is possible, some underfit streams are shown to

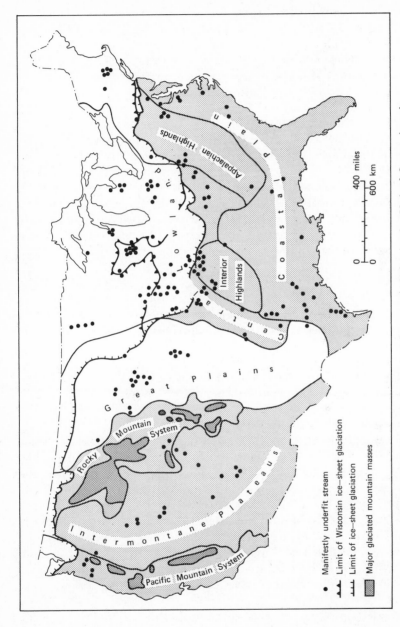

Figure 11-7 Manifestly underfit streams (*disks*) in the United States, in relation to natural regions and to former glaciation. (After Dury, 1964–1965, p. A30.)

Manifestly underfit stream

Limit of Wisconsin ice—sheet glaciation

Limit of ice—sheet glaciation

Major glaciated mountain masses

date back as far as early Pleistocene time; others received at least their latest conspicuous reduction in size at one or other of two very late transitions from cooler and moister to warmer and drier climate in Europe: those at the beginning of Allerød time and of Boreal time, respectively. Radiocarbon dates indicate that those were times, also, of relatively high levels of some pluvial lakes. It appears, then, that important changes in discharge have occurred repeatedly over a long time.

The peak discharges implied by underfit streams are too great to have resulted solely from reasonable increases in mean annual rainfall, or solely from reduced evaporation brought about by lowered temperature. Neither is the presence of frozen ground, preventing infiltration of rainfall on watersheds, adequate in itself. Dury postulated single rainfalls of exceptional intensity and duration, superposed on only moderately increased annual rainfall values, and aided locally by frozen ground. Although the variables involved have prevented translation of the theory into quantitative climatic terms so far, there exists in it a valuable potential for better understanding of glacial-age climates.

River Ice

The winter freezing of streams as a factor in their regimens at times of glacial maxima is recorded by cobbles and boulders, the largest nearly 2m in longest diameter, occurring in alluvium of various Pleistocene dates. Large "erratic" boulders were reported from Pleistocene alluvium in the Chesapeake Bay region by W. B. Rogers as early as 1875, and later by McGee. They were later examined intensively by Wentworth (1928), who found striations on many of them, and whose study is still the standard reference.

These clasts are distributed along stream valleys such as the Susquehanna, Rappahannock, James, Ancient Teays, and Tennessee. Apart from an occurrence in Texas, most lie within a belt about 600km wide, south of the glaciated region (fig 11-8). The striations can hardly be glacial, because the watersheds of nearly all these streams lie outside the glaciated region and the lithologies of the clasts are those of bedrock of the watersheds themselves. Most of the striated clasts are far coarser than the alluvium in which they are enclosed. Along these streams no striated stones have been found in alluvium known to be preglacial, nor in modern alluvium. Yet such stones are common in alluvium along large Arctic streams in which the action of river ice during the spring season is conspicuous.

This collection of data is taken to mean that the striations were made by river ice and that the stones were rafted downstream on ice blocks, at times when winter climates were more severe than they are today. Probably those times were glacial ages. The stratigraphic positions of the frag-

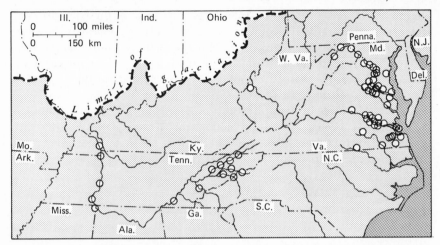

Figure 11-8 Distribution of observed striated cobbles and boulders in alluvium in central eastern United States. Concentration along certain streams results in part from intensity of search. Data from Wentworth, 1928; 1930; J. Petty (Sci. 71, 1930, 483). Not shown is occurrence in Texas (L. Reed, A.J.S., v. 15, 1928, 520).

ments are not fixed precisely within the Pleistocene, however, nor are former winter temperatures within the region concerned established beyond this inference: at times the rivers were seasonally frozen deeply enough to create ice pans capable of transporting boulders. Most of the rivers involved are subject to at least some winter freezing, with early-spring transport of ice pans. McGee noted that the maximum diameter of observed "erratic" boulders diminishes markedly from the Susquehanna on the north, through three intermediate rivers, to the Roanoke on the south, suggesting rapidly diminishing activity of river ice. Today average January temperature along the Potomac in western Maryland is $\sim -1°$ and along the Roanoke in eastern Virginia $\sim 3°$. The difference, $4°$, suggests a possible approach to glacial-age winter temperatures, but the southern limit of "erratics" on the Atlantic slope is not known, and would have to be established before an attempt to derive former winter temperature values could be made.

SUMMARY

Weathering profiles ("soils"), speleothems, alluvial stratigraphy, and stream-channel geometry are potentially useful in the reconstruction of parameters of ancient climates. Inferences from most of these features are still qualitative; only speleothems have thus far led to quantification. The response of stream regimens to climatic change is more sensitive in dry regions than in moist. With the probable exception of speleothems, none of the features named responds as sensitively as do the lacustrine features discussed in chap 12.

All these features bear implications of time, but only where radiometric dates have been extracted from them have rates of activity been established. Furthermore these rates, being dependent on local variables, cannot be applied to similar activity at other places and times.

CHAPTER 12

Fluctuation of Sealevel

THEORY OF EUSTATIC FLUCTUATION

Shifting of sealevel relative to the lands is clearly reflected in the geologic record and has occurred throughout geologic time. Many of the world's coasts bear evidence, both morphologic and stratigraphic, that such shifting has happened repeatedly within Quaternary time, and the records of tide gages indicate that changes are presently in progress. In order to emphasize that these changes are relative to the land, we speak of *relative sealevel,* or imply it, in circumstances in which the data available do not justify our being more specific. For example, it frequently happens that we are unable to distinguish between (1) crustal movement of a coast and (2) rise or fall of the level of the sea relative to a point at the center of the Earth. This second kind of change is necessarily worldwide, and is termed *eustatic.*

Several factors are capable of bringing about change in relative sealevel. We can neglect those which cause changes of very small amplitude and very short duration, and concentrate on those which cause changes capable of being preserved in the geologic record.[1] Factors of longer duration include the processes listed in table 12-A. In that table, (excluding 1A) (1) is distinguished from the other processes in that it is limited to localities or regions. Because of this, geologic study in some instances can establish (1) as the cause of a particular change by identifying deformation of a beach or other shore feature that was originally horizontal.

The six processes capable of causing worldwide eustatic changes in the level of the ocean are far more difficult, and in some cases thus far impossible, to isolate. Numbers (3) and (5) are considered negligible over the

[1] For a succinct review of all factors see Higgins, 1965; Kuenen, 1955, contains pertinent commentary. We must bear in mind that today's sealevel, widely used as a datum, is a variable level.

Table 12-A

Processes That Can Cause Change of Relative Sealevel

Process	Character of Change
1. Movement of crust involving a particular coast (May be tectonic or isostatic) 1a. Local compaction of sediment, especially peat	Local or regional

2.   Movement of crust beneath the sea (May be tectonic or isostatic) (Includes decanting of shallow continental seas)
 	
 3.   Sedimentation in the sea
 	
 4.   Creation of volcanic masses (flows, cones) beneath the sea
 	
 5.   Contribution of juvenile water to sea
 	
 6.   Thermal change in seawater
 	
 7.   Exchange of water substance between sea and land, involving:
    a. Glaciers
    b. Lakes
 	
 8.   Decanting of shallow seas into main ocean
	Eustatic

span of Quaternary time; (6) is likewise believed to be relatively small (Wexler, 1961, p. 870). Numbers (2) and (4) may not be negligible but are hard to evaluate. The effect of the latter may be offset by compensatory isostatic movement, but the effect of the former in changing the shape and capacity of the world ocean basin may be substantial. This factor should not be neglected, but hitherto it has been for lack of means of evaluating it.

This leaves us with (7). In it we have a process that can be estimated, at least as an approximation. The aggregate volume of all lakes is minute compared with that of glaciers, past and present; hence lakes are commonly neglected, leaving glaciers in the hydrologic cycle the chief consideration. Indeed, although other causes are present, movement of the crust and exchange of water substance in the hydrologic cycle (fig 16-1) are probably the two significant promoters of sealevel change within late-Cenozoic time.

According to the inventory in table 4-D, nearly 98% of the world's stock of water resides in the ocean, a little more than 2% resides on land in solid or liquid form, and barely a trace amount is contained in the atmosphere. The additional amount that the atmosphere could hold, or that the lands could hold in liquid form under reasonable assumptions of altered temperature are apparently small; hence the critical element is the exchange between water in the ocean and ice on the land. During a glacial age, with the volume of water substance in glaciers estimated to have been greater by a factor of 3 than that of today, the effect on sealevel would have been substantial (table 4-E). Since 1841, when this theory of glacial control of sealevel was first perceived by Maclaren, there has been little doubt that sealevel fluctuated as glacial and nonglacial climates succeeded each other. The sea is the ultimate reservoir; the amount of water it contains fluctuates with temperature. As the ratio of snowfall to rainfall increases during a glacial age, and as ablation is reduced, the ratio between water in the sea and ice on the land equilibrates at the lower temperatures then prevailing with reduced volume in the reservoir. During a period of higher temperatures than those of today, the equilibrium ratio results in increased volume in the reservoir.

Amplitude of Fluctuation

Deduction from Estimated Volumes of Land Ice. If the reservoir itself were not deformable, it should not be very difficult to estimate the amplitude of fluctuation of sealevel, both below and above its present position, by two methods. One method consists of deducing sealevels from the volumes of glaciers; the other consists of inference from geologic data.

In the first method the aggregate volume of glaciers during glacial ages and their volume during intervening warmer times are estimated. The difference between the two values is then converted into the thickness of a layer of water having the area of the world ocean. This procedure, detailed in Lawson, 1940, is followed in table 4-E.[2] All sea ice can be neglected, because it is floating and hence exerts no influence on sealevel regardless of its abundance.

There are three principal potential errors in this method. The least of these is that changes in the regimens of former glaciers probably were not contemporaneous, at least in detail. All or parts of some glaciers were shrinking while others expanded. Hence the outer limits of formerly glaciated regions are probably not isochronous, either individually or as a group.

A greater potential error lies in the estimation of average thicknesses and volumes of glaciers, particularly ice sheets, that no longer exist. Thus

[2] Several other calculations have been published; a recent example is in Donn et al., 1962.

far the profiles of such glaciers have been reconstructed by analogy with those of existing ice sheets, which for one reason or another may not be truly analogous. Furthermore the profile along one radius would have differed from that along another radius of the same ice sheet. As no attempts have been made to reconstruct these in detail, the volume estimates we possess at present are crude.

A third potential error, particularly along coasts, lies in our ignorance of the isostatic effect of thickening of the ocean by the addition of a layer of water to it. In table 4-E, column 7, it is calculated that if all existing land ice were to melt, a layer of seawater 65m thick would be created. The creation of this layer should cause sealevel to rise by 65m minus an amount resulting from subsidence of the sea floor under the weight of the added layer of water (chap 13). In the deep oceans this amount might well consist of one-third of the thickness of the added layer, but in the shallow, landward parts of continental shelves the layer would necessarily be thinner and isostatic adjustment, if any, would be less. This difference would modify the altitude at which the thickened ocean would meet the land. It could even result in an emerged strandline along a continental coast being higher than a contemporaneous strandline around an island in the deep ocean, as discussed by Bloom (1967, fig 7). Because of the uncertainties involved in such calculation, table 4-E does not include a column showing sealevels after isostatic adjustment. Research has not yet progressed to a point at which such a column would be meaningful.

We have, then, from table 4-E, column 7, an estimated raw difference of 132m between today's sealevel and that of a glacial age. That value must be reduced by the amplitude of isostatic subsidence of the sea floor under the weight of the 132m layer of water. If we assume that the factor to be applied is 1/3 (chap 13), the sealevel difference between the two times is not 132m but 92m. As noted above, the actual value may be greater. This deduced value applies to our model "glacial age," defined earlier. But in the Late Wisconsin glaciation, the one for which we have the most information, glaciers spread over a smaller area, possibly 10% less than in our model. Hence the volume of land ice was less and sealevel should then have stood less far beneath its current altitude. This would correspondingly reduce the sealevel numbers quoted above.

When we consider the higher sealevels of interglacial times we encounter a different problem: the interglacial states of the Antarctic and Greenland Ice Sheets. These glaciers contain nearly all the world's ice, equivalent, as we have noted, to a layer of ocean water about 65m thick. If an isostatic-subsidence factor of 1/3 were subtracted from this, we would get $\sim +43$m as the altitude of sealevel. However, it is very unlikely that the Antarctic Ice Sheet, at least, once it had formed, disappeared in any interglacial time. Because we do not know the extent to which either it or its counterpart in Greenland was reduced between glacial ages, we cannot at present make even a rough estimate of the altitude of an interglacial sealevel by deduction from the volumes of glaciers. As we note in the next section, the geologic record offers us little basis for estimate that is much better at present, but promises well for the future.

We are even less well informed with regard to the position of sealevel before the late-Cenozoic glacial ages. A major ice sheet existed in Antarctica about 4×10^6 years ago and possibly earlier [3] (Denton and Armstrong in Turekian, ed., 1971), whereas in Oligocene time the Antarctic climate was not favorable to large glaciers. Hence we think of the "preglacial" condition as having ended in Antarctica in later Miocene or Pliocene time. The transition from a nonglacial to a glacial status could have been slow and fluctuating, involving millions of years. At any rate, if we assume that the existing Antarctic and Greenland ice bodies together equal the volume of the world's land ice at the end of the "preglacial" condition, the "preglacial" ocean would have been thicker by 65m (table 4-E), and sealevel would have been higher by that amount, less a deduction resulting from the isostatic effect. Whether this calculation can be checked against geologic evidence acquired in the future remains to be seen.

Inference from the Geologic Record. Our problem here is to determine the amplitude of eustatic sealevel fluctuation with as much time calibration as possible. The pertinent geologic record consists, in a broad sense, of littoral sediments and other features above present sealevel and of terrestrial sediments and other features below it (fig 12-1); such occurrences are detailed elsewhere in this chapter. Direct observation has been aided in recent decades by echo-sounding profiles and by coring both on land and beneath the sea floor; C^{14} dates of fossil shells, driftwood, and other materials have aided materially in time correlation of littoral features.

Emerged marine strandlines and related features are present along many coasts and in places attain altitudes of hundreds of meters. Some, notably in the Mediterranean region, include sediments with warm-water invertebrate faunas; from such occurrences the strandlines were considered to be interglacial. Correspondingly high interglacial sealevels were then inferred, on the tacit assumptions that the implied sealevel fluctuation was eustatic in character and that the coasts during and since the rel-

---

[3] Fossils in Florida suggest that relative sealevel there stood near its present altitude between 4 and 7 million years ago (Webb and Tessman, 1967). If it is assumed that no subsequent crustal movement has occurred there, the sealevel can be considered eustatic, implying a volume of land ice much like that of today.

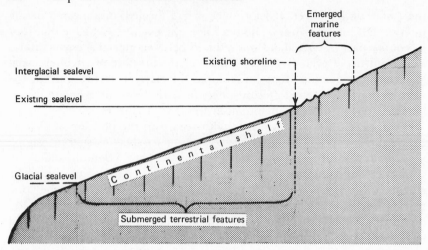

Figure 12-1   Idealized, dimensionless diagram of a nonglaciated coast, showing eustatically controlled positions of sealevel at various times. Noneustatic influences are neglected.

evant time had been stable. On those assumptions occurrences were correlated on a basis of altitude. A corollary of this concept of eustatic fluctuation on a stable coast was that the "preglacial" sealevel was much higher than that of today, and that through many subsequent fluctuations sealevel highs became gradually lower. Apart from its geophysical improbability (Kuenen, 1955, p. 201), this idea was shown to be inapplicable to Mediterranean coasts by the demonstration that strandlines there have been deformed by crustal movement. For a time subsequently it was claimed that although "tectonic" coasts, where post-strandline folding or faulting are demonstrated, are unsuitable for the correlation of strandlines by altitude, there are many "stable" or "relatively stable" coasts along which such study could be made. However, the demonstration of sea-floor spreading and the geomagnetic data supporting the concept of the displacement of continents, brought together cogently in the mid-1960's, raise doubt as to whether any coast is in fact stable. If the crust is acting as a series of moving conveyor belts, it would be prudent to avoid correlation of strandline remnants by altitude on any coast, and to rely rather on normal stratigraphic parameters as is done with strata in the continents (Flint, 1966). Any agreement of altitudes among stratigraphic correlatives on two or more coasts (and there seem to be a few examples of such agreement) might then be taken as evidence of crustal stability on those coasts, at least during the time since the strandlines were fashioned.

Summarizing evidence, some of which is detailed in a later section, on the depths of occurrence of littoral and related features, we can say that on a worldwide basis, many such features occur at depths of −100m or less, although some are deeper. Some pertain to glacial ages (mostly the last one) on independent evidence, and most of the remainder are assumed to be glacial because of their vertical positions. If we take these as a measure of glacial-age drawdown of sealevel, we must assume that no crustal movement has occurred, other than isostatic movements related to the weight of the water layer withdrawn and then returned.

An example is the outer edges or "breaks" of the continental shelves. Although at variable depths from one region to another, these are said to occur rather commonly at −130 to −140m in well-surveyed sectors (Shepard, 1963a, p. 266). The rather smooth surfaces of the shelves probably have resulted from effective erosion by surf, as has been inferred by various authors. However, we cannot confidently interpret the depth of −130 to −140m as measuring the eustatic drawdown of sealevel at the extremes of glacial ages until it has been established that the shelves have not been affected by movements of the crust. Net seaward downwarping, amounting to 30cm/km, on a shelf 100km wide would lower the shelf edge by 30m. The edge of such a shelf with a present depth of −130m would have a eustatic component of only 100m.

A similar argument applies to the positions of interglacial sealevels. In the case of an emerged beach with a present altitude of +40m, situated 100km inland from the coast, net landward upwarping of 10cm/km would require us to subtract 10m from the "eustatic" altitude of the beach, leaving a net altitude of +30m. Although hypothetical, these calculations illustrate the primitive state of our knowledge and the fact that all our directly measured sealevels are relative. Also they emphasize the importance of the tectonic factor, as yet commonly unknown, in the vertical positions of features indicative of former sealevels.

Another factor complicates our calculations still further. After subtraction of the tectonic factor, the remainder is not necessarily all eustatic. If, for example, we should decide to adopt a value of 100m for eustatic drawdown of sealevel below its position today, we must subtract from it a percentage x, representing isostatic adjustment to water weight, although the proper percentage is as yet undetermined. On this basis, then, the glacial-age portion of sealevel fluctuation would be 100m minus x. In view of the uncertainties, we can still improve little on the round value of −100m quoted by several researchers during the last half century. The interglacial-age portion is simply undetermined as yet, and will continue to be so until we have succeeded in separating the glacial-eustatic factor from the tectonic factor in the altitudes of interglacial littoral features.

CONTEMPORARY RISE OF SEALEVEL

Before considering more fully the geologic data of sealevel change, we should note the record of tide-gage readings on various coasts through the last century or substantial parts of it. The record varies from place to place, probably mainly because of crustal movement but secondarily and locally because of errors of various kinds (Kuenen, 1955). Data (not exclusively from tide-gage readings) from various coasts are summarized by Jelgersma (1961, p. 76), Wexler (1961, p. 869), and Gutenberg (1941, p. 730). After attempted elimination of supposed tectonic influence by selection of coasts in which readings are nearly uniform from point to point, the results seem to indicate contemporary rise of sealevel, with mean values ranging from <1 to >1mm/yr, and give some evidence of variation in rate within the past century. We must keep in mind, however, that there is no datum from which the results can be measured, and that consequently tide-gage data can be interpreted only on a basis of consistency, or lack of it, in the records of individual gages at points along a particular coast. Change of sealevel through time can be evaluated better from geologic data pertaining to the last 20,000 years or so.

RISE OF SEALEVEL SINCE THE LATE-WISCONSIN MAXIMUM

Nature of the Data

The variety of features of terrestrial origin which are submerged today is so large, their occurrence off continental coasts so general, and the glacial-age character of some of them so evident, that even before the advent of C^{14} dating it was clear that such features represent lowered sealevel related to the latest glacial age and probably also in part to earlier glacial ages. Most such features lie within the depth range of -100m; this fact constitutes the chief inductive basis for the long-held opinion, capably argued by Daly (e.g., 1934, p. 157–164), that glacial-age sealevels stood about 100m below the sealevel of today. Although the recognition of submerged terrestrial features is beset with difficulties (e.g., Kaye, 1959, p. 132) the number and variety of such features are impressive. A generalized list includes the following features.

1. *Submerged benches, beaches, bars, deltas, and other shore features,* represented both by morphology and by fossil-bearing sediments re-

covered in cores. These include the floors of lagoons and banks in the coral-reef region of the tropical Pacific (cf. Nugent, 1946), a shelflike strandline at ~ -100m in several parts of the Mediterranean (cf. Blanc, 1937, p. 627), a strandline at ~ -100m and ~ 650km long off the west coast of South Africa, benches off the east coast of North America (cf. J. Ewing et al., 1963) and the Long Forties Bank off eastern Scotland, with intertidal mollusks, at -70m.

Submerged deltas, especially those built of outwash entering deep water, have been tentatively identified off the south coast of Newfoundland and off the mouth of the Don River in the Black Sea. A massive delta fan, now submerged, occurs at the mouth of the Rhône at Marseille, and a submerged delta of the Yangtze River is thought to be graded to a sea'.evel at -50m. The general occurrence of large submerged deltas is discussed by Shepard (1963a, p. 273). However, a submerged delta does not prove rise of sealevel, because a bulky delta may have enough weight to cause crustal subsidence or may be built in a subsiding area.

2. *Relict sediments.* Normally the sediments on shallow sea floors decrease in grain size with increasing distance from the shore. Off eastern United States and in the northern Gulf of Mexico, however, ideal gradation is realized only in part. In some areas sediments far from shore are coarser than those inshore, and even include sand believed to be eolian, at depths as great as -50m. Most of the sediments on the Atlantic continental shelf of North America are relict (Emery, 1968.)

3. *Submerged stream valleys and alluvium.* Many shallow sea floors are cut by valleys whose forms suggest that they were made by land streams. Several of the best known are: (1) off the Hudson River, 140km long, ending at -70m; (2) off the Elbe, 500km long, ending at -80m; (3) off the Rhine (with the rivers of eastern Britain as tributaries), 720km long, ending at -90m; (4) off the Po, in the northern Adriatic Sea, 250km long, ending at -100m; (5) Bosporus and Dardanelles straits; and (6) on the Sunda Shelf, between Borneo and the Malay Peninsula, a valley 1000km long, ending at -90m.

The bedrock floors of large valleys in many parts of the world lie so far below sealevel that it is probable they were excavated at times when relative sealevel stood lower. They have been reported from Britain, South Africa, Australia, and the east coast of North America. Depths have little significance in detail because, for a given sealevel lower than today's, they would be influenced by the presence, width, and slope of an adjacent continental shelf. Those in New England (Upson and Spencer, 1964) attain an extreme depth of -120m which may well be attributable in part to glacial erosion. The bedrock floors of the valleys that form Delaware and Chesapeake Bays, both of which are in nonglaciated terrain, are at maximum depths of $-42m$ and $-60m$ respectively.

4. *Submerged eolian sand.* Along some coasts windblown sand, commonly related to beach dunes, extends below sealevel, proving submergence though not in itself establishing eustatic lowering of sealevel.

5. *Submerged fossil marine organisms,* of kinds that lived very close to sealevel, still at the positions at which they lived.

6. *Submerged fossil terrestrial organisms.* Many occurrences of deeply submerged freshwater peat and tidal-marsh peat (fig 12-6), tree stumps and even standing trunks, and bones and teeth of large terrestrial mammals (fig 12-2) have been reported from shelf areas off various coasts (e.g., A. Martin, 1959).

7. *Hiatus in radiometric dates.* Shelf sediments may consist of two sequences, each represented by a series of internally consistent dates on shells of littoral organisms or on other littoral features, with an obvious hiatus between the two series. It can be inferred that the hiatus represents a period during which the localities of the dated samples were emergent. Minimum depth of such emergence may match other evidence of drawdown of sealevel during the latest glacial age.

In this connection we note that lowering of sealevel is probably accompanied by lowering of water tables in coastal regions. This is observed in some areas of low relief. In Florida, at a point $\sim$75km inland, the core from a boring reveals a hiatus between two strata in a lacustrine fill, dated $\sim$8200 and $>$35,000 C^{14} years respectively. The hiatus was interpreted as reflecting the interval during which sealevel and water table were lowered during the Late Wisconsin glaciation (Watts, 1969).

Several of these sorts of data are combined in appraisals of deep borings in coastal areas. Examples are given in table 24-I.

Synthesis

Most of the sediments and features in the foregoing list pertain to marine transgression that has accompanied the deglaciation following the Late Wisconsin maximum. The sediments of this transgression were given the ancient name *Flandrian,* redefined by G. Dubois in 1924. Although applied at first to the European region, this name has been used increasingly in other parts of the world.[4] A thoughtful discussion of the

[4] Some authors have even employed the name to refer to all sediments, of whatever origin, that are believed to correlate with the Flandrian of Dubois. This usage has the disadvantage that any reasonably selected type section would be submerged and therefore would be studied only in cores—unless it is proposed to drop Dubois' meaning completely and redefine the term with no reference to its earlier meaning.

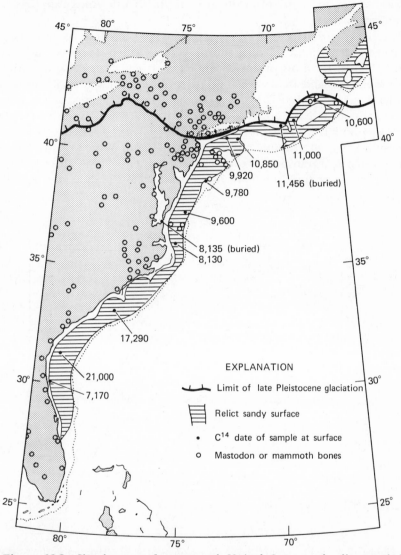

Figure 12-2 Sketch map of east-coastal United States and adjacent shelf, showing known localities of mammoth or mastodon bones and localities of C[14]-dated samples from the shelf. Samples are from the surface except where noted as buried. (After Emery, 1967.)

transgression in terms of sediments off the North American coasts is found in Curray, 1964.

Despite the variety of features listed in the preceding section, the results considered most reliable, and those most widely discussed, are based on C^{14}-dated fossil organisms of kinds that lived in very shallow littoral environments, and that were collected from localities at which they are believed to have lived, at known depths beneath present sealevel. With these it is possible to construct time/depth curves such as those shown in figs 12-3 and 12-4. The plots, representing former positions of sealevel, are subject to uncertainties such as range of depth of the organism, possible postmortem transport of the fossil to a higher or lower altitude before burial, range of the confidence limit attached to the date, and the like.

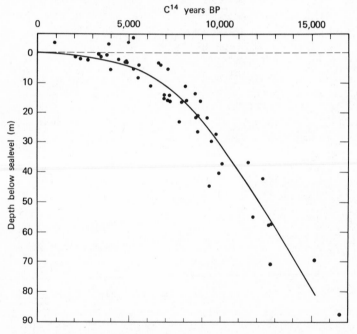

Figure 12-3 Submergence curve based on nearly 50 C^{14}-dated samples of organisms taken from growth positions judged to have been close to sealevel. Samples are from various depths along or off several coasts thought to have been "relatively stable." C^{14} ages are plotted against depths. Curve reflects progressive submergence, believed to be chiefly eustatic. Compare a similar curve for the Netherlands constructed by Jelgersma (in Sawyer, ed., 1966, p. 62) and the curve in Milliman and Emery, 1968, fig. 1); also fig 12-2. (After Shepard, 1963b, figs 1, 2.)

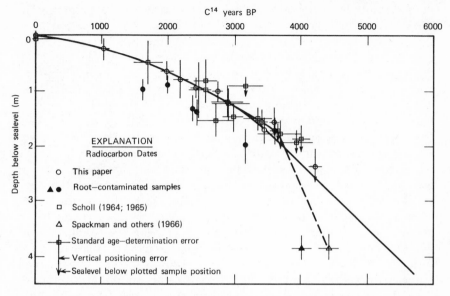

Figure 12-4 Submergence curve based on C^{14}-dated samples of peat, shell, and other materials from measured depths along the coast of southern Florida, believed to have been stable during the time period represented by the curve. Each sample is believed to approximate the sealevel of the time when it was formed. (Compare fig 12-3.) (Combines curves in Scholl and Stuiver, 1967, and Scholl et al., 1969. Control points for the right-hand part of the curve are given in the later publication.)

However, the results arrived at for various suites of samples spanning all or part of the last 15,000 years are rather similar; this suggests that any noneustatic element in the relative-sealevel change is minor, although differences of several meters at points on the curves are stressed by Newman et al. (1969, fig. 8).

Setting aside possible noneustatic sealevel rise, we note that these curves are asymptotic or nearly so; if accepted at face value they imply that glacier melting decreased continuously during the time periods they represent. On the other hand, some authors (e.g., Fairbridge, 1961, p. 156) have read from time/depth data a rise of sealevel marked by considerable fluctuation. Differences among the various curves result mainly, of course, from the C^{14}-dated samples selected for plotting, and opinions differ as to what coasts are "stable."

Is fluctuation probable? The record of fluctuation of glaciers and the curves of fluctuating climate through the last 15,000 years at least admit the possibility of corresponding fluctuation of sealevel. Re-examining the curves, we see that the envelope of plotted points in fig 12-4 permits local variations of a little more than 1m during the last 4000 years, with greater though poorly defined amplitude before 4000 BP. In fig 12-3, if the points plotted above existing sealevel are neglected, the envelope permits variation of ~5m at 4000 BP. Thus it is at least possible that some fluctuation did occur, although the huge volume of the world ocean militates against fluctuation of marked amplitude. Hopkins (ed., 1967, p. 463–465) saw in submerged strandlines in the Bering Sea evidence suggesting fluctuation, and thought the Bering Land Bridge was finally cut off ~10,000 BP. The Scandinavian and Laurentide Ice Sheets ceased to exist as such at some time not far from 7000 BP. Before that time their rapid melting may have tended to smooth out a sealevel curve that in the absence of large ice sheets might have responded more sensitively to the fluctuation of smaller glaciers. The role of the Antarctic Ice Sheet in influencing sealevel since 7000 BP is still almost unknown. When the curve in fig 12-3 has been extended another few thousand years into the past (cf. Shepard and Curray, 1967, fig 2) through the establishment of a substantial number of points, we will have a much better basis for judging possible fluctuation. Meanwhile the curves say that between 10,000 and 9000 years ago sealevel was rising at an average rate of ~80cm/100 yr, whereas during the latest 3000 years the average rate slowed to ~3.5cm/100 yr.

Sealevel During the Last Few Thousand Years. Opinions have differed as to whether sealevel has risen above its present position at any time since the Late-Wisconsin glacial maximum. The plot in fig 12-3 includes four points at positive altitudes, dated within the last 6000 years, although in a later plot (Shepard and Curray, 1967, fig 2; see also plot in Curray et al., 1970, fig 6, representing various opinions) those points do not appear. Several possible errors inherent in such points are well set forth by Shepard and Curray and by Jelgersma (in Sawyer, ed., 1966, p. 54). Although it seemed at first that the slightly higher temperatures of the peak of Hypsithermal time could reasonably have been accompanied by a sealevel higher than today's, it is now widely recognized that slight, temporary steepening of a curve like that in fig 12-3 could equally well be expected. The case for a positive sealevel rests on the stratigraphy and the accuracy of dating of samples from positive-sealevel features. It has been shown that some such features antedate the latest glacial age, that others have been locally uplifted, and that still others have resulted from storm-wave activity related to the existing sealevel. On the other hand, strandline data from the Bahamas (Lind, 1969) are held to indicate positive sealevel of postglacial date and eustatic origin.

It is too early to say that the sea has not risen eustatically above its position today, but the accumulated evidence seems to favor the view that it has not done so.

A special aspect of the study of recent sealevel is that a significant proportion of the dated samples come from layers of peat from former tidal marshes, cypress swamps, and the like. Peat is very compactable under the weight of overlying sediment, so that a body of peat formed, for instance, within a narrow portion of the vertical tidal range may become compacted to a lower position unless it is underlain by a very firm substrate.

INTERSTADIAL SEALEVEL

Little is known about the position of sealevel during the interstadial that preceded the Late Wisconsin glaciation. In the dacade 1950–1960 it was suggested that during that interstadial the sea stood higher than it does today. However, the fragmentary pollen record indicates a climate cooler than today's, implying that sealevel was lower. In the following decade an interstadial sealevel at around -10m was proposed; it figures in synthetic curves such as those of Curray (in Wright and Frey, eds., 1965a, p. 724; Milliman and Emery, 1968; Veeh and Chappell, 1970). The matter is still unresolved, but it will be settled eventually by the collection and dating of enough well-selected samples. It is important because of the light it could shed on the extent of the great ice sheets in the northern hemisphere during the interstadial. The extent is poorly known from direct evidence because the major Late Wisconsin expansion removed or concealed most of the evidence in the critical regions.

INTERGLACIAL SEALEVELS

Except along exceptionally unstable coasts, most of the strandlines of known or supposed interglacial age occur above the existing sealevel, in contrast with glacial-age sediments and strandlines, most of which are now submerged. For this reason the fund of knowledge of the emerged features is greater than knowledge of those which must be researched by bottom profiling and sampling, coring, and other seaborne techniques. However, much of the older literature on emerged strandlines includes correlations based on the assumption that no crustal movement has occurred; these must be recognized and considered separately from the reported data on the features themselves.

A useful guide to the literature is Richards and Fairbridge, 1965. In consulting any literature, however, one must bear in mind that although the presence of marine fossils *in situ* (not as transported particles) indicates former submergence, fossil indicators of sea*level* are normally not precise, because the depth ranges of some shallow-water marine organisms are not well known, and some of the known ranges are large. Water temperatures are normally inferred from the tolerances of the same forms or allied forms living today. Two other factors may need to be taken into account locally. One is the height of washing by surf during exceptional storms; on some Atlantic coasts this height exceeds +10m. Similar heights along ancient strandlines were inferred by Hörnsten (1964) from field measurements. The other factor is possible confusion of marine with nonmarine sediments. In central Florida an extensive surface mantle of sand, formerly thought to be of Pleistocene marine origin, is now interpreted as a concentrate of quartz, residual from lateritic weathering of underlying Pliocene phosphorite (Altschuler and Young, 1960). Such "pseudomarine" features might occur at the surface of other soluble rocks such as salt or carbonates.

Mediterranean Region [5]

Any discussion of the development of knowledge of late-Cenozoic strandlines begins logically with the coast that surrounds the Mediterranean and that continues beyond Gibraltar southwestward along the Atlantic coast of Morocco. The Mediterranean acquired its present general form beginning in later Miocene time, but may have been continuously a sea only since early in the Pliocene. Study of the strandlines began well back in the 19th century. Many of the numerous occurrences of emerged strandlines consist of cliffs and benches cut in bedrock, with associated patches and veneers of fossil-bearing, gravelly or sandy littoral sediment. These features exhibit various degrees of erosional destruction, determined by their relative dates of origin and by the local topography. In some areas morphology is lacking, and minimum positions of sealevel are marked only by the highest occurrences of fossil-bearing strata of known stratigraphic position. Such features are, of course, not strandlines, but we group them with the others. In some places such features reach altitudes as great as +900m; in others specific strandlines are known or inferred to pass beneath the sealevel of today. These differences result from late-Cenozoic crustal movements of various dates and commonly of localized character. Despite such local deformation, there are said to exist coastal segments, notably in Cyrenaica and in Atlantic Morocco, where

[5] General references: Mars, 1963; Gignoux, 1952; Norin, 1958, p. 7–10; see also Quaternaria, v. 6, 1962, p. 97–549.

emerged strandlines maintain their relative altitudes through considerable distances and so are not apparently deformed.

The sequence (of time-stratigraphic units, based on faunas) most commonly quoted in the literature before about 1950 (e.g., Gignoux, 1950) is:

Versilian (late-Pleistocene and Holocene)
Tyrrhenian (Pleistocene)
Sicilian (Pleistocene)
Calabrian (Pliocene or Pleistocene)

An Emilian unit, between the classical Calabrian and Sicilian, has been proposed. Some authors favored a Milazzian unit between Sicilian and Tyrrhenian; others favored a Monastirian unit following the Tyrrhenian; these units, however, have been subjects of much controversy. The confused state of knowledge of the sequence is reflected in table 17-B.

Most of the units were earlier thought to correspond with glacial ages. Later they were recognized, mainly through their faunas, as interglacial, except for the Versilian, which, like the Flandrian farther north, is supposed to span the time since the late-Weichsel low position of sealevel. Correlation of strandlines was based mainly on altitudes.

Many of the earlier beliefs are now recognized as adding up to a gross oversimplification. Faunal-stratigraphic and morphologic studies have established a good basis for correlation by stratigraphy rather than by altitude, and peculiarities of the faunal sequence are being attacked successfully. The mollusk fauna shows weeding-out of tropical forms in Pliocene time and their virtual elimination in the Quaternary. All these changes are regarded by Ruggieri (1967) as resulting primarily from change of climate.

However, even as early as Calabrian, "cold" immigrant mollusks from the coast of western Europe farther north appeared in the Mediterranean; this immigration was repeated, and despite the general decrease in tropical kinds, the Mediterranean also received "warm" immigrant mollusks from the west coast of Africa. The "cold" faunas were more at home in a deeper-water environment than were the "warm" faunas, which were littoral. According to Mars (1963) this state of affairs is explained by a hypothesis based on ocean currents through the Strait of Gibraltar, where a rock threshold at −400m separates the Mediterranean from the Atlantic. Today (see illustrated brief explanation in Longwell et al., 1969, p. 354), two currents flow through the strait, one above the other. Heavy saline water flows as a density current out of the Mediterranean, where evaporation is rapid, while lighter, fresher Atlantic water flows eastward above it, replacing the Mediterranean water lost by evaporation and outflow. The hypothesis suggests that in a glacial age lower temperature and reduced evaporation would raise slightly the level of the Mediterranean and reduce salinity to a value less than that of the Atlantic. The light, relatively fresh water would flow outward through the strait, while cold, heavier Atlantic water would flow inward underneath it, accompanied by a "cold," relatively deep-water fauna from the north. Interglacial conditions would resemble those of today, with a "warm" littoral fauna from the south entering the Mediterranean through the strait.

This hypothesis seems reasonable, but whether or not it becomes established, it emphasizes the value of considering as many factors as possible in attempts to explain faunal anomalies.

The present state of knowledge represents a large advance over that prevailing before mid-century. Strandlines are more numerous than had been supposed earlier. The stratigraphic positions of a number of them are known, relatively or approximately, by related marine faunas and, in northwestern Africa, by Stone-Age artifacts. Some strandlines or groups of strandlines are separated by evidence of deep marine regressions (table 24-I), most of them of unknown vertical extent, and some of them probably eustatic on a basis of related faunas. Although along some Mediterranean coasts deformation is evident, along others (such as Morocco, Cyrenaica, and Lebanon) the younger strandlines, at least, show little evidence of deformation, although they may have been lifted up epeirically.

The Sicilian and Calabrian clearly date at least as far back in time as the earlier part of the Pleistocene. The group of strandlines generally assigned to the Tyrrhenian date to the later part, although they represent two or three distinct marine transgressions, the latest of which seems to be marked by three or more individual strandlines. Within the reasonably near future meaningful correlations should become possible; meanwhile a table in Mars (1963, p. 82) gives one informed opinion as to the general sequence; another, rather similar table appears in Lumley, 1963, p. 576. Regressions reasonably attributable to glacial ages may number as many as five or six. A considerable number of Uranium-series dates are available (e.g., C. E. Stearns and Thurber, 1965), but pending the validation of dates obtained by that method (chap 15), it seems prudent not to quote them in a general review.

North Sea Region

The southern coast of the North Sea has been subsiding since before the beginning of Quaternary time. In consequence, borings in the coastal part of the Netherlands reveal a thick section of Rhine River deltaic sediments, in which are wedges of marine sediments (fig 12-5) that pinch out landward. The wedges record repeated transgressions of the sea, probably

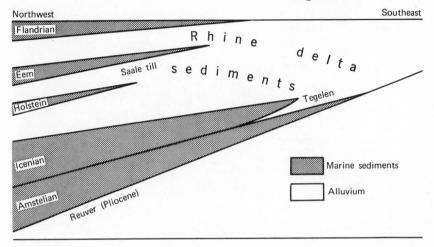

Northwest   Southeast

Figure 12-5   Idealized cross section through a part of the Netherlands (not to scale), constructed from borings, and showing five wedges of marine sediments intercalated in alluvium. (After Brouwer, 1956.) The uppermost wedge is shown in detail in fig 12-6.

of eustatic character and of interglacial age, superposed on the regional subsidence. Hence the stratigraphy is known but eustatic altitudes are not; the later half of the Flandrian transgression is known in detail (fig 12-6). Farther north, on the North Sea floor (fig 12-7), peat records the presence of several large, former swamps apparently of Boreal and Preboreal age, as determined by pollen stratigraphy (e.g., Jessen and Jonassen, 1935). The vertical positions of most are between −35m and −50m. A sample of Boreal age (table 24-C) was recovered from −37m; that depth fits the curve in fig 12-3 reasonably well, suggesting that during subsequent time epeiric movement of the northern North Sea floor may have been small. For a time during the Flandrian rise of sealevel, the Dogger Bank must have been an island. Interglacial marine sediments at various positions above and below today's sealevel occur at a number of localities around the southern British and Irish coasts (list in West, 1968, p. 277). Some are related to strandlines with preserved morphology. Interglacial sediments and strandlines are reported also from coastal points in western France and Spain, and buried interglacial sediments from lands south of the Baltic Sea (chap 24). Because of isostatic uplift, most of the marine sediments in northern Britain and Ireland and in Scandinavia are of Flandrian age.

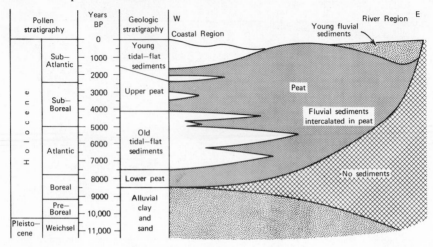

Figure 12-6 Schematic section through coastal western Netherlands (not to scale), showing complex wedges of marine (tidal-flat) sediments of Flandrian age, known from borings. C^{14}-dated pollen stratigraphy (table 24-C), is the basis of the time scale, to which the stratigraphic units have been fitted. (After Jelgersma, 1961, p. 85.)

Atlantic North America

On the east coast of North America the southern limit of glaciation, which intersects the coast near New York City, marks a boundary between marine sediments of Flandrian age on the rocky coast to the north, and marine features of presumed interglacial age on the low, sandy coast to the south. Owing to the presence of outwash as far south as Chesapeake Bay, however, the boundary is transitional rather than sharp. The marine features include gently sloping wave-cut cliffs, beaches, spits, barriers, and extensive areas of marine sediments, in places with fossils. The sediments are numerous (fig 12-8), consist primarily of quartz sand, and range in altitude up to ~+38m in Virginia and to more than 50m in Georgia. Most of the related fossils, chiefly mollusks, were collected from surface exposures and occur at less than +10m. They represent a fauna quite like that of today, living at temperatures similar to those now existing at the same places. Some fossils have been recovered from deep boreholes at higher altitudes, in places where fossils have been leached out of shallower positions. All samples of shells and wood that have been C^{14}-dated hitherto are more than 40,000 years old, suggesting though not proving that they are of pre-Wisconsin age. Buried cypress peat likewise

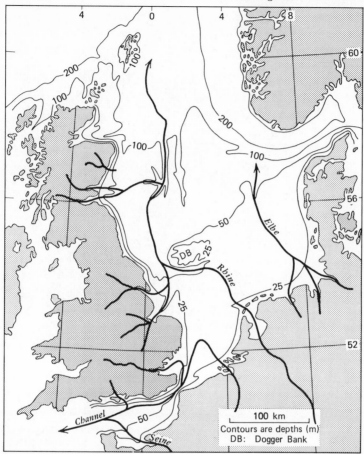

Figure 12-7 Sketch map of North Sea region, showing contours on the sea floor and positions of major rivers at latest time of lowered sealevel. Much generalized; the stream segments shown may not have been contemporaneous. (Streams mainly after R. G. Lewis, 1935, Geog. J., v. 86, p. 549.)

indicates temperatures like those at present. Sediments and morphology at the higher altitudes indicate greater chemical weathering and erosional modification than those at lower altitudes, indicating that probably at least two interglacial units are represented.

Emerged marine features on the Atlantic Coastal Plain were identified before the middle of the 19th century. However, most of the literature before the 1960's, like that on the Mediterranean strandlines, emphasized morphology and neglected stratigraphy, correlated shores by altitude,

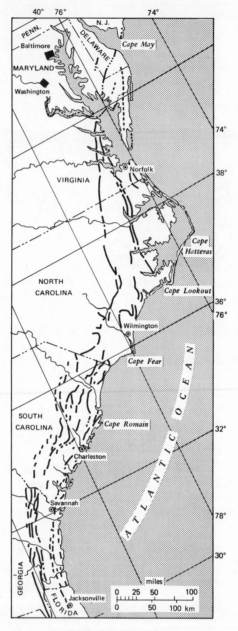

Figure 12-8 Representative emerged shoreline features of probable Quaternary interglacial age in coastal southeastern United States. Thick lines represent wave-cut cliffs, bars, and barriers of various ages. Some of the features

336

and regarded them as the products of successively falling sealevel.

The substitution of stratigraphy for morphology as the prime basis of interpretation began in 1960 in Virginia, and revolutionized the marine-interglacial history of the Coastal Plain (Coch, 1965; Oaks, 1965). Through detailed study of a long narrow belt at right angles to the coast, including the analysis of hundreds of borings, the project established a sequence of many relative-sealevel positions, each represented by offshore, beach and dune, and lagoon facies, some of the positions being separated by profound regressions. The transgressions reached various altitudes in an irregular pattern, and a simple interglacial seemed to be represented by a group of several sealevels. Subsequent studies in other states have followed the same procedure and have yielded comparable preliminary results, although firm correlations along the coast have not yet been achieved. Figure 12-8 shows that one of the emerged strandlines included an interglacial Cape Lookout, 85km inland from the present cape. Also some of the ancient shore features form angles with the present coast, indicating subsequent erosion.

Marine features continue through Florida. In the calcareous terrane of southern Florida episodes of solution seem to have alternated with marine transgressions represented by carbonate strata. This sequence is reasonably attributed to eustatic fluctuation of sealevel (Parker et al., 1955, p. 114). Marine sediments at positive altitudes are also present along parts of the Gulf of Mexico coast.

Near the mouth of the Mississippi River are broad terraces (chap 21) consisting of dissected alluvial fills and reaching an extreme altitude of nearly +100m. Traced downvalley, they pass beneath the recent sediments of the Mississippi delta. Loading of the crust by the weight of delta sediments is believed to have caused downwarping of the fills in the seaward direction, and upwarping is thought to have occurred in the upvalley direction. In theory the fills should grade upvalley into outwash bodies, but significant data from farther up the valley are lacking.

---

shown are composite, representing two or more strandlines of different ages at the same position. Available stratigraphic data are not sufficient to permit correlation through long distances. Compiled from various sources. *New Jersey:* P. MacClintock, 1943, J. Geol., v. 51, p. 460. *Delaware, Maryland:* R. R. Jordan 1964, Delaware Geol. Survey B. 10; unpubl. data. *Virginia:* Coch, 1965, fig 4; Oaks, 1965, fig 10. *North Carolina:* Doering, 1960, J. Geol., v. 68, p. 182; D. J. Colquhoun (unpubl.). *South Carolina:* D. J. Colquhoun, 1965, Atl. Coastal Pl. Geol. Assoc. Field Conf. 1965, text; Colquhoun and Johnson, 1968, Palaeo-geogr., Palaeoclimatol., Palaeoecol., v. 5, p. 105. *Georgia:* Hoyt and Hails, 1967, Sci., v. 155, p. 1542.

The floor of the Mississippi deep valley is as low as −105m and is filled with alluvium, much of which is thought to be of outwash origin. The deep valley is believed to have been cut during the Wisconsin Glacial Age (Fisk, 1944), but it may be older.

Along much of the Texas coast as well as on the coasts of New Jersey, Delaware, and Maryland large quantities of alluvium seem to have obscured and buried most of whatever interglacial marine sediments may have been present (cf. Flint, 1940b, p. 783).

On Bermuda, with very distinctive Quaternary stratigraphy, marine features are mainly calcareous rather than mainly quartzose, and have been a subject of study for many decades. The most recent study (Land et al., 1967; refs. therein) recognized a sequence of eolianites and fossiliferous marine limestones (table 22-C) as representing positive interglacial sealevels, and soils and surfaces of unconformity as representing negative, presumably glacial-age sealevels.

Figure 12-9 compares time/altitude curves of sealevel for three areas, each constructed from rather different data peculiar to the individual areas. They are very rough and possess little time calibration; however, they show similarities and appear to represent two interglacial ages.

Pacific North America

Like the Mediterranean region, the Pacific coast of North America has been obviously affected by tectonic movements within Quaternary time. A carefully studied example is the sequence of terraces, cut into bedrock and veneered with sediments including marine deposits, at Palos Verdes Hills near San Pedro, California (Woodring et al., 1946, p. 113–118). There are 13 benches, the highest reaching +400m. At least some of them are deformed, and the eustatic factor has not been separated from the tectonic. Similar benches occur along the coast of Baja California in Mexico.

Farther northwest, along the coast of Santa Barbara County, California, 17 marine strandlines ranging up to more than +300m have been identified (Upson, 1951). Those below +60m preserve morphologic details, and two, at ∼18m and ∼27m are associated with marine fossils compatible with a late-Pleistocene age. Borings through alluvium show three valley floors, graded to at least −60m and probably to −90m, that postdate the 27m strandline. Probably that strandline represents' a high sealevel of the last major interglacial and the deep valleys date from the succeeding glacial age. Some or all of the higher terraces may have reached their present altitudes by tectonic movement.

An example of the careful application of faunal data to Pacific strandlines is the study of Santa Barbara Island by Lipps et al. (1968). Good

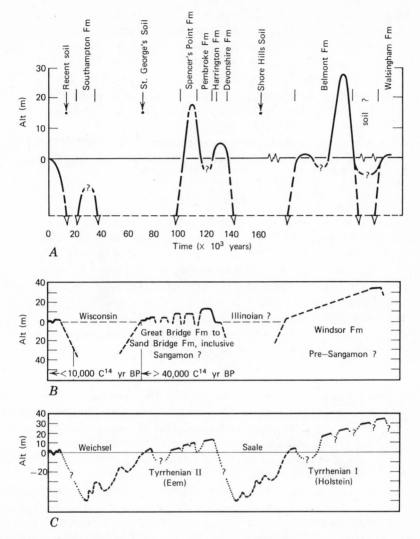

Figure 12-9 Generalized curves showing inferred fluctuations of relative sea-level through time in three areas. *A.* Bermuda (After Land et al., 1967, p. 1002). Time calibration, based in large part on U-series dates, is by those authors. *B.* Norfolk district, Virginia (Oaks, 1965, with unpublished changes). *C.* Southern Mallorca (After Butzer and Cuerda, 1962, p. 414). Correlations are by those authors.

ecologic data were obtained from faunas related to two strandlines at ~9m (Sangamon?) and ~40m (Yarmouth or Aftonian?).

One of the most complete, little-deformed late-Cenozoic marine sequences is that in Alaska, described by Hopkins, McCulloch et al. (Hopkins, ed., 1967, p. 47, 114), and developed mainly along the Bering coast. At least seven transgressions, some of them composite, are represented. They range in age from a few thousand to more than 2 million years, and the highest altitude reached by any is short of +100m. Some of the strata contain fossils. Most of the transgressions are believed to have been interglacials; stratigraphic relations of the latest of these suggest its age is Sangamon.

Other Coasts

On most other coasts Pleistocene marine features (with morphology preserved) are present locally but are less well known than those in Europe and North America. The literature on these is abundant but scattered; few titles deal with intensive studies on long coastal segments.[6] In some of them no attempt is made to distinguish between eustatic and tectonic factors; in others, such as Pacific South America, Japan, and South Australia, tectonic influences are fairly evident. Detailed review of these many occurrences is not likely to result in a useful synthesis until more sophisticated data have become available.

However, it is worth noting that there is agreement between the dates of emerged coral-reef terraces that are believed to represent transgressive maxima of eustatic sealevels on tectonically rising coasts as far apart as Barbados (lat ~6°S; long ~147°E) (Mesolella et al., 1969) and New Guinea (Veeh and Chappell, 1970). The maxima are dated approximately by U-series dates, mostly though not wholly on nonrecrystallized coral. Similarity of the two sequences of transgressive maxima, independent of their absolute altitudes which are mainly the result of uplift, is the basis of the belief that the maxima are eustatic. The time span involved is roughly the last 200,000 years plus or minus the errors involved in measurement.

Other data on Quaternary strandlines in Alaska are summarized in Heusser, 1960, p. 19–24. Chemekov (1961b) published details on the Pacific coasts of Siberia.

Stratigraphic, nonmorphologic evidence of four marine regressions and three transgressions on the coast of the island of Oahu, as revealed in core borings, led Lum and Stearns (1970) to correlate the transgressions with the interglacials recognized in continental United States. The eustatic nature of the changes is assumed.

---

[6] An intensive study of a short segment is cited in Hammen, 1963, p. 156.

EFFECT ON STREAM REGIMENS

The rising sealevel caused by change from glacial to interglacial conditions drowned the mouths of large streams in many parts of the world. In addition, at times when glaciers approached their maxima the falling sealevel laid bare vast areas of continental shelf, which were trenched and channeled by the extended rivers of those times. Along considerable stretches of some rivers the general effect of these slow fluctuations would have been twofold: (1) filling of valleys with sediment by slackened streams, and (2) erosion of valleys by accelerated streams. But this statement does not imply that fluctuations of sealevel alternately slacken and accelerate all streams. The response of a stream to a change of sealevel depends on the shape of its long profile, the volume and grain sizes of the load of sediment it is carrying, the configuration of the sea floor off the stream's mouth, the local stability of the crust, and other factors. It follows that no two streams will respond in the same way to fluctuations of sealevel. Accordingly it is unsafe to infer changes of sealevel solely from evidence of cutting and filling by streams, as has been done at times by writers in both Europe and North America. Evidence of changes of stream regimen may properly be shown to be in agreement with the scheme of sealevel fluctuations inferred from marine deposits, strandlines, and similar direct evidence, but in the present state of knowledge it does not constitute independent proof of movements of sealevel.

INFERENCE FROM BIOGEOGRAPHIC DATA

The similarity of faunas, especially mammal faunas, on pairs of land areas separated by shallow water implies the recent presence of connecting land bridges and thus supports the concept of lowered sealevels. Various land bridges thus inferred are discussed in chap 29. Because the depths of present submergence of most of them are less than the apparent amplitude of sealevel rise since the last glaciation, such bridges can be reasonably explained by sealevel movement alone. Other bridges may have been created or submerged wholly or partly by movement of the crust. Indeed, before any bridge can be fairly ascribed to lowered sealevel, the unlikelihood of crustal movement as the cause of present submergence must be shown. It is difficult if not impossible to exclude crustal movement entirely. In consequence a judgment of "probable but not proved" must be entered in the biogeographic case for temporarily lowered sealevel.

SUMMARY

No brief review could pretend to cover the scattered literature on geologic evidence of change of sealevel. However, not even a thorough review would solve the problem, which involves extremely varied sediments and land forms, some of which are submerged and therefore concealed, and many of which have been examined as a sideline to other kinds of research. Some of the unevaluated uncertainties in the results to date are:

1. No single standard altitude datum has been employed.
2. The precision of altitude measurement varies greatly.
3. The true vertical position of many strandline features with respect to the related former sealevel is not precisely determinable.
4. In many instances the eustatic influence is not or cannot be distinguished from the tectonic in evaluating a strandline lying above or below present sealevel.
5. Many individual studies pertain to short coastal segments, through which the influence of broad warping may not be recognizable. Even on long coasts the possibility of warping along axes parallel with a coast is not ruled out.

Although formidable, this list does not add up to a hopeless situation. On the contrary, it points to the need for more sophisticated study in which stratigraphy and faunal relations are emphasized over mere altitude. Such study has already begun; its overall results should be highly informative.

Meanwhile we can set down a few very tentative conclusions based on the foregoing discussion. The first three are little more than educated guesses.

1. Possible lowest position of relative sealevel during the last glaciation: -100m(?)
2. Possible highest position during the last interglacial: $+20$m(?)
3. Deduced amplitude of sealevel change through the last interglacial/glacial cycle:120m(?)
4. Emerged strandlines and marine sediments, probably referable to earlier interglacials, exist but are or may have been deformed and hence are not useful for measurement.
5. Hence the amplitudes of earlier sealevel fluctuations are not known. If proportional to the *areas* glaciated at various times they would probably not exceed the tentative amplitude stated above by more than about 10%.

CHAPTER 13

Glacial-Isostatic Deformation

DEGLACIATION AND CRUSTAL WARPING

It was mentioned in chap 2 that the rise of the Baltic coasts of Fennoscandia was caused by regional upwarping of the Earth's crust following removal from it of the temporary load consisting of the Scandinavian Ice Sheet. A similar explanation was advanced for the rise of the Great Lakes region in North America. Although in some areas gentle warping of nonglacial origin may play a part in the uplift, it is no longer doubted that the movement is primarily the result of the removal of temporary loads of glacier ice. The supporting evidence, modified after a summary by Gutenberg (1941, p. 750) is as follows.

1. In both Fennoscandia and North America the outer limit of the upwarped region parallels the limit of the latest glaciation.

2. In both regions the isolines (lines connecting points on any strandline that have been equally uplifted) are concentric to the area in which, according to independent evidence, the former ice was thickest late in the process of deglaciation. That area, of course, differed from the area of thickest ice before deglaciation began.

3. In both regions antecedent downwarping is suggested by the presence of marine deposits now bent up above sealevel, and in turn overlying subaerial deposits.

4. In both regions rate of uplift is of the same order of magnitude. During the last few thousand years the rate of uplift of Fennoscandia has slowed down from an earlier, much greater rate.

5. In both regions even the incomplete data obtained thus far show that gravity anomalies are negative,[1] and that they increase toward the

[1] That is, there are deficiencies of mass in these sectors of the globe. When in the future the plastic return flow in the upper mantle and the accompanying upwarping of the crust have been completed, presumably these deficiencies will have been wholly or largely eliminated.

central parts of the glaciated areas, indicating that crustal equilibrium in these regions has not yet been reached.

6. In other regions of former glaciation or former greater glaciation, such as Britain, Greenland, Svalbard, Novaya Zemlya, Siberia, Patagonia, and Antarctica, where postglacial upwarping is expectable, evidence of it has been observed.

So clearly is the direct relation between large glaciers and crustal warping now recognized that the presence of late-Pleistocene emerged marine deposits occurring widely throughout a high-latitude region may be taken as indicating the probability of extensive glaciation. In detail, however, the matter is complicated by the rapid rise of sealevel that has occurred throughout the latest major deglaciation; as a result the datum from which amplitude of upwarping could otherwise be measured has continually changed. A further complication, evident in some regions, is that tectonic movements unrelated to glacial loading occurred contemporaneously with glacial-isostatic deformation. The true glacial-isostatic movement must be isolated from these complicating factors and viewed apart.

THEORY OF WARPING

The theory of warping extends well back in history. Recent discussions include those by Andrews (1968a; 1968b), Broecker (1966), and Farrand (1962).

A large ice sheet is believed to constitute a load greater than the strength of material of the upper mantle beneath the crust. The crust subsides beneath the ice in basinlike fashion. Although the deformation is partly elastic, it is chiefly plastic; hence flow within the mantle transfers material outward, away from the basined area, in compensation for the extra load. With deglaciation, return flow within the mantle bulges up the crust, and eventually conditions are restored, or almost restored, to normal.

Recovery follows deglaciation progressively inward, but with a time lag, the relaxation time. Because of the lag it is possible to measure amounts and rates of recovery. Where areas were submerged during deglaciation, the surf at seashores and lake shores fashioned cliffs and beaches; recovery then deformed these traces of horizontal water planes. Those which are still preserved are inclined, standing highest where the former ice was thickest and postglacial upwarping greatest. The inclination of a warped strandline measures the differential uplift, in the direction of trend of the strandline, that took place between the time when

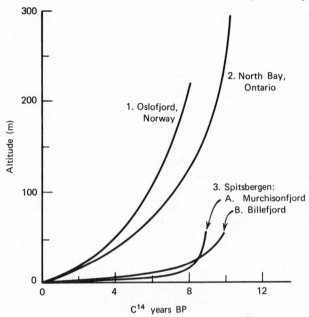

Figure 13-1   Curves of glacial-isostatic uplift, controlled directly or indirectly by C 14 dates. All show that the rate of uplift was rapid at first but decreased with time. (1: U. Hafsten, 1958, Norsk Geog. Tidsskr., v. 16, p. 83. 2: W. Farrand, 1962, p. 184. 3A: I. Olsson and W. Blake, 1962, Norsk Geog. Tidsskr., v. 18, p. 14. 3B: Feyling-Hanssen and Olsson, 1960, p. 123).

the strandline began to rise above the water surface and the present time. Because in some regions several successive strandlines are warped in this way, the traces of the strandlines constitute families of curves (fig 13-1) whose form depends mainly on rate of deglaciation. The time relations of such curves are controlled chiefly by C[14] dates, as is shown in an excellent example from East Greenland (fig 13-8), controlled also, in part, by stratigraphic positions of points on the curves.

The curves are smooth, a form probable if uplift is the result of lateral and upward flow of subcrust material.

The shapes of the curves show that, for any locality, rate of uplift was very rapid when the locality was deglaciated, and then decreased nearly exponentially. Where well controlled by C[14] dates, curves are displaced in time according to the moment when each area was deglaciated (Farrand, 1962, p. 196). Thus in North America, curves near the central area of the former ice sheet are younger than those in the peripheral zone; yet all are similar in form.

Apparently this means that, at any time, uplift beneath the ice sheet was relatively rapid, whereas uplift beyond the ice sheet was relatively slow, and that as the ice sheet diminished in area and thickness, the outer limit of relatively rapid uplift migrated inward as deglaciation proceeded.

Although there is no basic criticism of the concept that basining causes flow in the mantle, the disposition of the displaced material during glacial maxima is a matter of debate. It had been expected that such material would elevate the crust in an area peripheral to the ice sheet. But the evidence on this is wholly negative, even where shore features and stream terraces should reflect such movement and afford a basis for measuring it. Either the slopes of such a bulge are so gentle that they have not been recognized, or the displaced material is distributed in a different pattern. In addition, wide segments of the positions of the margins of former ice sheets along continental coasts are now submerged, making their reconstruction difficult.

An important question concerns the thickness of ice required to deform the crust. The answer is not known exactly, but we can say in general that there is no evidence that small glaciers have thus affected the crust. The small former ice cap of The Faeroes and that of Kerguelen Island in the southern Indian Ocean did not apparently bend their floors, although the larger, composite British ice caps did so. According to Daly (1938, p. 182) those former ice caps which caused crustal subsidence had diameters of more than 500km and thicknesses of more than 1km. However, these dimensions seem to be too large because the area of pluvial Lake Bonneville (longest diameter ~600km; average depth 145m) has been warped up since removal of the water load. This example is examined by Crittenden (1963); the problem of load is also discussed by Niskanen (1943).

Amplitude of upwarping depends on at least five factors: (1) density of ice (~0.9), (2) thickness of ice, (3) density of subcrust material (~3.3), (4) proportion of elastic to plastic deformation, and (5) degree of adjustment of the crust to the additional load. If then, we assume that adjustment is complete, maximum subsidence beneath a large ice sheet should equal roughly one-third of maximum ice thickness. Thus, if the Laurentide and Scandinavian Ice Sheets reached maximum thicknesses of 3km, basining and subsequent uplift in central areas should approximate 1km.

What effect does relief, consisting of highlands and deep depressions, exert on the deformation? Apparently the effect is at best small, because the isolines constructed upon warped shore features appear to follow broad regional trends regardless of local topography. This fact, as well as the failure of small glaciers to bend the crust, suggests that isostatic adjustment for differences of load does not take place locally but is distributed beneath a wide area.

The loads represented by proglacial lakes such as the early Great Lakes, and by long arms of the sea such as the Champlain Sea in the St. Lawrence Lowland, may have prolonged the effect of loading by glacier ice itself, delaying the earliest phase of crustal recovery. In treating the complex problem of loading off open seacoasts by increments of water accumulated during deglaciation, Bloom (1967) found some support for the concept that even rather shallow water should deform the crust by loading. However that may be, it is probable that the deep-sea floor subsides under the load of water returned to the ocean during a major period of deglaciation of the lands.

The form of glacial-isostatic movement may be complicated in some areas by tectonic movements unrelated to glacial unloading. Contemporary subsidence, for example, is occurring on the southern coast of the North Sea (summary in Jelgersma, 1961, p. 14) and is suspected along the coast of New York and New England (e.g., Fairbridge and Newman, 1968). Studies of glacially induced movement must take account of the possibility of such extraneous influence.

UPWARPING IN FENNOSCANDIA

In two parts of the world, Fennoscandia and eastern North America, studies of glacier-induced warping of the crust have been correlated across regions wide enough to give a general picture of the form of the uplift. That uplift is occurring or has occurred in a number of other glaciated regions has been established, but the data on them are not yet sufficient for broad syntheses.

Tide-Gage Records

The uplift in Fennoscandia is established on two lines of evidence: (1) tide-gage records showing that the land is rising differentially with respect to sealevel, and (2) instrumental measurement of former shorelines that are both emerged and deformed. The two lines of evidence taken together suggest nearly continuous movement since deglaciation of the Baltic region began more than 10,000 years ago.

The tide gages are placed by national governments at many points along the shores of the sea and of large lakes as well. The gages are fixed firmly to the rocky shore and have floats that show, and in some gages automatically record, changes in the position of the water surface with respect to the land. When the record of a large number of gages over a period of many decades is examined, and meteorologic and other extraneous effects are allowed for, it appears that upwarping in Fennoscandia is still

Figure 13-2 Contemporary uplift in southeastern Fennoscandia. Isolines represent rate of rise of land above sealevel in millimeters per year, determined from tide-gage readings. Dotted lines are interpolations; arrows indicate discrepancies between different national systems of observation. (After E. Fromm, Atlas över Sverige, 1953; Bergsten, 1954; E. Kääriänen, Fennia 89, 1963, p. 19. Compare data for North Sea and Britain in Valentin, 1954, pl. 15.)

in progress. Although century-old casual observations of emerging rocks and shoals had made it certain that uplift was occurring, it was in the present century that the inquiry was put on a sound, quantitative basis by Witting (1918). His results, amplified by later measurements, are shown in fig 13-2. The southern shores of the Baltic and North Seas lie beyond the outer limit of present uplift, which passes through Denmark and Latvia. Rate of uplift increases to nearly 1m/100 yr within a small area at the north end of the Gulf of Bothnia. The isolines delineate a rather flat-topped dome.

The current movement is causing lakes in the Baltic region to encroach upon southeastern shores and to recede from northwestern shores. The three connections between the Baltic and the North Sea (apart from the artificial Kiel and Göta Canals) are the Öresund, the Great Belt, and the Little Belt. Their minimum depths are, respectively, 7m, 11m, and 11m. Uplift of 11m relative to sealevel would therefore convert the Baltic into a lake or chain of lakes. According to Niskanen (1939, p. 24), uplift now in progress will result in emergence of the floor of the Gulf of Bothnia, so that about 8000 years hence the northern part of the Gulf will consist of a small lake unconnected with the sea. Apparently this prediction assumes no future rise of sealevel.

These calculations imply an aggregate relaxation time of many thousands of years, perhaps as many as 15,000.

Deformed Water Planes

The former position of the surface of the sea, or of a lake, referred to in the older literature as a *water plane,* is reconstructed from a variety of features. The most obvious of these are beaches, bars, wave-cut cliffs, and related strandline forms. Others include lag concentrates of wave-washed boulders, sediments with marine (or lacustrine) fossils (generally mollusks), and the surfaces of contact between marine and terrestrial sediments (fig 13-3) and between the foreset and topset strata in deltas. Isolines constructed on deformed water planes show that on a broad regional basis the deformation was domelike, similar to that inferred from tide-gage measurements. In Fennoscandia the deformation is measured on marine and lacustrine water planes (fig 13-4) that succeeded each other during the process of deglaciation. The central part of the dome lies in the region of the Gulf of Bothnia, where the ice sheet during deglaciation was thickest. Whereas the younger strandlines extend throughout the Baltic region, the older ones are present only in the southeastern part, because when they were made the rest of the region was still occupied by the ice sheet. Furthermore, as soon as the ice began to thin, recovery of the crust began. But actual displacement of the glacier margin had to occur, permitting lake water or seawater to occupy part of the region formerly covered with ice, before the making of shore features could start. Hence some recovery of the crust occurred before even the earliest strandlines were fashioned. We have no means of measuring directly the amount of this early recovery, but it must have been great, perhaps constituting as much as half of the total recovery. Accord-

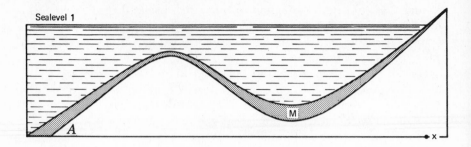

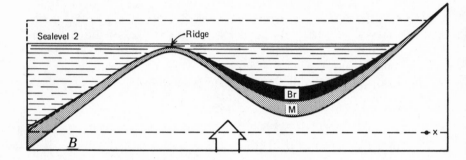

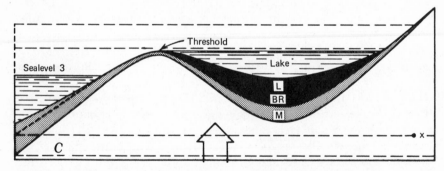

Figure 13-3 Idealized diagram showing three phases in the control of sedimentation in a coastal area subjected to uplift. Point x is the same in all diagrams. At first (A), marine sediment M accumulates over entire submerged area. B. Through uplift, a submerged ridge forms a shoal, causing deposition of brackish-water sediment Br in a bay. C. After further uplift the shoal becomes a threshold, retaining a lake, in which lacustrine sediment L accumulates. The L/Br surface of contact marks the time of emergence from the sea. (Adapted from U. Hafsten, 1960, Norges Geol. Undersøk. no. 208, p. 451.)

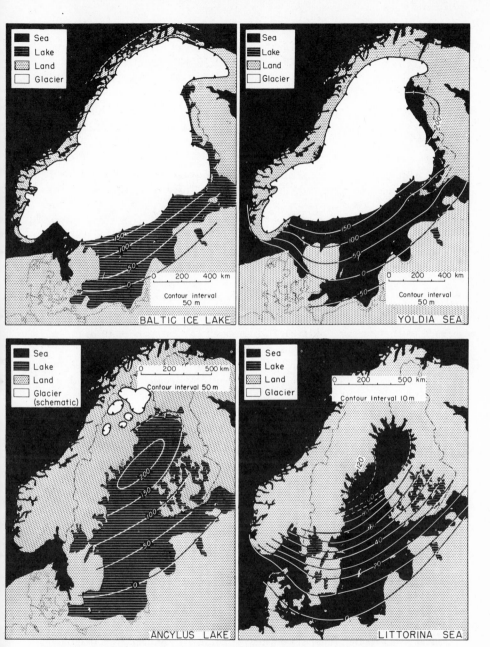

Figure 13-4 Successive water bodies in the Baltic Sea basin during the last deglaciation. Isolines (scaled in meters) show present positions of the related strandlines with respect to existing sealevel. (After Fromm, Atlas över Sverige, 1953, with addition of isolines in northern Norway after Andersen, 1965, from Marthinussen.) Before Baltic Ice Lake time the center of the ice sheet and the basin beneath it would have been farther southeast.

351

ing to Niskanen (1939), the measured portion of total recovery in Fenno-scandia is around 520m (270m before 8750 BP; 250m since then), with 210m more still to be expected, for a total of 730m. Gutenberg (1941, p. 762) cited similar values.

The domelike form of the uplift, with slope steepest on the flanks and diminishing both in the central part and near the periphery, appears in the maps in fig 13-4. The outer limit of deformation is a line that trends from southern Scandinavia northeastward through Leningrad. This form, as a whole, is not incompatible with the concave curves shown in fig 13-1, because the maps and the curves show two different things, albeit closely related.

When figs 13-2 and 13-4 are compared, it appears that the center of uplift has shifted northward during the last 7000 years. It is not known whether the shift is apparent only and an effect of slight errors in mea-surement. If it is real its cause is undetermined.

Geometry of Uplift Relative to Sealevel

Viewed in profile, the points of intersection of successive strandlines such as that of the Yoldia strandline with the Ancylus strandline (fig 13-5), are the result of both warping and submergence: warping of the basin of the Ancylus Lake, and submergence resulting at least in part from eu-static rise of sealevel to the Yoldia position. Because sealevel was rising throughout the progress of upwarping, today all the former strandlines pass below the sea in the southern part of the Baltic basin. The way in which this affected successive strandlines is shown step by step in fig 13-6.

In studies of the postglacial history of updomed regions, it is desirable

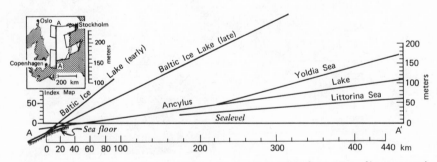

Figure 13-5 Deformed strandlines in southern Sweden, seen diagrammati-cally in vertical profile. Data have been brought into the plane of the profile from within shaded area on index map. (Simplified from E. Nilsson, 1953, fig 35; pl 4. Additional detail in E. Nilsson, 1968; comparable profile for Finland in Hyyppä, 1966.)

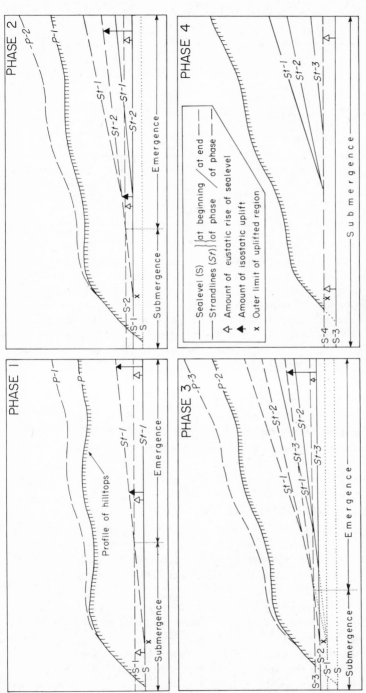

Figure 13-6 Effect of isostatic uplift and eustatic rise of sealevel on successive marine strandlines along a coast. Four postglacial phases are shown. Glacier margin has retreated toward the right, and lies to right of segment illustrated. A strandline formed at the beginning of any phase will have acquired, by the end of the phase, the profile shown. During phase 4 no isostatic uplift occurs. Vertical scale greatly exaggerated. Relations slightly distorted owing to representation of strandlines as straight lines instead of curves. (After an unpublished sketch by B. J. Andersen.)

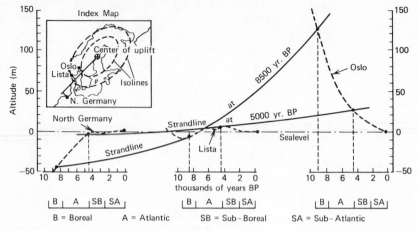

Figure 13-7 Reconstruction of sealevel positions for any desired times during postglacial isostatic uplift. Three curves, showing displacement of relative sealevel through time in three areas lying close to a radius of the domelike uplift, are placed in sequence. By means of vertical lines erected from the time scales, the position of sealevel in each area, for any desired time, is plotted, and the plotted points are connected by curve. From the data given here, curves for 8500 and 5000 BP have been constructed. (After a sketch by B. J. Andersen, based on B. J. Andersen, 1960, pl. 8.)

to reconstruct the position of sealevel at a particular place and time during the contemporaneous processes of uplift and rise of sealevel. This can be done graphically by the method shown in fig 13-7. The geometric relations of sealevel rise and crustal uplift are treated in detail in an excellent study by M. Okko (1967). A curve showing present altitudes of points, each formerly at sealevel, plotted against time, can be corrected (fig 13-8) for eustatic rise of sealevel during the time period by adjusting it to a standard curve of sealevel rise. An example, for the Helsinki locality, is given in Donner, 1969, p. 148. Similarly a curve showing inferred crustal movement, derived by subtracting a curve of observed relative sealevel from a curve of eustatic sealevel can be plotted (e.g. Kaye and Barghoorn, 1964). Conversely, rise of sealevel can be derived from isostatic-emergence data (Schofield, 1964).

Apart from the strandlines in Sweden represented in fig 13-5, analogous relationships in Finland are plotted and discussed in Hyyppä, 1966, and in Donner, 1964b.

The sequence and attitudes of Baltic strandlines as set forth here are those which represent the opinion of the majority of students. However, complete agreement on results is hardly to be expected. In some areas

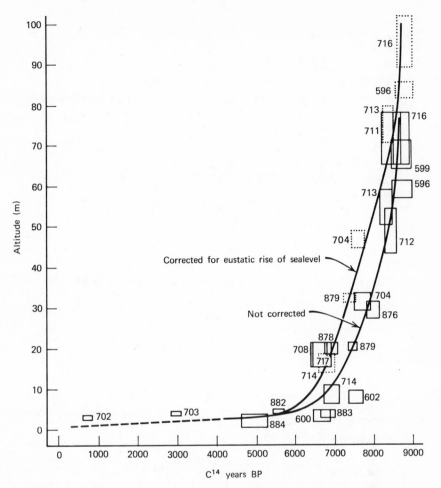

Figure 13-8 Curves in which C[14] dates of marine shells and driftwood are plotted against altitude of occurrence in a small area in East Greenland. Numbers are laboratory numbers of samples. In the curve corrected for eustatic rise of sealevel on the basis of a standard curve (fig 12-3), a few of the boxes representing confidence limits for samples are reproduced in skeleton form, as an aid to reading the curve. Rate of emergence was around 9m/100yr at first, decreasing nearly exponentially to 60cm/100 yr around 6000 BP. Thereafter it decreased to around 7cm/100 yr. (Adapted from Washburn and Stuiver, 1962.)

strandlines are so numerous and so closely spaced that piecing together a single one from discontinuous remnants is very difficult. In others some strandlines are so faint that they can lead to considerable error in determination of the positions of former water planes.

Regardless of minor differences of opinion, the domelike form of the uplift in the Baltic region and the close relation between that form and that of the ice sheet establish the glacial-isostatic cause of the movement. Because of this relationship the direction of maximum inclination of the profiles of warped strandlines (fig 13-5) at any place can be expected to approximate that of the local radius of the former ice sheet. This principle has been used in attempts to discriminate between local ice caps and a segment of a large ice sheet in areas where glacial features are either little known (e.g., Newfoundland; Flint, 1940a) or submerged (e.g., Svalbard-Barents Sea region; G. Hoppe et al., unpublished).

UPDOMING IN NORTHERN NORTH AMERICA

In eastern North America broad uplift of the glaciated region has occurred, but despite a number of excellent studies the resulting synthesis is less complete than is that for the Baltic region. The difference is due in part to the much greater size of the North American area, in part to the scantier distribution of strandlines within the region studied, and in part to the physical difficulties involved in examining in detail the little-inhabited region of Canada where the ice sheet was thickest during the closing phases of the last deglaciation.

For decades detailed data were confined to the southeastern sector of the glaciated region. Beginning around 1950, however, geologic exploration of central and Arctic Canada began to add new information about that vast region, and somewhat later the study of lacustrine strandlines of the Plains region of Canada began to yield results. It is already evident that the North American results are broadly comparable with those achieved in Fennoscandia.

Historical Review

As early as 1884 Upham recognized that the shorelines of the extinct Lake Agassiz are not horizontal, but he attributed this fact to the gravitative attraction of the ice sheet. A year later Gilbert observed that strandlines in the Lake Ontario basin were deformed, and soon afterward he inferred that the deformation had resulted from removal of the ice-sheet load. Before long Spencer inferred that what he called the focus of uplift of the Great Lakes region must lie in the region southeast of Hudson Bay.

The first attempt at broad regional synthesis was made by De Geer who, after several years' study of the postglacial uplift of Sweden, visited America in order to make comparisons. The map he published shows isolines on the imaginary surface of greatest uplift at each place throughout the sector from Newfoundland to Manitoba. Deformation of the shorelines of Lake Bonneville had been shown by isolines in Gilbert's classic monograph two years earlier. De Geer's synthesis showed that the deformation in North America was closely analogous to that in Fennoscandia.

We owe our knowledge of the warping of the classic Great Lakes region chiefly to the researches of Gilbert, Spencer, Leverett, Taylor, Coleman, J. W. Goldthwait, Stanley, and Hough. The data available up to 1915 were gathered into a summary by Taylor (Leverett and Taylor, 1915, p. 316-518) which, although amplified and modified in many ways, still stands as a monumental reference. There is also some information on the Hudson-Champlain depression (Woodworth, 1905; Chapman, 1937), although for New England in general, data are scanty (Hyyppä, 1955, with references; Tarr and Woodworth, 1903; J. W. Goldthwait, 1925; Leavitt and Perkins, 1935; Jahns and Willard, 1942).

The earlier information on eastern Canada, gathered mainly by J. W. Goldthwait, is mainly in publications of the Geological Survey of Canada, of which the most extensive deals with Nova Scotia (J. W. Goldthwait, 1924). Newfoundland data are given by Daly (1934) and by Flint (1940a, with references). Information on Ungava was summarized by H. C. Cooke (1930), and on Lake Agassiz by Johnston (1946). Farrand and Gajda (1962) discussed northern and eastern Canada as a whole, with a map; Andrews (1968a; 1968b) dealt with Arctic Canada, and Blake (1970) more specifically with the Queen Elizabeth Islands.

Tide-Gage Records

As in Fennoscandia, warping is inferred chiefly from tide-gage records and from deformed strandlines. As early as 1896 study of the tide-gage records of the Great Lakes convinced Gilbert that that region is being warped up toward the north at present. Subsequent study (Gutenberg, 1941, p. 739; S. Moore, 1948) confirmed Gilbert's opinion and extended it to include the Atlantic coast between New Jersey and Quebec. The average rate of warping (fig 13-9) seems to be a little less than 1mm/100km/yr, not quite as much as the rate in the Baltic region. As recalculated by Barnett (1966), warping at Churchill, on the west shore of Hudson Bay, amounts to 73cm/100yr. The movement is causing lakes to rise against their southwestern shores while receding from their northeastern shores.

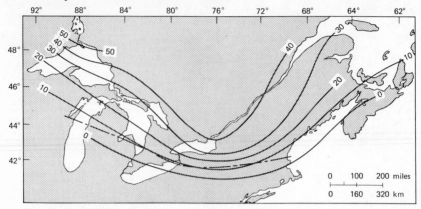

Figure 13-9 Contemporary deformation in eastern North America, calculated from tide-gage records. Continuous curves indicate amplitude (in cm/100yr) relative to a theoretical datum. The broken line marks the outer limit of warping of the Algonquin strandline and its eastward extension. (Gutenberg, 1941, p.'743. A plot employing additional information but based on sealevel, a rising datum, appears in S. Moore, 1948, pl. 1.)

Because of this, the shoaling of some Canadian harbors may in time become a serious problem.

Assuming that conditions in North America are analogous with those in Fennoscandia, Gutenberg calculated that the central part of the uplifted area, in the Hudson Bay region, should rise about 250m more before crustal recovery is complete. If no further change of sealevel occurred, such a rise would reduce Hudson Bay to a tiny inlet of Hudson Strait in the extreme northeastern part of its present area. The great Lakes region, being at a considerable distance from the center of the uplift, should experience smaller future movement. However, as noted hereafter, later estimates visualize a smaller future rise.

Deformed Water Planes

The succession of glacial Great Lakes, formed during the latest deglaciation, is described with sketch maps in chap 21. Numerous warped strandlines of those lakes combine with marine strandlines in the coastal region to give the information shown in fig 13-10. The figure shows the general parallelism of isolines with the peripheral zone of the ice sheet, the position of the outer limit of deformation, and the inward shift of the zone of most rapid deformation with time.

Outer Limit of Deformation. The outer limit of the region deformed at any time is determined from the vertical positions of strandlines. In the basins of Lakes Erie and Huron, for example, the strandlines of Glacial Lakes Whittlesey and Maumee (fig 21-6; table 21-F) are horizontal from their southern limits northward as far as Lake St. Clair and eastward as far as Ashtabula, Ohio. At these two localities they begin to rise, and continue to rise northward to the points where each of them ended against the glacier margin. A line connecting Lake St. Clair with Ashtabula is the outer limit of warping ("hinge line" in the older literature) for the lakes. Figure 13-11 shows, in idealized section, the origin of this relationship. The strandlines of the Glenwood phase of Lake Chicago, contemporaneous with Maumee and Whittlesey, are horizontal from Chicago north to Milwaukee, Wisconsin and Grand Haven, Michigan. From these places the strandlines rise northward until each comes to an end. Accordingly it is inferred that the Whittlesey outer limit of warping (fig 13-10, WH) crosses Lake Michigan through these two places.

Although this evidence indicates that there was no upwarping south of the Whittlesey outer limit of warping after Lakes Maumee and Glenwood Chicago had begun to form, it does not imply that upwarping of that region did not take place in the interval between the Late Wisconsin glacial maximum and the development of those lakes. The Whittlesey outer limit of warping therefore represents merely the outer limit of measurable warping in the Great Lakes region. Traced westward, this limit (whether or not it coincides in time of origin with Lake Whittlesey) trends northwesterly, passing north of Glacial Lake Dakota, whose strandlines apparently are not warped, but south of Lake Agassiz, whose strandlines (Elson in Mayer-Oakes, ed., 1967, p. 46) are warped even as far south as the lake outlet. On similar evidence the outer limit of warping is known to pass southwest of Glacial Lakes Souris and Regina.

Relation of Deformation to Sealevel. The surfaces of successive lakes formed during general deglaciation of a single basin fell rapidly when new, lower outlets were uncovered, and rose, less rapidly, when readvance of the glacier margin blocked an outlet. With the sea, however, the relations were different. During deglaciation eustatic sealevel appears to have risen continuously; if there were any reversals their amplitudes would have been small. Such a rise implies that the upper limit of occurrence of postglacial marine features across a broad region probably does not represent a single, isochronous water plane, but rather a composite, representing the joint effects of a locally rising crust and a generally rising sealevel. An example is the region around Hudson Bay.

The Bay is fringed by a belt, 50 to 350km wide, of marine sediments, including many beaches (fig 13-12), deposited in the proglacial/postglacial Tyrrell Sea (chap 22). The sediments are widespread at altitudes as much as 150m above the Bay and locally reach 285m, the high for the region.

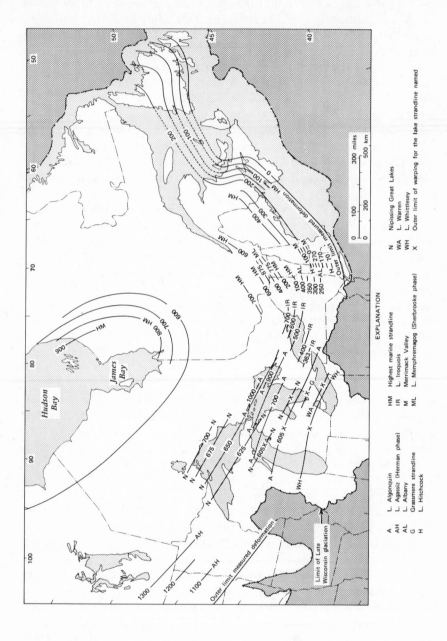

EXPLANATION

A   L. Algonquin
AH   L. Agassiz (Herman phase)
AL   L. Albany
G   Grassmere strandline
H   L. Hitchcock

HM   Highest marine strandline
IR   L. Iroquois
M   Merrimack Valley
ML   L. Memphremagog (Sherbrooke phase)

N   Nipissing Great Lakes
WA   L. Warren
WH   L. Whittlesey
X   Outer limit of warping for the lake strandline named

0   100   200   300 miles

0   200   500 km

360

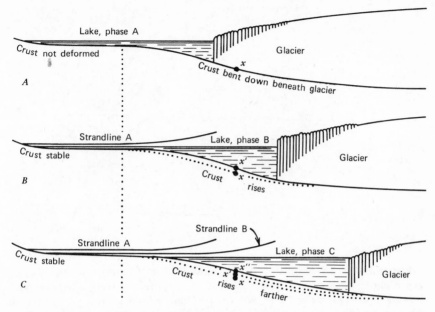

Figure 13-11 Idealized sections showing progressive upwarping of crust during deglaciation. *A.* Glacial lake (phase A) has formed between the receding glacier margin and higher ground to the left. *B.* Uplift has affected the area to the right of the vertical dotted line, bending the strandline of phase A to a higher position. Point *x* has been lifted to position *x′.* A new lake phase, B, has formed at a lower level. *C.* Renewed uplift has affected the area to the right of the vertical line, bending the strandline of phase B, and warping a little more the extreme right-hand end of A. Point *x* has been lifted to position *x″.* A third phase, C, has formed at a still lower level.

The upper limit of marine sediments represented by these altitudes has been called the *marine limit.* It is on the marine limit that the isolines in the Hudson Bay region (fig 13-10) are based.

Is this marine limit an isochronous surface? Apparently it is not, for

Figure 13-10 Deformation in eastern North America inferred from altitudes of strandlines, highest marine sediments, and related features dating from the Late-Wisconsin deglaciation. The curves are isolines; numbers are altitudes in feet above existing sealevel. In most of the lake sequences, only the highest and oldest lake level is plotted. Curves are provisional, as they are based on data of widely varying reliability, but they show both the general form of the upwarped region and inward migration of the zone of most intense deformation. (Selected and compiled from published sources and from data furnished by A. L. Bloom, H. W. Borns, and Carl Koteff.)

Figure 13-12 Strandlines (probably mostly storm beaches) exposed by emergence along the southwest coast of Hudson Bay during recent decades. View looking northwest near lat 55°50', long 87°00', in the vicinity of Ft. Severn. About 60 beaches are visible in the view. (Canadian Armed Forces.)

three reasons. First, invasion of the Hudson Bay region by the sea began around 8000 C^{14} years BP. Hence the measured uplift has been occurring through a period of about that length, with nonmeasured uplift having begun perhaps twice as long ago. However, owing to the pattern of deglaciation the measured uplift seems to have begun earlier in the southern part of the region than in the northwestern part. Second, sealevel has risen since the invasion began. Third, rate of deglaciation varied from one district to another. Where this rate was rapid the marine limit was "registered" earlier at any given point, and participated in greater uplift, than where the rate was slow.

Because of these factors the marine limit, although it possesses a dome-like form, probably does not give us a time datum from which rate of uplift can be calculated. The position of sealevel, however, can be calculated for a particular date by the method described in fig 13-7 or the one implied in fig 13-8. Isolines can then be plotted from that datum. Isolines thus related to a single position of sealevel, and recording uplift through a defined time interval, can be spoken of as *isobases,* a term introduced by De Geer but used by him in the sense in which we have been using isoline.

From a plot of isobases thus defined (fig 13-13), rate of uplift can be

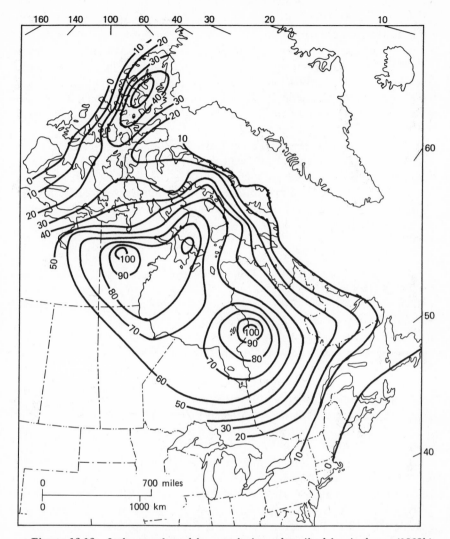

Figure 13-13 Isobases, plotted by a technique described by Andrews (1968b), from occurrences of postglacial marine sediments at 58 points in eastern and northern Canada where date and altitude of each occurrence are known. The curves are based on the sealevel of ~6000 BP, believed to be ~13m below that of today (Shepard, 1963b, fig 1). They show amplitude of emergence in meters ±5m, since that date. Three centers of uplift are evident, possibly reflecting the positions of comparatively thick residual ice bodies shortly before the date mentioned. (J. T. Andrews, unpublished.)

calculated. Andrews (1968b) used such data to estimate relaxation times for East Greenland, Arctic Canada, and Fennoscandia (1.7, 2.5, and 3.0 $\times 10^3$yr, respectively for a 10×10^3yr period of postglacial recovery.) and employed rates of uplift to calculate uplift still to be expected. His estimate of residual uplift for a part of the Hudson Bay region is 100 to 150m, a value that can be compared with Niskanen's estimate, already cited, of 210m for Fennoscandia. As for aggregate uplift, both measured and nonmeasured, we assume that the Laurentide Ice Sheet at its maximum was 3km thick in its central part, and that total recovery then amounts to around 1km. Of this we have measured 275m in the Hudson Bay region and expect 100 to 150m more, for a maximum of 425m. The balance, $\sim$575m, must be attributed to uplift, mostly very early, in the peripheral region, as much as 130m (400ft) of which has been measured (fig 13-10), with the remainder unaccounted for.

UPLIFT ELSEWHERE

Although many strandlines, most of them probably made by former lakes, are known to exist in northwestern Canada, there is little information on their attitudes. Much if not all of the Pacific Coast of Canada and southeastern Alaska was deformed following deglaciation. Postglacial marine sediments occur on Vancouver Island, with altitudes of <100m on the west coast, increasing to 180m in the Vancouver district. In extreme northwestern British Columbia similar sediments occur up to 120m. In southeastern Alaska postglacial marine sediments occur at many places, locally with a cold-water mollusk fauna; the highest altitude thus far reported is 90m.

Elsewhere around North American coasts we find deformation decreasing northward toward the Arctic Ocean, with an apparent outer limit of deformation near the northern shores of the Queen Elizabeth Islands. This limit, however, may relate to a deglaciation antedating the last one. On southern Baffin Island, amplitude of uplift decreases southeastward toward Davis Strait, suggesting that the eastern margin of the former ice sheet stood nearby.

Uplift near lat 72° on the coast of East Greenland (fig 13-8), measured by A. L. Washburn and Stuiver (1962), reflects deglaciation by the Greenland Ice Sheet. Upwarping in Iceland, amounting to at least 110m and dating from within the last 10,000 years, is shown on a map by Kjartansson et al. (1964, p. 128). Movement is recorded in various parts of Svalbard, and C^{14}-controlled curves (fig 13-1) for two fiords during the last 10,000 years have been published. Rate of uplift today is very small. Post-

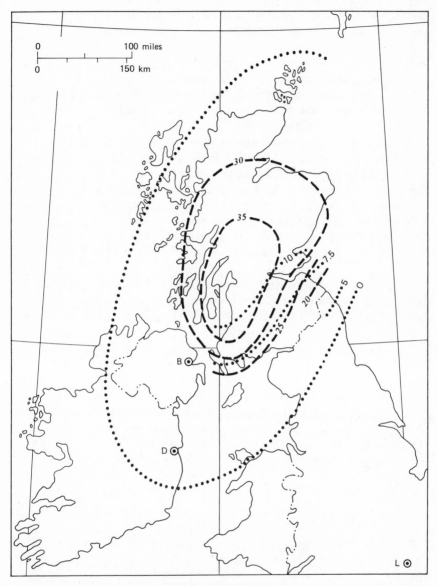

Figure 13-14 Isolines, in meters, on two deformed postglacial marine strand-lines in northern Britain. Dotted lines: ("100-foot" strandline, dating from around 8000 to 7000 BP.) Zero isoline from W. B. Wright (1937, fig 127); other data from Donner (1963, fig 4).

glacial marine features in Novaya Zemlya reach nearly 100m, and are reported along the coast of Siberia, with decreasing altitude eastward nearly to the Lena River.

In the British Isles (fig 13-14) postglacial uplift centered in the Highlands of Scotland, as indicated by the attitudes of two distinct strandlines, with the upper limit of marine features ~40m above today's sealevel for the older surface and ~12m for the younger. Other, less conspicuous strandlines are present, and there is some evidence that movement is still in progress. The subject is discussed, with references, in Sissons, 1962; Donner, 1963; Donner, p. 23, in D. Walker and West, eds., 1970.

In the Alps (Jäckli in H. E. Wright and Frey, eds., 1965b, p. 153) where the average thickness of the Late Würm ice mass is estimated at 500m, strandlines on which measurement of deformation could be based are lacking. However, it is suggested that renewed fault movement of recent date along old fractures, paralleling valleys, in which valley floors have moved up relative to interfluves, may be an expression of isostatic recovery.

In glaciated parts of the southern hemisphere, emerged marine features thought to be related to glacial-isostatic movement have been reported from Patagonia and from various parts of the Antarctic coasts.

SUMMARY

Deformed water planes occur wherever there is evidence of former ice sheets, or ice caps of substantial size. Their form is domelike, and stratigraphy and isotopic dates permit calculation and rate of movement. Amplitudes are compatible with minimum thicknesses of former glaciers estimated from geologic data.

CHAPTER 14

Stratigraphy

The sequence of Quaternary strata involves the same range of sediments, the same kinds of problems, and the same techniques as does the stratigraphy of older deposits, with the important addition that the direct and indirect effects of climatic change are more obvious than in older groups of strata, and that morphology (terraces, moraines, degree of erosional dissection), intensity of weathering, and soils play an important part in recognition and correlation. Sediments of marine, brackish, lacustrine, fluvial, and eolian origin are widely present in addition to those of glacial origin. In some regions quaternary volcanic rocks including layers of tephra are abundant, and in others warped, folded, faulted, and upheaved strata testify to the activity of Quaternary crustal movements.

For nearly a hundred years after the distinctive form and internal character of glacial deposits were recognized, such deposits were studied more widely than were the Quaternary strata of nonglacial origin. In the 20th century, however, knowledge of the latter increased rapidly, and the integration of nonglacial with glacial strata developed on a large scale. With the inception of techniques for coring the sediments of the deep-sea floor, Quaternary stratigraphy entered an important new phase.

Although the kinds of sediments formed during the Quaternary are represented also in older strata, Quaternary strata now exposed include a larger proportion of sediments of terrestrial and especially of glacial origin. Though erosion has dissected them, it has destroyed a relatively small proportion of their total bulk, so that they exist in abundance.

Terrestrial sediments of whatever age are more difficult to interpret than marine sediments, mainly because of greater lateral and vertical variability and smaller content of fossils. Many terrestrial strata are extremely thin and discontinuous. Some have unusual stratigraphic relations caused by rise and fall of sealevel and of lake levels, as well as by contemporaneous warping resulting from loading and unloading by ice sheets. Also they include erosional unconformities of unusual kinds. The interpretation of all strata rests basically on what William Smith in 1799 called the *order of superposition of the strata:* the deposits underneath are older; those above are younger. But this order is likely to be obscured in deposits controlled in part by local topography. For example, young Pleistocene terrestrial sediments may occur at altitudes lower than those of older sediments. This happens when alluvium covers the floors of valleys that have terraces mantled with older alluvium; and complication increases when loess indiscriminately mantles terraces of various ages. In such situations relative ages must be inferred from other features, perhaps outside the immediate area.

The abundance of small bodies of sediment representing former bays, lakes and ponds, dunes, and small streams, and of surfaces of unconformity representing deep, steep-sided valleys introduces great complications. Indeed the details of a single exposure of terrestrial strata may be so complex, lateral variation so abrupt, and variations in thickness so great, that there can be little assurance of continuity without continuous exposure. This situation is met by the maintenance of reference files of detailed descriptions and photographs.

Even where strata are continuous it is not always possible to correlate sections of till throughout a wide region on a basis of composition, which changes with changing character of the underlying bedrock. Conversely, in a region of uniform bedrock the individual layers of till in a sequence may be very similar and difficult to differentiate.

Another complicating factor is general lack of consolidation of most Quaternary sediments. Mass-wasting quickly conceals newly made exposures and vegetation soon covers them up. Hence an exposure cannot be revisited with assurance that its critical relations will still be visible.

The lateral variability and nonconsolidated state of many Quaternary strata have led to the use of special techniques for sampling, stratigraphic study, and mapping. These include the use of boring tools, notably the hand auger, for tracing the continuity of sedimentary units concealed by vegetation and by thin overlying deposits. Power-driven augers mounted on trucks, capable of boring rapidly through unconsolidated sediments to depths of 100m or more, have made it possible to determine subsurface stratigraphy where low relief and blanketing surficial material preclude the existence of many exposures. The raising of undisturbed cores more than 20m long from sea-floor sediments has revolutionized the study of Pleistocene marine stratigraphy. Core-boring of organic sediments from lakes and bogs has made possible the reconstruction of ecologic and climatic sequences, especially for the last 15,000 years. Finally, geophysical subsurface-exploration methods are used successfully in the recognition of buried topography and buried bodies of distinctive sediment.

Despite the special difficulties and the techniques employed to cope with them, we must not lose sight of the fact that interpretation of Quaternary strata follows the principles established, through a period of much more than a century, for strata of all ages.

OBJECTIVES OF QUATERNARY STRATIGRAPHY

The ultimate objective of the study of strata of any kind is threefold: (1) identification of units and their placement in sequence, (2) their lateral correlation, and (3) determination of *chronology* (dates and elapsed times) through as much of the sequence as possible. We shall deal with chronology in chap 15; here we are concerned with identification, sequence, and correlation. For these activities a descriptive scheme or framework is necessary, much as a catalog is necessary to a large library. The scheme must be one that can be understood and followed by specialists from various scientific disciplines, people who may be working on the same problem or in the same place at different times. The scheme must be unambiguous, adapted to the widest variety of stratigraphic situations, and easy to use.

Such a scheme has been set forth, in the form of a code embracing the whole geologic column, by a nonofficial North American group (American Commission on Stratigraphic Nomenclature, 1961). It has been adapted specifically to the Quaternary by the same group (Richmond and others, 1959). It is used widely (in principle, though not necessarily in full detail) in North American countries and has been adapted in a number of other countries.[1] A detailed recommendation for an adapted version, for use throughout Africa, is contained in Bishop and Clark, eds., 1967. The essence of any such code, in theory and in practice, is the recognition and complete separation of two different kinds of units. (1) Rock-stratigraphic units (and in "rock" we include all unconsolidated sediments). These are tangible, material bodies that possess recognizable, reproducible boundaries. (2) Time-stratigraphic units of rock. These are conceptual because they are limited above and below only by imaginary surfaces that represent divisions of time and that may or may not coincide with physical discontinuities within the rock. The distinction between these two kinds of units is essential to precise correlation, as will be shown presently.

[1] For one European scheme, emphasizing glacial stratigraphy, see Lüttig, 1964b.

Stratigraphic codes also recognize other kinds of units, as described below, and in all cases insist on sharp definition and careful application of stratigraphic terms. Stratigraphic units must be defined and described with reference to an existing *type section* at a specified locality, at which the unit so defined can be seen by others. Use of such a code eliminates the differences of terminology and differences of method that now exist among various disciplines in studying strata. Hence it improves reproducibility of description and recognition.

Relationship of Strata to Time

Early in the history of stratigraphy all the units recognized and described were rock stratigraphic. Two or more sequences of such units in different areas were (and in a great many cases still are) correlated on the assumption that the upper and lower surfaces of a rock layer in one place are contemporaneous with the corresponding surfaces of a layer in another place. The assumption is true only rarely, because very few rock units are *time parallel.* Probably the best example of a time-parallel stratum is a layer of tephra deposited during a single volcanic eruption; some such layers are deposited in no more than a few days. Layers of tephra are therefore eagerly sought as key horizons within the Quaternary sequence. Examples include the Bishop Tuff in California, the Pearlette Ash (possibly twofold) in central United States, the tephra from an eruption of the Glacier Peak volcano in Washington and from the later Mt. Mazama catastrophe in Oregon (e.g., Powers and Wilcox, 1964; Westgate and Dreimanis, 1967), and various layers of tephra in Iceland and beneath the floor of the North Atlantic Ocean. Some such layers are identified petrographically as related to a particular eruption.

Most strata, indeed, are *time transgressive;* that is, their boundaries cut across lines that represent moments of time. For example, the lower surface of a major layer of till (fig 14-1) is younger at its periphery than in the upstream direction, whereas with the upper part of the layer the reverse is likely to be true. The same applies to deposits made during a transgression of sea upon land and during the subsequent regression. In the same figure, the till layer forms a wedge projecting into a thick section of loess deposited during the glaciation responsible for the till. The base of the till is transgressive upon the loess underlying it, whereas the upper part of the loess is transgressive upon the till and was evidently deposited during deglaciation. Both contacts between till and loess, therefore, are time transgressive.

The dilemma caused by the almost universal time transgressiveness of strata is met by the creation of the time-stratigraphic units already mentioned, and by calibrating them approximately in terms of absolute time.

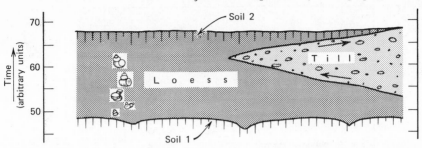

Figure 14-1 Idealized diagram showing two rock-stratigraphic units (loess, till), two soil-stratigraphic units (Soils 1, 2), and a faunal zone based on a homogeneous group of land snails. The faunal zone is stratigraphically coincident with the loess unit, which is also a time-stratigraphic unit, as is evident from its near-parallelism with the time lines (horizontal broken lines) established by C[14] dates. Although the base and top of the loess body are nearly time parallel, those of the till are time transgressive; the till was deposited by a glacier whose outer edge advanced (lower arrow) over loess and later retreated (upper arrow) through a period of at least 15 time units.

Two bases for calibration exist: radiometric dating (chap 15) and degree of evolutionary development of the fossils contained in the strata. Neither basis is precise, although radiometric dating is the most accurate, at least in Quaternary strata. Time-stratigraphic units, therefore, are defined by radiometric dates, by the fossils the strata contain, or by physical continuity with strata that do contain fossils. Hence many time-stratigraphic units correspond essentially with faunal or floral assemblages (table 14-A, column 4).

Kinds of Stratigraphic Units

Rock-Stratigraphic Units. The general terminology recommended for Quaternary strata is given in table 14-A, and examples appear in tables 14-B, 21-A, and 21-D. We have already defined as a category the rock-stratigraphic units shown in column 1. The *formation* is the basic unit; two or more of these constitute a *group,* and a formation can be subdivided into the smaller units named.

Time-Stratigraphic Units. Nearly all the units shown in column 1 are time transgressive and all are finite in extent. If we are to correlate them even approximately from one place to the next, let alone from one continent to another, we must relate them to a time sequence. This is the reason why time-stratigraphic units (column 5) were invented. We repeat that they are independent of the rock units and represent, by tangible thicknesses of rock, units of time contemporaneous throughout the world.

Table 14-A Terminology Recommended by ACSN for Quaternary Stratigraphic and Time Units. (After Richmond et al., 1959. Examples have been added.)

1	2	3	4	5	6	7
Rock–stratigraphic	Glacial–[b] stratigraphic	Soil–stratigraphic	Biostratigraphic	Time–stratigraphic	Geologic time	Radiometrically determined time
Group	Glacial or glaciation	Soil	Assemblage zone (marine invertebrates)	System	Period	(C14, K/Ar, or other) years BP
Formation[a]	Interglacial or interglaciation			Series	Epoch	
Member	Stadial or stade		Provincial mammal age (terrestrial mammals)	Stage or provincial stage	Age or provincial age	
Lentil	Interstadial or interstade			Substage or provincial substage	Sub-age or provincial sub-age	
Tongue						
Bed[a]						
Examples						
Wisconsin till	Wisconsin glaciation	Wisconsin soil		Wisconsin (Glacial Stage) (precisely defined)	Wisconsin Glacial Age	

[a] Or descriptive lithologic name such as till, outwash, gravel, loess.

[b] The term glacial–stratigraphic is substituted for the ACSN term climate–stratigraphic in order to divorce the term glaciation from exclusively climatic interpretation, and to permit its definition as given in chap 5. Not all glaciations are controlled by climate.

Just as terms such as formation and group are confined to rock units, so time-stratigraphic units are distinguished by exclusive terms. According to common usage at present, the Quaternary System is made up of a Pleistocene Series and a Holocene or Recent Series. Each of these in turn can include stages such as the Wisconsin Stage, and each stage can include substages.

Geologic-Time Units. The time involved in the formation of the strata embraced in a particular time-stratigraphic unit is a geologic-time unit (column 6); commonly it bears the same name as the corresponding time-stratigraphic unit. Thus the strata of the Wisconsin Stage were deposited during the Wisconsin (Glacial) Age, and those of a particular substage during a sub-age of the same name.

Some authors add the adjectival endings *an* or *ian* to the names of both time-stratigraphic units and the corresponding geologic-time units, to distinguish them still further from rock units. To others this refinement seems both unnecessary and cumbersome; with some geographic names it results in barbarisms such as *Wisconsinan* and *Twocreekan* (for Two Creeks). In this book we shall use the simpler form without adjectival endings, except for names that were first used or defined with such endings.

Biostratigraphic Units. Another category of stratigraphic units is the group of *biostratigraphic* units, which are characterized solely by their content of fossils. These are of several sorts. The unit most commonly mentioned in Quaternary stratigraphy is the *provincial mammal age*. This consists of the time represented by strata, in a particular region, characterized by a particular assemblage of fossil mammals. Examples are seen in table 29-A.

Soil-Stratigraphic Units. Ancient soils (chap 11) create a stratigraphic problem rarely met with in pre-Quaternary strata. They are not deposits, as are bodies of alluvium or loess, but are alterations of pre-existing material. To complicate matters further, the surface beneath which a soil develops normally truncates strata of more than one age (fig 11-6). Hence a soil is not a rock unit, and although it transgresses various strata, it may be nearly time parallel. In this respect it is analogous to an assemblage zone, characterized by distinctive fossils though it may cut across rock units. We call it a soil-stratigraphic unit (column 3), but give it a geographic name (e.g., Sidney Soil) like that given to a formation. Probably most soils are essentially time parallel.

Glacial-Stratigraphic Units. It is also necessary to be able to refer to glacial events, generally inferred from (time-transgressive) rock-stratigraphic units such as tills. Thus far, glacial-stratigraphic units (column 2) have been named principally in glaciated regions, and are expressed in terms of *glaciations* and *interglaciations,* either one of which can be sub-divided into *stades* and *interstades.* A glaciation, then, is a time of major expansion of glaciers, inferred from geologic data; a stade is a time of less marked expansion of glaciers within a glaciation. Thus we can speak of the Late stade of the Pinedale Glaciation. This implies no quantity of time expressed by a number, but rather the general time involved in that glacial event.

Geopolarity Units. A special kind of stratigraphic unit, restricted to volcanic rocks, is defined by the polarity of the Earth's magnetic field as reflected by the orientation and polarity of particles of magnetite and other magnetic minerals present in the rocks (Cox, 1969). The particles, as they formed and settled during solidification of magma, assumed ori-entations in conformity with the existing field, and since solidification they have retained a "memory" of that field, expressed by remanent mag-netism which can be measured. The polarity of the Earth's field has re-versed repeatedly during later Cenozoic time, and the reversals have been dated approximately by K/Ar methods, applied mainly to core samples. Many dated reversals have occurred at irregular intervals during the last four million years. The data on these make possible a time sequence based on polarity (fig 14-2). The large units in the sequence consist of *geopolarity epochs* (distinct from the epochs in the conventional geologic column) which represent long intervals predominantly of normal or re-versed polarity. These are subdivided into shorter intervals called *geopo-larity events,* during which polarity was the reverse of that which pre-dominated during the pertinent epoch. Epochs carry the names of persons; events carry the geographic names of type localities. Use of the geopolarity sequence in stratigraphic correlation is illustrated in table 20-A.

The geopolarity time sequence constitutes a worldwide framework into which dated strata of many kinds can be fitted, the common denominator being radiometric dates. The sequence is therefore a major factor in strat-igraphic correlation of Quaternary and even pre-Quaternary strata.

Correlation of Units

The correlation of strata from one place or region to another is accom-plished by five different means; these involve the use of fossils, artifacts, radiometric dates, temperature data, and rock-stratigraphic units.

Fossils. Of course, in a broad sense organisms are time transgressive, but in Quaternary strata they are less so than rock strata are. The empiri-cal experience of generations of geologists has shown that in a broad way assemblages of fossil animals and plants are the best available basis for time correlation. In the older strata this is partly the result of imperfec-

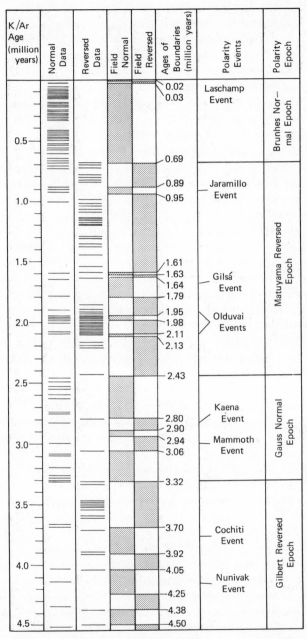

Figure 14-2 Geopolarity time sequence. Intervals when polarity of the geo-magnetic field was normal are shaded; intervals of reversed polarity are shown in white. (After Cox, 1969.)

tion of the stratigraphic record: much of the rock succession has been destroyed by erosion or remains buried far below the surface, so that many physical gradations and transitions are not apparent.

Nevertheless, in Quaternary strata correlation by means of fossils encounters special difficulties. Rates of change of Quaternary environments were generally more rapid than rates of evolution in Quaternary organisms. The same faunas may appear repeatedly in successive strata, and their transgression of time is commonly evident. For example, during and after the last major glaciation many large mammals in North America became extinct, whereas fewer extinctions occurred in East and South Africa. To the scientist of the future these facts could suggest correlation of the extinct North American fauna with the African fauna of today.

This difficulty does not imply that specific faunal and floral assemblages cannot be recognized within the Quaternary. On the contrary, Quaternary mammal faunas in western North America have been subdivided into the three assemblages shown in table 29-D, and mammal faunas in Europe have been subdivided similarly in several provincial regions. These subdivisions, like those in pre-Quaternary strata, are based on degree of evolution of the fossil organisms. We should note, however, that in Quaternary strata the occurrence of two successive strata, each stratum with a different assemblage of related fossils, can be shown very commonly to have resulted from nothing more than the invasion of an area by a new assemblage, resulting from change of climate or of some other environmental factor, without any evolutionary change in the organisms having occurred. Useful inferences can be drawn from such occurrences, which, however, have little to do with evolution.

Floral zones based on fossil pollen are discussed in a later part of this chapter.

Artifacts. In Europe, central and southern Africa, and elsewhere, attempts have been made to base correlation on the known succession of prehistoric assemblages of artifacts (*not* individual artifacts) some of them extending far down in the Quaternary sequence. The reliability of such correlation has been much debated by archeologists, among whom opinions differ. Assemblages of artifacts can be time transgressive; yet distinct stratigraphic sequences of such assemblages within specific regions have been identified, and if used with caution could become very useful for correlation. The accumulation of related C^{14} dates supports this opinion.

Radiometric Dates. The determination of dates expressed in numbers through C^{14}, K/Ar, and other radiometric measurements (chap 15) makes possible time correlation between two units that may lack fossils, be physically dissimilar, and be situated far apart, even in different continents. Stratigraphic sequences on a scale so minute as not to be correlatable in pre-Quaternary strata have been established and correlated by means of radiometric dates. Such dates constitute probably the most promising method of bringing a true time sequence to the stratigraphy of the Quaternary.

Temperature Data. On the assumption, not yet fully established but increasingly probable, that change of climate, particularly temperature, has been contemporaneous throughout the world (chap 16), evidence of temperature fluctuation can serve as a basis of time-stratigraphic correlation. The repeated fluctuation of temperature with time, discerned in deep-sea sediments, supports the assumption. Indications of temperature are implicit in glacial drift and in many fossil-bearing sediments, notably those with pollen sequences and marine invertebrates. Temperature correlation based on alternating glacial and nonglacial sediments has been and still is the mainstay of the stratigraphic classification of Quaternary strata. But as such correlation involves only rock-stratigraphic units, it is a time classification only in a crude way.

Rock Units. The development of Quaternary stratigraphy, like any other, begins with physical description. Although a very large number of rock units have been described, general time-stratigraphic correlation is still in the future. However, as indicated in a foregoing section, several kinds of rock units imply relatively high or relatively low temperatures. Thus the existing rock-stratigraphic classification implies in part an embryonic time-stratigraphic classification, if we assume reasonably that major temperature changes are broadly synchronous throughout the world.

The rock units most useful in correlation, as we have noted, are time-parallel units such as tephra deposited from a single eruption. Some tephra layers are datable by K/Ar and thus add further to their value as key horizons. Identification and correlation of tephra layers in western United States are discussed by Wilcox (p. 807–816 in H. E. Wright and Frey, eds., 1965a).

General Comments. Time-stratigraphic classification is more difficult to achieve in Quaternary than in older strata because (1) the best-known Quaternary strata are far more varied, far less continuous, and still very abundant, and (2) evolutionary differences among fossils are generally smaller and less conspicuous than their facies differences—the differences related to environment. Description and definition of the lithologic units can proceed, but much progress must be made before fully satisfactory time-stratigraphic units can be agreed on.

Table 14-B shows how a stratigraphic chart can be set up with data from two adjacent regions in which the sequences are rather well understood. Although in the table only the Wisconsin Stage is represented, text

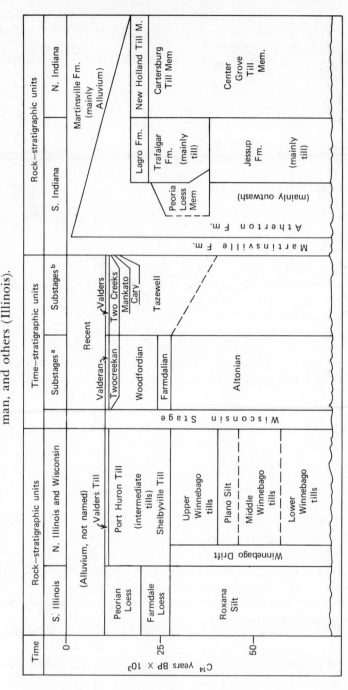

Table 14-B

Example of Correlation Between Two Adjacent States Through the Use of Time-Stratigraphic Units and C[14] Dates. (Adapted from W. J. Wayne, 1963, p. 14 (Indiana) and various publications by J. C. Frye, H. B. Willman, and others (Illinois).

[a] According to Frye, Willman, and others. Without the adjectival endings the names would read: Valders, Woodford, Farmdale, Alton.

[b] According to Leighton (1960, p. 550)

378

descriptions of many of the units are given in Wayne, 1963. As the stratigraphy is primarily glacial, no biostratigraphic units are named; however, further study of the nonmarine mollusks that are locally abundant could lead to the erection of a detailed biostratigraphic column. Such a column would probably play a large part in a chart relating to a nonglaciated region. Also some material exists for a soil-stratigraphic column.

In some regions the state of knowledge is such that time-stratigraphic units, if created, would be of little use. In them correlation, admittedly imprecise, is based on rock units.

TERMINOLOGY OF POST-PLIOCENE STRATA

History of Current Usage

The terminology of post-Pliocene strata, their equivalents and their subdivisions, had its inception about 1822 and evolved in conformity with growing knowledge. At present the major geologic-time divisions commonly current in the English-speaking world are:

Cenozoic (Kainozoic) Era
- Quaternary Period
  - Recent (Holocene) Epoch
  - Pleistocene ("Glacial") Epoch
- Tertiary Period   Pliocene Epoch and earlier epochs

The scheme now current is essentially of mid-19th century origin. Its evolution began with the introduction of the term *Quaternary* by Desnoyers as an addition to the standard designations *Primary, Secondary,* and *Tertiary,* all of which had been in use since 1760. Desnoyers (1829) applied the term to the strata that overlie the Tertiary rocks in the Paris basin. Soon afterward Reboul (1833) extended the term to include the strata characterized by living species of animals and plants, as distinguished from the Tertiary, whose fossil remains, he thought, "almost all belong to extinct species." [2]

Reboul's definition, we may note, was the first one to have a faunal basis. He subdivided the Quaternary as follows:

Quaternary period
- Historic epoch
- Anti [sic]-historic epoch

Six years later Lyell (1839, p. 621) introduced the term *Pleistocene,* applying it to strata whose content of fossil mollusks included more than

---

[2] That this belief was not soundly based was shown, across the English Channel, in that same year by Lyell (1830–1833, v. 3, p. 52–58), who was just then introducing the terms *Pliocene* and *Recent.*

70% of living species. Thus both Quaternary and Pleistocene were early identified with faunas.

The Glacial Theory, however, introduced a different basis of definition. In the same year in which that theory (chap 2) appeared, Schimper (1837) recognized a Glacial Period (*Eiszeit*). Shortly afterward Forbes (1846, p. 402) redefined the Pleistocene Epoch as equivalent to the "Glacial epoch"—"the time distinguished by the presence of severe climatal conditions through a great part of the northern hemisphere." Although Forbes thought of the "Glacial epoch" as a time of vast iceberg-laden seas rather than of great ice sheets, he was the first to define the Pleistocene in terms of climate.

In a paper read in 1854, Morlot (1856) subdivided the Quaternary ("Quartaire") into

$$\text{Quaternary period} \begin{cases} \text{Modern epoch} \\ \text{Second glacial epoch} \\ \text{Diluvian epoch} \\ \text{First glacial epoch} \end{cases}$$

and restricted it to the post-Pliocene. This classification, too, was glacial rather than faunal. Much earlier, Lyell (1830–1833, v. 3, p. 52) had introduced the term *Recent epoch,* defined as the time "which has elapsed since the earth has been tennanted by man" and recognized as younger than the Tertiary. As thus defined it included both the Pleistocene and the Recent (Holocene) of the classification commonly current today; likewise the definition was faunal. His later introduction of Pleistocene was really a substitution of Pleistocene for Recent, that is, synonymous with post-Pliocene. Forbes' subtraction of the Pleistocene from the Recent was long afterward agreed to by Lyell (1873, p. 3).[3] This reduced the Recent to the equivalent of postglacial.

Present usage is based essentially on the state of affairs dating from 1873. As defined by the U. S. Geological Survey. (Wilmarth, 1925, p. 45) the Quaternary System includes both Pleistocene and Recent, is characterized by animals and plants of modern types, and its time equivalent, the Quaternary Period, is commonly called the "Age of Man." The Pleistocene Series (time equivalent=Pleistocene or "Glacial epoch," also inaccurately called the "Ice Age" or "Great Ice Age") includes the extensive glacial strata of the northern hemisphere and contemporaneous rocks. The Recent Series includes the strata created since the disappearance of the former great ice sheets.

[3] The confusion resulting from the checkered history of these terms is succinctly stated by C. B. Schultz and Stout (1948, p. 571–574).

In France, a country without extensive glacial deposits, the tendency has been to define and subdivide the Quaternary on a basis of fossils. For example, Haug (1907–1911, p. 1776) regarded the Pleistocene Epoch as beginning with the appearance and intercontinental spreading of three new mammals: Bos [Leptobos], Elephas [Mammuthus (Archidiskodon)], the earliest true elephant, and Equus (one-toed horses of zebrine, caballine, and asinine kinds), and also with the appearance of man. This view was later supported by others (e.g., Kurtén, 1960, p. 3).

In German usage the classification is the same as that current in the English-speaking world, although the terminology differs. For Pleistocene Series Germans use Diluvium, and for Recent, Alluvium. This terminology had its origin not in Germany but in England, apparently with Mantell (1822, p. 274), who classified the surficial deposits into Diluvium (sediments laid down by agencies no longer operative, such as the Biblical flood) and Alluvium (sediments laid down by agencies still in force, such as existing streams). In German usage Diluvium (time equivalent = Eiszeit = Glazialzeit) and Alluvium (time equivalent = Postglazialzeit) are grouped together to form the Quartär (= Quartärzeit) in analogy with English usage.

Russian usage employs the term Antropogen as synonymous with Quaternary; within it are Pleistocene and Holocene.

Terms and definitions that were proposed in North America but never widely used are reviewed in Flint, 1957, p. 281.

The Pliocene / Pleistocene Boundary

The fixing of a satisfactory boundary between Pliocene and Pleistocene presents serious problems which have not yet been resolved. The most fundamental problem arises from the fact that Pliocene strata are mostly transitional into the Pleistocene, thus opening the way to a number of possibly arbitrary choices. Reviews of the matter are given in Flint, 1965 and R. C. Moore, 1949.

From the historical summary in the preceding section it can be discerned that from before the middle of the 19th century two distinct bases for defining the lower limit of the Pleistocene Series were in use. One consists of evolutionary differences between fossil organisms; it was implicit in the proposal by Reboul (1833), and is explicit in recent attempts to fix a boundary by means of planktonic microfossils. The other consists of evidence of climatic cooling, expressed mainly in glaciation. It began with Forbes' (1846) explicit equation of Pleistocene with Glacial. This definition has had many followers, and persisted in apparently good standing into the late 1960's. However, it is incompatible with two significant discoveries in high latitudes. One consists of evidence of migration of cold-water marine faunas and terrestrial floras into lower latitudes. The other consists of glacial sediments. Both are found in strata of indisputably Pliocene and even later Miocene age, radiometrically dated at as many as 10 million years BP. These discoveries show that Cenozoic glaciation is not exclusively Pleistocene and hence that Forbes' equation is not valid. Glacial climates are not a proper basis for fixing the Pliocene/Pleistocene boundary. This leaves us with the state of evolution of organisms, probably fortunately in that it is the basis on which all pre-Pleistocene stratigraphic boundaries are determined and so is consistent. If a nearly isochronous boundary can be found, well and good. Meanwhile no generally agreed-upon boundary exists. A realistic alternative to the continued emphasis on the base of the Pleistocene has been advocated by, among others, Dunbar and Rodgers (1957, p. 297), Flint (1947, p. 204), Hays and Berggren (unpublished), and West (1968, p. 225) is to eliminate Tertiary and Quaternary from the geologic column and to define Pleistocene in faunal terms, as was done by Lyell in 1839. Although this would do nothing to fix a boundary, it would make the existing uncertainty seem less serious.

Holocene (Recent)

Definition of the upper limit of the Pleistocene encounters difficulties as serious as those facing definition of the lower limit. Most scientists concerned with geologic stratigraphy draw a stratigraphic boundary between a Pleistocene Series and a Holocene (or Recent) Series, but there is no general agreement on its position. One of the more precise definitions places the base of the Holocene at the beginning of deposition of autochthonous continental sediments following deglaciation. Neustadt (p. 421 in Cushing and Wright, eds., 1967) proposed that the base of the Alleröd (table 24-C) be accepted as marking that event. Such a boundary, however, is time transgressive, and in a wide region the Alleröd sediments were glaciated during a major readvance. Many stratigraphers believe the base of a Holocene unit should be fixed by physical features dating from 10,000 years BP or somewhat earlier.

The concept of such a boundary stems from an attempt by Forbes (1846) to discriminate between glacial and postglacial. Forbes was concerned with a limited district, within which the boundary was distinct and perhaps nearly time parallel. Agassiz (1847, p. 233) took a broader view. In the same paper in which he elaborated the classic Glacial Theory he considered the problem of a boundary between glacial deposits and those of today, and predicted that it would be very difficult to solve. He was right, for throughout any wide region both glaciation and deglaciation are time transgressive. Hence any boundary based on rock-stratigraphic units is likely to be arbitrary, although with enough radiometric dates time boundaries could be drawn through strata representing many different facies. Even in the 20th century a Pleistocene/Recent boundary was defined vaguely, for example, as "the time since the retreat of the last ice sheet," and as the time when many large mammals in North America became extinct. Each of these events involved more than 10,000 years, an interval during which complex small-scale stratigraphic sequences were deposited in many areas.

Other proposed bases, although defined more sharply, are still necessarily arbitrary and some are of purely local significance. Among them are (1) the time when the shrinking Scandinavian Ice Sheet separated into two parts at Ragunda in central Sweden, (2) the close of the Lake Algonquin phase of the Great Lakes sequence (table 21-F), and (3) the beginning of the Flandrian eustatic rise of sealevel (Reade, 1872, p. 111; several subsequent authors). The latter event was contemporaneous everywhere, and has value locally, but its usefulness is limited by the difficulty of applying it stratigraphically. It is not recognizable in continental interiors, and along coasts the evidence, even where uncomplicated by crustal movements, is likely to be recognized only in borings below present sealevel.

Enquist (1918, p. 82, 94, 101, 102) proposed that the "Postglacial" in Fennoscandia began with the deglaciation of the Fennoscandian end moraines. This point is less arbitrary than some other proposals because a rapid temperature rise began around that time. A nearly identical postglacial time unit, beginning with the start of Preboreal time, is in use among pollen stratigraphers in northern Europe (table 24-C). A similar basis was applied to western North America by Antevs (1948, p. 176), who suggested a Neothermal time unit, occurring within the Pleistocene and having three subdivisions based on temperature criteria. For the same region Hunt (1953, p. 20) proposed a Pleistocene-Recent boundary at a stratigraphic change marking the onset of arid conditions after the latest glacial maximum. He thought that significant mammal extinctions were related to this boundary.

The foregoing discussion illustrates a real objection to assigning to the postglacial the value of a series, in that it has not yet been recognized as a time-stratigraphic unit. If the evidence were obscure, like the evidence of the onset of late-Cenozoic glaciation, we should be justified in generalizing our interpretation of it, but it is so detailed and so abundant that we cannot close our eyes to the gradual transition it records.

In view of the time-transgressive character of glacial and "deglacial" conditions, we favor, with Kay and Leighton (1933, p. 672), Dunbar and Rodgers (1957, p. 298) and West (1968, table 11.3), the concept that the Pleistocene should embrace all post-Pliocene time. Also we think it unwise to employ the terms glacial, postglacial, and recent in a formal stratigraphic sense. We prefer the use of local geographic names for stratigraphic units, freeing the three adjectives for their proper informal use. Events that occurred within any district while it was covered with glacier ice are referable to glacial time for that district; subsequent events are there referable to postglacial time. Since removal of the glacier cover was progressive, it follows that postglacial time began for some districts much earlier than for others, and that for most of Antarctica and Greenland it has not yet begun.

The term postglacial can properly be applied to events in regions that were never glaciated, but in which the indirect effects of glaciation are clearly marked, as for example the valleys of large streams draining glaciated regions. The cessation of outwash deposition and the beginning of trenching of the outwash or deposition of alluvium is an event that can be recognized in broad regions and that can be taken to mark the transition from glacial to postglacial time in a valley system. Where there is no direct connection between a glaciated and a nonglaciated district, such as is afforded by outwash, the term postglacial is hardly applicable to any feature in the nonglaciated district. For arid basins of western United States, where Pleistocene features record alternating episodes of moisture and dryness, terms such as Pluvial and Postpluvial have been suggested.

Summary

In this book we have used the term Quaternary in the title and at other points where the post-Pliocene is referred to. We have done this in order to facilitate understanding of the subject matter, inasmuch as Quaternary is the term in greatest common use. Nevertheless we dissent from common practice in that we favor the dropping of Tertiary System and Quaternary System from stratigraphic nomenclature. If this were done, the Pleistocene Series would include all post-Pliocene strata, as implied by Lyell in 1839. Further we believe, in view of the long time span of the succession of late-Cenozoic cold climates, that the Pliocene/Pleistocene boundary should not be based on climatic indications. Finally we think the terms recent and postglacial should be used only informally, and applied only within geographically restricted areas. We have little hope that these changes will come about soon, but we think they are soundly based.

POLLEN STRATIGRAPHY [4]

Palynology, the study of pollen and spores and their dispersal, has led to pollen stratigraphy, which has contributed much to our understanding of interglacial climates and vegetation, as well as to more detailed knowledge of climatic events in Europe during the last 15,000 years. Fossil pollen, identified in peat as early as 1885, soon began to be used in stratigraphy. Systematic pollen stratigraphy dates from studies made on Swedish peat bogs by von Post (1916).

From the point of view of the general science of the Quaternary, the principal objectives of the study of fossil pollen are reconstruction of the environments (especially the climates) in which former plants lived, and correlation of pollen-stratigraphic zones from place to place.

Technique [5] and Assumptions

When they flower, many trees and other plants shed pollen or spores in quantity. The grains, 0.1 to 0.01mm in diameter, are dispersed in various ways, often by the wind. Those grains which settle in lakes and bogs, and in the ocean as well, are protected against oxidation and are incorporated in accumulating sediments, especially *gyttja* (organic-rich lacustrine mud) and peat. The outer walls of pollen grains are remarkably durable both physically and chemically, and some grains can be identified not only as to genus but also to species and even subspecies. The grains are thus extremely useful microfossils, for they represent assemblages of plants that existed in the vicinity of the deposit at the times of deposition. Usually it is possible to draw qualitative inferences from pollen assemblages about former climates. In some instances one can even extract specific information from the occurrence of particular pollen species. For instance the tundra plant *Armeria sibirica,* whose pollen was collected from a stratum more than 12,000 C^{14} years old in southeastern Massachusetts, is today restricted to high-Arctic tundra in northern Canada, where the mean July temperature is lower by at least $10°$.

Peat and gyttja are very rich in pollen, while lacustrine inorganic sediments are less so. Pollen in deep-sea sediments is sparse, although, having been transported in part by rivers, it can represent in a qualitative way elements of the vegetation of entire watersheds.

Pollen grains have two substantial advantages over plant macrofossils,

[4] General references: Faegri and Iversen, 1964; West, 1968; Ogden, 1965; M. B. Davis in H. E. Wright and Frey, eds., 1965a, p. 377; M. B. Davis, 1969.

[5] Comments on technique in Mehringer, 1967, p. 132–146.

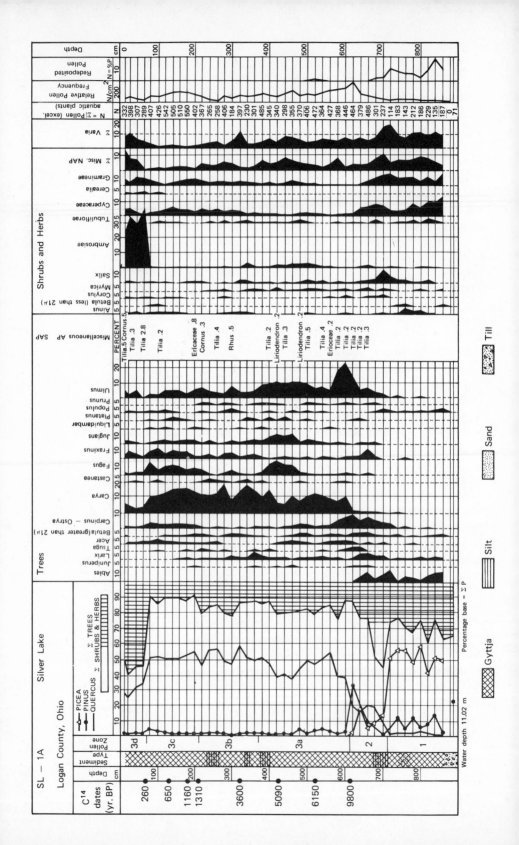

most of which represent the special environments of swamps and flood-plains where macrofossils can be preserved. Pollen grains can indicate the floras of uplands; they also lend themselves to quantitative treatment; they can be easily counted and dealt with by statistical methods.

The technology of collection and study was originally developed in northern Europe. The grains are extracted from sediments exposed at the surface or raised in cores from lake sediments or bog peat, and even from the sea-floor sediments described in chap 27. Lake sediments are better than bog peat, and basins of other kinds are better than kettle basins be-cause in some kettles persistence of ice delayed the beginning of deposi-tion of sediment. Cores are taken with great care to avoid contamination of one layer by another. In the laboratory samples from selected strati-graphic intervals are concentrated and examined under the microscope. At each interval generally 200 pollen grains (and in some studies thou-sands) are identified and counted. Usually the microfossils are divided into three parts, arboreal pollen (AP), nonarboreal pollen (NAP), and spores. Of these AP possesses the advantage that most trees shed more pollen and release it higher above ground than do shrubs and herbs.

The data resulting from identification and counting at any strati-graphic position can be expressed as percentages of total pollen or of total AP at that position, or as number of grains per cubic centimeter of sediment. The general plant assemblage that contributed the pollen at the relevant time is a *pollen spectrum*. Histograms of successive spectra can be connected to form a *pollen curve*. European scientists observed that pollen spectra collected at the surface of a bog agree qualitatively with the composition of the plant communities now living in the sur-rounding region.

The curves from a series of stratigraphic positions in a core are plotted in sequence against core depth (i.e., depth below the surface of the bog or below the lake or sea floor), to form a *pollen diagram* (figs 14-3, 14-4). Such diagrams, showing the percentage composition of the pollen "floras" of successive strata in a core, make it possible to perceive the sequence of development from one type of vegetation to another. They enable a pol-len analyst to draw inferences about changes in climate, other physical changes, and changes brought about by deforestation and other activities of man. The relation of a pollen "flora" at some horizon within a core to

Figure 14-3 Pollen diagram from a core, 9m long, raised from the floor of Silver Lake in central Ohio. All pollen types are plotted as percentages of total pollen. Deglaciation of the locality is recorded by the contact of gyttja on till. The flood of *Ambrosiae* (ragweed) in the uppermost 70cm of the core reflects clearing of forests by Europeans. (Ogden, 1966, pl. 1.)

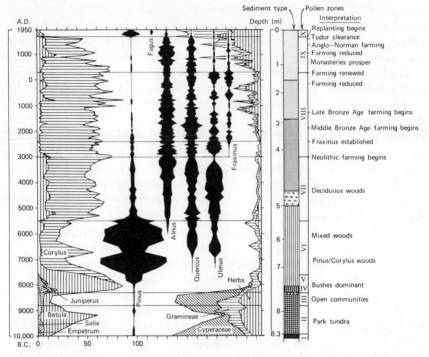

Figure 14-4 Pollen diagram of unconventional design from a core, 8.3m long, from Littleton Bog, County Tipperary, Ireland. Black and shaded areas show percentages of total pollen; along any horizontal line they add up to 100. Shaded areas represent bushes and woody shrubs (left) and herbs (right). Black areas in center are trees. Time calibration is based on eight C[14] dates from the core. The pollen zones at right are those shown in table 24-C. Note diagram is scaled, not to 100% but only in units of 10%. (G. F. Mitchell, p. 6 in H. E. Wright and Frey, eds., 1965b.)

the true flora of the region at the time of deposition of the pollen is not one-to-one, and in some cases may be far from being so. Particular species may be over- or underrepresented in the pollen spectrum for several reasons. (1) Some trees shed hundreds of times more pollen grains than others do; (2) some propagate the pollen more effectively; (3) transport mechanisms (wind velocities, streams, lake currents) can vary. Hence diagrams include distortions and their interpretation cannot be fully objective. On the other hand peat and gyttja are C[14] datable within the time range of the method, so that apart from the comparison of pollen diagrams, many pollen-bearing strata can be correlated through any distance by radiometric dates. Despite many technical difficulties pollen analysis has contributed material of great value to stratigraphy.

Some of the difficulties are minimized by major improvements in techniques. Such improvements include: (1) comparing the pollen spectra for a locality at various times in the past with the spectrum for the same locality at present—that is, with today's *pollen rain;* (2) feeding into the pollen diagram the effect of variations in rates of sediment accumulation, through the use of C^{14} dates; (3) correcting for differences in rates of pollen production by different species; and (4) preparing sediments in such a way that diagrams can specify absolute number of pollen grains per unit volume of sediment. This is an alternative to percentage calculation in which reduction of the count for one species automatically increases the proportion of other species.

Even with these improvements, however, pollen analysis cannot record change in the environment—perhaps least of all change of climate—with uniform fidelity. Probably such changes are recorded with least ambiguity at or near an ecotone (a boundary) such as that between tundra and subarctic forest, prairie and temperate forest, or savanna and tropical forest. In contrast, plant communities such as certain tropical forests, that cover large areas and consist of a wide variety of species, may not record climatic change in their central areas.

Even were these difficulties fully overcome, there remains the assumption that a former plant assemblage possessed the same internal interrelations and the same general environment (including climate) as the most nearly similar assemblage living now. Under this assumption the climate of the area in which the assemblage now lives is considered to have prevailed when and where the fossil assemblage lived. This broad inference would be modified by an experienced palynologist through giving, in each case, more weight to some factors and less to others.

Just as some glacial events are surges apparently unrelated to climate, so some changes of vegetation are independent of climatic causes. Two common examples are the inception and later changes of agricultural practice and the natural and artificial burning of vegetation. Hence all other likely causes of changes in vegetation recorded by pollen must be eliminated before changes in climate can be established with confidence.

Pollen Stratigraphy in Northern Europe

We mentioned in chap 2 the postglacial stratigraphy, developed for Scandinavian bogs, that forms the Blytt-Sernander sequence (table 24-C). The validity of the sequence, originally based on megafossils, was strengthened by von Post (1916), who pioneered in the statistical treatment of pollen in the bog strata. Later, Knud Jessen refined the sequence in Denmark by subdividing it into nine pollen zones (essentially "floral ages") numbered I to IX in the table, and based on pollen statistics. These zones have been identified throughout much of northern Europe, and their numbers are in wide use. Throughout the region C^{14} dates of zone boundaries have been found to be fairly consistent. This has made possible approximate time calibration of the sequence, as shown in the table. The sequence begins with deglaciation of the Salpausselkä and correlative moraines. It appears to indicate increase of temperature to a maximum, followed by decrease, although factors other than climate probably have affected the pollen data (Iversen, 1960). It opens the door to the possibility (though at present not to the reality) of correlation of climatic change between northern Europe and other parts of the world, even in regions in which the fossil pollen are from different species. In effect, pollen zones can be used as key horizons, despite the fact that such horizons are time transgressive. Radiocarbon dates, however, indicate that the time lag from place to place, defined by the boundary between two pollen zones, is so short that the problem of time transgressiveness is apparently not serious.

However, regional differences must be taken into account. At a Mediterranean station, for example, the pollen stratigraphy does not reflect the climatic cooling that is evident in northern Europe at the base of Zone IX (Beug, 1967a). The difference is interpreted, not as evidence that cooling did not occur so far south, but rather as a result of the comparative tolerance of the forest flora at the Mediterranean locality.

Pollen Zones I, II, and III are compatible, in their climatic implications, with geologic evidence of fluctuation of the margins of the Scandinavian Ice Sheet and of British glaciers. At sites situated beyond the limit of a glacial readvance, changes in pollen spectra reflect changes in the position of the ice-sheet margin.

From dated pollen zones plotted from various localities it is seen that during the last glacial age distinctive latitudinal belts of vegetation in northern Europe lay farther south than they do today (figs 23-7 and 23-8), and also that they differed somewhat in composition from the same zones today. It seems, moreover, that during deglaciation the climate of that region rose to a peak of warmth and dryness between about 5000 and 3000 BP, thereafter becoming cooler and moister. Earlier fluctuations of climate seem to be recorded by pollen (mainly as fluctuation in the AP/NAP ratio), as indicated in table 24-C.

Pollen Stratigraphy in North America

The development of pollen stratigraphy in North America is less advanced, at least as far as interpretation of former vegetation and climate is concerned, than it is in northern Europe. Apart from the very different sizes of the two areas and the smaller number of palynologists working in America, probably the chief reason is the far greater complexity of the American tree flora. Only about 30 species of trees exist in western Europe as a whole, and only 6 species of deciduous trees in northern Europe. In contrast, North America has more than 130 species, with as many as 30 species occurring within a single genus. The pollen of some of these species are very difficult to distinguish. For example, in eastern North America 13 species of the widespread genus *Pinus* are recognized, yet most of them cannot be identified from pollen.

Therefore it happens that fossil pollen of one species can be confused with that of another having different ecological affinities. The possible effect on interpretation of former climate is obvious.

Mainly because of such complexities, climatic changes in North America are reconstructed (table 14-C) in less detail than they have been in northern Europe. Although a postglacial maximum of warmth has been discerned in the record (marked, for example, by the expansion of prairie eastward into Ohio), it is not evident in all sequences. Also apparent only in certain sequences are earlier fluctuations indicated in the glacial record (Ogden, 1965). That these are either not present or unclear in some sequences may result from the comparative distances of such sequences from the margin of the ice sheet as well, perhaps, as from floral complexity.

Interglacial and Interstadial Pollen Stratigraphy

In northern Europe, where pollen-bearing strata have been studied extensively in relation to strata of glacial origin, the climate-stratigraphic terms interglacial (interglaciation) and interstadial (interstade) are used in a fairly precise sense (Zagwijn, 1960, p. 51). *Interglacial* implies a regional temperate climate, at its maximum at least as warm as the climatic maximum of Holocene time, at the same locality. *Interstadial* implies a regional boreal climate, even at its maximum lacking temperate deciduous forest. A typical continuous climate-stratigraphic sequence deriving, with changes, from Jessen and Milthers (1928, p. 334–375), in an area outside the limit of glaciation, is shown in table 14-D.

Although pollen-bearing strata of Wisconsin interstadial age are on record, North America has as yet yielded very little pollen information from interglacial strata. The European record is far better documented. Although it shows that the successive interglacials were alike in reflecting temperate climate, the vegetation of each was distinctive, not so much in the presence of distinctive plants as in the relative proportions of recurring plants. Because climate is only one of several factors (such as rates

Table 14-C

Vegetation Sequences in Parts of North America During the Last 14,000 Years. Boxes Represent Approximate Times of Deglaciation of Specific Regions. (Adapted from a chart by J. G. Ogden.)

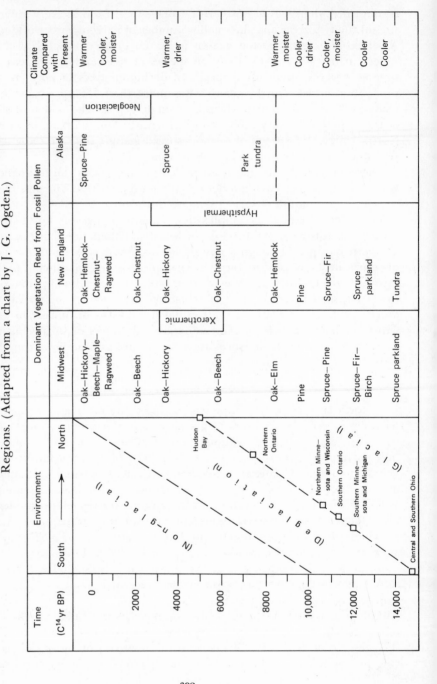

Time (C[14] yr BP)	Environment (South → North)	Midwest	New England	Alaska	Climate Compared with Present
0	(Non-glacial)	Oak–Hickory–Beech–Maple–Ragweed	Oak–Hemlock–Chestnut–Ragweed	Spruce–Pine (Neoglaciation)	Warmer
					Cooler, moister
2000		Oak–Beech	Oak–Chestnut		
4000		Oak–Hickory	Oak–Hickory	Spruce	Warmer, drier
6000	Hudson Bay	Oak–Beech (Xerothermic)	Oak–Chestnut (Hypsithermal)		
8000	Northern Ontario	Oak–Elm	Oak–Hemlock	Park tundra	Warmer, moister
10,000	Northern Minnesota and Wisconsin (Deglaciation)	Pine	Pine		Cooler, drier
	Southern Ontario	Spruce–Pine	Spruce–Fir		Cooler, moister
12,000	Southern Minnesota and Michigan	Spruce–Fir–Birch	Spruce parkland		Cooler
14,000	Central and Southern Ohio (Glacial)	Spruce parkland	Tundra		Cooler

Table 14-D

Expectable Sequence in Pollen-Bearing Sediments from the Apex of One
Glaciation to That of the Next. (Amplified from West, 1968, p. 303.)

Climate—Stratigraphic Units	Stratigraphic Character of Vegetation	Character of Climate	
Glaciation	Full glacial Early glacial	Cold	c
Interglaciation	Post temperate Late temperate Early temperate Pre—temperate	Temperate, s. l.	t
Glaciation	Late glacial Full glacial	Cold	c

and distances of migration and changing barriers to migration) that
could have caused such differences, no clear, major differences in climate
from one interglacial to the next have yet been established. The differences in pollen spectra, however, are sufficient to aid correlation, at least
within a single region.

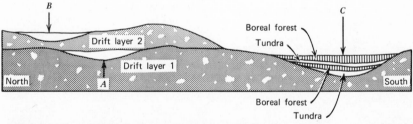

Figure 14-5 Idealized relations of pollen-rich bodies of peat or gyttja to layers of glacial drift in Europe. Body *A* demonstrates deglaciation, the minimum
amplitude of which can be measured if the border of drift layer 2 is known. If
not decapitated by the readvancing glacier, body *A* could record a complete
postglacial sequence, on the assumption that accumulation began as soon as the
site was deglaciated. Body *C*, similarly, could record a continuous vegetation sequence since the earlier deglaciation. Its pollen profile shows a tundra zone
sandwiched between two zones of boreal forest. (Not to scale.) In North America thus far, corresponding sequences are very poorly marked or not represented, possibly because of climatic differences. (See West, 1961.)

Special Applications

Pollen stratigraphy can be useful in ways that are not strictly stratigraphic. An example is the mapping of drift sheets in districts where drift is thin, discontinuous, or obscured by steep slopes or a thick forest cover. Figure 14-5, illustrating a sequence met with in Europe, shows that if the regional pollen sequence is well known, the presence of two tundra zones in body C should lead to the search somewhere north of locality C, for drift marking a glacial readvance. That drift is drift layer 2.

Another application of pollen stratigraphy is illustrated by the dramatically sudden replacement of spruce pollen by oak pollen in the Great Lakes region. Too rapid to be easily explained by plant succession, the change may have been caused by a regional change from northwest to southwest winds around 10,000 years ago (J. G. Ogden, in Cushing and Wright, eds., 1967, p. 124).

CHAPTER 15

Geochronology

CALENDAR TIME

Geochronology, the science of dating events in Earth history, began with guesses and estimates and has progressed to the establishment of more nearly accurate dates, mainly through measurements of radioactivity. It has not yet become possible to date any event with entire accuracy. Almost every method in use yields maximum and minimum dates between which, with stated statistical probability, the event occurred. The probable error may range from one year to thousands of years, depending on the dating method employed.

When we speak of dating we imply that we are dealing with sidereal time, calendar time, time expressed in units of equal length. Stratigraphy, though it establishes a sequence of events, tells us nothing of calendar time. Even time-stratigraphic units, which imply time synchrony of events at two or more places, afford no basis for measurement of calendar time. The calibration of a time-stratigraphic sequence, the placing of its component units within brackets of calendar time, is the role of geochronology. Taken together, therefore, stratigraphy and chronology [1] are the framework upon which Earth history is constructed.

EARLY ESTIMATES

As applied to the later Cenozoic part of Earth history, all estimates and determinations of the date of an event are arrived at by determining (1)

[1] The term chronology has been used by some geologists and by most students of prehistoric archeology as a synonym for stratigraphic sequence, without implication as to calendar time. This ambiguity should be generally recognized.

the rate of activity of some process in terms of units per year, and (2) the number of such units accomplished by the process since the event occurred; the second is then divided by the first. This theory is simple and finds its most nearly accurate application in measurements of radioactive decay. Before the application of radioactivity to dating events, the processes measured consisted of erosion of various kinds, deposition of sediment, and weathering, chiefly chemical. The results were not accurate because uncontrolled variability in rate invalidated (1); also in some cases (2) could not be measured accurately. These sources of error, when present, were never overcome.

Through making allowances for probable changes in rate, the measurements became estimates. Some of them proved to be of the right order when compared with C^{14} measurements, whereas others turned out to be invalid. The history of early estimates is given in the following brief review.

Delta Building. Attempts to date events through geologic processes were stimulated by the work of Croll (1875, p. 325), who deduced from certain relations between Earth and Sun that the last major time of low temperature had existed from 240,000 to 80,000 years BP. Thinking these values too great, geologists estimated the dates of inception of delta building when Swiss lakes were deglaciated, and obtained values of 12,-000, 16,000 (later reduced to 13,000), and 20,000 years on three major deltas. The inception of the modern delta of the Fraser River in British Columbia was estimated at 8000 BP on a basis of rate of advance of the delta front. Subsequently it was dated at >7300 <11,000 C^{14} yr. BP. Inception of the Bear River delta in northern British Columbia was estimated at 3600 years. The discrepancies among these dates do not necessarily reflect errors, for delta building began at different times at the various localities.

Retreat of Cliffs. Estimates of the time elapsed since the end of the Glenwood phase of Lake Chicago and since the inception of Lake Ontario (table 21-F) were based on recession of wave-cut cliffs. The first was far too small, the second too great. The discrepancy reflects the difficulty of allowing for variations in rate of cutting.

Stream Dissection. As early as 1888, Prestwich (1886–1888, v. 2, p. 534) estimated a duration of 15,000 to 25,000 years for the latest glaciation, and 8000 to 10,000 years for the duration of "postglacial" time (presumably in Britain), based on a guess as to the general extent of erosion. These values are close to those derived from C^{14} dates. An estimate, made in 1870, of the time elapsed since deglaciation of the escarpment south of Lake Erie, 12,500 years, is of the same order as C^{14} dates of contemporaneous events. Other estimates based on rate of valley erosion and volume of material eroded apply chiefly to the duration of interglacial events, but they are subject to errors arising out of the basic assumptions.

Retreat of Falls. Comparison of the rate of recession of Niagara Falls with the length of the gorge cut by the receding falls was long considered a reliable basis for dating a postglacial event. The event was the inception of the Falls, which occurred between the times of Lake Lundy and Lake Iroquois. The estimate, 20,000 to 35,000 years, was later reduced by the discovery of an error of 24% in the measurement of present-day recession. However, the estimate was completely invalidated by the discovery, through borings made for a bridge, that the middle segment of the gorge contains a thick fill of drift and hence must antedate not only Lake Lundy but also the latest glaciation of the district. The C^{14} date of Lake Lundy is around 12,600 BP (table 21-F).

Similar calculations of the recession of the St. Anthony Falls of the Mississippi River at Minneapolis, Minnesota, yielded 8000 and 12,000 years as the time elapsed since Lake Agassiz was drained. Measured variation in rate of recession during this period of observation invalidates those estimates also.

Weathering and Soil Development. Various estimates of the time elapsed since specific drift sheets were deposited have been made on a basis of weathering and soil development in the drift, as well as on depth of erosion. On this basis Penck (Penck and Brückner, 1909, p. 1169) calculated that if postglacial time in the Alps be taken as $=1$, then Riss-Würm interglacial time $=3$ and Mindel-Riss interglacial time $=12$. Taking a value of 20,000 years for postglacial time, based on the estimated time needed to build a delta into Lake Luzern and on other data, he thus derived 60,000 and 240,000 for the duration of the two interglacials named.

This is the basis of the widespread concept that the Mindel-Riss interglacial is "the great interglacial," having been much longer than the following one. If this concept is based only on Penck's data, its validity is doubtful. Not only was the estimated age of the delta later reduced substantially by recalculation, and not only do C^{14} dates imply that Penck's "postglacial" value is much too great, but the quantitative reliability of the weathering and soil criteria he used remains unevaluated. Several factors other than time may have contributed to the differences of weathering and depth of erosion observed by Penck. Opinions as to the relative length of the Mindel-Riss Interglaciation differ according to the kinds of data on which they are based. Meanwhile it would be prudent not to characterize that interglaciation as "great."

Similar estimates were made subsequently in Iowa, Indiana, and Bermuda, but various errors in the data used have invalidated them.

In summary, weathering and soil development constituted the chief basis, before the introduction of radiometry, for estimates of the duration of major units of the Pleistocene Epoch. Penck's conclusion that the "Ice Age" (=Pleistocene as a whole, as then understood and applied to the Alps), was "several hundred thousand years" long, is apparently the basis of a long-enduring belief that that epoch lasted about a million years. Radiometric dating has shown subsequently that ice ages were not confined to the Pleistocene, and that that epoch was substantially longer.

Other Geologic Processes. Estimates of time elapsed since the deglaciation of localities have been based also on rates of accumulation of peat and of precipitation of travertine by springs. One estimate of post-Hypsithermal time was based on the concentration of dissolved mineral substances in a lake without outlet, and on the rate of contribution of these substances by present-day inflow.

Several attempts were made to distinguish among glacial stades in regions of calcareous drift and moist to subhumid climate (e.g., Flint, 1949; Dreimanis, 1959). They achieved only partial success, apparently because rate of leaching varies with composition and texture of drift and with climate.

Chronology from the Earth's Celestial Geometry. If it were established that changes in the Earth's climates were controlled primarily by the combination of periodic variations in elements of the Earth's orbit, according to a hypothesis discussed in chap 30, it would become possible to date climatic events through at least part of Quaternary time. On the assumption that climates are controlled thus, chronologies have been deduced from insolation curves derived from measurement of the geometric variations (e.g., Zeuner, 1952, p. 142–145). The validity of such chronologies rests on two relationships. One is similarity of form between insolation curves and climate curves derived from other data (e.g., Broecker, p. 139 in J. M. Mitchell, ed., 1968); some authors have claimed such similarity, and perhaps it exists. The other relationship involves the quantitative competence of the geometric variations to have caused the climate changes. Such competence has been repeatedly denied. In the existing state of knowledge an opinion as to the validity of dates deduced from insolation curves would be premature.

DATING FROM RHYTHMIC NATURAL PROCESSES

Some natural processes possess a period or rhythm that leaves its impress on the sedimentary or other record. If the length of the period is known and the impress is distinct, simple counting will yield the length of the record, and, if the record extends to the present, absolute dates as well. Disadvantages include uncertainty as to length of period, errors in counting, and the fact that most records of rhythms are short. The chief chronologies developed on this basis are dendrochronology and varve chronologies.

Dendrochronology [2]

The study of the annual growth rings of trees has yielded a record that extends back through the last several thousand years. Many kinds of trees grow only during a part of each year; the fact is recorded by sheathlike layers of wood that appear as annual rings when the trunk is sectioned. The age of a newly felled tree is therefore determined by the number of rings it possesses. As no two growth seasons are exactly alike, successive rings, influenced by variations in climate, vary somewhat in thickness.

The relationships among tree rings can be measured and quantitatively tested, and are physiologically understandable. However, tree rings are not free from error. Errors up to as much as 10% in a single tree are possible, because as many as 10% of the trees in a stand can fail to show any ring at all in a year of poor growth. Hence for each area studied a mean curve must be constructed through comparison of the rings of a number of trees. The mean curve becomes a standard with which unknowns can be compared.

Within a single climatic province variations in ring thickness are fairly systematic in all trees. In southwestern United States the chief factor that controls tree growth is moisture, and the rings vary mainly with precipitation. In northwestern United States the controlling factor is temperature, and tree rings differ accordingly.

Within a single climatic province, therefore, systematic sequences of ring widths can be recognized in the standard curve. By matching these sequences between living trees and beams cut from trees felled at various times in the past, a continuous calendar can be made, extending well back in history, and validated by agreement with evidence of other kinds. Dendrochronology has been applied chiefly to the dating of prehistoric cultures in southwestern United States, through the use of house beams and other objects cut from wood. It is useful also as a climatic record, confirming other evidence as to the dating of unusually wet or dry periods. Its accuracy, at least within the province, is quite high.

Until about 1967 tree-ring dating was limited by a rather short time range, but in that year the completion of research on the bristlecone pine (*Pinus aristata*) in eastern California spectacularly extended the range

[2] General references: Stokes & Smiley, 1968; Fritts, 1966; Ferguson, 1968.

back to 7100 years BP. The extension led to comparison with the C^{14} chronology and confirmed that differences exist between the lengths of C^{14} years and tree-ring years (nearly=calendar years). This matter is mentioned again in this chapter.

Measurement of Varves

Rhythmites and Varves. The regular alternation, in sedimentary sections, of layers of different composition or texture, forming pairs or couplets, was attributed to an annual rhythm as early as 1769. Systematic counting of the couplets with the object of establishing a chronology began early in the 20th century. Since then couplets, or other regularly recurring groups of laminae of both organic and mechanical origin, have been identified, and the appropriate term *rhythmite* was applied by Bruno Sander to the sediment so layered, without implication as to thickness of pair or period of rhythm. The term is useful because in many sequences period is not established. The term *varve* (Swed. *varv,* a periodic repetition) was applied by De Geer to a couplet, regardless of origin, that represents an annual period.[3] Whether the period is annual or of some other length, it need only be known in order that the rhythmite be useful as a measure of absolute time. Rhythmites of lacustrine and marine origin have been identified in pre-Pleistocene strata (Bramlette, 1946; W. H. Bradley et al., 1942, p. 14). At least some of these are of organic origin, being related to the life cycles of organisms such as diatoms.

Organic Varves. A rhythmite deposited in a lake near Interlaken in Switzerland consists of thin couplets, each containing a light-colored layer rich in calcium carbonate and a dark layer rich in organic matter. Proof that the couplets are annual, and therefore varves, is established on organic evidence, first recognized by Heer (1865). The sediment contains pollen grains, whose number per unit volume of sediment varies cyclically, being greatest in the upper parts of the dark layers. The pollen grains of various genera are stratified systematically according to the season of blooming. Finally, diatoms are twice as abundant in the light-colored layers as in the dark. From this evidence it is concluded that the light layers represent summer seasons and the dark ones fall, winter, and spring. Counts of the layers indicate a record that is valid through at least the last 7000 years BP. Despite some obscurities the sequence agrees in many respects with that inferred from other kinds of evidence, so that the general inference that the couplets are varves is supported (Welten, 1944).

---

[3] According to Antevs, 1925, p. 1. In the original reference (De Geer, 1912, p. 242) the annual period is implied but not stated specifically in a definition.

Figure 15-1 Thick varves, deposited mechanically in the Baltic Ice Lake (fig 13-4), exposed near Uppsala, Sweden. The top of one of the dark-colored summer layers just touches the top of the pencil, 16.5cm long. Faint laminae, thinner than summer layers, are attributed to storms. At least 16 varves are visible; they become thinner upward, possibly because of retreat of the glacier that formed the northern shore of the lake. (R. F. Flint.)

Organic varves, deposited in stagnant seawater, are reported from a fiord in British Columbia. Their period is related to the season of bloom of a planktonic diatom.

Mechanically Deposited Varves. The Quaternary rhythmites most extensively studied contain couplets consisting of silt and clay deposited in proglacial lakes during the last deglaciation. Those in Sweden, Finland, Denmark, eastern North America, and Patagonia are best known, but such rhythmites occur in other regions as well. Generally the couplets range in thickness from a few millimeters to a few centimeters. They consist of a coarse member, dominantly silt, and a fine member, dominantly fine silt and clay (fig 15-1). In many couplets the coarse member is graded (*diatactic* in Swedish literature) and in some the fine member is graded likewise; some couplets, however, do not possess graded laminae.

Cores taken from a depth of 60m in glacier-fed Lake Louise, Alberta, showed couplets about 50mm thick, corresponding to the thickness of an annual layer of sediment estimated by sampling the load of the sole feeding stream. This relationship established the probability that these particular couplets are varves.

The environment of deposition is inferred from field data (e.g., Antevs, 1925, p. 33–44) and laboratory experiments (e.g., Legget and Bartley, 1953). Suspended sediment consisting chiefly of silt and clay was brought into a temporary glacial lake by streams from the melting glacier and from surrounding land during spring and summer. The silt settled early, but the clay particles remained in suspension, settling gradually during autumn and winter after melting had ceased and the surface of the lake had become frozen. This separation caused the observed gradation from the coarse member to the fine member of the couplet. The low temperature and hence relatively great density of the water was a factor in delaying settlement of the clay particles. With resumption of melting the next spring, new sediment entered the lake, and the coarse fraction, settling rapidly, produced the sharp contact between the top of the older couplet and the base of the younger one. Thus the couplet is a varve, the sediment year commencing with the spring.

The conditions described indicate why varves are particularly distinct in lake sediments of glacial origin. The finest sediment is segregated through being held in suspension during the melting season. However, varves do occur in nonglacial sediment. For example, 7522 varves were measured in such sediment in Ångermanland in northern Sweden (De Geer, 1940, p. 172). But despite such exceptions the majority of the known mechanical rhythmites are glacial-meltwater deposits; hence they record the near presence of glacier ice where and when they were laid down.

The salts in seawater, acting as electrolytes, are believed to flocculate suspended sediment, causing it to settle as homogeneous masses of particles of various sizes and reducing the distinctness of mechanical couplets. Apparently, however, flocculation is not confined to seawater. Mechanical couplets from at least one former glacial lake are not graded; they are believed to have been flocculated in both winter and summer, possibly through variations in the concentration of carbon dioxide dissolved in the lake water. Much remains to be learned about the origin of mechanical couplets.

Measurement and Correlation. (Agterberg and Banerjee, 1969.) Most studies of rhythmites are based on the assumption that the couplet period is annual, and have had as their objective the establishment of a chronology. The basic study was that of De Geer (1912; 1940) in Sweden. At closely spaced localities he measured the thickness of each couplet, and plotted the results at each locality as a curve of thickness. From the curve can be read relative thickness, the significant parameter because absolute thickness varies from one locality to another. Indeed varves are wedge shaped, thinning in the distal direction.

Curves from two localities are placed side by side and are moved up or down until a correlation appears. As in dendrochronology, curves from different localities are combined into a chronology longer than that obtainable from any one locality.

Correlation based on thickness assumes that during deposition the weather characteristics of any year, especially the summer season, were reflected, through amount of ablation, in the thickness of sediment deposited in the lake or lakes throughout a region whose width equals the distance between the two most distant localities at which a single couplet is recognized.

De Geer found that between curves obtained from localities only 1km apart he could get good correlation. He made measurements in Sweden, at approximately this interval, through a distance of nearly 1000km. He recognized individual couplets up to a maximum distance of more than 50km and showed that the layers lapped off on each other like shingles on a roof, extending farther north with increasing height in the section. From the resulting curves was compiled a Swedish varve chronology extending back, with one gap and one extrapolation, from A.D. 1900 through a period of nearly 17,000 years, on the assumption that the couplets are varves. Subsequent study, however, indicated that that chronology is reliable thus far only through approximately the last 12,000 years (Mörner, 1969; table 15-A).

De Geer's study was helped by the fact that the Baltic Ice Lake was wide and very deep, and received most of its sediment directly from the melting ice sheet. These conditions led to thin couplets that were uniform through rather long distances.

Varves belonging to a segment of De Geer's chronology were studied in Finland by Sauramo (1923; 1929), who showed that correlation based on thickness alone could lead to error and developed a more conservative method akin to that of ordinary stratigraphic correlation. The method was followed by later workers in Sweden.

For example, intensive study was made in Sweden by Järnefors and Fromm (1960) to check the original measurements; it resulted in validation of the De Geer sequence save for minor corrections. Much of the sequence was likewise remeasured from new samples by Nilsson (1960; 1968). He, too, validated De Geer's principal results, and applied his findings to the calibration of certain events, using the date of the youngest

Table 15-A
Dates of Events in Deglaciation of Southern Sweden. (Mörner, 1969.)

Event	Date (years BP)		Pollen zones
	From varves (Recalc. to BP scale)	From C[14] dates	
End of Younger Dryas; drainage of Baltic Ice Lake; inception of rapid deglaciation	10,163	10,000 to 9,950	III
Billingen (=Salpausselkä III) phase	10,250 to 10,350	(10,100 to 10,200)	
Interval of 100 to 200 years	10,450	(10,300)	
Skövde (=Salpausselkä II) phase	10,650	(10,500)	
Interval of about 250 years	10,890 .	(10,750)	
Taberg (=Salpausselkä I) phase			
Beginning of Younger Dryas	11,100	10,950 to 10,900	II
Allerød Interstade	11,950	11,750	Ic
Berghem Stade (Older Dryas)	(12,100)	11,900	Ib
Bolling Interstade		12,250	
Fjaras Stade		12,300	Ia
Agard Interstade		12,650	

Note. Parentheses indicate indirect dating.

noneroded varve as minimum for the locality. Mörner (1969) derived the dates shown in table 15-A; (see also fig 8–3); they are closely approximated by C[14] dates.

The younger Dryas unit (table 24-C) is represented in Norway by a single end moraine (the Ra Moraine) (B. G. Andersen, 1960), in Sweden by the three Central Swedish moraines formed within a period of 950 years (Mörner, 1969), and in Finland by the three Salpausselkä ridges, also formed, according to Sauramo, in 950 years. Opinions differ, however, as to the correlation between the Swedish and Finnish features.

The Swedish varve chronology (Nilsson, 1968, pl. 1; Mörner, 1969, pl. 2) implies rates of retreat of the ice-sheet terminus in the range 60 to 200 m/yr in Allerød time, 20 to 80m/yr in Younger Dryas time, and 225m/yr in Preboreal time. The maximum Allerök rate, 200m/yr, adds up to 100km in 500 years, a very rapid rate of deglaciation.

Dates within the varve chronology are in fairly good agreement with those determined by C[14]. Fromm (1938) analyzed the pollen and diatoms in the postglacial varves measured by Lidén in Ångermanland. A pollen horizon dated from the varves at 8300 BP was compared by Wenner (1968) with the C^{14} date, 8600 BP, of a comparable horizon in peat in the same area. Mörner (1969) compared several horizons between 12,000 and 7000 BP that have been dated by both methods (table 15-A).

De Geer subdivided his varve stratigraphy into four units, which he called Daniglacial, Gotiglacial, Finiglacial, and Postglacial. Although these names appeared frequently in the literature, they have become less useful than the names of pollen zones and the Blytt-Sernander units and are less frequently seen.

In summary, the Fennoscandian varve chronology developed by De Geer, Lidén, Sauramo, Nilsson, and others has stood up well as a means of dating events in Sweden and Finland through about the last 12,000 years. However, the validity of the older part of the sequence, that based on Danish strata, is still unclear. Research on rhythmite bodies in Denmark (Sigurd Hansen, 1940; 1965, p. 66) showed the common occurrence of laminations within a single couplet (fig 15-1), geneally in the coarse member. These were ascribed to redeposition of sediment after it had been stirred up by storms in shallow lakes. The glacial lakes south of and older than the Fennoscandian moraines were shallow, whereas those north of them were much deeper. Because the De Geer chronology interprets "storm laminations" as varves, Danish geologists did not accept the part of the chronology that antedates those moraines.

One could argue, however, that even though storm waves stirred the bottom of a shallow lake, the stirred-up fines would not settle until winter freezing of the lake surface made settling possible. On this basis one could accept the storm-lamina concept as explaining irregularities in a coarse member, yet believe that each fine layer must be a winter deposit, its top therefore being the top of a varve. As the Danish sequence begins close inside the limit of the Weichsel glaciation, De Geer's count of around 17,000 years elapsed since the beginning of deglaciation would be at least compatible with C^{14} dating of the presumably correlative Late-Wisconsin glacial maximum in North America.

In the Americas measurements in rhythmites have attained less success, possibly only because the work of the original investigators has not been followed up, as in Europe, by later investigators. Caldenius (1932), a student of De Geer, made a careful study in Patagonia, but no use of his work has been made by others in that region. Antevs (e.g., 1925) made studies in the United States and Canada. His measurements embraced rhythmites deposited in lakes that were narrower and shallower than the Baltic Ice Lake, and that received substantial sediment from land streams as well as from the ice sheet. A north-south distance of nearly 1000km was involved, but it included three gaps aggregating half the distance. The resulting chronology implies a period of 28,000 years from the Late Wisconsin glacial maximum to the time of deglaciation of the Cochrane district, Ontario; about two-thirds of this period is based on measurement, the remainder being interpolation. However, C^{14} dates of events within this sequence suggest a period of little more than 10,000 years. Possibly the cause of the discrepancy lies in the interpolations used. A modern study in Ontario, in part overlapping that of Antevs, is the work of O. L. Hughes (p. 535–565 in H. E. Wright and Frey, eds., 1965b), who established that the couplets under study are varves.

Varves as a tool in chronology are unlikely to compete successfully with the greater flexibility of C^{14} dates, but in areas where material datable by C^{14} is scarce or unavailable, varves may again come under study.

RADIOMETRIC DATING [4]

The discovery of radioactivity led to attempts to measure the time elapsed since the crystallization of bodies of igneous rock containing radioactive isotopes. For some decades the isotopes employed were members of the uranium and thorium series, with long half-lives, so that they were best adapted to measurements in ancient crystalline rocks. By the middle of the 20th century, however, other isotopes and a variety of techniques were under development, with results that made a great impact on the chronology of late-Cenozoic strata. Some have become routine and others are in various stages of development. Techniques applicable to geology are shown in table 15-B; others applicable to archeology are discussed in Brothwell and Higgs, eds., 1963. Our discussion is concerned principally with three methods of dating, involving respectively radiocarbon, potassium argon, and uranium- and thorium-series isotopes.

Radiocarbon Dating [5]

Developed by W. F. Libby between 1946 and 1949, radiocarbon measurement came rapidly into wide use for dating events within the last few tens of thousands of years. Radioactive carbon (C^{14}) is a nuclide created when neutrons derived from primary cosmic radiation bombard nitrogen (N^{14}) in the Earth's atmosphere. The C^{14} emits beta rays and disinte-

[4] General references: Broecker, W. S., p. 737–753 in H. E. Wright and Frey, eds., 1965a; Shotton, 1967b.

[5] References include Libby, 1955; 1961. Virtually all dates are contained in the journal *Radiocarbon*, published by the American Journal of Science.

Table 15-B

Dating Methods Pertinent to the Quaternary. (W. S. Broecker in H. E. Wright and Frey, eds., 1965a, p. 737.)

Isotope	Half–life (10³ yrs)	Method	Range (10³ yrs)	Materials	Likely Applicable to				
					Ocean Temp.	Sealevel	Glacier Extent	Arid Lakes	Pollen Sequences
C¹⁴	5.7	Decay	0.35 / 35–70	Organics–CaCO₃ Organics	+	+	+	+	+
Pa²³¹	32	Decay / Integration / Th²³⁰ normal	5–120 / 5–120 / 5–120	Red clay or *Glob.* ooze	+	0	0	0	0
Th²³⁰	75	Decay / Integration / Th²³⁰ normal	5–400 / 5–400 / 5–400	Red clay or *Glob.* ooze	+	0	0	0	0
		Growth	0–200	CaCO₃–(organics)	0	+	0	+	+
U²³⁴	250	Decay	50–1000	Coral	0	+	0	0	0
He⁴	—	Growth	No limit	Mollusks or coral	0	+	0	+	0
Ar⁴⁰	—	Growth	No limit	Volcanics	+	+	+	+	+
Cl³⁶	300	Growth	50–500	Ign. or met. rock	0	0	+	0	0
Be¹⁰	2500	Decay	100–8000	Red clay	+	0	0	0	0

grates, with a half-life of about 5730 years,[6] to N^{14}. Measurement of the C^{14} content of a substance makes possible calculation of the time elapsed since the active carbon it contains was created, if it is assumed that the flux of cosmic rays has not varied greatly during that time.

As rapidly as it is created, C^{14} oxidizes and thereby constitutes a minute fraction of the CO_2 in the atmosphere. Being soluble in water, atmospheric CO_2 exchanges with the hydrosphere. Plants metabolize CO_2 and animals absorb plant tissues, so that exchange with the biosphere also occurs. In consequence the concentration of C^{14} in atmosphere, hydrosphere, and biosphere is virtually uniform and represents equilibrium between rate of creation and rate of decay. At least this was the case before thermonuclear explosions added artificially made C^{14} to the atmosphere.

While an organism lives, its C^{14} maintains this equilibrium concentration, but with death metabolism ceases and radioactive disintegration reduces the specific activity of the C^{14} in its body substance. Specific activity is proportional to the total amount of C^{14} remaining and is, of course, a function of the time elapsed since death. Specific activity can be counted in carbon having the form of a solid or a gas. When specific activities are translated into years, counting errors increase with increasing age. Disintegration being a random process, each measurement involves a statistical uncertainty, conventionally expressed in such a way that the chances are 2 to 1 that the true value deviates less than the stated amount, and 19 to 1 that the deviation is less than twice that amount. A statistical uncertainty of <100 years is difficult to achieve. However, so small an uncertainty is not very significant in the evaluation of Pleistocene events that occurred more than 11,000 years ago, and in some events more recent than that.

The rather short half-life of C^{14} limits its theoretical use for dating to a span of eight or nine half-lives, say $\sim$50,000 years. Indeed, finite dates greater than about 50,000 years are made possible only by isotopic enrichment of samples before counting. Although a few finite dates greater than 60,000 years have been reported, they should be considered tentative until internal consistency appears in a group of several related dates in that range. At the near end of the span, material no more than 100 years old can be dated.

A C^{14} year does not necessarily equal a calendar year. C^{14} dates of wood from ancient bristlecone pine trees do not agree with the dates of

[6] This value, determined in 1962, supplants an earlier value of about 5568. To avoid confusion, however, published dates continue to be quoted on the basis of the older value. Such a date can be converted to the newer half-life value by multiplying it by 1.03.

the same wood as determined from ring counts. Similar discrepancies exist between C[14] dates and historical dates of wood from ancient Egyptian tombs, themselves subject to recognized uncertainties (e.g., papers in Olsson, ed., 1970). Nevertheless, curves or tables whereby C[14] years can be converted to calendar years are likely to become available in the near future. With its aid the discrepancies between varve chronology and C[14] chronology, both of which vary within modest limits, can be reassessed and perhaps reduced.

The substances most commonly dated include wood, charcoal, peat, shells, inorganic carbonates precipitated from saline lakes, and residual protein from bone, including human bone. Several sources of error in the sample itself, in addition to counting errors, are known. Among these is dilution of a substance with younger carbon, such as carbon introduced in the roots of growing plants while the substance lay below ground. Another source is the dilution of organisms, such as aquatic plants and shellfish, with ancient, "dead" carbon, while they lived in water containing bicarbonate derived from the solution of old rocks.

With increase in the number of samples dated, these and other errors are minimized. Each dated sample is evaluated in the light of its stratigraphic position, climatic implications, biogeochemistry, and other pertinent information, and each date must be readjusted or rejected as better samples are dated with improved techniques. Much expert judgment therefore enters into the evaluation of a new date.

It was C[14] dating that revealed the twofold character of the Wisconsin Glacial Age (Flint and Rubin, 1955), consisting of two glaciations separated by an interglaciation, and led to field studies that validated that history. C[14] dates also defined the Late Wisconsin glaciation in central North America as having occurred between about 25,000 and about 10,-000 BP, reaching its maximum at about 18,000. By fixing times when glacier termini overrode forests and incorporated wood in drift (fig 15-2), and uncovered other places in which peat then began to accumulate, radiocarbon has made possible a reasonably consistent calendar of glacial events. Dating by C[14] also permits time correlation of strata that are disconnected and separated by long distances. For example, glacial sediments are correlated with sediments of pluvial lakes, and glacial or postglacial sediments in North America with those in Europe, demonstrating that in these widely separated regions climatic changes occurred with broad contemporaneity. Despite its recognized errors, C[14] dating, within the short time it spans, makes possible worldwide correlation of comparatively recent strata and events with a precision as yet unattainable in older sedimentary rocks. In the field of archeology C[14] has made advances of prime importance in the dating and succession of prehistoric

Figure 15-2 Spruce stumps in growth position in soil developed in outwash gravel, upper Muir Inlet, Alaska. The age of one of them is 7025 ± 270 C^{14} years BP (R. P. Goldthwait, 1963). Apparently the trees were buried by outwash (upper left) generated by Neoglacial readvance. (F. D. Larsen.)

cultures. Up to the beginning of 1970 more than 40,000 C^{14} dates had been determined and published.

Potassium/Argon Dating

Although the span of radiocarbon dating is limited to a late part of late-Quaternary time, the potassium/argon (K/Ar) method is capable of yielding approximate dates that can cover the whole of the later Cenozoic with ease, provided only that igneous rocks containing primary biotite, muscovite, or amphibole are present and are related critically to strata of interest.

In such minerals one isotope of potassium, K^{40}, is radioactive, with a long half-life. Its decay products include Ca^{40}, the common isotope of calcium, and Ar^{40}, an inert gas, in the ratio of 89:11. Measurement of the ratio of argon to potassium in a suitable mineral measures the time through which argon has been accumulating, and hence dates the mineral at least approximately. In late-Cenozoic strata the samples used are mainly ancient flows of basalt and layers of tephra. The dates of two flows, for example, are respectively the maximum and minimum dates of any fossil-bearing sedimentary rocks, or perhaps glacial drift, that are sandwiched between them.

During the 1950's radiocarbon measurements gave great impetus to refinement of stratigraphy and correlation at the near end of the late-Cenozoic time span. Similarly, during the later 1960's potassium/argon dating began to reveal the antiquity of the series of glacial ages, showing that in high latitudes, cold times commenced well before the Pleistocene began. A related date is 3.3×10^6yr on sediments in France with a lower Villafranchian fauna. The value of all K/Ar dates for correlation has been enhanced through the use of the geomagnetic-polarity sequence.

Although few K/Ar dates, thus far, are younger than a few hundred thousand years, the method appears capable of yielding fairly reliable dates within a much shorter time span. If such dates materialize, their number and usefulness should be limited only by the stratigraphic distribution of datable lava and tephra. Table 15-C lists a series of K/Ar dates that are pertinent to the history of late-Cenozoic glaciation.

Meanwhile, attempts to date the Alpine glaciations through the stratigraphic relations of their supposed Rhine-terrace equivalents to volcanic rocks (Frechen and Lippolt, 1965, p. 28) have resulted in a sequence thought to range from Günz through the Mindel/Riss Interglacial, with dates ranging from more than 400,000 to around 150,000 years. The range agrees broadly, although not in detail, with that derived by Emiliani (1966b, fig 6) by extrapolation from C^{14} dates in sediment cores from the Caribbean Sea. One must bear in mind that series of dates, particularly long sequences, are subject to uncertainties of two kinds: uncertainties in stratigraphic correlation of the sediment under study, and uncertainties inherent in the dating techniques.

Uranium-Series Dating

Radiocarbon dating is limited to the last 50,000 years or less, while potassium/argon dating is limited, in terms of results thus far, to a time span antedating the last half million years or so. The need for filling the "datability gap" thus created is clear, because the gap is crowded with interesting events. This need led to experimentation with other radioactive isotopes. Promise was found in daughter nuclides of the uranium-decay series, not only because their half-lives are of the right length but also because their use is directly applicable to measurement of marine shells rather than being confined to volcanic matter. Hence these U-series daughters possess a large potential for geochronometry. Various applications of this method have been tried. One of them involves Th^{230} (io-

Table 15-C

Representative K/Ar Dates of Significant Stratigraphic Occurrences, Between 0.5 m y and 10 m y BP. Dates are minimum and both dates and stratigraphy are subject to correction or reappraisal.

Approx. K/Ar date (year B.P. $\times 10^6$)	Occurrence	Ref.
0.5	Late Trinil basalt, Indonesia. (Near *Pithecanthropus*.)	
0.7	Bishop Tuff, California. (Immed. underlain by till not firmly correlated.)	1, 1a
1.3	Bruneau Basalt, Idaho. (*Mammuthus* fauna.)	2
1.4	Basalt, McMurdo Sound, Antarctica. (Overlain by till of Ross Ice Shelf.)	4
1.5 to 1.9	Sutter Buttes, California. (Blancan fauna.)	2
1.6	Valros basalt, S. France. (Between Villafranchian and Astian/Plaisancian.)	2
1.7	Base of Bed I, Olduvai, Tanzania. (Fauna of Villafranchian affinity, with australopithecine.)	3
1.7, 2.0, 2.0, 2.1, 2.2	Basalt, Taylor Valley, Antarctica. (Between 2 glacial layers.)	4
2.47	Timaru Basalt, New Zealand. (76m above supposed Plio/Pleist. boundary.)	5
2.5	Marine sequence, New Zealand. (Lowest known indic. of cooling.)	6
2.5	Basalt, Lake Rudolf, Kenya. (Overlies strata with *Australopithecus*.)	7
2.7	Andesite, Wrangell Mts., Alaska. (Overlies tillites.)	10
2.7 3.1	Latite (2.7) and andesite (3.1), California. (Separated by Deadman Pass Till.)	8
3.1	Basalt(?), Iceland (Immediately overlain by tillite.)	9
2,8, 3.0, 3.1, 3.1, 3.1, 3.1, 3.2, 3.3, 3.6, 3.6	Basalt, Taylor Valley, Antarctica. (Overlain & underlain by evidence of glaciation.)	4, 4a
3.7, 3.9	Basalt, Wright Valley, Antarctica. (Underlain by evidence of glaciation.)	4
3.5	Basalt, Glenns Ferry Fm., Idaho. (Hagerman Fauna of Blancan age.)	2
3.6, 8.4, 8.7, 9.8, 9.9, 10.2	Andesite, Wrangell Mts., Alaska. (Interstratified with tillite.)	10
6.0 to 10.0	Basalt, Jones Mts., West Antarctica. (Related to tillite.)	11

References

1. Dalrymple et al., 1965.
1a. Gage, p. 401 in H. E. Wright and Frey, eds., 1965b.
2. Evernden et al., 1964.
3. Hay, R. L., 1963, Science, v. 139, p. 829.
4. Denton et al., 1970.
4a. Armstrong, R. L., et al., 1968, Science, v. 159, p. 187.
5. Mathews, W. H., and Curtis, 1966, Nature, v. 212, p. 979.
6. Stipp et al., 1967.
7. Patterson, B., and Howells, 1967, Science, v. 156, p. 64.
8. Curry, R. R., 1966, Science, v. 154, p. 770.
9. McDougall and Wensink, 1966, Earth and Plan. Sci. Ltrs., v. 1, p. 235.
10. Denton, G. H., and Armstrong, R. L., 1969.
11. R. L. Armstrong, unpublished.

nium), a daughter, several times removed, of U^{238}. It is based on the extent of radioactive disequilibrium between parent and daughter, induced by natural processes or by the introduction of parent or daughter from outside sources into a marine carbonate. If the system remains closed, in the course of time equilibrium is restored. But measurement, at an intermediate time, of the extent to which equilibrium has been restored can, in theory, give the geologic age of the carbonate.

However, the system does not necessarily remain closed. U^{238}, Th^{230}, or intervening daughters may move out of or into the system, either in the original marine environment or, more likely, through the activity of circulating ground water after emergence, and no clear indication that such activity has occurred is inherent in the carbonate. Hence the reliability of a date must be determined by comparison with dates derived otherwise, by evaluation of the geochemical environment of the sample, or through internal consistency in a series of samples in stratigraphic superposition. Even a "good" date may be accurate only to within ±25%. Geochemically, corals give fairly accurate dates, but the results of dating mollusk shells are so inconsistent as not to be dependable. As the occurrence of corals is restricted to the tropics and subtropics, their value is correspondingly reduced.

Many uranium-series dates have been obtained, mostly in the range 65,000 to 300,000 years. Because no good means of selecting the reliable ones has been devised, uranium-series dates are not quoted in this book, which cannot be corrected as easily as can articles in journals. Many of the dates thus far obtained, however, will be found in the paper cited in W. C. Bradley and Addicott, 1968; a discussion of certain discrepancies is given in Shotton, 1967b, p. 376.

Late-Cenozoic Time Scale

It would be very satisfying to be able to construct a time scale for the later Cenozoic beginning, say, in the second half of the Miocene, but the data now available are inadequate, as is reflected in the poor calibration in table 29-F. The principal difficulties lie in our inability, as yet, to fix a generally applicable Pliocene/Pleistocene boundary on the basis of organic evolution and on the great paucity of dates, within the Pleistocene, older than about 40,000 BP. Thus, although radiometry applied to geologic data has revealed a history that includes an unknown number of significant fluctuations of climate, we have only begun to be able to calibrate them in time. It would be of little value to estimate the length of Quaternary time as long as there is no agreement on the position of the base of the Pleistocene Series. The promise for the future, however, is very great.

CHAPTER 16

Late-Cenozoic Climates

It is hardly necessary to remind ourselves that the parameters of a former climate, such as temperature and precipitation, wind directions and velocities, are not measurable directly; they must be inferred from geologic, biologic, and isotopic data. The study of most such data for their climatic implications is still in its infancy; little integration on a broad scale is yet possible. We can call attention here to the features measured and to models that have been set up, but we cannot yet go very far with synthesis.

THE DATA OF FORMER CLIMATES

Nearly all the data from which inferences as to former climates have been drawn [1] involve fluctuation through time. We list them in table 16-A, noting that most are mentioned in other contexts elsewhere in this book.

Temperatures Inferred from Three Sources

The climatic parameter that has received the greatest attention is temperature. Inferences about former temperatures have come mainly, although by no means exclusively, from three sources: (1) the climatic snowline, (2) the ranges of organisms, and (3) oxygen-isotope ratios. The apparent relation between oxygen isotopes and temperatures of surface water of the ocean is discussed in chap 27. Although it was formerly thought that O^{18}/O^{16} ratios in shell carbonate indicated lowered temperatures in glacial ages, even in low latitudes, this is now uncertain. Until

---

[1] Cf. Schwarzbach, 1963, chaps 2-7.

Table 16-A
Data from Which Paleoclimatic Inferences Have Been Drawn

Source	Chief parameter [a]
Isotopic paleotemperatures	T
Cirque—floor surface, "glaciation limit", and similar features	T
Lower limit of frost wedging on highlands	T
Lower limit of gelifluction sediments	T
Altitude of treeline	T, P
Position of outer margins of glaciers	T
Eustatic fluctuation of sealevel	T
Outer limit of permafrost	T
Fluctuation of lakes in dry regions; stratigraphy beneath lake floors	P
Fluctuation of dissolution/precipitation relations in caverns in carbonate rocks	P
Distribution of active desert sand dunes	P
Ancient soils unstable under the existing climate	P
Changes in the regimens of streams	P
Fluctuation of ranges of organisms	T, P
Anomalies, especially disjunctions, in the distribution of organisms today	T, P

[a] T = temperature;   P = evaporation/precipitation ratio

existing uncertainties are clarified we do not know just how much temperatures, measured in this way, were changed in glacial ages. Fluctuation in the ranges of organisms is treated in chaps 14, 27, and 28. Such shifts are discerned in the record of land plants (table 16-B), land animals, and marine invertebrates, but the results in terms of temperature are mostly qualitative and yield little that is more definite than changes of a few degrees within the glacial-nonglacial cycle.

As for the snowline, we noted in chap 4 that the reconstructed glacial-age snowline is systematically lower than the existing snowline by some

Table 16-B

Temperature Changes Suggested by Fossil Terrestrial Plants

Time Unit	Region	Parameter	Departure from Existing Value at Locality of Occurrence (°C)	Source
Hypsithermal	Northern Europe	July mean	+2.0	J. Iversen, 1944, Geol. Fören. Stockholm Förh., v. 66, p. 477.
Younger Dryas (Fenno—scandian)	Central Europe	July mean	−6.0	J. Iversen, 1954,Danm, Geol. Undersög., ser 2, no. 80, p. 98.
Alleröd maximum	Northern Europe	July mean	−2.0 to −3.0	Idem. p. 97.
	Central Europe	July mean	−2.5	Firbas, 1949—52, v.1, p. 287.
Mindel—Riss(?) Inter—glacial age	Austrian Alps	Annual mean	+2.0 to +3.0	Klebelsberg, 1948—49.
Latest glacial maximum	Central Europe	Annual mean	−6.0	(Derived from altitudes of vegetation zones.)
Earliest glacial maximum	Western Poland	Annual mean	−4.0 to −5.0	Szafer, 1954.

1000 to 1500m in middle and low latitudes. This implies temperature lowered by 7° or more, depending on the lapse-rate value used in the calculation. Penck (1938a) adopted for maritime western Europe a value of 8°, the difference between the temperature at the existing snowline and the temperature at the altitude of the glacial-age snowline. Messerli (1967) derived a series of values for the Mediterranean region. For a thoughtful review of the uncertainties involved in deriving temperatures from snowline data see H. E. Wright, 1961a, p. 966.

One uncertainty involves the wintertime presence of a layer of cold air as much as 1km thick, overlying cold, snow-covered regions with high albedo. This happens today and results from radiational cooling. Similar conditions may have existed in central Europe at a glacial maximum, resulting in the exceptionally low temperatures suggested by ice-wedge casts (chap 10), and would have intensified the secular lowering of temperature (Flohn, 1953).

Information is not sufficiently precise in most cases to permit discrimination between mean annual and mean summer temperature. Snowline data pertain to summer temperature and so do at least some data on the ranges of plants. Oxygen-isotope ratios, on the other hand, probably pertain to mean annual temperature if they reflect temperature at all.

CHANGES IN PRECIPITATION

In the 19th century the opinion was expressed more than once that precipitation in a glacial age must have been greater than that of today, in order to account for the vast volume of water substance contained in the ice sheets. Since then opinion has been better informed (Brooks, 1949). Although the matter is not resolved, it has been repeatedly deduced that overall precipitation must have been less than that of today. The argument is based on lowered temperature with consequent reduced evaporation from the ocean surface, evaporation that today is greatest in the equatorial zone. Wind velocities in middle latitudes are supposed to have been higher than today's because of increased meridional pressure gradients. This should have increased the surface area of the sea, creating spray, and promoting evaporation. Yet such increased velocities would not have occurred in the equatorial zone, which did not participate in the increased pressure gradient. Considering these factors, Flohn (1953) estimated glacial-age evaporation at about 78 to 80% of that of today.[2] The concept has received support by local test and comparison of its consequences with geologic evidence. For example, from the fact that at glacial maxima glaciers on the inland slopes of the coastal mountains in Alaska expanded little beyond their present extent, Taber (1953, p. 332) inferred reduced precipitation, despite lowered temperatures. Manley (1951) derived a theoretical value of 80% of present values for precipitation on British highlands during glacial maxima. He thought this value correct within a narrow margin because higher or lower values conflict with the evidence. The higher firn limit resulting from a lower precipitation value would result in an area of snow accumulation inadequate to produce the glaciers recorded by the drift sheets; a lower firn limit would have resulted in cirques in southern British highlands where none exist. The deduction, therefore, is rather closely controlled by the known evidence. Again, the glacial-maximum firn limit rises from around 120m on the west coast of Scotland to between 500 and 600m in the Cairngorm Mountains 120km farther east. The steep rise, about 3.6m/km, is greater than the existing decrease in precipitation between the two districts would support; it demands glacial-maximum values smaller than existing values. A somewhat similar relationship exists between the Solway Firth and the northern Pennines in England.

A glacial-age precipitation map of Europe was attempted by Klein (in Klute, 1951, p. 277). It shows reduced overall precipitation and a steeper

[2] Discussion of recent changes and speculation as to their causes is found in Kraus, 1958.

decline from the Atlantic coast eastward than occurs today, but the values adopted, controlled by very few data, may be too small.

The marked general warming in the period 1925 to 1945 was accompanied in the northern hemisphere at least, by poleward shift of the prevailing storm tracks. The resulting invasion of high-latitude regions by warm, moist air masses brought increased precipitation to existing glaciers, but it also brought a greater amount of heat. During that period glaciers shrank very rapidly. These are matters of observation, but corresponding data for a comparable period of cooling do not exist because the latest full cycle of cooling antedates the instrumental record. However, cooling is apparently accompanied by a shift of the prevailing storm tracks toward the equator and by the general expansion of glaciers. Presumably precipitation then decreases in high latitudes, but despite the decrease, glaciers expand because of reduced temperatures. At the same time glaciers in lower latitudes expand also, as a result of both lowered temperature and increased precipitation.

This argument, based on the theoretical reverse of a measured set of conditions recently prevailing, is compatible with the concept that precipitation in and leeward of the chief glacier-covered regions was less during major glacial maxima, than it is in the same regions today. Even though we grant that glacial-age precipitation, overall, was less than today's by some specified value, that value cannot be translated into a proportional reduction of overall precipitation on the lands, because only about 20% of the water in the cycle falls on the lands (fig 16-1). Indeed it is expectable that in a glacial age the areal pattern of precipitation would have been different from that of today. Near the interface between the trade winds and the westerlies, precipitation could well have been greater than today's, as might also have been the case along the windward margins of ice sheets in meteorologically oceanic environments. This matter was the subject of a glacial-age model constructed by Flohn (1953), who on certain assumptions deduced equatorward shift of westerlies and trade winds such that in the northern hemisphere the precipitation/evaporation ratio increased in low-middle latitudes but decreased in a narrow zone close to the equator (fig 16-2). This would have narrowed the belt of equatorial rain forest.

Glacial-age precipitation values from place to place cannot be fixed by deduction, but must be inferred from the data of fossil organisms, snowline altitudes, pluvial lakes, and similar features. Yet even such data contain large uncertainties. One example is the pluvial lakes (chap 17), which were caused by lowered temperature as well as by increased precipitation. Another example is the apparent absence of trees in much of central Europe during the Weichsel Glacial Age. Probably this was not exclu-

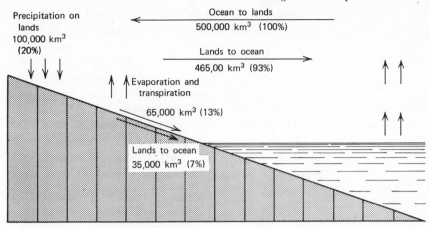

Figure 16-1 Annual volumes of water participating in the world hydrologic cycle today. (Cf. table 4-D.) In the cycle a water layer ~1.35m thick is evaporated from the ocean annually.

sively the result of climatic aridity. The effect of extremely low temperatures and of permafrost would have created physiological effects independent of those of precipitation deficiency.

The early notion that precipitation during glacial maxima must have been greater than that of today in order that ice could accumulate was refuted, for the Alps as a sample area, by Penck and Brückner. They

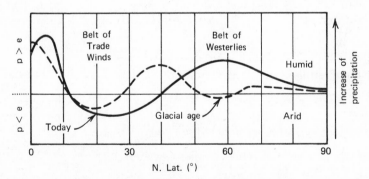

Figure 16-2 Model suggesting pattern of displacement of planetary wind belts in northern hemisphere as compared with today's pattern. The model shows a low-latitude zone in which the precipitation (p)/evaporation (e) ratio was less during a glacial age than it is today. (Adapted from Flohn, 1953, p. 270.)

argued that increased precipitation should have produced a glacial-age snowline with a steeper windward slope than it has, whereas actually the former snowline is nearly parallel with the existing snowline. This condition appears to be general throughout the world (chap 4). Indeed, ice can accumulate despite smaller precipitation values than those now measured, if temperatures are low enough to reduce ablation adequately.

Even if this is true, we are still faced with the problem of the volume of water substance in the ice sheets and the time involved in transferring it from the ocean to the lands. According to the estimate embodied in table 4-E, the lowered temperature in a glacial age caused net withdrawal from the ocean of a layer of water ~132m thick to build glaciers, the pluvial lakes being neglected because their aggregate volume was very small. The volume of that layer was ~47 $\times 10^6 km^3$, equal to about 3.5% of the volume of the ocean.

Could the known glacial-age ice sheets have been built within a length of time compatible with our meager data, if we assume that precipitation values were no greater than today's? We can approach the answer to this question with two crude tests. In the first (table 16-C), we note that (4) represents less than 3.2% of (3). Hence if ablation losses are neglected, only a little more than 3% of annual precipitation on the glaciated areas, at today's rate, could have built the glacial-age glaciers in 50,000 years, starting from a condition in which the extent of glaciers was equivalent to that of today. If we assume an initial no-glacier condition, which is very improbable, the percentage of annual precipitation becomes 4.8.

The foregoing calculation assumes uniform precipitation over the entire area ultimately glaciated. This is unrealistic because it does not accord with generally accepted ideas about how ice sheets form, and suggests a different approach. Climatic atlases indicate that today's annual

Table 16-C
Data for Testing Rate of Accumulation of Ice Sheets in a Glacial Age

	10^6 km³
1. Annual precipitation on World's land area (Nace, 1967)	0.1
2. " " on World's glaciated areas (=30% of (1))	0.03
3. Cumulative pptn. over 50,000 years (= 50,000 × (2))	1500.00
4. Estimated excess of solid H_2O residing on land in a glacial age over H_2O on land today (Table 4—E)	47.40

precipitation on the highlands in northeastern North America is about 2m, and in maritime Norway about 3m, for an average of 2.5m. Let us assume this amount falls entirely as snow on these highland areas and is so confined that it does not flow outward. If ablation losses are neglected, this much snow could build a body of ice 2.5km thick in 2000 years, over the area on which it falls. Other, more refined calculations for the Ungava region were made by Barry (1966, p. 42).

However, the snow would flow outward. Snowfall at the stated rate, occurring on only 2% ($=400 \times 10^3$km) of each of the areas known to have been covered by the Laurentide and Scandinavian Ice Sheets, compacting and flowing outward, could have formed those ice sheets in about 40,000 years. Ablation losses would, of course, increase the time, but snowfall on the remaining 98% of the areas would decrease the time by a far greater amount. These calculations, crude though they are, suggest that increase in overall precipitation rates is not required for the building of the glacial-age ice sheets within the time spans that were apparently available.

TREND OF TEMPERATURE THROUGH CENOZOIC TIME

Miocene and Pliocene Fluctuations

Let us return to temperature in former times, a parameter surrounded with fewer uncertainties than precipitation values are. We can examine the sequence of climatic changes having glacial implications, that can be read from the present record beginning with its earliest part. It appears at once from table 15-C that in Antarctica, Alaska, and Iceland, at least, glaciation has occurred repeatedly and perhaps continuously during the last ten million years, including most or all of Pliocene time. This fact reflects the result of secular cooling at high altitudes in high latitudes, that had been in progress since earlier Cenozoic time.

Figure 16-3 shows Cenozoic cooling inferred from land floras in three widely separated regions. Figure 16-4 shows comparable temperature curves constructed from data on marine mollusks in Pacific North America. Related to these charts are two others. One (Dorf in Nairn, ed., 1964, p. 19) shows latitudinal shifting of broad floral assemblages through Tertiary time. A second (Wolfe and Hopkins in Hatai, ed., 1967, p. 72) shows Tertiary climatic fluctuation based on the forms of leaf margins. In addition, data set forth by Wolfe and Leopold (in Hopkins, ed., 1967, p. 203) support the concept of rapid cooling around 15 million years BP, in later Miocene time. Apart from these data, Bandy (1968; Bandy et al., 1969) found in planktonic microfaunas evidence of cold episodes in late-

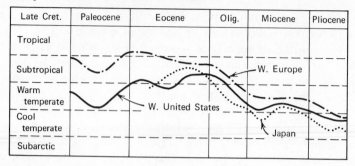

Late Cret.	Paleocene	Eocene	Olig.	Miocene	Pliocene

Tropical

Subtropical

Warm temperate

Cool temperate

Subarctic

W. Europe

W. United States

Japan

Figure 16-3 Irregular climatic cooling through Cenozoic time, inferred from fossil plants in three widely separated regions. (Japan from Tanai and Huzioka, *in* K. Hatai, ed., 1967, p. 93; other curves from E. Dorf, p. 20 in A. E. M. Nairn, ed., 1964.)

Miocene and mid-Pliocene time. Emiliani (1966b) noted, from a variety of data, similar episodes in northern and southern Italy, and Zagwijn (1967) found pollen evidence, in the Mio-Pliocene of the Netherlands, of fluctuation between warm-temperate and cool-temperate climate, possibly repeated once or twice. Rónai (1968) inferred from pollen data in Hungary a long cool interval in the late-middle Pliocene.

Having mentioned various inferred fluctuations of temperature, we must note a dissent by Axelrod and Bailey (1969) as to such inferences as are based on fossil plants. Those authors maintained that mean annual temperature alone is an inadequate measure of the thermal requirements of plants, and proposed that mean annual range of temperature is equally important. Failure to consider both factors, and to evaluate altitude and geographic position of a fossil flora, could lead to error in reconstructing a paleoclimate. Therefore, some inferred late Tertiary fluctuations of temperature may need re-examination.

Pollen study of Pliocene lignite at Wallensen im Hils, in northwestern Germany, showed fluctuations of temperature which, although less intense than the Pleistocene fluctuation, were regarded as forerunners of the latter (Altehenger, 1958).

Looking at the curves derived from these varied data, we can discern an overall downward trend in Cenozoic surface temperatures, amounting to several degrees. Superposed on the general trend is apparent fluctuation of several degrees' amplitude in Miocene, Pliocene, and Pleistocene time. From our very incomplete information on the extent and distribution of glaciation, we might surmise that at first, episodes of low temperature produced glacial responses in high parts of high latitudes. Large middle-latitude ice sheets seem not to have appeared until the Pleistocene, perhaps well after its beginning. If this was so, it may have been because of insufficiently high lands in or adjacent to those latitudes, or because secular temperature had not yet fallen quite enough to push the snowline down below the tops of middle-latitude highlands. In the latter case buildup of the Antarctic Ice Sheet might have been a factor in the lowering of temperature. Because of its polar position that ice sheet is rather stable today and very likely was so in late-Tertiary time. When at its fullest extent, bathed in the ocean around its entire perimeter, it must have contributed to refrigeration of lower latitudes through cold ocean currents. The establishment of mid-latitude ice sheets which were both large and metastable reinforced temperature changes brought about by other causes, and through repeated shrinkage and re-expansion created substantial changes in and near the territory they occupied.

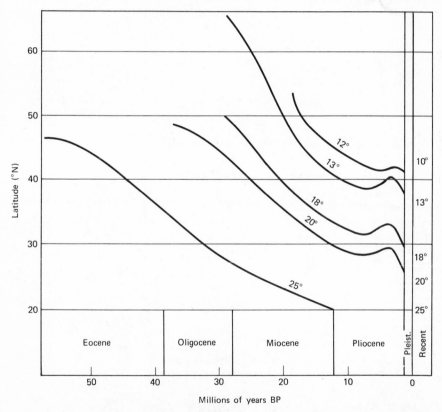

Figure 16-4 Latitudinal shift of February marine isotherms along the Pacific Coast of North America through most of Cenozoic time. (After C. W. Durham, 1950, G. S. A. B., v. 61, p. 1259.)

Quaternary Fluctuations

The distinction between Quaternary fluctuations and preceding ones is artificial, because we are clearly dealing with continual fluctuation, and because we do not know either the exact stratigraphic position or the date of the Pliocene/Pleistocene boundary within the continuum. Some of the more recent events listed in table 15-C are, of course, Pleistocene. But

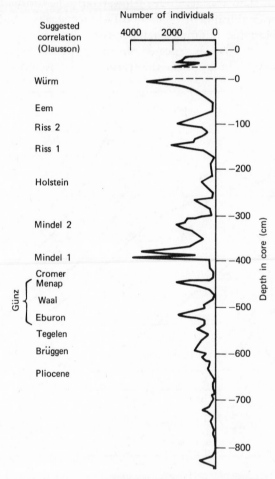

Figure 16-5 Curve showing changes in abundance of Foraminifera (total content of entire fossils) with core depth, in Cores 58 and 58B, eastern equatorial Pacific Ocean. Redrawn and tentatively correlated (to Alps stratigraphy) by Olausson (1961, fig 5) from curve constructed by Arrhenius (1952, fig 3.4.2). Curve has been reversed right for left.

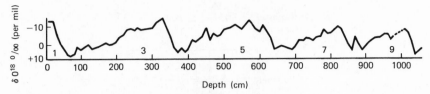

Figure 16-6 Curve showing fluctuation, through time, of the O^{18}/O^{16} ratio in surface water in the central Caribbean, as derived from a single foraminiferal species in a core. Odd numbers identify stages in which O^{18} was relatively low. (Emiliani, 1966a.)

the later fluctuations are best visualized in the curves developed from cores. Accordingly we shall examine several long sequences that extend through much or all of the Quaternary, and then look at shorter but more detailed sequences that include part or all of the Weichsel glaciation.

Long Sequences. Although some recorded sequences that include most of the Quaternary consist of single cores, one consists of two or more cores and two others consist of combinations of exposed sections and cores. Climatic information has been extracted from all of them.

One of the earliest published long-core analyses (fig 16-5) relates to the equatorial Pacific Ocean. The fluctuations of the curve are believed to be related to variations in supply of nutrients, in turn dependent on changes in climate. Another curve (fig 16-6) was plotted from the oxygen-isotope ratio in pelagic Foraminifera at many positions in a core from the central Caribbean Sea. The form of the curve through the last 70,000 years or so agrees broadly with curves from terrestrial data (fig 16-10). Because it is not certain how much of the fluctuation has resulted directly from temperature change and how much from change in isotopic composition of seawater (chap 27), it is not possible at present to derive from O^{18}/O^{16} ratios unambiguous conclusions as to temperature.

An interesting feature of the curve is that slopes representing the warming phase are steeper than slopes in the cooling phase. Similar asymmetry is seen in parts of the Bogotá curve in fig 16-11. If the asymmetry is both real and systematic, it suggests that warming is characteristically faster than cooling, and one could ask whether such a difference in rate of change originates in the source of heat or in factors in the response. An attempt has been made to relate the asymmetry to insolation changes arising from factors in the Earth's celestial geometry (Broecker and van Donk, 1970).

Because most of the curve in fig 16-6 is time calibrated only by extrapolation and the curve in fig 16-5 is not time calibrated at all, the two oceanic curves cannot be compared closely. The correlation in fig 16-5 is

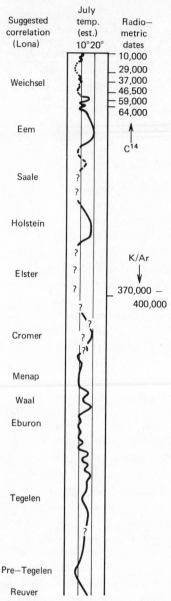

Figure 16-7 Curve showing fluctuation of temperature inferred from pollen stratigraphy in the Netherlands. Generalized from various cores and exposed sections. Time scale not uniform. (Zagwijn, 1963, tables 1, 2.)

arrived at simply by counting the cold phases from the surface downward. Although it may be good near the top, the reliability of such correlations decreases downward in any sequence that lacks either time calibration or recognizable key horizons.

In contrast with oceanic cores, most of the long sequences from terrestrial stations are based on the statistics of fossil pollen. An elaborate example is a curve developed from pollen data at a number of localities in the Netherlands (fig 16-7). In theory such data should provide a temperature record more sensitive in detail than the records pertaining to seawater. Even though the indicated amplitude of fluctuation of July temperature through about 12° is neglected, the variations still indicate variations of climate. The Netherlands curve, whatever its discontinuities, represents a nearly complete record of the Pleistocene. This is because it is almost continuous from around the base of the Cromer unit downward into the Reuver, known to be Pliocene on faunal and floral evidence other than its pollen content. The curve may well constitute a standard against which other European sequences can be compared.

One of those sequences is shown by the curve in fig 16-8, the well-known succession at Leffe in northern Italy, based on cores and surface exposures. The mammalian fossils keyed to it help to place its position in the Pleistocene sequence. It does not include the upper part of the Pleistocene, but bears a considerable resemblance to a low segment of the Netherlands curve.

The curve for Poland (fig 16-9) is much less detailed than the two preceding curves because it is based principally on plant megafossils, which cannot be sampled at many short intervals, as fossil pollen can be. Crude as the curve is, however, it shows substantial fluctuation of inferred temperature, and extends downward, like the Netherlands curve, into strata of unquestioned Pliocene age.

Two completely continuous European curves are those from Tenaghi Philippon in Macedonia and from Padul in southern Spain. The latter (fig 16-10) was constructed from pollen analysis, at rather long sample intervals, of a single core taken from a basin fill and is controlled through the last 50,000 years by C^{14} dates. The proportion of trees to shrubs and herbs, reflected in the pollen data, clearly indicates the temperature changes known to have occurred elsewhere in Europe, as is evident from the column of dates. With analyses of samples taken at much smaller intervals, a very detailed trace of these changes apparently could be obtained. Furthermore, as the bottom of the core does not reach the bottom of the pollen-bearing sequence, deeper coring should make possible extension of the curve into still lower Pleistocene strata. The curve from Macedonia is reproduced by Hammen (his fig 2 in Turekian, ed., 1971).

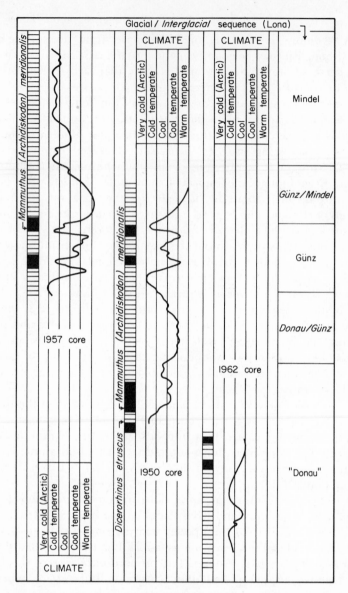

Figure 16-8 Curves showing climatic fluctuations inferred from kinds and abundances of fossil pollen in three closely spaced cores from an area at Leffe in northern Italy. Columnar sections show character of sediments, which are lacustrine: black = peat and lignite; shading = inorganic sediments, chiefly marl and clay. Arrows show stratigraphic positions of occurrence of specific fossil mammals. Aggregate length of cores, ~60m. Mean annual temperature at the coldest points on the curves is ~9° lower than that of today at the locality. (Lona, 1950, p. 169; 1963; Lona and Follieri, 1957, p. 93; Lona, unpub., 1964).

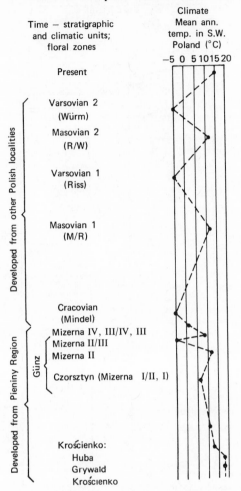

Figure 16-9 Composite chart showing time-stratigraphic and climatic units and floral zones for Poland (Szafer, 1953, p. 78–89; 1954, p. 122, 217–218). At right is curve drawn by R. F. Flint from temperature data cited by Szafer. For better comparison the vertical scale has been arbitrarily fitted to that of fig 16-5 and includes whatever distortion is inherent in that figure.

The longest of the European cores, with a length of 950m, comes from Jásladány, on the Hungarian plain 90km east of Budapest. It represents a sedimentary fill built into a subsiding basin. In it nearly the entire Quaternary and the upper and middle parts of the Pliocene are represented (Rónai, 1968; 1969). The sediments are marine in their Pliocene part and alluvial in the Quaternary part. The generally rich pollen content indicates

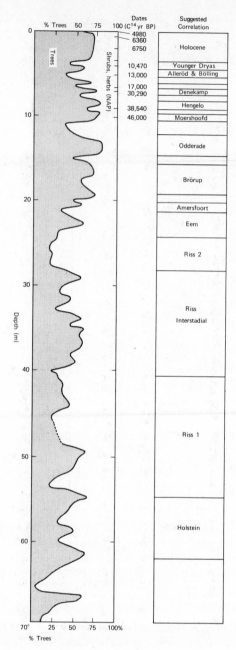

Figure 16-10 Curve showing proportions of AP to NAP in pollen samples from a core in peat and clay, Padul, Spain (lat 37°). The fluctuations are believed to reflect climatic changes. Positions of C[14]-dated samples and the authors' correlations are shown at right. (Florschütz, Menendez Amor, and Wijmstra, 1970, unpublished. Cf. Menendez Amor and Florschütz, 1964.)

a long, cool interval in the late Middle Pliocene, preceded and followed by warm climates. The Quaternary is subdivided by climatic implications into three parts, respectively warm, temperate, and cold. By extrapolation from the existing, measured rate of subsidence, Rónai estimated the length of the early warm interval at 800,000 years, and of the intermediate temperate interval and the late cold interval at 300,000 years each, for a total score of 1,400,000 for Quaternary time.

The Pliocene/Quaternary boundary is drawn, on a basis of mollusks, in the middle of a transition zone 110m thick, representing more than 200,000 years' time on the scale of extrapolation. In Rónai's opinion, the uppermost cold unit in the Quaternary probably represents the Riss, Riss/Würm, and Würm of the Alpine climatic sequence.

Three long cores from North American stations have been or are being analyzed. One, nearly 200m long, was raised from beneath the Great Salt Lake in Utah. As the sediment reflects changes in water depth (presumably of pluvial Lake Bonneville), the stratigraphy has been divided into "arid" and "pluvial" units. Although virtually uncontrolled by radiometric dates, the various parts of the core are estimated by Eardley and Gvosdetsky (1960) to total 800,000 years. The correlation by those authors is based on the assumption that a layer of tephra far down in the sequence is the Pearlette Ash of the eastern Great Plains region, known to be of Kansan age. But as this assumption, although possibly correct, has apparently not been established beyond doubt, correlation of the core sequence is not yet firm.

A second core, 265m long, penetrates the sediments of pluvial Searles Lake in southeastern California (fig 17-4). The stratigraphy embraced by the last 50,000 C^{14} years, well known from many industrial cores, reflects fluctuations of the pluvial lake that are compatible in climate and time with glacial events in the Great Lakes region (chap 17). When study of the long core, nearly six times as long as the industrial cores, has been completed (see partial results in G. I. Smith and Haines, 1964) it should be possible to extend such correlation downward, for comparison with the core sequences in Europe.

A third core, 200m long, from beneath the San Augustin dry lake in western New Mexico, has yielded a generalized pollen curve through approximately the upper half of its length (Clisby and Sears, 1956). The curve, most of which is not time calibrated, shows relative abundance of spruce and desert scrub varying with time. Although this relationship should have climatic significance, no clear resemblance to climatic curves better controlled by dates is apparent in the diagram.

Summing up for these three long cores, we can say that although it possesses excellent potential, North America has not yet furnished a core

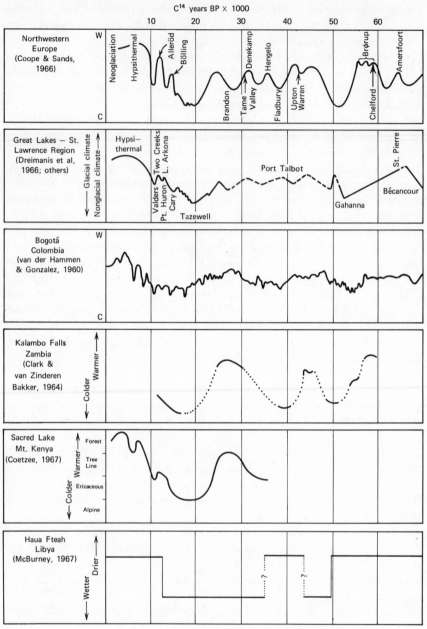

Figure 16-11 Climatic variations in various continents and latitudes within the last 70,000 years, inferred from terrestrial data. No sea-floor stations are represented. Further explanation in text. (Continued on p. 433)

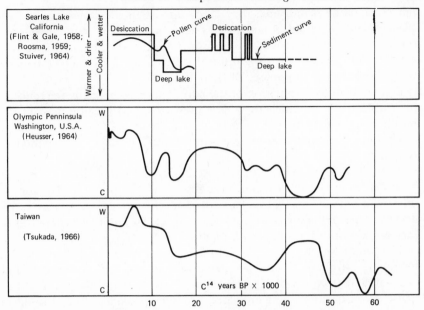

Figure 16-11 Climatic variations in various continents and latitudes within the last 70,000 years, inferred from terrestrial data. No sea-floor stations are represented. Further explanation in text. (Continued from p. 432)

sequence from which a continuous climatic record for the whole Quaternary has been extracted.

Shorter Sequences. Beginning in 1958 there began to appear curves representing fluctuation of one or more climatic parameters through some part of the last 70,000 years or so, at terrestrial rather than oceanic stations. Nine such curves are brought together in fig 16-11. All have at least some C^{14} time calibration, and this makes general comparison possible. The curves have been fitted to a uniform time calibration and the temperature calibrations of some have been replaced with simple "warmer" and "colder" end points.[3] Otherwise they are as constructed by their authors.

Most of the curves involve assumptions, either stated or implied by the authors. The basic data vary; they include pollen diagrams, glacial geology, and physical characteristics of lacustrine sediment. Yet despite these differences the curves are generally similar. Representing a wide though

[3] Tables 16-B and 16-D give selected data on the range of fluctuation of temperature at terrestrial stations.

not ideal range in latitude, they lend support to the concept that late-Pleistocene fluctuation of climate has been broadly contemporaneous, although the data are still inadequate to indicate whether differences between curves reflect true geographical differences or are merely artifacts resulting from the technology of their study.

Because detailed evaluation of the curves must be made by reference to the publications in which they were presented, and because most of the publications cited are rather readily available, the list that follows is short and is confined mainly to locality, source of data, and method of construction.

The *Taiwan* curve is based on a 12m core from beneath a lake at about 750m in central Taiwan, near 24°N lat. Temperature changes are inferred from altitude movements of a species or several species as reflected in percentage changes down the core. An amplified climatic history is given in a later paper (Tsukada, 1967).

The curve for the *Olympic Peninsula* is based on a core nearly 7m long, taken from peat and lacustrine sediments at Humptulips, a few miles inland from the Pacific Ocean near lat 47° and at alt 100m. Climatic reconstruction is again based on inferred changes of altitude of ecologic zones.

The *Searles Lake* curve pertains to a series of commercial cores from a dry lake in southeastern California, near lat 36°. It is based on data presented originally by Flint and Gale and later refined by Stuiver. The basic information consists of alternating deep-lake mineral sediment with fossils, and evaporites that indicate desiccation. The pollen curve, which agrees with the physical data, is the work of Roosma.

Haua Fteah is a large cavern on the Libyan coast, near lat 33°, with floor sediments that have been excavated to a depth of 14m. Through that depth environmental data were obtained from fossil mammals and from O^{18}/O^{16} ratios in marine shells brought into the cavern as food over a long time. The indicated temperature range through time is about 10°.

The *Sacred Lake* curve derives from a core about 11m long, taken from beneath a crater lake at 2440m on the side of Mt. Kenya, in Kenya and almost on the Equator. The curve is constructed from pollen data indicating vertical shifts of recognized ecological zones.

The *Kalambo Falls* curve is based on pollen data in a series of sediment samples from an important archeological site in northern Zambia at 8°30′S lat and at alt 1200m. Again the temperatures are inferred from amplitude of shift of vegetation zones.

The curve from *Bogotá* is drawn from the top 32m of a very long core in lake sediment of the Sabana de Bogotá, near 5°N lat in Colombia and at alt 2560m. Vertical shifts of ecologic zones, through a maximum amplitude of 1300m, constitute the basis for temperature inferences.

The *Great Lakes* curve is composite. In its present version the segment from 25,000 back to 70,000 C^{14} years BP derives from Dreimanis et al., whereas that from 25,000 BP to the present comes, through Flint and Brandtner (1961), from earlier published material cited by those authors. Parameters employed in determination of climate include presence or absence of glacial sediments in local stratigraphy, C^{14}-dated positions of the ice-sheet margin, and pollen data from local sections.

The curve for *western Europe* also is composite. It represents a wide region and is based on pollen analysis as well as data of other kinds. Its "older" part is more closely controlled by C^{14} dates than is the corresponding part of the Great Lakes curve. The labels to the right of the curve are the continental names of certain segments representing relatively mild climate. Those to the left are the names of localities in England that are significant for stratigraphic details and C^{14} dates. A very useful sequence of late-Quaternary climatic data for southeastern Europe and other lands around the eastern part of the Mediterranean Sea are assembled by Farrand (in Turekian, ed., 1971).

We note, in conclusion, that future curves showing late-Pleistocene climate will surely be far more refined than those we have just examined, and may indicate differences caused by differences in geographical position. However, the principal peaks and troughs seen in today's curves seem unlikely to undergo major change.

Oxygen-isotope Curves from Ice Sheets. The curves described in the foregoing paragraphs should be compared with the detailed curve (fig 16-12) constructed from data extracted from a core through the northwestern part of the Greenland Ice Sheet (Dansgaard et al. in Turekian, ed., 1971). The core, 1390m long and ending at the base of the glacier, represents a continuous sequence of annual layers of former snow. It was sampled at a very large number of points for $0^{18}/0^{16}$ ratio [expressed as the deviation $\delta(0^{18})^0/_{00}$ from a standard] in the ice. As explained in chap 27, this ratio, in high-latitude snow, depends mainly on the temperature of condensation at the time of ice formation. Hence changes in it should indicate general changes of climate, with the highest $\delta(0^{18})$ values marking the warmest climate. A curve representing a plot of $\delta(0^{18})$ values along the length of the core should therefore show sequence and relative amplitude of secular temperature change.

However, time calibration of the curve involves difficulties. Although annual incremental layers (essentially varves) are present in the ice sheet and are clearly visible in the upper part, farther down they are squeezed out by compaction and thus cannot be counted individually; yet the fluc-

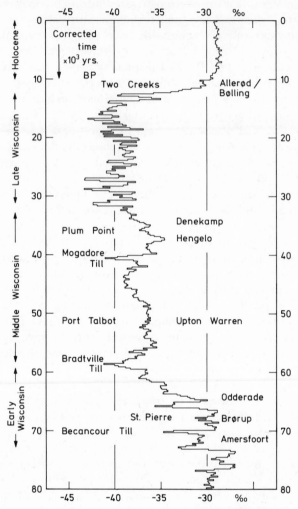

Figure 16-12 Curve representing a large part of the Camp Century, Green-land, ice core, in which δ(0¹⁸) is plotted against time, calibrated as explained in text. The curve compares well with the curves shown in fig 16-11; hence its authors have suggested, by means of labels, possible correlations with strati-graphic units elsewhere. (Dansgaard et al. in Turekian, ed., 1971).

tuations of $\delta(0^{18})$ values through groups of years are preserved. An early version of the curve (Dansgaard et al., 1969) was time calibrated by a mathematical treatment based on the Glen flow law (chap 3). The calibration is not linear, but when the top part of the curve (A.D. 1200 to A.D. 1969; fig 16-13) is compared with instrumental and historical records it appears to be accurate to within $\sim 3\%$.

The curve in fig 16-12 represents a large portion of the core and was time calibrated in a different way. Through Fourier spectra of mean $\delta(0^{18})$ values through the last 10,000 and 157,000 years in the time scale of the earlier version, 350- and 2000-year cycles of variation were discerned. On the assumption that the latter cycle continued downward, the lower part of the curve exceeding 45,000 BP was adjusted to fit the upper part. Despite the fact that much of its time calibration rests on an assumption, the curve as a whole bears a strong resemblance to the curves in fig 16-11, and because of the kind of data from which it was constructed, probably it is more accurate in detail than most of them. If details predicted by the Camp Century core curve show up in future curves based on quite different data, its peculiar worth will be firmly established.

A parallel study (Epstein et al., 1970) of an ice core from Byrd Station (near 80°S; 119°30′W), near the margin of the Antarctic Ice Sheet, resulted in a curve that, although less detailed, is compatible with that based on the Camp Century core. If compatibility persists through refinement of both curves, the probability that climatic change in the two polar hemispheres has been contemporaneous will be very high.

Temperature Ranges. Temperature values and ranges are still far too few to allow the continuous tracing of temperature through time in any terrestrial area. A number of spot values have been determined for various stations, mainly in Europe, but as some stations are in maritime climate and others are in continental climate, they are comparable only in a general way. Such values are assembled in Šegota, 1967; a representative group from that source is listed in table 16-D. We can compare the values with those in fig 16-6 if we bear in mind that the latter figure represents rather low-latitude oceanic stations, and that in it the temperature range is at any rate subject to adjustment. It is much too early to say whether the range of temperature fluctuation in continental regions will be the same for each glacial/interglacial cycle as is suggested for an oceanic region in fig 16-6, on a basis of $0^{18}/0^{16}$ ratios. However, the fluctuation of such ratios is supported qualitatively by the fluctuation, with time, of warm- and cold-water marine faunas, as shown in fig 27-5.

Synchroneity of Fluctuation. Looking at the curves in fig 16-10, we can see enough similarity to justify the tentative statement that climatic fluctuations at least as much as a few thousand years in length have been

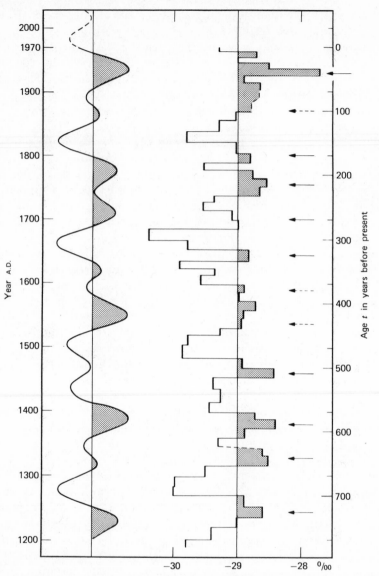

Figure 16-13 Curve developed from the top part of the Camp Century ice core, showing fluctuation of $\delta(0^{18})$ in the ice layers accumulated during the last 780 years. In terms of climate, low temperatures are to the left, high temperatures to the right. (Dansgaard et al. in Turekian, ed., 1971.) The curve is comparable with that for the Sierra Nevada developed by Curry (p. 40 in Schumm and Bradley, eds., 1969).

Table 16-D

Temperatures, Derived from Fossil Organisms and Other Indicators, for Various Stratigraphic Positions. (Selected from Šegota, 1967.)

Stratigraphic position	Region or locality	Mean annual temperature (°C) departure from today's value
Holocene	Europe	+1 to +3
	Southeastern Canada	+2.7
	Utah	+3.2
	Caribbean Sea	+1.5
Weichsel Glacial		
Younger Dryas	Germany	−6 to −8
Alleröd	North and Central Europe	−2 to −4
Older Dryas	Central Europe	−7
Bölling	Central Europe	−3
Late Würm max.	Central Europe	−10 to −15
Fladbury [a]	England	−6
Chelford [a]	England	−2 to −3
Eem (Riss/Würm) Intergl.	Alps	+2
(max.)	Netherlands	+2 to +3
	Czechoslovakia	+2 to +3
	Southeastern Europe	+2 to +3
	Toronto, Ont.	+2 to +3
Saale Glacial	Britain	>−9
Holstein Intergl.	Germany	+2 to +3
	Poland	+2
	Czechoslovakia	+3 to +4

[a] Localities indicated in fig 16–11.

broadly synchronous in North America and Europe, and probably also in other continents as well. Some events of smaller amplitude, such as Bölling and Alleröd, seem likewise to have been nearly contemporaneous, with possible offsets of a few hundred years. A graphic comparison (Heusser, p. 138, in Sawyer, ed., 1966) of pollen-stratigraphic data through the last 15,000 years for North Pacific America and southern Chile supports hemispheric contemporaneity of fluctuations having amplitudes of as little as 1500 years. However, we must exclude from this statement some glacial fluctuations of smaller amplitude, whether or not controlled by climate. On a scale of several tens of thousands of years we note broad intercontinental similarities. Even though we cannot yet pinpoint in time the beginning of the Weichsel Stage or the Wisconsin

Stage,[4] it looks as though the two were time correlatives. Each shows at least one major cold fluctuation around or a little before 50,000 BP and another, possibly more pronounced cold event between 25,000 and 10,000 BP. These two events are referred to in this book, respectively as the Early- and Late-Wisconsin (=early- and late-Weichsel) climatic events (and where appropriate, glaciations). In an earlier book (Flint, 1957, p. 341) the name "classical Wisconsin" was adopted provisionally for the later event, to distinguish it from the earlier one, which had just been recognized and whose date was then still uncertain. With advancing knowledge the need for the provisional term has passed, and it is not used in the present book.

In North America and in western Europe the two glacial events are separated by 20,000 years or more of fluctuating climate, of a character intermediate between that of the two glaciations and that of today. This rather irregular and composite event has been thought of as an intra-Wisconsin (Weichsel) interstade. In view of the rather low temperatures at Chelford and Fladbury in England (table 16-D; fig 16-11), during the intermediate period and during the correlative period in eastern North America (Dreimanis et al., 1966, p. 320) and in North Pacific North America (Heusser, 1964, p. 39), it seems likely that ice sheets were then rather extensive and sealevel therefore rather low.

The curves developed from some of the long sequences mentioned earlier in this chapter suggest that similar intermediate climates may have existed in pre-Wisconsin glacial ages as well. This is not positive, however, because the "long" curves as a group hardly permit climatic correlation far back of the Wisconsin, except by the dubious procedure of counting down from the top, or rather the near end, of the curve. Indeed it may be that instead of several distinct and rather similar glacials separated by distinct interglacials, we have merely a series of irregular fluctuations of various amplitudes, with the peaks and troughs of the largest ones representing the "glacials" and "interglacials" of the stratigraphic record. In either case the curves reproduced in this chapter yield no distinct evidence of periodicity. The matter is still speculative.

Synchroneity of climatic fluctuation, then, rests only on the data for Wisconsin and later time. The general similarities among the temperature curves (fig 16-11) and similarities between fig 16-12 and curves in Epstein et al., 1970 are a principal source of support for synchroneity. Other support is found in C^{14}-dated synchroneity of postglacial climate changes in northern North America with those in northwestern Europe and in

---

[4] The few dates in this range come from U-series measurements, the difficulties with which are mentioned in chap 15. The probable beginning of Wisconsin time is thought to be >70,000 <100,000 years BP.

parts of the southern hemisphere (e.g., Heusser, 1966, p. 300; H. Nichols, 1967) (table 26-F) and in synchroneity of climatic events in the instrumental record between A.D. 1800 and 1940 (e.g., Lamb, 1964). These matters are discussed in chap 30.

SUMMARY

Summarizing the rather diverse material discussed in this chapter, we can note that fluctuation of climate through time has been established principally from fossil organisms, oxygen-isotope ratios, and changes in the snowline. The chief climatic parameter indicated is temperature; its changes seem to have been greater in middle than in low latitudes. Range of fluctuation on a secular basis was probably not more than 7° and may have been less. Precipitation fluctuated, and may have diminished overall during glaciations, but locally may have increased. Building large ice sheets through normal operation of the hydrologic cycle poses no difficulties as to quantity of precipitation.

Glaciation occurred in the Miocene and Pliocene as well as in the Quaternary. Cold periods were more numerous than the four periods of the classical literature. The last (Wisconsin=Weichsel) cold period began probably >70,000 <100,000 years ago. Major fluctuations seem to have been synchronous throughout the world.

CHAPTER 17

Pluvial Features

Chapter 2 discussed the origin of the concept of former pluvial ages in regions not extensively glaciated. The features on which the concept was based at first were extensive lakes in areas now too dry to support lakes of such area and depth. The term pluvial, however, was originally applied to coarse alluvium in western Europe, and was later gradually transferred to the lakes and other features of dry regions with which we now associate it. In an effort to approach precision, we shall take *pluvial* (noun or adjective) to mean a climatic regimen of sufficient duration to be represented in the physical or organic record, and in which the precipitation/evaporation ratio results in greater net moisture available for water bodies and organisms than is available in the same area today or in the preceding regimen. Conversely, we shall take *nonpluvial* to mean a climatic regimen in which the precipitation/evaporation ratio is less than that of today, or distinctly less than that of a preceding or following pluvial regimen.

Because lakes were the earliest recognized and are among the most obvious of pluvial features, we shall devote most of this chapter to them. Problems of pluvials in specific regions are discussed in appropriate chapters.

PLUVIAL LAKES

Evidence of Pluvial Origin

In 1776, Vélez de Escalante (1943, p. 75) found shells near Utah Lake and reasoned that formerly a much larger lake had covered the area. Apparently this was the first recognized indication of what is now called Ancient Lake Bonneville. In the vast, arid Basin-and-Range region in western United States many former lakes, some of them hundreds of meters in depth and most lacking an outlet, are now known. As a group they were primarily pluvial. Supporting evidence includes C^{14}-dated synchrony of fluctuation of lakes in separate basins, and synchrony of deep-lake phases with glaciation. Local concentration of meltwater was not a major factor in the filling of most lakes because glaciers did not enter their watersheds, although in some basins meltwater played a part.

Whereas such data support the pluvial origin of the lakes as a group, they do not rule out nonclimatic influences in the creation or fluctuation of at least some lakes. Indeed, in other regions less well studied, where there are few ancient lakes, nonclimatic influences may be very difficult to rule out. Such influences include tectonic movement, the creation of dams consisting of lava flows, fans or avalanched debris, and shifts in the positions of the mouths of influent streams. Hence some supposed pluvial lakes may not in fact be pluvial, or at least not exclusively pluvial. Baringo Lake, on the floor of the Rift Valley in Kenya, has twice risen in response to the building of volcanic dams and twice fallen in consequence of breaching of its basin rim by faulting, all within the later half of Pleistocene time. Lake Rudolf has been similarly affected (Fuchs, 1950).

Lake Bonneville experienced a lowering of at least 100m because of erosion of its outlet, independently of climatic change. Because climatic controls can be distinguished from local controls only through meticulous field study, equation of lake fluctuation with climatic change must be made with caution.

However, the wide distribution, in arid and semiarid areas on all continents, of surely or apparently pluvial lakes, generally in latitudes lower than those of the extensively glaciated regions, supports the concept derived from the distribution and chronology of glaciation, that major Quaternary climatic changes were secular.

Temporal Relation to Glaciation

As we noted in chap 2, it occurred to Jamieson and to Lartet in the 1860's that the climatic conditions responsible for the building of ice sheets might also be accountable for the expansion of lakes, and the assumption that this was the case underlay much of the thinking about the matter. Lartet had found evidence of glaciation in Syrian highlands, and on the ground of general probability he correlated this with a former high stand of the Dead Sea, a saline lake.

Twenty years later similar lake deposits in Utah were found, so related to glacial drift as to indicate that a lake maximum was probably synchronous with a glacial maximum (Gilbert, 1890, p. 305–311). High-water lake phases occurred only slightly after glacial maxima. This discovery, made in the basin of ancient Lake Bonneville in Utah, is confirmed by an analogous relationship between lake sediments and outwash in the same basin.

Synchrony is established also by the relation of shorelines of Lake Russell and end moraines of former Sierra Nevada glaciers (Putnam, 1950; Gilbert, 1890, p. 311–315). Radiocarbon dates on sediments of various pluvial lakes indicate synchrony of high-lake phases and glacial maxima. From stratigraphic relations and radiometric dates, therefore, it appears that many pluvial lakes are related in time to glaciations.

By analogy with North American data it is argued that lakes in other dry regions such as North Africa, Western Asia, and Central Asia likewise fluctuated with glacial fluctuations. Apart from local exceptions where nonclimatic causes are involved, this is probably true. It is not necessarily true, however, that pluvial conditions close to the Equator were synchronous with glacial ages. Direct evidence on this matter is scanty, but one model (fig 16-2) relates such conditions to nonglacial times.

Relation to Climatic Parameters

Although climate is accepted as the primary control of the fluctuation of many lakes in arid regions, and although one can say confidently that the precipitation/evaporation ratio is the significant parameter, quantification of the ratio is another matter. Obviously the ratio of lake depth to lake area will affect evaporation, as will plant cover on the watershed. Attempts at quantification involve assumptions, and results are thus far no more than reasonable approximations.

In analyzing the various factors as they apply to Pluvial Lake Estancia in New Mexico, L. B. Leopold (1951) developed a model, based partly on the extent of lowering of the snowline, in which temperature was reduced and evaporation was decreased. At today's temperature 760mm of precipitation—more than double today's value—would be required to maintain the expanded lake. At present temperatures this much moisture would have brought pine forests down the mountains to the lake shore; yet there is no evidence that this occurred. Hence increased precipitation without reduced temperature cannot explain the lake. The method could be applied to other lakes as well.

It is improbable, however, that reduced temperature alone, without increased precipitation, could explain pluvial lakes. Although it may be that over some ice-sheet margins precipitation was less than it is today at the same places, pluvial lakes appear to imply increased precipitation, not necessarily of a large order and certainly varying from place to place. Lakes without outlet are sensitive to slight changes in evaporation or precipitation, as is evident in the recent record of the Great Salt Lake (table 30-A).

In an analysis of Pluvial Lake Lahontan in Nevada, Broecker and Orr (1958) calculated that with annual temperature 5° less, precipitation 80% greater, and evaporation 30% less than today's, the pluvial lake could be reconstituted. In analyzing the pluvial lake in Spring Valley, Nevada, Snyder and Langbein (1962) came to the comparable conclusion that with evaporation 30% less and precipitation 66% greater than today's, the lake could re-form. Under the calculated precipitation/evaporation ratio, precipitation did not exceed evaporation, in which case the district, and very likely the entire Basin-and-Range Province, would have attained only a semiarid condition in pluvial times, never becoming humid.

The aggregate volume of water stored in pluvial lakes was so small compared with that stored as ice and snow on the lands that it has commonly been neglected in calculations of the hydrologic cycle during glacial ages. The margin of uncertainty as to the aggregate volume of the former ice sheets (chap 4) is greater than the volume of the lakes.

Stratigraphy and Morphology

Although beaches, wave-cut cliffs, terraces, deltas, and other obvious features of shore morphology formed the basis of the earlier interpretations of pluvial lakes, attention has turned increasingly to sediments and especially to subsurface stratigraphy, in which cores have played an important part. The general relationships commonly encountered in lakes that have been studied intensively are shown schematically in fig 17-1. In more detail, facies changes from beach to lake-floor sediments are seen, and in the lake sediments are fossils of freshwater fishes and invertebrates, as well as pollen blown or washed in from surrounding slopes. Repeated fluctuation of lakes is indicated by overlapping of sedimentary facies, by

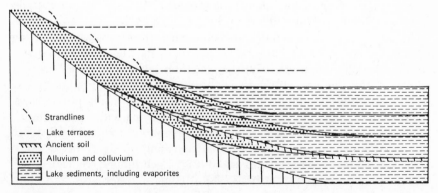

Figure 17-1 Schematic section of part of a pluvial lake basin now dry, showing common relationship between lake sediments, other sediments, ancient soils, and strandlines. Three former pluvial phases are indicated.

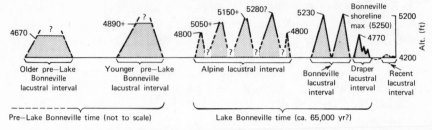

Figure 17-2   Fluctuation, with time, of Pluvial Lake Bonneville and its predecessors. Adapted from Morrison (1965, fig 2; Morrison in H. E. Wright and Frey, 1965a, p. 274).

interbedding of lake sediments with subaerial sediments, by buried soils, and by evaporites recording desiccation or near desiccation of lake floors. In some basins these fluctuations are as complex as those implied elsewhere by features indicating repeated advance and retreat of the termini of glaciers. The trace of fluctuating level through time can be constructed graphically from a pluvial lake whose stratigraphy is sufficiently well known (fig 17-2).

Lakes in North America

The Basin-and-Range region contains more than 141 closed basins (Hubbs and Miller, 1948, p. 148), mostly of tectonic origin. Most of them bear evidence of former lakes or expansions of existing lakes, but only about 110 former lakes have been identified thus far because some lakes flooded more than one basin. The former Lakes Bonneville and Lahontan together occupied more than 25 basins.

The distribution of former lakes is shown in fig 17-3; tabular data on the water bodies will be found in Snyder et al., 1964. The saline lakes that occur in these basins today are confined, with two exceptions having extraordinary sources of water, to the northern part of the region at and north of lat 39°. Meinzer (1922) suggested, on a basis of relative areas of the lakes, that the least dry parts of the Basin-and-Range province today are comparable with the driest parts of the province during pluvial ages. Southwestern New Mexico and southeastern Arizona had about as much lake area during the pluvials as southern Oregon has today. The distance between the two areas is 8° to 10° of latitude, and the difference between their mean annual temperatures is about 8°. Even though we neglect variations in area and depth of basins, local climatic differences, and other variable factors, we have a rough indication that temperatures in the southern part of the region were reduced at most by this amount. How-

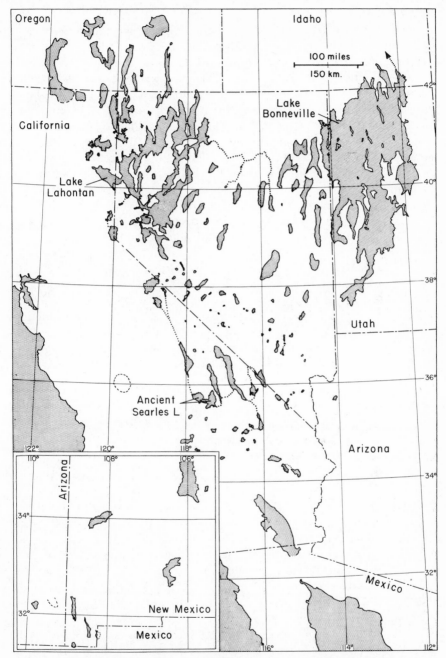

Figure 17-3 Sketch map showing pluvial lakes in western United States. Dotted lines represent overflow stream channels. (Compiled from several sources.)

447

ever, if, as is likely, the north-south gradients of temperature (hence evaporation) and (or) precipitation differed from those of the present, this value would be altered, probably toward a smaller one.

Lakes Bonneville and Lahontan. Lake Bonneville (tables 20-A, 20-B; Morrison in H. E. Wright and Frey, eds., 1965a, p. 273), by far the largest of the former lakes in western North America, occupied a number of coalescent intermont basins in Utah, Nevada, and Idaho. Today the only water bodies in those basins are the highly saline Great Salt, Provo, and Sevier Lakes. When at its maximum Lake Bonneville had an area of more than 50,000km² (nearly equal to that of the existing Lake Michigan) and an extreme depth of more than 330m. Fluctuations of the lake and its predecessors, as at present understood, are summarized in fig 17-2.

The basin-floor deposits record two deep lakes antecedent to Lake Bonneville (the lake recorded by conspicuous strandlines) and possibly extending far back into the Pleistocene. The relations of Lake Bonneville to ancient soils imply that that water body postdates a major interglacial, probably the Sangamon, and that it rose to high levels in two groups of fluctuations, separated by a time of low water. Two of the rises in the second group rose exceptionally high. One of these, the rise responsible for the "Bonneville" shoreline, overflowed, constituting a flood that left a conspicuous record of erosion and sediments (Malde, 1968). Although the high rises were originally believed to have been controlled by climate, it was later suspected that the Bear River, which today contributes heavily to inflow into the basin, was responsible. It seems likely that the river formerly detoured the basin and so made no contribution to the earlier lakes. Then it was diverted, by a lava flow, into the Bonneville basin, into which it brought a large increment of "new" water. This idea is interesting as an illustration of the mingling of nonclimatic with climatic influences on the level of a lake without outlet. If it is valid, Lake Bonneville was not at all times wholly "pluvial."

A core 200m long, from Saltair at the shore of Great Salt Lake, penetrates sediments possibly older than some of those exposed, and shows (Eardley and Gvosdetsky, 1960) variations in carbonate and clay-mineral content, as well as buried soils, invertebrate fossils, and fine-grained tephra. These features are believed, reasonably, to indicate several alternations of pluvial and arid climate. Correlation of the core stratigraphy with that derived from other data has been attempted (Eardley, 1968), as mentioned in chap 16.

At its maximum, the area of Lake Lahontan (Russell, 1885; Morrison and Wright, eds., 1967), the next largest lake, was less than half that of Bonneville, and the maximum depth about two-thirds that of the larger lake. The record of fluctuation of Lahontan is complex. It resembles that of Bonneville at least in its broad aspect; but as it has been less thoroughly studied, close correlation with the record of its sister lake is not yet complete. Apparently, unlike Bonneville, Lahontan did not overflow at any time.

Ancient Searles Lake. In southeastern California, a region very arid today, strandlines and floor sediments establish the existence of a former chain of large lakes, some of which were more than 200m in depth. One of these occupied a basin represented in part by the basin of the existing Searles Dry Lake, the site of a chemical industry whose core borings have yielded important data on the Searles Lake of pluvial times. At times that lake was the terminus of the chain; at others it overflowed into nearby basins. Lake-floor sediments include both silt/clay units and evaporites and reach depths of at least 265m below the existing dry lake. Generalized stratigraphy and inferred fluctuations of the lake appear in figs 17-4 and 17-5, and tables 20-A and 20-B. The two thicker "mud" layers represent deep-lake phases; the "salt" layers are evaporites and represent very shallow phases, possibly with complete desiccation at times.

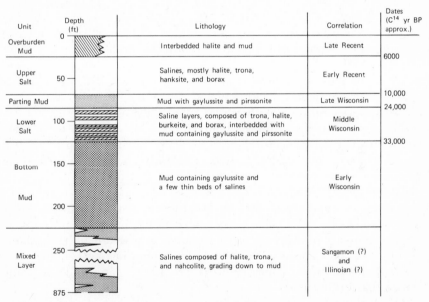

Unit	Depth (ft)	Lithology	Correlation	Dates (C¹⁴ yr BP approx.)
Overburden Mud	0	Interbedded halite and mud	Late Recent	
Upper Salt	50	Salines, mostly halite, trona, hanksite, and borax	Early Recent	6000
Parting Mud		Mud with gaylussite and pirssonite	Late Wisconsin	10,000 24,000
Lower Salt	100	Saline layers, composed of trona, halite, burkeite, and borax, interbedded with mud containing gaylussite and pirssonite	Middle Wisconsin	33,000
Bottom Mud	150 200	Mud containing gaylussite and a few thin beds of salines	Early Wisconsin	
Mixed Layer	250 875	Salines composed of halite, trona, and nahcolite, grading down to mud	Sangamon (?) and Illinoian (?)	

Figure 17-4 Subsurface stratigraphy of Searles Lake. (G. I. Smith in Morrison and Wright, eds., 1967, p. 296; based on data from Flint and Gale, Haines, Smith, and Stuiver.) From estimated rates of sedimentation, the age of the base of the Bottom Mud is considered to be 130,000 years, an unusually large estimate for the beginning of Wisconsin time.

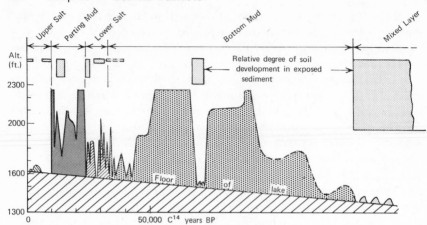

Figure 17-5 Inferred fluctuation of Ancient Searles Lake. Time calibration through latest 45,000 years is based on C^{14} dates. Earlier events are drawn to same scale based on extrapolation. (After G. I. Smith in Morrison and Wright, eds., 1967, p. 300.)

The "Parting Mud," the most closely studied of the deep-lake sediments, contains fresh-water mollusks that indirectly indicate temperatures around 4° lower than today's (if rainfall is neglected), and pollen that indicates expanded woodland around the lake. Many C^{14} dates show this "mud" layer was deposited beginning about 24,000 BP and ending about 10,000 BP. Deposition may, however, have ended as early as 12,000 BP, owing to the former CO_2 in the sampled lake carbonate having possibly not been in equilibrium with the atmosphere when it was precipitated. In either case, however, the dates cited nearly coincide with the Late Wisconsin glaciation of the southern Great Lakes region. They confirm the time coincidence of a pluvial with a glacial climate, at least for North America (Flint and Gale, 1958; see also Stuiver, 1964). Additional information (dates and other data) is supplied by G. I. Smith et al. (1967).

An assemblage of many C^{14} dates (Broecker and Kaufman, 1965) shows that during about the last 35,000 years the Bonneville and Lahontan basins experienced similar climatic fluctuation. The climatic sequence is similar to that in the Searles basin.

The Basin-and-Range region is not the only North American province that contained pluvial lakes. Many such lakes existed in the Great Plains province, notably in Texas, in basins caused by deflation, solution, and other nontectonic activities. One of these was Ancient Lake Lomax, about 30km in diameter and more than 15m deep, in western Texas. Its age as determined from fossil mollusks is Wisconsin (Frye and Leonard in Morrison and Wright, eds., 1968, p. 519).

The North American record of pluvial lakes is still little known, but future investigations promise a rich harvest of information on climatic changes. In most of the stratigraphy, ancient soils, and strandlines known thus far we have only the later part of the record, but longer cores from beneath the lake floors may supply some or all of the earlier part.

Lakes in Mexico and South America

Data on pluvial lakes in the Americas, south of the United States, are meager. Although such lakes are said (Jaeger, 1926) to have been abundant throughout the Mexican plateau, details are known only for the vicinity of Mexico City, where a thickness of at least 75m of lacustrine sediments, grading into stream deposits, is known from surface exposures and core borings. The section, which includes pedocal soils, appears to reflect at least three moist/dry alternations of climate, perhaps extending well back through Wisconsin time. Lake Chapala, south of Guadalajara, Jalisco, appears to have been higher in late-Pleistocene time, covering a far greater area than does the existing lake.

In the Atacama Desert in northern Chile, and adjacent dry parts of Argentina and Bolivia, saline and dry lakes are abundant (fig 17-6). Many of them are bordered by features recording former larger lakes (cf. Troll, 1928; Vita-Finzi, 1959). The basin of Lake Titicaca, at 16°S lat on the Peru-Bolivia border, is fringed by laminated clay as much as 100m above the present lake. From this sediment a former much larger lake (Lake Ballivián) is inferred. As the clay is overlain in places by outwash, it antedates at least one glaciation (Newell, 1946). Another pluvial lake (Lake Minchin) occupied confluent basins farther southeast in Bolivia.

Lakes in Asia

Aral-Caspian-Black Sea System. (Erinç, 1954; Pfannenstiel, 1951; summary by K. W. Butzer, 1958, p. 27.) The Caspian Sea is the world's largest lake. It is fed by the Volga and Ural Rivers, and its surface lies well below sealevel. Strandlines are present around its basin up to at least 50m above the lake. They are faint; possibly the water surface fluctuated so rapidly that it did not stand long at any one position. The distribution of both strandlines and fossil-bearing lake-floor sediments indicates that at one or more times (apparently glacial ages) the water rose high enough to become confluent with water in the basin of the Aral Sea, more than 600km to the east, and also to flood the Volga at least as far upstream as Kazan. As the lake surface fell, water from the expanded Aral Sea flowed into the Caspian through the Usboi, a remarkable mean-

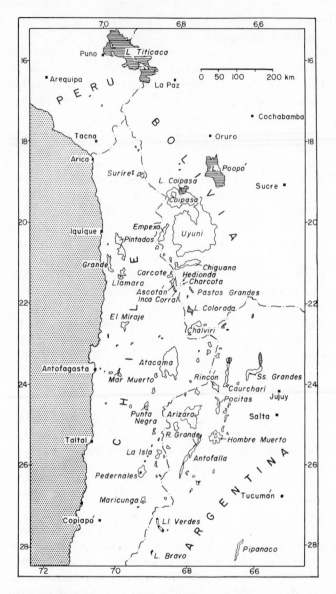

Figure 17-6 Sketch map showing basins in the Altiplano of Chile, Argentina, Bolivia, and Peru between lats 16° and 28°. Areas of dry lakes are open; wet lakes are shaded. Only the larger dry lakes are named. Basin-floor altitudes vary between 500m near the coast and about 3800m in the east. The map shows 113 basins, but additional, mostly very small basins occur within the area shown, and still other basins are present south of lat 28°.

452

dering channel 1km wide, 20m deep, and now dry. Similarly the Caspian was temporarily confluent, through the Manytch depression (now followed by the Baku-Rostov railway) 600km long, with the Sea of Azov and the Black Sea. The Black Sea was then a lake, with a surface well above its present surface. It overflowed at the Bosporus into the Mediterranean, whose level was then glacially low. Later the Black Sea lake was lowered and the Manytch depression became a river flowing into this lake from the Caspian.

These events were repeated, with some variations, more than once, but correlation with outside events is tentative. Table 17-A shows one recent correlation. Altitudes of the earlier features are not given, because tectonic movements that followed the Chauda-Bakinian sedimentation have complicated their meaning. It is considered probable that the Karangatian and Atelian sediments represent the last interglaciation and that possibly the two earlier regressions represent the glaciation next preceding it.

Unlike the lakes of western North America, the Aral-Caspian-Black Sea system of water bodies was augmented by large volumes of glacial meltwater, and so was pluvial only in part. The Aral Sea received increments from the glaciers of the Pamir Mountains and Tyan Shan in Central Asia, via the Oxus River (Amu Darya) and Syr Darya. The Caspian received the vast discharge of the Volga, which drained a large sector of the Scandinavian Ice Sheet. The Black Sea water body received much meltwater from the ice sheet via the Don and the Dnepr, and from Alps glaciers via the Danube. In addition, Caucasus glaciers contributed to the waters of the Black Sea and Caspian basins. Discharge from those water bodies through the Bosporus into the Mediterranean must have been large, because it included the drainage of most of central and eastern Europe.

Lake Lisan. The Dead Sea, an intensely saline lake fed chiefly by the River Jordan, is now about 80km long and 390m deep, and its surface stands at the remarkable altitude of −397m. As we have noted, Lartet early recognized pluvial features around the lake. Subsequent studies by Blanckenhorn, Picard (1943, p. 151–158), and Neev and Emery (1967; also Farrand in Turekian, ed., 1971) have established a history (fig 17-7) not unlike that of Ancient Searles Lake. Core borings have revealed, beneath the present lake, clay/silt units alternating with evaporites, and strandlines and ancient floor sediments have been mapped in the surrounding area. The Ancient Lake Lisan stood as high as −180m, was nearly three times the length of the Dead Sea, and was at least 190m deep. The greenish-gray, finely laminated floor sediments exposed above the present lake are very similar to those exposed around Searles Dry Lake. Radiocarbon dates on driftwood and inorganic carbonates, extrap-

Table 17-A

Quaternary Sequence Around the Black and Caspian Seas (modified from an unpublished chart by P. V. Federov, 1960).

Black Sea[a]	Caspian Sea	
	Stages	Features and regressions [a]
Nymphean terrace (1 to 2)	New Caspian stage	Late New Caspian terrace (−23)
		Derbentian regression
Thanagoriisk regression (−2 to −4)		Maximum New Caspian terrace (−22)
		Chelekenian regression
New Black Sea terrace (4 to 5)		Early New Caspian beds (−24)
Old Black Sea beds (−5 to −10)		Mangyshlak continental suite (−50)
New Euxinian beds (−10 to −40)	Khvalynian stage	Upper Khvalynian horizon (terraces: −16, −11, −2)
		Regression −50
Epi–Karangatian continental suite (Profound regression)		Lower Khvalynian horizon (terraces: 25, 35, 48)
Late Karangatian terrace (+12 to +14)		Atelian horizon: Continental suite / Marine ("Girkanian") beds
Early Karangatian terrace (+22 to +25)	Khazarian stage	Profound regression
Regression		Upper Khazarian horizon
Euxino–Uzunlarian terrace (+35 to +40)		Lower Khazarian horizon
Old Euxinian terrace (60)		
Regression		Urundzhikian horizon
Chaudian–Bakinian terrace	Bakinian stage	Upper Bakinian horizon
Lower Chaudian horizon		Lower Bakinian horizon
Tanaisian (Khaprovian) continental suite		Tiurkian continental suite
Gurian stage (Upper Pliocene)	Apsheronian stage (Upper Pliocene)	

[a] Numbers refer to altitudes, in meters, above or below present sea level (Black Sea) and above or below present level of Caspian Sea (alt. +28 m).

olated to the base of the clay facies, suggest that the latest pluvial began around 60,000 to 70,000 years ago, or longer if sediments transitional from underlying evaporites are taken into account. Postpluvial desiccation started around 20,000, with a brief recurrence of wetter conditions around 10,000 BP. The general climate/time agreement with pluvial lakes in western United States is good.

Other Asiatic Lakes. Because most of them have been seen only in re-

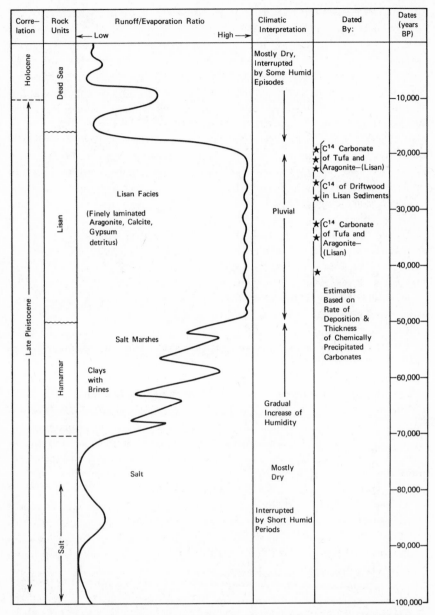

Figure 17-7 Inferred fluctuation of climate through time in the Dead Sea region, as interpreted by Neev and Emery (1967, p. 27).

connaissance, pluvial lakes in the rest of Asia are represented mostly by brief, spotty data, some of which are cited here. In central Turkey several basins without outlet contained more water during pluvial ages than they do now (Louis, 1938; Huntington, 1914, p. 564). Lake Burdur rose to an altitude 95m above its level of 1950 and, like Bonneville, overflowed. Strandlines mark three lower positions of the lake. Lake Tuz rose about 75m above its existing level but had no outlet. Konya basin is now dry, but strandlines and lacustrine sediments with fossils demonstrate a former lake. Lake Van has abandoned strandlines at 14, 30, and 60m above its existing surface. Lake Iznik has strandlines at 15, 45, and 55m above its 1949 level. These strandlines are not correlated with stratigraphic units elsewhere. In northwestern Iran, Lake Urmia, during one or more glacial ages, expanded to nearly twice its present area (Bobek, 1937, p. 165).

The basin of the two Kara Kul lakes on the Pamir Plateau in the Tadzhik SSR has two sets of abandoned strandlines, a comparatively fresh series less than 45m above the present lakes and an older series 45 to 97m above the lakes. These suggest two pluvial phases, which were held by Pumpelly (1905, p. 138) to be directly related to glacial maxima in the adjacent Trans-Alai Mountains.

Both lake-floor sediments and terraces establish the fact that Lake Issyk Kul in the Kirghiz SSR was greatly expanded at least once, and that this expansion coincided with an expansion of glaciers in the Tyan Shan (Prinz, 1909).

The deposits around the basin of Shor Kul, a salt lake in Sinkiang, 130km northeast of Kashgar, yield the following record. The lake expanded to more than 100m above its present level. Desiccation followed, during which the lake floor was covered with alluvium and in places channeled. Thereafter the lake re-expanded to about 30m above its present level. It is not apparent whether desiccation occurred before the present lake was established (Pumpelly, 1905, p. 210).

Sairam Nor, a lake in Sinkiang northeast of Alma Ata, was formerly 60m higher than now (Carruthers, 1913, p. 619). Lop Nor, a large, shallow lake at the mouth of Tarim River in central Sinkiang, has at least six strandlines around its basin, at altitudes up to 180m above its present surface (Huntington, 1907, p. 356). Between expansions the lake contracted to small proportions. Penck (1931, p. 13) thought it likely that the lowest part of the Tarim basin had been the site of a very large pluvial lake. The Turfan basin in eastern Sinkiang, 320km northeast of Lop Nor, has two abandoned strandlines, the higher of which is 60m above the basin floor (Huntington, 1907, p. 356). At least a dozen additional pluvial lakes in Sinkiang are probable. Strandlines as high as 133m above existing lakes are reported from basins in Tibet. Many basins with indications of former pluvial lakes exist in southern and southeastern USSR and in Tibet.

In Indian Tibet the Pang-gong Tso and Pangur Tso basins contained contemporaneous lakes during at least two glacial episodes, the later generation having been smaller than the earlier. In addition, distinct variations in lake volume have occurred since the climax of the latest glaciation (Hutchinson, 1939; Huntington, 1906).

Lakes in Africa

Basins in dry parts of Africa formerly and repeatedly contained large lakes. Beginning in the northeastern Sahara, in the Libyan Desert, we find in the Fayum deflation basin (alt −53m), now dry, evidence of former high-water phases that were fed by overflow from the Nile and hence were not in themselves pluvial lakes. In the Kharga basin farther south, layers of tufa seem to record pluvial phases, but the sequence is complicated by other factors.[1] Near Tibesti in the southeastern Sahara a caldera, now dry, contains the record of a former lake more than 350m deep. Dates from fossil shells, at two altitudes separated by 250m of lake sediment, range between 14,500 and 15,000 C^{14} years BP and are clearly compatible with the late Weichsel glaciation. Likewise the literature on the central and western Sahara contains references to lacustrine sediments, some with fossils implying moister climate, but the geology of the occurrences is not discussed in sufficient detail to make it clear which, if any, of the inferred lakes were pluvial.

Beyond the southern border of the central Sahara, the Lake Chad region appears to yield reliable evidence of a pluvial lake. According to Grove and Pullan (in F. C. Howell and Bourlière, eds., 1963, p. 230) the basin of Lake Chad displays a system of beaches and bars ∼50m above the existing lake. The former lake was 950km long, with an area of >300,000km², nearly that of the Caspian Sea. Apparently it overflowed via the Benue River to the Atlantic. If air temperature 5° lower than today's, based on glacial data from East Africa, is assumed, these authors visualized annual precipitation approaching 1600mm as necessary to balance evaporation losses, even if loss by overflow is neglected. Moreau (1963) thought that northward shift of the equatorial rain belt through about 5° latitude could have provided the necessary water. A much lower, younger strandline lies 6m above Lake Chad. Radiocarbon dates on tufa and other substances (Faure, 1966) indicate the presence of a lake

---

[1] Review of literature on the two basins in Butzer, 1958, p. 68–74; on the Saharan region as a whole in Faure, 1969.

~22,000 and ~9000 to 3000 years ago, but whether these represent one or two lake phases is not known. The fact that the water of Lake Chad is nearly fresh suggests that it has been reconstituted recently, after a period of desiccation. Other, compatible data on the Chad region are given by Servant et al. (1969).

In the southwest, the dry Makarikari basin in Botswana, in the extreme northern part of the Kalahari "desert," bears massive strandlines that define a lake ~34,000km^2 in area and as much as 45m deep. Other basins elsewhere in the Kalahari likewise contained lakes. In some there is evidence of post-lake dune building, the dunes being now inactive, fixed by vegetation. (A. T. Grove, unpublished.)

In the Eastern Rift Valley in Kenya, fluctuation of several lakes has been inferred from strandlines and other geologic features. In a review of studies of some of these lakes, Flint (1959a) called attention to the likelihood of recent movements of the crust and volcanic activity in this tectonically active belt, but thought some climatic influence was probable in the fluctuation of Lake Magadi and of a lake or lakes in the Naivasha sector. That opinion was substantiated and extended by later, detailed study of a core (Richardson, 1966) and strandlines (Washbourn, 1967), summarized in chap 26.

Still other pluvial lakes are reported from the highlands of southern Ethiopia. The scanty information about African pluvial lakes would be greatly enriched by data from core borings. A few cores now exist, and their detailed study can be expected to shed new light on a rather murky subject.

Lakes in Australia

Information about former lakes in Australia is not abundant and there are uncertainties as to the extent of climatic influence in the fluctuation of some of them. The dry interior of the continent abounds in dry lakes, about the history of which little is known. In South Australia ancient Lake Dieri (area ~100,000km^2; maximum depth >50m) is represented by at least four abandoned strandlines, embracing several dry basins including that of Lake Eyre. The ancient lake overflowed southward to the ocean. (David, 1950, v. 1, p. 616; E. D. Gill, 1961.) In northern Victoria the fluctuation and even the creation of some former lakes, represented by strandlines and shore dunes, has been held to be explicable by faulting, at least in large part, rather than by climatic variation (Bowler and Harford, 1966). However, other lakes in both northern and southern Victoria are believed to have fluctuated with climate in a sensitive manner, as developed in a detailed study by Bowler (1970).

Two lakes in which climatic influence is thought likely are Lakes Colongulac in western Victoria, and George, 30km northeast of Canberra. Shells in emerged sediments of Colongulac, now dry, date ~14,000 C[14] years, a time still glacial in North America and Europe. Lake George, now ~150km² in area, has two abandoned strandlines (at 13m and 30m). Galloway (1965) argued from various data that the high-level phases of both lakes coincided with lower, not higher precipitation values than those of today; if this was the case the lakes were pluvial only if their existence depended on lowered evaporation despite decreased precipitation. Although this opinion is opposed to that derived from study of North American lakes, knowledge of glacial-age climates is too imperfect to justify hostility toward any opinion that runs counter to general belief. Australian climates may have differed significantly from American ones.

Summary

Summarizing the climatic evidence afforded by lake basins in the dry belts, we can say that alternation of pluvial and dry phases occurred in North America, Asia, probably Africa, and possibly Australia. In major North American lakes pluvials were essentially coincident with glacial maxima. By analogy, such coincidence is expectable in other regions, though not necessarily universally. Support for a rather general coincidence of pluvials with glacials lies in the probability that growth of ice sheets and sea ice in higher latitudes was accompanied by displacement of the Arctic and Antarctic convergences toward the equator, traversing regions normally dry.

The problem of the number of pluvials and interpluvials, like that of the number of layers of glacial drift, is not yet solved. Nowhere are four glacial stages exposed in any single section of drift or at any one locality. The number of sections exposing as many as three drifts is small, though two drifts exposed at one place are fairly common. On the other hand the prospects for the presence of long, nearly continuous sequences in lake basins are better. Sections with alternating fresh-water lake deposits, evaporites, and in some places also erosion channels and alluvium should be present, having been preserved, except for small losses through deflation during dry phases. The stratigraphy of long cores taken from terrestrial basins is described briefly in chap 16. That more cores do not exist is a result mainly of the limited number of attempts made to obtain them; when more are at hand the relevant climatic history is sure to be extended.

OTHER FEATURES
IMPLYING CHANGES IN
EVAPORATION/PRECIPITATION RATIO

Some of the features, both physical and organic, listed in table 16-A, have been used as a basis for inference as to former pluvial and non-pluvial climates. Although these features are mentioned in chap 16, we should especially call attention here to anomalies in the ranges of living organisms as a possible key to former changes of climate, especially as regards changes in effective moisture.

One of the more interesting of such anomalies is the disjunct distribution of montane birds through a large part of the African continent.

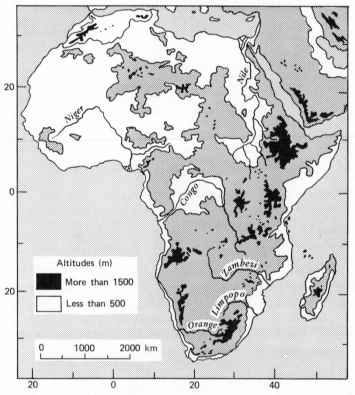

Figure 17-8 Sketch map of Africa showing areas above and below critical altitudes. (Adapted from Moreau, 1963.)

Much of southern, eastern, and north-central Africa consists of savanna and steppe ranging in altitude between a few hundred and a thousand meters. Standing as isolated "islands" above this terrain are many montane areas (fig 17-8) more than 1500m in altitude, receiving precipitation of more than 1000mm, and mostly clothed with forest. At the higher altitudes in such forests is a bird fauna believed to be incapable of adapting to a nonforest environment. Accordingly it is thought likely that the distribution of both forest and birds was formerly continuous (Moreau, 1963; Moreau in F. C. Howell and Bourlière, eds., 1963, p. 28). It is held that general reduction of temperature by 5° would have caused vegeta-

Table 17-B

Sequence of Quaternary Climates in Atlantic Morocco, Inferred from Fossil Faunas and Floras and from Physical Features, with Suggested Correlations to the Mediterranean Region. (Adapted from Biberson and Jodot, 1965.)

	MEDITERRANEAN REGION			ATLANTIC MOROCCO			
	Marine stages [a]		Terrestrial units	Marine cycles	Terrestrial cycles	Industries	Climate
Versilian		"Flandrian" (transgression) (regression)	Late Würm	Mellahian	Rharbian	Aterian	Today's climate
Versilian					Soltanian		Cool, wet. Forest
Tyrrhenian		"Neotyrrhenian" "Grimaldian" "Tyrrhenian 3" "Monastirian II" (transgression) (regression)	Early Würm	Ouljian	Presoltanian		Warm temperate; rather dry
Tyrrhenian							Wet temperate
Tyrrhenian		Eutyrrhenian "Tyrrhenian II" "Monastirian" Tyrrhenian s.s. (transgression) (regression)	Riss	Harounian	Tensiftian	Evolved Acheulean	
Tyrrhenian		Paleotyrrhenian "Tyrrhenian"; "Tyrrhenian 1" "Milazzian", "Milazzian 2" (transgression) (regression)	Mindel	Anfatian		Middle Acheulean	Warm, drier, savanna; valley cutting
Tyrrhenian					Amirian	Lower Acheulean	Moist—temperate; N. Atlantic mollusks Mediterranian; savanna Cool, wet. Forest & prairie
Sicilian	Upper	"Sicilian 2" "Milazzian" (transgression)		Maarifian			Warmer
Sicilian	Lower	Sicilian (regression)	Upper Villafranchian		Saletian	"Evolved pebble tools"	Cooler & possible wetter Warm temperate, like Gulf of Cadiz
Calabrian	Upper	Emilian (transgression)	Lower Villafranchian	Messaoudian		"Primitive pebble tools"	
Calabrian	Middle	Calabrian s.s. (regression)			Lower Villafranchian; Moulouyan		Cooler; possibly wetter Pliocene mammal spp.
Calabrian	Lower	Precalabrian of Selli Astian (upper part)		Moghrebian			Warm, moist, subtropical

[a] Quotation marks denote that definitions vary.

tion zones to descend the slopes to altitudes 500 to 700m lower than their present limits. This would have united the "islands" into a continuous belt (fig 17-8) through which free communication of both trees and birds could take place. Although not without its local anomalies and other problems, this explanation of today's isolated communities is appealing. The evidence, like much evidence of a biogeographic nature, is negative, consisting of absence of connections. As such it is not really able to stand alone, but it has value as confirmation of other data pointing in the same direction.

A study of another kind, along the Atlantic coast of Morocco, illustrates a different approach to the problem of pluvial climates in Africa. Biberson and Jodot (1965) made a conventional study of the stratigraphy of that belt and extracted from the sequence a considerable body of information pointing to repeated fluctuation of the evaporation/precipitation ratio as well as of temperature (table 17-B). The sequence is tied to the marine stages in the Mediterranean, because they constitute a succession that is geographically near. However, as the table shows, that succession is undergoing revision (chap 12); hence the authors were forced to include a multiplicity of terms. The upper part of the table shows relationships compatible with the concept that relatively moist conditions in northern Africa were at least broadly contemporaneous with glacial expansion in Europe.

CHAPTER 18

Late-Wisconsin Glaciers in North America

RECONSTRUCTION OF ENVIRONMENTS

As we look at even a small-scale map of North America in a glacial age (fig 18–5), it is obvious that the continent was then very different from its present aspect. Most of its northern half was buried beneath ice, its southern half was altered as to flora and fauna, and it was enlarged, especially on the east and south, by withdrawal of the sea. Once these primary differences are stated and we look for details, we at once encounter difficulties in reconstructing the former conditions. Our data must be drawn from sources of many kinds, and because they are too few to integrate into a nearly complete picture, we attempt to connect them with reasonable theory. Our synthesis is certainly incorrect in places; we construct it only so that it can be later reconstructed in surer fashion, just as we construct and reconstruct stratigraphic charts.

Our reconstruction represents the Late Wisconsin glaciation with its maximum about 18,000 years ago. Although that glaciation was somewhat less extensive than some of its predecessors, we have much more information about it because of its recency. Therefore we use it as a model, realizing that earlier glaciations may have differed from it in one way or another.

Although at the maximum, ice was continuous across North America from Atlantic to Pacific, the glaciers constituted two categories, differing in character and place of origin. The Laurentide Ice Sheet, a continuous body, occupied the eastern and central regions and the Cordilleran Glacier Complex occupied the western region. Together, but excluding Greenland, those two units covered an area of more than $16.2 \times 10^6 \text{km}^2$

463

(table 4-C) and constituted more than one-third of the world's glacial cover. With Greenland included they constituted nearly 40%.

CORDILLERAN GLACIER COMPLEX

Character and Extent [1]

The predominant characteristic of the Cordilleran Glacier Complex (figs 18-1 and 18-2) was the local origin of its ice. That is, the ice originated on high mountains and most of it remained there. In some areas, however, ice flowed downward to lower lands, forming piedmont glaciers and, in a central region, forming an extensive mountain ice sheet. This rather local character of the Cordilleran ice stands in strong contrast to the Laurentide Ice Sheet, which originated as local glaciers in the northeast and east and spread from those regions far to the west and south across lower lands.

The most conspicuous single element in the Cordilleran Glacier Complex was the least local part, the continuous interconnecting mass of valley and cirque glaciers, piedmont glaciers, and a mountain ice sheet, that centered in British Columbia and stretched northwest to the Aleutian Islands and south to Mount Adams near the Columbia River, a distance of 3760km. The glaciers formed chiefly in the Coast Ranges and Cascade Mountains and to a smaller extent in the Rocky Mountains farther east. Most of the lower but still chiefly mountainous country between these two great chains was buried beneath ice that at first flowed inward from the two chains. The piedmont glaciers, fed from eastern and western ranges, coalesced over this lower country so completely that the confluent mass formed a mountain ice sheet, the Cordilleran Ice Sheet. To a considerable degree the ice sheet was confined between the two high mountain chains, much as is the Greenland Ice Sheet today.

On the Pacific coast, glaciers flowing down the western slope of the Coast Ranges or through the great transverse valleys of the Fraser, Skeena, and Stikine Rivers entered the deep water immediately offshore, calved many icebergs, and in places perhaps formed ice shelves. The ice that flowed from the mainland of southern British Columbia coalesced with the local ice of Vancouver Island, so that at the maximum, directions of flow over the island were not radial but southwest. A great piedmont lobe protruded southward into the Puget Sound Lowland, while local glaciers on the high Olympic Mountains west of the Sound flowed

---

[1] Map references include Prest et al., 1968 (Canada); Karlstrom et al., 1964 (Alaska); maps on p. 232, 304, 343, 349 in H. E. Wright and Frey, eds., 1965a (western United States).

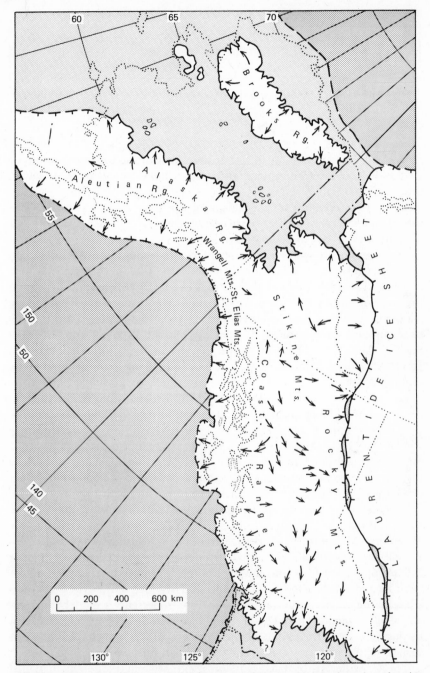

Figure 18-1 Sketch map of northern Cordilleran North America showing major glacial-age glaciers. Arrows indicate generalized directions of flow. Non-glaciated coasts are drawn at −100m isobath; glacier termini in oceans at −200m isobath. Compiled from Prest et al., 1968, and maps in H. E. Wright and Frey, eds., 1965a.

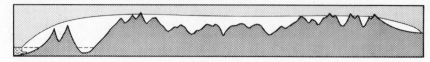

Figure 18-2   Cross-sectional model of a mountain ice sheet typical of the Cordilleran Ice Sheet.

downward to coalesce with it. Other lobes east of the Cascade Mountains terminated on the Columbia Plateau or in large valleys in northeastern Washington.

The rate of advance of the margin of the Puget lobe was rapid. A series of dates in the Seattle district shows that the glacier overrode that area at rates of 100 to $150m/C^{14}yr$.

Throughout the whole extent of the coastal mountains of Alaska large valley glaciers, originating in continuous snowfields, coalesced as piedmonts and possibly floating shelves that extended well beyond the line of the present coast. The ice was both thicker and more extensive on the seaward flank than on the interior side of the mountains, reflecting the predominant oceanic source of snowfall. On the Brooks Range, paralleling the Arctic coast of Alaska, the glaciers on the southern flank were better developed than those on the northern flank, indicating that the Gulf of Alaska was the chief source of snowfall for those glaciers too.

In Alaska the coastal mountains and the Brooks Range constitute two great prongs of the Cordilleran Glacier Complex, although apparently the glaciers of the northern prong did not coalesce with the northwestern end of the Rocky Mountains in Canada. Between the prongs the broad, low country drained by the Yukon River was not glaciated because of high summer temperatures and comparatively low precipitation values.

The width of the whole confluent mass from the west coast of Vancouver Island to Calgary, Alberta, beyond the eastern base of the Rockies, was nearly 900km. Farther north, from Yakutat Bay in Alaska to the Mackenzie Mountains in Yukon Territory, its width was not much less. In the latitude of Calgary the altitude of the glacier surface reached 2600m; only high peaks projected through it as nunataks. Outlet glaciers from it, and local valley glaciers as well, descended valleys on the eastern flank of the Rocky Mountains. Figure 18-3 represents a model similar to an east-west section through the confluent part of the complex. However, through much of the extent of the ice sheet, altitude of the actual upper surface of the ice sheet is not known and can only be suggested.

The Cordilleran ice reached its most extensive development in British Columbia. Glaciers diminished both northward into Alaska and Yukon and southward through western United States. Their distribution closely parallels that of the relatively few and small glaciers of today. The greatest extent of glaciers formerly, as now, was in regions where: (1) precipitation values were great, (2) precipitation fell chiefly as snow, and (3) summers were cool and short. In the Cordilleran region as a whole the combination of these conditions most favorable for building glaciers is found in coastal British Columbia and adjacent Alaska. It becomes less favorable northward toward northern Alaska and Yukon owing to decrease in total precipitation; southward into the United States it becomes less favorable owing to decrease in the second and third factors. Further details on Wisconsin-age glaciers in Alaska are given in Heusser, 1960, p. 13–19.

In addition to the confluent mass, the Cordilleran Glacier Complex included separate glacier groups and systems, each centering on a mountain range or other highland. Several of these were in Alaska. Many others, embracing at least 75 separate areas, lay in western United States (fig 18–11), the largest being the glacier complex, 400km long, on the Sierra Nevada in eastern California.

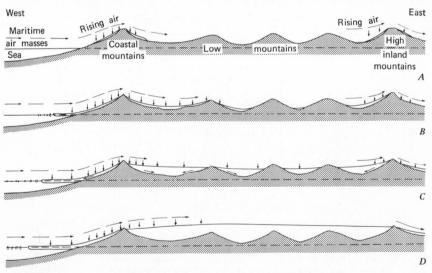

Figure 18-3 Vertical sections showing development of a glacier complex such as the Cordilleran. Diagrammatic; vertical scale much exaggerated. Nourishment is derived from maritime air masses moving eastward. Length of section about 800km. *A*. Development of glaciers in mountain valleys. *B*. Coalescence of valley glaciers to form trunk glaciers in intermont areas. *C*. Development of mountain ice sheet. *D*. Maximum ice sheet phase.

Growth

Growth of the Cordilleran glaciers accompanied gradual descent of the snowline as recorded by the altitudes of cirques. Table 18-A shows that in three Cordilleran districts descent of snowline and extent of glaciation increased with increase in existing precipitation values and with decrease in summer temperature. The relationship suggests that though absolute amounts may have differed, relative temperature and precipitation differences in the region were much the same during glaciation as they are today, and that though climatic belts may have shifted, their relative positions have remained unchanged.

Growth of the coalescent ice sheet centered in British Columbia occurred in four phases: (1) an alpine phase, in which small valley glaciers developed on the moist Coast Ranges and to a lesser extent on the drier Rocky Mountains farther east; (2) an intense-alpine phase, in which growth and coalescence of the valley glaciers led to the development of long trunk valley glaciers in major valleys; (3) a mountain-ice-sheet

Table 18-A

Comparison of Existing and Former Snowlines in Three Districts in Cordilleran Western United States

Parameter	Mt. Rainier National Park, Washington	Glacier National Park, Montana	Rocky Mountain National Park, Colorado
Altitude of highest peaks (m)	4,330	3,000	4,300
Mean annual precipitation (mm)	At Paradise (alt 1660) 5,000 At Longmire (alt 850) 2,000	2,500	760
Mean July temperature (°C)	12	18	19
Present altitude of snowline (m)	3,000	3,000	4,250
Altitude of former snow— line implied by cirques (m)	1,360	1,660	3,000
Difference between the two snowlines (m)	>1,660	>1,360	1,250
Relative extent of former glaciation	greatest		least

phase, in which further growth developed extensive piedmont glaciers along the flanks of the mountains, especially the Coast Ranges; this ice became so thick that it virtually buried the Coast Ranges; and (4) a continental-ice-sheet phase, in which, over the country between Coast Ranges and Rockies, the piedmonts coalesced into an ice sheet at least 2,300m thick over the valleys. This ice was confined between the enclosing ranges and had little movement. From this central reservoir, with a possibly domelike center near lat 53°, outlet glaciers discharged westward through valleys transecting the Coast Ranges, eroding them intensely (Armstrong and Tipper, 1948; Armstrong and Brown, 1954.)

The lofty mountains of Vancouver Island generated a separate glacier mass, perhaps of mountain-ice-sheet form, that became confluent with and was later overwhelmed by the ice of the mainland. In the zone of confluence its thickness reached 1200m. Indicators were carried eastward to positions near Vancouver, showing that temporarily, at least, the ice from the island dominated the mainland ice.

In this general sequence the transition from the third to the fourth phase, which is supported by evidence of reversal of direction of flow in the Coast Ranges, implies that the ice divide, the axis of outflow of ice, shifted from the mountain crest inward to a position over lower land. From the crest of the Coast Ranges the shift was eastward, but from the lesser crest of the Rockies it was westward. When the two axes of outflow coalesced, the Cordilleran Glacier Complex temporarily centered in an ice sheet.

The shift in the ice divide on the Coast Ranges could have occurred in this way (fig 18-3). The chief locus of precipitation then, as today, was on the western sides of mountains that faced the moist maritime air masses moving against them from the Pacific Ocean. Although accumulation probably was greater on the western side than on the eastern, discharge of ice by glacier flow was nevertheless far greater on the western side. This is because the average western slope is steeper and, more important, because on the west the ice reached the sea within a short distance, calved into deep water, and floated away. Hence, the ice that flowed westward had a short swift journey on a steep slope, but the ice that flowed eastward had a slower journey on a longer and much gentler slope and was not subject to loss by calving. The shift may have been favored to some extent also by wind-drifting of snow toward the east. Thus despite greater accumulation on its western side, the ice there drained away as fast as it accumulated, whereas on its eastern side, despite smaller accumulation, drainage was slower and terminal losses were less. East-flowing ice therefore accumulated and "backed up" in the valleys until its surface rose higher than most of the mountain summits.

Relation to Laurentide Ice Sheet

Precipitation on the Rocky Mountains, 500km to the east, probably was considerably less than on the Coast Ranges, and at any one time the glaciers must have been correspondingly smaller. There was no seawater to bring about calving; the early glaciers that flowed down the slopes of the Rockies coalesced to form piedmonts on both flanks of those mountains. While the western piedmonts merged with Coast Range ice to build the ice sheet over the Cordillera, the eastern piedmonts coalesced in some sectors with the Laurentide Ice Sheet on the Plains. We infer from indicators that ice from the Rocky Mountains never extended far eastward: in northern Montana 60km; in southern Alberta 80km, and in the latitude of Edmonton about 90km; at lat 58°40′ the Laurentide Ice Sheet brought drift at one time to within 8km of the base of the mountains.

These data show the effect of the strong rain shadow produced by the Rockies. Westerly winds yielded snow to the massive glaciers west of the Rockies and then, as they descended the east slope, were adiabatically warmed and precipitated little moisture. Although reduced as long as the Laurentide Ice Sheet was present, the rain shadow appears to have been continuously effective. The zone of local confluence (fig 18-1) of the two ice bodies, however, fluctuated repeatedly (Stalker, 1960b, p. 72).

Glaciers in the Far North

North of lat 60° the difference between the strong glaciation of the coastal mountains and the less strong glaciation of the mountains farther inland and to leeward becomes rapidly more pronounced. At Ketchikan, on the coast at lat 56°, annual precipitation is 3800mm, but it decreases rapidly inland. In consequence, glaciers generated on high mountains extended little beyond the inland bases of the mountains. Other factors in the decrease of glaciers toward the interior were increasing distance from the paths of air masses that move eastward and northeastward from the Pacific and the strong rain-shadow effect of the massive mountains along and near the Alaska coast. Mt. McKinley, the highest peak, reaches 6100m. At Fairbanks, in the interior, precipitation is only 280mm. Mainly because of deficient snowfall, then, lowlands and low mountains in northwestern Yukon and the interior of Alaska were not glaciated.

In contrast, the massive Brooks Range in northern Alaska was high enough to generate glaciers, despite contemporary precipitation values of only 250 to 300mm. Although at least 600m thick in places, these ice bodies coalesced only in part, and their northern termini failed to cover the Arctic coastal plain (e.g., Porter, 1966). Elsewhere in Alaska only a few isolated mountains generated glaciers.

Sea Ice in the Pacific-Arctic Region

Under existing climates floating sea ice (maps in Weaver, 1946), the greater part of which is frozen seawater together with snow, occupies most of the Arctic Ocean and extends southward into both the North Atlantic region and the Bering Sea. In winter, continuous or nearly continuous sea ice extends down nearly to lat 58° in the Bering Sea, and single ice masses extend much farther (fig 4–8). In summer the southern edge of continuous ice recedes north of Bering Strait. During glaciation the Bering Strait was wide, dry land. Near its coasts sea ice was probably thick and perhaps nearly continuous.

The distribution of glacial-marine sediments indicates that during glaciation ice shelves and bergs together dropped sediment over the floor of the Gulf of Alaska and the adjacent Pacific, from the coast of Washington northwestward at least as far as the Alaska Peninsula (Menard, 1953, fig 1). Evidently most bergs from the ice shelves of Cordilleran piedmont glaciers were carried westward past the Aleutians; only a few floated south.

Southern Limit of Continuous Ice [2]

South of the U.S.–Canada boundary, coalescent glaciers with local sources occupied the Cascade Mountains southward uninterruptedly nearly to the Columbia River. The west flank of the Cascade Mountains generated more ice than did the much drier east flank. North of lat 48° western glaciers were tributary to the massive Puget lobe, part of a thick piedmont glacier of which one lobe discharged seaward around the south end of Vancouver Island.

East of the Cascade Mountains the margin of continuous ice crossed the 49th Parallel and extended nearly 150km into the State of Washington, where it reached its farthest extent along the northern margin of the warm, dry Columbia Plateau. The ice there was essentially a wide, thick piedmont lobe fed from adjacent mountains. Farther east, across the width of the Rocky Mountain System, several such lobes occupied major valleys.

Highland Glaciers Outside the Continuous Ice

Within western United States south of the limit of continuous ice there were at least 75 separate areas of glaciers (tables 4-C and 18-B). Each area centers in a highland: a mountain range, group of ranges, or plateau. Most of the areas were occupied by valley glaciers singly or in groups,

[2] Crandell, p. 341 in H. E. Wright and Frey, eds., 1965a; Richmond et al., p. 231 in idem.

Table 18-B
Glaciated Areas in Western United States

Reference Number on Map, fig 18–4	Area	Highest Known Altitude of Highland (ft)	Type of Glacier	Number of Glaciations Recognized	General Altitude of Cirque Floors (ft)	Lowest Known Altitude Reached by Glaciers (ft)	Selected References
1	Juan de Fuca piedmont lobe		Piedmont gl	4		0	Crandell in Wright and Frey, eds., 1965a
2	Puget piedmont lobe		Piedmont gl	2		0	Idem
3	Olympic Mountains	7,954	Ice cap. Valley gl		4,500–5,000 in N	0	Idem
4	Cascade Mountains in Washington	14,408	Complex	3	4,500–5,000 in N	0	I.C. Russell 1900; Page, 1939
5a–c	Cascade Mountains in Oregon						
a	Mt. Hood district	11,225	Complex	2	4,500–5,000	200	Thayer, 1939
b	Mt. Jefferson district	10,500	Complex	3	5,500		
c	Crater Lake district	8,938	Complex	3	6,000		
6	Mt. Shasta	14,161	Complex	Obscured by volcanism	3,900		Diller, 1895
7	Klamath Mountains (many separate areas)	9,345	Valley gl	4	6,500	3,900	Sharp, 1960
8a–c	Sierra Nevada						
a	Lassen Peak district	10,457	Complex	Obscured by volcanism	7,000–7,500		Wahrhaftig and Birman, p. 305 in Wright and Frey, eds., 1965a
b	Mt. Elwell district	8,615	Valley gl		7,500	2,000 (W. flank)	
c	Principal area	14,495	Complex, ice sheet in northern part	9	7,500 in N / 9,000 at lat 38° / 11,500 in S	4,700 (E. flank)	
9	San Bernardino Mountains	11,485	Valley gl	2	10,300–11,300	8,700	Sharp, Allen, and Meier, 1959
10a, b	Okanogan and Spokane lobes		Piedmont gl	2+?		1,000	Flint, 1937a
11	Mountains in N.W. Montana and N. Idaho	10,438	Complex	2+?	5,500 in N / 7,000 in S	2,000	Dort, 1965a; Alden, 1953
12a–c	Bitterroot Mountains	10,131	Complex	3	6,000 at lat 47°30' / 6,500 at lat 46°		Lindgren, 1904, p. 51
13a–e	Clearwater Mountains	8,150	Complex		6,500 in NE / 7,000–7,500 in SW		Idem, p. 67
14a, b	Salmon River Mountains	12,078	Complex	2–3	8,500 in N / 9,000–9,500 in SW	4,000	Capps, 1940; Ross, 1938
15	Lookout Peak district	8,124	Valley gl				
16	Seven Devils Mountains	9,387	Valley gl, Cirque gl				
17	Wallowa Mountains	10,037	Valley gl, Ice cap	2	7,500	4,000	W.D. Smith et al, 1941
18	Elkhorn Mountains	8,922	Valley gl		7,500		
19	Strawberry Mountains	9,600	Cirque gl ?				
20	Owyhee Mountains		Cirque gl ?				
21	Steens Mountain	9,354	Cirque gl, Valley gl			9,000	W.D. Smith, 1927
22	Santa Rosa Mountains		Cirque gl				
23	Jarbidge Mountains and adjacent ranges	9,911	Cirque gl				Blackwelder, 1934

No.	Locality	Altitude (ft)	Glacier type	No.	Cirque/snowline altitude range (ft)	(ft)	Reference
24	Independence Range	11,000	Cirque gl	2			Idem
25	Pilot Range	10,704	Cirque gl				Idem
26a, b	Ruby–East Humboldt Mountains	11,356	Valley gl	2	9,000–9,500	6,100	Sharp, 1938
27	Shoshone Mountains	10,322	Cirque gl	1			Blackwelder, 1934
28a, b	Toiyabe Range	11,775	Cirque gl	1			Idem
29	Wassuk Range	11,303	Cirque gl	2+?			Idem
30	Sweetwater Range	11,646	Valley gl	2			Idem
31	White (Inyo) Mountains	14,242	Valley gl		12,000–12,500	9,300	Idem
32	Spring Mountain	11,910	Cirque gl	2		6,500	C.R. Longwell, unpubl
33	Snake Range	13,047	Cirque gl	2		9,300[a]	
34	Deep Creek Mountains	12,101	Valley gl		10,500+		Blackwelder, 1934
35	Tushar Mountains	12,173	Valley gl				
36	Glenwood Mountain	11,223	Valley gl, ice cap	3	11,000–11,300	8,500	Sharp, 1942
37	San Francisco Mountains	12,794	Valley gl	3	11,000	6,600	Flint and Denny, 1958
38	Aquarius Plateau	11,000	Ice cap	2		8,400	Hardy and Muessig, 1952
39a, b	Fish Lake Plateau	11,500	Ice cap	1	10,000–10,500		Spieker and Billings, 1940
40	Wasatch Plateau	11,000	Valley gl	2			Blackwelder, 1934
41	Stansbury Range	11,031	Valley gl	2		7,800	
42	Oquirrh Range	10,585	Valley gl	3		8,500	Idem
43a–d	Wasatch Mountains	12,008	Valley gl		8,500–9,000 in N; 10,000 in S	5,000	Atwood, 1909
44	Uinta Mountains	13,500	Complex	3	9,500–10,500		Idem; Bradley, 1936
45a–f	Yellowstone–Teton–Wind River highlands		Complex		9,000	6,600	
a	Madison and Gallatin Ranges	10,000	Valley gl, ice cap	5	10,000		
b	Snowy Range and Beartooth Plateau	12,850	Valley gl	3	9,000–9,500		Richmond, 1964
c	Teton Mountains	10,685	Complex		10,500		Horberg, 1938, p. 72
d	Absaroka Range	12,202	Valley gl		9,000–9,500		Horberg, 1940
e	Wyoming and Salt Creek Ranges	10,763	Complex	3	10,500	8,000	G.W. Holmes and Moss, 1955
f	Wind River Range	13,785	Valley gl	5			
46	Lost River Range	12,655	Valley gl		9,500–10,000		Dort, 1962
47	Lemhi Range	11,025	Valley gl		9,500–10,000		
48a, b	Beaverhead Mountains	10,960	Valley gl				
49	Centennial Mountains	10,211	Valley gl				
50	Snowcrest Range	10,546	Valley gl				
51	Tobacco Root Mountains	10,267	Valley gl				Jacobs in Schumm and Bradley, eds., 1969

[a] Anomolously low because of extraordinary local topographic conditions

Table 18-B *continued*

Reference Number on Map, fig 18–4	Area	Highest Known Altitude of Highland (ft)	Type of Glacier	Number of Glaciations Recognized	General Altitude of Cirque Floors (ft)	Lowest Known Altitude Reached by Glaciers (ft)	Selected References
52	Pioneer Mountains	9,210	Valley gl				
53	Red Mountain	10,150	Valley gl				
54	Anaconda–Flint Creek–Sapphire Mountains	10,475	Valley gl		7,000–7,500		Weed, 1902
55	Deer Lodge Mountains	8,789	Valley gl				
56	Crow Creek Mountains	9,432	Valley gl		7,500–8,000	6,300	Alden, 1932
57	Big Belt Mountains	9,478	Valley gl				Idem: Weed and Pirsson, 1896
58	Little Belt Mountains	9,000	Valley gl				Alden, 1932; G.R. Mansfield, 1908
59	Castle Mountains	8,606	Valley gl	2	8,000	5,000	Darton and others, 1906a, 1906b
60	Crazy Mountains	11,214	Valley gl	1	8,700–9,200		W.W. Atwood, Jr., 1937
61	Big Horn Mountains	13,165	Valley gl	3?	10,500–11,000	6,200	
62	Medicine Bow Mountains	12,005	Valley gl	3	10,000	7,500	
63	Park Range	12,220	Valley gl		10,500		Richmond in Wright and Frey, eds., 1965a
64	White River Plateau	12,001	Valley gl	5	11,000		Idem
65a–c a	Colorado Rockies Ranges with coalescent glaciers	14,255	Complex	4	11,000–11,500 at N; 12,000 at S	8,000	Ray, 1940; Capps and Leffingwell, 1904; Bastin and Hill, 1917 p. 57. Ives, 1938; Lovering, 1935
b	Mount Evans district	14,260	Valley gl	2	12,000	<9,000	Ray, 1940, p. 1877
c	Pikes Peak district	14,110	Valley gl	2		10,000	Idem, p. 1880
66	Grand Mesa	11,500	Ice cap	3		11,500	Retzer, 1954
67	La Sal Mountains	13,089	Complex	5	9,500–10,000		Richmond, 1962 Richmond, 1963
68a–c a	Sangre de Cristo Mountains (Northern) Sangre de Cristo Mountains	13,301	Valley gl	5	11,000 in W 10,500 in E		Ray, 1940
b	Blanca Peak district	14,290	Valley gl	2	11,000–11,500	8,800	Idem, p. 1882
c	Culebra and other ranges	13,546	Valley gl	2	11,000	8,500	Idem, p. 1887–1901
69	West Spanish Peak	13,623	Valley gl	2	12,000	9,000	Idem, p. 1885
70a	San Juan Mountains	13,725	Ice cap, Valley gl	5			Atwood and Mather, 1932; Richmond in Wright and Frey, eds., 1965a
70b	Canjilon Divide	10,000	[extension of San Juan Mountain glaciers]	1			H.T.U. Smith, 1936
71	Sierra Blanca	12,003	Cirque gl	2			H.T.U. Smith and Ray, 1941; Richmond, 1963
72	White Mts., Arizona	11,490	Cirque gl	1	10,600–10,800		Melton, 1961

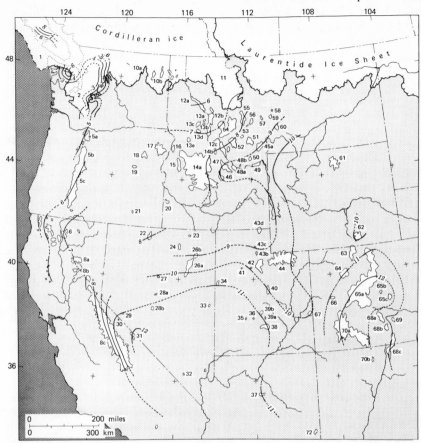

Figure 18-4   Glaciated areas in western United States and adjacent Canada, with much-generalized contours on mean altitudes of floors of lowest cirques, expressed in thousands of feet; they approximate the orographic snowline. Numbers on areas are keyed to table 18-B. (Contours in Washington adapted from Porter, 1964; elsewhere constructed from map and text data on cirques.)

but some were complex, even including ice caps. Distribution of these areas is indicated in fig 18-4, which shows extent of glaciation regardless of date; hence all the areas are larger, and some are considerably larger, than they would be if they represented only the Late-Wisconsin glaciation. The areas are generalized, some of them including smaller areas that were not ice covered.

The largest single area of former glaciation lies in the Yellowstone-Teton-Wind River highlands, which embrace several mountain ranges and plateaus. Each of the higher ranges supported numerous valley glaciers, which flowed out and in some places coalesced as piedmonts on lower terrain beyond. The Yellowstone Plateau was covered with coalescent piedmont ice that thickened and developed radial outflow from the plateau itself. The Beartooth Plateau to the northeast had an ice cap of its own.

The second largest individual area of former glaciation is in the Sierra Nevada. The central segment of the range bore an ice cap. Under the influence of nourishment from the west, the center of outflow of the ice cap shifted from the crest of the range to the western slope, from which ice flowed eastward up the valleys and across the crest. A dry climate prevented extensive glaciation of the eastern slope. The northern and southern segments of the range, being lower, bore valley glaciers only.

The former distribution of glaciers on the highlands of western United States is related principally to altitude, to latitude, and to the greatest source of moisture, Pacific air masses. The effect of altitude, which largely determined orographic precipitation on highland areas, is illustrated by mountain ranges in the southern part of the Basin-and-Range region, lying between the Sierra Nevada on the west and the Rocky Mountains on the east. There only ranges reaching altitudes of more than 3300m were glaciated. The effect of latitude is shown by the continuously decreasing altitudes of glaciated highlands from south to north. The effect of moisture source is revealed by the low-level glaciation near the Pacific coast compared with the higher level of glaciation in the Rocky Mountains.

The map, fig 18-4, shows these effects in descent of the level of cirque excavation toward the north and toward the Pacific. The regional snowline at a glacial maximum would slope in the same directions, but its slopes would be steeper and it would reach lower altitudes on the west and north.

Glaciers in Mexico and Central America. (Mexico: S. E. White, 1962; Lorenzo, 1959—existing glaciers only; Costa Rica: R. Weyl, 1956.) The three highest peaks in Mexico are the volcanic cones Ixtaccihuatl (5286m), Popocatépetl (5452m), and Orizaba (5675m), all near lat 20°. Although "Popo" is still active, apparently all three achieved essentially their present dimensions before the later Pleistocene. "Popo" has a summit firn field 1km in diameter, but drift, with end moraines, extends down to 4150m. "Ixta" has four small glaciers today; moraines occur as low as 3100m, and the inferred difference between existing and former snowlines is about 1900m. Orizaba, being highest and nearest the Gulf of Mexico from which the chief precipitation comes, has a large firn field draining into seven small glaciers. Details of its former glaciation are still unknown. Existing firn limits are about 5000m.

In Costa Rica, Cerro Chirripó (lat 9°30'; alt 3820m) was glaciated in the Wisconsin Glacial Age although it bears no glaciers today. The former ice reached down to at least 3400m and the indicated snowline then was ∼3500m.

Deglaciation

Throughout much of the Cordilleran Glacier Complex, deglaciation was characterized by thinning and terminal retreat, which transformed many ice caps into families of valley glaciers. This process of shrinkage was interrupted repeatedly by re-expansions, at least some of which are known to have been synchronous across regions of considerable size, and which seem to have been the result of changes of climate. The sequence of re-expansions, still known in only a fragmentary way, is discussed briefly in chap 20.

An exception to this rather common style of deglaciation is the main mountain ice sheet in British Columbia. Held in by high mountains on both sides and overlying a region of substantial relief, this body was affected by greatly reduced nourishment when general thinning set in. It appears to have shrunk away from the high ranges that confined it, with the last remnants occupying the center of its former area (Fulton, 1967).

LAURENTIDE ICE SHEET

Extent

East of the Cordilleran glaciers, northern North America was almost completely covered by ice, which when at its maximum extent constituted the Laurentide Ice Sheet (fig 18-5), named by G. M. Dawson in 1890 and defined by Chamberlin in 1895.[3] The area glaciated by the ice sheet, combining glaciations of more than one age, is believed to have totaled nearly $13.4 \times 10^6 km^3$ (table 4-C). However, the Late Wisconsin ice sheet was smaller, having possibly about 90% of the area quoted. The southern border of the glaciated region is clearly defined and is well known. The western border approaches the east base of the Rocky Mountains (fig 18-6) and in a few places touches drift of Cordilleran origin; but in respect to Late Wisconsin drift such places are few. The northern border lies north of the present coast except on Banks Island, where it lies inland. On the east, the ice on Ellesmere Island coalesced with the Greenland Ice Sheet, at least in the region north of lat 80°. Farther south the limit of glaciation is submerged. If it is assumed to coincide with the

---

[3] The history of these names and of the ice-sheet concept is outlined in Flint, 1943.

−200m isobath, the glaciated region included the banks off Newfoundland, Nova Scotia, and New England. Much of that area now submerged bears clear evidence of glaciation.

The extreme northeastern part of the glaciated region possesses mountainous relief and therefore may have resembled the Cordilleran Ice Sheet. It has been called the Ellesmere-Baffin Glacier Complex (B. G.

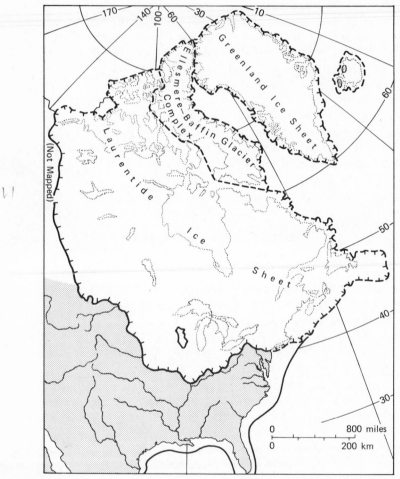

Figure 18-5 Sketch map showing extent and broad subdivision of glacial-age glaciers, regardless of age, east of the Cordillera. Where not known from geologic data, submerged limits of former glaciers are drawn speculatively along the −200m isobath; −100m isobath off Atlantic and Gulf Coasts suggests possible shoreline during a glacial age. The enclosed area in Wisconsin and Illinois is the "Driftless Area."

Figure 18-6 View southward along the upper limit of glaciation, Mayo District, Yukon Territory. Late-Wisconsin McConnell Moraine, built by the Laurentide Ice Sheet, winds around the east flank of Kalzas Plateau, a smooth nunatak. The moraine separates glaciated country (L) from nonglaciated nunatak, showing solifluction features in foreground. Twin hills (upper R) constitute a smaller nunatak, with moraine at their west base. Lat ~63°20′; long ~134°40′; alt ~1,800m. (U. S. Air Force photo; text ref: Bostock, 1966.)

Craig and Fyles, 1960). West of it the Parry Islands may have carried only disconnected ice caps during the Late-Wisconsin glaciation.

Directions of Flow

The directions in which the ice flowed are inferred mainly from streamline molded forms, asymmetric glacially abraded hills, indicators, striations, eskers, and end moraines. Some and possibly most of these features are a record of the latest movements in the areas of their occurrence, the glacier having erased evidence of earlier directions. As sequences of two or more directions of movement in some districts have

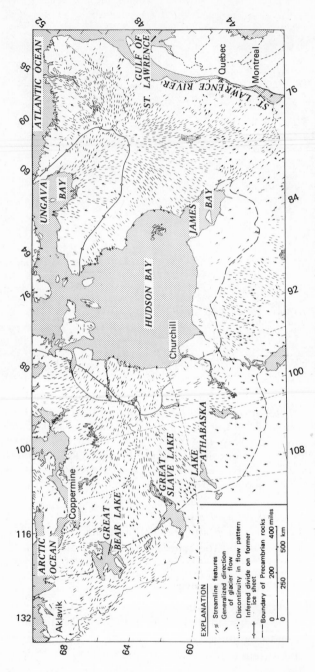

Figure 18-7 Streamline forms plotted from aerial photographs of a large part of Canada, from a map compiled under the direction of J. T. Wilson and checked locally on the ground. Also shown are an inferred ice divide and a line of discontinuity in the pattern of glacier flow. Arrows are mostly generalized from a variety of data. In part from B. G. Craig and Fyles, 1960. Probably most of these features were created at various times during the last deglaciation. For detail in other parts of Canada see Prest et al., 1968.

been established, sorting of the features throughout the region into a time sequence is full of uncertainty.

An example consists of the streamline features shown in fig 18-7. The time when these were molded is not known from evidence in the features themselves. However, their consistent relationship to end moraines and particularly to systems of eskers, both of which were made during deglaciation, suggests that they, too, are late features. Probably those near the periphery of the glaciated region are older than those in more nearly central positions.

Figure 18-7 shows two divides from beneath which ice flowed in both directions, as inferred from air photographs and locally from ground observations. An eastern divide, shaped like a U open to the north, coincides closely with an existing drainage divide. A western divide, trending southwest from the northwest end of Hudson Bay to near Churchill, Manitoba, does not coincide with a present-day divide. Instead it traverses rather flat country 150 to 300m in altitude. C^{14} dates indicate that the latest movement in both directions from this divide occurred not long before deglaciation of the region. Probably the eastern divide, as yet not dated, was in existence at an equally late time.

On the other hand, the similarity of the flow pattern seen in fig 18-7 (and in more detail in the Glacial Map of Canada, Prest et al., 1968) to an analog simulation made by field plotter (Mackay, 1965) suggests that the pattern overall may be related to the form of the ice sheet when at its maximum rather than in some later, reduced state. More study must precede any firm conclusion about the ages of the directional features.

Shifts in direction of flow during Wisconsin time are, nevertheless, detected from sand-size heavy-mineral indicators in the Great Lakes region. Ice that flowed southward into the basins of Lakes Ontario and Erie later flowed southwest along the axes of those basins. Ice that flowed southwest into the Lake Ontario basin later flowed southward, likewise along a basin axis (Dreimanis et al., 1957). Also the Glacial Map of Canada shows lines of discontinuity of direction of flow other than that shown in fig 18-7. One such line trends southeasterly from the head of James Bay, swinging south and southwest and disappearing near Sudbury, Ontario. It separates a southeast-flowing body from a southwest-flowing body. It is not known whether the bodies were contemporaneous or successive (Boissonneau, 1966–1968).

Origin and Growth

For the early history of the ice sheet, on which there is little direct evidence, we rely on analogy with the Scandinavian and Cordilleran glaciers and on deduction from reasonable assumptions as to climate. The main

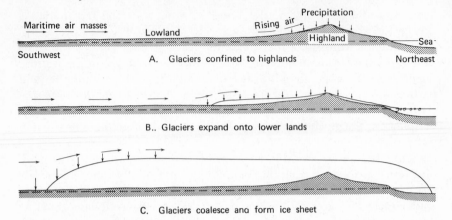

Figure 18-8 Idealized development of an ice sheet such as the Laurentide Ice Sheet. Only one highland is shown, although many must be visualized. Diagrammatic; not to scale.

body of the ice sheet was immigrant into most of the region it covered. Probably it developed originally as a glacier complex over wide areas of the Labrador-Ungava plateau (cf. Tanner, 1944, p. 178) as well as on the plateaus and mountains of Arctic islands such as Baffin, Bylot, Devon, and Ellesmere, where the highest peak exceeds 2400m. According to Barry (1966) the Labrador-Ungava region, with cold winters and short, cool summers, is meteorologically critical for the creation of an ice cover. He calculated that with today's mean summer temperature lowered by 3.3°, buildup of ice could take place, creating an ice sheet 2km thick in 20,000 years.

One can visualize a glacier complex, following secular lowering of temperature, expanding to form a continuous ice mass throughout much of northeastern North America (fig 18-8). We can think of the growth of that ice mass and its local influence on the wind system as the cumulative effect of small, persistent departures from the climatic conditions that characterize nonglacial times. Precipitation is brought to that region in moist, maritime air, mainly from tropical sources in the Atlantic and the Gulf of Mexico. It should have supplied a continuous source of nourishment, permitting the ice mass to expand toward the south and west. With its great albedo the ice would have chilled the air above it, thus constituting a meteorologic as well as topographic moving barrier to atmospheric movement and localizing the precipitation of snow.[4]

[4] This theory of origin is detailed in Flint, 1943; see also Leighly, 1949.

In analogy with the existing regimen over the Greenland Ice Sheet, presumably moist air rose over the high, cold margin of the glacier and over wedges of cold air extending in front of it, and precipitated snow over a wide belt of glacier surface. The precipitation was therefore of two kinds, cyclonic and orographic. Reflecting a large part of the incoming solar radiation, the ice sheet developed rather persistent high-pressure, anticyclonic conditions in the atmosphere above it. When the ice sheet reached its maximum extent, anticyclonic conditions over its surface may have been persistent enough to reduce the average number of traveling cyclones that could surmount it. The result would have been reduced snowfall on the ice, and this could have been a factor in limiting spread of the ice sheet toward the south and southwest, the directions from which relatively warm, moist air reached the glacier. It seems likely also that cyclones deflected to easterly paths along and south of the ice margin could have caused increased precipitation on a periglacial belt of territory.

We suppose that spreading of the ice sheet across the Dakotas and the Canadian plains was aided by snowfall from Gulf air masses moving northwest, by flow from better-nourished areas of the glacier, and by relatively low ablation values. During the spreading, Pacific moisture sources must have become gradually effective. Polar-continental air moving southeastward from the Mackenzie basin along the western margin of the growing ice sheet is likely to have been overridden by Pacific westerlies and northwesterlies, causing at least some snowfall. Polar northeasterlies, moving southwest from the Arctic Ocean, might have increased such snowfall by creating wedges of cold air over and near the ice sheet.

The formation of the Greenland Ice Sheet, described in chap 4, must have occurred while the Laurentide Ice Sheet developed. We can reasonably picture outlet glaciers from Baffin and Ellesmere Islands and from Greenland as entering Baffin Bay and Davis Strait, possibly forming ice shelves (Hattersley-Smith, 1960) that coalesced, perhaps even as far south as lat 60°. Ice shelves may have calved into the Labrador Sea, which at that time was probably covered with sea ice.

As the Laurentide Ice Sheet expanded, its active southern margin probably encountered and merged with plateau ice caps and valley glaciers already formed on local highlands such as those in eastern Quebec and Newfoundland as well as the Adirondacks, White Mountains, and Catskills. It is possible even that summer chilling induced by the presence of the growing ice sheet may have been great enough to create perennial snow or even a thin ice sheet in the cold continental region northwest of Hudson Bay, where today snow disappears during three summer months. Any such "advance glaciers" would have been overwhelmed by the Laurentide glacier and incorporated into it.

The position of the margin of the Laurentide Ice Sheet when at its maximum, like that of all ice sheets, was determined by at least three factors: (1) rapid calving on reaching a coast, (2) developing stability on reaching the line of a summer isotherm at which ablation balances rate of flow, and (3) reaching a topographic barrier, either localized or cumulative. Along the Atlantic Coast and much of the Arctic Coast of North America, factor (1) was dominant. Along the southern margin of the ice sheet, (2) and (3) were dominant, and along most of the western margin the control appears to have been exerted mainly by (3). The near congruence of the southern limit of glaciation by the ice sheet in different glacial ages suggests that summer temperature was similar during each glacial age. The other limits, controlled by coasts or by topography, should of course have been similar in each glaciation.

The various margins of the ice sheet are described briefly in a subsequent section. Because of the constraints mentioned, the ice sheet, as can be seen in fig 18-5, was broadly rectilinear rather than circular. Directions of flow in each marginal zone were approximately normal to the margin, although skewed locally by valleys and other topographic features beneath the ice. In the northeast, where the ice sheet proper merged into a glacier complex, mountain axes were dominant in controlling flow directions, and nunataks may have been numerous in places.

Thickness of Ice [5]

At one time or another during the Wisconsin Age the ice sheet overtopped all the highlands between New York and Labrador; hence in table 18-C the maximum heights are minimum values for local ice thickness, although not necessarily in Late Wisconsin time. Actual thickness, however, is not known. Farther west, the Lake Michigan lobe, at its maximum extent, was at least 580m thick at lat 44°, because its western margin reached 450m on the "Driftless Area" in Wisconsin, whereas the bottom of Lake Michigan is at −130m. The minimum estimate must be increased by the thickness of postglacial sediment on the lake floor and by the difference in altitude of the glacier surface between its western margin and its axis over the Lake Michigan basin. These corrections might double the minimum. In eastern South Dakota a probable maximum value, for a marginal locality and a time postdating the glacial maximum, is 500m; in Toole County, Montana, minimum thickness was 360m.

On the Canadian plains in southern Alberta, measurements at former

[5] Data in, e.g., Stalker, 1960b, p. 73; Westgate, 1968a, p. 58; Henderson, 1959b, p. 69; Flint, 1955b, p. 135.

Table 18-C

Highlands in Eastern North America That Were Covered by the Laurentide Ice Sheet at Some Time During the Wisconsin Glacial Age

Highland	Maximum Altitude (m)	Height above Surrounding Country (m)	
		Maximum	Average
Catskill Mountains, New York	1,274	1,270	900
Adirondack Mountains, New York	1,620	1,575	1,200
Mt. Washington, New Hampshire	1,905	1,500	1,360
Mt. Katahdin, Maine	1,596	1,425	1,270
Shickshock Mountains, Gaspé, Quebec	1,282	1,600[a]	850
Long Range, Newfoundland	803	1,050[a]	240
Torngat Mountains, Labrador	1,500	1,820[a]	900

[a]Surrounding country is in part below sealevel

nunataks give values between 300 and 700m (depending on distance from the margin) for Late Wisconsin ice and up to 900m for Early Wisconsin ice. Farther north (west of the Lesser Slave Lake) a minimum thickness of 600m was calculated. Crude as these values are, they support the concept that the ice sheet was substantially thinner in its western and southwestern parts than in its eastern part. The concept, supported further by the smaller inclination of the Lake Agassiz strandlines than that of the Lake Algonquin strandlines, is expectable inasmuch as the terrain along the western limit of the former ice sheet is more than 1000m higher than the terrain north of the Great Lakes. An ice sheet with an equilibrium profile therefore must be thicker in central Ontario than in Alberta, even though the effect of isostatic subsidence is neglected. If subsidence is taken into account the altitude of the Ontario region beneath the ice sheet would have been much lower and the difference in ice thickness between Ontario and Alberta therefore would have been considerably more. In the Mackenzie River valley at lat 65°, calculated minimum thickness increases to 1330m, but altitude is 1000m less than in southern Alberta localities. On Victoria Island, near lat 70°, minimum thickness is calculated at 545m (A. L. Washburn, 1947, p. 55).

In crude analogy with the existing Greenland Ice Sheet, the annual snowline might have lain 60 to 80km north, upstream, from the southern margin of the ice sheet.

Nonglaciated Areas

Two kinds of nonglaciated areas occur within the glaciated region. One consists of a highland partially submerged by the ice sheet—a nunatak in the strict sense. Several examples occur in Montana, Alberta, and Saskatchewan where the glacier was comparatively thin. Others, like the one in fig 18-6, occur farther north in western Canada. The other kind is an area between two outlet glaciers as they pass through mountains, generally near coasts. The commonly steep gradients of such glaciers result in comparatively thin ice, which, however, coalesces downstream at the bases of the mountains, isolating the nonglaciated higher slopes as a kind of nunatak. This geometry is common along the Greenland coasts today. Bare areas of this kind on Baffin Island, dating from former glacial maxima, have been inferred from the upper limit of distribution of glacial features and from the distribution, on interfluves, of rubbles of blocklike boulders, frost wedged from bedrock, that apparently were not touched by the Late Wisconsin glaciation (Mercer, 1956, p. 565). Similar nonglaciated areas in coastal Alaska and British Columbia (Heusser, 1955) and in Greenland (Iversen, 1952–1953) have been inferred from the occurrence, in lake sediments deposited soon after deglaciation, of pollen of hardy plants such as occupy nunataks in Greenland today. The plants are believed to have survived the last glaciation on isolated bare areas.

The rather common occurrence of isolated colonies of hardy arctic-alpine plants in districts much farther south led to a hypothesis that the latest ice sheet in eastern North America was so thin that it failed to overtop many highlands, that stood above the ice sheet as nunataks. Among them are the Torngat and Kaumajet Mountains in Labrador, the Long Range in western Newfoundland, the Shickshock Mountains on the Gaspé Peninsula, Quebec, the Keweenaw Peninsula on the south shore of Lake Superior, and the highest peaks in New York State and New England. The hypothesis was championed mainly by Fernald (1925), and was later supported by Lindroth (1963) on a basis of the distribution of living beetles. The idea found apparent confirmation, according to Coleman (e.g., 1921), in the supposed absence of glacially eroded surfaces and glacial deposits in these high places. In most localities, however, more detailed examination revealed geologic evidence of glaciation at much higher altitudes than had been recognized formerly [e.g., Ives, 1960b, p. 325, photo vii (Labrador); MacClintock and Twenhofel, 1940 (Newfoundland); Flint et al., 1942; Alcock, 1944 (Gaspé); R. P. Goldthwait, 1940 (Mt. Washington); Savile, 1961 (Queen Elizabeth Is.)]. In some cases, however, the higher glaciation antedates the Late Wisconsin.

In some areas botanical as well as geologic evidence fails to support the "nunatak hypothesis" (Wynne-Edwards, 1937; Rousseau, 1950.) A broad alternative view is that the expanding Laurentide Ice Sheet drove the arctic-alpine flora southward, and that during deglaciation the flora returned, ascended highlands, and persisted in those cool places after it had been replaced at lower levels by less hardy plants. More specifically, a refuge and migration route along a tundra corridor peripheral to the ice sheet has been suggested (S. T. Andersen, 1954).

Apart from the evidence of plants, there remain high areas such as the Long Range in Newfoundland and the Shickshock summits, where little-weathered glacial-erratic boulders demonstrate glaciation, but where, also, surficial rubbles consisting of frost-wedged blocks imply intense frost action while the areas were bare of ice. If it were demonstrated that such rubbles must antedate the Late Wisconsin glaciation (e.g., E. Dahl, 1955), correlation of the erratics with an Early-Wisconsin glaciation can be considered. Another possibility is that some rubbles may have been covered by nearly stagnant ice during later glaciation (e.g., Ives, 1966).

The Drift Border

Eastern Sector. Very broadly and neglecting detail, we can characterize the whole region of Laurentide glaciation as consisting of a central area of dominant glacial erosion, with regolith generally thin or absent, and a peripheral area of varying width, in which drift is generally in maximum thickness and abundance and end moraines are conspicuous. The peripheral area is conspicuous on the west and south, and on the east in the submerged area as far north as Nova Scotia. The remainder of the eastern sector and the northern sector, both submerged, are still unknown. Ever since late in the 19th century the outer boundary of the peripheral zone has been called the *drift border,* a convenient name that we shall continue to use.

Where glacier ice entered the sea, calving was conspicuous and ice shelves may have been present in places. Submerged end moraines, at least partly of submarine origin, parallel the southeast coast of Nova Scotia; they may mark the limit of the ice sheet in Late Wisconsin time, although at some earlier time or times the entire shelf was glaciated (L. H. King, 1969). Probably ice streams followed the submerged Cabot Strait and Georges Basin channels. Along the coast between New York and Newfoundland the weather may have been both foggy and stormy, because the Gulf Stream washed or at least approached the margin of the ice sheet; but the comparatively warm water may have prevented the formation of sea ice. In those sectors where the glacier was in contact with the sea, calving must have occurred. The configuration of the limit of glaciation near New York implies that the limit off Long Island must have been above sealevel, but just where, farther northeast, it reached the sea is not known. Data on the environment in the vicinity of Georges Bank are given by Emery et al., 1965.

Southern Sector. West of the Atlantic Coast the ice sheet reached its extreme extent in the Ohio-Mississippi basin, at lat 39° in the Wisconsin Age and 36°40' in the Illinoian Age, a lower latitude than any other ice sheet because an abundant source of nourishment lay to the south. The southern margin of the ice sheet was controlled both by summer temperature (fig 18-9) and by relief of the subglacial floor (figs 18-9 and 18-10). Areas topographically low imposed a lobate form upon the ice margin (fig 18-11). From the continental shelf to a point south of the eastern end of Lake Erie, the broad Hudson-Ontario lobe formed an irregular arc with its greatest extent near New York City where altitudes are lowest. West of it the even broader Great Lakes lobe occupied the low territory between the Allegheny Plateau and the Wisconsin highlands. Within the Great Lakes lobe can be discerned its individual components—the Lake Erie lobe and the Saginaw (Lake Huron) lobe, both with NE–SW axes, and the Lake Michigan lobe and the Green Bay lobe just west of it, with nearly N–S axes. The Lake Michigan lobe, however, is skewed toward the SW, possibly as a result of the Lake Erie-Saginaw ice mass thrusting against it. All these lobes are clearly the product of the channeling of flowing ice by the capacious bedrock valleys now occupied by units of the Great Lakes.

West of the Mississippi the Des Moines and James lobes were localized

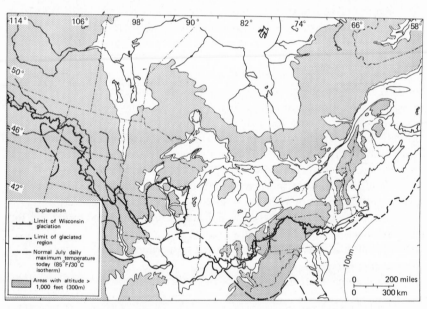

Figure 18-9 Relation of glaciated area to topography in eastern and central North America. (Data from Flint, 1943; Flint et al., 1945.)

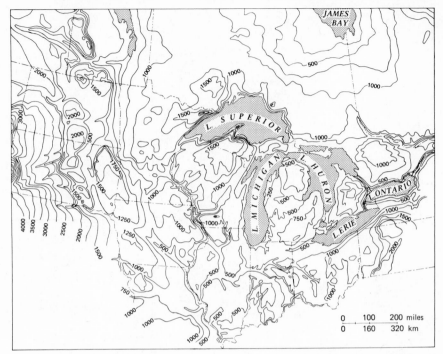

Figure 18-10 Bedrock surface beneath Quaternary sediments in central North America, inferred from borings and other subsurface data. (After Horberg and Anderson, 1956, fig 1; with modifications in South Dakota from Flint, 1955b, pl. 2.)

by broad N–S depressions now drained by the Des Moines and James Rivers. The absence of conspicuous lobes farther west on the Great Plains is consonant with the low relief of that region.

From Illinois to the Rocky Mountains the limit of glaciation strikes northwest, oblique to the eastward regional slope. Throughout most of this distance of nearly 2000km it parallels the isotherm of normal July daily maximum temperature (fig 18-9), and likewise climbs more than 1300m, reflecting the influence of altitude on the ice sheet.

The pronounced re-entrant in the limit of Wisconsin glaciation in Illinois and Wisconsin is caused by a comparatively high, plateaulike area (fig 18-10) that induced major lobation in that part of the ice sheet. The history of the area during earlier glaciations is a matter of debate. Early studies failed to find drift over its surface; it was therefore called the Driftless Area, and was classically mapped as shown in fig 18-5. Later

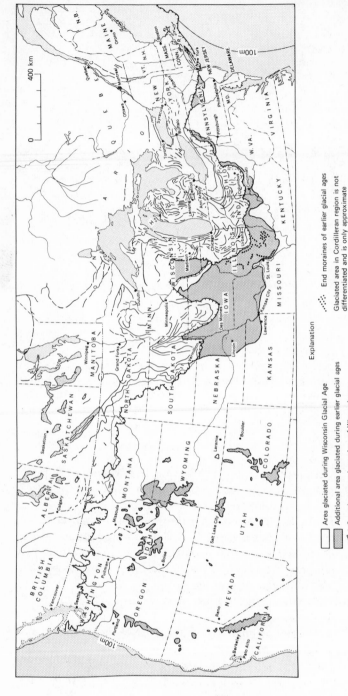

Figure 18-11 Extent of glaciation in northern United States and southern Canada, showing lobation induced by configuration of the terrain beneath the ice. At maximum extent of glaciers, shoreline may have stood near −100m iso-bath.

Explanation

☐ Area glaciated during Wisconsin Glacial Age

▨ Additional area glaciated during earlier glacial ages

〰 Conspicuous end moraines of Wisconsin age

∴ End moraines of earlier glacial ages

Glaciated area in Cordilleran region is not differentiated and is only approximate opinions have been divergent. The part of the area that lies in Illinois is still held to be driftless (Willman and Frye, 1969). The larger part that lies in Wisconsin has been claimed to have been mostly or entirely covered by ice during an Early Wisconsin glaciation (Black, p. 54 in H. E. Wright and Frey, eds., 1965a). The matter remains unresolved.

The age of the drift is not the same in all sectors of the southern drift border. From the Appalachians westward at least into Illinois, the border drift dates, or is supposed to date, from ~17,000 to ~20,000 BP, and belts of drift of lesser age lap off on it. Farther northwest, in some sectors drift 3000 to 5000 years younger runs close to the Late Wisconsin drift border or itself constitutes the most extensive Late Wisconsin drift. This pattern suggests that between 12,000 and 14,000 BP ice overran areas of slightly earlier glaciation. The explanation of the differences from one sector to another may lie in meteorologic factors; in any event it is not yet clear.

We must not leave the southern sector of the drift border without mentioning its conspicuous proglacial streams, recorded by long narrow valley trains. On the Atlantic slope the now-submerged Hudson and the Delaware and Susquehanna River valleys were the chief avenues of proglacial discharge. Farther west the huge Ohio-Missouri-Mississippi River system drained most of the southern sector of the ice sheet and deposited a great volume of outwash gravel and sand. These valley trains, as well as proglacial lakes along the former glacier margin, are delineated on large-scale maps (e.g., Flint et al., 1959).

Western and Northern Sectors. Continuing with our description of the limit of glaciation, we note that in southern Alberta there was little contact between the ice sheet and Cordilleran ice, although contact had been extensive during glaciations antedating the Late-Wisconsin glaciation. In the Northwest Territories, west of the Mackenzie River, the ice sheet was limited, not by a plains slope as it was farther south, but by abutting the main range of the Rocky Mountains, a barrier the glacier could not surmount.

The northern boundary of the ice sheet was irregular, with ice streams moving through major valleys and straits and calving along wide fronts. Probably some sectors were marked by ice shelves. Apparently the Arctic Ocean was covered with sea ice throughout the time of extensive glaciation. The sea ice, mixed with bergs created by calving, extended south into the Atlantic, depositing drift on the ocean floor and finally melting in the warmer water of the Gulf Stream. No comparable activity occurred in the Pacific because the Bering Strait did not exist. Through eustatic lowering of sealevel the strait was replaced by a wide land mass (fig 18-1), which prevented the movement of Arctic water and ice into the Pacific.

Partly for this resson the thermal and sedimentary histories of the Atlantic and Pacific Oceans during glacial ages were significantly different.

Deglaciation

Southern Region and Hudson Bay. We deduce reasonably that deglaciation resulted from regional increase of temperature that converted the economy of glaciers from positive to negative. The change may have occurred at somewhat different times in different parts of the continent, but the dates we possess (fig 18-12) suggest that such time differences were not great. Deglaciation involved not only abandonment of territory by retreating ice margins but also thinning of the ice bodies.

From the beginning of deglaciation (18,000 BP to ~15,000 BP, depending on the sector) to the final disappearance of the Laurentide ice *as an ice sheet* (apart from residual highland glaciers in Baffin Island, Labra-

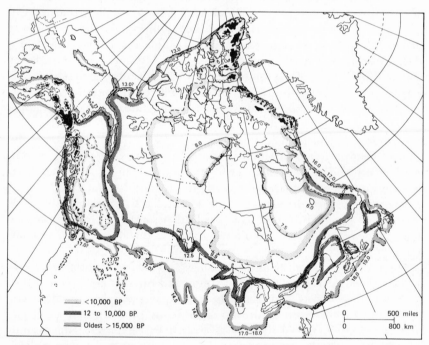

Figure 18-12 Speculative model of deglaciation by the Laurentide Ice Sheet from ~18,000 BP to disappearance of the ice sheet. Isochrons are based on C[14] dates as well as end moraines and other field data. Short arrows indicate significant readvance. (Prest, chap XII in Douglas, ed., 1970.) For more detail see Canada Geol. Survey Map 1257A, 1969.

dor, and elsewhere), the elapsed time was not much more than about 12,000 C^{14} years. Measured along a radius extending from the Hudson Bay region southward nearly to southern Ohio, where deglaciation began early, this time can be conveniently divided into four units: from the Late-Wisconsin Maximum to Lake Erie, about 4000 years; from Lake Erie through much of the Great Lakes history to the Valders readvance, about 3000 years; from the Valders event to the Cochrane readvance, about 3000 years; and from Cochrane to the final disappearance of the ice sheet proper, about 2000 years. Along this radius, therefore, deglaciation was an accelerating process in terms of area uncovered. Whether it was accelerating in terms of rate of conversion of ice to water is not determinable until ice thickness is known. An attempt, based on assumptions, was made by Farrand (1964).

Reserving the details of Great Lakes history for chap 21, we characterize briefly the main outlines of the ablation of a huge volume of ice. On the north and east, the earlier phases were characterized by much calving into the sea, especially in the deeper straits and channels between islands. On the west, where the ice sheet was flowing against the regional slope, deglaciation resulted mainly from regional thinning, stagnation, and building of ice-disintegration features. Probably thinning was promoted also by flow southeastward toward lower land in Montana and North Dakota (Stalker, 1960b, p. 74). The adverse regional slope also favored ponding of water in proglacial lakes. Together the lakes were of enormous extent, as can be seen on a map (Prest et al., 1968), although most details are still unknown.

On the south, deglaciation was mainly by retreat overland. From the drift border to the Great Lakes region the retreat was characterized by the deposition of lodgment till. The formation of the lakes accelerated recession through the inception of calving; yet at least two substantial readvances, climaxing at about 13,000 and 11,000 BP, occurred after the lakes had developed. As examples of the fast timetable we can cite retreat from the Lake Border moraines south of Lake Erie, mostly by calving, to the center of the Lake Huron basin, about 300km, followed by readvance of about 200km to build the Port Huron moraine, all within about 1000 years. In the following 2000 years the terminus retreated between 400 and 500km and then readvanced nearly the same distance (in the Lake Michigan basin) to the Valders maximum.

Gradual thinning of the ice sheet during deglaciation is demonstrated by increasingly pronounced lobation of successive drift borders (fig 18-11), especially well displayed in South Dakota (Flint, 1955b). End moraines were built, but they possessed generally less volume and continuity than earlier moraines. Shrinkage and thinning of the ice sheet should have weakened the anticyclonic effect over its surface and should have permitted more frequent passage of storms across it, increasing precipitation values. The southern periphery of the ice may have been ablated increasingly by warm rains coming from the Gulf of Mexico, in analogy with rain ablation of glaciers in Iceland today. A speculative reconstruction of climatic patterns during deglaciation is given by Bryson and Wendland (in Mayer-Oakes, ed., 1967, p. 271).

Parenthetically we note that various authors (e.g. Broecker et al., 1960) have emphasized, partly on evidence from marine sediments, the concentration of conspicuous secular temperature change in the single millennium that included 11,000 BP. Data are still too few for adequate evaluation of this interesting idea; the possibility remains that the change, although abrupt in some respects, was generally spread over a longer time.

The history of the Champlain Sea (the marine submergence of the St. Lawrence Valley) and of other marine transgressions along the Atlantic coast of Canada is summarized by Elson (1969). The essential data are set forth in fig 18–13, reproduced from that source.

North of the Great Lakes another readvance, the Cochrane, pos-

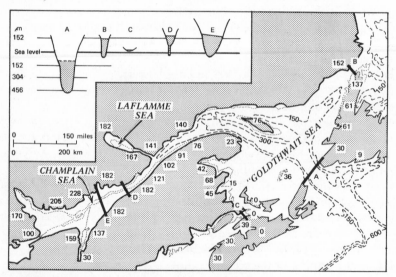

Figure 18-13 Areas of marine submergence during Late-Wisconsin deglaciation of southeastern Canada, with cross sections at five transects (*A-E*). Numbers are spot altitudes of marine limit in meters. Dashed lines are existing isobaths, in meters. After J. A. Elson (1969, p. 252.)

sibly in two phases, was achieved about 8300 BP. It may have been local, though possibly it may correlate with end moraines west and northwest of Hudson Bay. (Falconer et al., 1965; B. G. Craig, 1965b). Nevertheless, this is the latest readvance of the main ice sheet of which we have a record. Immediately following it there occurred a major event that led to the destruction of the Laurentide body—the incursion of the sea into Hudson Bay (Craig, p. 63 in Hood, ed., 1968). Calving had driven the glacier margin westward from the Labrador Sea through Hudson Strait. Shortly before 8000 BP seawater leaked into the Bay, and by calving made its way rapidly southward between the ice sheet and the Ungava Peninsula and westward around the margin of the ice. This event increased calving to a spectacular extent, and drove a wedge of seawater into the ice body. The wedge rapidly expanded into a wide body, the *Tyrrell Sea,* in which ice-rafted sediment was widely deposited. Calving soon separated the ice into two residual parts, both of them on higher ground, above the reach of calving. One part occupied the Ungava Peninsula while the other covered the region immediately northwest of Hudson Bay. Truncation of the former ice sheet by the Tyrrell Sea resulted in readjustments in rates and directions of flow in the two residual bodies, restoring equilibrium states (H. A. Lee, 1959, fig 6). The truncation seems to have been accomplished in little more than a thousand years. Hudson Bay was mostly cleared of ice by 7000 BP, setting off rapid isostatic uplift that shoaled the bay and reduced its area. It is thought that the residual glacier northwest of Hudson Bay disappeared before 6000 BP, and the body east of the Bay, lingering in the center of Ungava, at or soon after the same time (e.g., Henderson, 1959a, p. 68).

Eastern Region. Because of its more irregular topography, locally greater altitude, and land/sea relationships, North America east of the Great Lakes and Hudson Bay experienced some differences in the process of deglaciation. These included persisting local glaciers in areas high enough to maintain nourishment after deglaciation at lower altitudes had taken place and the effect of possible isolation of ice in the region south of the St. Lawrence Lowland.

Persisting local glaciers occurred on highlands in the eastern Arctic, for example Baffin and Ellesmere Islands. Probably the existing Barnes Ice Cap on Baffin Island represents continuous succession from the ice sheet (Løken and Andrews, 1966). Although it is believed that Newfoundland was overlapped by the Laurentide Ice Sheet, there is some evidence of one or more local ice caps, whose time relations to the larger ice body are not clear (cf. Tanner, 1944, p. 200; J. Lundqvist, 1965).

The Catskill and Adirondack Mountains in New York, the Green Mountains in Vermont, and Mt. Katahdin in Maine bear evidence of the existence of cirque glaciers after deglaciation by the ice sheet. In the White Mountains, New Hampshire, 11 cirques formed before the Late-Wisconsin ice sheet buried the mountains, but seem not to have been reoccupied during deglaciation (R. P. Goldthwait, 1970). The Shickshock Mountains in Quebec were occupied by cirque glaciers, possibly preceded by an ice cap, after general deglaciation of the vicinity. A local ice cap in Nova Scotia, postdating ice-sheet glaciation, was indicated by Hickox (1962), an ice cap in Maine was suggested by Borns (1963), and hypothetical residual ice bodies are indicated by Prest and Grant (1969, p. 12). Although earlier literature had reported evidence of northward flow of ice in southeastern Quebec, suggesting a regional ice cap in northern New England, the conclusion was shown (McDonald, 1967) to be invalid.

The New England – Maritime Provinces region constitutes a peninsula between the St. Lawrence River estuary and the open Atlantic. By about 14,000 BP deglaciation had uncovered the line of the present coast of Connecticut, by 13,000 ice was calving into the sea along the coast of Maine (Borns, 1963), and by 12,000 the margin of an active ice sheet lay along the southeastern slope of the St. Lawrence Valley and calved into the estuary somewhere northeast of Quebec City. Before 10,000 BP the Champlain Sea, an arm of the Atlantic, had replaced the ice sheet in the entire St. Lawrence Lowland, leaving the terrain southeast of it as a peninsula. Such a calendar demands rapid calving, at first probably along the Cabot Strait-St. Lawrence trough. During the period of rapid deglaciation of the St. Lawrence Valley, ice that formerly flowed southeast across the St. Lawrence would have been diverted northeastward down that valley. This change of flow direction would have reduced ice thickness over the New England – Maritime Provinces region, and may have been a large factor in a change in the glacial regimen there.

The change can be characterized as follows. From around 18,000 to less than 14,000 BP the glacier in coastal New England and Long Island was flowing actively and building successive end moraines. Through a brief intermediate interval moraine building was succeeded, in Connecticut at least, by the creation of ice-disintegration features (chap 8) in the terminal zone, with outwash heads forming at similar positions in adjacent valleys. This activity, suggestive of reduced rate of flow, was followed in turn by a nonsystematic pattern of kame terraces, eskers, and related features indicative of near or complete cessation of flow. The implication is that the ice between the St. Lawrence Valley and the Atlantic Coast finally thinned and disappeared by downwasting while the main ice sheet farther northwest continued with active flow and a moraine-building habit. This concept has not been validated by C^{14} dates simply because in New England an adequate number of dates from critical positions does not exist. It has been thought by some authors (e.g., Henderson, 1959a) that the ice body over the Ungava Peninsula, well after the creation of the Tyrrell Sea, finally disappeared in a similar manner, as suggested by the extensive esker system shown on the Glacial Map of Canada and by the pattern of temporary, ice-dammed lakes.

Cordilleran Glaciers. The central body of the Cordilleran Ice Sheet, overlying mountainous country between the higher Coast Ranges and Rocky Mountains, seems to have disappeared in part by downwasting, leaving few if any end moraines as a record of active flow during deglaciation. Most other Cordilleran glaciers were of the valley or piedmont type, whose recession was marked by readvances of termini with construction of end moraines. Some of these are dated and correlated, and are detailed in chap 20. In some of the many glaciers that still exist in the Cordillera, readvances have persisted down to within the past 200 years or so.

CHAPTER 19

North America Outside the Glacier-Covered Regions

When we seek to synthesize the environments—climate, soils, plant cover, and animal populations—beyond the glaciers at their maxima, we find little firm information. We can note at once that active deposition of outwash along major valleys was accompanied by the accumulation of loess over wide areas (chap 9), especially in central United States and along the Mississippi River extending southward almost to the Gulf of Mexico. Because it spread through a wide range of latitude, the deposition of loess is evidently not a sensitive indicator of climate. We can, however, extract a little, though not much, climatic synthesis from frost-action effects and from plants as represented by pollen. We shall, then, turn to these sources, leaving the fuller discussion of animals, most of which are climatically less sensitive, to chaps 20 to 22 and 28.

FROST-ACTION EFFECTS

Studies in Pennsylvania, along the Susquehanna River (Peltier, 1949) and in the northwestern part of that state (Denny, 1956) show that the Wisconsin ice sheet was fringed by a narrow belt characterized by conspicuous frost activity. The frost-produced features, now relict, are well developed through 15km or so from the drift border and then fade out toward the south. The surface, a dissected plateau of moderate to strong relief, is mantled by discontinuous frost-wedged rubble, particularly on slopes and valley floors. The mantles, possibly little more than 2m thick, are attributed to solifluction. Their profiles are long and smooth and

498

conceal minor lithologic irregularities. On valley floors the rubble merges with valley trains, augmenting the coarse, angular sediment fractions and progressively diluting the outwash fraction. The outwash sediments, augmented with rubble, form bodies that range in thickness from less than 10m to perhaps 30m and that have been dissected to form terraces. In the Susquehanna valley as many as four such units of Wisconsin age have been recognized. It is estimated that in some areas the volume of rubble implies lowering of hilltops by an average of 3m within the span of Wisconsin time.

When solifluction was active the ground beneath it was frozen, possibly to depths of 2 or 3m; however, the freezing could have been seasonal, without creating permafrost. Frozen ground is indicated also by block-fields, patterned ground, and frost-stirred ground (chap 10), developed at various places both outside and inside the drift border. The pattern of occurrence indicates that a climate cold enough to favor such features persisted for a time, perhaps as long as 4000 years, after the ice sheet had begun to shrink. Expectably, relict frost-action effects, including solifluction, are found much farther south of the drift border on Appalachian summits (fig 10-5) than in the lower country west of the Appalachians. Solifluction sediments have been identified at high altitudes at distances of more than 600km from the drift border, and features relict from a colder climate are reported (P. B. King and Stupka, 1950) from high altitude in the Great Smoky Mountains at about 35°30'. Wayne (p. 393 in Cushing and Wright, eds., 1967) showed the distribution of such effects in western Ohio, Indiana, and Illinois, as well as a synthesis of isotherms and vegetation zones. He inferred negative mean annual temperatures immediately beyond the glacier terminus in that region. A related fact concerns the diameters of spruce trees that grew in park tundra in central Ohio just before the glacial maximum. They are comparable with those in south-central Ungava today, where the mean July temperature is 11° cooler than it now is in central Ohio. To the north, in Wisconsin, Black (1964) showed the distribution of frost-action features, and inferred mean temperatures as much as 10° or even 15° below those of today on the basis of ice-wedge casts, relict features that imply permafrost. On the Great Plains, features such as relict frost-stirred ground and solifluction sediments occur, in places near the drift border, westward and northward at least through Alberta. Some of these, however, are older than Late Wisconsin. On the New England coast ice-wedge casts carry the same implication for a time shortly after the Late Wisconsin glacial maximum; features relating to the maximum itself are mostly submerged.

Thick layers of colluvium at moderate altitude in the Piedmont region of the Carolinas have been attributed to glacial-age frost action. However, Whitehead and Barghoorn (1962) consider increased rainfall during a glacial-age regimen a more likely cause of conspicuous debris-flow activity. Radiocarbon dates suggested Early Wisconsin as a probable time of origin. Similar though possibly less pronounced conditions might reasonably be expected at the Late-Pleistocene glacial maximum.

VEGETATION ZONES

The Problem of Reconstruction. In considering the distribution of vegetation during the Late Wisconsin maximum, not only beyond the Laurentide Ice Sheet but in the Cordilleran region as well, we must state at the outset that paleoecology, whether of plants or of animals, is more complex and its data more obscure than the glacial geology we have been describing. We are dealing here with *ecosystems,* natural dynamic systems that consist of the entire biotic community and its environment. An important factor in such environment is climate, but numerous other factors influence the distribution of biota. Consequently, with the small amount of information available, at best we can sketch only general outlines and at worst frame hypotheses to be established by current or future study.

A sound way to begin is to review the basic distribution of plants in North America today. Figure 19-1 shows that distribution in a greatly simplified manner. In general the vegetation zones form a belted pattern. In the west the belts trend generally north-south, parallel with the geologic structure as reflected in topography. In the east the belts trend more nearly east-west, generally parallel with isotherms, but also with a north-south pattern of northern forest on Appalachian heights, and of temperate forest, prairie, and steppe imposed mainly by decreasing humidity from east to west. Thus it is evident that the vegetation belts are controlled mainly by temperature and precipitation.

East-west belts may well have existed throughout the Cenozoic, but the north-south belts are mainly the result of the creation or uplift of mountains, mostly in the second half of that era. These events compartmented the Cordilleran region. Having adjusted to the compartments the vegetation, not only in the Cordillera but right across the continent, was then subjected to disruptions caused by temperature fluctuation, one result of which was the repeated spreading of glaciers over substantial parts of the territory. Thus environments, and with them entire ecosystems, have been very unstable. Various taxa survived glacial ages in high latitudes by "overwintering"—riding out the cold periods—in "refuges" that were not covered with ice. Figure 18-1 shows extensive nonglaciated areas in interior Alaska; probably there were other, smaller areas farther south.

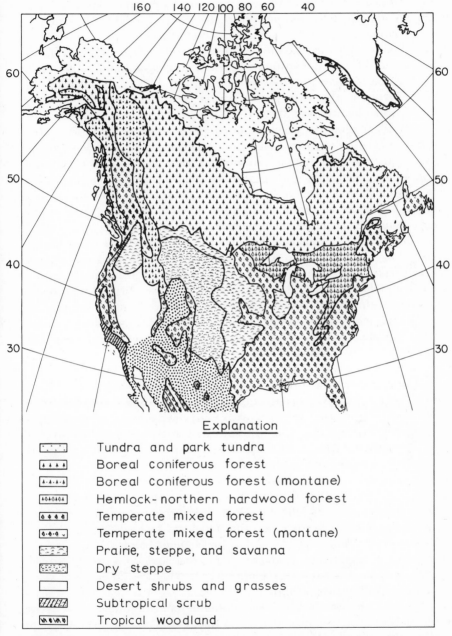

Figure 19-1 Natural vegetation zones in North America today, much generalized. (Compiled from various sources by R. F. Flint and prepared by L. A. Sirkin.)

Such areas are likely to have been "refuges." In middle latitudes survival was accomplished by shifting down the slopes of highlands into areas where climates were less severe.

On many highlands the kinds and proportions of taxa changed to various extents during the altitudinal shift of vegetation zones, under the influence of a complex group of controlling factors. Consequently our overall view represents broad generalization of a pattern so incomplete as hardly to be a pattern at all; yet in some parts of the country a tentative picture is emerging. It is being constructed by comparison of plant distribution today with that revealed by fossil plants, especially fossil pollen. With today's generalized pattern in mind, then, we can turn to the problem of reconstruction of the pattern that existed during the Late-Wisconsin Glacial Age. Our information is so spotty that we can do little more than mention the general results of study of two or three regions and assume that comparable relationships existed elsewhere.

Northwestern Cordilleran Region. In the mountainous coastal region extending from Oregon to Alaska (Heusser, p. 469 in H. E. Wright and Frey, eds., 1965a; Heusser, 1960), zones of vegetation lay at altitudes as much as 1500m lower than those they occupy today. It would not be correct to say that as temperature decreased, each zone migrated downslope intact. Rather, each taxon tended to abandon areas in which low temperature inhibited growth and to occupy new terrain at lower altitudes and at lower latitudes, where temperature was more favorable. The interrelationships among the plants themselves, particularly as regards tolerance of temperature and moisture changes, are so complex that downslope displacement differed somewhat among the members of a plant community. Indeed it appears that in many areas vegetation zones with the compositions they have today simply did not exist, so great was the disturbance of local environments. Yet migration is clearly evident in the relative abundances of key taxa in a series of pollen spectra. The overall pattern, then, was not different in style from that of today. Zones of vegetation formed a pattern that differed in detail but that still was related to mountains and other highlands. One marked difference consisted of tundra at the summits of highlands that today are covered by forest.

In northern and western Alaska the boundary between boreal forest and tundra migrated far to the south and east, and along the Arctic coast the tundra itself changed composition, reflecting the influence of colder summers (Colinvaux in Hopkins, ed., 1967).

Mammals recovered from placer alluvium and reworked loess, probably of Wisconsin age, in nonglaciated east-central Alaska include rodents such as the Siberian lemming, and also a beetle, that are characteristic of tundra. Their presence implies that when they lived, the treeline was at least 400m lower than today's. Forest returned to the district around 6800 C[14] years ago (Repenning et al., 1964). One could argue that such data do not compel the notion that in central Alaska the glacial environment differed from the present environment in a spectacular way. Indeed the glacial and nonglacial environments may have been broadly similar, differing mainly in respect to the position of treeline. Apart from fluctuation of the treeline, some temperature change is indicated by the freezing, thawing, and refreezing of silt clearly implicit in the record of ice-wedge casts.

It has been argued (Klein, 1965) that mammals now living in the territory fronting the Gulf of Alaska must have survived the glaciation in "refuges" (Heusser, 1960, p. 200–208) that included the Alaskan interior and small coastal areas that were not glaciated. Bison, for example, lived in central Alaska through the Illinoian, Sangamon, and Wisconsin Ages. Incidentally, all the vegetable matter recovered from carcasses of these and other mammals taken from frozen silt is that of plant species living today in Alaska.

Southwestern Region. (Martin and Mehringer, p. 433 in H. E. Wright and Frey, eds., 1965a; Wells, 1966; Wells and Berger, 1967; Wells, 1970; Malde, 1964.) The plant cover in southwestern United States, from western Texas to southern California, was very different from that in the Pacific northwest. Nevertheless the differences between the existing and Late Wisconsin patterns in both regions are rather closely parallel. Pollen data from alluvium, lake sediments, and archeological middens at about 50 localities show that vegetation zones were commonly lowered through at least 600m and in some areas perhaps much more (figs 19-2, 19-3). For example, in southwestern Nevada Mehringer (1967) inferred descent of the juniper/pine woodland by about 1000m. Such changes occurred under the influence of an augmented precipitation/evaporation ratio—in other words, of a pluvial climate. The spreading of those zones down the slopes of highlands converted much of the area now grassland into pine parkland; and the areas of desert vegetation, widespread today, were considerably reduced. Xerophilous woodland, however, was extensive in some areas. An interesting reconstruction of vegetation in southwestern Texas around 18,000 BP, based on wood-rat middens, is given by Wells (1966).

The transition from the pattern at the glacial maximum to desert, grassland, and Mexican woodland communities reached its peak about 12,000 BP; the pollen data reflect none of the fluctuations evident in some northern regions, notably northern Europe (table 24-C), very likely because Phoenix, Arizona is more than 1500km distant from the nearest point on the Late Wisconsin drift border. Nor is the postglacial warmth,

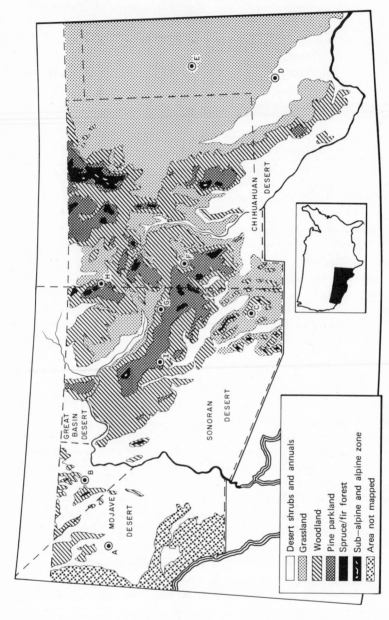

Figure 19-2 Vegetation zones in southwestern United States today. Lettered circles denote pollen localities. (After Martin and Mehringer, p. 438 in H. E. Wright and Frey, eds., 1965a.) Woodland in western Texas is actually less widespread than shown on this map.

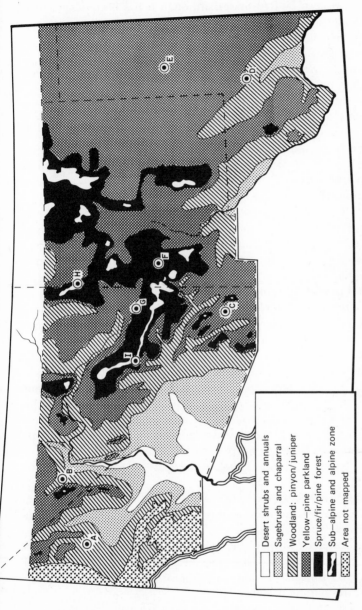

Figure 19.3 Vegetation zones in southwestern United States around the Late-Wisconsin maximum, reconstructed from pollen data chiefly from the localities denoted by lettered circles. (Martin and Mehringer, p. 439 in H. E. Wright and Frey, eds., 1965a.) The pattern implies that zones were displaced downward through a maximum distance of 1000m. Other authors (e.g., Wells, 1966) prefer a more conservative value, on the order of 600m, which would alter the map appreciably.

reflected in the Hypsithermal of more northern latitudes, evident in the data from southwestern United States, although it has been discerned in southern Colorado and northern Arizona. Martin and Mehringer suggested that in the southwest a change in seasonality may have blurred any effect that might otherwise have been registered.

In the southern Great Plains vegetation included much more woodland than it does today (Wells, 1970). Somewhat more precipitation than today's, at least in summer, is inferred from fossil mollusks. The classes and orientations of dunes on the Plains from southwestern Texas to northwestern Nebraska have led to the belief that around the glacial maximum, effective winds were NW, whereas subsequently they shifted to SW (e.g., H. T. U. Smith, 1965). In New Mexico and Arizona, loess deposited apparently during the Sangamon Interglacial Age was affected by the development in it of a pedocal soil under moister Wisconsin pluvial conditions.

In western Texas wood-rat middens include macrofossils of parts of trees indicating, more directly than by pollen data, dry-woodland vegetation as much as 800m lower than today's, at or soon after the glacial maximum (Wells, 1966).

In the central part of the semiarid Mexican Plateau the Wisconsin stratigraphic unit known as Becerra Superior has yielded a vertebrate fauna (Hibbard, 1955) including wolf, short-faced bear, mastodon, elephant, horses, camel, bisons including giant bison, antelope, ground sloth, glyptodont, gophers, a bird, frogs, toads, and turtles. This assemblage tells us little about the climate; it merely indicates the general nature of the glacial-age(?) fauna. Elsewhere in Mexico bog lemming, marmot, and masked shrew have been found at high altitude, without stratigraphic identification. Also not placed stratigraphically are pine vole and prairie dog, which, accompanied by the elephant and bison, are thought to have migrated into Mexico from the north at times (glacial ages ?) when moisture was great enough to create grassland, at least in corridors or closely spaced patches.

Eastern and Central Regions. Except for the Appalachians, most of the region that lay south of the ice sheet is one of low relief. Because of this we can expect that plant taxa displaced by glaciation would have shifted mainly toward lower latitude, rather than mainly toward lower altitude as in western North America. That does indeed seem to have occurred, but the extent of displacement and the proportions of various taxa that were involved are matters of debate.

Certainly in all the areas overspread by glacier ice, vegetation ceased to exist. Because boreal vegetation does show up in pollen spectra dating from 17,000 to 20,000 C[14] years BP, immediately beyond the drift border, evidently plant successions enforced by glacial climate kept pace with spread of the ice sheet. Rates of spread measured in Ohio, by the C^{14} dates of trees pushed over by the advancing ice margin and buried in till, range from $\sim$40m/yr to $>$100m/yr. During deglaciation, in places at rates even more rapid, vegetation evidently lost little time in repopulating areas of bare drift. Modern analogies exist. In postglacial time, on the moist coast of southern Alaska, bare drift has been covered with a succession of plants ending in mature forests of large spruce, all within a lapse of no more than 200 years. Despite the fact that in that oceanic climate conditions are exceptionally favorable for plant growth, this figure is impressive. Thus, although vegetation zones viewed stratigraphically are time transgressive, probably most of them are only weakly so. Apparently, then, plant communities quickly adjust to changes in the factors of climate and to glacial invasion. At a glacial maximum, probably a tundra community could have existed only near the margin of the ice sheet or on mountains of sufficient altitude.

Indeed, most palynologists now view the ice sheet as probably having been fringed by a narrow tundralike or park-tundralike zone, the trees present being mostly spruce. This zone seems to have continued southward as a long narrow offshoot along the crests of the Appalachians. At Cleveland, Ohio, about 24,000 BP, just before the Late Wisconsin invasion, that zone has been inferred from fossil insects (not pollen) as being open, rather barren country with very few trees, colder than today but not truly arctic (Coope, 1968). Terrain of this kind is established by pollen spectra from the Appalachians eastward, and also westward through Ohio. It is not yet reported from farther west, partly perhaps because most of the known pollen sequences developed from cores do not begin until well after the glacial maximum, at times when spruce forest had already been established. However, we are justified in *speculating* that a narrow tundra (in a broad sense) would have fringed the ice sheet westward to the foot of the Rocky Mountains. Eastward from the Appalachians, a much wider zone of "tundra," in which nonarboreal pollen is comparatively abundant, may have been predominantly grassland.

It was mainly in the narrow tundralike zone that the intense frost activity described in a preceding section took place. The close correspondence between a zone of tundra and a zone of frost action on the one hand, and the margin of the ice sheet on the other, favors the concept that the two zones were products mainly of low temperatures induced by proximity to the glacier, rather than primarily of secular cooling. However, we can add that the direct influence of the ice sheet in depressing temperature must have been confined largely to a narrow zone and did not extend outward through a wide belt of latitude. The numbers we

have quoted suggest a secular lowering of normal mean annual temperature by perhaps 5°, locally increased to 10° or even 15° in places along
the margin of the ice sheet. Measured cooling effects of small ice caps and
other glaciers in high latitudes on surface-air temperatures in summer
range from 1.5° to 4° (Havens, 1964).

When we attempt to assess the combined influence, on vegetation, of
secular and local lowering of temperature, not to mention possible
changes in precipitation, we run into difficulty arising mainly from two
circumstances: (1) the many kinds of North American forest trees (chap
14) compared with the impoverished European flora; and (2) the very
small number of detailed pollen studies that have been made. Because of
these obstacles the pattern of full-glacial vegetation zones is not yet clearly
apparent, and indeed even today's zones grade into each other by imper-
ceptible transition.

The spectrum of views as to the probable full-glacial pattern is very
wide. At one end is the concept (Braun, e.g. 1955) that only taxa close to
the ice sheet were affected, that the tundra and a spruce/fir forest constituted two rather narrow zones, and that south of these the temperate forest cover underwent little change. At the other end of the spectrum is the
view that nearly all the zones were displaced southward as recognizable
belts, with some shifts through as much as 1000km (e.g., P. S. Martin,
1958). The latter concept parallels one that has been considered by many
as valid for the European region, and that was adumbrated as early as
1860 by Charles Darwin, and developed by Hooker for American floras
in 1878. Both views are based on occurrences of pollen, but lack of
enough detailed studies and paucity of C^{14} dates for stratigraphic placement of sequences make final appraisal impossible. Research in the 1960's
(Whitehead, p. 417 in H. E. Wright and Frey, eds., 1965a; p. 237 in
Cushing and Wright, eds., 1967) demonstrated that various boreal species
lived in southeastern United States at times during the Pleistocene. Boreal forest or woodland existed in southern Pennsylvania, together with
abundant boreal vertebrates, although thus far this assemblage is known
only at a station at the rather high altitude of 500m. As the date of the
occurrence is 11,300 BP, probably the area was as cold or colder at the
glacial maximum.[1] Farther southwest, at Saltville in western Virginia (alt
520m), a partly comparable occurrence dating to 13,500 BP includes boreal mammals and a pollen spectrum thought to represent spruce parkland and prairie (C. E. Ray et al., 1967).

Full-glacial pollen spectra at low altitudes on the Coastal Plain, both
in Virginia and in North Carolina, show that spruce was dominant, and

[1] The detailed report on the locality (Guilday et al., 1964) is of special interest because of the convergence of data from both plants and animals as to former climate.

that pine occurred with it, but that at the more southerly locality spruce was less abundant. A spruce/pine assemblage existed in Georgia as well. In all these places, however, pollen of southern elements is present also. Such occurrences establish that in full-glacial time the climate was colder, but thus far they have furnished little quantitative information on the composition of the forests. It is one thing to show lowered temperature, but quite another to define and describe a former vegetation zone.

Therefore, when the ice sheet was at its maximum, boreal plants were indeed present, even at low altitudes, in forests from which they are absent today. Hence during glaciation forest composition did not remain unchanged. On the other hand the indicated presence of southern trees together with boreal ones suggests that instead of a boreal forest shifting southward as a whole, the pre-existing forest was invaded by boreal elements and became mixed, thus differing in composition from the boreal forest present in Canada today. Proof of the correctness of the suggestion is, however, still lacking. We might add that the long southward projection of the northern hardwood forest on the Appalachian summits today (fig 19-1) probably had a full-glacial predecessor in the form of tundra and boreal forest. Such a projection could have been a main avenue for the invasion of boreal elements into the southern forests of the time.

The Coastal Plain was widened by lowering of sealevel. If the sealevel of full-glacial time stood as low as -100m (fig 18-5), the Coastal Plain at lat 38° would have been wider than it is today by nearly 150km. Its increased area should have improved its effectiveness as an avenue for movement of both plants and animals, and the presence of the Gulf Stream close offshore should have been a source of warmth for at least its seaward part.

West of the Appalachians the zone of park tundra is believed to have been followed on the south and southwest by a wider zone of forest, dominated at least in its proximal part by spruce (H. E. Wright, 1968a). With the spruce were birch, alder, and tamarack, with local patches of temperate deciduous trees. Forest of this general kind extended southwestward through southern Iowa at least to northeastern Kansas, nearly 250km distant from the ice sheet in Iowa, and to southwestern Missouri, 400km from the drift border. At the Missouri locality a pollen spectrum rich in spruce is accompanied by a spruce log about 16,500 C^{14} years old, and bones of mammoth (Mehringer et al., 1968).

Boreal forest was present beyond the ice sheet in the Dakotas, because it is indicated in such areas at ~12,000 BP, well after deglaciation had begun. Probably it became more boreal in character toward the northwest, and could reasonably have continued to the Rocky Mountains. Whatever its extent and width toward the west, the forest merged southwestward into more open pine woodland or savanna, followed in south-western Texas and New Mexico by xerophilous woodland with pine (Wells, 1970). By analogy with the ecology of Illinoian time (Kapp, 1965), we can speculate that gallery stands of deciduous trees occupied major river valleys at least as far south as Oklahoma and northern Texas. Wendorf et al. (1961) believed that in the latter region open woodland characterized much of the terrain. If summer temperatures in the southern Plains were lower by 5° or more (Hafsten, 1964, p. 414 inferred annual values lower by 8° to 10°, based on pollen data), the summers could have been cloudier and the precipitation/evaporation ratio increased sufficiently, perhaps, to explain the differences in vegetation without any increase in yearly precipitation.

The description drawn from very scanty data, liberally added to by extrapolation and speculation controlled in part by what we think the climate would have permitted, is portrayed graphically in fig 19-4. We have

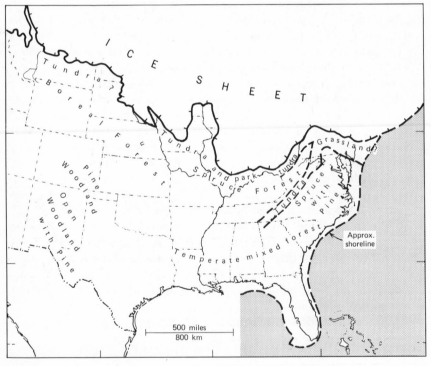

Figure 19-4 Sketch map suggesting possible pattern of vegetation in eastern and central United States at the Late-Wisconsin glacial maximum. Although controlled at a few localities, the sketch is very speculative. (Compiled from published sources.)

dared to write labels but we have drawn no boundaries, and we warn the reader that most of the pattern shown is only suggestive.

Summary. This very sketchy and speculative reconstruction of full-glacial vegetation in parts of North America leads us to the following tentative conclusion. The vegetation was unstable, continually readjusting to changes in climate and glaciers. Throughout the Cordilleran region, vegetation zones spread downslope to lower altitudes and also invaded lower latitudes, undergoing mixing during the process. East of the mountains, at least on the Plains and eastward from Indiana, the ice sheet was bordered by park tundra that likewise extended south along Appalachian summits. South of the tundra was forest dominated by spruce, extending from the Atlantic at least into the Dakotas. Southward, spruce diminished and the forest graded into temperate mixed forest that was open at all times to moist maritime air moving northward from tropical-Atlantic sources and hence tended to be conservative. Northward from northern Texas the forest was flanked on the west by open pine/spruce woodland and with deciduous forest along the rivers. Sketchy though it is, this pattern is what one might reasonably expect to have resulted from secular temperatures a few degrees below those of today, accompanied by the spread of an ice sheet to middle latitudes.

FAUNAS[2]

Present knowledge of North American faunas south of the extensively glaciated region under Wisconsin full-glacial climates is even more sketchy than knowledge of the flora. Much of the information comes from cavern sediments, difficult to correlate with the glacial stratigraphic sequence, so that C^{14} dates must be relied on heavily. If we assume that various faunal assemblages believed to be of "Wisconsin" age or known to date from the earlier phases of the last deglaciation are indicative of full-glacial faunas, we can speculate as follows, relying on data from the very few localities that have been carefully studied:

The park tundra and boreal forest in the eastern and east-central region were inhabited by, among other kinds, woolly mammoth, collared lemming, and caribou in the open parts and by mastodon, ground sloth, and moose in the more wooded parts. In addition there were horse, bison, musk-oxen of nonarctic kinds, snowshoe hare, weasel, marten and fisher, peccary, porcupine, mice, rats, voles, nonarctic lemmings, squirrels, woodchuck, bats, many birds, snakes, and invertebrates. Some of the bo-

[2] The following lists are generalized mainly from Hibbard and Taylor, 1960; Guilday et al., 1964; Guilday, 1971; C. E. Ray et al., 1967.

real mammals lived as much as 10° of latitude south of their present southern limits.

On the Great Plains in southwestern Kansas we find peccary, camel, badger, skunk, ground squirrels, prairie dogs, gophers, mice, shrews, and voles, as well as a variety of birds, turtles, a salamander, fishes, and invertebrates. The fauna implies reduced summer temperatures (Hibbard and Taylor, 1960). Farther south, in northern Texas, we find mammoth, mastodon, horse, camel, antelope, bison, tapir, peccary, deer, ground sloth, jackrabbit, and turtle living in open boreal woodland (Wendorf et al., 1961).

In desert and semidesert country in New Mexico, northern Mexico, and Arizona, the following fauna is found, distributed among five localities with many overlaps: horse, ass, tapir, pronghorn, bighorn sheep, shrub-ox, bison, elk, mountain goat, deer, bears, bobcat, jaguar and other cats, fox, wolf, coyote, skunks, weasel, ground sloth, marmot, prairie dog, squirrel, rabbit, southern lemming, desert rats, mice, and shrew.

In central New Mexico, a locality at 1700m with arid grassland today, a local fauna apparently dates from the late-glacial part of Late Wisconsin time. It includes two camels, elephant, horse, and a bear, all now extinct, as well as other forms that now live farther north where summers are both cooler and moister. This pattern suggests former grassland with sagebrush at the New Mexico locality (A. H. Harris and Findley, 1964).

Five localities in southern California contribute the following fauna of Wisconsin age. Not all the kinds have been found at all localities, although overlaps are common: horse, tapir, bison, pronghorn, shrub-ox, mountain goat, camels, peccary, two species of mammoth, mastodon, saber-tooth cat and other cats, skunk, badger, weasel, bears, fox, wolves, coyotes, ground sloth, and various small rodents. Not surprisingly, this fauna is similar to that in Arizona and the region adjacent and is probably representative.

These North American faunas, like their modern successors, differ from one region to another with differences in climate. None of the kinds is incompatible with the pollen data, nor with today's faunas in the same regions if we take into account the lower temperatures that prevailed at a glacial maximum. We must repeat, however, that our lists are meager, yielding only the most generalized and very incomplete sketch of animal life across the continent.

ENVIRONMENTS DURING DEGLACIATION [3]

After our crude sketch of full-glacial vegetation, we can try to summarize what the pollen record shows about the change in the pattern from full-glacial to the present. Table 19-A presents typical records for the northern part of the Cordilleran region, calibrated by C^{14} dates. Each column shows the vegetation, appropriate to latitudinal and other differences, changing with time. Apart from regional differences, the columns are similar. They start with comparatively low-temperature assemblages and reflect gradually increasing warmth that reaches a peak some 4000 to 6000 years ago, with subsequent reversion to slightly cooler, moister climate. This succession is reflected, in terms of temperature in one part of the north Pacific coastal region, in fig 16-10, col. 2.

Turning back to table 14-C, we can examine a typical sequence of deglacial changes for midwestern United States. Here again we can infer gradual warming to a peak around 5000 BP, followed by a modest reversion to cooler, moister climate. The sequence for Alaska is too much generalized to show the reversion evident in table 19-A; but the New England sequence shows it, as do also table 19-B and the sequences for northern Europe mentioned in chap 14. Not surprisingly the sequence, read upward through the period 14,000 to 6000 BP, broadly resembles the change we suppose occurred in dominant elements in the vegetation at the glacial maximum, if we could have followed them southward from the edge of the ice sheet. The generalized sequence for New England was developed from local sequences, each of which shows its own version of a postglacial succession imposed upon it by climatic change. The march of plant communities into terrain progressively vacated by the ice sheet illustrates again that colonization by plants was nearly as rapid as the deglaciation process itself, leaving little intervening time for bare ground to be affected by deflation, sheet erosion, or the cutting of gullies—processes that are not very evident in the deglacial record.

The first arrivals were pioneers, able to migrate rapidly. Some of them were specially capable of supplying nitrogen to the ground, thus promoting the development of soil. Such plants were followed successively by those that could migrate only at slower rates, but that were more stable. We can suppose that those elements of the fauna, particularly the mammal fauna, that did not become extinct during deglaciation, kept pace

[3] General references: Ogden, 1965; Heusser in H. E. Wright and Frey, eds., 1965a; Sirkin, 1967; Benninghoff, p. 70–78 in Bergstrom, ed., 1968.

Table 19-A Plant Succession Through the Last 13,000 Years, More or Less, in 11 Areas in Northern Cordilleran North America, Developed from Pollen Data. Vertical stripes indicate that no data are available. Assembled by C. J. Heusser (p. 481 in H. E. Wright and Frey, eds., 1965a).

C14 Years BP × 1000	Northwestern Alaska			Northern Alaska		Pacific Coastal Alaska			Pacific Northwest		
	St. Lawrence Island	Nome	Ogotoruk Creek	Umiat	Chandler Lake	Homer	Munday Creek	Montana—Ward Crks.	Granite Falls	Onion Flats	Liberty Lake
0							Sitka Spruce Hemlocks	Lodgepole Pine			
1	Grass–Sedge	Absent (?)	Absent (?)		Birch–Alder	Alder–Birch		Lodgepole Pine			
2	Birch–Alder Spruce	Absent (?)	Absent (?)	Absent (?)	Sedge Spruce	Sedge–Heath Spruce	Alder Sitka Spruce Hemlocks	Western Hemlock predominance	Western Hemlock predominance	Douglas Fir Oregon Oak	Yellow Pine predominance
3							Alder Sitka Spruce	Mtn. Hemlock Heath	Douglas Fir		
4											
5	Alder prominence	Birch–Alder Heath–Sedge Spruce	Birch–Heath	Alder prominence	Alder prominence	Alder maximum	Lysichitum maximum	Lysichitum maximum			
6	Birch–Heath	Birch–Heath	Birch–Willow Sedge–Grass	Birch Sedge–Heath Spruce	Birch–Heath Spruce	Alder maximum	Alder maximum	Alder maximum	Douglas Fir predominance	Oregon Oak maximum	Grasses Chenopods Composites
7	Grass–Sedge	Birch–Willow Sedge–Grass		Willow–Birch Sedge–Grass		Sedge	Sedge	Sedge			
8			Birch	Birch	Birch	Sedge	Sedge	Hemlock Sitka Spruce			
9	Birch	Birch		Birch	Birch	Willow–Birch Alder	Heath	Lodgepole Pine maximum	Douglas Fir	Douglas Fir	
10	Grass–Sedge			Sedge	Sedge	Birch–Willow Sedge		Alder–Willow Sedge	Lodgepole Pine maximum	Lodgepole Pine maximum	Lodgepole Pine maximum
11	Grass–Sedge Willow–Birch	Willow–Birch	Sedge–Grass Willow–Birch	Sedge–Grass Composites Willow–Birch	Sedge–Grass Composites Willow–Birch				White Pine	Sitka Spruce Fir	Yellow Pine
12											Grasses
13											

Table 19-B Postglacial Successions of Vegetation in Northeastern United States, Compiled and Correlated by L. A. Sirkin, 1967. Other areas and localities, including southern Quebec, are given in Lasalle, 1966, fig 16.

Approximate age (years BP)	Pollen zones (subzones)	Central New Jersey	Western Long Island	Martha's Vineyard, Massachusetts	Southern Connecticut	Central New England
1,000	Oak— (C) pollen Zone	C3b Oak, chestnut	C3b Oak, chestnut, birch		C3b Spruce, pine rise	C3b Spruce, pine rise
2,000		C3a Oak, alder	C3a Oak, holly, hemlock		C3a Oak, hemlock, chestnut	C3a Oak, hemlock
4,000		C2 Oak, hickory	C2 Oak, hickory		C2 Oak, hickory	C2 Oak, hickory
		C1 Oak, hemlock	C1 Oak, hemlock	C1 Oak	C1 Oak, hemlock	C1 Oak, hemlock
7,000	Pine— (B) pollen Zone		B2 Pine, oak			B2 Pine, oak
9,000		B Pine	B1 Pine	B Pine	B Pine	B1 Pine, spruce
10,500	Spruce— (A) pollen Zone		A4 Spruce, pine	A4 Spruce	A4 Spruce returns	A4 Spruce returns
		A2–A4 Spruce	A1–A3 Pine, spruce	A3 Pine, spruce	A3 Pine, spruce, oak	A3 Pine, spruce
12,500				A1, 2 Spruce, pine max	A1, 2 Spruce, pine max	A1, 2 Birch, spruce
	Herb— (T) pollen Zone	T Pine, spruce	T3 Pine, spruce, herb	Tb Spruce park	T3 Birch park–tundra	T3 Tundra
14,000			T2 Spruce park	Ta Spruce, birch alder park	T2 Spruce park–tundra	
15,000			T1 Park tundra	V3, 4 Tundra	T1 Tundra	Ice sheet
	Zone (W*)	W* Park tundra	W3 Park tundra	V1, 2 Park tundra		
17,000			W2 Near-tundra	?	Ice sheet	
			?	Ice sheet		
20,000			Ice sheet			
Modified from:	M.B. Davis in H.E. Wright and Frey, eds., 1965	Sirkin, Owens, and Minard, 1970 (unpublished)	Sirkin, 1967	Ogden, 1959	Leopold, 1956	M.B. Davis, 1958

Postglacial

Glacial

* W = Wisconsin.

515

with the movement of plants and took their places once again in communities that provided favorable habitats.

Brown and Cleland (p. 114 in Bergstrom, ed., 1968) set forth a persuasive argument concerning the way in which vegetation developed during deglaciation of the Great Lakes region. From pollen data, analyses of the stomach contents of fossil mammals, and the distribution pattern of Early Man in the period after about 14,000 BP, they visualized dilution of spruce-dominated forests with deciduous trees. The result was an irregular mosaic pattern of forest elements rather than a northward march of well-defined vegetation zones. Through this process the mosaic gradually gave way to a belted pattern that took form, in that region, in the neighborhood of 11,000 years BP.

CHAPTER 20

Stratigraphy and Climatic History of Western North America

THE CORDILLERAN REGION

The Cordilleran region of North America is primarily one of mountains with intervening basins. Many of the mountain ranges are high and some of the basins are deep. In consequence the Quaternary stratigraphy is dominated in mountain valleys by glacial sediments, and in basins by fills consisting of alluvium, the sediments of pluvial lakes, and (in coastal southern California) marine strata. Ancient soils developed in these various sediments play a part in correlation and in the recognition of former times of relatively warm, nonglacial climates in a sequence characterized by many climatic fluctuations. Likewise, layers of tephra ejected by Cordilleran volcanoes constitute stratigraphic horizon markers.

Until the middle of the 20th century, scientific study of these strata lagged behind that of ice-sheet glaciation of the country east of the Cordillera. Although the scientific explorers of the post-Civil War period recognized the effects of former glaciation in the mountains, and although both I. C. Russell and G. K. Gilbert, before the end of the 19th century, discerned stratigraphic sequences in the Basin-and-Range Region, systematic examination of glacial-stratigraphic successions in the mountains began effectively with the work of Blackwelder (e.g., 1931). After 1950 such study expanded rapidly. Stratigraphic names (Bull Lake, Pinedale) proposed by Blackwelder for units in the Wind River Mountains are in general use and have acquired almost as wide significance as have the terms Early and Late Wisconsin. Similarly, the glacial sequence in the Sierra Nevada is still based on that of Blackwelder, although it now includes several units recognized and inserted more recently. Relative ages of the drift bodies are determined mainly by statistical treatment of degree of weathering of granitic clasts. Modern studies, moreover, have led to correlations between glacial and pluvial-lake sequences, thereby tying more closely together glacial and pluvial events.

GENERAL CHARACTER OF
THE GLACIAL STRATIGRAPHY

The pursuit of stratigraphy in high mountains encounters special difficulties, for several reasons. The geologist must get about almost solely on foot and often faces logistical problems. Also, rapid erosion on steep slopes has removed much drift and in places has inhibited the development of weathering and soil profiles. Furthermore, in many highlands the bedrock is such that the drift is coarse-grained and permeable, a characteristic unfavorable for distinct soil development and for the preservation of embedded wood and other substances useful for C^{14} dating. Also glaciated areas are commonly not connected; hence correlation, impossible by continuous tracing, must be based on morphology, degree of weathering, relative extent and altitude of units, and radiometric dates. Finally, in mountain ranges such as the Sierra Nevada and the Wasatch, uplift and localized crustal movements have occurred since the beginning of glaciation. This may have affected the relative position of the snowline and the extent of glaciers.

Despite these disadvantages, a sequence of Wisconsin age, common to many highlands from the Rocky Mountains to the Pacific Coast, has been established. Correlation with the Wisconsin Drift of central North America, and with units within the Wisconsin drift, is based mainly on C^{14} dates, with consistent though far less precise support from relative degree of development of ancient soils. In basin fills additional support comes from assemblages of vertebrate fossils.

Selected principal sequences, excluding British Columbia and Alaska, are set forth in tables 20-A and 20-B, and other data are given in table 18-B. From the tables it appears that there are three recurrent elements which we can describe further. (1) One or more units of glacial drift older than a thick, very mature soil, commonly with reddish hues. Such drift lacks constructional topography. It is much dissected; in many areas it antedates conspicuous trenching of bedrock and consists only of scattered remnants, and is believed to antedate the last major interglaciation. (2) An extensive drift sheet, younger than any thick, very mature soil; it has subdued constructional topography and few surface boulders, is itself moderately weathered, and in places moderately well-developed soil separates it from overlying drift. It may consist of two units separated by a weakly developed soil. (3) A less extensive drift sheet, younger than (2), having pronounced constructional topography and a bouldery surface, and less alteration by weathering and soil development. It, too, may consist of two distinct units. The relationships of these three elements in one small area are seen in fig 20-1.

Both (2) and (3) together postdate the last major interglaciation and are correlatives of the Wisconsin Drift in central North America, (3) being Late Wisconsin and (2) being approximately Early Wisconsin. In addition, some sequences also include younger tills, in which tills of the Neoglaciation are prominent, as well as rock glaciers and protalus ramparts. Although a few sequences are known in far more detail than many

Figure 20-1 Sketch, made from a photograph, looking south into the mouth of Convict Canyon (lat ~37°35′), Sierra Nevada, California. End moraines of Tahoe and Tioga age are shown, as well as vertical positions of Sherwin and McGee Tills. (W. C. Putnam.)

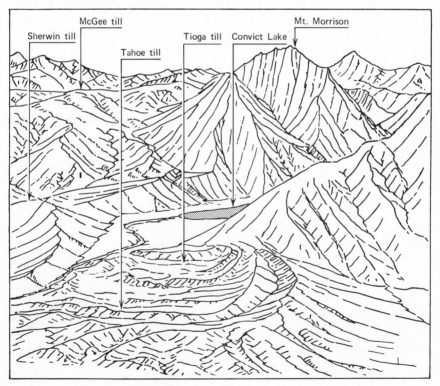

Table 20-A
Correlation of Late Quaternary Stratigraphy in Western United States.
(Birkeland et al., 1970.)

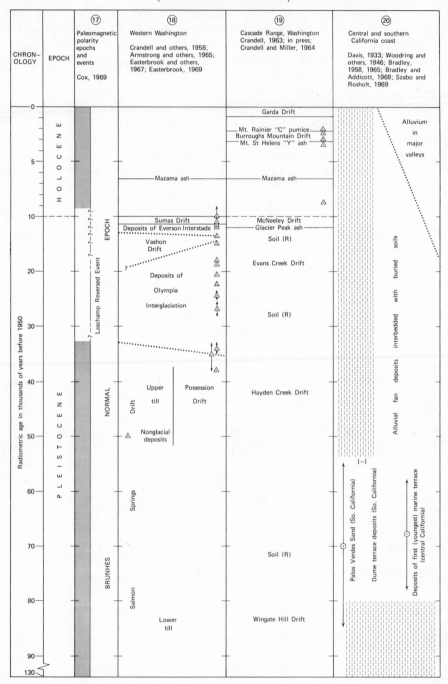

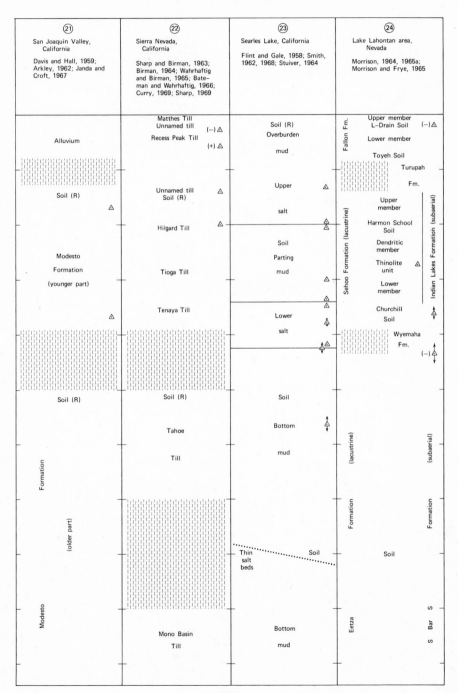

Table 20-A
Correlation of Late Quaternary Stratigraphy in Western United States. (Birkeland et al., 1970.) *(continued)*

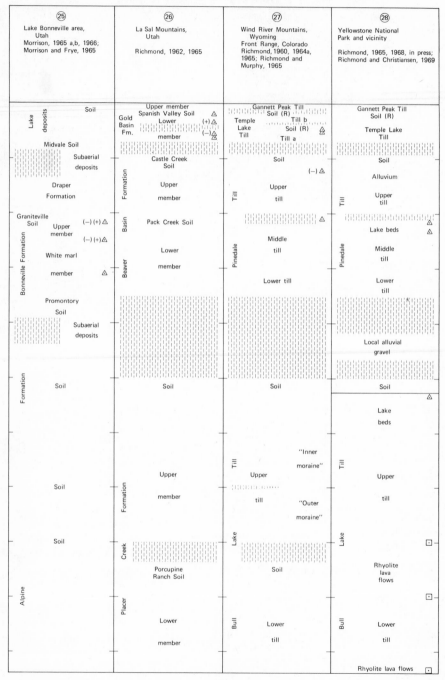

㉕ Lake Bonneville area, Utah — Morrison, 1965 a,b, 1966; Morrison and Frye, 1965	㉖ La Sal Mountains, Utah — Richmond, 1962, 1965	㉗ Wind River Mountains, Wyoming — Front Range, Colorado — Richmond, 1960, 1964a, 1965; Richmond and Murphy, 1965	㉘ Yellowstone National Park and vicinity — Richmond, 1965, 1968, in press; Richmond and Christiansen, 1969
Lake deposits — Soil	**Gold Basin Fm.** — Upper member — Spanish Valley Soil △ — Lower (+)△ member (−)△ △	**Temple Lake Till** — Gannett Peak Till — Soil (R) — Till b — Soil (R) △ — Till a	Gannett Peak Till — Soil (R) — Temple Lake Till
Midvale Soil			
Subaerial deposits	Castle Creek Soil	Soil	Soil
		(−) △	Alluvium
Draper Formation	Upper member	Upper till	Upper till
Graniteville Soil — Upper member (−)(+)△ (−)(+)△	Pack Creek Soil	Middle till	Lake beds △
White marl member △	Lower member	Lower till	Middle till
Promontory Soil			Lower till
Subaerial deposits			Local alluvial gravel
Soil	Soil	Soil	Soil △
			Lake beds
		"Inner moraine"	
Soil	Upper member	Upper till	Upper till
		"Outer moraine"	
Soil	Porcupine Ranch Soil	Soil	Rhyolite lava flows
			·
Lower member	Lower till	Lower till	
			Rhyolite lava flows ·

Left margin labels: Bonneville Formation; Formation; Alpine (column 25). Formation; Basin; Beaver; Formation; Creek; Placer (column 26). Till; Pinedale; Till; Lake; Bull (column 27). Till; Pinedale; Till; Lake; Bull (column 28).

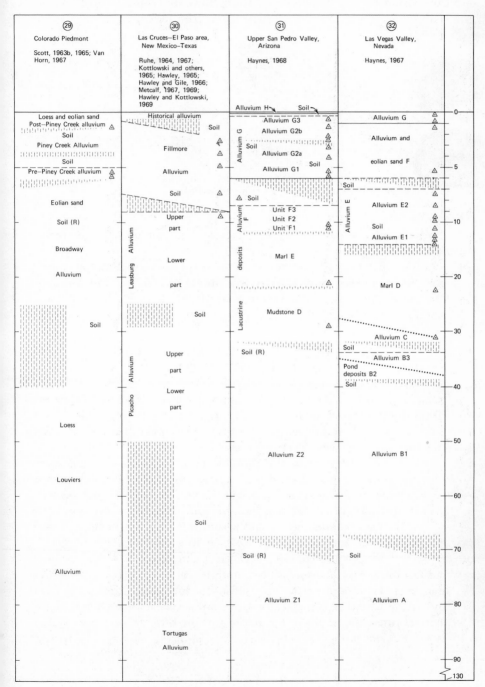

Table 20-A
Correlation of Late Quaternary Stratigraphy in Western United States.
(Birkeland et al., 1970.) *(continued)*

E X P L A N A T I O N

△ Radiocarbon date	⎫ arrows indicate uncertainty	——— Boundary between stratigraphic units; based on chronological data
○ Uranium–series date	⎭ of date where space permits	
		– – – Approximate boundary between stratigraphic units; close chronological control is lacking
(+)	Radiometric data indicate that this is the maximum age for the overlying unit	
(−)	Radiometric data indicate that this is the minimum age for the underlying unit	········ Units or events above and below line are partly of the same age
Soil	Buried soil, in places recognized in relict positions; (R) indicates a soil recognized only as a relict soil	▦ Erosion interval or time of nondeposition recognized between units

The references are those of the compilers; many of them are not listed in the present book. Additional data are given in table 18-B.

others, the overall pattern is as described above, without major discrepancies.

In only two areas are as many as three tills known, each deeply weathered, all in superposition, and of pre-last-interglacial age. The areas are the La Sal Mountains, Utah (table 20-B) and Glacier National Park, Montana (Richmond, 1957; 1962).

HYPSITHERMAL INTERVAL AND NEOGLACIATION

Much attention has been given to glaciations that have occurred within the last few thousand years, that is, since the disappearance of the Laurentide and Scandinavian Ice Sheets. The comparatively warm Hypsithermal interval, defined (Deevey and Flint, 1957) as "the time represented by four pollen zones, V through VIII in the Danish system" spans the period between ~9000 and ~2500 years BP. Following that period, a time of cooler climate, glacial readvances, and in places rebirth of small glaciers was recognized by Matthes, who called it informally the "little ice-age." Glacial maxima before and within this later time seem to cluster around positions shortly after 5000 BP, again soon after 3000 BP, and finally, and most conspicuously, between about 300 and 150 BP. Porter and Denton (1967) proposed substitution of the name Neoglaciation, first used by J. H. Moss in 1951, as a substitute for Matthes' informal term, and recognized that Neoglaciation overlapped somewhat on the Hypsithermal interval as defined. For use in this book we define Neoglaciation as *the rebirth and/or growth of glaciers following maximum shrinkage during the Hypsithermal interval.* The paper by Porter and Denton, as well as the more recent comprehensive paper by Denton and Porter (1970), documents Neoglacial advances, not only in the North American Cordillera but in other parts of the world. The widespread time coincidence of those advances affords good support for time coincidence, on a secular scale, of variations of climate. Incidentally, early Neoglacial cooling is reflected in the Sierra Nevada not only in physical evidence of glaciation but in the pollen record as well (D. P. Adam, p. 275 in Cushing and Wright, eds., 1967).

The columns in table 20-A that concern southwestern United States raise another question of terminology, involving the description of climate in southwestern United States during the last several thousand years. Antevs (1948) proposed a series of terms relating to "postpluvial" time in the Southwest. Of those *altithermal,* referring to a time of greater warmth and aridity than today's, and nearly equivalent to the "climatic optimum" of former usage in northern Europe, has continued in use. Although the existence, in the Southwest, of such a warm period has been questioned, most investigators in that region consider it a reality. Other terms, such as "megathermal," have been urged, and "thermal maximum" has been suggested to describe informally the highest point, or any one in a sequence of high points, on the postglacial time/temperature curve. *Altithermal* is not a rock-stratigraphic term and hence does not appear in tables 11-A, 20-A, and 20-B.

COMMENTS ON THE GLACIAL SEQUENCE

Although in most of the areas represented in tables 20-A and 20-B the relative ages of the units identified thus far is established, absolute dating and correlation of the units are poorly known. The tables indicate fairly the uncertainties, and like other tables prepared with less attention to detail, they will fill out and become firmer in time.

Table 20-B is noteworthy in that it includes a tentative calibration based in part on K/Ar dates, and in addition, calibration in terms of the paleomagnetic sequence. Dating of the Bishop Tuff (table 20-B, column 7) at 0.7 million years by the K/Ar process (Dalrymple et al., 1965) provided a minimum date for the underlying Sherwin Till, and thus intimated that glaciations had begun earlier than had been supposed. K/Ar dates indicate that the McGee and Deadman Pass Tills are far older than the Sherwin (fig 20-1), recording glaciation as much as 3 million years ago. Upfaulting with an amplitude of at least 1200m postdates the McGee Till.

Table 20-B

Correlation of Quaternary Stratigraphy in Western United States. (Birkeland et al., 1970.)

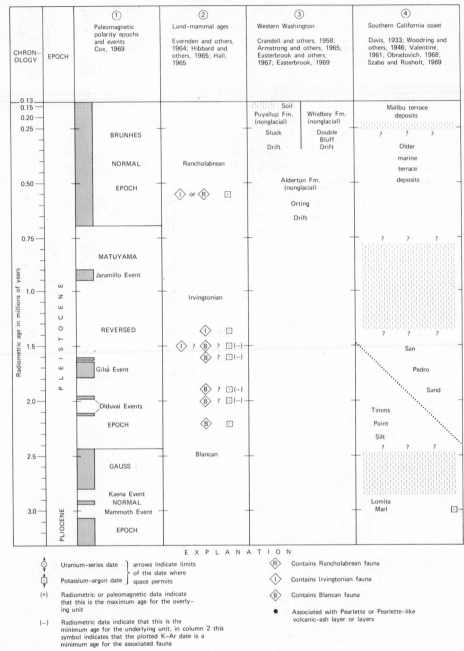

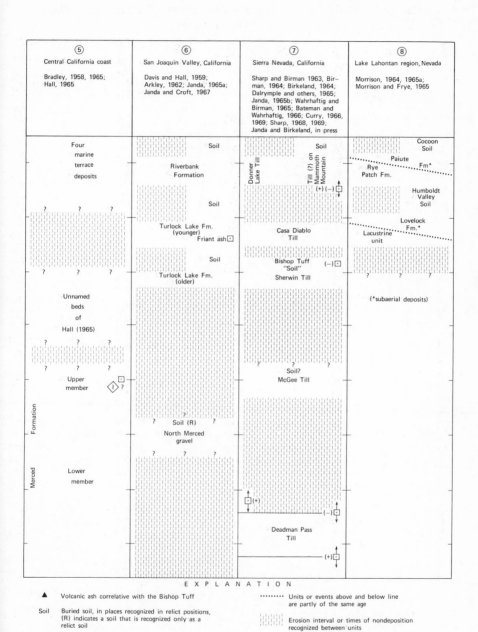

	(5)	(6)	(7)	(8)
	Central California coast	San Joaquin Valley, California	Sierra Nevada, California	Lake Lahontan region, Nevada
	Bradley, 1958, 1965; Hall, 1965	Davis and Hall, 1959; Arkley, 1962; Janda, 1965a; Janda and Croft, 1967	Sharp and Birman 1963, Birman, 1964; Birkeland, 1964; Dalrymple and others, 1965; Janda, 1965b; Wahrhaftig and Birman, 1965; Bateman and Wahrhaftig, 1966; Curry, 1966, 1969; Sharp, 1968, 1969; Janda and Birkeland, in press	Morrison, 1964, 1965a; Morrison and Frye, 1965

EXPLANATION

▲   Volcanic ash correlative with the Bishop Tuff

Soil   Buried soil, in places recognized in relict positions, (R) indicates a soil that is recognized only as a relict soil

————   Boundary between stratigraphic units, based on chronological data

— — —   Approximate boundary between stratigraphic units, close chronological control is lacking

··········   Units or events above and below line are partly of the same age

   Erosion interval or times of nondeposition recognized between units

Soil   This indicates that the time interval was characterized by both erosion and soil formation; times of both are uncertain

527

Table 20-B
Correlation of Quaternary Stratigraphy in Western United States. (Birkeland et al., 1970.) *(continued)*

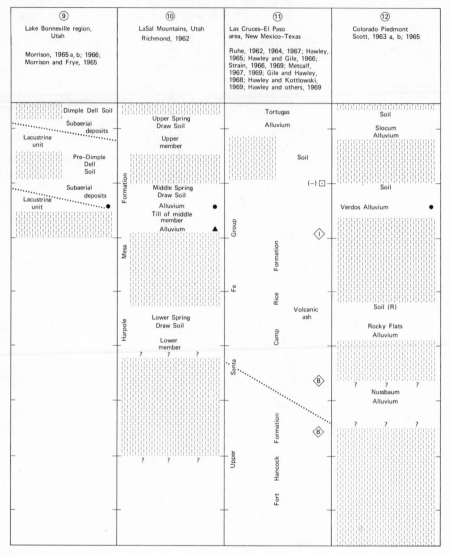

⑨ Lake Bonneville region, Utah — Morrison, 1965 a, b; 1966; Morrison and Frye, 1965	⑩ LaSal Mountains, Utah — Richmond, 1962	⑪ Las Cruces–El Paso area, New Mexico–Texas — Ruhe, 1962, 1964, 1967; Hawley, 1965; Hawley and Gile, 1966; Strain, 1966, 1969; Metcalf, 1967, 1969; Gile and Hawley, 1968; Hawley and Kottlowski, 1969; Hawley and others, 1969	⑫ Colorado Piedmont — Scott, 1963 a, b; 1965
Dimple Dell Soil; Subaerial deposits; Lacustrine unit; Pre-Dimple Dell Soil; Subaerial deposits; Lacustrine unit ●	Upper Spring Draw Soil; Upper member; Middle Spring Draw Soil; Alluvium ●; Till of middle member; Alluvium ▲; Lower Spring Draw Soil; Lower member; ? ? ? ; ? ? ? (Formation / Mesa / Harpole)	Tortugas; Alluvium; Soil; (−) ⊡; ⟨I⟩; Volcanic ash; ⟨B⟩; ⟨B⟩ (Group / Fe Rice Formation / Camp Rice Formation / Santa Fort Hancock Formation / Upper Fort Hancock Formation)	Soil; Slocum Alluvium; Soil; Verdos Alluvium ●; Soil (R); Rocky Flats Alluvium; ? ? ? ; Nussbaum Alluvium; ? ? ?

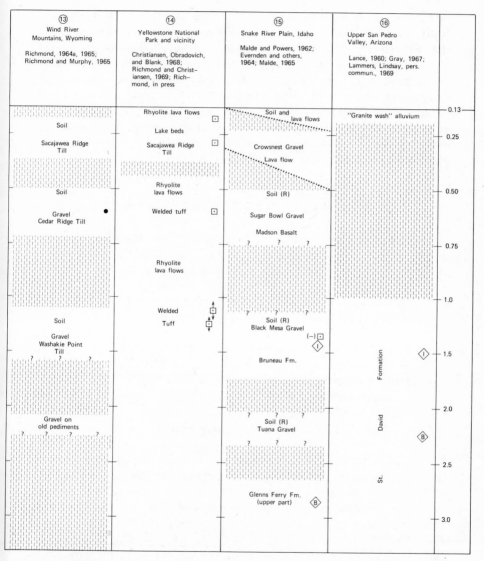

The references are those of the compilers; many of them are not listed in the present book.

Correlation of any part of the Cordilleran sequence with the glacial stratigraphy of central North America by any means other than radiometric dating involves difficulties. C^{14} dating has established the essential equivalence of major climatic events in the two regions through about the last 30,000 years and their probable equivalence through the last 50,-000 years. But the only area thus far found in which drift of Cordilleran origin approaches contact with drift of Laurentide origin lies near the eastern base of the Rocky Mountains in Montana. There Bull Lake Drift is reported (Richmond, 1960) to merge with Laurentide drift of apparently Early-Wisconsin age, whereas Pinedale drift merges with a younger drift as yet not correlated except by inference. As far as it goes, this information is compatible with the relevant C^{14} dates.

NONGLACIAL TERRESTRIAL STRATA

Nonglacial sediments in the southern part of the Cordilleran region [1] are of both terrestrial and marine origin. The terrestrial sediments include abundant basin and valley alluvium (chap 11), as well as deposits made in pluvial lakes (chap 17; correlated locally with glacial drift), at tar seeps, and in caverns. The cavern sediments have special interest because of their local content of pollen, mammal remains, and artifacts.

Among the better-known sequences of basin alluvium is the fill beneath the San Joaquin "Valley" in central California (table 20-B, column 6), where Quaternary sediments are as much as 1500m thick. Beneath a part of the region Quaternary alluvium overlies Pliocene marine sediments conformably, and locally contains a Blancan mammal fauna. Fossils occurring higher in the alluvium are of Irvingtonian age.

Local accumulations at tar seeps are important for ecology and for the regional history of vertebrates. The most famous are the Rancho La Brea beds, exposed in the city of Los Angeles and consisting of a mass of sand, silt, and clay thoroughly impregnated with tar that seeped upward from underlying oil-bearing strata and hardened as a result of slow evaporation. The deposits are filled with the bones of fossil mammals and birds and the remains of coniferous trees. The place is the site of an active oil seep to which animals came, perhaps thinking it a pool of water, and became trapped in the tar. Carnivorous animals fed upon the carcasses of the victims and in turn were caught. The bones of hundreds of individu-

---

[1] See a comprehensive review by Wahrhaftig and Birman (p. 299–340 in H. E. Wright and Frey, eds., 1965a).

als have been collected, and many thousands more are believed to be still in the tar (Jahns et al., 1955, chap 3, p. 55–57).

The rich fauna with 39 species of mammals (including *Canis, Smilodon, Camelops, Bison, Equus, Mylodon, Nothrotherium, Mammut,* and *Archidiskodon*), as well as 60 kinds of birds (Stock, 1956), is upper Pleistocene, apparently spanning the Illinoian, Sangamon, and Wisconsin ages. Radiocarbon dates of some of the material fall within the range 40,000 to 4500 BP. (Ho et al., 1969). Similar seeps at McKittrick and at Carpinteria, California, contain a somewhat similar fauna and a flora suggesting a cooler climate.

MARINE STRATA

The coastal region of southern California includes mountain ranges and basins, in both of which are marine sedimentary rocks as much as 1800m thick, of Quaternary age. These strata have been deformed, intensely in some areas, and the deformation, although spread through late-Cenozoic time, was intense as late as mid Pleistocene. Overlying the marine strata is a variety of local continental sediments. The sequence has been studied intensively, in part because it is petroliferous.

The Los Angeles and Ventura sedimentary basins contain particularly significant sections (table 20-B, column 4; Woodring et al., 1946) because in them marine Pleistocene sediments overlie marine Pliocene sediments with continuity. The Pliocene-Pleistocene boundary has been drawn tentatively at the base of the Santa Barbara Formation in the Ventura basin and at the base of the Lomita Marl and Timms Point Silt in the Los Angeles basin, at the position of a marked discontinuity in the invertebrate fauna. Also strata above the boundary contain *Equus.* The discontinuity is evolutionary and seems to have no climatic implications. Although certain mollusks occurring above the discontinuity had given rise to an inference of climatic cooling, they occur in associations that imply water depths as great as 180m. Hence the inferred low temperature may be related to depth rather than to climatic change. This idea was developed fully in a discussion by Valentine (1961, p. 421), who also indicated that major provincial differences preclude correlation of these marine faunas with faunas in Europe. Despite the known existence of other strata in California, he concluded that the Pliocene-Pleistocene boundary in California has not yet been determined satisfactorily.

COASTAL TERRACES[2]

The California coast and its Santa Barbara Islands are marked by flights of terraces, most of them somewhat deformed. Each terrace consists of a wave-cut bench, in places with a fossiliferous marine cover, and commonly with at least a partial cover of alluvium. These features mark former positions of sealevel relative to the land. Although overlying alluvium precludes precise measurement of altitudes, the terrace forms occur up to +400m or more. As many as 13 terraces have been identified within a single segment of the coast.

Stratigraphic positions of some of the terraces are shown in table 20-B, columns 4 and 5. At least one terrace seems to be Late Pleistocene, is little if at all deformed, and may be entirely a product of a eustatically controlled position of sealevel. However, most are older. One group cuts and therefore postdates the San Pedro Sand of Early Pleistocene age; where stratigraphic controls are lacking, a Pliocene age for a high terrace is not excluded. Some terraces vary in altitude along the coast. Because of this, and because of the high altitudes of some units, crustal movement is believed to have played an important part in the present configuration of these features.

PUGET SOUND-VANCOUVER REGION[3]

The coastal region that includes the lowland around Puget Sound, and the Fraser Lowland on both sides of the Strait of Georgia in southwestern British Columbia, provides a rich and complicated sequence of Quaternary strata. Glacial sediments extending from sealevel to high altitude combine with a succession of glacial and nonglacial sediments, including marine sediments that seemingly extend far back into Pleistocene time. The record shows that valley glaciers in the Olympic, Cascade, and other adjacent mountains coalesced locally with lowland ice of the Cordilleran Ice Sheet, which later shrank to the condition of piedmont and valley glaciers. Detailed study since the early 1950's replaced the original reconnaissance with the sequence shown in tables 20-A, 20-B, and in more detail in table 20-C.

The oldest known drift yet discovered is the Orting Drift, a thick suc-

---

[2] Wahrhaftig and Birman, p. 329 in H. E. Wright and Frey, eds., 1965a; Emery, 1960, p. 5–8; Woodring et al., 1946.

[3] References include Crandell, p. 341–353 in H. E. Wright and Frey, eds., 1965a; J. E. Armstrong et al., 1965.

Table 20-C

Stratigraphic Sequences in the Puget Sound–Vancouver Region. (After J. E. Armstrong et al., 1965.)

Glacial–Geologic Units		C^{14} years BP × 1000	Fraser Lowland: British Columbia (Armstrong et al, 1965)	Fraser Lowland: Washington (Easterbrook, 1963)	East–Coastal Lowland Vancouver Island (Fyles, 1963)	Puget Lowland (Crandell, 1963; Easterbrook et al., 1967)
Fraser Glaciation	Sumas Stade	10–, 11–	Sumas Drift	Sumas Drift	Capilano sediments	
	Everson Interstade	12–, 13–	Whatcom glaciomarine deposits; Newton stony clay; Cloverdale sediments (Capilano sediments)	Bellingham glaciomarine drift; Deming sand; Kulshan glaciomarine drift		
	Vashon Stade	14–, 15–	Surrey Drift	Vashon Drift	Vashon Drift	Vashon Drift
	Evans Creek Stade ?	17–, 19–, 21–, 23–				Evans Creek Drift
Olympia Interglaciation		25–, 29–, 33–, 36–	Quadra sediments		Quadra sediments	Kitsap formation
Salmon Springs Glaciation			Semiamu Drift (age unknown)		Dashwood Drift (age unknown)	Salmon Springs Drift; Possession Drift; Whidbey Fm; Double Bluff Drift
Puyallup Interglaciation; Stuck Glaciation; Alderton Interglaciation; Orting Glaciation		(Not calibrated)	?	?	?	

cession of two or more tills separated by alluvial and lacustrine sediments, together probably representing oscillation during a single glaciation. This drift is oxidized, though not deeply. It seems to have acquired its cover of overlying strata rather soon after it was deposited. It is followed by the Alderton sediments, a sequence of alluvium and lacustrine sediments, with significant colluvium resulting from ancient volcanic mudflows from Mt. Rainier. Pollen indicates that the climate, at least at one time, resembled that of today.

Next in succession is the Stuck Drift, consisting of till, outwash, and, locally, ice-margin lake sediments. The overlying Puyallup Formation is much like the Alderton. Pollen occurring immediately above the Stuck Till indicates the cool, moist climate expectable during deglaciation. That climate was followed, as recorded in other exposures, by warming to about today's level and later by cooling. The Puyallup sediments are weathered to an extent consonant with a time lapse much longer than that represented by existing postglacial time.

The Salmon Springs Drift consists of two or more layers of till interbedded in a sequence of sand, gravel, and other sediments. Pollen separated from a layer of peat indicates a climate cooler and moister than today's. Probable correlatives of this drift in various parts of the lowland are shown in table 20-C. The succeeding Olympia sediments are thought to correlate, at least in part, with the interstade immediately preceding the Late Wisconsin drift, as suggested by stratigraphic position and C^{14} dates. They consist principally of lacustrine fine sediments and alluvium. Pollen indicates a cool, moist climate, on the whole cooler than today's. During at least one glaciation of Salmon Springs age glacier ice in the Olympic Mountains attained the status of an ice cap. As indicated in the next paragraph, the boundary between Olympia sediments and the succeeding Fraser glacial sediments seems to be conspicuously time transgressive.

The Fraser Drift is broadly equivalent to the Late-Wisconsin drift in the region east of the Cordillera. Its deposition seems to have been complicated by the presence of both local valley glaciers and the Puget Lobe of the Cordilleran Ice Sheet. The valley glaciers climaxed and retreated through some distance before the larger ice body reached its greatest extent. The time lag between the two climaxes may have amounted to 3000 years or even more. The Evans Creek Drift represents the climax of valley glaciers in the mountains east of the Puget Lowland, and in the Olympic Mountains as well, where the glaciers were much less extensive than they had been in Salmon Springs time.

The Vashon Drift was marked by the climax of the Puget Lobe around 15,000 BP. Till was deposited beneath the lobe, while in the sea beyond it thick, extensive glacial-marine sediments accumulated. Marine sedimentation continued, building deposits having a maximum thickness of at least 160m, through the phase of deglaciation. Indeed these marine sediments, built beyond the retreating ice margin, are the Everson sediments. Overlying till and outwash (the Sumas Drift) in the eastern part of the Fraser Lowland indicates a readvance that climaxed between 11,000 and 10,000 BP. Thereafter the ice melted out of the lowlands. In nearby mountains drastic deglaciation was followed by the deposition of at least two drift bodies of Neoglacial age.

ALASKA-YUKON REGION

Early Glaciations

The glacial sediments thus far identified in Alaska can be divided conveniently into those indicating early (i.e., mainly pre-Pleistocene) glaciations and those indicating later (mainly or wholly Pleistocene) glaciations. In one early category are tillites and glacial-marine strata, both occurring in the coastal belt that fronts on the Gulf of Alaska. Tillites occur at various places in the St. Elias, Wrangell, and Chugach Mountains. First described by Capps (1916) and later studied by Denton and Armstrong (1969), these sediments are numerous. They are interbedded with ancient basalt flows, through which many of them have been dated by K/Ar. The dates range from ~3 million back to at least 10 million years BP; hence the tillites range through Pliocene and possibly late-Miocene time, the span within which these mountain ranges formed and rose to great height. The tillites are overlain locally by four tills of lesser age, together embracing the last 1.6 million years and possibly the last 2.7 million years.

Deformed glacial-marine strata with known local minimum thickness of 3000m occur in the coastal belt adjacent to these mountains, in the Lituya Bay, Malaspina, and Yakataga districts, and on the Trinity Islands and Middleton Island. The strata in the latter place are described in detail by D. J. Miller (1953). Mollusks collected from glacial-marine strata at various places range from earlier Pleistocene down through Pliocene and possibly into upper Miocene. These occurrences are consistent with inferences of late-Miocene and Pliocene cooling, mentioned in chap 16, based on evidence from fossil floras.

Glaciation in Alaska therefore began early, perhaps at first only on high coastal mountains with glaciers descending their flanks and entering the Gulf.

Later Glaciations

Of the four tills mentioned as overlying the tillite sequence, the young-est probably correlates with the Kluane Glaciation (table 20-D) as indi-cated by C^{14} dates; correlation of the three others still lacks a basis (Denton and Armstrong, 1969). However, glacial-geologic sequences, their younger parts controlled by C^{14}, have been established in several parts of Alaska. Some of those which appear to extend well back of Wisconsin time are represented in table 20-D. Tables with more details, including short sequences, are given by Péwé et al. (p. 360 in H. E. Wright and Frey, eds., 1965a); and Hamilton (p. 204 in Schumm and Bradley, eds., 1969). Also the Quaternary Map of Alaska (Karlstrom et al., 1964) in-cludes a correlation table. From these tables it seems evident that both Early and Late Wisconsin glaciations, separated by an interstadial, are represented in Alaska, and that an unknown number of pre-Wisconsin glaciations are also present. The history of vegetation in Alaska, through these climatic fluctuations, is discussed by Colinvaux (p. 207 in Hopkins, ed., 1967).

Vertebrates of Illinoian age, found in Alaska (mostly near Fairbanks) include hunting dog, wolf, fox, a large cat, mastodon, mammoth, elk, giant elk, moose, caribou, large bison, various musk-oxen, mountain sheep, and horse. In addition clear evidence of the presence of beaver has been found. (Péwé and Hopkins, p. 267 in Hopkins, ed., 1967.)

Late Wisconsin and Neoglacial stratigraphy in the Muir Inlet area on the coast of southeastern Alaska, where marine as well as glacial sedi-ments are involved, is outlined by Haselton (1966).

Reconnaissance (Bostock, 1966; Vernon and Hughes, 1966; Hughes et al. 1969) in various parts of Yukon Territory produced the sequence shown in table 20-D, right. It seems likely that in that sequence the McConnell Glaciation approximates the Kluane Glaciation, but the posi-tions of the pre-McConnell glaciations are still unknown. The character and morphology of the Reid Drift suggest an Early Wisconsin date, but this is speculative.

Marine Strata; Bering Bridge

The coasts of Alaska, particularly western Alaska, expose marine sedi-ments of late-Cenozoic age. These represent a series of transgressions, some of which are at least in part the result of eustatic fluctuation of sea-level. The integrated sequence is given in table 20-D, where it is mostly not fitted to a time scale. Hopkins (p. 50 in Hopkins, ed., 1967) placed the Krusensternian transgression in postglacial time, the Woronzofian in the mid-Wisconsin interstade, and the Pelukian in the Sangamon Inter-

Table 20-D

Correlation of Selected Quaternary Sequences in Alaska and Yukon Territory. Nonmarine Units are Glacial-Geologic Units.

WESTERN ALASKA (Marine transgressions) Hopkins et al., 1965	Time: Years BP × 10³ (Scale changes at 10 × 10³)	SEWARD PENINSULA Hopkins, 1963	CENTRAL BROOKS RANGE (Glacial-geologic units) Detterman et al., 1958; Porter, 1966	COOK INLET Karlstrom, et al., 1964	SOUTHWESTERN YUKON TERRITORY Denton & Stuiver, 1967	SOUTHWESTERN YUKON TERRITORY Rampton, 1969	EASTERN YUKON TERRITORY Bostock, 1966 Hughes et al., 1969
Krusensternian	0 / 2 / 4 / 6 / 8	Mount Osborn	Fan Mtn., glac'ns (2) Alapah Mtn. glac'n / Itkillik glac'ns	Alaskan glac'ns — Tunnel / Tustumena	Neoglaciation / Slims nonglacial	Neoglaciation	Neoglaciation — Kluane L. seq.
Woronzofian	10 / 20 / 30	Salmon Lake	Anivik Lake / Antler Valley / Anayaknaurak / Banded Mtn.	Naptowne glac'ns — Tanya / Skilak Lake / Killey / Molschorn	Kluane	Macauley	Mc Connell
Pelukian	40 / 50 / 60 / 70	Nome River	Sagavanirktok glac'n. and Anaktuvuk glac'n.	Knik / Eklutna	Boutellier nonglacial / Icefield / Silver nonglacial / Shakwak	Mirror Creek	Reid
Kotzebuan		Iron Creek		Caribou Hills			Pre—Reid { Klaza / Nansen
Einahnuhtan				Mount Susitna			
Anvilian							
Beringian (Pliocene in part?)							

(Not calibrated)

Neoglaciation Tentatively correlated with (Positions unknown; not correlated)

537

glaciation. He assigned older strata to time positions based in part on K/Ar dates; the greater part of the oldest, the Beringian, is older than 2.2 million years.

On the Arctic Coast the thick Gubic Formation, of Quaternary age, is as yet too little known to be subdivided formally. However, it contains a number of distinctive faunules in widely spaced areas. These may mean that the Gubic consists of several thin units, each of a different age (Mac-Neil, 1957).

The marine sequence in western Alaska is closely related to the history of the Bering Strait land bridge. According to Hopkins (p. 457 in Hopkins, ed., 1967) Beringia, the broad land of which a large part today lies submerged beneath the shallow Bering and Chukchi Seas, stood above sealevel during early- and mid-Tertiary time. As a land mass, it formed a barrier against the mingling of faunas of the Pacific and Arctic Oceans, but afforded a path for the interchange of Asiatic with American land faunas and floras. Late in Miocene time Beringia was briefly cut by a seaway. During most of Pliocene time it was again a land bridge for the passage of mammals. Near the end of the Pliocene, between 4.0 and 3.5 million years ago, Beringia was again cut by the sea. Throughout the Quaternary, land and sea alternated as sealevel fell and rose eustatically in the succession of glacial and interglacial ages. Flooding occurred during at least six episodes. Today an emergence of only 46m would re-create the bridge; 100m of emergence would bring much of Beringia out of water.

Other Sediments

In the broad lowland drained by the Yukon River in interior Alaska, Quaternary sediments consist mainly of alluvium, colluvium, and loess. The alluvium is not a unit; in places it is so exposed as to show a stratigraphic sequence. Near Fairbanks, for example, two bodies of gravel (not outwash) are overlain by sediments possibly of Illinoian age. The nonglacial origin of these gravels emphasizes the fact that during glacial ages most of the interior of Alaska was not marked by glaciers. Streams originating in glaciated mountains, however, did carry some outwash sediments to the Yukon River. Such sediments were a source of localized sand dunes and of more widespread loess, most of it of late-Pleistocene age, and some of it thick in places. Some loess has been reworked by creep and solifluction, with the result that in many valleys such material has become thick. A large number of the Quaternary fossil mammals recovered in Alaska were carcasses rather than skeletons, enclosed in reworked loess that had been continuously frozen since soon after emplacement of the carcasses.

Permafrost, which is present in most of Alaska, contains within itself a record of sequence of events. Frozen ground, ice-wedge casts, (chap 10) and their stratigraphic relations together record freezing during Illinoian(?) time, deep thawing in the Sangamon Interglaciation, two principal episodes of freezing in Wisconsin time separated by some thawing, and shallow thawing in the Hypsithermal, followed by the colder Neoglacial climate. Where pollen records exist, they confirm this sequence. Details are given by Heusser (1960, p. 181–189), who cited episodes of glacier advance in the Glacier Bay district, 90km northwest of Juneau, at about 800, 1500, 2500, 4200, and 7000 BP. He notes that the two oldest of these advances occurred within Hypsithermal time as defined, and not within Neoglacial time, and explains them as the result, in part, of increases in humidity evidenced independently by pollen data. On mountains more than 4500m high in a coastal environment, he observed, "rising humidity will undoubtedly affect glaciation, even under hypsithermal conditions."

The climatic record in Alaska over the last 13,000 years or so, as read from pollen-bearing sediments, is given in table 19-A.

MEXICO AND CENTRAL AMERICA

Glacial Stratigraphy. Literature on the Quaternary of Mexico and Central America is meager and centers on glaciation of high mountains and on sediment fills in basins. Glaciation is areally very small; that reported in the literature is confined to two recently active volcanic cones in central Mexico and to a highland in Costa Rica.

The only glacial-stratigraphic sequence yet published is developed on the high cone Ixtaccihuatl (S. E. White, 1962). There four principal drift bodies and a possible older drift are identified by degree of weathering and morphology and are named. In addition several young recessional moraines are identified. Correlations with the sequence in the Rocky Mountains are suggested. If these are accepted, both Wisconsin and Neoglacial drift bodies are present. Literature is cited on the proposition that in this part of Mexico glacial-age climates were moister and warmer than those of nonglacial times, which were drier and cooler.

The neighboring cone Popocatépetl bears evidence of glaciation, assigned to the Wisconsin Age, in the form of erosional features, end moraine, and alluvial terraces (S. E. White, 1951). Pre-Wisconsin glaciation may not have occurred because of late date of upbuilding of the volcanic cone.

The glaciation in Costa Rica mentioned above appears to be confined to Cerro Chirripó, a high point on the Cordillera de Talamanca. It has been studied (Weyl, 1956) only in reconnaissance, and because of fresh morphology and slight weathering is inferred to be of Wisconsin age.

Nonglacial Stratigraphy. Probably the basin fills of Mexico and Central America embody a more complete and more sensitive historical record than do the glaciated mountains, but few have been investigated. The best known sequence is that of the Cuenca de México in the vicinity of the Capital. The upper part of the fill, representing somewhat less than 150m of sediment, is described by Mooser et al. (1956). It consists of fine-grained alluvium and lacustrine layers, locally with caliche, and is believed, despite very sparing control by C^{14} dates, to include both Wisconsin and pre-Wisconsin units. The data, however, do not seem to preclude the possibility that the entire known section is of Wisconsin age. Climates during that time are discussed by Lorenzo (1967).

Another basin, the Guadiana Valley in the State of Durango, contains a sequence (Albritton, 1958) parts of which have been correlated tentatively with units in the Cuenca de México and with the Neville, Calamity, and Kokernot Formations in Texas (table 11-A).

CHAPTER 21

Stratigraphy of Central
North America

For convenience of discussion we shall consider central North America as embracing the country between the Appalachian Plateau and the Rocky Mountains, including the Great Lakes region and the Canadian Provinces between the Canadian Shield and the Rockies. In terms of stratigraphic knowledge this region constitutes a reasonably distinct unit. It differs strikingly from the Cordilleran region, and the differences are reflected in the glacial drift. The country consists mainly of plains and rolling lowland; hence the mantle of glacial drift was nearly continuous where deposited and has remained so despite subsequent erosion. The bedrock, mainly rather flat-lying sedimentary strata, includes abundant claystones and carbonate rocks; as a result the till layers constructed from them are rather fine grained. Exposures of drift are mainly in valley sides and in artificial cuts along major roads. Correlation through wide areas is less difficult than in a mountain region. Also, partly because the drift is generally less permeable to ground water, fossil material is better preserved, facilitating correlation by fossil content and by C^{14} dates.

We shall outline the stratigraphy as systematically as the state of knowledge permits, but we cannot avoid a somewhat uneven treatment. Tables 21-A, 21-C, and 21-D show regional sequences and principal stratigraphic names; many purely local names are not included.

PRE-WISCONSIN STRATIGRAPHY

The classical glacial-stratigraphic sequence (table 21-A) and the history of its nomenclature (table 21-B) constitute the basis of our discussion.

Table 21-A

Tentative Correlation of Pre-Wisconsin Stratigraphy in Central North America.

GLACIAL-STRATIGRAPHIC UNITS[b]	INDIANA (After WF[a], p. 67) — Glacial-strat. units: stades	INDIANA — Rock & soil-strat. units	ILLINOIS (WF[a], p. 46) — Glacial-strat. units: stades	ILLINOIS — Rock- & soil-strat. units	EASTERN NEBRASKA (Neb. G. Surv. B. 23, 1965, p. 4; WF[a], p. 190) — Rock & soil-strat. units	EASTERN KANSAS (Kan. G. Surv. B. 189, 1968, p. 60; WF[a], p. 206) — Rock & soil-strat. units	NORTHWESTERN TEXAS (WF[a], p. 206, Frye & Leonard, 1963) — Rock & soil-strat. units	TIME STRATIGRAPHIC UNITS
Sangamon Interglaciation		Sangamon Soil		Sangamon Soil	Sangamon Soil	Sangamon Soil	Sangamon Soil	Sangamon Interglacial Stage
Illinoian Glaciation	Richmond Abington Centerville	Butlerville Till (Loveland loess)	Buffalo Hart Jacksonville Liman	Buffalo Hart Till Roby Silt Jacksonville Till & outwash Mendon Till Petersburg Silt	Loveland loess Crete Sand, etc. Beaver Creek loess & alluvium Grafton loess & alluvium Santee Till	Loveland Fm Crete Fm	Eolian sands Intermediate Terrace alluvium	Illinoian Glacial Stage
Yarmouth Interglaciation		Yarmouth Soil		Yarmouth Soil	Yarmouth Soil	Yarmouth Soil	Yarmouth Soil	Yarmouth Interglacial Stage
Kansan Glaciation	Colombia Garrison Creek Interstrade Alpine	Cloverdale Till (Cagle loess)		Till & outwash Proglacial silt Sankoty Sand	Clarkson Till Sappa loess, etc. w/ Pearlette Ash Grand Island sand, etc. Walnut Creek loess, etc. Red Cloud alluvium Cedar Bluffs Till Nickerson T. Atchison Sand	Sappa Fm w/ Pearlette Ash Grand Island Fm Cedar Bluffs Till Nickerson Till Atchison Fm	Tule Fm and Hardeman Terrace alluvium	Kansan Glacial Stage
Aftonian Interglaciation		Afton Soil		Afton Soil	Afton Soil	Afton Soil	Afton Soil	Aftonian Interglacial Stage
Nebraskan Glaciation				Till, outwash	Iowa Point Till Fullerton loess, etc. Holdrege sand Elk Creek Till Seward loess, etc. David City sand	Fullerton silt Holdrege sand Nebraskan Till David City Fm	Nebraskan Terrace alluvium (Red R.) Blanco Fm.	Nebraskan Glacial Stage

[a] H.E. Wright and Frey, eds., 1965a. [b] Glacial-stratigraphic units are the climate-stratigraphic units of the American Commission on Stratigraphic Nomenclature.

Table 21-B

Origin of Glacial and Interglacial Stage Names in Central North America

Unit	Named by	Original Reference	Type Locality or Region	Remarks
Wisconsin	T.C. Chamberlin	J. Geikie, 1894, Great ice age, 3d ed., p. 763.	State of Wisconsin	Originally named East Wisconsin. Its name was later shortened (Jour. Geol., 3, 1895, p. 270–277).
Sangamon	F. Leverett	Jour. Geol., 6, 1898, p. 176.	Sangamon County, Illinois	Adopted from Sangamon Soil of Worthen (1873).
Illinoian	F. Leverett	T.C. Chamberlin, 1896, Jour. Geol., 4, p. 874.	State of Illinois	When first proposed, the name was written Illinois, later changed.
Yarmouth	F. Leverett	Jour. Geol., 6, 1898, p. 239.	Yarmouth, Iowa	Sections at two wells with soil and peat.
Kansan	T.C. Chamberlin	J. Geikie, 1894, Great ice age, 3d ed., p. 755; Jour. Geol., 3, 1895, p. 271.	Northeastern Kansas (Standard sections are near Afton Junction, Iowa)	This name was applied in 1894 to the drift now called Nebraskan, but was transferred (Jour. Geol., 4, 1896, p. 872) to the drift sheet to which it is at present applied.
Aftonian	T.C. Chamberlin	Jour. Geol., 3, 1895, p. 270.	Afton Junction, Iowa	First described by Bain (Iowa Acad. Sci., Proc., 5, 1898, p. 93–98). Type sediments may be Kansan outwash.
Nebraskan	B. Shimek	Geol. Soc. Am., Bull. 20, 1909, p. 408.	State of Nebraska	Not well exposed in Nebraska; formerly known also as sub–Aftonian and pre–Kansan.

Note: Glacial units are shown in roman type; interglacial units in *italic type*

The sequence recognizes only four major glaciations (some of them multiple) but no proof has been found that more than four did not occur. Stratigraphy in North America, unlike that in Europe, New Zealand, and elsewhere, is very poor in sections reflecting transition from pre-Pleistocene to Pleistocene strata. Thus in most parts of the continent there seems to be an important hiatus at the base of the lowest known glacial layers.

As indicated in chap 11, strongly differentiated mature soils are not developed in Wisconsin drift, but do occur in Illinoian, Kansan, and Nebraskan drift. The youngest such soil therefore is a convenient key horizon, very probably correlative with the similar soil present in the Cordilleran region. Because that soil is generally ascribed to the Sangamon Interglacial, the discussion that follows consists of two parts, pre-Wisconsin and Wisconsin stratigraphy.

Nebraskan Glacial Stage

Glacial sediments of Nebraskan age are accompanied by questions as to correlation and even as to identification, largely because they are poorly known in detail. This in turn results from their relatively great age and widespread concealment beneath younger sediments. Nebraskan drift not covered by younger drift has not been identified anywhere with certainty, although it may possibly occur west of the Mississippi River in northeastern Iowa. Because it is generally buried, this drift has been identified mainly from scattered exposures and from borings.

As we have noted, however, in many places identification leaves much to be desired, partly because it dates from early decades of the 20th century when it was naively supposed that each of the four major glaciations was a single event represented by a single layer of till. Since that time the Wisconsin and Illinoian drifts have been shown to be complex, and the Kansan drift is rapidly taking on a similar character. However, the practice of labeling as Nebraskan any till found underlying till believed to be Kansan still persists. The entire array of "Nebraskan" occurrences needs careful revision on a basis of lithologic and other studies designed to identify tills on criteria other than mere ordinal position in a sequence of layers. Such revision would result in removal of some occurrences from the Nebraskan category. Probably some of these would be recognized as Kansan, and just possibly some might prove to be older than the Nebraskan of the present classification. Two tills of Nebraskan age have been identified in Nebraska, and the complexity of a Kansan-Nebraskan sequence in northeastern Kansas is demonstrated (Dort, 1966).

Figure 21-1 indicates present knowledge (not very precise) of the distribution of Nebraskan drift. Best developed in southwestern Iowa, northern Missouri, eastern Nebraska, and northeastern Kansas, that drift consists of massive clay-rich till locally 60m thick, with lenses of stratified material. Buried outwash is present extensively in northern Missouri, and in Nebraska and Kansas as the David City and Holdrege units. In those states Nebraskan loess exists as well, as the Seward and Fullerton units.

Drift thought to be Nebraskan is reported from Alberta, Minnesota, Wisconsin, Illinois, Ohio, and Kentucky, but in no instance is identification entirely firm. In Kentucky, likewise, are scattered large erratic boulders of crystalline rocks, possibly a lag concentrate either from till or from ice-rafted sediment deposited in proglacial lakes. The boulders have been said to be Nebraskan, but correlation is speculative. In Illinois, Nebraskan drift may include not only till but also a buried valley fill of

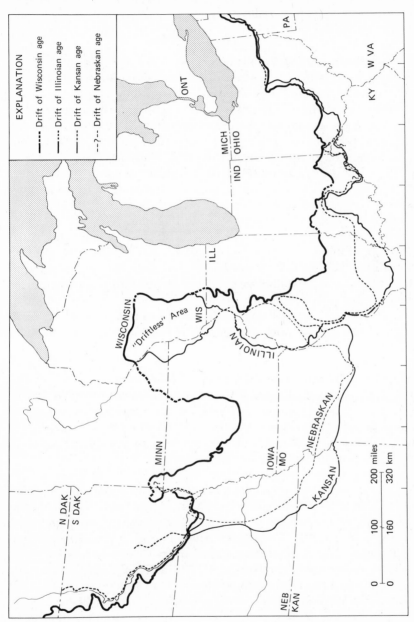

Figure 21-1 Known and inferred outer limits of four glacial stages in central North America. (Modified from Flint et al., 1959.)

sand, 90m thick. In Kansas a correlative body of sand and gravel contains a mollusk fauna implying cool-temperate climate. Peat containing a sub-arctic flora underlies till, said to be Nebraskan, in southern Minnesota; it implies a climate cooler than today's.

Blancan-Villafranchian Problem. No sequence displaying a Pliocene-Pleistocene transition well documented by fossils is known in central North America. Nor does the Nebraskan drift yet show a significant relation to strata bearing fossil mammals. The position of the Nebraskan Glaciation in the Pleistocene succession is obscure. In the Plains region (and in the Cordilleran region as well) the three faunal ages based on mammals (fig 21-2; chap 29) include one, the Blancan, which in current classification spans part of the Pliocene as well as part of the Pleistocene. Within it, K/Ar dates suggest there may be a lapse of at least 1.5 million years between the Rexroad local fauna, widely considered Pliocene, and the faunas accepted as Pleistocene. Within this time, faunal information is still too scanty for the fixing of a Pliocene-Pleistocene boundary based on faunal evolution.

A critical aspect of this matter concerns the affinity of the Blancan fauna with that of the Villafranchian in Europe. The name Blancan is derived from the Blanco Formation. This is a body of terrestrial sand, silt, and clay with a type locality at Mount Blanco in northwest Texas; it overlies unconformably Pliocene Ogallala alluvial sediments which, generally, in the Great Plains region have a Pliocene mammal assemblage equivalent to that of the Plaisancian/Astian stage in Europe.

The mammal fauna of the Blanco Formation (there is also a mollusk fauna) was first believed to be Pliocene, but as knowledge increased it was found to have associations with faunas of other stratigraphic units elsewhere in the Great Plains region and at localities as far west as Arizona, California, and Idaho. This led to the setting-up of the Blancan Provincial Mammal Age, applicable to a wide area and classified as Pleistocene.

According to Colbert (Colbert et al., 1948, p. 628; McGrew, 1948, p. 549) the Blancan fauna is characterized by the appearance of new genera, especially *Plesippus* and *Nannippus* (horse), *Borophagus* (dog), and *Procastoroides* (beaver), all of which are non-immigrant forms derived from autochthonous stocks. Extinction of as many as 25 mammal genera is indicated between the Pliocene Hemphillian Mammal Age immediately preceding, and the Blancan Age. Some Hemphillian genera, of course, persisted into the Blancan, and some immigration occurred in Blancan time. All in all, there is a close affinity between Blancan and Villafranchian faunas (Colbert et al., 1948, p. 630), but the Hemphillian-Blancan break is not really sharp. Faunal lists for several typical localities are given by Taylor (1960, p. 14), and a list of known Blancan genera is given by Evernden et al. (1964, p. 150).

Since at least as early as 1937 (J. R. Schultz, 1937) the faunal affinity between Blancan and Villafranchian has been recognized and does not seem to be seriously questioned (see discussion in C. B. Schultz and Stout, 1948, p. 571, with an extensive bibliography). However, as far as its relationship to major time-stratigraphic units is concerned, the Blancan Age has been considered, by various authors, to be (1) Pleistocene, (2) Pliocene, (3) intermediate between Pliocene and Pleistocene but neither the one nor the other, and (4) inclusive of parts of both Pliocene and Pleistocene. These varied opinions are possible mainly because the Blancan and Villafranchian faunas belong to wholly different provinces separated by a vast distance and connected, as far as we know, only by the pathway over the former land bridge at the Bering Strait.

There is at least some indication in Blanco strata and Blancan faunas of the climatic fluctuation that seems to characterize the basal Pleistocene in Europe. According to Evans (Colbert et al., 1948, p. 617), lacustrine strata and evidence of wind activity in the Blanco Formation record a relatively moist climate preceded and followed by drier ones. McGrew (Colbert et al., 1948, p. 551) said the Grandview and Hagerman local faunas (both Blancan) include small boreal mammals that imply a cold, moist climate at some time late in the Blancan. J. R. Schultz (1937, p. 94) listed evidence of cooling, such as immigration of the boreal vole *Mimomys* into southwestern North America. Hibbard (Colbert et al., 1948, p. 551) characterized the Rexroad fauna in Kansas as having a "warmer" character, an opinion Taylor (1960) shared. The cold fluctuations discerned by Lona (1950) at Leffe in the Italian Alps appear to have begun in earliest Pleistocene time, whereas most of the evidence of cool climate in the Blancan Age is not closely fixed in the stratigraphic sequence. That which is fixed approximately appears to be Blancan, and indeed Taylor (1960, p. 15, 19) suggested a Nebraskan glacial age for the lower part of the Ballard Formation in Kansas, which he placed in the later part of the Blancan Mammal Age.

It appears that even if the Blancan and Villafranchian Ages were defined more precisely than they are at present, they would probably not prove to be exact counterparts in time. Nevertheless as far as present knowledge of their faunas alone is concerned they are approximate counterparts. Although some authorities in Europe place the beginning of the Pleistocene at the beginning of the Villafranchian Age, others visualize the possibility of subdividing the Villafranchian into a Pleistocene part and an earlier Pliocene part, based on faunal differences. However, the Villafranchian faunas are not yet well enough known to make such subdivision possible. There, for the present, the matter rests.

Aftonian Interglacial Stage

The sediments described originally as Aftonian consist chiefly of gravel and sand and very likely represent outwash, both Nebraskan and Kansan. However, soil developed in Nebraskan till or outwash and overlain by Kansan till, as well as evidence of erosion at a corresponding position, have long been accepted as Aftonian Interglacial features. Aftonian soil, in most occurrences decapitated so that only the B horizon remains, occurs in Nebraska, Kansas, Iowa, Missouri, and other states. In some exposures the soil is accompanied by peat, and in one exposure possible decomposed loess overlies Nebraskan till, representing Aftonian weathering.

Interglacial gravel and sand occurs in the southern part of the Canadian Great Plains. Termed "Souris gravel and sand" in southwestern Manitoba, and possibly correlative with the "Flaxville gravel" in eastern Montana, this sediment is thought to postdate the earliest Pleistocene glaciation of the Plains. For that reason it is mentioned here under the Aftonian rubric, although its true correlation is unknown (cf. Klassen, 1969, p. 8).

Fossil mollusks in southwestern Kansas indicate a semiarid climate in Aftonian time, with winters warmer than today's, resulting in greater evaporation despite mean precipitation comparable with the present (Berry and Miller, 1966; see also Taylor, 1960). Summers were mild. The mammals of the region, represented by the Sanders local fauna, give compatible climatic indications (Hibbard, 1956).

Kansan Glacial Stage

Kansan drift extends beyond younger drift sheets in a broad belt of country west of the Mississippi River, probably including southeastern South Dakota and possibly including the Maunsell Till in southern Alberta. In Illinois, Indiana, and Ohio it is wholly overlapped by Illinoian drift, but in northern Kentucky it is locally at the surface. Throughout the central region it consists mainly of clay-rich till similar to and, in most places, not easily distinguishable from Nebraskan till, but in some areas conspicuous bodies of outwash are present. The drift is generally 10 to 30m thick and in places exceeds 45m. It preserves no constructional topography, and because in some districts it is confined to broad interfluves, valley cutting in those areas is believed to postdate the drift.

The presence in the lower part of the Cloverdale Till, in southern Indiana, of fragments of local bedrock and inclusions of red soil derived from it suggest that the ice sheet of Kansan time was the first to reach that region (Gooding, 1966). The two apparent lobes of Kansan drift in Illinois (fig 21-1) are based on very scanty data, and may actually have been coalescent. Drift that may be Kansan is exposed in southern Alberta, overlain by several younger tills.

The Kansan drift includes three known tills in Nebraska and two in Kansas. Weakly developed soil zones in Kansas and in Missouri further indicate multiple glaciation in Kansan time. Probably it was Kansan ice that made the first radical alterations in the drainage pattern in Ohio, Indiana, and Illinois (fig 9-3). In Kansas and Nebraska upper Kansan loess and alluvium, together constituting the Sappa Formation, in many places expose a thin layer of fine tephra, the well-known Pearlette Ash. This layer, possibly the product of a single eruption originating in the Cordillera, occurs within a region extending from South Dakota to Texas and is a useful key horizon. Kansan loess, overlain in the same exposure by Loveland, Farmdale, and Peorian loess layers, is reported (L. L. Ray, 1965, p. 59) from near the Ohio River downstream from Owensboro, Kentucky. The Kansan loess constitutes indirect evidence of the former existence of a Kansan valley train, erosion remnants of which are presumably buried beneath younger outwash.

The Kansan environment in southern Indiana, indicated by snails in loess, included a boreal mixed forest. Mollusks from sediments of late Kansan age in western Kansas and others in northwestern Texas imply temperature lower than today's, with prairie and woodland but not extensive forests, and with lakes and ponds. Climate was more equable than today's, with cooler, moister summers, although annual rainfall may not have been much greater. Gallery forests grew along the larger streams; savanna characterized the interfluves. Certain fossil rodents indicate that summer temperatures were lower in Kansas than in Texas (Getz and Hibbard, 1965).

In Nebraska fossil mammals of early Kansan age indicate a plains environment with a climate perhaps slightly moister than today's. Late-Kansan mammals include mammoth and musk-oxen (suggesting tundra), and *Parelephas,* horse, bison, moose, beaver, and giant beaver, a group suggesting boreal forest. Mammals of late-Kansan age or older in northern Texas indicate a savannalike environment under a maritime climate with a suggested precipitation of ~750mm, 20% greater than today's. In contrast with existing streams, many streams then were perennial.

Yarmouth Interglacial Stage

The Yarmouth Interglacial Stage is represented in most of the central states by widespread soil developed in Kansan drift and other sediments, and varying in character with regional variation in the climate. Much of the Yarmouth Soil (in northern Missouri constituting part of the Ferrelview Formation) is of accretion-gley character. The Yarmouth Interglacial is represented further by peat at four localities in Iowa, at two in Illinois, and at one in Indiana. Comparison of the soil with its modern equivalents suggests that the Yarmouth climate was much like the present one, though possibly slightly warmer and drier (Simonson, 1954). Many of the minor stream valleys in northern Missouri originated on the exposed surface of the Kansan drift, presumably in Yarmouth time. Pollen in the peat in Illinois, chiefly fir, pine, and tamarack, implies a climate cooler than today's, and suggests that possibly a transition to the succeeding Illinoian is represented. A similar possibility exists at the Indiana locality, where a cool, moist climate is indicated by snails (Wayne, 1958). In summary, a Yarmouth soil is widespread and is compatible with the concept of a nonglacial climate. In contrast, peat between Kansan and Illinoian drift layers implies a cooler, moister climate and therefore may belong in the Illinoian rather than in the Yarmouth.

Mollusks in alluvium along the Cannonball River in southwestern North Dakota indicate an age that is probably Yarmouth though possibly younger. Their ecology is that of grassland with gallery woodland, under a climate much like that of today (Tuthill et al., 1964).

The Borchers fauna in southwestern Kansas comes from strata that immediately overlie the Pearlette Ash and that are confidently referred to the Yarmouth. Mammals in the fauna indicate a climate as warm as or warmer than that of the present. This and many other local faunas in the fossil-rich sequence in Meade County, Kansas (fig 21-2) have yielded a great wealth of information useful for the reconstruction of Pleistocene ecology.

It was suggested by Deevey (p. 648 in H. E. Wright and Frey, eds., 1965a) that the Yarmouth is unrecognizable away from its type area. However, the data summarized above indicate that the Yarmouth Soil is at least as well developed and widespread as the Aftonian Soil. It has been implied that the Yarmouth Interglacial, if correlative with the Elster/Saale in Europe, should have been exceptionally long and therefore well marked. However, the concept that the Elster/Saale Interglaciation was unusually long (chap 15) is open to question.

Illinoian Glacial Stage

The Illinoian drift is best developed in its type region, Illinois, where it is the surface drift throughout a wide area. Eastward, its belt of outcrop narrows across Indiana and Ohio and is nearly continuous into the Appalachian region. On the north shore of Lake Ontario, near Toronto,

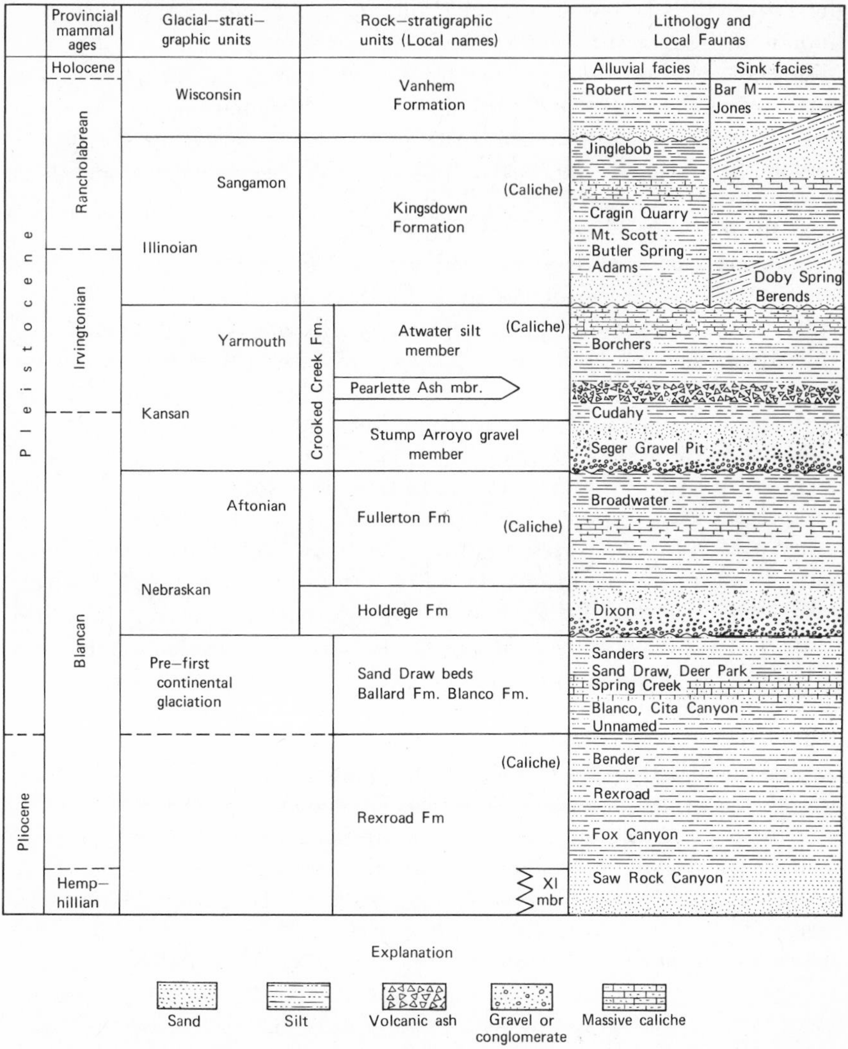

	Provincial mammal ages	Glacial–strati–graphic units	Rock–stratigraphic units (Local names)	Lithology and Local Faunas	
				Alluvial facies	Sink facies
Pleistocene	Holocene		Vanhem Formation	Robert	Bar M / Jones
	Rancholabrean	Wisconsin			
		Sangamon	Kingsdown Formation (Caliche)	Jinglebob / Cragin Quarry / Mt. Scott / Butler Spring / Adams	Doby Spring / Berends
	Irvingtonian	Illinoian			
		Yarmouth	Atwater silt member (Caliche) [Crooked Creek Fm.]	Borchers	
		Kansan	Pearlette Ash mbr.	Cudahy	
			Stump Arroyo gravel member	Seger Gravel Pit	
		Aftonian	Fullerton Fm (Caliche)	Broadwater	
	Blancan	Nebraskan	Holdrege Fm	Dixon	
		Pre–first continental glaciation	Sand Draw beds Ballard Fm. Blanco Fm.	Sanders / Sand Draw, Deer Park / Spring Creek / Blanco, Cita Canyon / Unnamed	
Pliocene			Rexroad Fm (Caliche)	Bender / Rexroad / Fox Canyon	
	Hemp–hillian		Xl mbr	Saw Rock Canyon	

Explanation

Sand   Silt   Volcanic ash   Gravel or conglomerate   Massive caliche

Figure 21-2 Stratigraphy of late-Cenozoic local faunas in northwest Texas, adjacent Oklahoma, western Kansas, and north-central Nebraska. Revised, with aid of C. W. Hibbard, from a manuscript by M. F. Skinner and Hibbard, and Hibbard and Taylor, 1960. Applicable C[14] dates: Robert local fauna, ~11,100 BP; Bar M l.f., ~21,000 BP; Jones l.f., ~27,000 BP.

the York Till is believed to be of Illinoian age. The great complex of Illinoian drift, including minor areas in eastern Iowa and southeastern Minnesota, is of northeastern origin. Farther west, Illinoian till occurs in northeastern Nebraska and probably also in the adjacent part of South Dakota. Certain occurrences of drift west of Lake Superior are possibly Illinoian. In southern Alberta the Brocket Till, underlying several tills of Wisconsin age, may possibly be Illinoian.

In Illinois and Indiana the Illinoian Stage includes three distinct till layers, so related as to suggest that their episodes of deposition were separated by only short periods of time. In Nebraska only one till is identified, but three layers of loess suggest that Illinoian ice approached that state three times. In Ohio two till layers are known. The Illinoian is the earliest drift that still retains constructional topography. In the Liman Substage in Illinois both probable moraines and probable kames of substantial size are present. The loess related to the Illinoian till is the Loveland loess (or silt), widespread from Indiana westward into Nebraska and Kansas. Illinoian outwash is present throughout the region, in some places preserved as terraces standing higher by 15m or more than adjacent outwash of Wisconsin age. This is consistent with the relatively great areal extent of Illinoian drift. In the southern part of the Great Plains the Illinoian Stage is represented mainly by alluvium, commonly in terraces, in major valleys.

In southwestern Kansas and adjacent Oklahoma the climate of the Illinoian maximum, implied by pollen in the Kingsdown Formation (Kapp, 1965) is thought to have been characterized by annual temperature about 5° lower than today's, and by cooler and moister summers; nevertheless the climate remained subhumid. Vegetation was pine savanna with some spruce and with deciduous trees along streams. The vegetation there resembled that of the Late Wisconsin maximum in northwestern Texas (chap 19). Fossil mammals of the Illinoian maximum in southwestern Kansas, likewise, indicate cooler summers characterized by more effective moisture, although not necessarily with more rainfall. Many of the fossil species today live much farther north or northeast, or at higher altitudes (Hibbard and Taylor, 1960). The mollusk fauna of southwestern Kansas and adjacent Oklahoma, known from five localities, gives climatic indications compatible with those sketched above (B. B. Miller, 1966). The extensive Illinoian local mammal fauna (Guilday, 1971) recovered from Cumberland Cave in western Maryland differs from Appalachian local faunas of Wisconsin age in that it includes both northern (though not arctic) and western elements.

In late Illinoian time the summers became warmer and drier, spruce almost disappeared, pine diminished, and juniper, *Artemisia,* and chenopod/amaranths became abundant. The character of the vertebrate and mollusk faunas changed correspondingly. The changes represented the beginning of the transition toward the succeeding interglacial climate.

Sangamon Interglacial Stage

The stratigraphic name *Sangamon* derives from a widespread soil recognized and named in Illinois as early as 1873. The soil occurs through the central region from the eastern Great Lakes to western Texas. It ranges in character from Prairie soil and accretion-gley in Indiana, Illinois, Iowa, and eastern Nebraska, southwestward through Chernozem and Chestnut, to Desert soils, with reddish hues common, as the climate changes from humid to arid. West of the Mississippi especially, the Sangamon Soil is exposed so commonly that it constitutes a much-used key horizon. It is developed in both till and loess, and in parts of southern Iowa it merges with the Yarmouth Soil, so that the two soils appear at first sight to form a single unit.

Besides the soil, the Sangamon Interglacial is represented by significant sediments, locally well exposed. A pair of exposures near Richmond in southeastern Indiana have yielded pollen from between Illinoian and Wisconsin tills. The sequence common to both exposures, developed by Kapp and Gooding (1964), is as follows:

Local zone	Vegetation	Climate
VI	Spruce, fir, pine	Cold, moist
V	Transition	Cooler, dry
IV	Oak/hickory maximum	Warm, drier
III	Transition	Warming
II	Pine	Cool, drier
I	Spruce, pine	Cool, moist

In this succession we see continuous change from the end of the Illinoian Glaciation through an interglacial maximum to the beginning of the Wisconsin Glaciation. In Zone IV the climate was at least as warm as today's. The forest of Zone VI was overridden by the advancing Wisconsin glacier.

In another exposure of Sangamon sediments, near Martinsville, Indiana, lacustrine mollusks indicate the cool climate of the early Sangamon, possibly representing Zone I and perhaps Zone II of the foregoing sequence.

The most famous locality of Sangamon sediments in the central region is the Don Valley brickyard near Toronto, where the lacustrine Don Formation (Karrow, 1967, p. 24), at least 8m thick, is exposed. This unit overlies disconformably the York Till, of probable Illinoian age, and does not record the basal part of the interglacial pollen sequence. The base of the Don Formation contains pollen of a deciduous forest like that of today (Terasmae, 1960) and therefore seems to represent, climatically, a middle part of the interglacial. One species of tree represented by pollen (the sweet gum, *Liquidambar*) and other species represented by macrofossils do not live in the Toronto area today, but occur much farther south. From these it is calculated that mean annual temperature was then higher by $2°$ to $3°$. Besides plant macrofossils there are mammals consisting of bison, deer, a short-faced bear, giant beaver, and woodchuck; these however are not sensitive indicators of climate. Pollen from a higher horizon shows gradual change from hardwood trees to conifers. Thus it reflects the slow onset of boreal climate, probably the one leading to the Wisconsin Glacial Age. Farther north, the Missinaibi beds (chap 22) may belong to the Sangamon Interglacial.

Southwestern Kansas and adjacent Oklahoma afford a glimpse of Sangamon (and later?) ecology in the southern Great Plains. The information comes from pollen, mollusks, and vertebrates in the upper part of the Kingsdown Formation, at two horizons that appear to represent early and late Sangamon respectively. Early Sangamon vegetation included sagebrush, composites, and grass, indicating semiarid open grassland, with warmer winters and possibly cooler summers than today's. The fauna (Cragin Quarry local fauna) included dire wolf, sabertooth cat, giant jaguar, mountain lion, columbian elephant, a ground sloth, horses, camels, antelope, rabbit, gopher, and lizards.

In latest Sangamon or earliest Wisconsin time the region was pine savanna, rather like the southern Ozark region or the Coastal Plain. This implies greater warmth and rainfall than today's, with relatively warm winters. The vertebrate fauna (Jinglebob local fauna, figs 21-2, 21-3) bears out the inference. It has been analyzed in terms of three groups. The first consists of residual elements of an earlier glacial-age fauna, including more northern kinds such as bog lemming, masked shrew, prairie vole, and two jumping mice. The second consists of Coastal Plain elements that immigrated from the south across a distance of possibly 800km. These include Columbian mammoth, certain shrews, a vole, a jumping mouse, rice rat, and box turtle. The third consists of kinds such as ground sloth, short-faced bear, and the giant *Bison latifrons*, not closely indicative of climate.

This pattern shows that significant climatic change occurred within the Sangamon itself, even though we cannot yet trace it in detail. The two faunal horizons are separated by massive caliche, possibly representing the B horizon of a pedocal soil; the caliche layer emphasizes a change in the moisture regimen, the exact nature of which is not known. Also the

Figure 21-3 Environment in southwestern Kansas and adjacent Oklahoma in latest Sangamon or earliest Wisconsin time, represented by the Jinglebob local fauna. (C. W. Hibbard.)

pattern illustrates the mixing of faunas as climates change, continually creating new combinations. It leads us away from the rigid concept formerly held, that floras and faunas shifted intact from one latitude to another as climate changed, much as a complete stage set is changed between the acts of a play.

Stratigraphy that occurs between two thick tills near Fort Qu'Appelle, Saskatchewan, is certainly interglacial and probably Sangamon. Fluvial sand and gravel ~30m thick contain driftwood and vertebrate bones, including *Bison latifrons,* woodland musk-ox, moose, horse, columbian(?) mammoth, camel, and turtle.[1] This is a temperate-climate assemblage, suggesting savanna and open-grassland environment.

[1] Ehsanullah Khan, unpublished.

Central Gulf Coastal Plain [2]

The sequence of sediment layers and depositional plains along the Gulf Coast and its relation to alluvium farther up the Mississippi River is best treated as a concrete problem; hence we digress from our stratigraphic treatment to a regional one. Along the coast we find a succession of four late-Cenozoic sedimentary units, which antedate the existing river floodplains (fig 21-4) but each of which was deposited in a similar manner. In present outcrop each of these four units consists of alluvium, coarse in the basal part and finer toward the top. The exposed upper surfaces of the units bear traces of meanders comparable with those of the Mississippi River today. As subsurface data show, each grades seaward through estuarine into marine facies. Because of subsidence, each thickens in the same direction, forming a wedge. Aggregate thickness of the wedges at the outer edge of the continental shelf is at least 1200m and possibly as much as 3000m.

Built by the Mississippi and other rivers, each unit, with its broad flat surface, added to the width of the Coastal Plain as it was deposited. After each episode of deposition trenching occurred, leaving the latest sediment as a pair of terraces along rivers but producing little or no topographic differentiation on the seaward slopes of interfluves. The stratigraphic/morphologic units are named as in table 21-C. Their dips diminish with decreasing age, owing to subsidence in the seaward direction and actual uplift in the landward direction. Thus all parts of all the units have been deformed.

The concept that the origin of these features lies in eustatic fluctuation of sealevel was introduced by Fisk (e.g., 1944, p. 69). According to that idea the sediments were deposited during rises of sealevel to interglacial maxima, and their trenching occurred during regressions to glacial minima. Foraminifera in the offshore marine facies suggest unusually deep-water positions and hence, by inference, high sealevels; intervening coarser sediments are correlated with low sealevels. Although not established firmly on independent evidence such as glacial origin of much of the alluvium, the concept has been accepted provisionally. However, Saucier and Fleetwood (1970, p. 886) held that along the Ouachita River, a major tributary, alluviation was greatest during late interglacial and early glacial episodes.

The alluvial unit next older than the four bodies described [3] is the Citronelle Formation, of Pliocene or early Pleistocene age. The four al-

[2] Bernard and Leblanc in H. E. Wright and Frey, eds., 1965a, p. 137–185; Fisk, 1944; 1951; G. E. Murray, 1961, p. 521.

[3] Doering (1958, p. 782) suggested a different correlation.

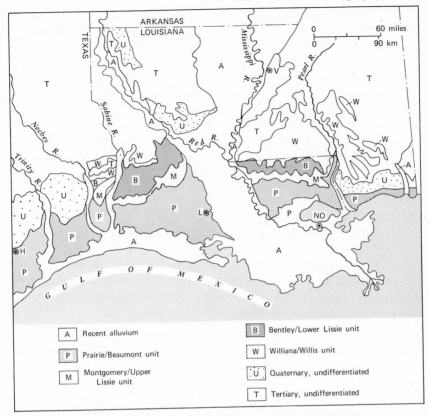

Figure 21-4 Geologic sketch map showing Quaternary sedimentary and terrace units between the Mississippi Delta region and Houston, Texas. Name used in Louisiana is given first, followed by name used in Texas. The name Citronelle is used in Mississippi. Cities: NO=New Orleans; V=Vicksburg; L= Lafayette; H=Houston. (Simplified from Bernard and Leblanc in H. E. Wright and Frey, eds., 1965a, fig 2.)

luvial bodies are Quaternary. It is not known whether more than four cycles of fluctuation, not yet clearly identified, are present. Until the sequence is more fully studied it seems unwise to try to correlate its units with specific interglacial ages. We note, however, that Saucier and Fleetwood (1970) recognized in the Ouachita River Valley a terraced body of alluvium (the Deweyville terrace and alluvium) that is younger than the Prairie, the youngest of the four mentioned above. In their opinion the Prairie and Deweyville units are related respectively to the Sangamon Interglacial and to the period between ~30,000 and ~13,000 C[14] years BP.

Opportunity for further study lies in the Mississippi Valley between the terraces in Louisiana and the well-studied glacial sequence upstream, a sector in which knowledge is surprisingly scanty. The late Wisconsin outwash that partly fills the valley appears to have older counterparts in the Missouri/Illinois segment at least (e.g., Rubey, 1952; Robertson, 1938; Shaw, 1911b; Matthes, 1931). Erosion remnants are found most commonly in the mouths of tributary valleys, where outwash fills were built into them, ponding some of them. Although sediments of Illinoian and possibly even greater age have been recognized tentatively, these have not been identified in the thick fill studied by Fisk farther down the valley.

The features described embody a problem not unlike the European problem of the terraces and sediments along the Rhine. The problem of the sediments along the Mississippi cannot be solved without close study of the terraces and sediments between the limit of glaciation and the region of the delta. No such study has yet been published.

Farther west along the Texas coast, other alluvial bodies of Quaternary age occur (table 21-C). They too are of alluvial/marine origin, the sediments being related to Coastal Plain rivers. Correlation of the units with those near the mouth of the Mississippi is still uncertain.

Red River Alluvium

A problem analogous in some respects to that of the Lower Mississippi River is that presented by alluvial terraces along the Red River in Texas. Although tributary to the Mississippi, the Red River was not appreciably affected by glaciation and did not carry outwash. Through its Texas-Oklahoma segment the Red River Valley contains a succession of alluvial

Table 21-C
Correlation of Late-Cenozoic Stratigraphy, Gulf Coast (Compiled)

Standard section		TEXAS		LOUISIANA		MISSISSIPPI		ALABAMA		FLORIDA	
Series	Group or Stage	Formation	Member	Formation	Member	Formation	Member	Formation	Member	Formation	Member
Rec- ent											
Pleistocene	Houston	Beaumont Lissie Willis	Terrace alluvium	Prairie Montgomery Bentley Williana	Terrace alluvium	Loess; Terrace alluvium		Terrace alluvium		Silver Bluff Pamlico Wicomico Okefenokee	Alluvial- and marine units
	--?--	--?--	--?--								
Pliocene	Citronelle	Goliad				Alluvium and deltaic sediments		Alluvium and deltaic sediments		Alluvium and deltaic sed- iments; Caloosahatchie Marl in south	

fills, each forming a paired terrace. According to the description by Frye and Leonard (1963), each fill represents aggradation during the later part of a glacial age, whereas deep erosion occurred during the early part. The time sequence of the fills is inferred from the relative intensity of development of the soils on them and from fossil mollusks. Each fill is said to be less bulky than the next older one, so that terrace altitude decreases with decreasing age. Apparently no correlation has yet been made, along the lower Red River, between the sequence described above and that within the Mississippi River Valley.

WISCONSIN STRATIGRAPHY

General Characteristics

The Wisconsin drift is the surface drift throughout a wide region (fig 21-1). It varies greatly in thickness and somewhat in physical character. In some areas it differs from the older drifts in that the till is distinctly richer in cobble and boulder sizes. Most of these are of rocks derived from the Canadian Shield, suggesting that in that region climate during or preceding erosion by the Wisconsin ice sheet differed from that of earlier glacial ages in promoting intense frost wedging. Also, in major valleys Wisconsin drift includes outwash fills which, because of their youth, have been preserved more extensively than have pre-Wisconsin valley trains.

The topography of Wisconsin drift is constructional, with little postglacial modification. End moraines, built in offlapping fashion by readvances during general retreat of the glacier margin, are numerous. The profiles of soil zones in the surface of Wisconsin drift or within distinct units of it are only weakly developed.

The Wisconsin is the only one of the glacial stages that is datable by C^{14}; thousands of samples from that stage have been dated. Indeed it was through discrepancies among groups of dates that the Wisconsin drift began to be subdivided on a realistic basis. The lower limit of the Wisconsin Stage is the top of the Sangamon Stage; in many exposures it is very distinct. The upper limit, as we noted in chap 14, still lacks a generally accepted definition. But the subdivisions within the Wisconsin have been the subject of radical change. Chamberlin (1883) proposed a twofold classification, as did Leverett (1899), who based the difference between his Early and Late Wisconsin on morphologic "freshness" and on trends of end moraines rather than on intensity of weathering. In 1929 Leverett classed Wisconsin drift into three groups, Early, Middle, and Late, all of them falling within the Late Wisconsin as defined in the present book. Soon afterward (Kay and Leighton, 1933), these subdivisions were replaced by four different ones, Iowan, Tazewell, Cary, and Man-kato, to which F. T. Thwaites added Valders in 1943. Thereafter, when C^{14} dating was applied to the stratigraphy, the Tazewell, Cary, Mankato, and Valders were shown to belong to strata less than about 25,000 years old (Flint and Rubin, 1955), while others were considerably older al-though still post-Sangamon. The strata less than about 25,000 years old were provisionally called "classical Wisconsin drift" (Flint, 1957, p. 341) in order to distinguish them as a group from the older strata also of Wis-consin age, although then mostly without names. The provisional term proved useful, but being no longer needed, it is not used in the present volume.

Continuing with the checkered story of Wisconsin-drift terminology, we note next that the Iowan till in Iowa was shown with probability to be nonexistent (Ruhe et al., 1968). The presumed Iowan till appears to be much-eroded Kansan and Nebraskan till; the overlying "Iowan loess" is C^{14}-dated as of Wisconsin (mostly Tazewell) age. In addition, C^{14} dates indicate that Cary drift and Mankato drift overlap in time, and may be contemporaneous or nearly so.

As a result of these relationships, Wisconsin nomenclature is undergo-ing a series of drastic changes. The State Geological Surveys of Illinois and Indiana have adopted local time-stratigraphic units (cf. table 14-B). Publications of some other groups and individuals have tentatively made use of *Early Wisconsin* and *Late Wisconsin,* and in some cases *Mid Wis-consin* as well. Because consistent names have had to be adopted for the purpose of discussion in the present book, the following names are used provisionally in the sense of time-stratigraphic names and are defined in terms of C^{14} dates:

Late Wisconsin	25,000 to 10,000 years BP
Middle Wisconsin	55,000 to 25,000
Early Wisconsin	>55,000

As this scheme is wholly a matter of convenience and is not proposed for general adoption, no apology is needed for the obvious fact that terms such as *early* connote time. They have long been in the literature, they favor no particular part of the continent, and they represent a useful (al-beit not exact) analogy with names in the region of the Alps, where the widely known subdivisions of the Würm Glacial Age are possibly broadly correlative.

In most sectors the areal outer limit of the Late-Wisconsin drift is fairly well defined. Through considerable distances that drift overlaps earlier Wisconsin drift bodies, concealing their margins. In such areas the earlier drift is sparsely exposed in valleys that cut through the overlying drift. In other sectors, such as southern Ohio, northwestern Illinois, and parts of Wisconsin, older Wisconsin drift extends beyond the Late Wisconsin drift boundary. Viewed broadly, the limits of the various Wisconsin drifts seem to be rather similar. The best single source of details of these occurrences, still few in number, consists of papers in H. E. Wright and Frey, eds., 1965a. Various other occurrences of pre-Late Wisconsin sediments in northern United States and southern Canada are probably or possibly Wisconsin rather than pre-Wisconsin, and await further study before their places in the sequence can become known.

Present Correlations

Some of the sequences developed thus far in various parts of central North America are gathered together in tables 21-D and 21-E. Like all correlation charts they represent opinion as well as fact, are certainly incomplete, and are probably defective in many places. Therefore they are no more than reports of progress and must be regarded as very tentative. The time calibration of their lower parts is for the most part poorly controlled, and considerable readjustment is expectable. Despite these qualifications, such charts are far better than none at all. In fig 21-5 an attempt has been made, following the style of fig 2 in Flint, 1955a, in the early days of C^{14} dating, to chart the "shapes" of the glacial invasions in Wisconsin time. The abscissa represents distance (not to scale) between the Hudson Bay region and the Ohio River. The "wedge-shaped" units are left open at their northern ends because there is little information about the extent of deglaciation between each two glacial expansions, although minimum extents are known in some cases. One deglaciation, apparently within the range of fig 21-5, may have uncovered the country back to the southern part of the Hudson Bay Lowland (McDonald in Hood, ed., 1968, p. 97). Because the lower "wedges" of readvance are poorly known, and because data from a wide east-west territory are represented in a single plane, the relative lengths of the lower wedges are not a suitable basis for inference concerning relative distance of readvance with time.

Great Lakes [4] and Other Water Bodies

Glacial Great Lakes. The Great Lakes occupy parts of an extensive system of large stream valleys, controlled by rock structure and dating from some unknown time in the later Cenozoic. The valleys were repeatedly enlarged by glacial scouring and were isostatically depressed under the weight of the ice sheet. Although remnants of pre-Wisconsin lake sedi-

[4] General references: Hough, 1963; 1966; Calkin, 1970.

Table 21-D

Tentative Correlation of Wisconsin Stratigraphy in Central North America. Categories of Terms Not Uniform. Time Scale in Lower Part Becomes Decreasingly Reliable Downward. Compare table 14-B

C14 years BP × 10³ (approx). Note scale change at 15,000.	NORTHWESTERN TEXAS (CLIMATIC) (Wendorf et al., 1961)	EASTERN KANSAS (Kan. G.S.B. 189, 1968)	EASTERN NEBRASKA (Nebr. G.S.B. 23, 1965)	ILLINOIS & WISCONSIN (H.E. Wright & Frey, eds., 1965a; Bergstrom, ed., 1968, p.33)	INDIANA (Wayne, 1965; Gooding, 1963)	OHIO (Various sources)	SOUTHERN ONTARIO (Dreimanis, et al., 1966; Dreimanis, 1967)
10	San Jon Pluvial	Bignell Fm	Bignell loess and alluvium	Valders Till / Two Creeks peat			Leaside Tills
		Brady Soil	Brady Soil	Port Huron Drift		Ashtabula Till	Halton Till / Wentworth Till / Cary/Port Huron Int. / Port Stanley Till
15	Monahans Interpluvial		Peoria Silt	Lake Border Drift	Lake Border Drift	Hiram Till	Lake Erie Interstade
20	Tahoka Pluvial / Cooke T. alluvium	Peoria Formation		Bloomington Till	Bloomington Till	Lavery Till	Catfish Creek Drift
				Shelbyville Till	Shelbyville Till	Kent Till	
30	Rich Lake Interpluvial		Hartington Till / Farmdale Soil	Farmdale silt and peat / Upper	Connersville Silt		Plum Point Interstade
				Winnebago Drift		Mogadore Till / Gahanna Drift?	Southwold Drift
40			Peorian Loess	Plano Silt (Upper)		Sidney Interval	
	Terry Pluvial	Gilman Canyon Formation	Gilman Canyon Formation	Plano Silt (Middle) / Roxana Silt	Fayette Drift (?) (Center Grove Till?)		Southwold Drift
50	Ambrose Terrace alluvium (Red R.)				New Paris Silt?		Port Talbot 2 Interstade
60				Plano Silt (Lower)	Whitewater Drift	till	Port Talbot 1 Interval / Bradtville and Sunnybrook Tills
70							Scarborough Fm

Suggested Temperature Change (Colder ← → Warmer)

Table 21-E

Tentative Correlation of Wisconsin Stratigraphy in Southern Canada East of the Rocky Mountains. Time Scale Before 20,000 BP is Suggestive Only

Time Terms used in this table	SOUTHERN ALBERTA (Westgate, 1968, p. 64)	WEST CENTRAL SASKATCHEWAN (Christiansen, 1968, p. 335)	SOUTHWESTERN MANITOBA (Klassen, 1969, p. 3)	HAMILTON—GALT DISTR., ONTARIO (Karrow, 1967, p. 48)	TORONTO DISTRICT ONTARIO	CHAMPLAIN LOWLAND (McDonald, 1968)
			Assiniboine Valley sediments			St. Narcisse mor.
						Champlain Sea seds
Late	Manyberries Ash			Halton Till	Leaside Till (upper)	Highland Front mor.
	Oldman Drift	Battleford Formation		Wentworth Till	Leaside Till (lower)	Till
	Etzikom Drift			Port Stanley Till		
	Pakowki Drift		Lennard Till	Catfish Creek Till	Meadowcliffe Till	
	Wild Horse Drift				Seminary Till	Gentilly
Middle?		"Older Till"	Unnamed middle deposits			
	Elkwater Drift?		? – ? – ?			
			Minnedosa Till	Canning Till?	Sunnybrook Till	
Early			Roaring River Clay			
			Shell Till			St. Pierre Interstade
						Bécancour Till

C14 Years BP × 10³ (approx). Note scale change at 15,000.

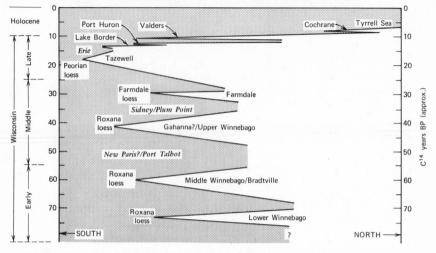

Figure 21-5 Diagrammatic sketch showing glacial invasions of the territory between the Appalachians and the Mississippi River during Wisconsin time. Represents both wedges of sediment and trace of supposed ice-sheet margin through time. Glacial sediments and loess in vertical letters; nonglacial sediments in slant letters. Time divisions at left are arbitrary. Pattern becomes poorly controlled toward base and is partly conjectural. (Compiled.)

ments exist, the magnificent sequence of glacial lakes whose successors are the modern Great Lakes dates from the Late Wisconsin deglaciation. Early members of this sequence were ponded between high ground on the south and the ice sheet on the north. As the ice receded the lakes widened; during glacial readvances the lakes narrowed. By degrees the water bodies became confined to the basins they occupy today.

The record of the former lakes is extensive. It consists of wave-cut cliffs and beach ridges that indicate shorelines now abandoned, as well as deltas, bars of various kinds, and broad lake-floor sediments. At any one former water level these features end abruptly or gradually on the north and east. This northern and eastern limit indicates the approximate position of the margin of the ice sheet at the time when the lake stood at that level. In general successively lower strandlines extend successively farther north and east, and the lowest ones extend right around the present lakes, showing that when they were made, the ice sheet had wasted entirely out of the Great Lakes basins. Abandoned channels that served as outlets are known for most of the lake phases described hereafter, but in the hilly, forested country north of the present lakes probably not all the outlets that must have existed have been identified.

The sequence of lakes depended basically on three factors: (1) retreat and readvance of the glacier margin, dominant in the earlier history; (2) deepening of outlet channels by outflowing streams, whose discharges varied greatly from time to time; and (3) elevation of outlet channels by isostatic upwarping as the crust was relieved of its load of ice, a factor significant in creating the Nipissing phase.

The sequence of lakes as presently understood is shown in table 21-F, and maps of representative lake phases in fig 21-6. The table indicates that as long as the ice margin, whether retreating or readvancing, stood within the Great Lakes region, the glacier/lake system was unstable and changes were frequent and rapid. In contrast, after about 10,000 BP deglaciation had freed most of the lakes region of ice and the system became more stable. The chief changes then were brought about by isostatic recovery of the crust, which altered altitudes of lake outlets.

Glacial Lake Agassiz. Closely related to the Great Lakes in time and mode of origin was Lake Agassiz (reference: Elson, Zoltai, and others in Mayer-Oakes, ed., 1967), whose sediments cover an aggregate area of ~500,000km², mainly in Manitoba, Ontario, and Saskatchewan, but also extending into Minnesota and the Dakotas. The area of the lake itself, however, was at no time greater than ~200,000km²; maximum depth exceeded 200m. History of the lake during deglaciation punctuated by readvances is read from beaches, lake-floor sediments, deltas of influent streams, and outlet channels, calibrated by C^{14} dates. The lake surface fluctuated repeatedly under the control of alternating retreat and readvance of the ice sheet. The lake endured through about 5000 years, from ~12,500 to ~7500 BP. During an early part of this time its outlet was southward into the Mississippi River drainage. Later outlets were toward the east into Lake Superior, and finally, after a brief confluence of Lake Agassiz with Lake Ojibway-Barlow (chap 22), northeastward into the Tyrrell Sea.

Other Lakes. (Map by Lemke et al. in H. E. Wright and Frey, eds., 1965a, p. 17). Late in the process of deglaciation of the James lobe in eastern South Dakota, the proglacial Lake Dakota (Flint, 1955b, p. 123–127) was ponded between the ice and high ground. The Lake lengthened northward after the manner of Lake Chicago and drained southward over a broad area of drift to the Missouri River. When at its maximum the lake was nearly 200km long.

Lake Souris was a small, nearly contemporary lake in northwestern North Dakota, dammed between the receding glacier and a drift-covered bedrock escarpment. Lake Regina (Johnston and Wickenden, 1930), nearly 500km long, occupied the basin of the South Saskatchewan River and drained into Lake Souris, which in turn discharged into Lake Agassiz.

Four proglacial lakes (from east to west Glacial Lakes Glendive, Jor-

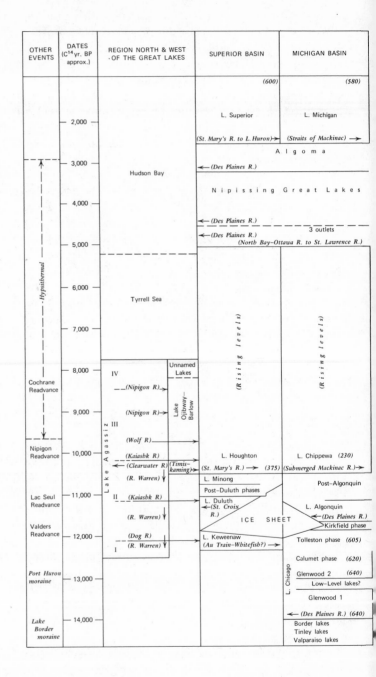

OTHER EVENTS	DATES (C¹⁴yr. BP approx.)	REGION NORTH & WEST OF THE GREAT LAKES		SUPERIOR BASIN	MICHIGAN BASIN
				(600)	*(580)*
	— 2,000 —			L. Superior	L. Michigan
				(St. Mary's R. to L. Huron)→	*(Straits of Mackinac)* →
				A l g o m a	
	— 3,000 —	Hudson Bay		←*(Des Plaines R.)*	
				N i p i s s i n g G r e a t L a k e s	
	— 4,000 —				
				←*(Des Plaines R.)*	
					3 outlets
	— 5,000 —			←*(Des Plaines R.)*	
				(North Bay–Ottawa R. to St. Lawrence R.)	
Hypsithermal	— 6,000 —	Tyrrell Sea		*(Rising levels)*	*(Rising levels)*
	— 7,000 —				
	— 8,000 —	IV	Unnamed Lakes		
Cochrane Readvance		_ _*(Nipigon R.)*→	Lake Ojibway– Barlow		
	— 9,000 —	*(Nipigon R.)*→			
		III			
		*(Wolf R.)*_			
Nipigon Readvance	— 10,000 —	*(Kaiashk R)*_		L. Houghton	L. Chippewa *(230)*
		←*(Clearwater R)*	*(Timiskaming)*→	*(St. Mary's R.)* →	*(375)* *(Submerged Mackinac R.)*→
		(R. Warren) ↓		L. Minong	Post–Algonquin
Lac Seul Readvance	— 11,000 —	II	*(Kaiashk R)*	Post–Duluth phases	
				L. Duluth	L. Algonquin
Valders Readvance		*(R. Warren)* ↓		←*(St. Croix R.)*	←*(Des Plaines R.)*
				I C E S H E E T	Kirkfield phase
	— 12,000 —	*(Dog R)*		L. Keweenaw	Tolleston phase *(605)*
		I	*(R. Warren)* ↓	*(Au Train–Whitefish?)* →	
					Calumet phase *(620)*
Port Huron moraine	— 13,000 —				Glenwood 2 *(640)*
					Low–Level lakes?
					Glenwood 1
					←*(Des Plaines R.) (640)*
Lake Border moraine	— 14,000 —				Border lakes
					Tinley lakes
					Valparaiso lakes

Table 21-F
Sequence and Mutual Relations of the Late-Wisconsin Great Lakes. [a]
(Compiled by R. F. Flint, Aleksis Dreimanis, and J. A. Elson, from various sources.)

HURON BASIN	ERIE BASIN	Niagara Gorge segments	ONTARIO BASIN	ST. LAWRENCE, L. CHAMPLAIN-HUDSON VALLEY	DATES (C[14] yr. BP approx.)	Position of sea level (ft.) relative to today's, according to smooth curve of Curray (1964, p. 180)
(580)	(573)		(246)			
L. Huron			L. Ontario		— 2,000	
(St. Clair–Detroit R.) →		Upper				
p h a s e					— 3,000	
(St. Clair–Detroit R.) → (595)		Great		St. Lawrence		
p h a s e				River	— 4,000	
	Lake	Gorge				
(St. Clair–Detroit R.) →	Erie				— 5,000	
(St. Clair–Detroit R.) → (605) →						
					— 6,000	— 20
(Rising levels)	(Rising levels)	Whirlpool	(Rising levels)	Estuary (St. Lawrence Valley below Montreal)	— 7,000	— 30
		Rapids			— 8,000	— 50
				Lampsilis Lake		
		Gorge		(St. Lawrence Lowland) / Laflamme Sea (St. John Lowland)	— 9,000	— 80
L. Stanley (165)					— 10,000	— 115
(North Bay–Ottawa R.) →				Mya phase (boreal)		
phases					— 11,000	— 150
(Fossmill) → (450?)	Early		Early L. Ontario	Hiatella phase (subarctic)		
L. Algonquin	L. Erie		(St. Lawrence R.) → (−25)			
(St. Clair–Detroit R.) (605)		Lower	Post-Iroquois phases →	L. Vermont	— 12,000	— 165
(Trent R.) → (560)	(Niagara R.) →	Great	L. Iroquois (330)			
Early Lake Algonquin (605)	(440)	Gorge	(Mobawk R?) → (330)	Intermediate lakes		
←? Lakes Grassmere (640) & Lundy (620) (Mohawk R.) →						
← L. Warren 1–3 (690–675) and Wayne (655) → ?	Lake Whittlesey				— 13,000	
← L. Saginaw 2						
← (Grand R.) (695) ← (Ubly Channel) (738)			Lakes			
(<615?) →	L. Ypsilanti →		Vanuxem →			
← (Grand R.) L. Arkona 1–3 (710–645)			← Hall	L. Albany (Hudson Valley)		
L. Saginaw	Lake Maumee		Newberry →		— 14,000	
← (Grand R.) (730)	← (Imlay Chan.)–3 (790)		Watkins →			
	← (Grand R.) –2 (760)					
	← (Wabash R.) –1 (800)					
	Pre-Maumee lakes					

Additional left-margin labels (Champlain Sea column, St. Lawrence–Champlain Lowland; Champlain Valley) appear vertically between the Ontario Basin and St. Lawrence columns.

[a] Numbers in corners of boxes are altitudes, in feet, of lake surfaces in the nondeformed region. Arrows show directions of lake outlets. All elements subject to readjustment.

567

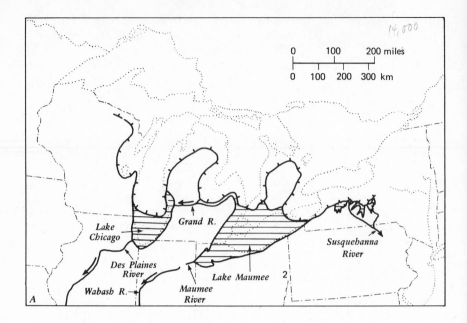

_14,000_

Figure 21-6 Sketch maps showing phases of the glacial Great Lakes at three times during deglaciation. _A_ At the time of Lake Maumee 2. _B_ At the time of the Port Huron Readvance. _C_ At the time of Lake Chippewa.

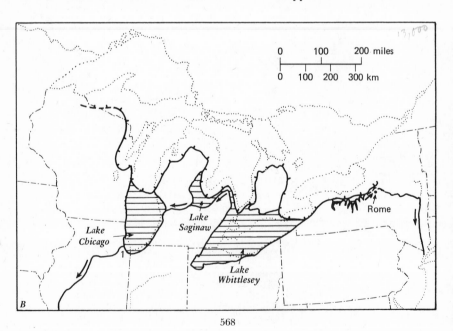

_13,000_

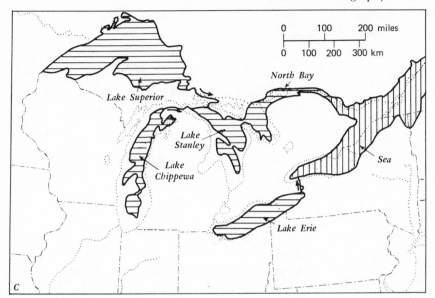

dan, Musselshell, and Great Falls) occupied the valleys of north- or east-draining rivers when the Wisconsin ice sheet was at or near its maximum extent—although whether that extent represents Early or Late Wisconsin ice is uncertain. Along the receding margin of the ice sheet farther northwest in Canada a number of proglacial lakes, some of them very large, existed. The largest, Lake McConnell (B. G. Craig, 1965a) spanned nine degrees of latitude. Among its modern successors are Lake Atha-baska, Great Slave Lake, and Great Bear Lake.

Two Creeks Section. (Reference: Thwaites and Bertrand, 1957; West, 1961). One of the most famous localities in North American stratigraphy is an exposure in a wave-cut cliff back of the modern beach of Lake Michigan at Two Creeks, Wisconsin. Discovered in 1907, the stratigraphy consists, in ascending order, of (1) till, (2) varved silt and clay, (3) a thin peaty layer (forest litter), (4) stratified fine sediments with mollusks and stems, branches, and leaves of trees, mostly spruce, and (5) till. A number of correlative exposures were found subsequently throughout a wide region. Layers (3) and (4) were early referred to as a "forest bed" and later were termed "peat." Neither term is wholly appropriate, but the inferred history is fairly evident. The lower till (1) is apparently of Port Huron age. During the post-till deglaciation Lake Chicago flooded the area, depositing varved sediment (2). The lake surface then subsided and a spruce forest developed, with birch, pine, and other trees (3). Glacial (Valders) readvance raised the lake level and drowned the forest; sediment with forest debris (4) was deposited. Finally the Valders ice arrived at the locality, burying the sedimentary sequence beneath till and continuing southward some 150km to its maximum extent near Milwaukee. Wood from the Two Creeks section was first C^{14}-dated at $\sim$11,400 BP, and subsequently several other dates from this and other exposures were obtained. Still later Broecker and Farrand (1963) measured several samples with as great precision as possible, and obtained a resulting average of 11,850 BP.[5] It has been suggested that the Valders readvance into the Lake Michigan basin was a surge and was therefore not related to climatic change.

Environments During Deglaciation

Vegetation. For the changes in vegetation that accompanied deglaciation we can refer to table 14-C, a sketchy picture that reflects the incompleteness of the available pollen data. In the region south of the Great Lakes the tundra that apparently fringed the ice sheet at its maximum had given place at some time before 14,000 BP to spruce parkland. Thereafter, with rising temperature and shrinkage of the ice sheet, the region east of the prairie underwent continual change, the forest becoming one dominated by oak beginning around 8000 BP. This crude generalization does not imply that a sequence of vegetation zones identical with today's followed each other northward as the ice receded. Rather we should visualize individual taxa as spreading northward, each in accordance with its own capability of adjustment to the environmental factors in the newly exposed territory. Hence probably no assemblage precisely resembled any assemblage now living, although many may have been similar in some respects to those of today. As far as the effect of glacial readvance on vegetation is concerned, the Valders readvance did not replace spruce forest with tundra. At two stations, one in Wisconsin 16km beyond the limit of readvance (West, 1961) and the other in Minnesota some 300km south of the limit (Jelgersma, 1962), the effect consisted of increase in the proportion of shrubs within a spruce forest.

West of the Great Lakes, environmental inferences are drawn from pollen at few localities. The sequence from a basin in drift of Cary age in central Iowa (P. H. Walker, 1966) begins at $\sim$13,000 BP, with conifer (mostly pine) woodland, giving way after 10,500 to a mixed forest of birch and hardwoods, and after 8000 to prairie. About 3000 BP the prairie was invaded by oak, thought to reflect a reversion to moister climate.

[5] Although this date can be accepted readily, the suggested correlation of Two Creeks with Bölling instead of Alleröd in Europe does not appear compelling; apparently the correlation remains an open question.

Farther north and west, localities in northeastern and southwestern South Dakota have furnished pollen evidence that spruce forest existed early in the period of deglaciation (Watts and Bright, 1968). At the former locality the sequence shows warming, with mixed deciduous forest and savanna, to a peak between about 8000 and 4000 BP, characterized by prairie with scattered trees. This condition was followed later by a moister climate.

Northwest of the Great Lakes, on the Plains between lat 48° and 51°, the pollen record is burdened with uncertainty as to the distances through which pollen could have been blown, in that open country, to sites of deposition. The record begins with a time around 12,000 BP, long after deglaciation had begun. Pollen-bearing sediments overlying the latest drift start with spruce forest or with open vegetation dominated by spruce. Between 10,500 and 8000 BP, depending on latitude, this changes to prairie or to parkland dominated by oak. In eastern North Dakota the prairie persists to the present; farther north where temperatures are lower, within the last 4000 years or so the prairie was invaded by deciduous trees and locally pine, creating the condition that existed up to just before modern cultivation (McAndrews et al. in Clayton and Freers, eds., 1967; Shay in Mayer-Oakes, ed., 1967; J. C. Ritchie in idem). As for the country between the western margin of Lake Agassiz and the Rocky Mountains, almost nothing is yet known of the vegetation during deglaciation, although an abundance of lake basins in the surface of the drift favors the prospects for sampling.

Mammals. The data on fossil mammals related to the period of deglaciation are very meager. The barren-ground caribou (*Rangifer arcticus*) is known from two localities in Michigan, although unfortunately at neither one is stratigraphic position well established. The chief interest in this species lies in the fact that it indicates the presence of tundra or park tundra where and when it lived. The mammoth is known also, but from a single locality near Flint, Michigan, where associated pollen indicates open spruce forest and the date is 11,400 C^{14} years. A similar environment is indicated for a woodland musk-ox from southwestern Michigan, dated at ~13,000 BP. Other taxa in the same time range include giant moose, caribou (*R. tarandus*), and whitetail deer, all from forest environments, though not necessarily with dominant spruce.

Also present in abundance during the period 15,000 to 9000 BP was the mastodon, which has been found in great numbers, probably owing to the good preservation afforded by a bog habitat. Associated pollen indicates a forest environment, although not exclusively spruce forest. The disappearance of mastodon around 9000 BP led Dreimanis (1968) to suggest that the extinction of these animals was hastened by their having been trapped in disappearing spruce forest, cut off from the boreal forest farther north by a growing belt of drier pine and hardwood forest in the northern Great Lakes region. In the Appalachian region, by 11,000 BP deciduous forests had replaced or were replacing spruce and pine, and boreal mammals were replaced by temperate kinds.

CHAPTER 22

Stratigraphy of Eastern North America

GENERAL CHARACTERISTICS

The Appalachian region, New England, and eastern Canada taken together constitute a natural region different from much of the central region described in chap 21. On the whole relief is greater, slopes are steeper, topographic texture is finer, and the bedrock includes more sandstones and other rocks rich in quartz, and fewer rocks rich in primary carbonates. Likewise prevailing climates are different. As a group these differences result in drift that is generally thinner and less continuous, coarser in texture, richer in quartz, less calcareous, more permeable, exposed with less frequency, and more altered by mass-wasting. End moraines are less bulky and more lobate in detail, and in some areas fewer. Also, because a much larger proportion of the terrain is forested, the tracing of moraines and other morphologic features is more difficult. When we add that much of the drift border lies under water, it is not hard to understand why glacial stratigraphy has made slower progress in the eastern than in the central region. The correlation chart (table 22-A) is even less firm than that in table 21-A, and it is noteworthy that the more detailed sequences are based on areas, such as those on the Coastal Plain, whose characteristics are exceptional to those cited above. The chart lacks even the tentative C^{14} time calibration applied to table 21-D. The comparatively small number of C^{14} dates is the result, in part, of the permeability of much of the drift, permitting destruction by oxidation of buried wood and similar organic matter.

573

APPALACHIAN REGION

The map, fig 18-5, shows a pronounced though broad re-entrant in the drift border extending from eastern Ohio into Pennsylvania and New York. The re-entrant marks the sector in which the ice sheet crossed the high Appalachian region, at altitudes exceeding 700m. The apex of the re-entrant likewise approximates the boundary between ice of the Ontario lobe and ice of the Erie lobe. In eastern Ohio the drift border tends to parallel the sandstone-capped, scarp-like margin of the Allegheny Plateau, which it overran.

In the western part of the Appalachian region the stratigraphy (G. W. White, 1969) has been traced continuously eastward from the central region and so is well tied to the Ohio sequence shown in table 21-A. Its study is greatly aided by abundant exposures in coal-strip mines. The sequence (table 22 A) begins at the base with till known to be pre-Illinoian, succeeded by Mapledale Till, twofold and probably Illinoian, with a thick soil developed in its surface. The Mapledale Till forms a narrow belt outside and parallel with the border of the Wisconsin drift, which here consists of a belt of till, the Titusville Till, of Early Wisconsin age. The latter till, with a soil developed in its surface, is overlain in offlapping fashion by four Late-Wisconsin tills, of which the lowest, the Kent Till, is C^{14} dated at around 23,000 BP. Textural differences have been useful in the correlation of these tills.

Farther east, in adjacent New York (Muller in H. E. Wright and Frey, eds., 1965a, p. 99) and northern Pennsylvania (Leverett, 1934), firm correlation with the sequence in Ohio has not yet been achieved. An outer fringe of drift present in places may be pre-Wisconsin, Early Wisconsin, or both. At Otto, New York, drift of Wisconsin age is underlain by an organic zone containing coniferous pollen that suggests a climate cooler than the present one (table 22-A). Originally placed at the close of the Sangamon (MacClintock and Apfel, 1944), the organic zone was later thought to be Early Wisconsin both because of its ecology and because of a C^{14} date of ~64,000 years. Its position is in abeyance pending thorough pollen study; so great a finite C^{14} age, also, might actually be infinite.

It is believed that an equivalent of the Kent Till is present in southwestern New York, but in that region both nomenclature and correlation of Wisconsin drift bodies are much disputed (see Muller, op. cit.) except for the Valley Heads drift, which is probably younger than basal Late Wisconsin and older than about 12,000 BP. That drift is represented by an end moraine that has been traced across much of western New York.

Table 22-A
Stratigraphic Sequences (Glacial Drift) in Eastern United States

Standard sequence	N.E. OHIO / N.W. PENNSYLVANIA	SOUTHWESTERN NEW YORK	NEW JERSEY/EASTERN NEW YORK/N.E. PENNSYLVANIA	LONG ISLAND, CONNECTICUT, RHODE ISLAND	CAPE COD, MARTHA'S VINEYARD, NANTUCKET	BOSTON DISTRICT, MASSACHUSETTS
(references)	G.W. White, 1969; Shepps et al., 1959	Muller in H.E. Wright & Frey, eds., 1965a, p.106	Leverett, 1934; MacClintock & Richards, 1936; MacClintock, 1940; Peltier, 1959; Lockwood & Meisler, 1960; Connally, unpubl.	Shafer & Hartshorn in H.E. Wright & Frey, eds.,1965a,p.113; Donner, 1964; Kay, 1960; MacClintock & Richards, 1936; Fuller, 1914	Kay, 1964a, 1964b; Schafer & Hartshorn in H.E. Wright & Frey, eds.,1965a,p.113; Ogden, 1963; Woodworth & Wigglesworth, 1934	Kay, 1961; Judson, 1949
Wisconsin — Late	Ashtabula T; Hiram T; Lavery T; Kent T	Valley Heads drift; A correlative of Kent Till	Bridgeport readvance; Luzerne readvance; Rosendale readvance; Wallkill m; New Hampton m; Pellet's Island m; Augusta m; Ogdensburg/Culver's Gap drift; Wisconsin drift of Leverett, Peltier, & others (in part) — *Lake Albany seds.*	Lake Hitchcock sediments; New Haven Clay; Hamden Till; Minor moraines in S. Conn Harbor Hill m; Charlestown m; Ronkonkoma m	Ellisville m; Stratified drift, outer Cape Buzzards Bay m; Sandwich m; Vineyard m; Nantucket m	Lexington Drift of Judson
Wisconsin — Middle						
Wisconsin — Early	soil; Titusville T	Organic sediments at Otto and Gowanda?	Wisconsin drift of Leverett Peltier, and others (in part)?	Hempstead Gravel; Montauk Till; Manhasset Fm; Herod Gravel; Lake Chamberlain Till?	Early Wisconsin m.; outwash proglacial sediments	weathered zone; marine clay; Till in drumlins
Sangamon	soil		Cape May Fm (marine phase)	Jacob Sand; Gardiners Clay		Marine(?) clay; weathered zone
Illinoian	Mapledale T (?)		Illinoian till; outwash in Delaware Valley		Late Illinoian till; m; Early Illinoian? till; m	Outwash; till; marine(?) clay
Yarmouth					weathered zone	
Kansan	Slippery Rock T ("pre-Illin.")		"Pensauken Fm"/ "Bridgton Fm" complex (in part younger) and "Jerseyan" drift	Till; Jameco gravel } "pre-Illinoian"	Kansan drift	Till (position uncertain)
Aftonian				Mannetto gravel	Aquinnah cg. Marine— and related seds	
Nebraskan					Nebraskan drift	

The valley of the Allegheny River contains bulky outwash. A comparatively old, high-level valley train is said to be Illinoian, with one or more Wisconsin bodies at lower altitude. The valley of the Susquehanna River likewise contains valley-train terraces, the oldest extensive unit being of probable Illinoian age.

In central and western New York, north of the border of the Wisconsin drift as currently mapped, evidence of earlier glaciation is partly morphologic. Small valleys tributary to the Cayuga Lake valley near Ithaca, broader and deeper than the postglacial valleys in that district, were cut into the side of the main valley after it had been glaciated at least once. Then they were themselves glaciated and filled with Wisconsin till. In widening the Cayuga Lake valley, Wisconsin ice planed away the downstream segments of some of these interglacial valleys and left them hanging conspicuously above the main valley.

A similar valley in Chittenango Falls State Park implies a rather long interglacial time. In the same region gravel said to be of Wisconsin age contains cobbles of conglomerate which itself contains erratic pebbles; the relationship implies earlier glaciation.

EASTERN PENNSYLVANIA, NEW JERSEY, AND NEW YORK

Southeastward through northeastern Pennsylvania and New Jersey information continues to be scanty. Reconnaissance (Leverett, 1934) across that region, partly confirming earlier work, established the presence of Wisconsin, Illinoian, and pre-Illinoian drift, including many outwash valley trains. The Illinoian drift mainly forms narrow lobes projecting down valleys beyond less-lobate Wisconsin drift. Some exposures of the till are weathered reddish. Leverett's pre-Illinoian drift, in many places identified only doubtfully, consists of thin patches of much-weathered till and thinly scattered erratics outside the areas of Illinoian drift.

Farther north, in the Catskill Mountains region, two distinct drift sheets occur. Northeast of the mountains a little-eroded drift shows conspicuous end moraines, including one at its border. On the southwest is a thicker drift with a more subdued surface, less distinct end moraines, and conspicuous postglacial gullies and fans. Although both drifts are presumably of Wisconsin age, agreement on their correlation has not been reached.

Along the valley of the Delaware River, Peltier (1959) and Lockwood and Meisler (1960) identified Illinoian as well as Wisconsin outwash (fig 22-1), the latter covered thinly with loess. Leverett (1934) concurred with earlier workers in recognizing Illinoian and an older drift called Jerseyan in New Jersey; on a basis of semiquantitative treatment of weathering in gneiss clasts, MacClintock (1940) concluded that the Jerseyan drift is significantly more weathered than the Illinoian and therefore probably older.

Bodies of weathered gravel in New Jersey, earlier named Pensauken, Bridgeton, and Beacon Hill, were thought by MacClintock and Richards (1936) to be remnants of a formerly extensive complex of alluvium (in part outwash) of Illinoian and older ages. The distribution of the sediments suggests that each unit was deposited first along pre-existing valleys, that the valleys were gradually filled up, and that the alluvium was then spread over the intervalley areas of the Coastal Plain until it formed a coalescent, nearly continuous spread whose surface was a broad alluvial plain sloping seaward.

The stratigraphic relations of the alluvial complex suggests that between the times of alluvial filling and lateral spreading, streams deeply re-excavated the valleys, some of them to depths below present sealevel.

The topographic position and degree of weathering of the highest and oldest alluvial unit, the Beacon Hill gravel, are such that it has always been regarded as pre-Pleistocene. Possibly it is correlative with the Brandywine Gravel of Maryland. The Bridgeton and Pensauken Formations are younger. The latter is thought to be of Illinoian age at least in part. However, as the complex contains conflicting elements (striated erratic cobbles at some localities; a warm-climate flora at one locality) not only different ages but also different origins seem to be involved.

Higher in the sequence the fossiliferous marine phase of the Cape May Formation belongs in a Sangamon time of positive sealevel; a temperate climate is indicated by the fossils.

Virtually all workers have recognized Wisconsin drift in New Jersey, but Early and Late Wisconsin, if both present, have not been differentiated. In the latitude of New York City, proglacial Lake Passaic was related to Wisconsin ice at and near its maximum extent.

Through the first 100km or so of Late Wisconsin deglaciation, five or more end moraines were built in the Wallkill Valley, west of and parallel with the Hudson River. These moraines represent the period before proglacial Lake Albany (Woodworth, 1905), occupying the Hudson Valley, entered the Wallkill. The Hudson Valley itself seems to have contained a lake during all or most of the Late Wisconsin deglaciation. The dam that created the lake may have been the end moraine, now breached, that crossed the valley at the Narrows between Staten Island and Brooklyn, but postglacial submergence has obscured the critical area. Varved clay occupies the Hudson Valley and adjacent areas from Hack-

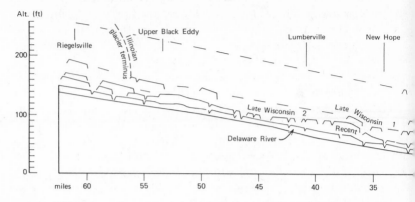

Figure 22-1 Profiles of five terraces along Delaware River. The four highest represent dissected outwash trains of various ages. The lowest consists of postglacial, finer-grained nonglacial sediment covering older sediments. Steepening

ensack, New Jersey discontinuously northward beyond Albany, New York. Not only does marine submergence now extend through the entire distance, but the valley lies within the region of isostatic uplift; as a result the level of the lake was repeatedly readjusted. Although kame terraces formed in some parts of the valley, probably deglaciation took place mainly by calving into the proglacial lake, which endured until some time after the Albany district had been deglaciated. Indeed it has been proposed that lake water persisted into the time when the Hudson/St. Lawrence divide had been deglaciated and Lake Vermont (table 21-E) had come into existence north of it.

While the glacial Great Lakes were evolving farther west, the northern slope of the Allegheny Plateau south of Lake Ontario was being deglaciated. The north-draining valleys which today contain the Finger Lakes were filled with meltwater to form the successive lakes Watkins, Newberry, Hall, and Vanuxem, followed by the larger lakes Wayne, Warren, and Lundy (table 21-F). The creation of Niagara Falls caused the lake water in northwestern New York to fall to the level of Lake Iroquois.

During one or more earlier deglaciations the Hudson, instead of being ponded, carried outwash sediments across the continental shelf then emerged, depositing them as a large delta at a position represented today by depths of −75m to −750m. The delta, defined by a reflection survey (Ewing et al., 1963), seems unlikely to have been active during the Late Wisconsin deglaciation because it is not compatible with a lake in the Hudson Valley.

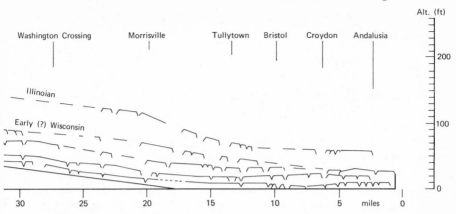

of gradients near Morrisville occurs where the valley widens abruptly and the proglacial river built large fans. Apparent reverse slopes near Bristol are an artifact of the projection employed. (Peltier, 1959, fig 24.)

THE GLACIATED COASTAL REGION

Long Island and the Cape Cod district are parts of the Coastal Plain. For this reason the character of their drift is rather similar throughout a distance of nearly 400km, and differs both internally and morphologically from the drift in the region of older rocks farther north. Most of the Pleistocene strata are exposed in wave-cut cliffs.

The Pleistocene sequence on Long Island overlies a nucleus of Cretaceous sediment. The basic study (Fuller, 1914) established the stratigraphic units; later studies assigned different ages to some of them (table 22-A), but no new units were added. The Gardiners Clay, an interglacial(?) marine unit with a mollusk fauna and pollen, both indicating a temperate climate, is not well defined and may consist of two or more units (Upson, 1970); neither is its stratigraphic position established firmly. Horizons correlated with it elsewhere are not all necessarily of the same age; even a unit of Yarmouth age is not excluded. Likewise the Manhasset Formation, assigned by later students to the Wisconsin, may possibly be older than the Early Wisconsin bracket in which it is currently placed.

As table 22-A shows, the section in the Cape Cod district (mostly exposed on Martha's Vineyard) includes a longer sequence, in which Kaye (1964) recognized each of the glacial stages of central North America, including two Illinoian tills and both Early and Late Wisconsin drift. A rather long section in the Boston area likewise includes two pre-Wisconsin tills, a probable correlative of the Gardiners Clay, and both Early and Late Wisconsin drift (Kaye, 1961). Interglacial sediments in cores, from the shelf ~100km off Long Island and ~70 km off Nantucket Island, contain pollen that indicates temperate hardwood forest, and so are probably interglacial. However, stratigraphic correlation is indefinite (Livingstone, 1964).

Although the great majority of exposures in southern New England show a single till, probably Late Wisconsin, exposures of two tills in superposition have long been known. Some of these pairs are lodgment and ablation tills representing a single glaciation. Others are clearly two lodgment tills of at least slightly different ages. At one locality in Connecticut two superposed tills (Hamden Till above; Lake Chamberlain Till below) are distinctive as regards lithology, fabric, and striation directions, though the physical disconformity between them is not marked by weathering (Flint, 1961). A till older than the surface till (although not certainly equivalent to the Lake Chamberlain Till) is now known at many localities, especially in western Connecticut. The rather moderate extent of weathering in its surface suggests that the underlying till is early Wisconsin, although that age cannot be said to be established. Overlying the Late Wisconsin (surface) till is widespread stratified drift, consisting of minor bodies of outwash and a great variety of drift of ice-contact character.

Two end moraines, each with extensive outwash, together cover most of the surface of Long Island and continue eastward through the Cape Cod region. Although not C^{14} dated on Long Island, these moraines are widely considered to be of Late Wisconsin age.

Figure 22-2 indicates their spatial relationships. The Ronkonkoma-Vineyard-Nantucket line is the older and probably lies at or rather close behind the outer limit of Late Wisconsin glaciation. In the absence of unequivocal evidence, it is generally assumed that these moraines date from around 20,000 to 18,000 BP. However, possibly the Vineyard moraine includes drift of earlier and later age as well. Behind this first line is the Harbor Hill-Charlestown-Buzzards Bay-Sandwich moraine. The first line displays two interlobate angles, the second line three. Indeed the "forearm" of Cape Cod consists of thick stratified drift deposited in an interlobate position. Behind the second line are discontinuous younger moraines. This is the case also in coastal Rhode Island and Connecticut, where two thin morainal lines are evident despite discontinuity. The trends of these lines appear to be unrelated to the existing shoreline, reflecting the probably emergent condition of the belt during deglaciation. Minimum dates of about 14,000 BP for these moraines are estab-

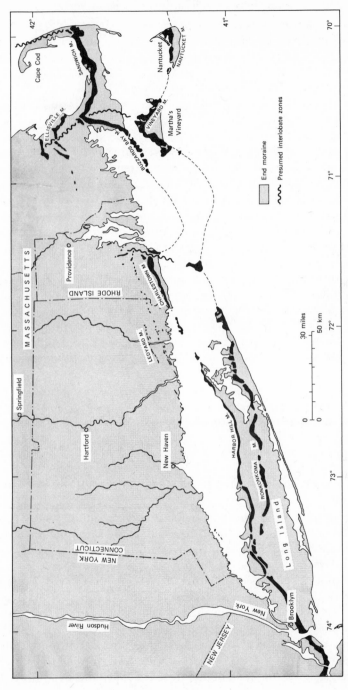

Figure 22-2 Sketch map showing known end moraines between the Hudson River and Cape Cod. (Compiled from published and unpublished sources.)

581

lished at two localities in coastal Connecticut.

Farther inland, end moraines have not yet been found. Instead kame terraces and other ice-contact features are abundant. These features indicate that after deglaciation of the coastal belt the ice sheet became thin and separated into narrow tongues that occupied valleys.

At many places in New England local lakes formed during deglaciation, and fine-grained sediments including varved silt and clay were deposited in them. Such sediments in New Haven, Connecticut, contain, as megafossils, leaves of the boreal plants *Dryas integrifolia, Saxifraga oppositifolia,* and *Salix arctica?* By far the most extensive of the lakes, probably ranking in size with Lake Albany, was Lake Hitchcock (Jahns and Willard, 1942) in the Connecticut River valley. Held in south of Hartford by a massive accumulation of drift, the lake extended progressively northward, probably by calving, until its proximal end lay in the northern parts of Vermont and New Hampshire. More than 4,000 rhythmite couplets have been counted or estimated. The date of inception of the lake is unknown; probably it was well after 14,000 BP. Although the lake is thought to have been still in existence as late as ~10,700 BP, the date when it was destroyed by draining is not closely known either.

Because much of the coastal glaciated region northeast of Cape Cod is submerged, little stratigraphy has yet been developed from that region. Sections including two or more tills are reported from as far northeast as southern Newfoundland, but correlations are still uncertain or nonexistent. Postglacial marine sediments, themselves emerged isostatically, overlie the latest till in many parts of the region as far south as Massachusetts, where amplitudes of eustatic rise of sealevel and isostatic uplift are equal.

ST. LAWRENCE RIVER REGION

The stratigraphy along the St. Lawrence River (Gadd, 1960) begins with the Bécancour Till, thought to represent a pre-Sangamon or a very early Wisconsin glaciation. The till is overlain by nonglacial sandy alluvium with a layer of peat, the St. Pierre peat, a sequence regarded as representing a nonglacial interval, either Sangamon or early Wisconsin. Plant megafossils, mainly parts of trees, in the peat imply a climate cooler than today's, and are more than 40,000 C^{14} years old. Next in sequence is a body of varved proglacial-lake sediments, named Deschaillons, deposited when the ice sheet, advancing up the St. Lawrence valley, created a temporary dam which held in a wide lake. In time the advancing ice covered the entire area, obliterating the lake, and depositing the

Gentilly Till. That till is not known to be broken by any surface of un-conformity or by nonglacial sediments and is directly overlain by sedi-ments of very late Wisconsin age. Hence it is thought to span the Port Talbot, Plum Point, and post-Farmdale intervals represented in central North America and sketched in fig 21-5. This concept implies that the re-advances in the latter region, through a time interval of forty or fifty thousand years, occurred while the St. Lawrence region remained contin-uously beneath the ice sheet. If correlative with the Gentilly Till, the Malone Drift (table 22-B), farther up the St. Lawrence, carries the same implication.

These tills, respectively, are overlain by drift deposited during Late-Wisconsin deglaciation, and that drift, in turn, is overlain by sediments of the Champlain Sea (chap 18). Summary accounts of the littoral and bottom sediments of this temporary water body, which extended into the basin of Lake Ontario, are contained in Coleman, 1941, Karrow, 1961, and Johnston, 1916; see also Gadd, 1961. Wagner (1967) gave a detailed

Table 22-B

Stratigraphic Sequences in Eastern Canada. (For Adjacent Vermont see Stewart and MacClintock, 1969.)

Stratigraphy	Time	SOUTHERN HUDSON BAY LOWLAND/NORTHERN ONTARIO & QUEBEC	ST. LAWRENCE VALLEY WEST OF MONTREAL	ST. LAWRENCE LOWLAND
		McDonald in P.J. Hood, ed., 1968 Hughes, 1965	MacClintock & Stewart, 1965	Gadd, 1960 McDonald, 1967
Holocene	6 8 10 12 (C14 years BP × 10³)	Tyrrell Sea sediments Lake Barlow—Ojibway seds. — Till of Cochrane Re-advance	Champlain Sea sediments Ft. Covington Drift	Sediments of St. Narcisse mor. Champlain Sea sediments Sediments of Highland Front moraine
Pleistocene Wisconsin Late Middle Early (Not to scale)		Till Proglacial—lake sand & silt (positions uncertain) Till	Malone Drift (position uncertain)	Gentilly Till Deschaillons varved seds. (Tills and intertill sediments in Sharbrooke district)
Sangamon		Missinaibi nonglacial sediments?		St. Pierre peat?
		pre—Missinaibi till		Bécancour Till?

list of invertebrate fossils, many of which are cold-water taxa; in addition
Beluga, a whale, occurs in the sediments. Common species of mollusks include *Yoldia arctica, Macoma balthica,* and *Hiatella arctica,* all of which
are tolerant of low temperature and low salinity, expectable in view of the
large volume of meltwater discharged into the narrow sea.

On the north side of the valley Foraminifera from Champlain Sea sediments occur in sediments of the St. Narcisse moraine; this testifies to the
presence of an active ice margin ending in the sea. Emergence was
brought about by isostatic uplift; a result was the modern St. Lawrence
River, trenching the blanket of marine sediments that covers the valley
floor.

NORTHERN ONTARIO AND HUDSON BAY

Much of the vast area of central and northern Canada is known only
in reconnaissance, and some is not known at all. We shall emphasize here
the stratigraphy of the region between the Great Lakes and Hudson Bay,
from which a sequence (table 22-B) has begun to emerge.

Nonglacial sediments, of interglacial or interstadial character, have
long been known throughout a wide region around James Bay. These
Missinaibi sediments (McDonald in Hood, ed., 1968) are, however, best
known from the watershed of Moose River, tributary to the head of the
bay. Overlain and underlain by till, they appear in three facies: alluvium,
peat, and marine sediments. The peat, including flattened branches,
twigs, leaves of small trees, and pollen, indicates boreal forest dominated
by spruce, much like that living in the region today, together with areas
of bog. This ecology, and the existence of a marine facies, suggest interglacial conditions with the sea (an ancient Hudson Bay) near its present
level. However, it has been suggested that the Missinaibi beds are of Wisconsin interstadial age, perhaps correlative with the St. Pierre peat. The
matter is unresolved pending the appearance of further data.

Overlying the Missinaibi sediments are two tills separated by the sand
and silt of a proglacial lake, the latter implying glacier ice in, but apparently not south of, the Hudson Bay region. This might represent an interstade between two Wisconsin glaciations.

Centered farther south in the region south of James Bay, and overlying
till that is apparently the younger of the two post-Missinaibi tills, are the
sediments of Lake Barlow-Ojibway. The history of that temporary lake is
outlined in Hughes in H. E. Wright and Frey, eds., 1965b, and Boissonneau, 1966–1968. Lake sediments, including varved silt and clay as well

as littoral sands, cover an area of more than 175,000km² north of the Great Lakes and south of James Bay; their distribution is plotted on the Glacial Map of Canada. The lake, with a maximum depth approaching 250m, was ponded between the ice sheet and ground that at the time sloped gently northward. (Subsequent isostatic uplift reversed the slope in the southern part of the lake area.) Part of the area of Lake Barlow-Ojibway is underlain by till derived in large part from sediments of the lake. This relationship reflects readvance of the ice-sheet margin through more than 100km, to the vicinity of Cochrane, Ontario, near lat 49°.

The Tyrrell Sea (chap 18) closely followed the deglaciation after the Cochrane readvance, although marine sediments are actually in contact with the lacustrine sediments only in an area southeast of James Bay. The sediments of the Tyrrell Sea are fossiliferous, with assemblages of mollusks not unlike those of the Champlain Sea.

ARCTIC CANADA

Nonglacial sediments, surely or probably Pleistocene, occur at a few localities in Arctic Canada and are summarized by B. G. Craig and Fyles (1960). The only unit known to be extensive is the Beaufort Formation, which crops out along the Arctic coastal fringes of the westernmost Arctic islands through an overall distance of nearly 700km. It consists of rather coarse fluvial sediments at least 60m thick, apparently the remnants of an ancient coastal plain that antedated deep dissection. Reconnaissance shows that the lower part of the Beaufort Formation locally contains pollen of spruce, pine, fir, and hemlock, certain hardwood trees, and birch, willow, alder, and herbs. This assemblage indicates a climate warmer than that expectable of an interglacial thermal maximum at a high latitude. Pollen in the upper part of the formation contains spruce and pine but apparently no hemlock nor hardwoods; here a climate cooler than the preceding one is indicated, although still not as cool as that of today. The stratigraphic position is so imprecise that although the sediment is thought to be Pleistocene, it is not certainly so.

In at least three other localities in the islands nonglacial sediments are identified. One, on Banks Island, overlies till that in turn overlies the Beaufort Formation. Megafossils include small trees and beaver-gnawed sticks. The pollen shows spruce, pine, birch, alder, and tundra plants, indicating partial forest and a climate warmer than today's. Data from the other localities are even more scanty. Farther west, in the delta of the Mackenzie River, pollen indicates former vegetation like that of today;

presumably it is interglacial. Farther west, in Alaska, much more information on former nonglacial climates in Arctic latitudes is available (chap 20).

Knowledge of the Quaternary stratigraphy of Greenland is confined to a few small disconnected coastal areas, mostly at the heads of fiords; generalization is not possible. The known stratigraphy is confined to the latest glaciation. Examples of areal studies include A. L. Washburn and Stuiver, 1962; Flint, p. 90–210 in Boyd, 1948. See also Weidick, 1971.

ATLANTIC COASTAL PLAIN

Returning to temperate latitudes, we examine the stratigraphy of the Atlantic Coastal Plain south of the glaciated region. In chap 12 we described the Coastal Plain briefly in connection with emerged marine strandlines (fig 12-8), and contrasted it with the glaciated region north of New York City. Southward through New Jersey, Pennsylvania, and the northern part of the Chesapeake Bay region the coastal belt is characterized by conspicuous sandy and in places gravelly alluvium, apparently including substantial amounts of outwash from the Delaware and Susquehanna Rivers. Marine sediments, mostly finer grained, are present as well, particularly in areas away from the principal rivers. Farther south along the coast marine sediments are more conspicuous, extending farther inland and to higher altitudes, although estuarine fills and alluvium are present along major streams. Along the Gulf Coast near and west of the Mississippi River, alluvium is dominant.

Along the lower Mississippi the alluvium is thought to date from times of major deglaciation and deposition of outwash (table 21-C). A similar relationship along the submerged Delaware River downstream from the segment shown in fig 22-1 was thought likely by Jordan (1964) after detailed study of fluvial and marine sediments in Delaware. However, the marine sediments there are possibly interglacial, relict from a time or times of higher sealevel and a different regimen of the Delaware River.

Outwash of Wisconsin age along the Susquehanna River upstream from the Coastal Plain is described by Leverett (1934), and along the submerged continuation of the river beneath Chesapeake Bay by Hack (1957). Here the discrepancy between the fluvial sediments of glacial age, all at negative altitudes, and the marine sediments of interglacial age extending up to positive altitudes is very apparent.

Preliminary study of a section exposed in the city of Washington, D.C., yielded pollen and spores from which a sequence of climatic changes was inferred, changes possibly embracing the Yarmouth, Illinoian, Sangamon, and Wisconsin Ages (Knox, 1969). As the only C^{14} date, near the top of the section, is infinite, the sequence is not calibrated and no firm correlation is yet possible.

Although little of the Coastal Plain Quaternary has been mapped in detail, the generalization that from Virginia southward through Florida Quaternary sediments are preponderantly marine is probably true. Until 1960, however, field study was hampered by a pervasive concept that morphology, represented particularly by "terraces," was the key to Quaternary history in the coastal belt (Flint, 1940b). "Terraces" and strandlines were correlated on a basis of altitude, and formations were even defined on a similar basis, regardless of possible crustal movement and of change of altitude caused by compaction or erosion since deposition. Some of the formations named in the older literature are effectively invalid because their definitions involve altitude. Again, the morphology of a strandline is emphemeral as a surface feature, because it is destroyed by reworking or is buried beneath younger sediments during subsequent marine transgression.

In a detailed study in southeastern Virginia, Oaks and Coch (1963) abandoned both the morphologic approach and the ill-defined stratigraphic names. They based their study primarily on stratigraphic relations, many of which were established only by examination of the subsurface.

The resulting sequence (fig 22-3) includes at least 8 formations of Quaternary age, chiefly marine. Most of these are divided into two or more facies comparable with the facies being deposited today. Table 22-C shows that sequence, and in addition sequences in South Carolina (in large part employing the old stratigraphic names) and in Florida based mainly on stratigraphy rather than on morphology. From Virginia southward the sedimentary units seem to be chiefly interglacial. The glacial ages are represented by surfaces of unconformity, and by lag gravels in channels, that indicate cutting by streams graded to lowered sealevels. In Florida the sediments are mainly calcareous; north of Florida they are mainly quartzose.

The Florida sequence in table 22-C is weakened by differences of opinion as to the stratigraphic positions of various units. These cannot be resolved in a broad review, but the differences are clearly brought out in the references cited in the table, and by Richards (p. 133 in H. E. Wright and Frey, eds., 1965a).

The often-used term *Pamlico Formation* is widely identified with fossil-bearing marine sediments related to positive sealevels of the Sangamon Interglaciation. It was first applied when stratigraphic knowledge was in a very primitive state, and it now appears that the unit is so poorly de-

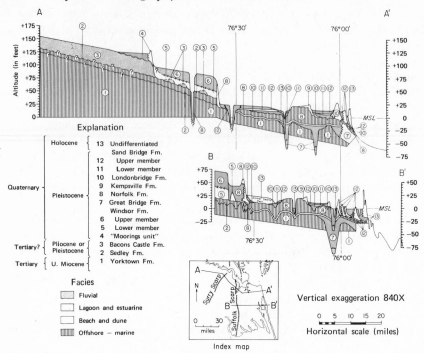

Figure 22-3 Section across southeastern Virginia normal to the coast, showing late-Cenozoic formations and facies. (R. Q. Oaks and N. K. Coch, unpublished).

fined that it is not truly useful. Its current meaning is hardly more than "mild-climate marine sediments less than about 10m above existing sea-level." Oaks and Coch (1963) avoided the term for this reason, although it still appears in the Carolina and Florida sequences. The fossils, mainly mollusks, indicate a climate as warm as or slightly warmer than that of today.

Apart from interglacial sediments of marine origin, a few occurrences of terrestrial sediments are known. Summarized by Whitehead (in H. E. Wright and Frey, eds., 1965a, p. 425), all are at <~30m and are of probable Sangamon age. At Washington, D.C., several pollen-analyzed sections combine to suggest a nearly complete interglacial sequence, from cool to warm to cool. At the thermal maximum (represented by a fossil cypress swamp with huge tree stumps) the temperature was at least as warm as today's. The Horry (pronounced *O-rée*) Clay, exposed in both Carolinas, yields a partial record of an interglacial climate with vegetation like that

Table 22-C

Selected Stratigraphic Sequences Beneath Atlantic Coastal Plain and in Bermuda. Correlations, Where Shown, Are Tentative.

Correlation	SOUTHEASTERN VIRGINIA N.K. Coch; R.O. Oaks (in part unpublished) Local sequence	SOUTH CAROLINA (Colquhoun et al., 1968; Colquhoun & Johnson, 1968, p.124) Local sequence	Suggested correlation	BERMUDA (Land et al., 1967, p.997) Eolianite	BERMUDA Marine limestone	BERMUDA Soil	SOUTHERN FLORIDA Parker et al., 1955, p.115–116; DuBar, 1958
Holocene	Dismal Swamp peat and other seds ———Regression———	Recent seds ——Regression——				Recent soil	Surficial sand & alluvium; Lake Flirt Marl (Recent)
	Sand Bridge Fm (2 mbrs)	Silver Bluff "terrace"	—Wisconsin II	Southampton Formation			Pamlico Sand
		Princess Anne "terrace"	—Wisconsin I			St. George's Soil	
	———–Regression?——— Londonbridge Fm	——Regression——			Spencer's Point Formation		Miami Oolite / Anastasia Fm / Key Largo Ls
Sangamon	————–Regression?——— Kempsville Fm[a]	Pamlico Fm, Horry Clay, & Ladson Fm	Sangamon Intergl.	Pembroke Formation	Harrington Formation		Ft. Thompson Fm
	Norfolk Fm[a]	Talbot Fm		Devonshire Formation			
?	Great Bridge Fm ———Regression———	——Regression——	—Illinoian Gl.			Shore Hills Soil	(Pleistocene)
?	Windsor Fm (2 mbrs)	Wicomico Fm	—Yarmouth Intergl.		Belmont Formation		
?	———–Regression?——— "Moorings unit"	——Regression—— Okefenokee Fm	—Kansan Gl. —Aftonian Intergl.	Walsingham Formation		Soil (?)—partly re-worked by Belmont erosion	
Pleist. or	——Bacon's Castle Fm ——(Regression)——	Coharie Fm					
Pliocene	Sedley Fm ——Regression——						Caloosahatchie Marl (Pliocene)
Miocene	Yorktown Fm	Duplin Marl					Tamiami Fm / Bone Valley Fm? (Miocene)

(Correlation with other areas has not been attempted)

[a] U.-series dates on corals lie within the range 62,000 to 86,000 BP (U.N. Goddard and N.K. Coch, unpublished).

589

at present. It indicates a warming trend ending with a marine transgression ("Pamlico"?). The Ladson Formation, mainly marine, near Charleston, South Carolina, includes pollen-bearing sediments in which pine, oak, and hickory are dominant, implying a climate differing little from today's. Similar climatic indications are given by pollen from organic clay exposed near the mouth of the Neuse River in coastal North Carolina (Whitehead and Davis, 1969). All these sections, then, are alike in indicating a Sangamon(?) climate which at its thermal maximum was little warmer than that of today. They can be compared with the section of the Gardiner's Clay on Long Island (Donner, 1964a) apparently indicating the middle and later parts of the Sangamon Interglaciation.

Also very likely of Sangamon age, a channel fill consisting of alluvium at Vero Beach, Florida has yielded an impressive array of terrestrial and freshwater fossils (Weigel, 1962). These include insects, fishes, birds, amphibians, reptiles, and mammals. Many of the taxa, including more than a third of the mammals, are extinct. The list of mammals collected includes ground sloth, wolf, bear, panther, saber-tooth cat, mastodon, mammoth, horse, tapir, camel, and bison. Neither the animals nor the associated fossil plants apparently indicate a climate markedly different from that now prevailing.

Table 22-C includes the sequence from Bermuda mentioned in chap 12. The units are not correlated, even tentatively, with those of the mainland because no reliable basis of correlation is yet known. It is thought, however, that the stratigraphy was controlled primarily by fluctuation of sealevel.

ENVIRONMENTS DURING DEGLACIATION

In summarizing the environment (chiefly vegetation) during the deglaciation of northeastern America, we should begin with fig 19-3, the map suggesting the vegetation existing when the Late Wisconsin ice sheet was at its greatest extent. The sequence of subsequent changes in New England is outlined in table 14-C, and is discussed by Davis in H. E. Wright and Frey, eds., 1965a, p. 391; by Ogden, 1965; and by Deevey, 1958.

In stratigraphic terms the herb zone (tundra or tundra-like grassland), which characterized the region beyond the ice sheet, followed the retreating ice margin through at least some distance inward. In southern New England that zone persisted from deglaciation of the present coastal region ($\sim$14,000 BP), gradually changing to park tundra through immigration of spruce. By about 11,500 the cover included abundant spruce and was progressing toward a closed forest. However, between $\sim$9,500 and $\sim$7,500 BP spruce was gradually replaced by pine, and this was later followed by hemlock and deciduous trees. (Farther northeast the timetable was somewhat different, but the scarcity of data hardly justifies details.) Around 5000 BP a change marked by reduction of hemlock occurred. This might have meant a drier climate, though possibly it might be related to fire or even the activity of man.

The earlier part of this history is reflected in fig 22-4, in which pollen data, related to time of deglaciation at seven localities, are arranged in south-north sequence. Changes in vegetation are correlated between localities as indicated by broken lines.

Farther west, in central New York, north of the Valley Heads moraine, pollen from a peat body ($\sim$12,700 BP) indicates spruce/tamarack forest with little fir or birch. When that forest existed, coastal southern New England was still wholly or largely treeless, while most of the St. Lawrence Lowland (Lasalle, 1966) still lay beneath the ice sheet, whose margin abutted the north slope of the Appalachian highlands. The exposed land was a treeless expanse of grass and shrubs (essentially tundra). However, the immigration of spruce soon began; this led to the development of park tundra, which in turn evolved toward spruce forest. That vegetation was later replaced successively by pine and by deciduous trees.

Various difficulties have impeded detailed studies of pollen in northern Canada, so that there are few data to be summarized. A compact review must await the reduction of local data by palynologists.

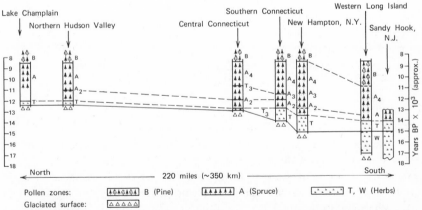

Figure 22-4 Cross-section of late-glacial vegetation zones through a belt that includes the Hudson Valley. Subzones indicated by letters are those summarized in Sirkin, 1967: [Sub]zones W and T represent the herb zones. Subzones A_2 through A_4 characterize the spruce zone. Zone B represents the pine zone, included to show the upper boundary of the late-glacial. The still-younger Zone C (mainly deciduous trees) is not included. Not all subzones for T and A as defined by Deevey, 1958 are recognized at each locality. Data from Connally and Sirkin, 1967, unpubl.; 1969, unpubl.; Sirkin, 1967; 1968, unpubl.

CHAPTER 23

Late Weichsel Environments in Europe

INTRODUCTION

During the latest maximum of extensive ice sheets, the general aspect of Europe, from the Atlantic to the Urals and the Caucasus and from the Arctic Ocean to the Mediterranean, was very different from that of today. Secular cooling coupled with the advent of a major ice sheet and the formation of other extensive glaciers resulted in a very different pattern of vegetation, and the fauna changed accordingly. At the same time sealevel stood lower relative to the land, exposing shelf area, much of which, however, was overspread by glaciers.

The latest maximum of extensive ice sheets occurred, as it did in North America, at ~17,000 to 20,000 C^{14} years BP; yet by 7000 BP the ice sheets had vanished and by 5000 most other glaciers either had disappeared or had shrunk to small size. In stratigraphic terms the latest maximum was the Weichsel Glaciation; the corresponding glaciation in the Alps was the Late Würm Glaciation (chap 24). Although in strict language the name Würm has nothing to do with a continental ice sheet, in most parts of the world other than North America that name has been identified commonly with the last expansion of such ice sheets in Eurasia. In the present discussion we avoid such loose usage and apply the name Weichsel to the last major ice-sheet glaciation in Europe, a glaciation generally believed to correlate with Wisconsin glaciation in North America. We apply the name Würm only to the region of the Alps, not reached by a continental ice sheet.

Events in Europe were similar in some ways to those in North America. Between lat 50° and lat 70°, maritime air masses built up an ice body on the Scandes, the Scandinavian mountain chain; it gradually spread over the whole northern part of the continent. Elsewhere glaciers formed in those highlands where altitude and precipitation were adequate to bring about net accumulation of snow. Chief among the glaciated highlands was the mass of the Alps which, at the glacial maximum, was covered by a complex mountain ice sheet. Glaciers formed likewise in high parts of the Pyrenees, Carpathians, and Apennines, and on highlands in the Balkans, central Europe, France, the Iberian Peninsula, and Britain.

At earlier glacial maxima, conditions were similar. Glaciers formed on the same group of highland areas and spread along much the same paths. The chief difference lay in the extent of earlier glaciers. For example, at the glacial maximum that immediately preceded the last interglacial the area of the Scandinavian Ice Sheet was substantially greater than that of its late-Weichsel successor.

SCANDINAVIAN ICE SHEET

Extent and Thickness

The map, fig 23-1, shows the relations of the Scandinavian Ice Sheet during the last glaciation and also at the earlier time or times when glaciation was most extensive. The difference in extent of ice at various times is substantial, but is not known quantitatively because parts of the western and northern sectors of the limit of the region glaciated at both times are now submerged and are therefore hidden from view. The southern boundaries, being mainly land boundaries, are known best, but even these become less certain toward the east and north. According to the most recent maps the Scandinavian Ice Sheet, when at its greatest known maximum, was coalescent with ice spreading from the Ural Mountains, and through that with a Siberian Ice Sheet farther east. On the southwest the Scandinavian ice body merged with glaciers of British origin. Off the west and north coasts of Norway we have placed its margin at the 200m isobath, an arbitrary position that we consider reasonable. Also we have extended the margin along that isobath to the northeast tip of Novaya Zemlya, a highland that was itself a source of local glaciers. To the north two highland archipelagoes, Spitsbergen and Zemlya Frantsa Yosifa, likewise generated local glaciers, essentially small mountain ice sheets, the outer limits of which are little known.

Because the northern limit of the Scandinavian Ice Sheet is submerged and uncertain, throughout the 20th century it has been commonly drawn

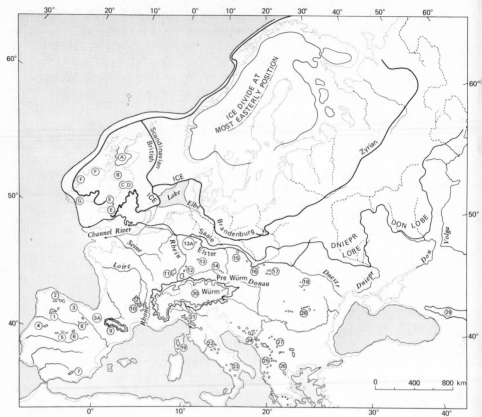

KEY TO LETTERS SHOWING CENTERS
OF OUTFLOW OF GLACIER ICE IN
BRITAIN AND IRELAND

A   Scottish Highlands
B   Southern Uplands of Scotland
C   Cumberland highlands
D   Pennine Chain
E   Mountains of Wales
F   Mountains of Connemara and Donegal
G   Mountains of southern Ireland

KEY TO NUMBERS SHOWING SEPARATE
AREAS OF INDEPENDENT GLACIATION
IN CONTINENTAL EUROPE

SPAIN AND PORTUGAL

1   Sierra de Pena Negra
2   Sierra de Picos
3   Penas de Europa
3A   Sierra de Aralar
4   Serra da Estrela
5   Sierra de Gredos
6   Sierra de Guadarrama
7   Sierra Nevada
8   Sierra Cebollera

FRANCE

9   Pyrenees Mts.
10   Auvergne and Cevennes Mts.
11   Vosges Mts.

GERMANY

12   Schwarzwald
13   Bayerische Wald
13A Harz Mts.
14   Bohmer Wald
15   Riesen Gebirge

POLAND AND CZECHOSLOVAKIA

16   West Beskiden Mts.
17   Tatra Mts.
18   Carpathian Mts.

CORSICA

19   Mountains of Corsica

SWITZERLAND AND ADJACENT
COUNTRIES

20   The Alps

ITALY

21   Etruscan Apennines
22   Roman Apennines
23   Southern Apennines

BALKAN COUNTRIES

24   Dinaric Alps
25   Pindus Mts.
26   Mt. Olympus
27   Rhodope Mts.
28   Transylvanian Alps

U.S.S.R.

29   Caucasus Mts.

rather close to the mainland coast of Europe. However, the possibility that in the last glaciation the ice sheet spread entirely across the floor of the Barents Sea to merge with the ice over Spitsbergen and probably also with that over Zemlya Frantsa Yosifa was visualized by Emery (1949, fig 1). Since about 1965 closer study has been given to this idea (e.g., Schytt et al., 1968). If indeed such spreading occurred in the last glaciation, it is even more likely to have done so at one or more earlier Quaternary maxima. Apart from direct evidence that offers it support, the concept is attractive for two reasons. (1) Nearly the whole area of the Barents Sea is less than 400m deep today and would have been even shallower at a glacial maximum. (2) Relatively warm maritime air masses that nourish glaciers in Spitsbergen and Norway today, and that nourished much larger glaciers in those areas in the past, would have been present to foster expansion of Scandinavian ice northward beyond the line of the existing coast. Under such circumstances the merging of glaciers in the far north would have been broadly similar to the accepted merging of Scandinavian with British glaciers in the southwest. The lines in fig 23-1 are drawn so as to take account of both the conventional and the "expanded" Scandinavian Ice Sheet.

Another uncertainty as to the limit of glaciation concerns the area represented today by the Shetland Islands. B. N. Peach and Horne (1879) demonstrated from the bearings of striations at 320 localities that those islands had been overrun by the Scandinavian Ice Sheet from NE to SW, and that later they were covered by radially flowing local glaciers. These facts imply that when the earlier set of striations was made, the Skagerrak, in places more than 500m deep today, was filled with ice, which therefore extended locally beyond the 200m isobath, taken as our general limit of glaciation in areas now submerged. Apparently the Shetland striations were made during the latest glaciation. We can suppose then that during any earlier, more extensive glaciation likewise, the Skagerrak was filled with ice.

Thickness of the Scandinavian Ice Sheet is not well known. It is thought to have been as much as 3000m locally over the Sognefjord (Hol-

---

Figure 23-1 Positions of glaciers in Europe at specific times, and chief centers of spreading in British Islands. Thick lines: limits of Weichsel/Würm glaciation, undifferentiated; thin lines: limits of earlier glaciations, labeled where identified. Reliability varies greatly. Coast is drawn conventionally at the −100m isobath. Drainage in North Sea−English Channel region is shown as it is believed to have been in Weichsel time. Compiled from many sources, including Valentin, 1958, map 4 (North Sea region) and Kaiser, 1960, pl. 1 (parts of southwestern Europe).

tedahl, 1967, p. 194). Most if not all mountain summits seem to have been overtopped by ice, and the 3000m value is quoted repeatedly. The thickest part of the ice body stood slightly west of the present Gulf of Bothnia, as indicated by isostatic data (chap 13). That radial outflow occurred from that vicinity across the mountains to the Atlantic is recorded by erratics of eastern origin occurring high on the mountain slopes, as well as by striations.

The possibility that ice-free areas existed on some interfluves on the western flank of the mountains must be considered, but many authorities think such nunataks are unlikely. The question is reviewed by R. Dahl (1963; see also Löve and Löve, eds., 1963, passim). The ice sheet covered the narrow continental shelf and calved into the ocean, as indicated by abundant glacial-marine sediments extending far off the present coast (Holtedahl, 1959).

Although in a broad sense the ice, when at its maximum, flowed radially outward from a center west of the Gulf of Bothnia, flow was influenced by topography. Pre-existing deep valleys that dissected the western flank of the mountains in Norway channeled the ice effectively and were greatly deepened and basined in consequence. Topography also shaped the southern margin of the ice sheet. Its influence was especially conspicuous in Germany, where the terrain reached by the ice sheet at earlier maxima had greater relief than that in southern Britain to the west or in Russia to the east. From west to east across Germany the glaciated region ends at the northern flanks of highlands such as the Sauerland south of the Ruhr Valley, the Erzgebirge, the Riesengebirge, and the West Beskiden Mountains. At the Weichsel maximum the ice sheet terminated farther north in a region of smaller relief, but even here, according to Bülow, the limit of glaciation in places coincides with the southern ends of preglacial topographic lows. Likewise in Poland the most extensive pre-Weichsel ice sheet terminated along the northern flank of the Sudetes and Carpathian Mountains, ponding lakes against the bases of the mountain slopes.

Origin and Growth

Scandinavian Mountains. The Scandinavian mountains are the backbone of the Norwegian-Swedish peninsula, reaching nearly from end to end of its 1000km length. Crest altitudes are commonly 1200 to 1500m, and in southern Norway exceed 2500m. The mountains are carved from a broad erosion surface upheaved late in Cenozoic time. Their trend is oblique to the paths of maritime air masses that approach the Norwegian coast from the southwest and cause precipitation ranging from 750mm in the north to more than 3000mm in places in the south. Precipitation varies roughly with crest altitude. Much of it falls as rain, but there is enough snowfall, coupled with cool, cloudy summers, to maintain ice caps and valley glaciers with a combined area of more than 5000km².

Phases of Growth. The Scandinavian Ice Sheet had its origin in these mountains, for drift of mountain provenance occurs east, south, and west of the mountain axis. If a small reduction of mean temperature, with or without change in precipitation amount, were to occur now, the result would be expansion of the existing glaciers into a partly coalescent complex. Probably this is what happened at the beginning of each glaciation.

If we assume initial temperature reduction, accumulation should have increased, particularly on the windward western slope of the mountains, where glacier ice should have been most abundant at first. However, indicators, streamline molded forms, and other geologic evidence show that toward the glacial maximum, ice was thickest in the region east of the mountains. This apparent anomaly, analogous to the development of glaciers in British Columbia (chap 18) is readily explained by regional topography. On the short steep western slope, flow should have been rapid. Within a short distance the ice reached deep water, where rapid calving prevented great extension in the seaward direction. In contrast, ice flowing eastward followed a longer, gentler slope and hence flowed more slowly, and terminal loss was less because calving was unimportant. Therefore east-flowing glaciers could expand and coalesce on the Bothnian lowlands into piedmont glaciers and eventually into an ice sheet (fig 23-2) that buried the mountains.

The accumulation of ice east of the mountain crest implies shifting of

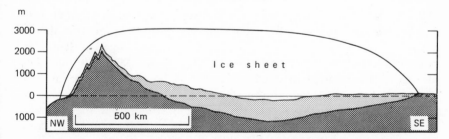

Figure 23-2 Cross-section sketch along a NW-SE line from the coast of Norway north of the Sognefjord to a point near Minsk, USSR, showing estimated profile of the Scandinavian Ice Sheet at its maximum in Weichsel time. Profile is adapted from Antarctic Ice Sheet, as in Robin, 1964. Bedrock surface beneath ice is shown both as at present and after isostatic subsidence. Profile of ice after subsidence would have been lower also, but is not shown because amplitude of upbuilding by snowfall is not known.

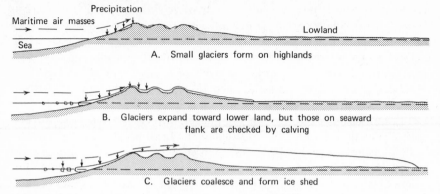

Figure 23-3 Idealized development of an ice sheet such as the Scandinavian Ice Sheet. Length of section about 800km. Diagrammatic; not to scale.

the ice divide. At least in central and northern Scandinavia, the center, or rather axis, of radial outflow of ice shifted from the mountain crest to a position that at one time was a minimum of 160km and possibly as much as 400km farther east, overlying much lower terrain (fig 23-3). The change may have been aided by drifting of snow to leeward under the influence of westerly winds and, more importantly, by nourishment from more frequent southeasterly winds (Hoppe and Liljequist, 1956), as indicated hereafter. In southern Norway, however, a local center of outflow seems to have been located over the highest part of the mountains, at least during deglaciation.

Evidence in Denmark confirms the inference that the ice divide shifted eastward. The distribution and stratigraphic relations of indicators in that country show that during pre-Weichsel glaciations ice invaded Denmark first from Norway and later from the Baltic region, and that at least once during deglaciation the center of outflow again shifted westward to the Scandinavian mountains (Madsen et al., 1928, p. 98, 88). Similar shifts are implied in northern Germany by indicators in the drift sheets (Richter, 1953) and also by striations (Schulz, 1967). The similarity of the record of successive glaciations demonstrates that the ice sheet formed repeatedly in the same manner; however, at the end of the latest glaciation it may have disappeared so quickly that flow from the mountains immediately to the north did not reestablish itself over Denmark (Wennberg, 1949).

In northern Fennoscandia successive positions of centers of outflow, related to the last deglaciation, have been discerned in streamline molded forms and in striations (Hoppe and Liljequist, 1956, p. 47–58). Successive directions of flow clearly evident in southern Finland (Virkkala, 1960) are not, however, obviously related to shifting of an ice divide. Further evidence of eastward shift of the ice divide is that the domed-up lake and marine strandlines formed during and after the uncovering of Fennoscandia by the latest ice sheet are related to a center in the area of the Gulf of Bothnia (fig 13-4).

Nourishment. Cyclonic air masses approaching from the southwest nourish Scandinavian glaciers today and probably did so during the latest glaciation (figs 4-8 and 4-9). One cyclonic path led northeastward, skirting the Norwegian coast, and caused snowfall on the western slope of the ice sheet. Another path lay along the southern margin of the ice. But as there are no high mountains on that path, air masses moving along it tended to hold the greater part of their moisture until they encountered the ice sheet. Hence there was much accumulation along the southern slope of the glacier. Accumulation on the southeastern part probably was augmented by snowfall from southeast winds. Snowfall on the southern slope would have occurred in spring and summer more than in winter when frequent outbursts of cold air from over the ice sheet would have checked the moving cyclones. The result would have been colder and drier winters in the lowlands of central and eastern Europe than those that now prevail. At all seasons the ice-covered Alps would have acted as a barrier against the transport of maritime air from the Mediterranean region. Indeed this lack of a tropical-maritime source of moisture may have been a principal reason why Eurasian ice sheets were smaller than the Laurentide body. The latter was supplied mainly from the Gulf of Mexico and other tropical-Atlantic water, whereas most of the ice in northern Eurasia seems to have been nourished from the North Atlantic.

As air masses moving eastward along the southern flank of the ice sheet lost moisture by abundant precipitation, nourishment diminished toward the east, as is shown by evidence that the eastern part of the glacier was thin. For example the Timan Hills, standing hardly more than 200 m above surrounding lowlands southeast of the Barents Sea, created a reentrant more than 450km long in the margin of the combined Scandinavian-Siberian Ice Sheet, at the glacial maximum. Similarly, at an earlier glacial maximum, hills between the Dnepr and Don Valleys, standing only 75m to 150m above the surrounding country, created a reentrant about 450km in length.

In the north the storm track that follows the Norwegian coast continues eastward along the Arctic coast of Europe and Siberia. Snowfall related to this track maintains the existing glaciers on Novaya Zemlya and perhaps also those on islands farther north. Atlantic air masses following this same path may have nourished, albeit scantily, the northern flank of the Scandinavian Ice Sheet.

Asymmetry. The ice sheet was asymmetric. Its maximum southeast radius was nearly 1300km, reaching nearly to Moscow,[1] whereas its west and northwest radii were little more than 300km long. Very likely the explanation lies in the relation of the ice sheet to the deep sea and to the movement of moist air masses that nourished the ice. Because of the short, steep slope on the west side of the mountains, rapid discharge of ice on that side resulted in migration of the ice divide eastward to an equilibrium position. The principal centers of outflow in the southern part of the glacier received abundant nourishment and clearly were able to maintain a flow of ice toward the east. Such flow, combined with precipitation from air masses traveling eastward along the southern flank of the ice sheet, is adequate to explain the rather great southeastward extent of the glacier.

The Laurentide Ice Sheet was generated in highlands and grew westward and southward because maritime air, approaching it first mainly from the southwest and later mainly from the south, nourished the parts that faced in those directions. Lowland centers of radial flow, controlled by snowfall, therefore migrated west and south. In contrast the Scandinavian Ice Sheet, also generated in highlands, grew eastward despite nourishment by westerly winds. This occurred because deep water prevented expansion toward the west, and because the Scandinavian mountains were neither high enough nor extensive enough to create an effective rain shadow in their lee. It is not probable that centers of radial outflow existed in the eastern part of the ice sheet, because nourishment in that region was substantially less than in the western region; apparently it diminished rather steadily toward the continental interior.

The simple concept of a single lowland center or axis of outflow, implied by radial streamline forms, striations, and eskers, influenced early opinion in North America, where an analogous pattern was visualized for that continent. But the North American pattern was different in that the Laurentide Ice Sheet expanded *toward* its sources of moisture, while the Scandinavian glacier, checked at its western margin by a water barrier, had to expand in the opposite direction (fig 23-3).

No water barrier, however, impeded expansion of the ice sheet toward the southwest, in the North Sea region. Except for the deep immediately off the south coast of Norway the water is shallow. With lowered sealevel during glaciation, the North Sea floor, when not actually covered by the glacier, was partly or wholly emergent. Nowhere could the water have been deep enough to float a thick ice sheet. This explains why Scandinavian indicators occur in Britain and in the Netherlands. Certainly during

---

[1] In an earlier glaciation this radius was about 2000km, reaching nearly to Volgagrad.

two glaciations, and probably during the last one, the main ice sheet spread southwestward toward the British Isles, a distance of 650km from the Scandinavian mountains, and coalesced for a time with the much smaller, independent British glaciers. Coalescence is supported also by the presence of the "Channel River" (chap 9), which would not have existed had the Rhine been free to flow northward across the North Sea region.[2] Although the "Channel River" existed during earlier glacial maxima, its existence during the latest maximum is uncertain. On its northeastern side the Scandinavian Ice Sheet in Weichsel time may have coalesced, through a short sector, with Ural Mountains ice; this is uncertain because the critical area is now submerged.

The borders of major drift sheets are traced, as a whole or in part, in fig 23-4. The two older borders (Elster and Saale) trend along the northern bases of central European highlands. One of them, the Saale, penetrates far down the basins of the Dnepr and Don Rivers and the headwaters of the Volga. The younger border (Brandenburg and supposed correlatives) describes a great arc from western Denmark through northern Germany, Poland, and northern European Russia. In its western part this arc approximates today's July sealevel isotherm of 20°, but eastward it swings northeast, crossing isotherms with lower values. The trend reflects decreasing precipitation with increasing distance from moisture sources, as does the trend of the Laurentide drift border on the North American Great Plains.

The earliest Pleistocene marine deposits of eastern England include sediments derived from the southeast and believed to represent one wing of a delta of the Rhine, built at a time when part of the North Sea floor stood above the sea. Possibly later blocking of the North Sea basin by the Scandinavian Ice Sheet diverted the Rhine westward and so created the "Channel River," a segment of which is now the Strait of Dover. This event would have been similar to the early Pleistocene integration of the Missouri River by the Laurentide Ice Sheet.

Viewed on a north polar map (fig 4-8) the Scandinavian and Laurentide Ice Sheets and adjacent glaciers are seen to be grouped around the North Atlantic Ocean. Such a grouping of glaciers which at the last maximum constituted in area a large proportion of all the glacier ice in the northern hemisphere, is not difficult to explain. In the North Atlantic region heat from low latitudes is brought farther north, in air masses and in ocean currents, than anywhere else in the hemisphere. This makes possible abundant precipitation from warm moist air masses.

[2] The "Channel River" existed at a time or times when the Strait of Dover was an isthmus. Shotton (1962) argued that the isthmus existed through at least most of Quaternary time including interglaciations.

SEPARATE GLACIERS

British Glaciers [3]

Glaciers formed repeatedly on British highlands, nourished by precipitation from relatively warm maritime air moving eastward from over the Atlantic. At principal glacial maxima they took the form of large ice caps, merging in a complex pattern. At times they coalesced with the Scandinavian Ice Sheet.

Principal Centers. The chief centers or groups of centers, from which ice flowed outward are named in fig 23-1, identified in fig 23-4, and summarized in the following list.

1. The Scottish Highlands in central and northern Scotland, with summits 1000 to nearly 1350m; today the snowline lies barely above the highest summit. These highlands constituted a cluster of individual centers, most of which lost their identity during a gradual merging of Scottish ice, although islands such as Skye and Mull seem to have maintained local outflow during the greater part of the time.

2. The Southern Uplands of Scotland, south of the Edinburgh-Glasgow lowland, with summits exceeding 750m.

3. The Cumberland highlands ("Lake District"), with summits at more than 900m.

4. The Pennine Chain, with summits at nearly 900m.

5. The mountains of Wales, including several groups scattered between the Bristol Channel and the Irish Sea, with summits between 750 and more than 900m.

6. Mountains of Connemara and Donegal in western Ireland, with summits at 600 to 750m.

7. Mountains of southern Ireland, including several groups from Kerry on the west to Wicklow on the east; summits are at 600 to 750m. Although coalescent with the main British ice during at least one glacial age, these highlands, in the last major glaciation, generated many separate glacier systems.

Although altitudes are comparatively low the snowline is also low, owing to a maritime climate with pronounced summer cloudiness. According to Manley (1951), who attempted a reconstruction of glacial climate in Britain, a very small temperature reduction would again bring glaciers to the higher summits.

[3] General references include Charlesworth, 1957, p. 749–787; W. B. Wright, 1937; Donner, 1957, 1962; G. F. Mitchell, 1960; Penny, 1964; Shotton, 1962, 1967b, 1967a.

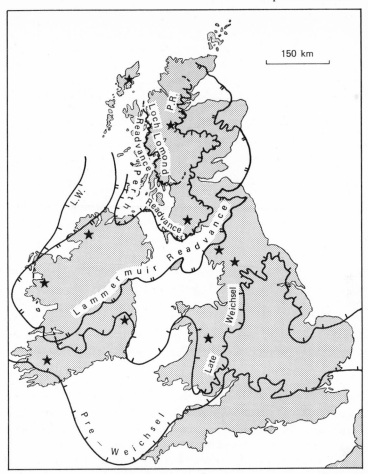

Figure 23-4 Extent of glaciation at various times on British Islands as understood in 1969. Much generalized in some sectors. Data compiled from many
sources, including Charlesworth, 1956; Donner, 1957; Synge and Stephens, 1960;
Mitchell, 1960; Penny, 1964; Shotton, 1967a, 1967b; Sissons, 1967, p. 125.) Stars
show representative centers of outflow of local ice, at one time or another.

These centers are identified mainly from distributions of indicators
and secondarily from striations and streamline forms. Identification is
complicated by the general coalescence of ice from various centers and by
fluctuation in the relative strengths of ice streams from different centers
as the glaciation waxed and then waned. The ice streams were controlled
by topography, having been guided by major valleys and by lowlands
such as the Irish Channel.

Probably British glaciers formed early in each glaciation, enlarging into piedmonts and small ice sheets, which merged into nearly continuous ice. The coalescent ice covered or nearly covered the Outer Hebrides and occupied the continental shelf from southern Ireland nearly to the Shetlands. In places it may have terminated in shelves. The ice in Scotland grew so thick that for a time the ice divide shifted eastward, as is attested by the transport, westward across highland areas, of drift from lower country on the east. This is, in miniature, what happened in Scandinavia. But ice thickness could not have been far greater than the heights of the mountains, for otherwise the centers of radial outflow would not have been so closely related to highlands as is inferred from striations and indicators. Evidence of this kind shows further that ice thickness over the several centers, and energy of outflow from them, varied from time to time, producing drift relations that are complex in the extreme. The various ice masses shouldered each other and jockeyed for position. Ice flowing westward from Scotland, England, and Wales at times not only filled the Irish Sea basin but crowded against eastern Ireland, depositing drift foreign to that country.

Relation to Scandinavian Ice. In at least two glaciations the Scandinavian Ice Sheet invaded eastern England, spreading drift from Norway, Sweden, and the floors of the Baltic and North Seas through as much as 60km inland from the coast. British ice merged with Scandinavian ice and, in eastern England during one glaciation at least, later replaced it.

As British drift of Weichsel age contains no primary Scandinavian components, Scandinavian ice at the maximum of that glaciation evidently failed to reach Britain, just as it failed to attain the limits reached by its predecessors in continental Europe. But it did make contact with British ice off the east coast. This is inferred from indicators which show that British ice was abruptly turned both northward and southward, parallel with the present coast from the Shetland Islands to Norfolk, a circumstance that could only have been caused by an opposing ice sheet acting as a barrier.

The glaciation labeled late Weichsel in fig 23-4 seems to have covered most or all of the till believed to have been deposited earlier in Weichsel time, although earlier outwash, dated around 42,000 BP, is known in western England. In this respect the relative extents and stratigraphic relations of these British drift bodies are not unlike those of drift bodies of Wisconsin age in central North America.

Deglaciation. Late Weichsel deglaciation was marked by at least three re-expansions of British glaciers (table 24-C): the Lammermuir (~14,000 BP), Perth (~12,000 BP), and Loch Lomond (~11,000 BP), corresponding fairly well with late re-expansions of the Scandinavian Ice Sheet. Details of glaciation on British highlands at these times, together with snowline fluctuations, are discussed by Manley (1959). The Lammermuir Readvance, developed mainly from Scottish and northern Irish centers, was however accompanied by the expansion of local glaciers in the Pennines, Wales, and southern Ireland, not shown in fig 23-4. The Loch Lomond Readvance followed a marine transgression over northern British coasts (fig 13-14), and in turn was followed by a second transgression that reached its peak in Hypsithermal time. The transgressions were complicated by local isostatic uplift that accompanied deglaciation, and do not in themselves imply fluctuation in the eustatic rise of sealevel. During the general rise the valley of the "Channel River" was invaded and submerged, and thus Britain was cut off from the continent at some time before 7000 BP. Ireland is thought to have been cut off from Britain at around 8000 BP. The history of these land bridges involves many problems posed by the present and former distribution of floristic and faunistic elements on adjacent lands. Data on the history of these bridges are given by Shotton (1962) and Mitchell (1960).

The Faeroes

The Faeroes (Grossman and Lomas, 1895), an archipelago of 26 islands, are plateaulike, with a general altitude of about 300m, above which mountains rise to nearly 900m. Scandinavian ice did not reach The Faeroes, which were covered by a local ice cap that probably extended outward over the shallow shelf surrounding the islands, into deep water. The upper surface of the ice reached at least 500m above present sealevel and may have stood even higher. Striations show that flow was radially outward. The movement should have been vigorous, for the islands are well situated to have received abundant snowfall. At some time, probably during deglaciation, local valley glaciers were present instead of the ice sheet, as recorded by numerous well-developed cirques, most of them at very low altitudes.

Iceland and Jan Mayen Island

Iceland (Thorarinsson, 1937; Hoppe, 1968; Einarsson et al., p. 312 in Hopkins, ed., 1967), a high and mountainous island with extreme altitudes of more than 1650m, attained a part of its height and many of its major topographic features during the late Cenozoic, in consequence of crustal movements. Although much glacier ice exists today, particularly in the south, during one or more glaciations probably most of the present land and at least some territory now submerged were covered with ice. Striations indicate that late in the last glaciation a composite ice sheet flowed radially out from the central part of the island, coalescing with a lesser ice sheet on the northwestern peninsula. Principal nunatak areas, which themselves carried cirque glaciers, were in the north, northwest, and northeast, that is, on the lee side of the island away from the source of snowfall. Ice thickness varied greatly, ranging from about 350 to 900m, and probably averaging more than 600m. Around parts of Iceland the glacier seems to have terminated in an ice shelf.

Some 500km northeast of Iceland lies Jan Mayen Island (Kinsman and Sheard, 1963), a volcanic mass 50km long at sealevel and with a cone summit 2325m in altitude. Several glaciers now occupy the slopes of the cone, and at least three glaciations have occurred within the last 5000 years. Yet the record of earlier glaciation is dimmed, as in some of the Aleutian Islands, by recent volcanic activity. Even as recently as the time since the last Interglaciation the island may have had a very different form and greater extent. However, climatic conditions that would increase glaciers in Greenland and Spitsbergen would also bring glaciers to Jan Mayen. Hence this island, whatever its form, probably would have been glaciated under such conditions if it possessed sufficient altitude to intersect the snowline.

Spitsbergen, Bear Island, and Zemlya Frantsa Yosifa [4]

Although Spitsbergen today is covered with glaciers extensively, formerly it was probably covered completely, with ice joining together the islands of the archipelago and extending over shelf areas far beyond the present coasts. On the west coast, ice may have been 1000m thicker than it is now. We have already noted the possibility that Spitsbergen glaciers coalesced with the Scandinavian Ice Sheet. That they coalesced with glacier ice centered on Bear Island, 240km to the south, is probable. That island, with a summit altitude of 536m, bears no ice today but was a former local center of radial outflow. It is connected with Spitsbergen by a broad shelf now scarcely 75m below sealevel. During glaciation the shelf would have been emergent; on it Spitsbergen ice could have deployed, merging with Bear Island ice.

The 130 islands that constitute the Zemlya Frantsa Yosifa archipelago are almost completely covered with ice today. It is generally believed that ice was even more extensive at former times, but very little information is available.

The Alps [5]

The massive mountain chain of the Alps, about 1050km long and with altitudes reaching from 4000m to more than 4500m, bore a more exten-

[4] Klebelsberg, 1948–1949, p. 570–578.

[5] The classic reference is Penck and Brückner, 1909. A better organized account is Heim, 1919–1922, v. 1, p. 197–344. Klebelsberg, 1948–1949, p. 660–716 is an excellent

sive body of ice, in glacial ages, than any other highland on the continent south of the Scandinavian Ice Sheet. Because of its size and height it influenced European climates significantly. When at its maximum, the former ice body covered an area estimated at 150,000km², in Switzerland, Austria, Italy, France, and Germany. Many outlet and valley glaciers spread out on surrounding lowlands (fig 23-5), reaching altitudes of 500m on the north and as little as 100m on the south side. The snowline stood as much as 1200m below its existing position.

The Late Würm glaciation began with valley glaciers, which enlarged and overspread divides between them. Thus were formed mountain ice sheets which, over deep central valleys, reached 1500m in thickness and through which only the higher peaks projected as nunataks. In radially oriented valleys glacial erosion was intense. In and beyond the mountain region bulky valley trains were built by proglacial streams, and loess was deposited extensively.

As the Alpine chain consists of many individual ranges, the glacier mass, when near its maxima, was a glacier complex with a number of local centers of outflow. However, shifting of centers away from range crests did occur; geologic evidence of flow southward across the Simplon, St. Gotthard, and other passes implies at least temporary northward migration of a major ice divide.

In the northern marginal part of the chain, at the last major maximum, the snowline stood about 1200m lower than now; at earlier maxima it was 1300 to 1400m lower than now. At the Late Würm maximum the southern, Mediterranean flank of this great glacier complex was better nourished than the northern, continental side. Perhaps this explains why southern ice tongues then reached the positions held by their predecessors before the last major interglacial, whereas northern tongues fell considerably short of the earlier positions. The difference might be the result of uplift of the Alps during the interglacial. A number of readvances punctuated and followed the Late Würm deglaciation (chap 24).

Other Highlands in Continental Europe [6]

Wherever European highlands reached above the snowline during glacial ages, ice formed on them. Thirty-one glaciated areas and groups of areas are shown in fig 23-1. Nearly all were glaciated, chiefly by valley glaciers and cirque glaciers, at the time of the latest glaciation. In some of them direct evidence of earlier glaciation exists. A representative description of one of these areas, the Bayerischer Wald, is that by Ergenzing-

summary. More recent discussions of specific districts include Jäckli, 1965 (Switzerland) and Aubert, 1965 (Jura Mts.).

[6] Data on these areas are brought together in Klebelsberg, 1948–1949, p. 716–740.

Figure 23-5   Map of southern Alps (mainly in Italy) and Po Valley during the Late Würm glacial maximum. Piedmont glaciers are spreading over the

er (1967). Apart from the list in fig 23-1, local glaciers may have existed on the Erzgebirge, Thüringer Wald, and Rhön. The evidence in these highlands is very obscure; if there were glaciers on any of them, they probably antedated the last interglaciation.

DEGLACIATION

The general pattern of deglaciation is suggested in figs 23-1 and 23-4, and is amplified in fig 23-6. The Brandenburg moraine and its correlatives are generally considered to represent the extreme position of Late-Weichsel ice, at roughly 20,000 years BP. Approximate dates of younger moraines, built during deglaciation in Denmark and southern Sweden under strong control by topography, are given in fig 23-6. Continuation

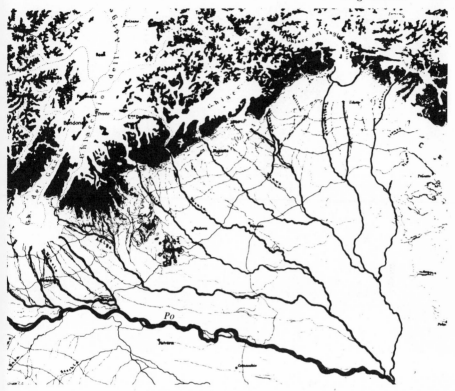

margin of the alluvium in the lowland south of the mountains. (Drawn by
Bruno Castiglioni for Atlante Fisico-Economico d'Italia, 1940.)

of these moraines toward the east is shown in Woldstedt, 1969, p. 6 (West
Germany), Gellert, ed., 1965, pl. in pocket (East Germany), and Roszko,
map, p. 74 in Galón, ed., 1968 (Poland).

Deglaciation seems to have been more rapid on the drier southeastern
flank of the Scandinavian Ice Sheet than on the oceanic western flank.
Readvances of significant amplitude occurred around 15,000 to 14,000,
13,000, 12,000, and 10,500 BP in southerly sectors, but were not necessarily contemporaneous in all sectors of the ice-sheet margin. The latest
of these events followed deglaciation of unknown amplitude, and resulted
in building the Salpausselkä, Central Swedish, and Ra moraines in southern Scandinavia. Probably the indicator fans in Finland (fig 7-18) were
developed mainly around that time.

With rise of the snowline the zone of ablation widened and the ice
sheet thinned rapidly. As a result, ice of Baltic origin in southern Sweden

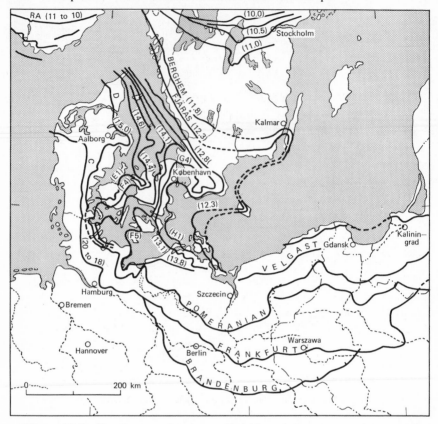

Figure 23-6 Sketch map showing principal end moraines and other lines marking recessional positions of the margin of the Scandinavian Ice Sheet in Denmark, Sweden, and Germany, with labels according to Danish/Swedish and German terminology. Moraines are identified by name or number (E1, etc.) Other numbers are calculated approximate dates, in years x 10^3 BP. Positions and dates in Denmark/Sweden from Mörner (1969); positions in Germany from Woldstedt, p. 257 in Rankama, ed. (1967).

and western Denmark formed a conspicuous lobe, which for a time flowed locally north where, earlier, Norwegian ice had flowed south. Again, in parts of northern Germany and Denmark, thinning resulted in stagnation and, in places, separation of peripheral parts of the glacier. During deglaciation successive water bodies, both lacustrine and marine, fringed the wasting ice sheet. In the east, lakes were ponded by ice in the heads of the valleys that normally drained northwestward into the Barents Sea, the White Sea, Lake Onega, and Lake Ladoga. In the south, in the Polish and German sectors, a whole series of small lakes fringed the ice margin. In time they were replaced, in the Baltic Sea basin, by a succession of lakes and arms of the sea, controlled as in North America by a combination of crustal warping and changes of outlet brought about by deglaciation.

Baltic Water Bodies [7]

Baltic Ice Lake. After the southern margin of the shrinking ice sheet had abandoned the innermost end moraines in northern Germany and eastern Denmark, a succession of small ice-dammed lakes formed along what is now the southern part of the Baltic basin. As the glacier terminus retreated these integrated to create the Baltic Ice Lake (fig 13-4). This large water body spilled out through a succession of channels to the sea, still far below its present level. At least three such outlets led westward to the arm of the sea north of Denmark. A later one led northeastward across the Lake Ladoga district in Finland to the White Sea, which then, owing to depression of the crust beneath the ice sheet, extended farther south than it does at present. The outlets were opened by deglaciation and closed by crustal uplift or by glacial readvance, as were the outlets of the earlier Great Lakes in North America. In consequence the lake level rose and fell. Thus are explained the several strandlines of the lake.

In the Baltic Ice Lake was deposited a part of the varved silt and clay described in chap 15. As the ice melted, the lake widened northward. Readvance of the glacier to the Ra, Central Swedish, and Salpausselkä moraines occurred while the lake existed.[8] At that time the glacier, with a terminus perhaps no more than 50m thick in the Finnish sector, built into the lake much deltaic sediment. This material merges northward into bulky eskers, testifying to a major system of glacial drainage in thin ice. Shortly after the moraines were built, deglaciation in southern Sweden exposed a new outlet so low that the lake was drained down to the level of the sea, still much lower than it is today.

Yoldia Sea. The sea therefore entered the Baltic basin, replacing the lake and constituting the Yoldia Sea named for its inhabitant, a mollusk originally named *Yoldia arctica* but later renamed *Portlandia arctica*, which lives today only in arctic waters. This mollusk, and indeed the entire fauna of which it was a member, lived in the cold sea water off southern Norway while the Ice Lake occupied the Baltic basin. With the draining of the lake and the entrance of the sea across southern Sweden into the Baltic basin, the *Yoldia* fauna migrated eastward. These mol-

[7] General references include Hansen, p. 61–65, Lundqvist, p. 177–181, and Donner, p. 234–244, all in Rankama, ed., 1965.

[8] Relations of the lake to the Salpausselkä are explained in M. Okko, 1965.

lusks, however, flourished chiefly in the western part of the Baltic region; farther northeast the melting ice diluted the salt content of the water to a point unfavorable for them. The Yoldia Sea lasted in the Baltic basin through nearly 1000 years, fashioning a number of distinct strandlines along the continually rising coast.

Ancylus Lake.[9] The connection of the Baltic with the ocean across southern Sweden was gradually raised above sealevel as crustal uplift continued. This again converted the Baltic into a lake, the Ancylus Lake, named after the mollusk *Ancylus fluviatilis*. At first the lake drained westward through a channel across Sweden. The northwestern shore of the lake was the shrinking ice sheet. The lake existed for nearly 2000 years, while the ice margin in northern Sweden shrank so far back that it ceased to be in contact with the lake, also receding from its western shore because of crustal uplift. Early in this time continuing uplift raised the outlet to the level of the Öresund, the channel that now separates Denmark from Sweden. As the original outlet was raised still more, the outlet of the Ancylus Lake was shifted to the Öresund.

Littorina[10] **Sea.** The general rise of sealevel that accompanied deglaciation gradually submerged the Öresund and converted the Ancylus Lake into an arm of the sea. There were some short-lived transitional phases, but the sea at its greatest extent is named the Littorina Sea after the marine snail *Littorina littorea* (the common periwinkle) abundant in its sediments. The confluent water outside the Baltic basin has been called the Tapes Sea, after the mollusk *Tapes decussatus*. The *Littorina* fauna implies distinctly warmer water than does its marine predecessor, the *Yoldia* fauna; the implication agrees with pollen evidence set forth in chap 24. It is, moreover, a saltwater fauna, implying a wide, deep opening to the ocean.

After having endured more than 3000 years, the Littorina Sea merged imperceptibly into the present Baltic Sea, in which an earlier *Limnaea* phase and a later *Mya* phase are recognized. The stratigraphic transition from *Limnaea* to *Mya* implies that the Baltic water was becoming increasingly brackish, as its connection with the ocean was narrowed and shoaled by uplift.

Marine Transgression

Marine transgressions such as those represented by the Yoldia and Littorina Seas occurred also along the Atlantic and Arctic coasts of Europe.

[9] A review by Fredén (1967) reveals in an interesting way the large number of researchers who have been concerned with the problems of this lake through a period of more than 80 years.

[10] This spelling is correct (Bequaert, 1943). The spelling *Litorina* still persists, however, even in some recent publications.

During deglaciation the retreating ice was replaced by water in most fiords and other valleys along the outer coasts of Norway, Finland, and European Russia. Conspicuous, fossil-bearing marine sediments cover lowlands along these coasts. An example is the White Sea-Lake Onega lowland, just east of the boundary between Finland and USSR. When the ice sheet stood at the Salpausselkä, its northeast margin fringed that lowland, bathed in sea water that extended far south of the present Barents Sea coast.

Late Phase of Deglaciation

Around 9500 BP the Scandinavian ice body was still an ice sheet, albeit thin and possibly stagnant through much of its periphery. It stood over the country west of the Gulf of Bothnia and was firmly connected with a network of glaciers in the mountains west of it. By 8000 BP the lowland ice sheet had disappeared and glaciers persisted only in the mountains. Shortly thereafter increasing warmth had raised the tree limit in the mountains to 200m above that of today, and probably also had done away with most of the remaining glaciers.

ECOLOGY [11]

European Vegetation Today

The simplest way to visualize environments in Europe during the last glacial age is to compare them with those of today. Figure 23-7 shows the broad distribution of vegetation in Europe at present. The pattern is controlled principally by temperature and moisture, which in turn are determined by latitude, position with respect to sources of moisture, and major topographic features. In terms of area the principal control by temperature is exerted by differences of latitude. In this respect the vegetation of western and central Europe forms a pattern consisting of E-W belts that succeed each other from the coast of the Barents Sea to the Mediterranean. The belt of (1) arctic tundra is succeeded in turn by (2) boreal forest, (3) northern coniferous and broadleaf forest, (4) temperate forest, and (5) Mediterranean vegetation.

These belts, however, are distorted by massive highlands which alter the climate in their vicinities and introduce local patterns based on altitude. The Scandes, Alps, Pyrenees, and other highlands impose patterns based on the trends of the highlands. In eastern Europe and Asia Minor the E-W pattern of vegetation is distorted by another factor, decrease in

[11] See the summary treatments, emphasizing vegetation, by Hammen in Turekian, ed., 1971, and emphasizing climate, by H. E. Wright, 1961a.

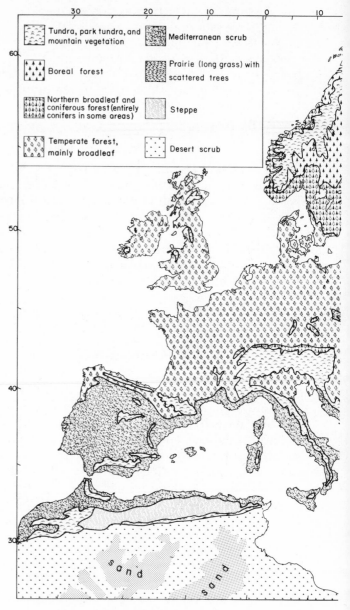

Figure 23-7 Sketch map showing generalized distribution of vegetation in

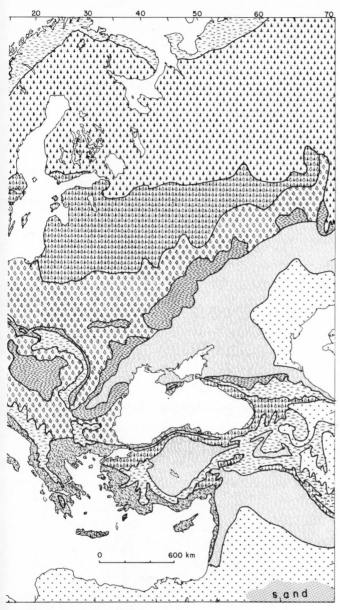

Europe at present. (Compiled from various sources.)

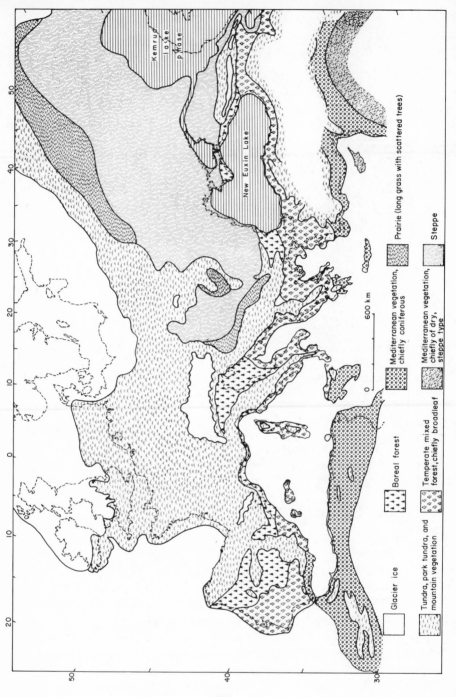

Kemrud Lake

phase

New Euxin Lake

600 km

Glacier ice

Tundra, park tundra, and mountain vegetation

Boreal forest

Temperate mixed forest, chiefly broadleaf

Mediterranean vegetation, chiefly coniferous

Mediterranean vegetation, chiefly of dry, steppe type

Prairie (long grass with scattered trees)

Steppe

precipitation toward the east. Western forests are succeeded by prairie, which occupies an irregular NE-SW belt, followed by belts of steppe and of desert brush. This pattern, like that farther west, is distorted in the vicinity of high mountains such as the Caucasus, Taurus, and Zagros ranges.

The pattern as a whole is fully adjusted to a predominantly temperate environment in which glaciers are negligible and in which prevailing westerly winds bring moisture from the Atlantic and distribute it, with decreasing intensity, eastward across the land mass. Winters are cold except along the Atlantic coastal fringe and in the Mediterranean region, and summers in the latter region and in central and eastern Europe as well, are continental with relatively high temperatures and minimum cloudiness.

Ecologic Zones at a Glacial Maximum

When we compare this overall pattern with that at the time of the Weichsel/Würm maximum as reconstructed in part from fossil pollen and in part by deduction, we find a striking difference. Figure 23-8 presents one interpretation of the glacial-age pattern. Another (Frenzel, 1967, p. 213) diverges from it in some respects, but the divergences are not major. Most of them arise rather from differences in the drawing of boundaries across gradational sequences. Neither reconstruction can be accurate in detail because data are not abundant, but together they present a picture that is broadly acceptable.

As we have seen, the northern half of the land mass was covered with ice and the principal mountains carried either ice caps or abundant valley glaciers. Vegetation was therefore confined to little more than half the area it occupies now. Like that of today, the vegetation consisted of tundra (in the broadest sense) grading southward into forest and eastward into steppe. There was far more tundra, prairie, and other vegetation with few trees or none at all, and far less forest than at present.

Tundra.[12] Tundra and park tundra (with much birch) fringed the glaciers. Along the margin of the Scandinavian and British glaciers such vegetation formed a wide belt. Beneath it was permafrost, inferred largely

[12] A good summary of ecologic conditions is given by Shotton, 1962; see also West, 1968, p. 292.

Figure 23-8 Sketch map showing generalized distribution of vegetation in Europe at the Weichsel/Würm maximum. Sealevel is at −100m; the Black Sea and the Caspian Sea are interconnected lakes. (Adapted from an unpublished updated version by J. Büdel, 1969, of the sketch map in Büdel, 1949.)

from the presence of ice-wedge casts. Mean annual temperature values are probably 10 to 12 degrees lower than those of today. Winters were extremely cold, with cool, cloudy summers in the west, grading eastward into warmer, more continental summers. The climate of central Europe was especially severe because that region was barred from warmer Mediterranean influence by the presence of the ice-capped Alps. This situation had no counterpart in eastern and central North America.

Northern France and the ice-free parts of England were subject to an intense frost climate. Although frost cracks are less abundant than in central Europe, effects such as frost-stirred ground and frost-wedged rubble are common, and gelifluction sediments are so widespread that gelifluction must have been well-nigh universal.

Throughout Europe much of the tundra, and prairie as well, was underlain by loess, then accumulating actively along and in the lee of rivers that were depositing outwash sediments. Even rivers of nonglacial origin seem to have been aggrading conspicuously, possibly because of active gelifluction on their watersheds.

Treeless vegetation covered the tops of the higher mountains wherever glaciers were not present; at lower altitudes and on the lower mountain tops it graded into park tundra and in places downward into boreal forest or prairie. In this form treeless vegetation extended down the Apennines into southern Italy, through Dalmatia into Greece, and over the Carpathians and the Caucasus. According to Bonatti (1966), steppe vegetation in the northern Mediterranean region implies that the climate was semiarid.

Differences between the climates on the two sides of the Alps were marked. The north side was covered with tundra under a severe frost climate. On the south the belt of tundra was narrow; within a short distance southward it graded into cold steppe and boreal forest, maintained by the less-cold climate of the lower altitudes in the Mediterranean region.

Toward the east the belt of tundra north of the Carpathians narrowed and extended northeastward across European Russia. It graded southward into park tundra and cold steppe with scattered trees. South of this, in turn, from the Hungarian plain eastward around the southern flank of the Carpathians, lay a belt of tundra developed on loess, with gallery forests along major rivers. This belt was flanked on adjacent mountain slopes and widely along its southern margin by park tundra and cold steppe with scattered trees. This vegetation extended along the Adriatic coasts and through the Balkans to the Black Sea.

Among the tundra animals were reindeer, musk-ox, mammoth, woolly rhinoceros, arctic hare, various lemmings, tundra vole, and other rodents, with moose, giant "elk", and carnivores such as cats and hyena in the more wooded areas. Toward the east, also, were saiga antelope and *Bison priscus*. In the meadows around the fringes of the Alps and in the higher parts of other European mountains were chamois, mountain sheep, alpine hare, marmot, and cave bear. Such tundra mammals as reindeer and mammoth were abundant as far south as southwestern France. At many localities throughout Europe fossil mammal faunas occur mixed. Such mixtures are expectable under changing climatic conditions, with pioneer taxa entering an ecozone while other taxa, formerly common in the zone, linger on. Horse, cattle, and bison, not mentioned above, are not peculiar to any one of these zones, but occur scattered through all of them.

Forest. In western and southern Europe the tundra was succeeded by patches of forest or by areas of prairie, marked by long grasses with groves of trees. Forest became nearly continuous only at the latitude of the Mediterranean, where it formed a belt along the coast, ranging in character from boreal to temperate assemblages. The Mediterranean region as a whole, however, is believed to have been much drier than it is today. Forest extended from Greece around the southern coast of the Black Sea (then a lake) and back, westward, to the Crimea. Fossils from the European forest environment near the Weichsel/Würm maximum include deer, goats, pig, beaver, bear, wolf, fox, lynx, badger, marten, and wolverine. This collection, of course, was differentiated somewhat according to forest type.

Steppe. In eastern Europe and Asia Minor the tundra on the north and forests on the west were succeeded by steppe, in places with intermediate strips of prairie. A simple belt of steppe extended from the Hungarian plain east of the alps around the end of the Carpathian Mountains, and so northeastward, as a band as much as 800km wide across European Russia, reaching from near Moscow to the Caspian Sea. At that time the Caspian was greatly enlarged, extending nearly 400km north of its existing northern shore. Across this vast belt of terrain covered with short grasses the channels of major rivers were fringed with gallery forests. The steppe was inhabited by such mammals as ibex and other antelopes, horse, ass, and an assortment of rodents including marmot, suslik, jerboa, and various mice.

Latitudinal Shift of Zones. Another look at the two vegetation maps makes it clear that at the Weichsel/Würm maximum, temperate forests now concentrated in Europe were growing as far south as the African coast of the Mediterranean. The displacement in latitude of the belt of tundra amounted, in lowland areas, to as much as 20°, whereas the displacement of temperate forests in lowland areas was more nearly 10°. The basic displacement can be attributed mainly to secular decrease of temperature, whereas the additional displacement of the tundra belt can reasonably be attributed in large part to the presence of great bodies of ice adjacent to the tundra on the north. At the glacial maximum the belt of forests was compressed between the tundra on one side and the steppe and desert of the Saharan region on the other.

Changes During Deglaciation

As in North America, amelioration of climate and shrinkage of glaciers were accompanied by appropriate shifts of ecologic zones.[13] Such shifts occurred rapidly, as is indicated by agreement among C^{14} dates of specific pollen zones at widely separated localities (table 24-C). Pioneer elements of the tundra assemblage occupied new terrain almost as quickly as it was uncovered by melting ice, and these were succeeded by other assemblages appropriate to the postglacial climate. Fluctuations that resulted in glacial readvances met with immediate responses in the character of local floras. Temperatures, as indicated by pollen data, rose to a peak at some time or times between about 7000 and about 5000 years BP. Another conspicuous although less warm peak was reached around 900 BP.

The ecology of interglacial times is generally assumed to have resembled that of today. Hence the appearance of fossils that indicate a climate substantially different from that of today may pertain to an interglacial maximum that has not yet been reached by the postglacial climate. One example is mentioned by Ložek (p. 385 in Cushing and Wright, eds., 1967; Ložek 1969). He stated that terrestrial mollusks in interglacial strata in Czechoslovakia indicate substantially greater precipitation than now falls there, as well as a more oceanic climate with milder winters and relatively wetter summers, though with increasing dryness toward the end of each interglacial. Another possible example is the mild-climate strata near Toronto, mentioned in chap. 20.

[13] Representative discussions of these changes include Lüdi, 1955, for the northern Alps; P. Müller, 1957, for the southern Alps; Beug, 1967b, for southern Europe; West, p. 48 in Rankama, ed., 1967, for Britain; various authors (p. 46–61, 120–123, 185–188 in Rankama, ed., 1965) for Scandinavia; and Hammen et al., 1967 and de Jong, p. 359–370 in Rankama, ed., 1967, for the Netherlands. A detailed account of re-immigration of lake fishes and marine mammals is given in Segerstråle, 1957.

CHAPTER 24

Stratigraphy of Europe

INTRODUCTION

Any attempt to summarize the Pleistocene stratigraphy of Europe within a brief space would be an act of temerity even for a European. For a North American it is little short of foolhardy; yet some sort of summary, however sketchy, is essential. Although full of interest, the undertaking involves a voluminous literature published in many languages and expressing a variety of opinions as to the significance of critical evidence. Furthermore, the record of human cultures extends well back into the Pleistocene of Europe and plays a significant part in stratigraphic interpretation. Where tested objectively, the data of prehistoric cultures have been found to agree with data from fossils and physical features, suggesting that cultures are no more steeply time transgressive than are the related faunal and floral zones, and that assemblages of artifacts are therefore useful as stratigraphic horizon markers. Notwithstanding, the merit of assessing the stratigraphic sequence solely from the point of view of geologic criteria is that the strengths and weaknesses of specific correlations stand out more clearly. Such an assessment is attempted here, and it is left to better qualified persons to weigh the cultural evidence against the geologic, where discrepancies exist.

Also, the present summary is independent of any theory of the cause of Pleistocene climatic changes. Some theories impose on the sequence of changes rigidities that impair objectivity and lead to attempts to fit the stratigraphic facts into some preconceived pattern rather than let the facts determine what hypotheses are permissible.

Finally, brief treatment is hindered by the lack of a terminology of glacial and interglacial stages that can be applied with confidence to all parts of Europe. Because of this lack, use of the terminology applied by

621

Penck and Brückner to the region of the Alps alone has been extended by authors to all parts of Europe without, however, a secure basis of stratigraphic data. Such extension has been deplored repeatedly, but it continues in use, apparently because it is convenient. The convenience, however, is false, for the use of terminology that is not well supported by the stratigraphic facts tends to stifle inquiry and to delay improvement of the stratigraphic column. In some countries geologists aware of this danger have developed stratigraphic names restricted to local sequences. Progress lies in this direction, and in the present chapter we have adopted a policy of caution.

Northern Continental Europe [1]

We can begin logically with the large northern part of Europe that was overspread by the Scandinavian Ice Sheet and with the belt immediately south of the ice. There the stratigraphy is comparatively well known and is fairly well correlated from one sector to another. The stratigraphic sequence in the west is represented by the columns for the Netherlands and for northern Germany and Denmark in tables 24-A and 24-B. The units shown there are essentially cold and temperate climatic units except that for the Netherlands a sequence of rock-stratigraphic units has been established, and this is included. For this western region the sequence at present consists of 11 major units of post-Pliocene age. The youngest unit, the Weichsel, apparently spans at least the last 70,000 C^{14} years, but the time spans of the older units are still poorly known. The cold and temperate units alternate. The latter are identified or supported by pollen data; among the former, the top three are represented by glacial drift whereas the lower units are identified principally on a basis of fossils. Possibly in the cold times of pre-Elster date the ice sheets did not extend as far south as the southern shores of the North Sea and the Baltic, and so left no direct records.

Subdivision of the Pleistocene. In table 24-A these units are grouped into Lower, Middle, and Upper Pleistocene in more than one way because there is no generally accepted standard. Such terms are normally built into the geologic column with definitions based on fossils. In this case, however, in their apparently earliest use the names (in the form Alt-, Mittel-, and Jungdiluvium) were based mainly on physical stratigra-

---

[1] These general references are suggested: Fennoscandia: Rankama, ed., 1965; Belgium: Zagwijn and Paepe, 1968; Netherlands: de Jong in Rankama, ed., 1967; West Germany: Woldstedt in Rankama, ed., 1967; East Germany: Gellert, ed., 1965; Poland: Galón, ed., 1968; European USSR: Gerasimov, ed., 1965; Gromov and Nikiforova, eds., 1969; Goretskiy and Kriger, eds., 1967; Markov, Lazukov, and Nikolaev, 1965; Markov and Velitchko, 1967; Moskvitin, 1970.

phy and only secondarily on fossils, and were applied to a very restricted area in the Bavarian Alps (Soergel, 1924). Their purpose seems to have been convenience in grouping units having climatic implications rather than the creation of provincial faunal or floral ages. They do not seem ever to have been strictly defined, and so in stratigraphic tables their boundaries differ from one author to another. These variations are reflected in certain columns in table 24-A. The Lower Pleistocene includes relict taxa of Pliocene character both among mammals and among plants, whereas the Middle Pleistocene does not. The Upper Pleistocene contains many more boreal mammals than does the Middle Pleistocene.

Pre-Elster Nonglacial Sediments. At various localities in Denmark interglacial sediments occur as erratic glacial boulders either incorporated in or beneath till. Some contain early Pleistocene fresh-water mollusks; others contain a flora, possibly Pliocene material that was redeposited in Pleistocene sediments. It is thought probable that all are of Cromer age.

In Carpathian Poland the work of Szafer (1954; 1953) on pollen and macroflora bridge part of the wide gap between the Pliocene, there represented by the Krościenko Floral Age, and the sediments of the earliest known glaciation (Podlas) in Poland. Temperature fluctuations are discerned in the sequence (fig 16-18), which, however, is possibly not yet complete.

Among pre-Elster sediments that are thought to represent cold-climate episodes are layers exposed between the Rhine and Maas Rivers near Brüggen and Venlo. A "Brüggen Glacial" unit is thought to be among them. Details are given by Boenigk (1970).

Elster Drift. The three highest cold units, represented by till at the surface in one district or another, were named by Keilhack after three German rivers, Elster, Saale, and Weichsel. Of these units the Elster has the smallest extent at the surface, being widely covered by the succeeding Saale drift. In Weichsel time the ice sheet failed to achieve its earlier, Saale extent in any sector. In this respect the relations are rather like those between the Illinoian and Wisconsin drift layers in North America east of the Mississippi River. The Elster drift continues eastward through Poland, where it is referred to the Kraków (also known as the South Polish) glaciation. Farther east, where it is called the Oka drift, it passes beneath the succeeding Dnepr drift. Little is known of its occurrence in Russia, if indeed it is present at all.

Elster till, locally reaching 30m in thickness, is deeply oxidized and leached and is so extensively dissected that in some districts it consists of isolated separate patches. Its surface expression is erosional rather than morainic; it lacks closed basins. Locally the till has been eroded to form a residual boulder pavement, overlain by Saale drift. In some districts there

Table 24-A
Selected Pleistocene Stratigraphic Sequences in Europe

Britain (East Anglia) (Stages)[d]	Netherlands (Rock–stratigraphic units)	Netherlands, North Germany, and Denmark	Germany: Rhine Terraces	Poland
Weichsel (Devensian)	*Kreftenhaye Fm*	Weichsel	Younger Low T. / Older Low T. **(Upper)**	North Polish (Baltic): Leszno, *Brørup*, Szczecin
Ipswich	Schouwen Fm (M) / Asten FM Eem Fm (M)	*Eem* (M, in part)		*Eem* (M, in part)
Gipping (Wolstonian)	Drente Fm — Veghel and Urk Fms	Saale: Warthe, *Treene*, Drente	Krefeld Middle T. / Lower Middle T.	Middle Polish: *Kujaw interstade*, *Masovian interstade*, Drente
Hoxne	Eindhoven Fm / Holstein Fm (M) / Emmen Fm	*Holstein* **(Middle)**		*Masovian*: Uppermost, Upper, Middle, Lower
Lowestoft Stade / Anglian *Corton Interstade* / Gunton Stade	Potclay Fm	Elster: II, I	Middle Middle T. / Upper Middle T.	South Polish (Kraków): II *interstade*, I
Cromerian / Beestonian / Pastonian	Sterksel and Enschede Fms	"*Cromer*": II *Bilshausen* and other, I sediments		*Cromer*; Mizerna IV, III/IV, III
Baventian (M)		Menap	Younger Main T.[b]	Podlas; Mizerna II/III
Antian (M)	Kedichem and Harderwijk Fms	*Waal*		Mizerna II
Thurnian (M)		Eburon **(Lower)**	Older Main T.[b]	
Ludhamian	Tegelen Fm	*Tegelen*		Mizerna I/II, I
Waltonian / (Older Red Crag) (M)	Icenian sequence (M) / Kiesel oolite / Merksem Fm (M)	Brüggen	High terraces	
Coralline Crag [Pilo.]	Reuver Clay [Pilo.]	Kaolin sand, etc. [Pilo.]	terraces [Pilo.]	Krościenko, Grywald, & Huba floras [Pilo.]
West, 1967, p.286; Shotton & West in George et al., 1969	Heide & Zagwijn, 1967; Zagwijn & Paepe, 1968; Woldstedt, 1969, p.32	Woldstedt, 1969, p.32; Gellert, 1965	Jong, p.408 in Rankama, ed., 1967; Woldstedt, p. 288 in idem.	Rühle & Mojski, 1965, pl. 3a; Różycki, S.Z., 1967b; Szafer, 1953; 1954

Note: Interglacial and interstadial units are shown in *italic* type. (M) = marine unit. Correlations are suggested only, especially in lower part. [a]Authorities disagree as to the existence of a widespread drift layer in this position. [b]May be considerably younger (Montfrans & Hospers, 1969). [c]Essentially the classical correlation. For an alternative scheme see text. [d]Correlations in lower part of column are uncertain and provisional.

Table columns under "**S o v i e t   U n i o n**" span the European, Provincial mammal ages, and Sea of Azov Region columns.

	European	Europe & Siberia Provincial mammal ages	Sea of Azov Region — Strata		Loess	Alps[c]
Valdai	Ostashkov / *Mologo–Sheksna* / Kalinin	Upper	Loess / *Veselo–Vosnesensk Soil* / Loess	Steppe / / Desert brush	Younger Loess with *Soils*	Würm — Lower Terrace gravels
	Mikulino — *Mga* (M)	Paleolithic	*Risklinenniy Soil*	Wooded steppe	*Soil*	R/W — *Moosburg gravels*
(Interglacial sequence)	Moskva / *Odintsovo* / Dnepr	Khazarian	Loess	Steppe / Woodland	Older Loess / *Soil* / Older Loess	Late stade / Riss / Early stade (Upper terrace gravels)
(Glacial sequence)	*Likhvin* — North and Old Euxinian transgressions (M)	Singilian	*Sea of Azov Soil*	Steppe / Sea of Azov Soil	*Soil*	M/R
(Glacial sequence)	Oka?[a] — Chaudian and Bakinian transgression (M)	Tiraspolian	Loess	Steppe	Older Loess	Mindel — Late stade / Early stade (Younger cover gravels)
	Apsheron and Gurian beds (M)		*Platovo Soil*	Steppe	*Soil*	G/M
	Upper Kvialnikian beds	Tamanian	Platovo Terrace "Scythian" Clay	Desert brush / Woodland		Günz — Older cover gravels
	Khaprovian and Ergenian beds; Upper and Middle Akchagylian beds (M); *Lower Kvialnikian beds*		*Margaritovka Terrace*	Desert brush		D/G
		Khaprovian	↑ Khapry Terrace ↓			Donau — Donau gravels
	Lower Akchagylian beds	Moldavian				Biber? — "High gravels"?
			Zone of Terra Rossa			
	Gromov, V.I., Vangengeim, E.A., Kind, N.V., Nikiforova, K.V., & Ravsky, E.I., unpubl.	Flerow & Sher (Flerow in Turekian, ed., 1971)	Dobrodeev, 1969		Woldstedt, p.288 in Rankama, ed., 1967	Wolstedt, p.269 in Rankama, ed., 1967; Penck & Brückner, 1909

Table 24-B
Selected Weichsel Stratigraphic Sequences in Europe

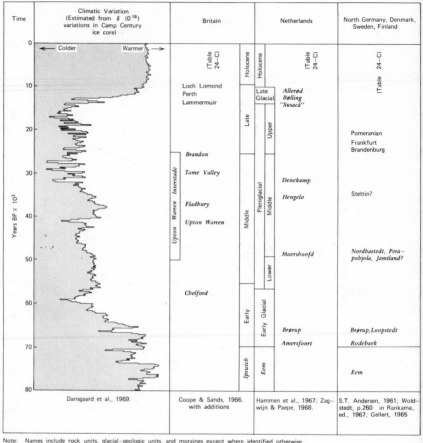

Time	Climatic Variation (Estimated from δ (0¹⁸) variations in Camp Century ice core)	Britain	Netherlands	North Germany, Denmark, Sweden, Finland

Note: Names include rock units, glacial–geologic units, and moraines except where identified otherwise.
Interstadials and interglacials are in italic type.

is no till at all; the drift is represented merely by scattered erratics of Scandinavian origin.

Along the northern flanks of highland areas in Germany, lake deposits occur along the margin of Elster drift. They contain zones of intense crumpling and faulting that indicate episodes of temporary advance, when the glacier margin rode forward over the lake floors. At least two re-expansions of the ice sheet have been inferred from evidence of this kind. The lakes formed where the ice sheet blocked north-flowing streams. In the valleys of some north-flowing rivers are remnants of al-

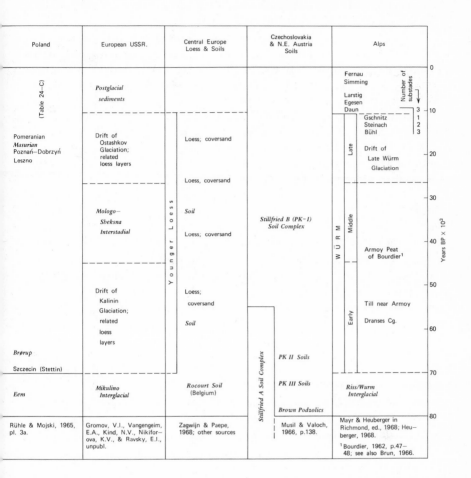

Poland	European USSR.	Central Europe Loess & Soils	Czechoslovakia & N.E. Austria Soils	Alps	
(Table 24–C)	*Postglacial sediments*			Fernau Simming Larstig Egesen Daun	Number of substades
				Gschnitz 1 Steinach 2 Bühl 3	
Pomeranian *Masurian* Poznań–Dobrzyń Leszno	Drift of Ostashkov Glaciation; related loess layers	Loess; coversand		Late Drift of Late Würm Glaciation	
		Loess, coversand			
	Mologo– Sheksna Interstadial	Younger Loess Soil Loess; coversand	*Stillfried B (PK–I) Soil Complex*	Middle WÜRM Armoy Peat of Bourdier[1]	
	Drift of Kalinin Glaciation; related loess layers	Loess; coversand Soil		Early Till near Armoy Dranses Cg.	
Brørup			Stillfried A Soil Complex PK II Soils		
Szczecin (Stettin)	*Mikulino Interglacial*	*Rocourt Soil (Belgium)*	PK III Soils	*Riss/Wurm Interglacial*	
Eem			Brown Podzolics		
Rühle & Mojski, 1965, pl. 3a.	Gromov, V.I., Vangengeim, E.A., Kind, N.V., Nikifor– ova, K.V., & Ravsky, E.I., unpubl.	Zagwijn & Paepe, 1968; other sources	Musil & Valoch, 1966, p.138.	Mayr & Heuberger in Richmond, ed., 1968; Heu– berger, 1968. [1] Bourdier, 1962, p.47– 48; see also Brun, 1966.	

luvial fills, now forming high, much-dissected terraces related to such ice blockades. Two distinct glacial advances within the Elster glaciation are inferred also from the distribution and provenance of erratics in the drift. Pollen data give indications of treeless vegetation, possibly tundra, in Poland.

Holstein Interglacial Sediments. [2] At localities in the Netherlands, Denmark, northern Germany, and Poland the Holstein Interglaciation is

[2] Summaries and detailed data in S. T. Andersen, 1963; Kempf, 1966.

represented by marine and (or) alluvial and lacustrine sediments. In places these sediments occur between Elster till and Saale till. The name Holstein is derived from the German region in which such sediments are abundant. During that interglaciation the Holstein sea transgressed more than 100km, in some valleys, over what today is land. The water body occupied the North Sea and Baltic basins and was essentially an ancient counterpart of the Littorina Sea. Partial sections, no one of which represents the entire interglaciation, indicate a climate changing from cold-temperate to temperate. Among marine mollusks are *Portlandia* (*Yoldia*) *arctica, Arctica islandica,* and *Saxicava arctica;* also less boreal forms such as *Littorina littorea.* Foraminifera related to the thermal maximum indicate temperatures like those of the North Sea today. The terrestrial facies of the Holstein Interglaciation contains the "warm" mullusk *Paludina diluviana* and the water fern *Azolla filiculoides.* The pollen record in Denmark indicates a transition from pioneer to more stable vegetation under a temperate, oceanic climate. Deep weathering of Elster drift is another feature of the Holstein Interglaciation. The Holstein Interglaciation in its type region is double, with cold-climate features sandwiched between fossil-bearing temperate-climate sediments (Ducker, 1969, p. 46).

Except in long borings through undisturbed sediments, the pollen record from base to top of an interglacial sequence is rarely complete, although an interglaciation can be pieced together from two or more incomplete though overlapping parts. According to Turner and West (1968), four recognizable zones occur in each interglaciation in northwestern Europe. These are (1) *pre-temperate,* a boreal zone featuring birch and pine, (2) *early-temperate,* with mixed-oak forest dominant, (3) *late-temperate,* marked by late immigrants, and (4) *post-temperate,* another boreal zone repeating zone (1). If these zones are recognizable in all the interglaciations they are obviously very useful for correlation.

Saale Drift. The distribution of the extensive Saale drift is generally well defined, though in places it is not differentiated clearly from the Elster. Despite considerable mass-wasting and dissection, the Saale drift retains some morainic topography, including subdued end moraines. Sections near Leipzig consist of three Saale tills separated by alluvial and lacustrine sediments. These and other sections, plus end moraines, lead to the belief that the Saale glaciation was complex and can be subdivided into two chief parts, called, in western Europe, Drenthe and Warthe. In Poland and USSR other names are applied to them. In a number of places the Drenthe and Warthe tills are separated by interstadial sediments, the Treene sediments, once thought to be a unit of interglacial rank. According to that earlier view, later abandoned, the Warthe till was thought to occupy a position at the base of the Weichsel whereas the

Drenthe represented the entire Saale.

In places massive bodies of outwash and ice-contact stratified drift are present in the Saale sequence. Also, along North German rivers such as the Weser and Lehne, and along the Rhine and Maas in the Netherlands as well, are remnants of a sand-gravel fill known as the "Middle Terrace." Its sediments were derived from the south, upstream. Locally it contains a "warm" vertebrate fauna near its base and a "cold" mollusk fauna near its top; its northern end, at the Saale drift border, includes glacial-lake rhythmites and is overlain by Saale till. Formed in much the same manner as the Minford Silt in Ohio, the fill is supposed to represent a transition from the Holstein Interglacial to the Saale Glacial maximum, as the climate cooled and rivers were dammed by the expanding ice sheet. Remnants of an older and higher fill, doubtless formed in an analogous manner, are similarly related to the Elster till. Finally, a younger fill, the "Low Terrace," is related in the same way to the Warthe till.

In Poland the Middle Polish (Saale) drift is subdivided into three glacial units (with tundra) separated by two interstadial bodies of alluvium, one of which contains horse, cat, cave bear, rodents, and Acheulean implements. In European Russia the same drift consists of two glacial units (Dnepr, Moskva) separated by an interstade (Odintosovo) marked by temperate broadleaf forest.

Eem Interglacial Sediments. The most recent interglaciation, the Eem, is named for a small stream in the eastern part of the Netherlands. Originally the name was applied only to marine strata, but it was later extended to embrace the entire interglacial sequence. It is represented by alluvium, peat, weathering zones, and a conspicuous marine facies. The latter is traceable from the Netherlands through Denmark, northern Germany, Poland, and the Baltic countries, to the USSR, where Baltic seawater spread nearly 200km southeast from the Gulf of Finland and extended northeastward as a strait that connected the Baltic with the White Sea and thus converted most of Fennoscandia into an island (fig 24-1). Marine mollusks indicate temperatures higher than those of today, and thus imply an ice-free Arctic Ocean and probably complete deglaciation of Scandinavia. In the USSR, sediments of Eem age carry the local name of Mikulino.

In some districts, as in Friesland on the southeast coast of the North Sea (Sindowski, 1958, p. 172) entire transgression and regression facies are known. It appears that the peak of transgression postdated the thermal maximum read from pollen data. The fauna in western Europe is distinctive, testifying to water as warm as or warmer than that in the southern part of the North Sea today. The index fossil is an extinct clam, *Tapes senescens*. The Eem Sea, like the Holstein Sea, was analogous to the Lit-

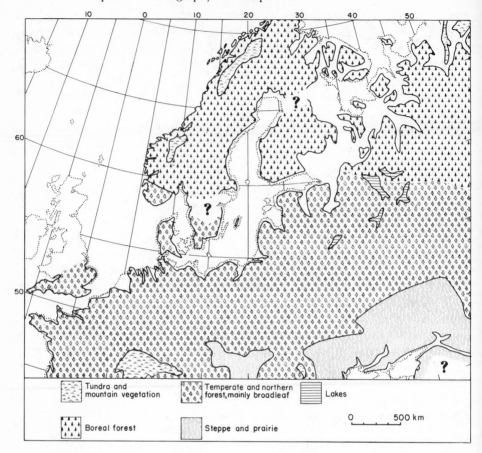

Figure 24-1 Sketch map showing approximate coastline and generalized distribution of vegetation in part of Europe in the middle part of the Eem Interglacial Age. (Adapted from Gromov and Nikiforova, 1969.)

torina Sea of recent date. Near the top of the Eem sequence, as for instance in a boring at Skaerumhede in northern Denmark, the fauna records a marked decrease in water temperature and by inference the approach of another ice sheet. At Husum in Schleswig both Eem and Holstein marine sediments, separated by Saale outwash, occur in a single section.

Pollen profiles show all or parts of the sequence *cold→warm→cold*. When at their maximum, temperatures may have been comparable to those of the postglacial thermal maximum of a few thousand years ago.

Pollen diagrams of the Eem sequence differ from those of the Holstein as to the proportions of the various kinds of plants present, so that in many localities the two units can be differentiated independently of the identification of an intervening till.

In Poland the pollen record begins with spruce forests, succeeded in time by broadleaf forests with larch. The mammal fauna included Merck's rhinoceros, *Bison priscus,* cave bear, and various rodents. Farther west, in central Europe, mammals included beaver, wolf, lynx, cave bear, weasel, marten, rhinoceros, the elephant *Palaeoloxodon antiquus,* horse, ox, bison, giant elk, deer, moose, and wild pig.

Weichsel Stratigraphy. Across much of northern Europe, strata of Weichsel age give a first impression of simplicity, implying a single expansion and fluctuating shrinkage of the Scandinavian Ice Sheet. This first impression arises from the relatively great extent of late-Weichsel drift, which, in western Europe at least, conceals older Weichsel strata beneath. From detailed study of strata (in part cores from borings) in the Netherlands and Denmark, it became evident that the widespread surface drift, now known to date from within the last 25,000 years, belonged to a Late-Weichsel glaciation and that earlier cold events of post-Eem date could be recognized. The earlier climates were inferred mainly from windblown sand, frost-action effects, and pollen data, stratigraphic positions were evident in cores, and time calibration was made possible by C^{14} dating. As a result the record of Weichsel climatic changes in the Netherlands-Denmark region is outstanding (table 24-B; fig. 16–10; Hammen in Turekian, ed., 1971). Reinforced by many compatible data from nearby Britain; it constitutes an unusually firm sequence. It includes, as table 24-B shows, a number of short fluctuations during the Late Weichsel deglaciation; still other, Holocene fluctuations are set forth in table 24-C. We have placed quotation marks around the name "Susacá" in the Netherlands sequence, because the name is derived from an occurrence in the Colombian Andes rather than from a local occurrence of interstadial climatic indicators. In fact this climatic fluctuation thus far has been recognized only in Colombia, eastern Africa, and Spain (Hammen and Vogel, 1966). The use by Netherlands scientists of the names Pleniglacial and Late Glacial is given in table 24-B in the Netherlands column, but these names are not used widely in other countries, perhaps because it is realized that conventional geographic names are preferable.

The stratigraphic position of the Stettin (in Poland, Szczecin) unit is not agreed on. In Germany the unit consists only of an end moraine, which was overridden by the ice sheet at some time before the Brandenburg glacial maximum, and which is tentatively assigned to late Weichsel.

Table 24-C
Pollen-Stratigraphic Sequence in Northern Europe Through the Last 14,000 Years

C[14] YEARS BP	GLACIAL EVENTS CONTINENTAL EUROPE	GLACIAL EVENTS BRITAIN	BALTIC WATER BODIES	CENTRAL FRANCE (HIGHLANDS)	SOUTHWEST GERMANY	ENGLAND[b]
1000				Fields, pastures, forest of beech, oak	Fields, pastures, forest of beech with spruce, oak	Fields, pastures, oak forest
2000				Beech–fir with oak, alder		
3000				Beech–fir forest with oak,	Beech forest	Woodland of oak, lime, alder, hazel
4000				lime, ash	(Clearances)	(Clearances)
5000			Littorina Sea	Linden–oak forest with elm	Lime with elm, ash, oak	Elm decline / Oak with lime,
6000						alder, etc.
7000			Ancylus Lake	Hazel–oak forest with elm	Lime, hazel with elm, etc. / Hazel with elm, lime, etc.	Hazel with pine, elm, oak
8000						
9000			Echineis Sea / Yoldia Sea	Birch–pine forest with oak	Pine	(With hazel) / Birch forest
10,000	Ra and Central Swedish moraines / Salpaus–selkä I,II	Loch Lomond Readvance	Baltic Ice Lake	Open forest of birch, pine	forest	Juniper stage / Tundra
11,000		Perth Readvance			Open birch forest. Juniper–Hippophai	Open birch forest / Tundra
12,000			?	Pioneer vegetation of juniper, Artemisia, etc.		
13,000	East Jylland (Pomeranian?) Readvance	Lammermuir Readvance	Ice lakes in central Skåne		Pioneer vegetation of Artemisia, etc.	
14,000						
Representative references	J. Donner, 1966: Comment. Biol. Soc. Sci. Fennica, v.29, no. 9. N.–A. Mörner, 1969: Sveriges Geol. Undersök., ser. C, no. 640.	R. West, 1968. p.263	E. Hyyppä, 1960: Int. Geol. Congr., 21st, Excur. Guide C 35. J. Donner, 1966: Comment. Biol. Soc. Sci. Fennica, v.29, no. 9.	G. Lang & W. Trautmann, 1961: Flora (Jena) v.150, p.11–42. G. Lemée, 1956: Soc. Bot. France B. v.103, p.83–94.	F. Firbas, 1949: Spät– und nacheiszeitl. Waldgeschichte Mitteleuropas, Jena, v.1. A. Bertsch, 1961: Flora (Jena) v.151, p.243–280. H. Müller, 1962: Geol. Jb, v.79, p.493–526.	H. Godwin 1956: History of the British flora, Cambridge. W. Pennington, 1964: Roy. Soc. London Phil Tr., sec B, v.284, p.205–244.

[a] Compiled by Johs. Iversen and R.F. Flint in 1969.   [b] For sequence in Ireland see fig 14–4.

NETHERLANDS	BLYTT-SER-NANDER-& OTHER NAMES	POLLEN ZONES	WESTERN DENMARK	SOUTHERN SWEDEN (INCLUDING SCANIA)	CENTRAL POLAND	SOUTHERN FINLAND	C[14] YEARS A.D./B.C.
Fields, pastures, heaths. Forest of beech, oak	Sub–Atlantic	IX	Fields, pastures, heaths, beech forest (transition)	Forest of spruce or beech with pine, oak Local fields, pastures Spruce, beech, pine, oak (Clearances)	Fields, pastures; forest of pine, hornbeam, spruce	Pine– spruce forest	A.D. —1000 —0
Woodland of oak with lime, beech, ash, hazel (Clearances) Elm decline	Sub–Boreal	VIII	Browsed forest of lime, ash, oak (Clearances) Elm decline	Oak with pine, lime, ash (Clearances) Elm decline	Pine, oak, lime, hornbeam (Clearances)	Birch–pine forest with lime,	—1000 B.C. —2000 —3000
Lime with elm, ash	Atlantic	VII	Lime, with elm, oak, (Climax forest)	Lime, with elm, oak pine	Lime, pine with elm, oak	elm, hazel Birch–pine forest with hazel, elm	—4000 —5000
		VI	(transition)	(transition)			—6000
Pine– hazel forest	Boreal	V	Hazel– pine	Pine– hazel	Pine forest with hazel, elm	Pine forest with birch	—7000
Birch– pine forest	Pre– Boreal	IV	(+ pine) Birch forest Juniper stage	(+ pine) Birch forest Juniper stage	Pine forest	Birch forest Park tundra	—8000
Park tundra	Younger Dryas	III	Park tundra	Tundra	Pine park tundra	Tundra	
Pine forest Birch forest	Alleröd	II	Open birch forest	Birch park tundra	Pine forest Birch forest	Ice	—9000
Tundra	Older Dryas	c	Tundra	Tundra	Birch park tundra		—10,000
Birch park tundra	Bölling	b	Birch park tundra	Ice			
Tundra	Oldest Dryas	a	Tundra		Tundra		—11,000
Polar desert			Polar desert				—12,000

(Left margin spanning labels: "Postglacial of pollen stratigraphers"; "Hypsithermal"; "Late Glacial"; "Full glacial (Pleniglacial)")

T.v.d. Hammen, 1951: Leidse Geol. Med., v. 17, p.71–183 W.v. Zeist, 1956; Palaeohistoria, v.4, p.113–118. T.v.d. Hammen et al., 1967: Geol. en Mijnb., v.46, p.79–95.	Blytt, 1876; Sernander, 1910.		K. Jessen, 1935: Acta Archaeol. (Copenhagen) v.5, no.3. J. Iversen, 1954: Danm Geol. Undersög. ser.2, no.80, p.87–119; Ibid. ser.4, no.3.	M. Fries, 1958: Acta Phytogeogr. Suecica (Uppsala), no. 39, 64 p. B. Berglund, 1966; Opera Bot. (Lund), v. 12, nos. 1,2.	K. Wasylikowa, 1964: Biuletyn Peryglac., v.13, p.261–417. M. Ralska–Jaciewiczowa, 1966: Acta Palaeobot. (Krakow), v.7, no.2	J. Donner, 1963: Acta Botan, Fennica (Helsinki), no.65.	

[c] Incorporates the correction factor of S. Th. Andersen (1967) based on variation in the abundance of pollen produced by different taxa.

In Poland, where it seems better documented, the unit consists of a variety of sediments, with fossil mammals, assemblages of Mousterian artifacts, and pollen indicating park tundra. It is assigned to the early Weichsel. Similarly, in European USSR Kalinin drift is thought to be early Weichsel, antedating a mid-Weichsel Mologo-Sheksna interstadial identified from pollen data. The Interstadial is followed by Ostashkov drift, believed to be Late Weichsel.

Mid-Weichsel interstadial sediments are reported (Korpela, 1969) from northern Finland as far north as 67°, where they are encountered in borings through surface till at 12 localities. The date of one sample is ~45,000 years, and the vegetation is of subarctic character. Obviously at that time deglaciation had progressed to some latitude north of the Arctic Circle, a fact that implies a very great reduction in the volume of the Scandinavian Ice Sheet. The sediments form the basis for a *Peräpohjola Interstade*. Somewhat similar occurrences are reported from localities in middle and northern Sweden. Dates on samples from four localities are infinite (>35,000 to >40,000 BP). The sediments are time-stratigraphically termed *Jämtland Interstade* (J. Lundqvist, 1967). Three superposed tills exposed near Lindköping are thought by Gillberg (1969) to be possibly representative of three glaciations of Weichsel age.

Apart from these occurrences, however, the mid-Weichsel interstade is elusive. Various occurrences of pre-Brandenburg sediments are known, but their lower stratigraphic limits are not fixed closely. Notable among such units is the Rixdorf sequence, exposed in a gravel pit at Neukölln (formerly Rixdorf), a suburb of Berlin. The name Rixdorf is applied to a whole group of deposits of this date in the Berlin district. The rich mammal fauna collected (not all at one locality) from this horizon includes groups suggesting more than one climate. Mammoth, woolly rhinoceros, musk-ox, reindeer, and Arctic fox suggest a boreal climate; the presence of *Palaeoloxodon antiquus, Dicerorhinus kirchbergensis,* lion, hyena, beaver, deer, bixon, ox, moose, and horse indicates varying degrees of warmer climate. The assemblage has been interpreted as two faunas, a "warm" one, and a later "colder" one immediately preceding the approach of the Brandenburg ice. Bones of other animals are present but they seem to have been reworked from much older deposits. The sequence is overlain by till of Brandenburg age. As these characteristics do not resemble those of the Eem sediments, a mid-Weichsel position seems likely.

The late-Weichsel drift, like the Late Wisconsin drift in America, possesses conspicuous constructional topography that includes end moraines (fig 23-5) and many closed basins, and that is little altered by mass-wasting or dissection. It is not mantled with loess in Germany and Denmark, but in Russia thin loess is present.

North of the Pomeranian moraine in East Prussia are many occurrences of lake sediments that are said to lie between two tills. They contain mollusks whose climatic implication is indefinite. These are the "Masurian interstadial" deposits, seemingly implying a post-Pomeranian deglaciation, though their true position is doubtful. They might be equivalent to the Bölling unit shown in table 24-C.

The principal "key horizon" in the deglaciation sequence is the ice-sheet margin represented by the Salpausselkä and the Central Swedish and Ra moraines in Fennoscandia, dating from 11,000 to 10,000 BP. No later, general ice-sheet readvance is known on the Continent, and this possibly parallels the emerging facts on the Laurentide Ice Sheet in Canada. Nevertheless climatic fluctuation is clearly evident in the pollen stratigraphy (table 24-C).

Interglacial and interstadial sediments in Fennoscandia are very infrequently reported and are probably little represented, owing to thorough glacial erosion of generally hard-rock terrain. In Denmark marine sediments of Holstein age are known at three localities. Others of Eem age occur along the coast below the existing sealevel (Hansen, p. 38 in Rankama, ed., 1965), whereas in Norway Eem sediments are known only as fragments of shells embedded in till (B. G. Andersen, p. 112 in idem).

Although interstadial sediments are little represented there, Fennoscandia has an unsurpassed record of the sequence representing the last 13,-000 years of Quaternary time. As stated in chap 14, pollen-stratigraphic study began in Sweden and has been carried to a fine point in all the Fennoscandian countries. Table 24-C is self-explanatory. It represents only the regions for which well-established, accurately dated pollen sequences are available, although less well-controlled sequences for other countries exist. The nine pollen zones indicated in the table originated with Knud Jessen in Denmark, but in 1953 were formally adopted in the regions shown. Local zone designations, archeologic stratigraphy, and the "De Geer" stratigraphic sequence are given conveniently in an informative table by T. Nilsson (1961). Although the units of the Blytt-Sernander scheme, in the column adjacent, are based on megafossils, they are in good agreement with the pollen zones and the names are still in use. We need to point out, however, that the Blytt-Sernander names end downward with the Pre-boreal. The zones below it were applied at various later times. Alleröd and Bölling are two localities in Denmark, and *Dryas* is the name of a small plant common in Scandinavian pollen spectra from the tundra belt.

Some of the difficulties involved in creating worldwide pollen-stratigraphic units are discussed by Hafsten (1969).

The table makes it possible to follow progressive changes in vegetation during and after deglaciation, with maximum warmth in Atlantic time, followed, in the Subboreal, by cooler, moister climates. Details are given by Fries (p. 55 in H. E. Wright and Frey, eds., 1965b).

The climate has not returned to the warmth of Atlantic and early Subboreal time (Sawyer, ed., 1966). Outside northern and central Europe the same rise to a temperature peak, followed by cooling, is discernible in pollen zones and other evidence, but the subdivisions shown in the table are rarely identifiable. Hence the term Hypsithermal was proposed (Deevey and Flint, 1957) to designate the time when temperatures were higher than they are today. It is useful because it is general, spanning about 6500 years, and because it can be applied anywhere in the world.

Minor fluctuation of climate in postglacial time is recorded by the *recurrence horizons* found in some peat-bog sequences in northern and central Europe. These are the tops of zones of oxidation of the peat. The oxidation occurred during protracted intervals of lowered water table. The unoxidized peat that overlies each recurrence horizon indicates the renewal or upward growth of the peat that resulted from rise of the water table during a moister and perhaps cooler climate. The concept is supported by appropriate differences between the assemblages of plants below and above a recurrence horizon. At least five such horizons have been identified rather widely (table 24-D). The features are C^{14}-dated from pairs of samples, respectively from just above and just below the top of the oxidized zone; resolution is generally less than 100 years, although

Table 24-D

Dates of Principal Recurrence Horizons Identified in Europe[a]

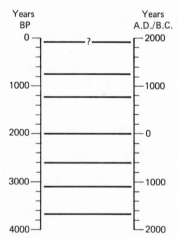

[a] Adapted from Bennema, 1954, p. 72, after Granlund and other sources.

in one of the surfaces shown in the table it is about 200. The dates, of course, indicate the beginning of moister conditions following the oxidation; they tell us nothing about the duration of the drier, oxidizing environment.

Not only fossil pollen but also fossil marine invertebrates demonstrate the rise to the Hypsithermal and the subsequent decline.

The term (*clearances*) in table 24-C is based on conspicuous increase in NAP, including weeds that accompany the destruction of forest. This increase reflects the clearing of forest by Neolithic people.

Pre-Elster Stratigraphy. We return now to the rather scanty information available on horizons older than Elster drift. The most detailed known sequence is in the Netherlands, a country that occupies a key position, being coastal and also marginal to former glacial invasions. Both marine and terrestrial strata, both of them fossil bearing, are present. The marine sequence reflects a transition from a warm, Pliocene mollusk fauna to a colder fauna.[3] The terrestrial sequence begins at the Pliocene Reuver Clay with a warm-temperate flora. The overlying sediments (Pleistocene, pre-Tegelen) include alluvium with a flora indicative of cold climate. The "black-bones" bed with Villafranchian and derived Pliocene faunal elements, that underlies the estuary of the River Schelde, is believed contemporaneous.

Next in succession is the Tegelen Clay, a partly deltaic alluvium, with upper Villafranchian mammals, fresh-water mollusks, and a temperate flora that becomes colder toward the top. Overlying this is a sequence of sediments, the Eburon, Waal, and Menap, with floras indicating cold, temperate, and cold climate respectively. At the top of this pre-Elster sequence is the Cromer, another temperate unit. Actual till of Elster age occurs in the Netherlands, but only as inclusions in Saale till; hence it does not prove that the ice sheet invaded the country in Elster time.

Table 24-E shows how guide fossils have been used by Netherlands geologists in erecting the stratigraphic sequence, rather as horizon markers than as indicators of climate.

Apart from the Netherlands sequence, we mention two other successions among a number of pre-Elster occurrences in Europe. One is in Carpathian Poland (fig 16-8), where alternations in the flora of successive strata imply rather early Pleistocene changes of climate. The other, at Bilshausen near Göttingen in Germany, just south of the glaciated region, is an interglacial layer whose stratigraphic position and flora indicate a Cromer age. Its pollen diagram shows a cold→warm→cold succession. Its fossil mammals include giant elk, moose, a deer (*Capreolus*), and a rhinoceros (*Dicerorhinus etruscus*).

[3] Details are given in Voorthuysen, 1950; Lagaaij, 1952.

Table 24-E
Guide Fossils, Including Plants, Mollusks, and Mammals, for Pleistocene Stratigraphic Units in the Netherlands [a]

	Arvicola bactonensis	*Mimomys intermedius*	*Mimomys pliocaenicus*	*Viviparus diluvianus*	*Viviparus glacialis*	*Azolla filiculoides*	*Azolla tegeliensis*	*Arvicola* (recent type)	*Microtus sp. sp.*	*Anancus arvernensis*	*Mammuthus (Archidiskodon) planifrons*	*Mammuthus (Archidiskodon) meridionalis*	*Loxodonta (Hesperoloxodon) antiqua*	*Mammuthus (Mammuthus) primigenius*
Weichsel								x						
Eem														
Saale								x	x				x	
Holstein	x	x				x		x					x	
Menap, Cromer, Elster	x	x	x	x		x								
Tegelen	x			x	x	x						x		
pre—Tegelen										x	x			

[a] Van der Vlerk and Florschütz, 1953, p. 3.

Britain and Ireland [4]

Lower Pleistocene. The most continuous and best-known Lower Pleistocene sequence in Britain occurs in East Anglia. As in the Netherlands, the British Quaternary sequence begins with marine strata and nonglacial terrestrial sediments. Along the North Sea coast of England, in Norfolk, Suffolk, and Essex, a succession of fossil-bearing littoral-marine sands (long known as "Crag") and estuarine silts and clays are facies variations in a subsiding basin. They are gently inclined northward so that a long continuous sequence is not exposed at any locality and good exposures of even limited extent are few. In a review of the stratigraphy of the Crags, Boswell (1952) emphasized that although the sequence contains zones (some of them defined only vaguely) of indigenous fossils, chiefly mollusks, there are present in the Crag sediments many shell fragments derived secondarily from older parts of the sequence. The Crags also contain vertebrate fossils, both in littoral-marine sediments and in a nonmarine conglomerate that unconformably overlies the Coralline Crag and is generally placed in the basal part of the Red Crag.

Summarizing the data from these fossils, Boswell showed that in the

[4] West, p. 1–87 in Rankama, ed., 1967; West, 1968; Mitchell, 1960; Shotton and West, p. 155 in George, ed., 1969; Walker and West, eds., 1970.

lower part of the Red Crag sequence (table 24-F) boreal gastropods increase markedly in comparison to their numbers in the underlying Coralline Crag, and at the same time southern kinds diminish. Furthermore the lower part of the Red Crag contains teeth of both zebrine and caballine horses and of *Archidiskodon meridionalis,* unknown in the Coralline Crag. Boswell therefore drew the Pliocene-Pleistocene boundary at the base of the Red Crag. An earliest-Pleistocene date for these fossils is further indicated by the associated occurrence of mastodon, which died out in Europe early in the Pleistocene.

The climatic implications of the presence of Foraminifera and Bryozoa support this boundary and strengthen correlation across the North Sea. Marine invertebrates likewise are a basis for correlation of the Red Crag with the Calabrian marine unit in Italy (table 24-I).

The study of pollen and lithostratigraphy in various layers of a core bored at Ludham, Norfolk and other cores led to the recognition of "climatic stages" (West, 1963, p. 152), most of which are included in table 24-A, column 1. Several of them are marine or estuarine. Two (Thurnian and Baventian) represent vegetation consisting of cold oceanic heaths, whereas two others (Ludhamian and Antian) represent temperate forests.

Elaboration of the Quaternary stratigraphy of East Anglia is currently in a transitional state, as is evident from comparison of the references cited in this section of the present chapter; hence the sequence in table

Table 24-F

Lower Pleistocene Marine Stratigraphy and Floral Ages in East Anglia[a]

	Floral ages	Marine rock units	
Pleistocene	Baventian Antian Thurnian	? Weybourne Crag	
		Norwich Crag (s.l.)	
	Ludhamian	*Scrobicularia* Crag	
		Newbourne— Butley Crag	
		Red Crag	
Pliocene		Boyton Crag Gedgrave Crag	Coralline Crag

[a] West, 1963.

24-A is provisional. The later references do not group the units into Lower, Middle, and Upper Pleistocene brackets, although such grouping was attempted in West, 1963, p. 152. In that reference, also, the units were labeled *temperate* or *cold,* according to the evidence of fossils. Although cold units occur as far down in the sequence as Thurnian, the oldest known till is the Cromer Till (= Norwich Brick-earth). It occurs within the Anglian Stage rather than in the Cromer Stage. This confusing fact arises in part from the history of terminology and of stratigraphic correlation in the region.

Middle Pleistocene. The Middle Pleistocene sequence as visualized by West (1963) begins with the Baventian Stage. That is succeeded by units that include Cromer peat, estuarine silt, and beach sand. Contained vertebrate fossils imply a generally temperate climate, and pollen shows this gradation through time: pine/birch→mixed oak forest→conifers, the cool-temperate-cool sequence of interglacials.[5] These units are overlain by the Cromer Till (North Sea Drift of older usage), the stratigraphic relations of which are not clear. By some the Cromer Till and the Lowestoft Till above it are together correlated with the Elster Till of the Continent. In East Anglia the two tills record different flow directions in the glacier. Whether that glacier was the Scandinavian Ice Sheet, as suggested by erratics from Norway, is not quite certain because the erratics might have been picked up by the ice from older drift in northern Britain. At any rate the Cromer and Lowestoft Tills are separated by the Corton Beds, a marine unit of uncertain climatic significance. The Lowestoft Till was known in the older literature as Chalky-Jurassic Boulder Clay. In parts of southern England it represents the most extensive of the glaciations.

At Hoxne (pronounced *Hoxon*) in Suffolk, and in other parts of East Anglia, numerous bodies of lacustrine sediment occur between Lowestoft Till and the succeeding Gipping Till. Their pollen records show a cold→temperate→cold climatic succession, evidently an integlacial, with widespread forests during the warmest middle part. In southern England correlative littoral and estuarine marine sediments occur along the coast at various altitudes.

According to one correlation the Gipping Till is basal Upper Pleistocene. A distinctive, chalky till, it is the surface drift throughout much of East Anglia and is correlative with the Saale Till of the Continent. Nonglacial sediments, apparently within it, suggest it may be twofold. It is followed stratigraphically by sediments of the Ipswich unit.

Ipswich interglacial sediments, named for gyttja and related sediments

[5] The flora of the "Cromer Forest Bed" succession is discussed by Duigan (1963, p. 180) as to its stratigraphic implications.

near Ipswich, Suffolk, include marine deposits occurring above sealevel on the south coast, and also terrestrial deposits with a pollen pattern resembling that of Eem sediments in Denmark. Although the pollen profiles of the Cromer, Hoxne, and Ipswich interglacials are distinctive, plant taxa changed little from one to the next. Rather, because of climatic changes of varying duration and intensity, taxa migrated through varying distances and occupied regions of western Europe through varying lengths of time.

Weichsel. Following the Ipswichian sediments is drift earlier called Newer Drift but more recently called simply Weichsel because its correlation with the Weichsel of the Continent is obvious. In Britain as on the Continent, early-Weichsel drift is elusive; however, early glaciation is indicated by indirect evidence. Formerly, the outermost Weichsel drift (fig 23-4) was thought to be early Weichsel, but a late-Weichsel date (<30,000 BP) for a segment of the western part of it, and a date near 18,000 BP for a locality in eastern Yorkshire (Penny et al., 1969) were established subsequently.

Be that as it may, as a whole the Weichsel sequence in Britain includes at least one early-Weichsel cold episode and a principal Late-Weichsel cold episode with pronounced late fluctuations. Between them is a period of less-cold though at no time temperate climate—an interstadial in time-stratigraphic terms—in which glaciers were much reduced in area and temperature fluctuated repeatedly through a reduced range as indicated by both floras and faunas. Significant data, for example, have been obtained from insect faunas (e.g., Coope and Sands, 1966) as well as from fossil pollen. The fluctuations are evident in the temperature curves in fig 16-11, column 9 and in table 24-B. These curves, and the subdivisions implicit in them may well become standards to which other, less well known sequences in Europe will be fitted as knowledge of them increases.

South of the British ice sheet, whose margin bisected England along a NE-SW line, frost-action effects are abundant. Ice-wedge casts, polygons, and involutions are observed in many places (fig 24-2), and in addition widespread gelifluction is indicated. Tundra (in the broad sense) was widespread.

Late-Weichsel events in Britain are complicated by the existence of several independent highland centers of spreading of glaciers, mentioned in chap 23. Deglaciation after the late-Weichsel maximum was interrupted by at least three re-expansions of principal British glaciers to the successive Lammermuir, Perth, and Loch Lomond positions, with corresponding activity on lesser highlands. Subsequent climatic fluctuation, well documented by pollen data from lakes and bogs, is summarized in table 24-C.

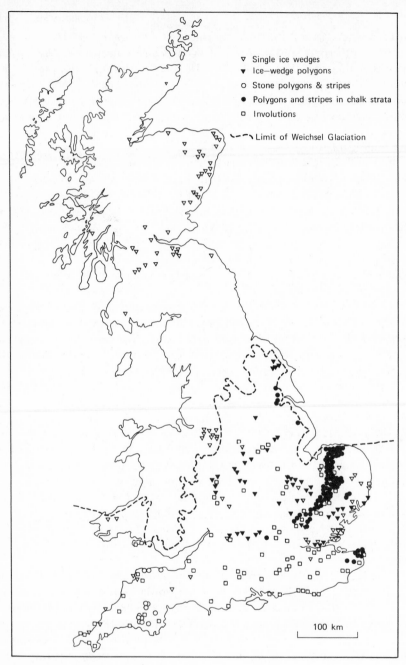

Figure 24-2 Frost-action features of Weichsel age in Britain. Compiled by R. G. West (1968, p. 275).

The Pleistocene sequence in Ireland includes sediments correlated with the Lowestoft, Hoxne, Gipping, Ipswich, and Weichsel of the British column. Much of the drift was deposited by glaciers that originated in Ireland, although at certain maxima ice of Scottish origin invaded eastern Ireland. Certain beaches, occurring above sealevel in southern and eastern Ireland, are considered interglacial.

Iceland

Iceland is one of the few lands from which very early glaciation has been reported. In northern Iceland a thick sequence of Pliocene and Pleistocene marine strata includes 10 tillites and several ancient lava flows. On certain assumptions, K/Ar dates and paleomagnetic correlation suggest that ages of the tillites range between ~ 0.7 million and possibly as much as 3 million years ago (Einarsson et al. in Hopkins, ed., 1967). Elsewhere on the island are other occurrences of till, apparently younger. Likewise two interglacial units have been tentatively identified, with marine fauna in one and fauna and flora in the other suggesting climates about like those of today (Thorarinsson, p. 377 in Löve and Löve, 1963; Schwarzbach, 1955). No integrated column has yet been published. Sequences representing the time since the last glacial maximum are discussed by V. Okko (1955) and by Einarsson (p. 355 in Löve and Löve, eds., 1963).

The Alps Region [6]

The Alps region was the site of the birth of the Glacial Theory and was also the locus of a pioneer, unified study of Pleistocene glacial stratigraphy, that of Penck and Brückner (1909). In the absence of anything comparable from a lowland region, the latter work became a standard as well as a classic. It presented a sequence of four major glaciations named Günz, Mindel, Riss, and Würm (after four streams in the foreland of the northwestern Alps) separated by three major interglaciations referred to, for an obvious reason, as G/M, M/R, and R/W. This neat scheme was a tempting standard for correlation, and workers took uncritical advantage of it to so great an extent that it achieved worldwide renown.

Despite the high quality of Penck and Brückner's work, the Alps region is not well adapted to be a standard even for Europe, let alone for the world. Because of the height and steep slopes of the mountains erosion has destroyed much of the sedimentary record. Likewise its drift bodies are isolated, lacking contact with those related to continental glaciation farther north, except for related loess sheets and perhaps also for

[6] Review papers include H. E. Wright, 1961a, p. 959–966; Movius, 1960, with discussion; Richmond, ed., 1968, embraces several regional papers; see also Burri, 1963.

outwash in the valley of the Rhine. Furthermore, the classic study of Penck and Brückner was concentrated mainly on certain valleys of the northern watershed, and could not properly represent the Alps region as a whole. Today stratigraphic studies in various lowland areas of Europe have outstripped in significance the synthesis of data from the Alps, and only since about 1960 has a new wave of more detailed, critical study of the Quaternary of the Alps taken form.

In much of the region the sequence has been constructed primarily from outwash, some bodies of which grade upvalley into till. Penck and Brückner recognized four main bodies of outwash, of which the two older are represented only by small remnants with little topographic expression, whereas the two younger are represented by extensive, terraced valley trains. Their relative ages are fixed by weathering profiles, by extent of erosion of bedrock, by relationships with interbedded sheets of loess, and by morphology. At one place or another, however, each outwash mass is traced directly into till. Some of the older bodies are multiple, implying more than one episode of aggradation and suggesting major fluctuations of glaciers at those times.

The two younger outwash units are related to end moraines which, in the next-to-youngest unit, are so greatly modified by mass-wasting that they retain little constructional topography. In the youngest unit, however, end moraines are both massive and little altered, and constitute a sequence that extends upstream to the heads of some valleys, implying repeated glacial readvances following deglaciations of mostly unknown extent. In this respect the sequence resembles many of those in the North American Cordillera.

Two decades after the publication of Penck and Brückner's work, Eberl (1930, p. 305–308) discovered pre-Günz outwash (without till) in the northern Alps. The discovery implied an early and perhaps complex glaciation to which the name Donau was given. Although for two more decades little attention was paid to this early glaciation, it was later confirmed on the southern flank of the Alps, in the following way. Although Penck had been unable to identify Günz drift on the south flank, that drift was later recognized at several localities by Nangeroni (e.g. 1950; Venzo, 1965). A sequence of sediments from Italian Alpine localities yields a composite section extending from upper Pliocene through Mindel (fig 16-7). Detailed studies of its fauna and flora, including pollen studies, reveal three cold phases that are said to antedate the drift identified as Günz. These are correlated with the Donau glaciation and are separated from the Günz (also represented by "cold" pollen) by sediments with pollen indicative of a warmer climate (fig 16-7; Venzo, 1955). The floral evidence consists mainly of high-Alpine conifers at relatively low altitudes. The sequence below the top of the "cold" strata correlated with the Günz has a Villafranchian mammal fauna (chap 29; Lona and Follieri, 1957, p. 97).

Areal extent of the various drift bodies indicates that in general the Riss was the most widespread glaciation, the Mindel as extensive in some places but less so in others, the Würm less than Riss or Mindel, and the Günz apparently least of the four. The Donau is recognized at so few places that little can be said of its extent, which, however, appears to have been small. The existence of pre-Donau outwash on a high terrace in the northern Alps has been claimed, and is tentatively recognized by Woldstedt (1969, p. 68) as representing a Biber Glaciation (I. Schaefer, p. 13 in Richmond, ed., 1968).

The Mindel and Riss units, and probably the Würm as well, consist of at least two cold phases. The evidence consists of multiple tills, multiple outwash valley trains, and pollen data mainly from the southern flank of the Alps.

Reasoning from indirect evidence, Penck and Brückner (1909, p. 1161, 1168) believed that the M/R interglacial was much longer than the others, and labeled it the "Great Interglacial." This belief has never been validated firmly, and moreover is weakened by various kinds of negative evidence. It is discussed briefly in chap 1.

Interglacial and Interstadial Stratigraphy. Occurring between bodies of drift are weathering zones, surfaces of erosion, and nonglacial sediments with fossils. Thick zones of mature weathering, commonly with red iron oxides, are developed in drift of Günz and Mindel age and according to some geologists, in Riss drift as well. Drift generally agreed to be of Würm age is weathered much less thoroughly. Where two drift bodies are in contact, the interface is one of erosion; in some occurrences the erosion surface possesses substantial relief.

Interdrift sediments consist commonly of alluvium, but also include peat or lignite, lake, and cavern deposits. Although many such occurrences are known, few are securely correlated because few are overlain or underlain by drift of known stratigraphic position. Those containing pollen [7] are less helpful for correlation than they might be because within any one altitude range the plant assemblages are rather alike. Also they resemble the assemblages that live at the same altitudes today; indeed most of the species found as fossils are still living. These facts suggest that between glaciations maximum temperatures were similar both to each other and to that of today, within a range of 2° or so. In most lo-

---

[7] Locality lists and brief partial summaries appear in Lüdi, 1953, p. 132–181; 1957; Klebelsberg, 1948–1949, p. 696–704; Venzo, 1955, chart following pl. 12; Woldstedt, 1954–1965, v. 2, p. 183.

calities the pollen shows one or more parts of a standard sequence that can be climatically interpreted as cool → warmer → cool. The actual plant groups vary, of course, mainly with altitude of the locality.

A classical interglacial occurrence, the Hötting breccia (Penck. 1921), an indurated colluvium exposed near Innsbruck, seems at first to be an exception to the generalization drawn above. It has yielded 42 plant species, some of which are distinctly southern and do not now live in the Alps. The implied former mean annual temperature was warmer than the present by about 2° and the implied snowline was higher than it is now by about 400m. This temperature and snowline, however, are not exceptional, for they were attained again as recently as Hypsithermal time. Hence the climate of Hötting time can be said to resemble broadly that of the present. The Hötting sediment has been assigned, without proof, to the M/R interglacial, although by some it is referred to the R/W.

Würm Stratigraphy. In the confused history of the Alps succession probably the greatest confusion has arisen over the sequence and nomenclature of Würm strata.[8] The causes include the proposals of various schemes, with agreement on none, lack of a sound basis in the Alps region itself for some of the units created, and inclusion in the sequence of names based on occurrence outside the region. Part of the trouble therefore is substantive and part is conceptual, involving nomenclature. The principal substantive weakness is that sediments representative of glaciation early in Würm time, say before 50,000 BP, have not been clearly and definitely recognized, and that opinions differ as to whether certain drift bodies are of Würm or Riss age. Penck originally visualized a mid-Würm interstadial, the Laufen, but in 1920 abandoned the concept in favor of that of a single major glaciation, a view advocated in more recent time by Büdel. A different concept accompanied by a different nomenclature arose from study by Soergel (1919) of the succession of loess layers and ancient soils in central Europe outside the Alps proper. Loess units to which Soergel had given local numbers were correlated by others with glacial layers in the Alps and the numbers were prefixed by W (for Würm); thus the terms WI, WII, and WIII came into use in Alpine stratigraphy. The numbers were used in a different sense by still others.[9] Meanwhile Woldstedt (1956b), applied the terms Early, Middle, and Late Würm to stratigraphy in northern Germany. Still other names in

---

[8] Summaries in Woldstedt, 1954–1965, v. 2, p. 176–180; H. E. Wright, 1961a, p. 959–961.

[9] An example consisting of five Würm "cold" units and based mainly on a variety of data from outside the Alps is Bourdier, 1961–1962, p. 279–291. The climatic sequence here is probably valid; we call attention to it only in the context of its terminology.

the literature include Old Würm and Young Würm, and also Old Würm and Main Würm separated by an interstadial.

In this uncontrolled state of affairs we are left with a wide choice of names. Pending unification of opinion among people experienced in Alpine stratigraphy, we must adopt names in order to be able to refer to the sequence in various parts of our present discussion. Having in mind the agreement in C^{14} dates through the last 30,000 years or so among climatic events in northern Europe, North America, and the Alps region and the probability that still earlier events also were contemporaneous, we can make a reasonable choice. For our purpose alone and without implying a proposal for more general use, we temporarily employ these names:

Late Würm	Strata dating between the Holocene and 25,000 BP, more or less.
Middle Würm	Strata, probably mainly interstadial, dating between 25,000 BP and 50,000 BP, more or less.
Early Würm	Strata dating between about 50,000 BP and the end of the last interglaciation.

This nomenclature is defensible only to the extent that it employs names rather than numbers, does comparatively little violence to any existing usage, and is reasonably compatible with nomenclature in other regions where the sequence is better known. The name Würm ought to be retained both because of its priority and because of its extraordinarily wide use—including, of course, its use in correlations whose strength is doubtful.

From table 24-B it appears that we have a fairly good sequence back through the last 20,000 years or so, represented best by end moraines, extents of drift sheets, and ecology derived from pollen data, although critical C^{14} dates are much less abundant than they are in sequences in northern Europe. Fluctuations of glaciers and also fluctuations of climate have been referred to by a number of European authors as "oscillations." This term is generally used without definition, although Movius (1960, p. 356) restricted it to a short climatic warming, excluding its use for episodes of cooling. The term is little used in North America.

The record of Alpine Würm stratigraphy older than Late Würm is very spotty. Middle-Würm interstadial sediments are represented, only possibly, by peat layers at Armoy, near Thonon, south of the Lake of Geneva, and near Hörmating in the northern foreland. Although they are said to exist near Armoy, Early Würm glacial sediments have not been identified with certainty, possibly because of later Würm erosion and deposition, which were substantial. Despite this highly unsatisfactory representation, it is nevertheless probable by analogy that Early-Würm glaciation occurred and was followed by a long interval of greater warmth; the record of those events will probably be found in time.

Indeed the Early Würm may be represented by drift hitherto referred to the Riss. G. M. Richmond (8th INQUA Congress, 1969, Rept. in course of publication) considered the Alpine Riss as younger than the soil representing the last interglacial. He equated the Mindel with the major glacial preceding that soil. If his correlation is correct, Mindel would equate with Saale and pre-Mindel glacial units would be moved downward accordingly. Because of this alternative possibility it is doubly undesirable to apply the Alpine stratigraphic names outside the Alps region.

Detailed discussion of the stratigraphy of a large segment of the Pyrenees, with correlations to the sequence in the French Alps, is given by Alimen (1964; see table 24) and by Taillefer (p. 19–32 in INQUA, 1969a).

NONGLACIATED REGIONS

In contrast to the glaciated regions, where the stratigraphy is dominated by drift alternating with interglacial sediments, the stratigraphic record of nonglaciated regions is read from outwash, loess, frost-action features, alluvium and stream terraces, marine sediments, ancient soils, and cavern sediments. Flora and fauna are quite commonly present in some of these categories of sediments, and artifacts play a larger part in the stratigraphy than they do in North America. Our brief discussion will concentrate on loess and the soils developed in them, alluvium along certain streams, the Villafranchian, and the marine strata of Mediterranean Europe.

Loess and Soils [10]

The areal distribution of loess in Europe is actually somewhat greater than that shown in the map, fig 9–16, owing to discoveries made since that map was compiled. Most European loess dates from or near glacial maxima, and much of it was derived from outwash valley trains. Pollen spectra from loess, however, reflect the near absence of trees in the tundra that prevailed in northwestern Europe, in contrast with the presence of trees, probably as gallery forests, in the steppe of southeastern Europe. Active deposition of loess took place under severe climates, possibly drier than those of today, and in regions where the vegetation was mostly tree-

---

[10] The most recent general study is in INQUA, 1969b.

less, consisting of tundra, prairie, and steppe. The character of the mammal faunas appropriate to those environments has already been mentioned.

In Hungary and the Ukraine region of European Russia loess is thick and constitutes a well-nigh universal cover. In one district only has the relation of loess to till been demonstrated. The Dnepr (Saale) till of the Dnepr Lobe is enclosed in loess without weathering at the contact. This particular loess is therefore of Saale age. Expectably it antedates a thick mature soil and is overlain by two loess sheets separated by a soil of Chernozem type, suggesting that twice during the last major glaciation the ice sheet entered the Dnepr drainage basin and created an outwash valley train. In Hungary 11 loess sheets, separated by soil zones, have been observed in a single section, but have not yet been correlated firmly.

The specific stratigraphic distribution of loess in Europe is shown in condensed form in tables 24-A and 24-B. It has been investigated in most countries and proves to be rather similar right across Europe from France and southern England to the USSR.[11] The Weichsel is represented by "Younger Loess," in many places including both Late and Early subdivisions. Saale and Elster are represented by "Older Loess," the Saale in places twofold. The ancient, buried soils recognized thus far are related to intervals of nonglaciation. Soils of intra-Weichsel, Eem, and Holstein age are widely present and even a Cromer soil has been reported. Probably still-older loess layers and soils have been extensively removed by erosion and hence have not yet been identified. In the North Sea coastal belt the Younger Loess includes as many as four members, some of them accompanied or even replaced by eolian sand.

The soils range from Podzolic types in areas that were forested during nonglacial intervals to Prairie and Chernozem types in less humid grassland areas.

Loess/soil stratigraphy has been studied extensively on the Danubian Plain in eastern Austria and adjacent parts of Czechoslovakia and Hungary. Typical relations in the upper part of the regional sequence (figs 24-3 and 24-4) include both single soils and "complexes" of two or more soils developed mainly in layers of loess (e.g., Demek and Kukla, eds., 1969, pl. 32). Local rather than glacial/interglacial names have wisely been attached to the soils. In Czechoslovakia soils have been identified by letters and numbers. In Austria place names have been applied; both are shown in table 24-B. Thus we have interglacial-soil names such as Still-

---

[11] Detailed description of the loess sequence in a wide region of Czechoslovakia in Macoun et al., 1965; Ložek, 1964, esp. p. 64; Demek and Kukla, eds., 1969. Distribution map and stratigraphy of loess and ancient soils in Ukraine, USSR, are given by Veklitch in INQUA, 1969b, p. 145. Older loess in Ruske, 1965.

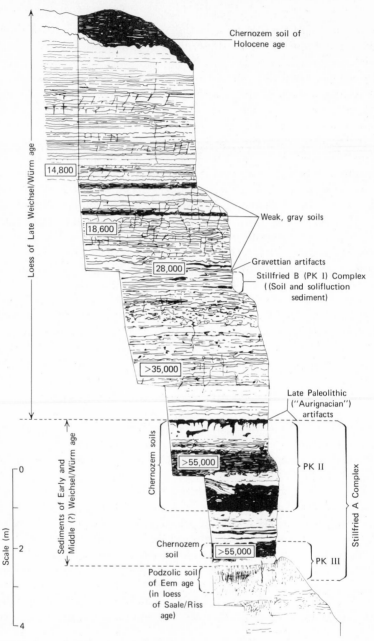

Figure 24-3 Detailed section exposing a nearly complete sequence of Weichsel/Würm loess and soil units. Numbers in boxes are C[14] dates. The exposure, in a brick-clay pit at Dolni Věstonice (Unter-Wisternitz) in southern Czechoslovakia, extends about 4m deeper than the base of the illustration. Adapted from Klima et al., 1961, with unpublished data from Vojen Ložek, 1969.

Figure 24-4 Face exposed in brick-clay pit near Brno, Czechoslovakia, only about 40km from the locality of fig 24-2 and with approximately the same stratigraphic sequence. Height of face about 20m. The steeply sloping grooves were made by excavating equipment. (Vojen Ložek.)

fried and Krems. Unfortunately the literature includes two widely repeated names, "Paudorf" and "Göttweig" that were applied to soils supposed to date from mid-Weichsel times, and then were extended to refer to intervals of deglaciation in the Alps sequence. It was learned later that both soils are of pre-Weichsel age, each representing a different interglacial. It would preclude further confusion if both names were abandoned.

Farther west, near Regensburg in Bavaria, a sequence similar to those along the Danube in Austria has been given detailed study. With an exposed thickness of 24m, the succession consists of loess and gelifluction colluvium representing glaciations, and buried soils representing nonglacial climates. Brunnacker (1964b) tentatively identified the glacials and nonglacials from Würm back through pre-Günz (see also Brunnacker, p. 96 in Morrison and Wright, eds., 1967).

Before leaving the topic of ancient soils we must emphasize that although we have discussed them in connection with loess, soils occur in every other kind of Quaternary sediment. They vary in facies in accordance with the local climate at the time of formation and, as in North America, it is observed that all soils truly red in color are no younger than the youngest interglaciation.

Alluvium and Stream Terraces

The loess sequence is intimately related to a sequence of stream terraces along the Danube (Fink, p. 179 in H. E. Wright and Frey, eds.,
1965b) which reach as much as 150m above the river and represent several phases of Pleistocene filling and cutting as well as Pliocene events.
Loess layers can therefore antedate or postdate one or more terraces and
thus contribute, along with the fossils contained in the loess, to the construction of a firm stratigraphic sequence.[12]

Another flight of terraces with complex stratigraphy including loess occurs along the River Somme; some of the richest exposures are in the vicinity of Amiens. There are at least four distinct terraces, each cut into
bedrock. Overlying them are sediments that include alluvium, loess, gelifluction colluvium and other frost-action features, and ancient soils. The
strata contain abundant fossil mammals and also artifacts. All the frost-
action features and the loess pertain to glaciations, whereas most of the
alluvium and the soils pertain to nonglacial climates. The fossil mammals and artifacts belong to both, as indicated in the generalized section,
fig 24-5. The relation of the various sediments to the rock benches is a
matter of debate. At present the historical sequence is developed wholly
from the sediments without regard to the benches.

Rhine River Sequence.[13] Apart from loess layers and ancient soils in
the nonglaciated region, terraces in the Rhine Valley have been given
much attention. Terraces occur along many rivers in Europe, but the
Rhine is the only river that connects the area of the Alps glaciation with
that of glaciations of the North Sea floor. The river, more than 1000km
long today, was more than 1600km long during times of glacial low
sealevel. In the Alps region the terraces are of the fill type and consist
of outwash sediments. However, the hope that the Rhine terraces would
afford stratigraphic correlation between Alps and North Sea for each
glacial age has not been fulfilled. The chief reason is that the Rhine,
a river apparently of Pleistocene age, flows through a long segment
between Basel and Frankfurt where intermittent tectonic activity has
introduced major discontinuity into the record except for the low
terraces of Weichsel age. Another complication results from the fact that
one unit, the Middle Middle Terrace, is cut by at least three younger
units in the latitude of Köln and Solingen.

Some of the terraces have been correlated with the glacial-geologic sequence by means of loess layers and ancient soils, and turn out to belong

[12] Besides the Fink paper mentioned above, see also Klima et al., 1961; Ložek,
1961–1962; Musil and Valoch, 1966; Prosek and Ložek, 1957.

[13] Woldstedt, p. 271–273 in Rankama, ed., 1967; Zeuner, 1959, p. 82–91.

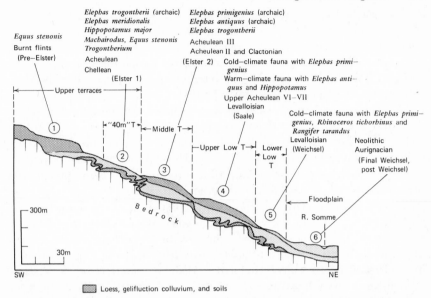

Figure 24-5 Schematic profile and section of one side of valley of the River Somme near Amiens, showing Quaternary sediments, soils, and stream terraces. Fossil mammals and artifacts found in sediments beneath the several terraces, together with correlation, are listed above. Section after Bourdier, modified by Alimen in Rankama, ed., 1967, p. 101.

to glacial ages (table 24-A). The correlation, on such evidence, of three pairs of two terraces each with Weichsel, Saale, and Elster drifts, respectively, supports the inference read from the drift succession that each of the principal glaciations consisted of at least two stades.

Many other large streams in Europe are flanked by terraces closely related to sediments that contain fossils and artifacts. From all of them contributions to the continental stratigraphic sequence have been made.

Southern Europe

Terrestrial Sediments. Among the most critical parts of the Quaternary stratigraphy of southern Europe are the sediments, both terrestrial and marine, of the earlier Pleistocene. While southern and central Italy were still submerged beneath the sea in late Pliocene and early Pleistocene time, northern Italy and adjacent parts of France and Spain were gradually emerging in such a way that terrestrial sediments were being laid down in parts of the northern region as early as the late Pliocene. Some sediments were deposited along the coast; others accumulated in local basins created by complex warping and faulting. Part of the terrestrial sediments in northern Italy form the basis of the Villafranchian (pronounced *Villafrankian*) Stage, named for Villafranca d'Asti, a town south of Torino. The name refers also to a mammal fauna, the record of which is centered in southern Europe but extends north to the Netherlands and Britain, far eastward, and southward to Africa.

Two clusters of studied localities in France, one in the south-central region and the other in the southeast, and another cluster in northern Italy, have furnished the bulk of the existing information [14] based not only on mammals but also on floras, which in places are rich.

The Villafranchian spans a long time, the limits of which have not yet been determined. It has been considered as extending upward to include the Tegelen, and downward either to the base of the Pleistocene (not defined) or into the Pliocene (Astian or lower). It includes faunas of both "cold" and "warm" character. At the risk of overgeneralization we say that the sequence appears to include at least three and perhaps four units, by no means sharply defined, based on faunal changes. Viret, a French authority, characterized the difference provisionally as in table 24-G. Table 24-H attempts climatic significance and correlation. Lüttig (1959, table 40) visualized four units in the Arno River watershed in northern Italy: (1) Stellicione beds (warm), (2) Ricasoli beds (cold), (3) Ville beds (warm), and (4) Sansino beds (cold). He regarded (1) as Pliocene and the others as lower Pleistocene.

Definition of the Villafranchian is in part a matter of semantics. Villafranchian in the wide sense includes both some Pliocene and some Pleistocene, and will have to be studied intensively before first-rate definition and subdivision are achieved. Meanwhile we may note that it is from Villafranchian fossils that Pleistocene was separated from Pliocene on the basis of the first appearance in it of certain forms including (1) the first true elephants, represented by *Archidiskodon planifrons,* an Asiatic species usually associated with the last of the mastodonts; (2) the first zebrine horses, usually associated with surviving *Hipparion;* (3) the first bovid (*Leptobos*); (4) the first *Dicerorhinus etruscus,* a cursorial rhino. With these are the large carnivore *Machairodus* and relatives, and the small *Megantereon;* also *Ursus etruscus, Crocuta perrieri,* and *Canis arvernensis,* together with certain rodents (cf. table 29-A).

The importance of the Villafranchian faunas is enhanced by their occurrence in strata in Germany, Hungary, Romania, southwestern USSR, southern Asia, China, and Japan. Somewhat related faunas are known in

[14] Summarized by Flint in H. E. Wright and Frey, eds., 1965b, p. 509–514. Details in references therein; especially Viret, 1954; Bonifay, 1964; Lüttig, 1959.

Table 24-G

Villafranchian Faunal Units and Their Characteristics[a]

Units	Faunal Characteristics (Names of some taxa have been modernized)	Significant localities
(Fourth)	Introduction of *Dicerorhinus kirchbergensis*	Tegelen [a]
(Third)	*Archidiskodon meridionalis* Probable extinction of mastodonts in Europe	Leffe[c], Senèze[b]
(Second)	Coexistence of *Anancus arvernensis* and *Archidiskodon meridionalis*	Saint—Vallier[b], Olivola[c], Chagny[b], Chillac[b]
(First)	*Zygolophodon borsoni; Tapirus;* absence of *Elephas*	Perrier[b]

[a] Netherlands   [b] France   [c] Italy

[a] According to Viret (1954, p. 183).

Africa.[15] Some of the alluvial strata in northern USSR (table 24-A) contain Villafranchian faunal elements, but correlations are still tentative, possibly because the stratigraphy has not been studied with sufficient intensity across a wide region. It appears in the table that steppe and woodland vegetation (based on pollen) correlate with cold episodes, in contrast with desert brush, which correlates with warmer episodes. Irregularities in the sequence suggest that possibly some stratigraphic units are still unidentified.

Marine Sediments. A significant fact about the Villafranchian in Italy is that some part of it (not the lowest part) grades by facies change into the marine Calabrian. This relationship affords a starting point for correlation of marine with terrestrial sequences in the Mediterranean region.

Perhaps the oldest known Quaternary marine strata in that region are exposed at Le Castella in the Province of Calabria, on the south coast of Italy. There a Plio/Pleistocene sequence more than 1600m thick is known. In it the base of the Pleistocene strata was determined (Emiliani et al., 1961) on a basis of temperature-dependent faunal characteristics, notably the appearance, in deep water, of the foraminifer *Anomalina baltica,* and in shallow water, of the mollusk *Arctica (Cyprina) islandica.*

[15] Literature is cited by Flint in H. E. Wright and Frey, eds., 1965b, p. 509.

Table 24-H Tentative Correlation of Plio-Quaternary Vertebrate Faunas in Southern Europe with Climate and Stratigraphy[a]

Successive Faunas	Localities (All in France)	Faunal Characteristics	Indicated Climate	Suggested Correlation with Alps Stratigr. Sequence	Series Subdivisions
Quaternary faunas	(Numerous localities)	Locally, cool-climate forms	Temperate; at times cold	Postglacial / Würm / Riss	Upper Quaternary
	Lunel–Viel	Abundant, reflecting conspicuous immigration of Quaternary taxa	Growing warmer	M/R Interglaciation	
	L'Escale (in part)	Impoverished fauna	Cold	Late Mindel	Middle Quaternary
Transitional faunas	Sinzelles (in part)	Becoming abundant	Cool–temperate	Early Mindel	
	Durfort	Very meager	Temperate	G/M Interglaciation	Lower Quaternary
	Sénèze (in part)	Becoming impoverished	Cold	Late Günz	
Villafranchian faunas	Saint–Vallier	Rich	Cool–temperate	Early Günz	
	Perrier–Roccaneyra	Rich	Temperate; dry	interglaciation?	
	Perrier–Etouaires	Rich, with conspicuous immigration of Villafranchian taxa	Cool–temperate	Pre–Günz cold episodes	Upper Pliocene
	Vialette	Meager	Cold		
Pliocene Faunas	Montpellier	Rich	Warm–temperate		Middle Pliocene

656

Both are boreal taxa. The basal Pleistocene stage is the Calabrian. A similar Plio/Pleistocene transition is exposed at Castell'Arquato, east of Piacenza in northern Italy, and also at localities in the Romagna district in north-central Italy. A "standard" marine sequence for this part of the Mediterranean region is set forth in table 24-I. We are concerned here not with changes of sealevel and altitudes of ancient strandlines (chap 12) but with the succession of marine strata and their implications for climate. In the table the "classical" sequence, current for several decades, is given because of its frequent appearance in the older literature. Actually it was based on remarkably few data, and it has been replaced, tentatively, by a more modern sequence based on more information.

The Calabrian and Sicilian faunas include cold elements, but include so many taxa of temperate affinity that they are not strictly "glacial." Also the entire sequence boasts only one truly warm fauna, the Lusitanian or *Strombus bubonius* fauna, characteristic of the Eutyrrhenian unit. Probably these odd facts are related to the geometry and currents of the Mediterranean basin. The general mechanism (Ruggieri, 1967; see also Ruggieri, p. 141 in H. E. Wright and Frey, eds., 1965b) involves the

Table 24-I

Succession of Quaternary Marine Strata in Mediterranean Europe

		"Classical" sequence of authors	Sequence according to Bonifay (1964)	Versilian sequence from core boring near Pisa, Italy. Depths (m) below sealevel. (Blanc, 1937.)
Holocene		Flandrian	Versilian (c)	⌐ 1 Peat with fir, pine, spruce, alder ├ 11 Eolian sand
Pleistocene		Tyrrhenian	Neotyrrhenian (w) Eutyrrhenian (w)	├ 12 Marine sediments with *Purpura* fauna and seeds of *Vitis vinifera* ├ 27 Brackish–water silt
		Milazzian	Paleotyrrhenian (w)	├ 30 Peats and organic clays with fir and pine ├ 60 (18,500 C[14] yr?)
		Sicilian	Sicilian (typical)	├ 61 Marine sands with *Vitis vinifera* ├ 73
		Calabrian	Emilian Calabrian (c)	├ 74 Lacustrine silt and clay and eolian sand; land snails
Plioc.		Astian	Upper Pliocene	└ 90

Note: (c) — Includes "cold" elements. (w) — Includes "warm" elements.
Other units contain either mixed or nonindicative faunas.

incursion into the Mediterranean of "cold," rather deep-water taxa of northern Atlantic origin during glacial times and "warm," littoral taxa of southern Atlantic origin during nonglacial times. Data from sources other than invertebrate faunas, however, are still too scanty to permit firm correlation of the Mediterranean marine units with the glacial/interglacial succession. Versilian sediments are clearly equivalent to the Flandrian of the North Sea coasts (Blanc, 1937). On the west coast of Italy they extend from existing sealevel down to at least -90m and record two (Early and Late Weichsel?) episodes of cold climate and lowered sealevel. The Neotyrrhenian and Eutyrrhenian units of Bonifay probably belong to the Eem interglaciation, and the Paleotyrrhenian unit, somewhat deformed, may be of Holstein age. Correlation of the older units must await additional data.

CORRELATION WITH NORTH AMERICA

It is not yet possible to make really firm correlations between Europe and North America except as between the late and middle parts of the Weichsel/Wisconsin units. In those units similarities in both the climatic character and sequence of the events represented afford a good basis for correlation, supported by a general agreement among C^{14} dates. The early Weichsel/Wisconsin is not in the same class because it is less well controlled, but at least there seem to be no apparent conflicts. The reliability of an Eem/Sangamon correlation seems to be fair. On both continents it directly underlies the last glacial complex and records extensive marine transgression. If there were any radiometric dates to support it, the correlation could probably be called good (table 24-J).

Table 24-J
Reliability of Correlation of Principal Glacial-Stratigraphic Units Between Europe and North America

Northern Europe	Reliability of correlation	North America
Weichsel	(Good)	Wisconsin
Eem	(Fair)	Sangamon
Saale		Illinoian
Holstein	(Increasing uncertainty)	Yarmouth
Elster		Kansan
(Older units)		(Older units)

Table 24-K
Stratigraphic Sequence in European Russia Fitted to the Radiometrically Dated Sequence of Geopolarity Units and Correlated with the Classical Alps Sequence [a]

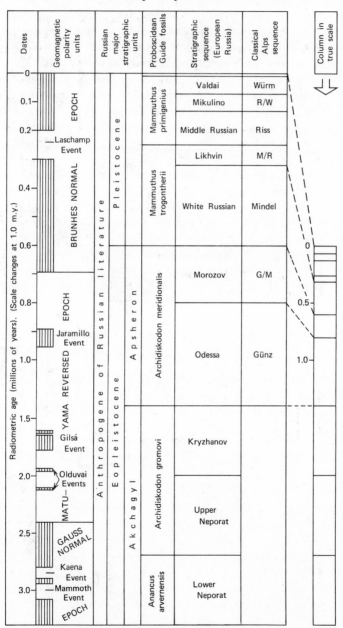

[a]From Gromov et al., 1969, p. 54, with minor adjustment and true-scale column added. For more detail in the Russian sequence see table 24-A.

Glacial-geologic units of pre-Eem or pre-Sangamon age are still not controlled by reliable radiometric dates, and there is nothing in physical characteristics nor content of fossils that compels correlation. Provisionally, therefore, we are obliged to consider correlations as uncertain.

Probably all the units listed in the table date from within the last 1 million years. Among older units the only entities that are potentially correlatable are the Villafranchian (European) and Blancan (American) provincial mammal ages. These lend themselves to K/Ar dating; the available dates indicate that the Blancan Age began substantially earlier than the dated Villafranchian Age (Evernden et al., 1964). However, owing to the fact that Villafranchian stratigraphy is poorly known and even less well controlled by dates than is the Blancan, good correlation is not yet possible.

Intercontinental correlation in the near future is most likely to be aided by use of the geomagnetic-polarity sequence. In a first attempt we can compare tables 24-K and 20-B, both of which were fitted to that sequence. The latter table includes only a modest number of radiometrically fixed points; the former contains very few. Indeed its stratigraphic units, in the vast length of time that lies beyond the reach of C^{14} dating, are placed mainly by means of fossils and other conventional techniques used in comparison with stratigraphy in western Europe. The tables are therefore sketchy, but what they do show is that apparently only a minor proportion of the last 3 million years or so is represented by the classical sequence of glaciations and interglaciations. This proportion recalls the early view of Hershey (1896), who visualized a long interval of erosion (in North America) in post-Pliocene, "preglacial" time, an interval that was at least as long as the aggregate time occupied by the glaciations. A similar view was held by LeConte, and later by J. E. Eaton.

CHAPTER 25

Quaternary of Asia, Australia, and Pacific Islands

The line that separates Asia from Europe extends from Novaya Zemlya southward along the crest of the Ural Mountains, onward to the Caspian Sea, and thence westward to the Black Sea. The Asiatic part of the Eurasian Continent is so vast that within it, despite the vigorous efforts of field scientists in the USSR, few areas are known in detail, many data are based only on reconnaissance, and some areas in the northeast are still virtually unexplored. Furthermore, literature on China published since about 1950 is little known in Western countries. In consequence our discussion of Asia must be more sketchy and of much lower quality than our discussion of the European region. We shall begin with Siberia— essentially Soviet Northern Asia—and then continue with China and other territories southeast, south, and southwest of it.

SIBERIA

Extent of Former Glaciers

General Distribution. In fig 25-1 we have plotted areas in Siberia that are believed to have been glaciated, distinguishing between maximum glaciation and glaciation mainly of Zyrianka age, since the last major interglacial. The plot is a compromise among various opinions and is very tentative. It shows that the Scandinavian Ice Sheet, when at its maximum, was confluent across the Urals with ice of Siberian origin, which in turn extended eastward to the shore of the Laptev Sea. East and south of that confluent ice mass, all Pleistocene glaciers were confined closely to

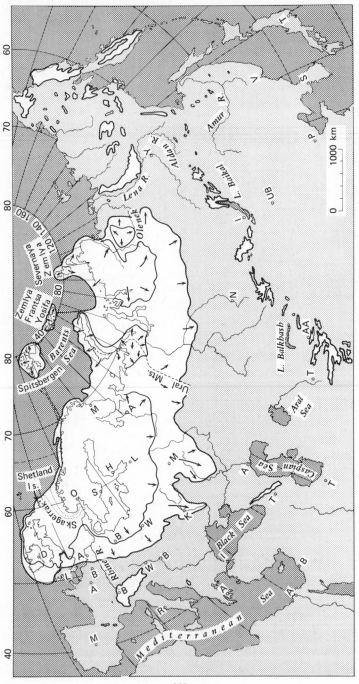

mountains, despite the fact that at some places in extreme northeast Siberia such glaciers reached the sea. Even the ice sheets and ice caps of northern Asia and Arctic islands were likewise centered on highlands; however, they are believed to have been thin. The largest single ice body, the Siberian Ice Sheet, was smaller than its Scandinavian neighbor and failed by several degrees of latitude to reach as far south.

The overall extent of glaciers in Siberia diminished from west to east, except in districts close to the Pacific Ocean, a local source of moisture. These relationships reflect the strong influence of a continental climate and imply that at glacial maxima, as today, most of Siberia was removed from sources of abundant moisture. Almost the only areas capable of supporting glaciers in Siberia today are Arctic-maritime; the mountains of the interior, though high, receive too much heat in summer to permit the existence of more than a few small glaciers. The relation of the former Siberian glaciers to sources of nourishment is discussed by Flint and Dorsey (1945).

With lower glacial-age temperatures, particularly in the summer season, glaciers were built on Arctic highlands by moisture from the Atlantic and, in early phases, from parts of the Arctic Ocean that were not yet frozen. Even today, despite the Ural Mountain barrier, Atlantic air masses penetrate into the interior of northern Asia. In extreme eastern Siberia glaciers received limited moisture from Pacific Ocean sources.

Along the polar front, where tropical air masses come into contact with cold polar air, activity should have been greater than now as a result of increased temperature differences between polar regions and low latitudes. At first, therefore, Siberian glaciers should have enlarged rapidly. But with complete freezing of the surface of the Arctic Ocean and growth

---

Figure 25-1 Sketch map of Europe and northern Asia, showing supposed boundaries of principal Pleistocene glaciers.

⌐﹁ Boundary (where known) of area glaciated in latest glacial age (not shown in eastern Siberia). Includes Weichsel, Kalinin and Ostashkov, Zyrianka and Sartan glaciations.

⌐﹁ Boundary of maximum glaciated area regardless of age. (Mainly Saale = Dnepr age).

·····... Alternative boundary on assumption that Scandinavian Ice Sheet covered floor of Barents Sea.

Arrows suggest, very tentatively, general directions of glacier flow in relation to the two boundaries. (Not shown in western Europe). *Note:* Boundaries in Siberia are indefinite. Compiled from many published sources, and from unpublished material for Norway supplied by Björn Andersen. Because some sources are not compatible, compilation is not of uniform quality, and in places is hypothetical.

of the Scandinavian Ice Sheet, precipitation in Siberia should have diminished, and the ice sheet there may have begun to shrink before the Scandinavian glacier had attained its maximum. Thereafter, probably just enough Atlantic air reached Siberia to maintain glaciers in equilibrium or slow decline during the remainder of a glacial age. The region was cold, and ablation values should have been continuously small.

Siberian Ice Sheet. The boundary between European Russia and Siberia is the long north-south line of the Ural Mountains, nearly coinciding with the meridian of 60°E, and the northern continuation of this highland in the Arctic island of Novaya Zemlya. The Urals reach extreme altitudes of about 1500m. The high northern part of Novaya Zemlya, at present covered by an ice sheet, reaches an altitude of about 1000m.

East of the 90th meridian and separated from the Urals by the broad lowland of the Ob' River is a series of three more highlands: on the south the Putorana Mountains (1500m), in the middle the Byrranga Ridge (1000m) forming the Taymyr Peninsula, and on the north the Severnaya Zemlya archipelago (450m) in the Arctic Ocean. Today this group of highlands receives greater annual snowfall than any other area of comparable size in the USSR. At the greatest glacial maximum (presumably the Saale maximum in European terms) not only all these highlands but the intervening lowlands were covered by a single ice sheet, with an area of more than 4,000,000km^2 but with a thickness, measured at the Urals, of only about 700m. Nowhere was it thick enough to bury the highland summits; probably it sloped away from each highland center of outflow and never possessed a simple, single domelike surface.

West of the Urals the ice sheet merged with the Scandinavian Ice Sheet, which in that region was thin because it, too, was poorly nourished. But as the Siberian glacier began to wane, the better-nourished Scandinavian ice expanded into the vacated region.

At the last major glacial maximum the Siberian Ice Sheet, like the European glaciers, was much less extensive than at earlier maxima. Though ice formed and flowed radially outward from each highland area, ice from the Urals and Novaya Zemlya reached southward only to lat 64°, where its margin is marked by a conspicuous end moraine. On the east it reached the Yenesey River only near the Arctic coast, where it appears to have coalesced with a Severnaya Zemlya-Byrranga ice mass. At that time the Putorana Mountains constituted a separate area of glaciation.

Central Siberian Plateau. The Central Siberian Plateau, 600 to 1000m in altitude, lying between the Yenesey and Lena Rivers, had a thin ice cap some 300km in greatest length. The Yenesey Hills, to the southwest, may have harbored small glaciers also.

Northeastern Siberian Mountains. The mountains between the Lena

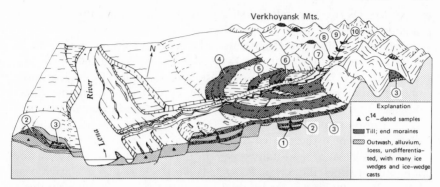

Figure 25-2 Block diagram showing end moraines, terraces, till, and other sediments between the Verkhoyansk Mountains and the Lena River, near lat 66°, long 126°, in the Republic of Yakutia, northeastern Siberia. In places altitudes exceed 2000m. Till and end moraines of Zyrianka and Sartan Glaciations are shown, together with small end moraines of Holocene age. Glacial sediments are numbered in order of decreasing ages. Not to scale; width of diagram is measured in tens of kilometers. (Adapted from unpublished sketch by N. V. Kind and V. V. Kolpakov.)

River and the Pacific Ocean, attaining heights of 2000 to 4500m, supported a complex of valley, piedmont, and ice-cap glaciers, the extent of which is still not well known, although in places they have been mapped in some detail (fig. 25-2; Chemekov, 1961a). Whatever their actual extent, the glaciers are likely to have been thin because of deficient precipitation.

The influence of Pacific moisture sources is evident on the Kamchatka Peninsula, where glaciers reached present sealevel on the east coast but on the west coast descended only as far as 300m above sealevel.

Arctic Islands. All the major island groups in the Arctic Ocean, not only Spitsbergen and Zemlya Frantza Yosifa but also Novaya Zemlya, Severnaya Zemlya, the Novosibirskiye Ostrovi and Ostrov Vrangelya were completely glaciated at one or more times.

At their greatest maximum, glaciers from Novaya Zemlya and Severnaya Zemlya coalesced with glaciers on the mainland to form the Siberian Ice Sheet. Coalescence was made possible by the fact that the intervening water areas are shallow. The sea floor, whatever its degree of emergence may have been at that time, differed little in general from the low plains along the Arctic coast today.

Although little more than 200m high, the Novosibirskiye Ostrovi generated a thin ice cap 250km in diameter; its southern margin reached the mainland coast. Probably the De Long Islands to the northeast were glaciated also. Ostrov Vrangelya, 600m high, was the center of a smaller ice cap that apparently failed to spread to the mainland coast.

Southeastern Siberian Highlands. The highlands west of the Sea of Okhotsk, between the Amur and Lena Rivers, are 1200 to 2500m in maximum height. They carried discontinuous glaciers fed by high-altitude snowfall brought by westerly storms and by easterlies from the Sea of Okhotsk. The many highlands in the region of Lake Baykal supported a glacier complex much like that of the Yellowstone-Teton-Wind River highland region in the United States.

The massive Altai mountain system in the border region of Russia and Mongolia includes the Sayan Range (3500m), which had a glacier complex at least 300km in length and extending down to an altitude of 900m. This complex resembled those in the Sierra Nevada and Cascade Mountains in western United States. Details are given by Seliverstov, p. 117 in Saks, ed., 1966.

Glacial Stratigraphy

The huge size of Siberia, the fact that much of it is clothed with close-growing subarctic forest, and its inhospitable climate together explain why only two major glaciations and an interglaciation have thus far been recognized in the region. General data are given by Strelkov (1965). Table 25-A presents a recent stratigraphic scheme with correlations to the European sequence.[1] The European Weichsel is represented in Siberia by the Zyrianka and Sartan Drifts[2] separated by Karginsky Interstadial sediments; the correlation is said to be fairly firm. The European Eem is represented by both marine and nonmarine strata, and the Saale by two drift layers separated by interstadial peat and soil. Correlation of all these units to the larger European units is strengthened by similarities within the mammal faunas, the characteristics of which are sketched in the table.

Glacial Lakes; Marine Sediments on the Arctic Coast. Emerged marine sediments identified with both glacial and interglacial ages occur along the Arctic and Pacific coasts of Siberia, in places extending more than 300km inland (Troitskiy, 1966; 1969). Probably the emergence of such sediments is the combined result of crustal warping and fluctuation of sealevel. The region marginal to the Siberian Ice Sheet on the south, between the Ural Mountains and the Yenesey River, was the site of a proglacial lake or lakes when the glacier was at its last maximum. The impounded water, held by a broad glacier dam in the drainage basins of

---

[1] See also Zubakov, 1965; 1969; Gromov and Nikiforova, eds., 1969; Markov, Lazukov, and Nikolaev, 1965; Markov and Velitchko, 1967.

[2] Together termed Valdai Drift by some authors. In this usage Valdai=Weichsel.

the Ob' and the Irtush Rivers, was controlled by a spillway at the head of the Tobol River that discharged into a route now dry, leading to the Aral "Sea" east of the Caspian. Probably the development of this lake or lakes paralleled the sequences of lakes formed in Fennoscandia and in central North America during shrinkage of the ice sheets in those regions.

Environments in the Nonglaciated Regions

The vast region outside the rather limited glaciers spans as much as 25° of latitude. The belts of vegetation today, like those in North America, are latitudinal in a general way except where conspicuous highlands interrupt the pattern. At the last glacial maximum the pattern was similar, but was displaced toward the south (fig 25-3). Where it was not covered by glacier ice the Arctic Coast was flanked by a broad belt of tundra, which graded with long transition into cold steppe. The major highlands, where not covered with glaciers, were clothed in montane grass and herbs. Toward the southeast the steppe graded into woodland, with small areas of forest in Pacific Coastal areas receiving sufficient rainfall. Although generalized, the pattern, reconstructed from pollen data, presents a consistent and reasonable picture. It differs from the corresponding map of Europe (fig 23-8) in that it reflects a much drier, more continental climate except along the Pacific Coast. Maps that are more comprehensive areally, and that show more detail for more glacial ages (although on a small scale) are seen in Frenzel, 1968 (see also Frenzel, 1967).

The Siberian vegetation, thus dominantly treeless, supported mammal faunas in which grazers predominated. On the tundra were mammoth, musk-ox, reindeer, and lemming, while overlapping these and extending far southward through the steppe were woolly rhinoceros, elephants, horse, bison, antelopes, cattle, and various rodents.

SOUTHWESTERN ASIA

The former glaciers of Turkey (Birman, 1968; H. E. Wright, 1961b), Syria and Lebanon (Kaiser, 1963, p. 128, 147), Iran, and the Caucasus region were confined entirely to mountains and were fed by westerly winds bringing moisture from the Atlantic, the Mediterranean, and the Black Sea. The snowline rose eastward from western Turkey into Iran, but was 1200m to possibly as much as 1800m lower than today's position. In Turkey a single glaciation, consisting of two maxima separated by a deglaciation and correlated with the Würm, has been recognized. Data on glacia-

Table 25-A

Table 25-A

Tentative Pleistocene Stratigraphic Sequences in Siberia and North China and Partial Correlation with Europe [a]

Europe	Siberia		
		PHYSICAL STRATIGRAPHY	REPRESENTATIVE FAUNAS [c]
Weichsel		Marine sediments Sartan Drift; loesslike silt Karginsky Interstadial sediments; soil Zyrianka Drift; loesslike silt	Mammuthus primigenius, Coelodonta, Bison priscus Saiga, Rangifer, Ovibos, Dicrostonyx, Lemmus *(North)* M.primigenius, Coelodonta, B. priscus, Poephagus, Spirocerus, Procapra, Bos, Allactaga, Ochotona *(South)*
Eem		Kazantzevo Interglacial sediments; marine sediments; Chernozem soil	
Saale		Taz Drift; Sanchugov sediments (marine), loesslike silt Peat & soils of Messo Interstade Samarovo Drift; loesslike silt	M.primigenius, Coelodonta, Saiga tatarica, Bison priscus Ovibos, Rangifer, Alopex, Lemmus, Dicrostonyx *(North)* Mammuthus trogontherii, Bison priscus
Holstein	"Diagonal sands" (W.Siberia)	Alluvium, peat Tobol marine sediments	Palaeoloxodon antiquus, Ursus spelaeus *(West)*
Elster	"Blue loam Series" (W.Siberia)	Alluvium and lake sediments (N. reg.); alluvium in S. reg.	Mammuthus trogontherii, Dicerorhinus, Alces latifrons, Praeovibos, *(North)* Archidiskodon, Coelodonta, Equus sanmeniensis *(Transbaikal)* Equus cf mosbachensis *(West)* Palaeoloxodon namadicus, Equus sanmeniensis, Alces latifrons, Trogontherium cf. cuvieri *(East)* Equus sanmeniensis Gazella sinensis Sinocastro *(Transbaikal)*
	Ferruginous seq. (Baikal reg.)	Saardakh sequence (Lower Lena R. region)	Archidiskodon(?), Equus san— meniensis *(East)* Allohippus cf robustus, Para— camelus gigas, Elasmotherium, Antilospira cf. gracilis *(West)*
	Kyzylgir seq. (Altai reg.)	Betekei sequence (N.Kazakhstan)	Paracamelus praebactrianus, Trogontherium minus, T. cu— vieri
Pliocene		Karabulakh sequence (Zaisan region)	

[a] A more detailed chart appears in Pei, 1939.

[b] Kurtén and Vasari, 1960.

[c] Cf. Flerow & Sher (Flerow in Turekian, ed., 1971).

[a] Modified from V. I. Gromov, E. A. Vangengeim, N. V. Kind, K. V. Nikiforova, and E. I. Ravsky, p. 5–33 in Nikiforova, ed., 1965; Chang, 1968.

PHYSICAL STRATIGRAPHY	REPRESENTATIVE FAUNAS	
Panch'iao Alluvium ——— Panch'iao erosion ——— Alluvial and la—custrine silt / Malan loess s.s.	Mammuthus primigenius *(North)*	H
Alluvium, lacustrine sediments	Palaeoloxodon namadicus, Coelo—donta antiquitatis, Equus hemionus, Spirocerus, Crocuta crocuta, Megaloceros, Spirocerus	Upper Pleistocene
Upper Lishih Huangti — ———Chingshui erosion——— Loess Reddish brown soil Loess	Equus, Cervus, Miospalax fon—tanieri	Upper Pleistocene
Lower Lishih Huangtu — **Reddish clay** — **Zone C** — Alluvium	Equus, Nictereutes, Ursus angus—tidens, Siphneus cf. wongi, Lagurus simplicedens, Sus lydekkeri, Palaeoloxodon takuna—gai, Dicerorhinus kirchbergensis.	Middle Pleistocene
Loess (orange), reddish brown soils	Equus sanmeniensis, Dicero—rhinus? kirchbergensis, Para—camelus gigas, Spirocerus peii, Megantereon, Lagurus simpli—cedens	Middle Pleistocene
Choukoutien cavern sediments: [=upper Elster?][b] Upper Sanmen	Palaeoloxodon cf. namadicus., Equus sanmeniensis, Trogontherium cuvieri, Lepus wongi, Hyaena sinensis, Megantereon, Dicerorhinus kirch—bergensis, Euryceros.	Middle Pleistocene
Wucheng Huangtu — **Zone B** — ——— Huangshui erosion ——— Loess (red) with 6 soils Lower Sanmen Lower part of Tingtsun alluvium	Nihowan fauna: Proboscihipparion sinensis, Equus sanmeniensis, Coelodonta, Bison palaeo—sinensis, Nictereutes sinensis, Megantereon, Prosiphneus. Archidiskodon planifrons, Para—camelus	Lower Pleistocene
Red clay — **Zone A** — Local lake sediments; alluvium	Elephas, Equus, Bos, Paracamelus, Cervus, Ovis	Lower Pleistocene
——— Fenho erosion ———	Sinotherium Hipparion	Plio.

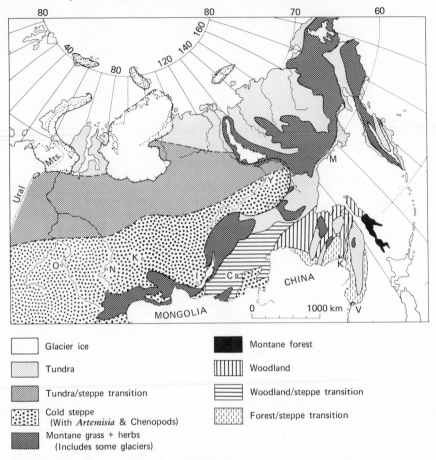

Figure 25-3 Natural vegetation zones in Siberia, around the maximum of the Sartan Glaciation, reconstructed mainly from pollen data. (Replotted on the base used for fig 25-1, from Giterman et al., 1968, p. 219).

tion of the Taurus Mountains, and of the Zagros Mountains on the Iran-Iraq boundary are given by H. E. Wright (1961b). Farther north and east in Iran the volcanic cone Kūh-I-Savalān (lat 38° 16′, long 47° 50′; alt 4240m) carried glaciers. The higher mountain groups of the great El-burz Chain (highest peak is Demāvend, 5671m) between Tehran and the Caspian Sea had valley glaciers that reached down to 1800m. The Zagros Mountains, dominated by Zardeh Kūh (4280m) west of Isfahan were gla-ciated, as was probably Takht-I-Shāh (lat 33° 20′, long 49° 22′; alt 4300m). Equally high mountains west, south, and east of Isfahan may also have had glaciers, even though the snowline rises toward the southeast.

At one time the main chain of the Caucasus (Great Kavkaz) was ice covered almost continuously through a distance of nearly 650km. The chain is high, with many parts of the crest exceeding 4000m, and culminates in Mt. Elborus (5600m). In the western part of the chain the snowline, now at 2700m, descended to 1400m and glaciers reached as low as 500m; in the arid eastern part snowline altitudes were and are much higher.

In the Little Kavkaz south of the main chain, five separate areas of glaciation have been recognized. These are on the higher parts of this chain, dominated by Mt. Aragats, (4095m) and Mt. Kiambil (3710m). In the Caucasus region two glaciations are inferred, but little is known about their correlation. In Syria, Mt. Hermon (2500m) and Mt. Lebanon (3070m) carried glaciers.

In the Zagros Mountains pollen data extend back to at least 40,000 years (H. E. Wright, 1968b; Zeist, 1967), and thus into the Middle Weichsel of the European continental sequence. The mountains were covered with an *Artemisia* steppe at the height of the last glaciation; the climate was therefore cool but dry. Changes in vegetation during deglaciation are known. Alluvial stratigraphy of parts of Iran is summarized by Vita-Finzi (1969).

The Quaternary stratigraphy (entirely nonglacial) of a part of Israel is given by Picard (p. 337–366 in H. E. Wright and Frey, eds., 1965b).

CENTRAL AND SOUTHEASTERN ASIA

The high mountain chains of central Asia form an irregular trident facing east. The long teeth of the trident are the Himalaya on the south, the Kunlun Shan in the center, and the Tyan Shan on the north. Between the first and second is the high plateau of Tibet; between the second and third is the great Takla Makan desert basin. The three mountain chains meet in a complex knot of ranges in the region in which Afghanistan, Pakistan, Sinkiang, and southern Asiatic Russia come together. The short handle of the trident extends westward into Afghanistan as the Hindu Kush and related mountains.

Northeast of the Tyan Shan another great chain, the Altai, with subsidiaries such as the Khangai and Kentei Mountains, reaches into Mongolia and onward almost to Lake Baykal. East of the Kunlun Mountains, in Sinkiang, Kansu, and eastern Tibet, are many high ranges separated by desert basins and plateaus. East of the Himalaya are the high ranges of southeastern Tibet, northern Burma, Sikang, and Yunnan.

All these chains are made up of numerous individual ranges, of which some are well known but others have hardly even been explored.

The ranges of the trident are high. The Himalaya, culminating in Mt. Everest (8730m), are the highest, but the Tyan Shan, Kunlun Shan, and some of the mountains in eastern Tibet, Sikang, and Yunnan have summits that exceed 6000m. Many of the higher mountains carry glaciers today. Throughout the entire region most of the higher mountains formerly supported numerous valley glaciers, many of which coalesced over the divides, but rarely did the coalescent masses reach the condition of mountain ice sheets.

Because of continental climate the existing snowline is high. It is domelike, standing highest in the central region and descending toward the comparatively moist Himalaya and the comparatively cold northern ranges of the Altai. The snowline of the greatest glacial maximum was arched likewise, rising from less than 4200m in Kashmir to more than 5700m in parts of Tibet and then falling northward to 5000m in the eastern Pamirs. Much farther north, in the Siberian Altai, it reached down to less than 2100m.

In the Pamirs, where the three mountain chains come together, glaciation was intense. Great valley glaciers as much as 250km long were formed; some of them coalesced at the mountain bases to form huge piedmonts. In some districts the glacier complex may have approached the condition of a mountain ice sheet.

An extensive compilation of what is known, particularly with regard to snowline altitudes, was elaborated by Wissmann (1959). Recent data on the Hindu Kush in Afghanistan are given by Grötzbach and Rathjens (1969). Nevertheless there remain many areas on which information is hardly more than conjectural. Wissmann calculated that lowering of the snowline ranged between 250 and 1100m, the amount depending chiefly on local precipitation.

Detailed studies of glacial stratigraphy have been few. One of them (Alešinskaja and Bondarev, 1969) equates glacial with local pluvial climates, as was done earlier in western United States. The Issyk Kul basin in Kirgiziya lies at the northern base of the Tyan Shan. Lacustrine sediments in the basin, supported by a few radiometric dates and pollen studies, record a deep pluvial lake, the latest high phase of which was coeval with the latest glaciation of the mountains, an event more recent than $\sim 26,000$ C^{14} years BP. Earlier climatic fluctuations are recognized, and mammal faunas are listed.

A second detailed study, in which several scientists have participated, is that of the long section exposed in the right bank of the Aldan River in Yakutia, about 300km upstream from the river's mouth. A summary is published by Boyarska (Bojarskaja, 1969). The sequence includes several layers of fine-grained alluvium and loess, with many fossil mammals and some pollen. Climatic fluctuation is indicated, but as opinions differ on interpretations within the section, the stratigraphic positions of the various units are not yet fixed firmly.

The mountains that flank the Kashmir basin in the northwestern Himalaya are the site of the "classic" glacial sequence in which four glacial-drift units were identified by De Terra and Paterson (1939), during a study based on earlier work by Dainelli. In nearby Swat Kohistan, Porter (1970) recognized a somewhat different sequence of three principal units (table 25-B); it resembles the upper parts of successions in western United States.

Study of fossil plants in the Kárewa lake sediments in the Kashmir

Table 25-B

Comparison of Glacial-Drift Units in Western United States and Swat Kohistan. Numbers in Parentheses Indicate Number of Recognized Substages.[a]

	CATEGORY I		CATEGORY II		CATEGORY III
	Sharp—crested, bouldery moraines; weak to moderate soils		Broad, smooth moraines with few surface stones; moderate to strong soils		Strongly weathered drift; moraines, if present, are subdued; depositional morphology generally poorly preserved; strong to very strong soils
Northeastern Sierra Nevada[a]	Tioga (2—3?)		Tahoe (2?)		Donner Lake and Hobart
Central Rocky Mountains[b]	Pinedale (3)	Time break	Bull Lake (2)	Greatest apparent time break	pre—Bull Lake
East—Central Cascade Range[c]	Cle Elum (3—5)		Kittitas (2)		Horse Canyon
Swat Kohistan	Kalam (3)		Gabral (2)		Laikot

[a] Birkeland (1964)
[b] Richmond (1960; 1965)
[c] Porter (1966; 1969)

[a] S. C. Porter, 1970.

Table 25-C

Sequences in Southern and Southeastern Asia

Northern Punjab, Sub–Himalayan India (Sahni and Khan, 1959, in part after W.D. Gill, 1951, and other authors.)				Upper Burma (Movius, 1949)	Java (Hooijer, 1956; Movius, 1949)
Stratigraphy			Paleontology	Stratigraphy	Mammal faunas
Pleistocene — Upper	Terraces and alluvium		*Equus asinus, Hyaena crocuta, Rhinoceros karnuliensis, Hystrix*	Red gravel and sand Pagan silt	
Middle			*Palaeoloxodon namadicus, Hexaprotodon palaeindicus, Equus namadicus, Bos namadicus, Rhinoceros dec–canensis*	Main Terrace Nyaungu Red Earth	Ngandong
	Upper Boulder Cg.			High Terrace	
				(Mogok Karst Fauna)	Trinil (700,000 BP?)
Lower	Siwalik series — Upper	Lower Boulder Cg.	*Equus sivalensis, Equus cautleyi, Bubalus platyceros, Rhinoceros platyrhinus, Rhinoceros siva–lensis, Hypselephas hysudricus, Bison sivalensis, Camelus siva–lensis, Hemibos acuticornis* and *Hemibos triquetricornis.*	Lateritic Gravel Uru Boulder Cg.	Djetis (2,000,000 BP?)
		Pinjor Zone		Upper Irrawaddy beds	Kali Glagah Tjidjoelang
		Tatrot Zone			
Pliocene	Middle	Dhok Pathan Zone	*Proamphibos kashmiricus, Hip–parion theobaldi, Hipparion antelopinum, Leptobos falconeri, Camelus sivalensis, Archidiskodon planifrons* and *Meryco–potamus dissimilis.*	Lower Irrawaddy beds	

basin, recorded in a series of papers by G. S. Puri, led to a synthesis of vegetational history and showed that the district was elevated by as much as 600m during the latter part of the Quaternary.

In the sub-Himalayan Siwalik Hills in the northern Punjab region of India, the thick Siwalik Series (table 25-C) consists of sediments built up by streams draining the rising Himalaya (W. Gill, 1952). Within this series three faunal zones are of interest: the Dhok Pathan Zone with a late Pliocene mammal fauna, the Tatrot, and the Pinjor. The Tatrot Zone, overlying the Dhok Pathan unconformably, has a Villafranchian fauna with *Archidiskodon,* impoverished in comparison with the (subtropical) Pliocene fauna below, and implying a cool-temperate climate (summarized in Movius, 1944, p. 16; Thenius, 1959, p. 289).

The fauna of the succeeding Pinjor Zone includes also *Leptobos* and *Equus,* and on other evidence it implies a climate warmer than that of the Tatrot. Here we have a suggestion of climatic fluctuation at the beginning of Pleistocene time. The higher part of the sequence consists of alluvium, loess, and terraces. Although not yet related firmly to the glacial sequence in the Himalaya, that stratigraphy does include vertebrate fossils and human industries (Movius, 1949).

Nonglacial Stratigraphy. The greater part of southern Asia, from the western Himalaya eastward to the Pacific, lacks altitude sufficient for glaciation. In river valleys and basins of that vast region beginnings have been made at constructing stratigraphic columns. Based mainly on alluvial stratigraphy, the component units are identified through fossil mammals, physical stratigraphy, and morphology, and are correlated between regions on these characteristics and on similarity of sequence (table 25-C). The Pleistocene Series is divided into Lower, Middle, and Upper parts, based mainly on fossil mammals.[3] The Lower Pleistocene is characterized by a Villafranchian fauna whose leading forms are *Archidiskodon planifrons, Equus,* and primitive cattle. The Lower Pleistocene faunas shown in the table, as well as the faunas associated with the Upper Irrawaddy Beds in Burma are Villafranchian. Furthermore, related faunas occur in strata in Ma Kai Valley and Yuanmo basin in Yunnan (Hooijer, 1956; Viret, 1954, p. 185) and in the Lower Sanmen beds near Peiping (table 25-A; Movius, 1944, p. 50).

On a basis of its Villafranchian fauna the Lower Pleistocene of southern Asia appears to correlate with European strata lying between the base of the Pleistocene and at least the top of the Tegelen. Closer correlation is not yet possible.

The Middle Pleistocene Narbada fauna of northwest India, Mogok Karst fauna of upper Burma, and Trinil fauna of Java differ among themselves (though all include *Loxodonta [Hesperoloxodon] namadica*) but also differ conspicuously from the Villafranchian. The Upper Pleistocene faunas, again, have a distinctive complexion.

Apart from faunal studies, beginnings have been made in the development of physical stratigraphy in Cambodia and Thailand (Takaya, 1967, p. 571) and in Ceylon, with tentative correlations with Thailand and north India (Takaya, 1968, p. 330).

China, Mongolia, and Korea

The highland of eastern Tibet, Sikang, and northwestern Yunnan lies at altitudes between 4000 and 6000m with at least one peak (Minya Konka) exceeding 7500m. Unlike western Tibet, which lies in the rain shadow of the Himalaya, this highland is not shielded from the moist southerly monsoon winds and is therefore not particularly dry. The climate is of high-alpine rather than steppe character, with much precipitation at high altitudes. Wissmann (1959) published a large map showing present and former glaciation. He (Wissman, 1937; see J. G. Andersson, 1939, p. 47) thought coalescent glaciation likely, represented it on a map, and discerned four glaciations. Richardson (1943; see also Wu, 1948) inferred that glaciers, including ice caps, had been virtually continuous above an altitude of about 4000m and had descended in some valleys to 3000m. Mei-Ngo (1960) recognized two glaciations of Yun Ling Shan

[3] Developed in Dietrich, 1953, and Hooijer, 1962. Takai (1952) reviewed the fossil mammals of the whole of eastern Asia.

(5900m), with end moraines at 3200m and 2800m respectively.

The Tsinling Shan, an east-west range in central China south of the Wei River between Lanchow and Sian, reaches altitudes of 3000 to more than 5400m. The higher groups of peaks are cut by cirques, some of which are as low as 4200m. Glaciation has been reported from the summit of the Datsin Shan (2800m) northeast of the great bend of the Hwang River in Suiyuan west of Peiping. In Outer Mongolia cirques at about 2400m occur in the Khangai Range (lat ~47°N, long 101° 30'E; alt 3000m).

The Lushan Range, in east-central China between Nanchang and Changsha, has been the subject of much debate. It has such land forms as cirquelike niches, hanging valleys, U-valleys, and moraine-like topography. On it also are diamicts interpreted as till. This assemblage has been taken to indicate glaciation of the range, despite altitudes that barely exceed 1800m (J. S. Lee, e.g., 1947). This interpretation was challenged by Barbour (1935) who thought an early Pleistocene glaciation, with the topographic details obliterated by erosion, a less likely explanation of these features than peculiar structural relations affected by extensive mass-wasting. Kozarski (1961) identified the diamicts as gelifluction sediments, but the overall question of former glaciation remains unanswered. According to Kobayashi (1965b), the currently recognized sequence of glaciations in China consists of a 1st (Poyang), 2d (Taku), 3d (Lushan), and 4th (Tali) glaciation. Each is considered to have been coeval with a pluvial phase, and each two are thought to be separated by an interglacial. We note that use of the name Lushan appears to refer to the controversy mentioned above; if so, the validity of the glaciation to which it is attached might be questioned.

Nonglacial stratigraphy in north China is summarized in tentative form in table 25-A. Much additional information, with faunal data, is given by Chang, 1962.

In northeastern Korea (fig 25-4) the Seturei Range (2540m) at the northeast end of the Kaima Plateaus bore small glaciers, as inferred from 18 cirques occurring at about 2000m and moraines as low as 1700m. The implied low snowline resulted from large precipitation values caused by the winter monsoon (Sasa and Tanaka, 1938). Farther southwest in Korea the Shan Alin, with peaks exceeding 2400m, was likewise glaciated (Berkey and Morris, 1927), probably under much the same climatic conditions as those in the Seturei Range.

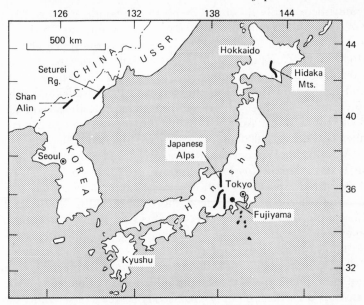

Figure 25-4   Sketch map showing mountain areas of Quaternary glaciation in Japan and Korea.

JAPAN AND TAIWAN

Glaciation

In describing the Quaternary of Japan we begin with glaciation and continue with nonglacial stratigraphy. The glaciation of Japan was confined to the higher parts of high mountains. Although it is the highest peak in Japan, the volcanic cone Fujiyama (3748m) bears no evidence of glaciation. As with other cones of recent date, the explanation probably lies in the destructive effects of post-glacial volcanic activity. Only two mountain groups (fig 25-4) show glaciation: the Japanese Alps (lat 36°) on the island of Honshu and the Hidaka Mountains (lat ~43° 30′) on Hokkaido. The former, whose highest peak reaches 3180m, consist of three ranges that in plan form an inverted Y. They are marked by discontinuous cirques having a pronounced NE aspect. The altitudes of their floors suggest a glacial-age snowline at~2700m, about 1200m below the calculated existing snowline, which passes above the mountain tops. Small end moraines occur downslope from many of the cirques, indicating the former presence of cirque and short valley glaciers. The glaciers were nourished mainly in winter by northwest winds, which today bring 800 to 2000mm of annual precipitation to the mountains. In the vicinity of the cirques and on lower peaks that bear no cirques rock glaciers and protalus ramparts occur, as well as block fields, patterned ground, and other frost-action features. Kobayashi (1958, p. 54) attributed all the glacial and frost-action features to the very late Pleistocene, and suggested that they represent three phases of cold climate. Radiocarbon dates of layers of tephra in a cirque within the Japanese Alps bracket two end moraines between ~27,000 and 15,750 years BP, suggesting that the glaciation was of Late-Wisconsin age (Kobayashi and Shimizu, 1966).

The Hidaka Mountains (max. 2040m) on Hokkaido bear 30 cirques, mainly with NE aspect and with floors at 1450 to 1600m; their relatively low altitude reflects their latitude, 8° north of the Alps. "Periglacial" features also are present in wide variety (Kobayashi, 1955). Three sets of end moraines occur (Horie, 1965, p. 179), suggesting two glacial episodes, but little is known of their dates or correlation.

Related to the glacial record, although more satisfactory from the point of view of climate, is the stratigraphic record of pollen, which, however, extends back through only the last 11,500 years or so. Shortly before that date mean annual temperature in central Honshu was 5° to 7° lower than it is today, and postglacial temperature fluctuations were rather like those in northwestern Europe (Tsukada, 1967).

The nonglacial stratigraphy of Japan is complex because Cenozoic tectonic activity has created so many sedimentary basins that there is little continuity of strata. Explosive volcanic activity has resulted in many layers of tephra, which have been exploited as key horizons as well as for radiometric dates. Magnetic-reversal stratigraphy likewise has been brought to bear on the sequence. Several correlation charts, mainly involving alluvial and volcanic strata, have been constructed, of which one, in adaptation, appears here as table 25-D. Other tables appear in Kobayashi, 1965a. Sakaguchi (1961; correlations on p. 504) made a special study of the stratigraphy of the Japanese Holocene, and a study of marine sediments along the coast, with correlation chart, appears in Takai and Tsuchi, 1963. We may note that the occurrence of elephants of various kinds in Japan at times during the Pleistocene implies a land connection with the Asiatic mainland. Whether the connection was continuous or discontinuous through time, it is likely to have been situated at the Korean Strait, where the distance is least and where minimum water depth today is between 100 and 200m.

Taiwan

In southeast-central Taiwan (lat 24° 10' to 24° 30'N, long 121° 00' to 121° 20'E) a NE-SW range of mountains more than 40km long culminates

Table 25-D
Quaternary Stratigraphy of Japan According to Ikebe [a]

Time (10⁴ years)	Magnetic-polarity units	Epochs	Stratigraphy — KINKI	Stratigraphy — South	Stratigraphy — KANTO	Leading mammal fossils	Fossil hominids and artifacts
0	BRUNHES NORMAL EPOCH	Holocene	Umeda formation	Yurakucho f.	Numa coral	Paleoloxodon naumanni; Mammuthus primigenius	Jomon; Hist. & Yayoi; Homo sapiens; Point; Microliths; Hand axes
1							
5		Pleistocene — Late	Lower Terrace (Itami)	Tachikawa Loam / Tachikawa Terrace; Musashino Loam / Musashino Terrace; Shimosueyoshi Loam			
10		Pleistocene — Middle	Middle Terrace (Uemachi); Higher Terrace (Meimi); ∿∿ Rokko movement (culminat.) ∿∿	Shimosueyoshi f.; Tama Loam; Byobugaura f.; Naganuma f.	Narita f.; Semata f.; Yabu f.; Jizodo f. (NARITA G.)		
50	E (Jaramillo event)	Pleistocene — Early — Upper		Kasamori form.; Nagahama f.; Chonan f.; Kakinokidai f.		Stegodon orientalis; Tomistoma	
100	MA — TUYAMA REVERSED (Jaramillo event)	Lower	Azuki; Pink — OSAKA Group; Katada formation	Kokumoto f.		Stegodon akashiensis; Archidiskodon proximus; Archidiskodon shigensis	"Nipponanthropus akashiensis"?
150	(Olduvai event)	"Earliest"	Askashi f.; Yellow — KOBIWAKO Group; Gamoh f.	Umegase f. — Ichijiku formation			
200	GAUSS NORMAL E.	(Lowermost); Pliocene — Late	Sayama f.; ANGE Group			Stegodon sugiyamai; Metasequoia flora	PLEISTOCENE / PLIOCENE
				Otadai formation — KAZUSA Group			

[a] Simplified from Ikebe, 1969.

in the peak Tung Shan (3997m). The summit of the range is a narrow plateau indented along its northern and eastern margins by 35 cirques, whose floors average 3600m. Moraines extend down to 3300m. (Kano, 1934–1935). Precipitation today is 2500mm. The implied glacial-age snowline, some 900m higher than in the Japanese Alps, reflects the difference of 12° of latitude between the two areas, despite greater precipitation in Taiwan.

Pollen study in central Taiwan (Tsukada, 1966) has led to a stratigraphy controlled by C^{14} dates back to ~35,000 BP. From it was developed the curve seen on the left in fig 16–10.

PACIFIC ISLANDS

We have included Japan and Taiwan with the Asiatic mainland, with which they are grouped in much of the literature, but we have still to deal with other islands, especially Hawaii, New Guinea, and the Galápagos, the first two because of their former glaciers and the last because of its pollen record.

Hawaii. Mauna Kea (4180m), a volcanic mountain on the Island of Hawaii, was glaciated down to about 3200m by a small ice cap believed to date from the Wisconsin Age. The snowline of glacial maxima is inferred to have been some 600m lower than the existing snowline at 4250 to 4550m. Wentworth and Powers (1941) saw in the glacial sequence four sedimentary units which they interpreted as drift. Three of the units they suggested were pre-Wisconsin. However, H. Stearns (1945) contended that only the youngest of the four is of glacial origin, consisting of till and outwash. He assigned it to the Wisconsin, and regarded the next-oldest unit as fanglomerate and the two oldest units as volcanic-explosion debris. The question whether Mauna Kea was high enough in pre-Wisconsin time to generate glaciers remains unanswered. In view of the extremely recent age of the Hawaiian cones, pre-Wisconsin glaciation may not have occurred.

Apparently the only published pollen study in Hawaii is that made by Selling (1948). The resulting pollen stratigraphy is confined to the last few thousand years, dated only in a relative way because the study antedates the radiocarbon method. However, within their short span the Hawaiian diagrams are rather similar to those from western Europe.

Galápagos. The Galápagos Archipelago, on the equator at long 90°, off the coast of Ecuador, has been an object of scientific interest since Darwin visited it in 1835, and is currently being investigated through fossil pollen. A preliminary paper (Colinvaux, 1968) described the coring of the sediments of a crater lake, 6m deep and without outlet, on Isla San Cristóbal (Chatham I.). Unpublished results supported by C^{14} dates suggest that through at least 40,000 years prior to ~10,000 BP the lake was dry, and that since 10,000 BP it has contained water. These suggestions, if confirmed, will raise interesting questions about the climatic history of the islands, which seems to have been similar in some respects to that of equatorial East Africa (chap 26).

New Guinea and Borneo. Although close to the equator and clothed mainly with rainforest, the islands of Borneo and New Guinea possess highlands that have been glaciated. One of the highlands on New Guinea,

Table 25-E
Critical Data on Pleistocene Glaciation of Peaks in New Guinea and Borneo [a]

New Guinea

Highland	Position	Alt (m)	Snowline alt (m)		Lowest evidence of glaciation (alt, m)
			Existing	Glacial —age	
Carstensz Toppen	4° 05'S 137° 09'E	5,030	4,650	3,485	2,120
Juliana Top	4°39'S 140°18'E	4,702	4,545	3,545	2,921
Mt. Wilhelm	5° 46'S 144°59'E	4,680	4,666	3,560	3,303

Borneo

Mt. Kinabalu	6°03'N 116°32'E	4,077	>4,077	3,710	3,210

[a] Data from Koopmans and Stauffer, 1968; Reiner, 1960; John Chappell, unpublished.

Carstensz Toppen, even today supports a small ice cap. According to Dozy (1938) the atmosphere surrounding the glaciated summits in New Guinea is so moist that a very small reduction of temperature would re-create the former glaciers. One valley glacier was 16km long and descended to 2000m. Table 25-E embodies the critical data, with two published references, on some of the summits known to be glaciated. Other summits in New Guinea that are believed to have been glaciated include the Salawaket Range and Mt. Giluwe, both near or exceeding 4000m.

AUSTRALIA

Glaciation

Although comparable with the conterminous United States of America as to area, Australia differs from it as to topography and climate. Its northern coastal region is tropical, its southern coastal belt is warm temperate, and its vast interior region is arid. Mountains are few; the highest peak attains only 2215m, and there are no existing glaciers. In consequence Quaternary environments were quite different. Glaciation and

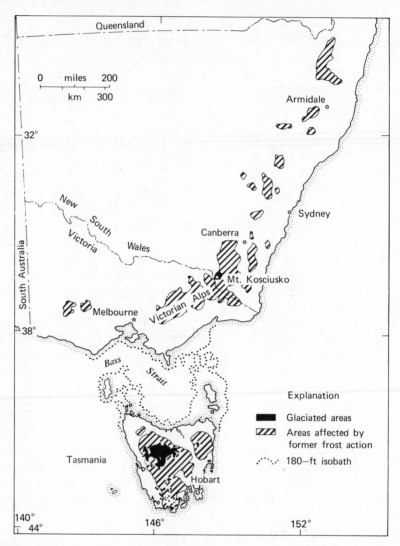

Figure 25-5 Sketch map of southeastern Australia, showing areas covered by Pleistocene glaciers and areas covered by mantles of gelifluction or solifluction sediments.

frost action were sharply restricted, and as climates fluctuated the changes related to them were concerned with water tables, lakes, stream regimens, and vegetation.

Former glaciation (fig 25-5) of the mainland was confined to a single group of confluent cirque and valley glaciers in the Snowy Mountains. The ice covered a combined area of barely 52km² and never attained the status of an ice cap (Galloway, 1963). The orographic snowline stood at about 1970m. As can be seen in fig 25-5, mantles of gelifluction and solifluction sediments were far more widespread, in the Snowy Mountains extending down to ~1000m. Block fields and patterned ground also occur in limited areas. Frost-action effects in Victoria were described by Talent (1965), and those in the Snowy Mountains by Costin et al. (1967).

In Tasmania, reaching to nearly 44° S lat, glaciation was far more extensive, although scattered and confined to high areas (J. N. Jennings and Banks, 1958; Derbyshire et al., 1965; Derbyshire, 1968; Galloway, 1963). Most of the ice bodies were clusters of cirque and valley glaciers, but the Central Plateau (lat ~42°; max. alt 1600m) bore an extensive ice cap. The outer limit of the ice is fixed at but few points because deep dissection and dense forests make field study difficult. Glaciation, however, was locally intense, the former ice was at least 300m thick in some places, and individual end moraines as much as 240m thick are known. Ice was thicker on the western and northwestern flanks of the Plateau than it was elsewhere, implying that related moisture-bringing winds were about the same as today's winds. Contemporaneous gelifluction or solifluction extended down at least 600m below the highest parts of the plateau.

Although only one glacial stade, marked by fresh little-weathered drift, is firmly recognized thus far, several exposures on the southeast flank of the plateau reveal drift that is strongly oxidized and drift containing clasts that are completely decomposed. On the northern flank a friable till overlies a lithified till. Such occurrences as these suggest that a stratigraphic sequence is present, at least in places.

Wood from the fresh drift, said to be from an end moraine near Queenstown, was dated at ~26,000 C¹⁴ years BP. If the occurrence is as stated, this date represents the initial advance of the King outlet glacier of the Central Plateau ice cap, an event that would correlate with the Late Wisconsin in North America. Similarly, according to Costin et al. (1967, p. 990), the glaciation of the Snowy Mountains occurred at least 15,000 to 20,000 C¹⁴ years ago, as estimated by the dates of peat overlying frost-action features correlated with the glaciation.

Nonglacial Features

Evidently glaciation played but a small part in the response of Australia to climatic changes. There can be little doubt that this was a consequence of generally low latitude and scarcity of substantial highlands. Zones of weathering in the stratigraphic sequence, pluvial lakes, and extensive dunes are among the conspicuous nonglacial features that are closely related to former climates. Apart from areas along the eastern half of the south coast, few occurrences of late-Cenozoic marine strata occur in Australia. From this it can be inferred that most of the continent as we now know it has stood above sealevel during that time. During the last ten or even twenty million years, therefore, opportunity for long-continued deep weathering has been great. In Western Australia, where late-Cenozoic climates have been generally warm, lateritic weathering is common, and in places is exposed to depths as great as 30m. The inception of weathering may date from Pliocene or even Miocene time. In places two zones of lateritic weathering are exposed, the younger one thinner and separated from the older zone by a surface of unconformity. Zones of lateritic weathering are exposed on the coast of Victoria and in the great dry interior region.

Analysis and tracing of ancient weathering zones play a relatively important part in field research into the late-Cenozoic history of Australia because there are comparatively few other features to claim the attention of scientists. Not only are ferruginous lateritized materials present, but materials cemented by secondary calcium carbonate and by secondary silica also are widely exposed.[4] In the wide interior region two and even three successive surfaces of Tertiary age are identified in places, each surface marked by zones in which one or more of these substances have accumulated. Probably the activity that led to the creation of each surface was tectonic, although the possibility of climatic change has not been excluded.

Pluvial lakes occur in Australia, particularly southern Australia, but as noted in chap 17 few firm conclusions as to former climate have been drawn from them. The most conspicuous indicators of climate are the striking and extensive systems of dunes that cover much of the great arid central region (fig 25-6). The dunes are of the longitudinal type, and within the region they cover, annual rainfall values vary from a little

---

[4] Some Australian scientists routinely apply to such materials the apt terms *ferricrete*, *calcrete*, and *silcrete*, long ago introduced by Lamplugh (1902). Lamplugh defined the terms as sediments (*materials* might be more appropriate) cemented respectively by secondary iron oxide, secondary calcium carbonate, and secondary silica. These terms are employed likewise in southern Africa, and some believe they deserve wider use.

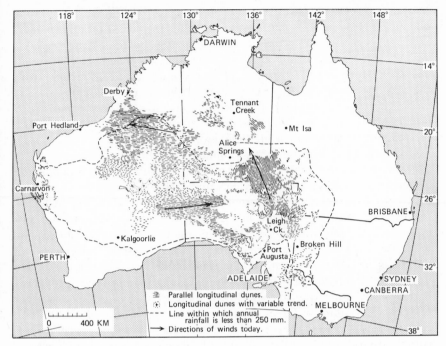

Figure 25-6   Sketch map showing longitudinal dunes in the central region of Australia and their relation to rainfall today. (Dunes after J. N. Jennings, 1968).

more than 250mm down to as little as 120mm. In some areas desert brush grows sparsely in the sand, but over large areas the dunes are bare. As the figure shows, the general wind system today forms a counterclockwise pattern. In the southern winter, southern Australia lies within the belt of westerly winds, whereas in both winter and summer the northern part of the dune region lies within the belt of Southeast Trades.

Hence both the winds and the precipitation of today are compatible with the presence of dunes and with their pattern; no major change seems required to explain the dunes. In this way Australia differs from some large dune areas in North America and Eurasia. However, little is known about the dunes in detail, and we cannot exclude the possibility that different groups of the dunes may have originated at different times. Like those in North Africa, most of the groups of Australian dunes lie in the lee of bodies of alluvium and probably were derived from alluvium. However, no evidence yet found points clearly to an interglacial origin. It is not even known how far back in time the process of dune building began.

In the extreme southwestern and southeastern parts of Australia, in the belt of semiarid country that nearly surrounds the arid interior, are many occurrences of lunettes (chap 9). Most lunettes are related to dry lakes and are relict from former climatic environments. They are rich in silt and clay, which were transported in sand-size aggregates in analogy with the clay in the North American paha. However, most lunettes also include quartz sand and even tiny pebbles, and some lunettes consist predominantly of quartz sand. The latter are related to dry lakes having quartz-sand beaches, which grade lakeward into silt.

Clay lunettes have been rather commonly attributed to the deflation of lake floors exposed during dry times. On the other hand, sand lunettes are believed by Bowler (in Mulvaney and Golson, eds., 1970) to have resulted from deflation of beach sediments while the beach was being actively replenished by surf and hence while the lake existed. In Bowler's opinion most clay lunettes were built during change from a cold climate with low evaporation to a warmer one, in which seasonal efflorescence of salts assisted deflation as long as the water table lay close below the lake floor. Stratigraphy of some lunettes shows quartz sand systematically overlain by clay, indicating a cold climate that later changed to a warmer one. In the Willandra Lakes, on the western edge of the Riverine Plain in New South Wales (chap 11), C^{14} dates suggest that such a change occurred around 15,000 BP, a reasonable date in terms of mid-latitude chronology in the northern hemisphere.

Elaboration of a stratigraphic sequence for any part of Australia is beset with difficulties. Sediments with marine invertebrates are confined to discontinuous coastal areas and are not abundant. Sediments with vertebrate fossils are patchy and few; knowledge of these is reviewed by Woods (1962). The study of fossil pollen is troubled by the fact that *Eucalyptus,* Australia's most common tree, is represented by hundreds of species, many of which have very similar pollen grains. Many weathering zones and ancient soils are distinctive, but means of dating them firmly are not yet available. Difficulties in interpretation of layers of alluvium like those in the Riverine Plain region are set forth in chap 11. Some of the obstacles mentioned or implied above arise from the comparative warmth and aridity of Australia. Water tables are low, and many Quaternary sediments are coarse, so that oxidation can be expected to have destroyed much datable fossil material.

The longest published sequences are derived from weathering zones and soils and are based on fluctuation of the precipitation/evaporation ratio. However, they are neither time calibrated nor correlated with sequences overseas. Examples include the qualitative curves developed by Whitehouse (1940, pl. 14) and by van Dijk (1959, p. 36), and the climatic sequence by Jessup (1961, tables 1, 2). These show several fluctuations between aridity and relative humidity but remain unevaluated. Van Andel et al. (1967) identified, off the northwest coast, sediments dating from ~19,000 to ~15,000 C^{14} years BP and indicating marked lowering of sealevel, cooler water offshore, and savanna vegetation on the adjacent land. They correlated the sediments with the last glaciation and inferred a climate substantially more arid than today's. They suggested that these conditions would be met by northward shift of the northern limit of the belt of westerly winds through 5° to 10° of latitude.

Climatic data inferred from fossils and other evidence are cited by E. Gill (p. 461 in Fairbridge, ed., 1961). A review of the climates of Australia today with an attempted reconstruction of former climates (Gentilli, p. 465 in Fairbridge, ed., 1961) assembles many interesting data from the literature.

Valentine (1965) analyzed a littoral marine invertebrate fauna at Port Fairy, Victoria, that reflects water temperature 1° to 2° higher than today's and that is probably interglacial. Inferred sealevel at the time the fauna lived stood at ~+5m. He noted that C^{14} dates on Quaternary marine fossils in Victoria are either infinite or <5000 BP, thus suggesting lowered sealevel during the last glacial age. Earlier Pleistocene stratigraphy is shown in table 25-G.

NEW ZEALAND

The New Zealand Alps, parallel with and west of the axis of the South Island, were widely and intensely glaciated (fig 25-7). In the higher areas, culminating in the Tasman Mountains (Mt. Cook, 3766m), many glaciers exist today. The former glaciers constituted a complex of valley and cirque glaciers and a few small mountain ice sheets. Altogether the complex was more than 600km long and nearly 75km wide. On the plains east of the mountains the ice spread out as piedmonts, and on the west it reached tidewater through deep fiords and possibly terminated in a floating shelf. Glacial erosion was intense, reflecting high precipitation (today approaching 2500mm at the higher altitudes) that resulted from the ideally maritime position of the New Zealand Alps. Stewart Island has a small area of glaciation on its highest part, at ~970m.

The North Island was nearly free of glaciation. The volcanic cone Mt. Ruapehu (2798m) whose crater is filled with glacier ice today, was formerly glaciated more extensively, with development of cirques having floors at ~2400m. Mt. Egmont (2610m), another volcano on the west

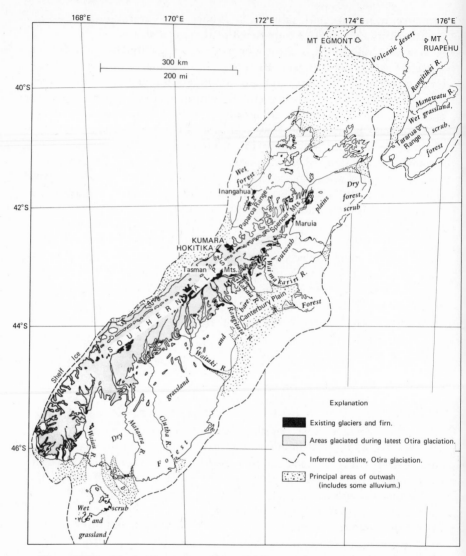

Figure 25-7 Glaciated areas, existing glaciers, and vegetation pattern at last glacial maximum in New Zealand. (Glacial data from an unpublished map by R. P. Suggate; vegetation from Wardle, 1963, p. 15.)

coast, has small glaciers. Farther south, the Tararua Range (1560m) was slightly glaciated.

The snowline (R. Willett, 1950) was lowered, on the South Island, by at least 900m below that of today. Outwash is well developed and is especially abundant on the South Island, where in places it constitutes successions of massive fans. Frost-action effects, chiefly in the form of gelifluction sediments, are widespread on parts of the South Island and occur likewise on the North Island. At one point such sediments extend below present sealevel, and in some places more than one gelifluction layer is present. The sediments are assumed to date from the Otira glaciation (table 25-F), when forests had disappeared from higher slopes, and frost

Table 25-F

Composite Pleistocene and Holocene Sequence, South Island, New Zealand [a,b]

Holocene	Dun Fiunary	
	Jacks Stream	
	Ferintosh	
	(Thermal maximum of Hypsithermal ?)	
	Birch Hill	
Pleistocene	Otira	3d. advance
		2d. advance
		1st. advance
	Oturi	
	Waimea	
	Terangi	
	Waimaunga	
	Waiwhero	
	Porika	
	Early glaciations	

[a] Compiled from Bowen, 1967; Gage, 1961; McGregor, 1967; and Suggate, 1965.

[b] "Warm" units are in italics.

wedging and movement of regolith were thereby favored.

Thin layers of loess occur on both islands; as many as three units have been identified. On the South Island, at least, they are believed to have been derived mainly from outwash.

At the height of the last glaciation Cook Strait (table 28-A), today ~100m deep, did not exist and the two islands were joined. During the postglacial rise of sealevel the strait was re-created prior to 9900 years BP.

From local sequences developed by various geologists a tentative consolidated glacial-stratigraphic sequence for the northern part of the South Island has been developed (Suggate, 1965), consisting of four major glacial units and three units representing warmer climate. In the Ben Ohau Range in the Southern Alps, McGregor (1967) recognized four glacial advances, each represented by a moraine, of Holocene age. Putting these together, we get the composite column in table 25-F. The "warm" units are represented by marine sediments, some of them fossiliferous, along the west coast and by wave-cut cliffs and benches. Interglacially red-weathered regolith of three different ages is identified on the North Island (Te Punga, 1964). Although not correlated with the sequence given, it implies warm, seasonally moist climate, such as might have prevailed at times when the belt of Southeast Trades lay south of its present position.

The early glaciations indicated at the base of the table consist of till interbedded with tectonic conglomerates of Early Pleistocene and possibly Pliocene age, folded and uplifted, at localities in the western part of the South Island. Such are the tills of the Ross glaciation of Gage (1961) and of the Old Man Group (Bowen, 1967). Considerable diastrophic movement occurred after the deposition of these sediments. The Kaikoura Orogeny began in the Pliocene and uplift continued throughout the Pleistocene. This ample elevation is regarded by some as the cause of glaciation on the South Island.

The pattern of vegetation during the Otira glacial maximum is represented on a map by Wardle (1963) and is shown in simplified form in fig 25-7, based mainly on data from pollen. Wardle believed that elements of today's flora survived the glacial climate in the far south and also in the northern part of the North Island. The altitudinal zonation of forest trees today is described in Wardle, 1964, and the climatic interpretation of Pliocene and Pleistocene fossil plants is discussed by Couper and McQueen (1954). Table 26-F gives pollen data pertaining to very late Quaternary time.

An important element in the nonglacial stratigraphy of New Zealand consists of continuous fossiliferous marine sequences exposed in the Wanganui basin and near Frankton, both on the North Island, and represented in a long boring near Timaru on the South Island (table 25-G).

Table 25-G

Late-Cenozoic Marine Sequence on the North Island, New Zealand, and in Southern Australia [a]

New Zealand units			Australian (Victoria) units		
Correlation	Series	Stage	Rock units		Stage
Holo—cene		Postglacial			
Pleistocene / Upper	Hawera	(Glacial— and inter — glacial stages)			
Pleistocene / Middle	Wanganui	Castlecliffian	Whaler's Bluff Fm.	Werrikoo Mbr.	Werrikooian
Pleistocene / Lower	Wanganui	Okchuan	Whaler's Bluff Fm.	Werrikoo Mbr.	Werrikooian
Pleistocene / Lower	Wanganui	Nukumaruan	Whaler's Bluff Fm.	Werrikoo Mbr.	Werrikooian
Pleistocene / Lower	Wanganui	Hautawan [a]	Whaler's Bluff Fm.		
Pliocene	Wanganui	Waitotaran		Maretimo Mbr.	
Pliocene	Wanganui	Waipipian		Maretimo Mbr.	
Pliocene	Wanganui	Opoitian			
Mioc.	Tara—naki	Kapitean			

Note. Foramimiferal zones are given in Jenkins, 1967, fig 2.

[a] In part equivalent to Waitotaran.

[a] C. A. Fleming, 1962; other references in Flint, p. 505 in H. E. Wright and Frey, eds., 1965b.

Both marine invertebrates and pollen indicate climatic cooling in Waitotaran time, with at least some fluctuation thereafter. This inference is the basis of the Pliocene/Pleistocene boundary shown in the table and K/Ar-dated at or somewhat earlier than 2.5×10^6 years.

Biogeographic data and problems are discussed compactly by C. A. Fleming (1962a). It is not known whether the near-absence of a fossil land fauna in New Zealand is the result of long-continued isolation or merely of the effective dissolution of fossils under the moist climate of a largely forested region.

CHAPTER 26

Quaternary of Africa, South America, and Antarctica

AFRICA

General View

Africa is the world's warmest and second-largest continent. It possesses few mountains; indeed most of it is plateau, great areas of which lie more than 900m above sealevel. Apart from the Rift Valley System, which formed in Miocene time and has been subject to active faulting and volcanism ever since, the African Continent has not undergone great changes since mid-Cenozoic time. With the exception noted, therefore, climate has probably played a major part in late-Cenozoic changes in weathering and soils (comments in Flint, 1959a), lake levels, eolian activity, possibly the regimens of streams, and the distribution of plants and animals.

Changes of all these kinds have occurred; they are real, at least in a qualitative sense. But their interpretation in terms of climate has failed as the basis of a consistent, understandable picture of glacial-age climates in Africa. The failure to achieve such a picture stems from deficiencies and defects, four of which seem the most important:

1. The small number of researchers and the small number of radiometric dates for a continent whose area is nearly four times that of the conterminous United States.

2. Uncritical interpretation of former climate from certain kinds of features.

3. A widespread assumption that all changes must have been synchronous and areally regular. This has led to widespread correlation

of local sequences with generalized "standard" sequences.

4. Failure to construct an adequate climatic model for glacial-age Africa.

Probably the last of these is the most significant, in that the existence of a meaningful model would act to support or invalidate item 3 above.

The prime basis for a model is an understanding of African climates today. Those climates differ greatly from the more regular climatic system that prevails in, say, middle-latitude North America. The difference arises in part from the fact that Africa is bisected by the equator and lies mainly between the subtropical belts of high pressure, which strongly influence its climates. Local topographic influences on climate are less important than in Eurasia and the Americas, because mountains are few; hence African climates depend chiefly on the broad atmospheric-circulation pattern.

Theoretically Africa lies within the two trade-wind belts on either side of the equator. The symmetry of this arrangement is, however, distorted by the presence over tropical Africa, and in summer over Asia, of an area of low pressure set up by greater heating of the land than of the sea surface. It is distorted further by seasonal shifting of this equatorial pressure trough over Africa, as well as by shifting of the belts of subtropical highs. The result is a seasonal system of winds and precipitation.

The zone of greatest heating, coinciding with the overhead position of the Sun, shifts from the southern hemisphere in January to the northern in July. The pressure trough follows, lagging about a month behind. The subtropical highs shift in the same directions, but less far. The inflow of maritime air, occurring as discrete air streams associated with the subtropical highs over the ocean, brings moisture onto the continent, precipitating it according to season. The precipitation is both orographic and convective, according to locality, temperature, and time. In southern Africa the traveling weather systems common in middle latitudes are weakly developed and are confined mainly to the far south and to the winter.

The seasonal character of the climates of equatorial and southern Africa is best seen in a comparison of the January and July air-circulation patterns. In January (fig 26-1) the pressure trough lies close to the equator and often merges with a monsoon low centered near 20°S over Rhodesia. Winds are strong. Three air streams dominate the circulation:

1. Along the west coast from the Cape to the mouth of the Congo the southeast trade-wind air stream is deflected so as to parallel the coast, giving rise to southerly and offshore winds. Southerly winds activate the cold Benguela Current in the Atlantic Ocean; the current flows north

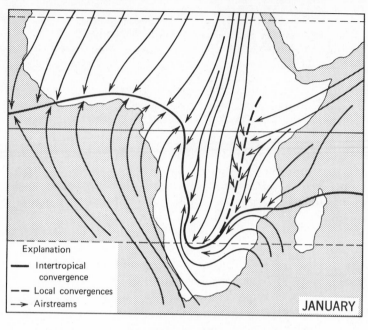

JANUARY

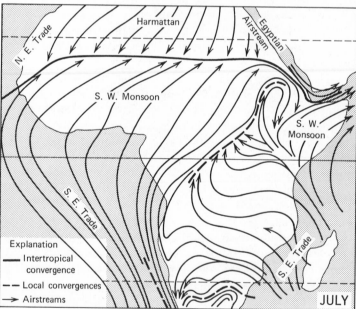

Explanation

—— Intertropical convergence

-- Local convergences

→ Airstreams

JULY

Figure 26-1 Mean zones of convergence and air streams over equatorial and southern Africa in January and July. From Flint (1959a), after Wellington; Brooks and Mirrlees.

along the west coast of South Africa. Offshore winds drag surface water westward away from the coast; this water is replaced by cold water welling up from moderate depth close inshore. Over the adjacent land the wind system, aided by the slope of the ground toward the coast, causes a general descent of air from the plateau in the interior to the coast, and cloudless skies result. Except over the extreme southern tip of the continent, the same is true in July. In consequence much of southwestern Africa ranges from semiarid to intensely arid.

On the southeast coast, in Natal and southern Moçambique, the southeast trade-wind air stream is accelerated by monsoonal indraft. Being onshore and having passed over ocean water about 5° warmer than the water off the west coast, it is quite unstable and precipitates rain on southeastern Africa, particularly on highlands. This is the cause of the rainy season (November to March) in that region.

2. Farther north along the east coast during the northern winter, an air stream that begins as the Asiatic northeast monsoon and merges into the trade-wind system is drawn south across the equator by monsoonal indraft. This air stream brings rain, chiefly convectional, to Kenya; most of the rain falls in November, immediately following the overhead passage of the Sun. This air stream contributes similarly to the southern summer rains in southern Tanganyika and Rhodesia.

3. Meanwhile the third air stream, associated with the flow from subtropical-high cells over Eurasia, crosses the equator at the height of summer and penetrates southern Africa as far as the Tropic of Capricorn. As this air is mainly of continental origin, it is comparatively dry; rain is scarce except on the high, cool East African mountains.

The intertropical convergence zone separating the southeast trade-wind air stream from the two northern air streams, and the convergence zone separating the two northern air streams from each other, are shown in fig 26-1. It is evident that most of the precipitation in southern Africa originates in the Indian Ocean.

By July the equatorial pressure trough and the intertropical convergence have migrated well north of the equator. The southern subtropical high-pressure belt likewise has shifted northward closer to the Tropic of Capricorn. The westerly wind system of the southern hemisphere, with its traveling cyclones, brings rain to the southern tip of the continent. With northward migration of the sun, the interior of southern Africa becomes cooler than the adjacent Indian Ocean and has developed a high-pressure system characterized by descending and outflowing air rather than by indraft of maritime air masses; this results in a dry season. Farther north, however, the equatorial east coast escapes the high-pressure

system, and, being also open to the southeast trade winds without inter-
ference by the island of Madagascar, receives rainfall from air masses flowing in from the Indian Ocean. On the west coast dry southwest winds
prevail as in January.

The distribution of mean annual rainfall resulting from this circulation pattern is shown in generalized form in fig 26-2.

In summary, the climates that characterize the various regions of the
continent are seasonal, but the seasons throughout the continent do not
coincide. Southwestern Africa experiences little seasonal change in wind
direction and is dry throughout the year. The western part of the coast
of South Africa has a Mediterranean climate with a winter rainy season
(May-August). The rest of South Africa, with most of Rhodesia and
Moçambique, has a pronounced summer rainy season and a dry winter
season. Much of equatorial eastern Africa has two rainy seasons, each
intermediate between summer and winter. These accompany the passage
of the overhead Sun, centering in April-May and October-November,
respectively. The former season, because it partly coincides with the
northeast monsoon, is the wetter of the two.

Extreme northern Africa is a different case. In the northern winter rain
is common. The Mediterranean coast lies within the belt of westerly
winds, while the northeast trades are south of the Tropic of Cancer. In
summer, with the intertropical convergence close to that tropic, the Mediterranean coast is swept by the northeast trades and precipitation is very
slight. This relationship is like that in middle America and southern
Asia.

With the exception of the Mediterranean belt, then, Africa is marked
by a complex climatic pattern. Under today's temperatures the distribution of rainfall over the continent is the prime factor that defines the existing zones of vegetation. There was no ice sheet to influence them. Nevertheless the glacial data from Africa's few high mountains provide a fact
of basic significance. The regional snowline was lower than today's by
amounts consistent with those in Europe. This justifies the inference that
glacial-age temperatures were lower than today's by a few degrees, and
hence furnishes a basis for the construction of a glacial-age climatic
model. A rudimentary model was attempted by Zinderen Bakker (p. 125
in Bishop and Clark, eds., 1967) and a rationale for "pluvials" in northeastern Africa was outlined by Flohn (1963). According to these models,
during a glacial age the belt of annual swing of the intertropical convergence was shifted southward, with effects on precipitation that differed
from one area to another. If this is what actually happened, it could account for much of the current confusion.

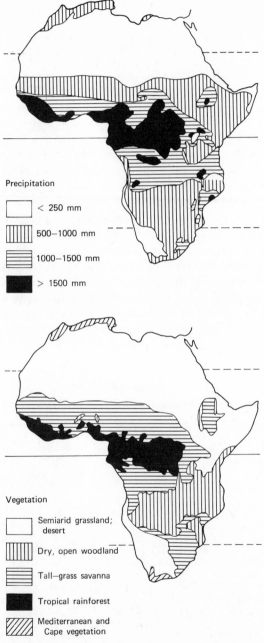

Precipitation

☐ < 250 mm

▥ 500–1000 mm

▤ 1000–1500 mm

■ > 1500 mm

Vegetation

☐ Semiarid grassland; desert

▥ Dry, open woodland

▤ Tall–grass savanna

■ Tropical rainforest

▨ Mediterranean and Cape vegetation

Figure 26-2 Precipitation values and vegetation belts (generalized) in Africa today. (After J. D. Clark, 1965.)

Glacial and Frost-Action Features

Our principal interest in the glacial and "periglacial" features of Africa lies in the fact, already noted, that they provide critical temperature data of basic importance for African climate. Also they have begun to yield stratigraphy that should eventually be comparable with that of Europe. These features are confined to high mountains in eastern Africa and Morocco.

The High Atlas Mountains in Morocco reach an extreme altitude of 4,165m and carry perennial snow today. Former glaciers were confined to the rectangle formed by lat 30°45′ and 31°25′N and long 7°15′ and 8°50′W. They occupied valleys whose heads are higher than 2900 to 3500m, depending on position and aspect. The largest glacier, more than 5km long, reached down to 2600m. End moraines are recognized but their stratigraphy has not been elaborated.

The rather scanty data on the former glaciers in eastern Africa are assembled in table 26-A. Of the glaciated highlands in the Kenya-Uganda-Tanzania region, close to the equator, numbers 2, 3, 4, and 5 are volcanic cones. The most complete single reference on the entire group is Osmaston, 1965. The highlands examined thus far display deglacial stades or

Table 26-A
Glaciated Areas in Eastern Africa

Highland	Lat. (approx.)	Long. (approx.)	Alt (m) (approx.)	Existing Glaciers	Lowest Altitude of Former Glaciers
1. Ruwenzori (Margherita Peak), Uganda	0°24′N	29°54′E	5119	Down to 4400 m	2000 (E) 3600 (W)
2. Sattimma, Aberdare Range, Kenya	0°19′S	36°38′E	3600	No	
3. Mt. Elgon, Uganda	1°08′N	34°33′E	4315	No	3400
4. Mt. Kenya, Kenya	0°10′S	37°18′E	5501	Down to 4900 m	3550[a]
5. Kilimanjaro, Tanzania	3°05′S	37°22′E	5897	Down to 4550 m	3575
6. Mt. Badda, Ethiopia	7°55′N	39°23′E	4133	No	
7. Simien Mountains, Ethiopia	13°14′N	38°25′E	4575	No	
8. Mt. Guna, Ethiopia	11°43′N	38°17′E	4206	No	Probably glaciated
9. Amba Farit, Ethiopia	10°53′N	38°50′E	3952	No	Probably glaciated
10. Mt. Chilalo, Ethiopia	7°50′N	39°10′E	3861	No	Probably glaciated

[a] Glaciated area ~430km^2

substades marked by moraines, of which the later members show good agreement among the highlands, thus implying that climatic change was a controlling factor. Radiometric dates, if and when they become available, could furnish a good check on this agreement. As yet there is little information on major glaciations previous to the last one, beyond the recognition of two drift sheets differentiated by weathering, on Mt. Kenya. The lowest former snowline inferred from Mt. Kenya data stood >1000m below the apparent existing snowline (Osmaston, 1965, table 6.12), if possible changes in precipitation are neglected.[1] This value is comparable with those on other highlands in low and middle latitudes, and argues for comparable reduction of temperature, at least at high altitudes, during glacial ages.

Nivation cirques, limits of frost wedging, and other features in the Lesotho Mountains, at lat 30°S, suggest a glacial-age snowline at ~3000m (G. Harper, unpublished), an altitude compatible with that cited above.

Intense frost shattering of bedrock occurred on the coast of Libya, and led to a semiquantitative climatic evaluation by Hey (1963). He concluded that in both Early and Late Weichsel times, even with no change from today's precipitation values, the district must have experienced a fall of mean temperature, sustained through the winter months, amounting to several degrees. This conclusion is compatible with the snowline data from African mountains.

Fluctuation of Lakes

In chap 17 we reviewed the data on former lakes in the desert region of northern Africa, and mentioned former lakes in Botswana and in the Eastern Rift Valley in Kenya. Among the latter, lakes in the Naivasha-Nakuru sector were studied by E. Nilsson (e.g., 1963), and as noted in chap 17 were re-examined by Washbourn (1967), and by J. Richardson (1966) who inferred from core data from the Naivasha basin that a larger, deeper lake than today's was in existence ~9000 BP and persisted until ~5600 BP. Then it shrank and possibly disappeared. About 300 BP the lake was reconstituted and thereafter underwent repeated fluctuation to the present time. This history implies a drying climate after ~5600 and a wetter one after ~3000, a sequence that fits, broadly at least, the relatively dry Hypsithermal followed by the wetter Neoglaciation discerned in middle north latitudes. R. L. Kendall (1969), after study of cores from the floor of the northern end of Lake Victoria, on the equator, indicated the following climatic sequence:

[1] Data on highlands in Ethiopia are summarized in Butzer, 1964, p. 311.

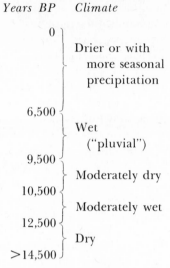

Years BP   Climate

If not viewed too closely the two records are in broad agreement. Both indicate a "pluvial", climate until roughly 6000 BP, a date at which climates in middle latitudes had been growing generally drier (and warmer) for at least three millenia. The presence of these lakes in Holocene time, both at or near the equator and also in the basin of Lake Rudolf at 3°N, suggests that the regional climate was not in phase with Holocene climates in Europe, and if so that this part of Africa experienced a pluvial or pluvials of a different kind from those of middle latitudes. More radiometric dates on more cores should help clear this matter up. Meanwhile a generally similar sequence is visualized by Servant et al. (1969) in the Chad region, far to the northwest, possibly suggesting a broader extension of these climatic conditions.

Windblown Sand

Windblown sand, predominantly in the form of longitudinal dunes, is concentrated chiefly in two parts of Africa, the great Saharan region ($\sim$8 $\times$ 10^{6}km^2) in the north and the Kalahari region ($\sim$2 $\times$ 10^{6}km^2) in the southwest. Both include inactive and active dunes, and in parts of the Sahara, at least, stratigraphic evidence of two or more generations of dunes exists. Both areas include alluvium, and in both it is suspected that the dune sand was derived originally from alluvial sources. The times of inception of intense eolian activity are poorly known but are supposed to be later Tertiary.

It is obvious that the extensive areas in which eolian sand is now covered or partly covered by vegetation were bare (and hence drier) at former times when the sand was active. In semiarid western Rhodesia, on the edge of the Kalahari region, parallel ridges of sand, some' of which are >100km long, occupy an area of at least 14,000km². Now covered with woodland, they are believed (Flint and Bond, 1968) to be remains of longitudinal dunes reduced to stumps by sheet erosion. If they are, a dry period of dune building was succeeded by a period of humidity sufficient for the inception of woodland and with rainfall intensity sufficient to support sheet erosion. However, no radiometric dates exist. Literature on the Kalahari region is summarized in Flint, 1959a, p. 363.

The climatic history of the Saharan region implied by dunes and by abundant traces of streams that no longer flow is a matter of debate. The evidence is conflicting, and the accumulated study of that huge area by a comparatively small number of scientists is not yet large enough to resolve the conflict. Many data are conveniently assembled by Monod (p. 119–229 in Howell and Bourlière, eds., 1963; see also Heinzelin, p. 292 in idem). They show a majority equating pluvials with European glaciations, but with some favoring a pluvial = interglaciation relationship. Although the conflict cannot be resolved from existing data, we note that the terms *pluvial* and *nonpluvial* connote fluctuations of climate possibly larger than those which may actually have occurred. Furthermore, there is evidence (e.g., J. D. Clark, p. 602 in Bishop and Clark, eds., 1967; Butzer, 1958) of moister climate in parts of northern Africa between ~9000 and ~5000 to 4000 BP, sometimes referred to as a "subpluvial." This evidence recalls the Holocene high lakes in northern East Africa, mentioned above, and suggests that we may be dealing with a different sort of "pluvial" from those evident in western North America and indeed in Africa itself.

Fossil Plants and Vertebrates

It is well established by pollen data that at former times the altitudes of floral zones on various African highlands have fluctuated. A few C^{14} dates indicate that on some highlands at least, vegetation zones were substantially lower at the time of the Weichsel glaciation in Europe (e.g., fig 16-10; Coetzee, 1967) than they are today. However, opinions differ as to the chief cause. Zinderen Bakker (1966) held that this was reduced temperature. On the other hand Livingstone (1967) maintained that increased precipitation was a more likely chief cause. The difference springs mainly from different interpretations of the ecology of the pollen of grasses. We agree that the matter is not yet wholly resolved.

It is unfortunate that the greater part of the pollen record thus far developed comes from relatively moist areas where fossil pollen has not been destroyed by oxidation, whereas the record of dunes and lakes comes in large part from drier areas. In this fact we again see the possibility already mentioned, that in parts of Africa "pluvial" climates may have been of more than one kind in terms of the chain of causes.

In connection with the established altitudinal shift of floral zones on African highlands we note the accordance with it of the inference by Moreau (chap 17) that montane forests were formerly more nearly continuous than they are today. That inference, however, was questioned by R. L. Kendall (1969).

Fossil pollen shows that at the times of Late Acheulean and Aterian cultures (both of Weichsel/Würm age), Mediterranean flora invaded the North African desert widely, implying precipitation >600mm.

As we have noted elsewhere, fossil land animals, mammals in particular, afford less resolution in the reconstruction of former climates than do plants, and the Quaternary mammals of Africa (general reference: H. B. S. Cooke, 1968) are no exception. Furthermore, because of their peculiarly African character they are difficult to correlate closely with European taxa, in part because throughout most of later Cenozoic time Africa was isolated from Europe and Asia and there was little or no communication. Hence application of the term *Villafranchian* to African faunas is inappropriate unless it is written "Villafranchian." It is thought likely that some of the faunal elements so designated in the literature are Pliocene. No evidence of wide population shifts like that in Europe and North America has been recognized in Africa, probably because the environmental changes in Africa were less profound. The living African faunas were divisible into three rather distinct groups: North African, East African, and South African.

Stratigraphy

With a few exceptions African Quaternary stratigraphic sequences are short, or undesirably tied by correlation to other sequences, or based in part on assumptions. Some of these disadvantages have come about through the erection of a "standard" sequence for East Africa that was based on climate rather than on objectively defined strata. Local sections in and far away from East Africa were correlated with it, and must be re-evaluated before they can be employed in syntheses. Critical analyses of this "climatic stratigraphy" will be found in H. B. S. Cooke, 1958 and in Flint, 1959b. Stratigraphic correlation sank to so low a state that an international conference met to establish new principles, and a lengthy volume (Bishop and Clark, eds., 1967; see esp. p. 375) resulted. After a decade or so of wide use of those principles it should be possible to erect sound regional sequences. Table 1 in H. B. S. Cooke, p. 180 in Bishop and Clark, 1967, represents an interim attempt for southern Africa. The volume just referenced contains much stratigraphic information, including sequences for the Rift Valley in Kenya, the Olduvai Gorge, and other critical areas.

Artifacts occur abundantly in Quaternary strata at many African localities. Although they are not, as individuals, a basis for correlation, assemblages of artifacts are so used, and their sequence is compatible with the sequence based on fossils and radiometric dates. Many African correlation tables include cultures routinely, as exemplified in the time-calibrated generalized table 26-B, and also in table 26-C, an example of marine sequences along the coast of northern Africa. Table 26-D gives a cultural stratigraphy for three broad regions of Africa. It is not yet possible to give a long and detailed correlation table apart from the cultural sequence because the geologic strata are compartmented into a large number of basins, many of them small. Such a table must await far more field study than has been accomplished to date. Meanwhile, out of many excellent local studies we mention those of Bishop (p. 293 in H. E. Wright and Frey, 1965b) on the Western Rift Valley in Uganda, which contains the Kaiso Beds with a Villafranchian-like fauna; Biberson and Jodot, 1965 on the Atlantic Coast of Morocco (chap 17); Rognon (1967, p. 492–530) on the Atakor Massif in the central Sahara, and Chavaillon (1964) on stratigraphy in the northwestern Sahara.

Table 26-B
Generalized late-Pleistocene Cultural Stratigraphy of Africa[a]

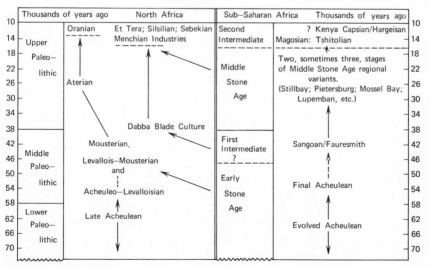

[a] J. D. Clark, 1965.

Table 26-C

Provisional Correlation of Marine and Terrestrial Strata in Atlantic-Coastal Morocco [a]

Mediterranean Marine Stages	Moroccan Stages		North African cultures	North African Proboscidean faunas	Major subdivisions	
	Marine	Terrestrial				
		Rharbien	Néolithique	Elephas africanus		
Versilian	Mellahien	Soltanien	Capsien Epipaléolithique Atérien		Late Pleistocene and Holocene	
Neotyrrhenian	Ouljien					
		Présoltanien	"Moustérien de tradition acheulienne"	E. atlanticus— maroccanus		
Eutyrrhenian	Rabatien	Tensiftien	Acheuléen évolué		Middle Pleistocene	
Paleotyrrhenian	Anfatien	Amirien	Acheuléen moyen Acheuléen ancien			
Sicilian II	Maarifien		Acheuléen primitif Pebble Culture finale	E. iolensis E. atlanticus		
		Saletien	Olduvaien	E. meridionalis		
Emilian	Messaoudien	Regreguien	Pebble Culture	E. recki	Early Pleistocene	
"Typical Calabrian"	Moghrebien	Moulouyen	Pebble Culture archaique	E. africanavus Anancus osiris Stylohipparion Lybitherium	Pliocene	
"Atypical Calabrian"			Sans industries			

[a] After Choubert, 1965.

SOUTH AMERICA

Climate, Vegetation, and Glaciers

Although it, too, is crossed by the equator and is affected by the annual swing of the intertropical convergence, South America has thus far presented fewer problems in Quaternary climate than has Africa. The difference results in part from the presence of the Andes Cordillera, running down through western South America from end to end of the continent. South of lat 20° this mountain chain is swept by westerly winds at all seasons; farther north it is affected by southeast trades and by monsoonal

Table 26-D
Cultural Stratigraphy for Three Broad Regions of Africa [a]

Absolute Dates B.C.	Absolute Dates BP	Cultures: Sub-Saharan Africa	Cultures (Stone Age)	Cultures: North Africa	Cultures (Palaeolithic)	Main Localities: South Africa	Main Localities: East Africa	Main Localities: North Africa	Main Pleistocene Divisions
		Wilton, Smithfield, Tshitolian, Elmenteitan	Late Stone Age / Inter-mediate / Second	Capsian, Later Oranian, etc.	Epi-Palaeolithic	Matjes River, Oakhurst, etc. Cave of hearths Kalambo Falls	Njoro, etc. Ishango, Naivasha	El Mekta La Mouillah, etc. Haua Fteah	Post Pleistocene
8,000	10,000								
		Magosian, Lupembo-Tshitolian		Oranian Et Tera	Upper Palaeolithic	Rose Cottage Cave Howieson's Poort	Magosi Gambles Cave	Taforalt	
		? Kenya Capsian	Middle Stone Age	Aterian and		Mumbwa	Enderit Drift, Malewa Gorge, Olduvai Bed V	Berard, Dar es Soltan, Mugharet el Aliyia, Sidi Mansur, Kharga Oasis	
18,000	20,000	Lupemban, Pietersburg		Levalloisian		Bambata, Mufo			
		Stillbay, Mazelspoort		Dabba Blade Culture		Boskop	Singa	Wadi Gan, El Guettar	
28,000	30,000					Florisbad	Dire Dawa	Ain Fritissa, Jebel Irhoud, Haua Fteah, Hajj Creiem, Sidi Zin, Sidi Abderrahman	Upper Pleistocene
		Sangoan and Fauresmith	First Intermediate	Levallois–Mousterian and Acheuleo–Levallois	Middle Palaeolithic	Broken Hill	Eyasi Beds		
38,000	40,000					Hopefield	Nsongezi (M/N Horizon)		
48,000	50,000					Montagu Cave of Hearths Kalambo Falls	Isimila	Lac Karar, Tihodaine, Adrar Bous	
		Later Acheulean Stage	Earlier Stone Age	Late Acheulean	Lower Palaeolithic	Cornelia	Olorgesailie, Kariandusi, Kanjera, Olduvai Bed IV		
						Vaal Younger Gravels	Olduvai Bed III?	Temara, Rabat, Ternifine, Sidi Abderrahman	
		Earlier Acheulean Stage		'Middle' Acheulean			Olduvai Bed II		
	490,000	Chellean Stage		'Early Acheulean' (Clacto–Abbevillian)		Kromdraai, Swartkrans, Sterkfontein Extension, Klipplaatdrif	Olduvai Bed I, Laetolil, Omo	Yayo, Ain Brimba	Middle Pleistocene
	1,230,000	Oldowan		Oldowan		Basal Older Gravels, Makapan, Sterkfontein, Taung	Kaiso, Kanam	Ain Hanech, Ain Boucherit	Lower Pleistocene
	1,850,000								

[a] After J. D. Clark, p. 20 in R. A. Lystad, 1965, *The African World*, F. A. Praeger, London.

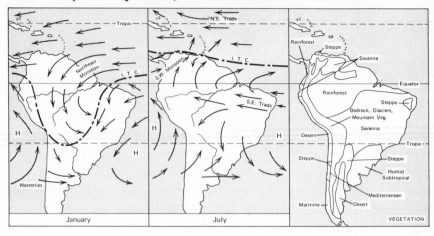

Figure 26-3 Mean positions of intertropical convergence and air streams over South America in January and July (Garbell, 1947, p. 108–109), and distribution of vegetation types (After Knoche and Borzacov).

winds (fig 26-3). These winds, in varying degree, bring oceanic moisture to the highlands, mainly in the local winter season, and maintain snow and glaciers in favorable places from Cape Horn to the Caribbean Sea. North of the equator moisture is brought from the northeast, but in the south it comes from the west. The existing snowline rises from <1500m in the far south to a maximum of ~6000m near lat 20°, north of which, in conventional fashion, it declines to <5000m in the equatorial belt. The glacial-age snowline as measured by cirque floors parallels the existing one at distances of ~500m to >1000m below it.

The distribution of glacial-age glaciers along the Cordillera (fig 26-4) varies roughly with snowline altitude. In the far south a mountain ice sheet more than 200km wide and in its axial part 800m to >1200m thick calved into the Pacific along the coast of Chile. South of lat 50° it formed a piedmont lobe that calved into the Atlantic. Farther north, however, the ice sheet became narrower. At ~43° the western margin of the glaciated region emerges from the sea and rises toward the north along the Chilean piedmont. On the Argentine side the ice formed a series of small piedmonts east of the Cordillera. At some point north of 38° the eastern and western margins withdrew into the Cordillera, and north of 30° the continuous area of glaciation breaks up into groups and clusters of what were small ice caps, valley and cirque glaciers, mostly at 4500 to 5000m. These continue northward, their spacing depending chiefly on summit altitudes. The most northerly ice body (Notestein, p. 620 in Cabot, 1939)

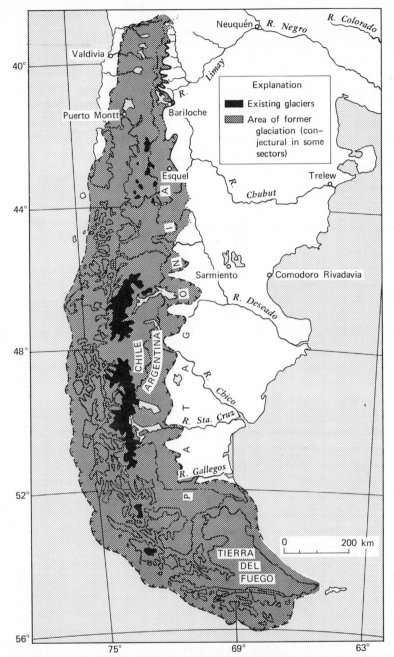

Figure 26-4 Existing and former glaciers in southern Argentina and Chile. Limit of former glaciers in Chile: compiled from Caldenius (1932), Heusser (1966, fig 1), and Feruglio (1949–1950, fig 227); in Argentina: north of 43° from Flint and Fidalgo (1964, 1969); south of 43° from Caldenius (1932) with local refinement by Fidalgo and Riggi (1965).

capped the Sierra Nevada de Santa Marta, facing the Caribbean in Colombia. The remarkable narrowness of the ice sheet and ice caps reflects
the rain-shadow effect of the high Cordillera. Throughout much of the
region east of the Andes and south of lat 35° annual precipitation is less
than 250mm. The eastward rise of cirque floors (fig 4-5) reflects the
warm, dry climate of the country to the east, which must have promoted
rapid ablation.

The eastern limit of glaciation south of 39° is distinct; none of the low
mountains east of it have been glaciated. The drift border between lat
41° and the Strait of Magellan was mapped in reconnaissance by Caldenius (1932) and between 39° and 43° was improved and detailed by Flint
and Fidalgo (1964; 1969). Several authors have asserted that local, low-
altitude glaciation affected the warm, dry country between the Andes and
the Atlantic. That the assertion is not supported by geologic features is
argued in the papers last cited and more fully by Polanski (p. 458 in
H. E. Wright and Frey, eds., 1965b). The error arose in part from interpretation of volcanic mudflows as till and of boulder lags left by erosion of
mudflows as glacial erratics.

Less is known of the glaciation of the Chilean flank of the Cordillera.
Early, basic data are given in Brüggen, 1950. Segerstrom (1964) summarized, with a map, data on Quaternary sediments but did not plot limits
of glaciation. The abundant rainfall, locally reaching nearly 3000mm and
generally comparable with that of the Scandinavian Mountains, has led
to effective erosion and to forest growth, both of which have impeded
field study of glacial drift. The abundant presence of volcanic-mudflow
sediments likewise has resulted in exaggerated interpretation of the limit
of former glaciation. However, wood and organic sediments immediately
underlying till at two localities near Puerto Montt (fig 26-4) are 15,000 to
16,000 C^{14} years in age. If, as it seems to be, the underlying drift (an end
moraine) is only a little older, it could be compared with Late-Wisconsin
drift in North America. Other dates from Chile are cited by Heusser
(1966, p. 279–281), with discussion of the possibility of glaciation of pre-
Late-Wisconsin age. Loess occurs locally overlying drift in both Chile and
Argentina.

Developed stratigraphic sequences are extremely few. The single one
that extends back through and possibly beyond the last major glaciation
pertains to the Argentine flank of the mountain ice sheet between 39°
and 43°. There three glaciations, possibly stades of a single major glaciation, from oldest to youngest named Pichileufú, El Condor, and Nahuel
Huapi (Flint and Fidalgo, 1964; 1969), were recognized and differentiated on a basis of weathering, as mentioned in chap 11. The same units
appear to be present throughout much of the region south of 43°. Near

lat 46° Fidalgo and Riggi (1965) described four layers of till and erected a stratigraphic sequence that included nonglacial units.

On the Chilean flank of the Andes near lat 41° end moraines, with C^{14} dates of ~18,000 and ~16,000 BP, evidently correspond with the Late-Wisconsin deglaciation in North America, whereas another moraine, on less firm evidence, may correlate with Early Wisconsin drift (Heusser, 1966, p. 280).

In the vicinity of La Paz, Bolivia, four glaciations (perhaps of the rank of stades) are described (Dobrovolny, 1962), named from oldest to youngest: Patapatani, Calvario, Milluni, and Choqueyapu. The bracket of time they represent is not known. Movement of the Andes occurred in post-Milluni time.

Another sequence is that on the Sierra Nevada del Cocuy, a part of the Cordillera Oriental at lat 6°N in Colombia (Gonzalez et al., 1965). It consists of four successive bodies of glacial drift so related to a local pollen sequence partly dated by C^{14} that it is known to span the last 14,000 years.

In Andean Peru three glaciations, estimated to have occurred within the last 13,000 years, are mentioned (without supporting data) by Lanning and Patterson (1967, p. 50). Lemon and Churcher (1961) found evidence in Ecuador of a wetter climate at some unspecified time in the Quaternary, but thought the coastal belt might have been as arid then as it is today.

The most striking unit of late-Cenozoic nonglacial stratigraphy consists of the Tehuelche-type gravels. These sediments constitute several veneers, of different ages, of pediment gravel, as well as of the products of its redeposition. Tehuelche-type gravels are everywhere thin, but they cover an enormous aggregate area of pediments, mostly dissected, in eastern Argentina south of about lat 42°, where precipitation averages ~175mm, and are traceable uninterruptedly to the Atlantic coast. The sediments consist mainly of rounded and polished pebbles of the most resistant rocks of the Andes and highlands to the east, in a matrix of reworked fine-grained tephra and nonvolcanic fines, commonly cemented with calcium carbonate. Tehuelche-type sediments underlie glacial drift; the ages of all bodies are believed to be Pleistocene.[2]

Other Quaternary pediment gravels occur almost continuously along the eastern base of the Andes at least as far north as 30°. In a detailed study of pediments, alluvium, loess, and volcanics in the Mendoza district, Polanski (1963, pl. 2) developed a stratigraphic chart. Quaternary

[2] Summary with references in Caldenius, 1940; detailed study in one area in Fidalgo and Riggi, 1965.

pediments occur also in southeastern Brazil. They were described by Bigarella and Andrade (p. 433 in H. E. Wright and Frey, eds., 1965b), who linked them with glacial/nonglacial climatic changes.

In connection with the Quaternary of Brazil, Damuth and Fairbridge (1970) reasoned from the occurrence of glacial-age arkosic turbidites in many sediment cores off the northeast coast of Brazil that much of that country was semiarid or arid during Late Wisconsin time.

Loess not only occurs near the mountains but also continues eastward to the sea and extends at least as far north as the Río de la Plata. It is not primarily related to outwash; most of it is desertic. Beneath wide areas it is not clearly distinguished from alluvium. An extensive field of sand dunes, largely inactive, covers part of the region between Buenos Aires and the Andes, but nothing is known of its age or origin. In the many coastal cliffs, however, sections of sediment as thick as 40m are exposed; from them was developed the Pampean Faunal Age (table 26-E; Frenguelli, 1950). The sediment is nonstratified silt and fine sand, probably alluvial, with fossil vertebrates (many extinct) including toxodonts, sloths, insectivores, glyptodonts, carnivores, and rodents. No plants are known. These fossils and the sediment itself suggest no climatic change; they seem compatible with a climate like that of today. Descriptive treatment of the Holocene part of the Quaternary sequence in Argentina is the work of Castellanos (1962).

Along the entire coast of southern Argentina are many emerged strandlines and bodies of marine sediment with invertebrates that have contributed to the marine faunal ages in the table (Feruglio, 1949–1950, v. 3). Probably most or all of them antedate the last glaciation. On the coast of Chile at La Serena (lat ~30°) a marine sequence spanning the interval from upper Pliocene through Holocene is developed from marine-invertebrate fossils occurring in a flight of terraces (Herm and Paskoff, 1967) somewhat like those on the coast of southern California.

Table 26-E
Late-Cenozoic Stratigraphy in Coastal Argentina [a]

	Rock - stratigraphic units		
		Nonmarine	Marine
Holocene		Querandí Fm	LaPlata Fm
Pleistocene	Pampean Group	Luján Fm	
		Buenos Aires Fm	Belgrano Fm
		Ensenada Fm	
Pliocene		Chapadmalal Fm	

[a] (Harrington, p. 153 in Jenks, ed., 1956, after Ameghino

Studies of pollen stratigraphy in southern South America are confined mainly to southern Chile, where the moist climate favors preservation of pollen grains. Table 26-F shows results from three localities, one of them in Tierra del Fuego. Although the sequences are short, they are reasonably similar in their climatic implications to Holocene pollen sequences on other continents. The differences among them may be caused in part by lack of close control by C^{14} dates; if and when that cause is eliminated, any residual differences may hold interest for their bearing on secular climatic history.

ANTARCTICA

Antarctic and Subantarctic Islands

Table 26-G summarizes glacial information about the larger islands in Antarctic and subantarctic waters (fig 3-14). From it appears the fact that virtually all the islands now have glaciers; all or nearly all were formerly more extensively glaciated, some completely so.

Macquarie Island south of New Zealand presents a peculiar problem (Mawson, 1943, p. 45–54, 76–83, 92). Surrounded by deep water, this island has been completely glaciated. Striations and indicators show, however, that the former ice did not spread out from a center on the island but crossed the island from farther west where now there is only deep water. Not only were rock fragments carried to the east coast from outcrops on the west coast; they were lifted 363m in the process. Therefore the glacier must have been large. The present mass is regarded as a residual horst left by profound postglacial block-faulting movement in a land mass most of which is now deeply submerged.

Antarctic Continent

According to the evidence of Tertiary fossil plants, before Miocene time the Antarctic Continent (fig 3-14) seems to have enjoyed a temperate climate. Indeed up to about the beginning of the Oligocene the continent appears to have been attached, in its southwestern part, to Australia. By late-Miocene time, however, ice had appeared in West Antarctica and was already rather extensive. This is inferred from tillite in the Jones Mountains (near the coast of West Antarctica at lat 74°), containing erratics derived from the Ellsworth Mountains and other areas well to the east; the tillite is $\sim 10 \times 10^6$ K/Ar years old. In the McMurdo Sound region east of the Ross Ice Shelf an outlet glacier of the ice sheet in East Antarctica existed at least 4×10^6 K/Ar years ago. Data from a core in sea-floor sediments indicate that bergs from calving Antarctic ice

Table 26-F

Tentative Correlation of Climatic Changes Through the Last 16,000 Years, Inferred from Pollen Stratigraphy at Three Sites in Southern South America, with Similar Data from Elsewhere [a]

Millennia BP	Europe	North Pacific North America	New Zealand	Kenya	Colombia	Fuego-Patagonia	Laguna de San Rafael, Chile	Llanquihue, Chile	Pollen Zones
0	Sub-Atlantic	warmer	drier		cold, wet 14°	cool, moist	warmer	warmer 14 – 16°	
1		cool, wet	colder, wetter	cooler			cooler, wet	cooler, wetter 12 – 13°	VIII
2									
3	Sub-Boreal	warm, but cooler drier	drier	cooler	cooler	dry	cool, but warmer	cool, but warmer, drier 14 – 15°	VII
4				warmer	warm relatively dry 16°				
5		(cooler)		cooler	relatively cold, wet 14 – 15°	warmer, drier	cooler, wetter	cooler, wetter 13 – 14°	VI
6	Atlantic	warmer, wet	warm, wet	warm, wet	warmer, wet 17 – 18°	cooler, wetter			
7		drier / warmer, wet		warm, wet	relatively warm, dry 14 – 17°	warm, dry	warmer, drier	warmer, drier 15 – 16°	V
8	Boreal			warmer, wetter			cool, wet		
9	Pre-Boreal	cool, wet	cold		relatively cold, dry 10 – 14°	warmer		warmer 13 – 15° / cool, wet 11 – 13°	IV
10	Younger Dryas	colder		colder	cold, dry 9 – 10°	cooler, wetter		colder, wetter 10 – 11°	III
11	Allerød	warmer		warmer	warmer, dry 14°	warmer, drier			
12	Older Dryas			colder		colder, wetter, warmer, drier		warmer, drier 14 – 15°	II
13	Bølling	cold		warmer	cold, wet 6 – 9°				
14	Oldest Dryas			colder, drier		cold, wet		cold, wet 11 – 13°	I
15									
16									

Postglacial / Late-glacial (Millennia BP)

Table 26-G
Chief Sub-Antarctic and Antarctic Islands: Glacial Data [a]

Name	South Latitude (approx.)	Longitude (approx.)	Area (km²)	Highest Alt (m)	Existing Glaciers	Former Glaciation
Auckland Is.	50°32'	166°13' E	606	>6600	No	Partially glaciated
Macquarie Is.	53°30'	158°57'E	230	430	No	Entirely glaciated
Heard Is.	53°10'	74°35'E	350	1818	Ice cap	Yes
Kerguelen Is.	49°25'	69°53'E	3413	1854	78	Extensive ice cap
Crozet Is.	46°30'	51°00'E		>1500	Yes	Yes
Bouvet Is.	54°30'	3°30'E		900	Yes	Glaciers now cover the island
South Georgia Is.	54°-55°	36°-38°W	3750	2787	Yes	More extensive than now
South Sandwich Is.	56°-59°	26°15'W		1360	Yes	(Actively volcanic)
South Orkney Is.	61°00'	45°-46°W	400	900	Yes	Extensive
South Shetland Is.	61°-63°	54°-63°W		2060	Yes	

[a] The best general reference is *Discovery Reports*, published in many volumes by Cambridge University Press. Ref. on Kerguelin Is.: Bellair, 1965; on existing glaciers: Mercer, 1967.

were depositing sediment even before then—at least 5×10^6 years ago (chap 27). However, although calving ice does imply glaciers, it does not compel the belief that an ice sheet was then present. At any rate these and other data (Denton et al., 1970), although still fragmentary, leave no doubt as to the great antiquity of glaciation of the Antarctic Continent.

Much of the information on glacial history since about 4 million years ago is drawn from the McMurdo Sound region and the Ross Ice Shelf. It concerns three groups of glaciers: (1) the Ice Sheet of East Antarctica and its outlet tongues, (2) the Ross Ice Shelf, and (3) local valley glaciers east of the Shelf. The history of glacial episodes as at present understood is summarized in table 26-H, which shows that information is still fragmentary. A sequence developed somewhat earlier (Grindley, 1967, p. 583) for a region farther south likewise distinguished between events connected with the Ice Sheet and those connected with the Ross Ice Shelf; the time span involves the last million years or more. Still earlier, Péwé (1960) identified in the Taylor Valley, McMurdo Sound region, four successive glaciations by an outlet glacier, from which fluctuations of the East Antarctic Ice Sheet could be inferred.

As yet little can be said as to the mutual time relations of fluctuations of the three groups of glaciers. It seems clear, however, that Ross events have not been synchronous with Taylor events (table 26-H). The Ross I event, bracketed between ~9500 and ~35,000 C^{14} years BP, is compatible in age with the Late Wisconsin and correlative events in the northern hemisphere. It is likely that the four Ross fluctuations have been controlled by fall and rise of sealevel caused by growth and decay of former

Table 26-H

Schematic Correlation Chart and Chronology of Glacial Events in the McMurdo Sound Region, Southern Victoria Land, Antarctica [a]

Taylor Glaciations (Ice Sheet in East Antarctica West of Taylor and Wright Valleys)	Ross Sea Glaciations (Ross Ice Shelf)	Alpine Glaciations
Taylor I	4450 yrs. BP (L–627; Marble Point) 5900 yrs. BP (L–462; Hobbs Glacier) 6100 yrs. BP (Y–2401; Hobbs Glacier) 9490 yrs. BP (Y–2399; Hobbs Glacier)	Alpine I 12,200 yrs. BP (I–3019; Hobbs Glacier)
	Ross I	
	34,800 yrs. BP (no laboratory number given; Cape Barne) >47,000 yrs. BP (Y–2641; Cape Barne; same locality as sample dated 34,800 yrs. BP) >49,000 yrs. BP (Y–2642; Cape Barne)	Alpine II K/Ar dates; 2.1 to ~0.4 m.y. (Walcott Glacier area)
Taylor II	Ross II	
	Ross III	
Taylor III K/Ar dates; 1.6 to 2.1 m.y. (Taylor Valley)	Ross IV K/Ar dates; 3.1 to 1.2 m.y. (Walcott Glacier area)	Alpine III K/Ar dates; 2.1 m.y. (Taylor Valley)
Taylor IV K/Ar dates; 2.7 to 3.5 m.y. (Taylor Valley) and 3.7 m.y. (Wright Valley)		K/Ar dates; 3.5 m.y. (Taylor Valley)
Taylor(s) V		

Note: The original table includes sources of the C[14] dates.

ice sheets in the northern hemisphere, through the mechanism described briefly in chap 3. By that mechanism the Ross ice would expand during northern-hemisphere glaciations and shrink during interglaciations and hence would be in phase with the northern changes of climate.

Fluctuation in thickness of the East Antarctic Ice Sheet, however, is a different case. Such fluctuation is indicated clearly by the Taylor events, which as shown in the table were not synchronous with the Ross fluctuations. Unfortunately the controls under which it occurred are still obscure.

The pole-centered Ice Sheet is a very conservative body and is not likely to have fluctuated by more than a minor proportion of its mass since it came into being. This opinion is supported by measurements of thermoluminescence of carbonate rocks (Ronca and Zeller, 1965) from Taylor Valley. This parameter is temperature sensitive, and seems to indicate that average temperature in the valley has not exceeded 0° in the warmest month during at least the last 1,400,000 years. That time span would embrace all the classical interglacial ages recognized in the northern hemisphere, but represents only part of the history of the Ice Sheet, for as we have noted, K/Ar dates and deep-sea cores indicate that that great body has been in existence through at least 4 or 5 million years.

As a final note we mention the presence of end moraines and discontinuous patches of till on the very restricted areas of the continent that are not covered with ice. Major parts of such areas, however, consist of bare, glacially eroded bedrock. In contrast, on the floor of the adjacent sea glacial sediments are nearly continuous.

It has been noted (Warnke, 1971) that although the volume of glacial-marine sediments inferred from sea-floor cores raised from around the Antarctic Continent must be very large, the Antarctic glaciers and icebergs today are nearly devoid of rock debris. The discrepancy suggests a change, in or through time, in rate of transport of drift by Antarctic glaciers, but if such a change occurred, the cause has not been isolated.

Submarine ridges, such as the Pennell and Iselin Banks extending across the mouth of the Ross Sea, are believed to be a series of submerged end moraines. Some of them stand so high that icebergs in the shoal water above them run aground. Core samples from the bottom off the Antarctic coast have been identified as marine till. This material mantles the sea floor throughout a zone many hundreds of kilometers in width surrounding the continent. Probably much of it was deposited from ice shelves and icebergs.

CHAPTER 27

Stratigraphy of Deep-Sea Floors [1]

POTENTIAL OF SEA-FLOOR STRATIGRAPHY

The sediments beneath the floors of the deep oceans are in most areas less than 200m thick, a fact that was mysterious when first established but that finds a reasonable explanation in the theory of sea-floor spreading. Although according to that theory the stratigraphy of the sediments should not include a record of ancient geologic events, it seems likely that the late Cenozoic is well represented. If this is the case perhaps the most continuous long record of late-Cenozoic climatic fluctuation can be expected from the sea floor. Much of the existing information is based on cores, some of which are as much as 20m long, and in which even the thinnest laminae are virtually undisturbed by the coring process. Other data are the result of geophysical exploration.

The supposition, many decades old, that the strata beneath the deep sea are continuous and are the product of slow, wholly vertical sedimentation upon a floor of low relief was found to be incorrect. We now know that beneath most parts of the oceans relief is about as great as on the lands and that in many places slopes are steep. We know also that bottom currents create channels and transport much sediment, that slump occurs on many hillslopes, and that wide plains are created by the accumulated deposits of innumerable turbidity currents. This knowledge tells us that surfaces of unconformity can be expected in core sequences, and that to minimize such breaks and to avoid the occurrence of undesirable turbidites, cores should be sited on nearly level tops of plateaus and on the sides of hills that have very gentle slopes. A surface of unconformity is rarely detectable by inspection but it can be inferred by discrepancies between thicknesses of sediment and radiometric dates.

[1] General references: M. N. Hill, ed., 1962–1963, v. 3; Arrhenius, 1967.

716

Core sediments are mixtures of particles derived from the lands and those formed in the ocean itself. The sediments of terrigenous origin are mainly *lutite,* which consists of particles, mostly of clay size, derived from the land during the general process of erosion. Most lutite is brought by rivers; a much smaller proportion is brought by winds. Accompanying it is dispersed or concentrated fine-grained tephra. In some regions in high and high-middle latitudes glacial-marine drift, some of it very coarse, is a conspicuous terrigenous element. In contrast with lutite, the sediment of oceanic origin consists chiefly of the shells and other hard parts of very tiny, single-celled marine animals such as Foraminifera and radiolaria and plants such as diatoms and coccolithophores. These two elements, the lutite derived from the lands and the oceanic material consisting essentially of microscopic fossils, can exist in any proportion.

The proportion depends in large part on the rate of deposition of lutite, which varies with both locality and time. Because two-thirds of the drainage area of the world's lands empties into the Atlantic, lutite accumulates more rapidly on the floor of the Atlantic than on that of the Pacific; hence a Pacific long core is likely to reach much farther down through the stratigraphic column than does an Atlantic core of equal length. Average rates of sedimentation range from much $<1\text{cm}/10^3$ yr in some open environments to $>50\text{cm}/10^3$ yr in some narrow seas. It was shown (Broecker et al., 1958), by means of C^{14} dates applied to an Atlantic core, that at the core site sedimentation of lutite was more rapid during the Late-Wisconsin glacial sub-age than in nonglacial times, by a factor of 3 or 4. Causes of the increased rate probably include rapid mechanical weathering, glacial erosion, enhanced discharge through proglacial streams, exposure of wide continental-shelf areas, and transport by winds of increased frequency and volocity.

As early as 1908 it became apparent that ocean-floor sediments are layered. Much later it developed that a good deal of the layering results from conspicuous changes in the proportions of the two elements we have described. This raised interesting questions as to the causes of the changed proportions. For example, an increase in the proportion of microfossils to lutite might be absolute, having resulted from a change in the oceanic environment (increased warmth or increased nutrients) that favored a larger population of micro-organisms. On the other hand the increased proportion might be relative—the result of mutual dependence within a closed system—having been caused by dilution of a constant population of micro-organisns by more rapid deposition of lutite. The two possibilities have different implications as to climate. Properly studied, cores from carefully selected sites provide many kinds of critical information about late-Cenozoic history. Such information includes rates of sedimentation, rates of erosion of the lands, temperatures of surface and bottom seawater, extent of sea ice, volumes of glacier ice on the lands, and paleomagnetic data. Our present interest lies mainly in paleotemperatures, in the correlation of strata from one core to another, and in the time calibration of core stratigraphy.

PALEOTEMPERATURE

Relation of Sea Temperature to Air Temperature

In the study of sea-floor stratigraphy as in the study of the lands, paleotemperature is a primary objective. We want to know both the amplitude and the chronology of former fluctuation of temperature. In this inquiry the relation between sea temperature and air temperature is important. The sea receives its heat almost entirely by absorbing heat radiated from the Sun and loses heat by radiation and convection to the atmosphere and by evaporation. Over the sea as a whole, loss of heat must balance gain, although heat exchange within different parts of the sea occurs through the medium of ocean currents. As a general result we would expect that in any area of the ocean/atmosphere interface the average annual temperature of the surface water of the ocean would approximate that of the contiguous atmosphere. The temperature equilibrium established between surface seawater and the air above it should mean that secular changes in climate should be reflected in changes in the organisms living near the surface of the deep sea, especially in low latitudes far removed from the direct influence of glaciers. When we recall that the sea-floor sediments in vast areas of the ocean consist mainly of the shells of pelagic Foraminifera, and that these animals are sensitive to variations in water temperature, the connection between such sediments and climatic change becomes obvious. The core sediments with their fossil faunas are manipulated in many ways, with good effect.

Attempts to apply similar temperature study to shallow-water mollusks in coastal areas have not met with great success because local temperature differences brought about by longshore currents and by coastal configuration may be great enough to mask the effects of more general temperature changes.

To return to deep-sea sediments, three lines of research, all having the objective of determining temperatures and therefore successive climates, have been followed. These concern (1) areal and stratigraphic distribution of taxa, (2) productivity, and (3) oxygen-isotope ratios.

Areal and Stratigraphic Distribution of Taxa

Faunal zonation in samples from the deep-sea floor was first recognized by Philippi in 1908. Schott (1935) mapped the general areal distribution of planktonic Foraminifera now living in the Atlantic Ocean. Because the same species occur in Pleistocene sediments beneath the sea floor, it is possible to draw climatic inferences from such sediments by measuring variations in the frequencies of species that are sensitive to temperature. An early example was the identification by Stubbings (1939) of four "cold" strata and three intervening "warm" strata in six short cores from the Arabian Sea. The basis consisted of two index Foraminifera. Stubbings thought the "cold" units represented glacial times, and on assumptions as to rate of sediment accumulation calculated the age of the latest one as 17,000 BP. Analogous results were obtained by others (mentioned in Phleger, 1951; list of species by latitudes in Emiliani and Flint, p. 906 in Hill, ed., 1962–1963, v. 3). The advent, in the late 1940's, of techniques for obtaining long, undisturbed cores, and of the C^{14}-dating technique almost at the same time put this use of stratigraphic paleoecology on a more sophisticated basis. Foraminifera of benthic habitat likewise reflect temperature change, but the changes observed in planktonic taxa are more marked.

Long stratigraphic sequences that reflect climatic variations have been elaborated through the fine measurement of frequencies of temperature-dependent taxa. Among such measurements are those of Lidz, 1966, Ericson et al., 1961, Bandy et al., 1969, Geitzenauer, 1969 (based on coccoliths), and Zhusé, 1961 (based on diatoms). A list of micropaleontologic studies of cores from Atlantic and Mediterranean waters is assembled in Emiliani and Flint (p. 911 in Hill, ed., 1962–1963, v. 3). Where data are adequate, the results permit correlation through considerable distance (figs 27-1 and 27-7). A study of Red Sea cores showed two "cold" and two "warm" zones (Herman, p. 325 in Morrison and Wright, eds., 1968).

The taxa to be counted must be selected with care to insure a high and uniform degree of temperature dependence. It is claimed (Lidz, 1966) that the *Globorotalia menardii* group of Foraminifera, used as a unit for counting, is undesirable because its four varieties differ from each other as to temperature significance.[2] Other more refined methods include counting and statistical treatment of whole assemblages of taxa instead of single taxa, and smoothing and broadening the temperature base by determining relative abundances of low-latitude to middle-latitude species.

Coiling Directions. A faunal parameter apart from relative abundance of species consists of dominance within a species of individuals possessing

---

[2] Comparison of Atlantic with Pacific cores in terms of this foraminiferal group suggests a rather complex time relationship of temperature fluctuation in the two oceans, but with generally colder water in the Atlantic (Ericson and Wollin, 1970). Acceptance, however, must await clarification of the temperature significance of taxa within the group.

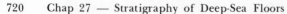
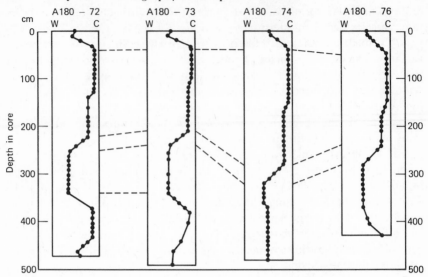

Figure 27-1 Temperature variations estimated from changes in relative abundance of "warm" (W) and "cold" (C) planktonic Foraminifera, and their correlation among four cores raised from the equatorial Atlantic Ocean. Numbers at top identify the cores. The curves represent all of Wisconsin time and an unspecified length of pre-Wisconsin time. (Ericson and Wollin, 1956, p. 120.)

a right- or left-coiling habit. In some species such dominances are consistently distinct from one region to another (fig 27-2) and their occurrence in various parts of a core sequence can be correlated from one core to another (Ericson and Wollin, 1956, fig 9). The dominances observed are empirical. Some of them change through depth (=time) in a given group of cores. Probably the changes result from changes in environment, but the controlling cause is not known. Changes in temperature, in salinity, and in polarity of the Earth's magnetic field are possibilities.

Faunal Displacements in the Pacific. Study of "warm" and "cold" microfaunas is applied not only to cores but also to sections, exposed in the continents by uplift and erosion, that represent deep-water marine environments. Living planktonic Foraminifera that occur also as fossils through later tertiary strata are useful as surface-water temperature indicators through late-Cenozoic time, when compared through a wide range of latitude. Bandy (1968a, 1968b) recognized in the Pacific three broadly occurring microfaunas, to some extent gradationally related yet distinct, as follows:

1. Polar fauna: Dominated by *Globigerina pachyderma* (left coiling), plus a group of other distinctive species of that genus.

2. Transitional fauna: Dominated by G. *pachyderma* (right coiling), plus *"Globigerina" dutertrei* and other distinctive taxa.

3. Tropical fauna: G. *menardii* complex, plus *Sphaeroidinella dehiscens* and other taxa.

Traced through late-Miocene, Pliocene, and Quaternary time, these three faunas form a pattern somewhat as suggested by the generalized model in fig 27-3. Comparison of the present latitudinal limits of these faunas with the water temperature obtaining within those limits permits the inference that temperature changes of as much as 12° have occurred

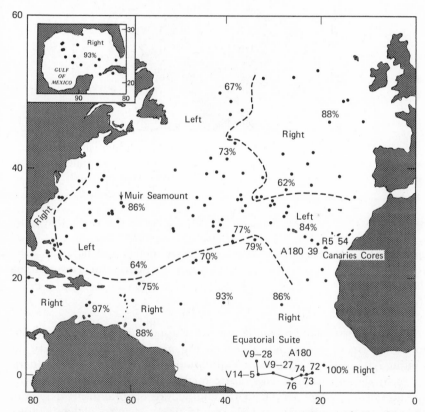

Figure 27-2 Modern distribution of predominant right- and left-coiling in the foraminifer *Globorotalia truncatulinoides* as reflected in percentages in the uppermost layer in cores (dots) from the Atlantic Ocean. The distribution is fairly consistent (Ericson, p. 834 in Hill, ed., 1962–1963, v. 3).

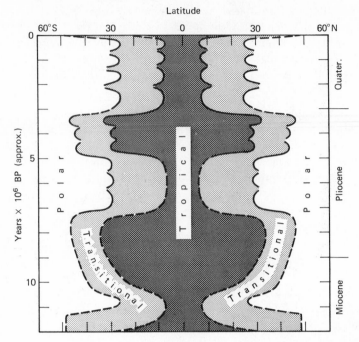

Figure 27-3 Suggested model showing latitudinal displacements of plank-tonic foraminiferal faunas in the Pacific through late-Cenozoic time, on the assumption of synchroneity between the polar hemispheres. (After Bandy, 1968a).

in some regions. The inferred major cold intervals shown in the figure fall in late-Miocene, mid-Pliocene, and Quaternary time, in fairly good agreement with intervals based on other parameters and noted in chap 16. The earliest known cold interval based on planktonic Foraminifera dates at least as far back as 13×10^6 years BP (Bandy et al., 1969).

Productivity

In many parts of the ocean, temperature appears to be a dominant factor in the size of population of pelagic organisms or, in other words, the productivity of those organisms. Populations of planktonic Foraminifera, for example, increase as temperature increases. This being true changes, in successive sea-floor strata, in the abundance of foraminiferal tests or their chemical equivalent in CO_2 should reflect changes of temperature of the surface water and by inference changes of climate. Measurements of abundance have given consistent results (e.g., Piggot and Urry, 1942, fig 3 and refs.).

In certain other oceanic areas the presence of nutrients seems to be the chief factor in productivity. In cores from the equatorial Pacific (fig 16-5), layers rich in calcium carbonate are correlated with times of cold climate, when the circulation of cold water of Antarctic origin was accelerated. Such water wells up along the equator, bringing nutrients to tropical surface waters. Here too the results of measurement are consistent (Arrhenius, 1952; p. 708 in Hill, ed., 1962–1963, v. 3). As can be seen in fig 27-7, correlation in the equatorial Pacific is in opposite phase to that in the equatorial Atlantic, where layers rich in calcium carbonate are assigned, on a basis of temperature, to interglacial rather than to glacial times. Because oceanic circulation conditions in the two regions are quite different we can accept both correlations.

In papers in press, W. F. Ruddiman and H. D. Needham and others analyzed mechanically deposited sediments and microfaunas in long cores from the Atlantic Ocean. The results suggested several pronounced cold episodes, dated by means of geopolarity to within the last 700,000 years. Earlier temperature fluctuation seems to have been less pronounced. Possibly the later group of episodes are represented in the Pacific curve, fig 16-5. Also, speculatively, they might be identical with the glacial ages in North American and European terrestrial stratigraphy. If they were, the classical glaciations of middle latitudes by large ice sheets would together have represented only a minor part of Pleistocene time.

Oxygen-Isotope Ratios

A means of estimating former temperature that is perhaps more sensitive than measuring abundances of taxa is based on isotopic geochemistry. It consists of measurement by mass spectrometer of the ratio O^{18}/O^{16} in the calcitic tests of planktonic Foraminifera. That ratio depends on two factors prevailing where and when the shells were secreted: (1) the oxygen-isotopic composition, and (2) the temperature of the ocean water. Variation of the ratio with temperature results basically from differences of vibrational energy between molecules of H_2O and CO_2 containing the lighter and heavier isotopes. The variation in the composition of the ocean results from the higher vapor pressure of the H_2O^{16} molecule, which causes it to evaporate preferentially from the ocean surface (fig 27-4).

The two factors operate in the same direction. During a glacial age water temperature diminishes, and with it the difference in vibrational energy between the two molecules decreases, so that $CaCO_3$ secreted at such a time is relatively rich in O^{18}. At the same time the water has become residually enriched in O^{18}, owing to net loss of ocean water by evaporation and storage of the water in glacier ice. This change, too, is

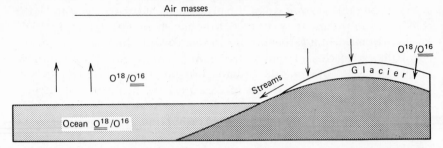

Figure 27-4 Oxygen-isotope ratios in the hydrologic cycle during a glacial age. The lighter isotope is preferentially evaporated from the ocean, transported to lands, and built into glaciers, leaving a residually greater concentration of the heavier isotope in ocean water.

reflected in the isotopic composition of $CaCO_3$ then being secreted. During an interglacial age, with higher water temperature but also with dilution of the ocean by meltwater enriched in O^{16}, conditions are reversed and are reflected in the isotopic composition of carbonate shells.

The O^{18}/O^{16} ratios in shell carbonate are measurable, and their variations in different taxa through time are comparable (fig 27-5). In the

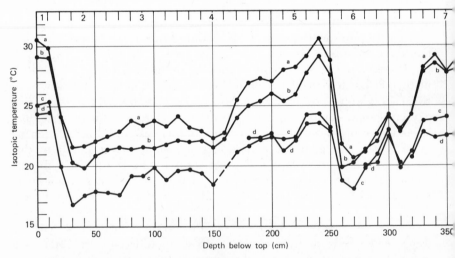

Figure 27-5 Fluctuation of O^{18}/O^{16} ratio with depth in a core, and hence through time, in each of four foraminiferal taxa, showing close similarity of the curves. From Core A179-4, tropical Atlantic Ocean. Numbers above the curves identify temperature units. (Emiliani, 1955, p. 549.) Ordinate is scaled in de-

layers of sea-floor sediment recovered in cores and controlled broadly by C^{14} dates, the ratios correspond with broad fluctuations of temperature recognized on the lands. Thus they appear to reflect glacial and interglacial times, as was shown first by Emiliani (1955) in cores from the tropical Atlantic and the Caribbean. This relationship seems well established. However, in each fluctuation of the ratio O^{18}/O^{16} in shell carbonate, we do not know what proportion to attribute directly to temperature change, the remainder being attributable to an accompanying change in isotopic composition of the ocean water. In order to estimate the temperature difference, for the water of the tropical Atlantic Ocean, between the Late Wisconsin glacial maximum (say 16,000 to 18,000 BP) and today, it is necessary to estimate both the aggregate volume of glacier ice on the lands at that time and its isotopic composition.

The basic calculation is that of Emiliani (1955), who estimated that ~70% of the O^{18}/O^{16} ratio measured in shell carbonate is attributable to temperature, and arrived at an adjusted value of 5° for the temperature difference, in tropical Atlantic waters, between Late Wisconsin time and today. Other calculations were made by H. Craig (1965), Olausson (1965) and Shackleton (1967), resulting in estimates of little or no temperature change in the tropical Atlantic. Dansgaard and Tauber (1969) measured the average O^{18} content of the ice in the Greenland Ice Sheet that

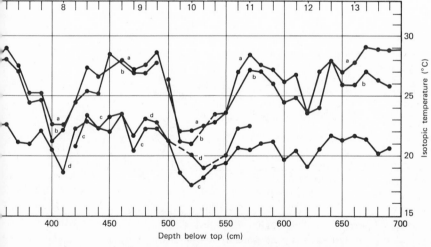

grees of isotopic temperature on assumptions as to volume and isotopic composition of glacial-age ice sheets. Taxa: (a) *Globigerinoides rubra,* (b) *Globigerinoides sacculifera,* (c) *Globigerina dubia,* (d) *Globorotalia menardii.*

formed during the Late Wisconsin maximum, and calculated that not more than 30% of the variation measured in Foraminiferal tests could be attributable to temperature change. Duplessy et al. (1970) obtained data indicating to them that isotopic ratios in Foraminifera reflect solely variations in the isotopic composition of the ocean. Therefore the opinions cited (based in part on assumptions) tend toward minimizing temperature changes near the surfaces of tropical oceans. They are surprising for two reasons. First, they do not correspond with evidence (chap 4) that glacial-age snowlines were lowered not much less at the equator than in middle latitudes. That fact would seem to demand a proportional though not necessarily equal change in the temperature of tropical-Atlantic surface seawater. Second, these opinions do not explain the substantial qualitative agreement between the O^{18}/O^{16} fluctuations measured in shell carbonate and the fluctuations in abundance of "warm" and "cold" organisms (figs 27-6 and 27-7).

A promising path out of the conflict of data and opinion seems to lie in re-examination of the assumptions regarding the isotopic compositions of the former ice sheets. It has been supposed that these were similar in composition to those of the existing Greenland and Antarctic Ice Sheets, but this is not necessarily the case. The former ice sheets reached much closer to the tropics and might have been built from snow preferentially richer in the heavier isotope than was the diminished precipitation that fell farther north and east (cf. Emiliani, 1970; 1955, p. 543).

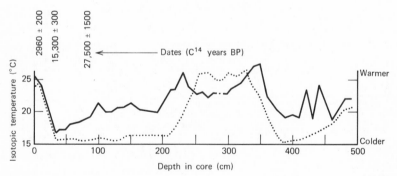

Figure 27-6   Curves showing fluctuation of two parameters in Core A180-73, equatorial Atlantic Ocean, through core depth. Continuous line: O^{18}/O^{16} ratio expressed on the left ordinate as isotopic temperature (Emiliani, 1955, fig 4). Dotted line: relative numbers of warm- and cold-water planktonic Foraminifera, scaled on the right ordinate (Ericson et al., 1961, fig 25). Three C^{14} dates provide partial time calibration. Similarity of the curves is obvious.

Although the isotopically measured fluctuation between glacial and nonglacial climates in Indian Ocean cores agrees with that in Atlantic cores (Oba, 1969), in Pacific cores it is much smaller (Emiliani, 1955, p. 560). The faunal fluctuation found by Zhusé (1961) in the diatom faunas of northwest Pacific cores was comparably small. The difference between the cores from the two oceans may lie partly in the accessibility of the Atlantic to cold Arctic as well as Antarctic water, because temperature of bottom water in the open ocean anywhere is conditioned mainly by that of surface water in the polar regions. Another factor may be the smaller size of the Atlantic as a reservoir.

Ice-Rafted Detritus

Glacial-marine drift (chap 7), widespread on ocean floors, is a direct indicator of cold climate. It was first identified and studied stratigraphically in a series of 11 cores raised in 1936 during a crossing of the North Atlantic by the cable-laying ship *Lord Kelvin* (Piggot and Urry, 1942). Four layers characterized by abundant drift, the thickest >30cm thick, were found beneath a layer of ooze with "warm" Foraminifera. Between the drift layers were strata of ooze with "cold" Foraminifera, and beneath the whole sequence was thick ooze with a "warm" fauna. From a modern viewpoint this succession looks rather like Holocene overlying Wisconsin with its long metastable interstadial, in turn overlying a substantial part of the Sangamon. Uranium-series dates on one of the cores (Piggot and Urry, 1942, p. 1195), although obtained by a method now known to encounter difficulties, look fairly reasonable through ~72,000 years.

Whatever the timetable, the alternations of sediment resulted from fluctuation of the southern statistical limit of sea ice and icebergs that drifted from the Greenland Sea and Arctic Ocean and that dropped drift onto the sea floor as far south as ~30° on the west side and ~45° on the east. An excellent description of the sequence is given by Bradley et al. (1942); erratic material constitutes as much as 30% of each "cold" layer. Later Conolly and Ewing (1965) found it possible to correlate zones of glacial-marine drift with zones already designated on a faunal basis as glacial.

Similar sediments are known to occur in the northeastern Pacific and in the Antarctic zone of the Southern Ocean as well (Lizitzin, 1960). According to Goodell et al. (1968), glacial-marine sediment occurs at the bottom of a core, >25m long, from lat ~65°S. Its extrapolated age is ~5 × 10^6 years. These authors concluded that calving from the Antarctic Ice Sheet began at least as long ago as that extrapolated date, thereby confirming the great age of the ice sheet (chap 26).

Time Calibration of Temperature Fluctuations

Adequate time calibration of strata revealed in cores has been slower in developing than laboratory research on the strata themselves has been. In a quarter century of study of long cores direct dating of strata is still rather limited, although much progress has been made on dating strata by indirect methods. We have mentioned the Uranium-series dates obtained on the *Lord Kelvin* cores. Although these look reasonable in the light of today's knowledge, the method has been little used because of inherent disadvantages.

With the advent of C^{14} dating, the method was applied easily and rather widely to the upper parts of cores, with resulting maximum dates in the range of 30,000 to 40,000 years. Such dates are useful for correlation and for determination of rates of sedimentation and of faunal change, but they are severely limited to the upper parts of cores. Dates of horizons farther down in cores have been estimated by extrapolation, based on rates of sedimentation determined by C^{14} dating, and assuming in most cases that rate of sedimentation was uniform. In one group of cores in the equatorial Pacific (including Core 58; fig 16-5) chronology was attempted by extrapolation from a C^{14}-controlled top part. This was done by measurement, at closely spaced intervals along the core, of the content of titanium oxide in the lutite fraction. The extremely fine-grained TiO_2, delivered by rivers, is assumed to become uniformly distributed through the ocean before its very slow settlement rate brings it to the bottom; rate of settlement is supposed to be nearly constant. The resulting calibration gave an age of some 1.7×10^6 years for the base of the core (Arrhenius, 1952, p. 24–29, 199), a good result for a pioneer attempt, if a much later geopolarity correlation (discussed in a later section) is accepted.

The use of direct K/Ar dating is confined to rare layers of nearly pure tephra, reasonably interpreted as the product of single eruptions. Although tephra has been identified in Atlantic cores from the region south of Iceland, such sediments are not sufficiently abundant to have been significant in dating and correlation of cores.

The most flexible and widely employed means of time calibration of cores is the use of geopolarity units. Because the prime use of such units is stratigraphic, we shall discuss them in the following section.

CORRELATION OF STRATIGRAPHY BETWEEN CORES

Correlations between cores (fig 27-7) have been made by means of microfaunal similarities, by similarities in O^{18}/O^{16} curves and productivity

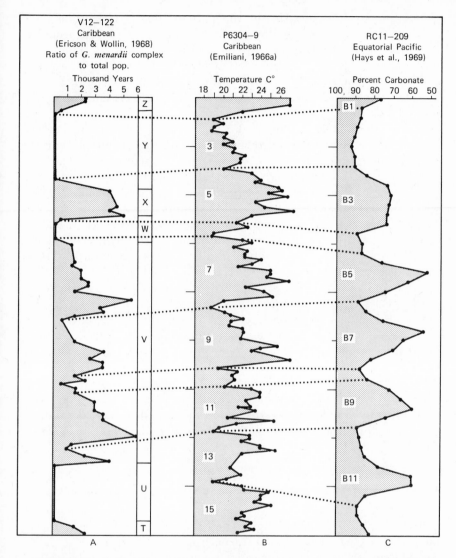

Figure 27-7 Comparison of key parameters in 3 cores. *A*. Curve based on relative abundance of Foraminiferal warmth indicators. *B*. Curve based on 0^{18}/0^{16} ratio. *C*. Curve based on productivity of Foraminifera. In this curve productivity decreases left to right; hence curve *C* is broadly inverse to *A* and *B*. Letters and large numbers denote stratigraphic units. (After Hays et al., 1969; dotted lines added.)

curves, and by C^{14} dates (in the top parts of cores). The use of these methods, however, is eclipsed by the application to cores of the geopolarity units (fig 14-2) established in continental sequences, an application that also extends a radiometric-dating framework from lands to ocean and facilitates or actually provides correlation of strata. The framework already extends back through the last 4.5 million years and can be expected to extend farther.

The magnetic-polarity framework was first applied to cores by Opdyke et al. (1966) by measuring the polarities of grains of detrital magnetite at intervals along seven cores from the western Antarctic Ocean. This permitted the core sequence, which had already been divided into four faunal zones, to be fitted to the terrestrial geomagnetic "standard," after which the terrestrial K/Ar dates could be applied to the faunal zones in the cores. It was found incidentally that evolutionary changes (including extinctions in the core faunas) coincide rather closely with magnetic reversals, opening up the possibility of a cause-and-effect relationship (Hays et al., 1969, p. 1495).

Later similar correlation was applied to 15 cores from the equatorial Pacific (Hays et al., 1969). Cores were examined both for stratigraphic ranges of dominant microfossils and for productivity as measured by the percentage of calcium carbonate in unit thicknesses of core. The positions of calcium-carbonate maxima (fig 27-8) proved to be similar to those in Core 58 (fig 16-5), for which we described the calibration by means of TiO_2 content. The geopolarity scale was then applied to the 15 cores, whose carbonate maxima could be correlated visually (fig 27-7), with those of Core 58. The geopolarity scale thus carried to that core implies that the sediment at the base of the core is somewhat more than 2 million years old.

PLIOCENE/PLEISTOCENE BOUNDARY

In the days when Cenozoic glaciation was believed to have been confined to the Quaternary Period, various proposals were made to define the Pliocene/Pleistocene stratigraphic boundary on a basis of the first appearance of glaciation or even of a merely cold climate. When the demonstration of Pliocene and even Miocene glaciation invalidated such a basis, attention turned more fully to the search for faunal changes that could be accepted as the boundary. Attention was focused particularly on microfaunas that could be widely recognized in cores as well as in strata exposed in the continents. Especially favored were planktonic organisms that ought to bear a close relation to prevailing air temperatures. As an

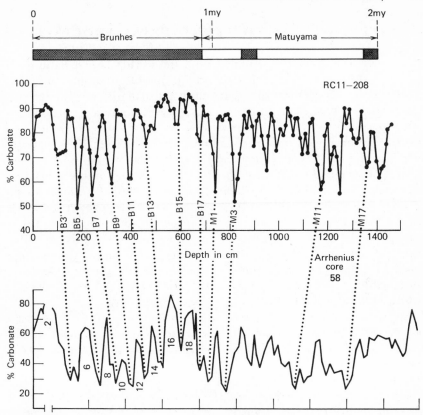

Figure 27-8 Correlation of cores in the equatorial Pacific with the geopolarity scale. Geopolarity in Core RC11-209 was measured directly. Carbonate maxima in the two cores were correlated by inspection, making possible rough time calibration of Core 58. (After Hays et al., 1969.)

example, one of the resulting proposals (Bandy and Wade, 1967) suggested a Pliocene/Pleistocene boundary at the horizon of disappearance of *Globorotalia tumida menardii* (right coiling) and the near-disappearance of discoasters, star-shaped calcareous micro-organisms of doubtful affinity. The date of that horizon seems to be $\sim3 \times 10^6$ years BP. In the same study the Miocene/Pliocene boundary was tentatively fixed at the horizon of appearance of the foraminifer *Sphaeroidinella dehiscens,* around 11×10^6 years BP. These changes pertain to a single core from the tropical western Atlantic. It remains to be seen, of course, whether these faunal changes are sufficiently widespread to be serviceable as standard boundaries.

INFERENCES AS TO TERRESTRIAL PROCESSES

Because cores consist of sediment from the world's largest basin of deposition, the ocean floor, their study should produce information on processes active over the continents and particularly on the relative importance and rates of such processes. As we have noted, cores furnish data, by inference, on air temperatures, on the relative amount of ice stored temporarily on the lands, and on the relative strength of planetary winds as indicated by rate of equatorial upwelling of cold oceanic water. Layers of ice-rafted detritus indicate glacial conditions on the lands. Conspicuous accumulations of turbidite, whether or not these form abyssal plains, record active turbidity currents and by inference lowered sealevel, exposed shelves, and accelerated slump activity on continental slopes. Minute study of the distribution of lutite, by components, on the Atlantic floor indicated to Biscaye (1965) that eolian sediment derived from northern Africa was not abundant compared with alluvium, and that such sediment is confined to the basin east of the Mid-Atlantic Ridge, which seems to have baffled transport toward the west.

As we noted earlier, the presence beneath the ocean floor off northeastern Brazil, of abundant arkosic lutite of Late-Wisconsin age led to the hypothesis that northeastern South America was semiarid during glacial ages. In theory the pollen that is present, though not abundant, in deep-sea sediments should provide useful information about contemporaneous conditions on nearby continents. However, the source of such pollen is difficult to fix. Pollen is transported by streams as well as directly by winds, and transport through a large river may involve long distances with much mixing en route. Subsequent further transport by ocean currents is also likely. As a result little progress has been made in firm inference based on pollen extracted from cores.

CONCLUSION

The study of deep-sea sediment cores is still in its infancy, but already it has demonstrated its basic importance. Microfaunas have established temperature changes of substantial magnitude through late-Cenozoic time. Such changes began at least 13 million years ago, in the Miocene. Their maximum amplitudes seem to have been rather similar, although degree of resolution is not yet very good. It is not known whether the temperature fluctuations are periodic. Core data suggest that at times of lowered temperature atmospheric circulation (and with it the wind-driven oceanic circulation) became more rapid. Most of these results are compatible with data derived from continental sources.

CHAPTER 28

Quaternary Fossils I:
General, Plants, Invertebrates

ASPECTS OF THE USE OF FOSSIL MATERIAL

Stratigraphy and Correlation

In foregoing chapters we have referred repeatedly to fossils as indicators of former climate in a particular region and as characteristic of particular strata. In chaps 28 and 29 we shall try to set down the evolutionary significance and stratigraphic position of fossils and the basis of climatic inference drawn from them. The exposition involves some repetition and some overlap, but conveniently brings the basics of Quaternary fossils (though by no means all the details) into one place. We shall emphasize fossil mammals (chap 29) because in chap 14 and several subsequent chapters we have given attention to fossil pollen, and in chap 27 we have dealt with marine invertebrates.

The study of Quaternary fossils has developed as part and parcel of the general study of fossils in the geologic column as a whole. Individual occurrences were reported, each one at best including taxonomic identification, areal occurrence, and stratigraphic position. From the accumulated reports a pattern began to be discerned. As the sequence of Quaternary strata was followed from base to top, the organisms they contained were found to differ in a rather systematic way. The differences resolved themselves into two chief kinds, evolutionary and environmental. Evolutionary differences formed a broad, gradual pattern of change from base to top, not very pronounced to be sure, but still recognizable. Environmental differences seemed to be superposed on the evolutionary change. They constituted a zigzag or alternating pattern marked by little or no change

733

in any single taxon but by appreciable change in assemblages of organisms. The assemblages carried with them appreciable environmental implications. The inference drawn from this dual pattern of change was obvious. Through the last 3 million years or so, as in earlier geologic time, organic evolution was pursuing its course, but in some taxa at least its direction was clearly affected by repeated changes in environments. Except for planktonic and benthic microfaunas, the majority of the Quaternary fossils on record come from middle latitudes. Hence our knowledge of environmental changes during the Quaternary mainly concerns middle-latitude regions.

Nearly all this knowledge, apart from pioneering aspects, dates from the 20th century and the latter part of the 19th. It is based on observations of the relation of fossils to the order of superposition of the Quaternary strata, like those begun by William Smith a century earlier, on much older strata of marine origin. The Quaternary study, however, is more complicated because most of the terrestrial strata are very discontinuous. Indeed some of them are wholly limited to spot occurrences in caverns, small bogs and lake basins, and other highly localized areas and cannot be traced in the field from one locality to another.

Quaternary paleontology, then, is based pragmatically on the occurrence of particular assemblages of fossils in particular strata. Of course facies is involved, although as yet Quaternary literature includes few descriptions of facies change through the various marine to the various terrestrial environments. Instead pollen, plant megafossils, terrestrial vertebrates, terrestrial invertebrates, littoral invertebrates, shelf organisms, and pelagic and benthonic microfossils are generally dealt with separately. Nevertheless stratigraphic occurrence remains the observational basis of all special studies of evolution, environment, and former climate inferred from fossils.

The use of fossils in correlation needs no special comment. Chapter 14 contains comments on correlation, and chap 24 notes the subdivision of the Pleistocene in Europe on a basis of fossil mammals. Some stratigraphic units in the Quaternary of Britain have been defined on a basis of fossil pollen (West, 1968). Provincial faunal ages based on mammals have been defined for the terrestrial Quaternary of North America, although these are not the principal means of correlation on that continent. Subdivisions based on mammals and used in various parts of the world are shown in tables appearing in chap 29.

We turn now to brief summaries of the present state of knowledge of Quaternary evolution, environment, climate, and biogeography derived from studies of fossils.

Evolution

The fossil record indicates that a considerable number of genera and a large number of species of mammals have arisen since the beginning of Quaternary time, that is, during a period of 3 million years, more or less. The philosophy of the appearance of new forms runs like this. During Earth history, at times of very slow physical change ecologic conditions are stable and so, in consequence, are faunas. Competition is not keen; it extends little beyond permitting any one taxon to perfect itself slowly in its particular niche. However, should empty niches become available, taxa able to occupy them are likely to evolve at a rapid rate. Thus rates of evolution depend to a considerable degree on rates of change in the environment and hence ultimately on geologic processes.

The later Cenozoic—essentially post-Oligocene time—witnessed rather rapid changes in the continents as compared with the earlier Cenozoic. Broad uplifts and the localized rise of new mountain ranges occurred. The resulting compartmentation of lands created many barriers, which in themselves induced local ecologic contrasts in climate, patterns of vegetation, and other environmental characteristics. Glaciation began, and with it lowered temperature, changes of sealevel causing the exposure of wide shelf areas, and the appearance of many lakes and marshes.

Most new species are thought to arise because faunas become isolated by the creation of barriers that interfere with genetic mixing. Therefore we could expect the new or different mountains, seas, deserts, forests, glaciers, and lakes of late-Cenozoic time to lead to speciation, provided these barriers endured through a long enough time. However, because real differences between species are hard to establish, the exact number of truly Quaternary species of mammals is not a matter of general agreement. According to Kurtén (1968, p. 254), of 119 mammal species now living in Europe and adjacent Asia, six date back to the late Pliocene; the rest appeared at various Quaternary times, six of them in the Weichsel Age. Similarly, only ~10% of the existing species of plants in northwestern Europe extend farther back in time than the Miocene Epoch (E. B. Leopold, p. 215 in P. S. Martin and Wright, eds., 1967).

We have emphasized late-Cenozoic compartmentation of the land rather than reduction of temperature because it does seem to have been the more fundamental influence, at least on terrestrial organisms. Specific adaptations to low temperature did occur, but mainly these affected fleshy parts and were not necessarily visible in fossil skeletons. Examples are the development of wool (a thermal insulator) and blubber (an insulator and a source of energy in winter), in elephants, rhinoceroses, and musk-oxen, of blubber in polar bears, and of food- and water-storage structures in some camels. Apart from such adaptations, the majority of the Quaternary changes in mammals is not related to low temperature, at least in an obvious way. The repeated changes of temperature in late-Cenozoic time could, however, have stirred and mixed populations through many enforced changes of range, and could thus have induced evolutionary changes of other kinds.

Two groups of mammals, elephants and bears, are reasonably well known through Quaternary fossils, although there are serious gaps in both records. Figure 28-1 sketches their evolution, mainly in Eurasia. The first elephant, *Archidiskodon planifrons,* entered Europe from Asia or Africa. It and its direct successor, *A. meridionalis,* were comparatively unspecialized, living in both woodland and more open country. Later in the Quaternary specialization occurred. The line split into two parts, mammoths

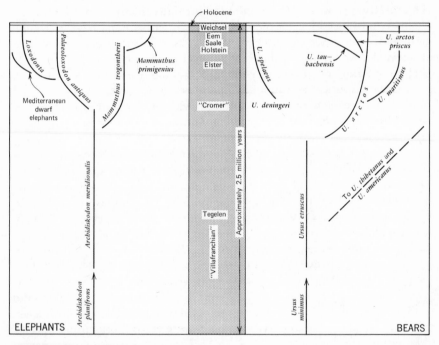

Figure 28-1 Apparent evolutionary pattern of bears and elephants through Quaternary time. Bears after Kurtén, 1957; elephants compiled from data in Kurtén, 1968. [Aguirre (1969) visualized a different pattern of elephant evolution, in which lineages diverge at much earlier times.] Time scale is only suggestive.

and elephants. The mammoths specialized to steppe and tundra grazing and died out near the beginning of the Holocene. The elephants specialized mainly to woodland browsing. They gave rise to some short-lived forms that became isolated on various islands, evolved to dwarf size, and died out early. Of the two main lines, one died out early in Weichsel time and the other still survives.

From a Villafranchian stem, *Ursus etruscus,* the Asiatic black-bear line branched off and much later branched again, giving rise to the Asiatic and American black bears. The main stem then developed two branches, the cave-bear branch and the brown-bear branch. The cave bears (*U. deningeri*) of the lower Middle Pleistocene evolved, about "Cromer" time, into the well-known *U. spelaeus,* which existed in great numbers but died out not long before the end of Late Weichsel time. The brown bears are represented chiefly by *U. arctos,* which includes the Alaska brown bear and grizzly bear as well as Eurasian varieties. The *arctos* line gave rise, probably in late Middle Pleistocene time, to the polar bear, *U. maritimus,* with its specialized sea-ice habitat.

Other groups show rather similar history. Among the rhinoceroses the long-lived Villafranchian form *Dicerorhinus etruscus* persisted until mid-mid Pleistocene. Thereafter two principal forms existed. *D. hemitoechus* was a steppe dweller, whereas *D. kirchbergensis* (Merck's rhinoceros) preferred woodland and savanna environments. The woolly rhinoceros *Coelodonta antiquitatis* entered Europe from the east in Saale time, inhabiting tundra and grassland. All three kinds (woodland, steppe, and tundra) persisted well into or through Weichsel time.

The introduction into Europe of unmistakably cold-adapted mammals occurred slowly and gradually, beginning apparently in Elster time. The Quaternary evolution of mammals is further discussed in Zeuner, 1958, p. 377–390.

When we compare the Quaternary history of plants with that of mammals, we find a considerable number of living species of trees that survived all the changes since the beginning of the Quaternary. Clearly the environmental changes were more rapid than the pace of plant evolution through adaptation. Indeed, the evolution of plants in Europe through later Pliocene and earlier Quaternary time was expressed in large part by progressive extinction of species that were exotic or thermophile or both, while more local species of European character survived. The changes, represented by sequences in northern Italy, the Netherlands, Poland, European Russia, and the Caucasus, are graphically shown by Grichuk et al. (1965, p. 324; also Zagwijn, 1960; Szafer, 1953, 1954). They are considered to have been primarily an effect of climatic cooling experienced repeatedly. Today, in western Europe, only about 20% of the plant species present in latest Pliocene time are still living there.

A compact history of the flora of Britain through the Pleistocene is given by West (p. 1 in D. Walker and West, eds., 1970); data for the Netherlands are given in van der Vlerk and Florschütz, 1950. Postglacial floral changes in northern Europe (table 24-C) are detailed for Denmark by Iversen (1960).

Environment,[1] Biogeography, and Climatic Change

During late-Cenozoic time plants and animals were conspicuously and repeatedly displaced from the areas in which they lived. The chief agency of change was changing climate, a basic element in the environments of organisms. As emphasized in preceding chapters, when a glacial age set in, the displacements were downward on the sides of highlands and equatorward in broad regions of lesser relief. Such changes of latitudinal or altitudinal range are independent of pre-existing evolutionary trends in individual taxa; yet the stresses of competition inherent in displacement and mixing surely affected the directions of evolution in many forms. Had the displacements lasted through longer periods than they seem to have done, probably their effect on evolution would be more evident.

The changes induced by a cooling climate were accomplished only rarely by actual migrations of individual plant or animal species. Rather the changes resulted from slow, progressive spreading of populations toward more hospitable areas. Such movements involved the invasion of areas already inhabited, creating communities with new combinations of organisms. With a warming climate, some elements of a cold-loving community, living beyond the margin of an ice sheet, would follow the glacier as it shrank, while other elements would remain, to become part of a community that favored more temperate conditions and that might well include immigrants from other regions. Such repeated mixing goes far to explain why faunas and floras of the successive glacial and interglacial ages were not exactly alike even though the successive climates could have been similar.

Biogeography.[2] As we have seen in other chapters, some biomes [3] (the regional complexes of communities characterized by distinctive vegetation and climate: tundra, broadleaf forest, and the like) shifted through 10° to more than 20° of latitude and through more than 1000m of altitude. Such changes are read from biota of many kinds: pollen, plant macrofossils, marine microfauna, other invertebrates, and vertebrates, and

---

[1] Background material on distribution of land animals today: Udvardy, 1970.

[2] Examples other than those mentioned below are described in Deevey, 1949.

[3] The forces that control plant communities are examined by Gleason and Cronquist (1964); a wider treatment of general ecology is that by Odum (1962).

are compatible with the inorganic record of glacial, pluvial, and other climatic events not directly connected with organisms. In a broad sense there are three possible effects of climatic changes on established biota: adaptation, migration, and extinction. Hitherto we have mentioned the shifting of biomes almost as though the Earth's surface were quite featureless. Actually irregularities and differences—mountains, seas, rivers, and the like—greatly complicate the movements of organisms and are major elements in the reconstruction of biotic responses to climatic change.

An example of the biogeographic influence of mountains is found in glacial-age Europe. The morphology of western Europe, as we noted in chap 23, is such that lowland plants, when forced by oncoming cooler climate to spread southward early in Pleistocene time, encountered mountain barriers. As a result, taxa that could not surmount or avoid the barriers became extinct. This seems to be the chief explanation of Quaternary impoverishment of the western European tree flora. Hickories, sassafras, magnolias, sweet gums and *Liriodendron* (tulip) were destroyed, although they live on in eastern North America where no barrier was present on the south.

An example of biogeographic barriers of another kind consists of the present-day distribution of arctic and subarctic plants through the lands that surround the Arctic Ocean. The distribution is anomalous; it includes disjunctions in the ranges of certain plants, which are apparently not explainable in terms of today's geography. The existing distribution implies diffusion eastward and westward from a center that embraces Alaska and adjacent Siberia, rather than from the south. According to the theory of Hultén (1937), in preglacial time these plants occupied a continuous circumpolar belt. Each time the Laurentide, Scandinavian, and Siberian Ice Sheets formed they destroyed large segments of the circumpolar flora, but that flora was conserved in Beringia, the glacial-age land area that consisted of the emerged Bering Sea floor and Alaskan and Siberian lands adjacent to it. Hence Beringia, free of continuous glaciers, constituted a refuge in which boreal plants could survive during a glacial age. During deglaciation, plants that had harbored in the refuge diffused into regions newly exposed from beneath ice sheets, and repopulated them. Such disruption and repopulation occurred with each succeeding glacial age and led to the present distribution pattern of arctic flora, believed to consist of the hardiest survivors of the repeated glacial disruptions. Although widely accepted in principle, this theory encounters a number of difficulties, especially as regards the times when the disruptions occurred. One informed opinion (Löve and Löve, 1967) holds that during the Weichsel glaciations at least, local areas in Norway and other boreal lands functioned as plant refuges in addition to the region of Beringia.

A theory of analogous disjunction of a continuous biome, in this case montane forest in Africa, is described in connection with fig 17-8. As noted there, biogeographic theories of this kind are based on negative evidence, yet they represent reasonable explanations of anomalous, large-scale patterns. A minute examination of the distribution of Carabid beetles led to the hypothesis that Newfoundland had not been continuously covered with glacier ice during Wisconsin glaciations (Lindroth, 1963). The data from glacial geology are not adequate to permit a strong denial of the hypothesis, which should be given recognition pending the accumulation of further geologic data.

In partial analogy with the Beringia refuge, the continental shelf of eastern North America, widely emergent during glacial ages, should have been both a refuge and a corridor for migrations of both plants and animals.

The warming trend in Hypsithermal time apparently resulted in the elimination of many local refuges. Thus, rise of the upper limit of trees eliminated alpine meadows on many hills, as is graphically shown in fig 28-2 for highlands in central Alaska, where, following disjunction of their range, small mammals were extinguished by this process.

The field of biogeography is bristling with problems that arise from observed disjunctions, some of which suggest climatic change while others do not. Study of such problems rarely leads to results that are unequivocal, partly, no doubt, because the basic evidence is more likely to be negative (a gap in a range) than positive. Lack of space precludes our going further into this fascinating field, and we recommend to the reader reviews such as that in Deevey, 1949, and texts on zoogeography such as that of Darlington (1957).

Climatic Inferences from Biota. How are data on former climate developed from fossils? The reasoning is based on comparison of areal ranges. In the largest possible collection of fossils from a single locality or district the existing range of each taxon (usually genus rather than species) is determined. If the fossil locality lies within the ranges of all similar taxa now living, the fossils constitute no evidence of climatic change. If it lies outside the range of a taxon, the difference in critical climatic parameters (such as mean annual precipitation or mean temperature of coldest of warmest month) between the present range and the nearest part of the former range is taken as the amplitude of change, with time, in the parameter measured (table 28-A).

The foregoing procedure is based on two main assumptions: (1) Climate exercises the primary control on the distribution of the taxa exam-

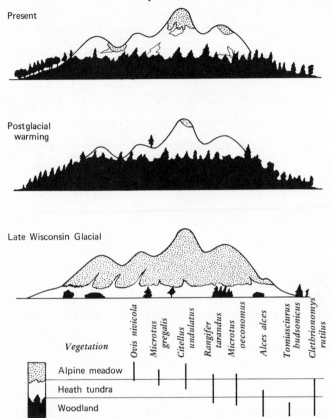

Figure 28-2 Diagrammatic illustration of vegetation changes that could have resulted in the extinction of three species of small mammals from the Yukon-Tanana upland. Vegetation is divided into three zones: alpine meadow, heath or shrub tundra, and woodland. Ranges of dominant mammals in these zones are shown below. (After Guthrie, 1968.)

ined. (2) The nearest living relatives (if involved) are subject to the same limiting factors as were the fossil taxa. These assumptions are reasonable, especially if the fossils are not very ancient, and are accepted because the method gives consistent results. For example, in some cases similar results are extracted independently from pollen and from mammals. The procedure must be employed with extreme care. Indeed the best results are obtained when, instead of just a qualitative list of the fossils present, quantitative study of an entire community is undertaken. Small mammals

Table 28-A
Temperature Changes Suggested by Fossil Terrestrial Plants

Time Unit	Region	Parameter	Departure from Existing Value at Locality of Occurrence	
			(°C)	Source
Hypsithermal	Northern Europe	July mean	+2.0	J. Iversen, Geol. Fören. Stockholm Förh., v. 66, 1944, p. 477.
Younger Dryas	Central Europe	July mean	−6.0	J. Iversen, Danm. Geol. Undersög., ser. 2, no. 80, 1954, p. 98.
Alleröd maximum	Northern Europe	July mean	−2.0 to −3.0	Idem, p. 97.
	Central Europe	July mean	−2.5	Firbas, 1949—52. p. 287.
Weichsel glacial maximum	Western Poland	Annual mean	−4.0 to −5.0	Szafer, 1954.
Mindel—Riss (?) Interglacial	Austrian Alps	Annual mean	+2.0 to +3.0	Klebelsberg, 1948—49.
Earliest glacial maximum	Central Europe	Annual mean	−6.0	(Derived from altitudes of vegetation zones).

prove to be more sensitive to minor changes of climate than are large mammals.

Although in some instances climate can be inferred from a single plant species (the tundra plant *Salix herbacea* indicates proximity to the ice sheet during the last deglaciation of northern Europe), only rarely can this be done with a single species of mammal. A whole fauna must be considered in order to reduce errors resulting from wide climatic tolerances of various species. For example, woolly mammoth and reindeer are not climatic indicators unless they occur in large numbers, for they are found at localities and in association with other organisms that are definitely not of glacial character. This reflects the fact that the ranges of most organisms overlapped during glacial and nonglacial times as they do today. Hence it is not surprising to find, in a single stratigraphic horizon at the same locality, a mixture of animals characteristic of tundra or steppe with animals common to temperate forests.

Although comparison of past ranges with present ranges on the principle of uniformity gives results that are generally consistent, there are exceptions. The climatic tolerances of some groups of mammals seem to have been greater in the past than is suggested by the distribution of their nearest existing relatives. Musk-oxen, for instance, connote a high-arctic climate if we consider only the range of the one existing genus, *Ovibos*. Musk-oxen are not uncommon in the Pleistocene of central North America and even as far south as New Mexico, but they include other genera which, for all we know, may have been well adjusted to temperate climates and hence do not necessarily imply arctic climates at latitudes of 30° to 40°. Indeed contemporaneous plants and small mammals imply that toward the lower end of that latitude range climates even at glacial maxima were not arctic.

Again, whereas *Hippopotamus* today is confined to the tropics and subtropics, its common occurrence in Pleistocene sediments in central and western Europe as far north as England does not justify the inference that climates in that region were formerly tropical. This genus, a grazer, is reported to range on equatorial mountains in Africa today, to altitudes at which water temperature never exceeds 3°. It is possible, also, that its occurrences as fossils in Europe represent species whose tolerances may have differed from those of the one species that survives today. Conversely, the cougar (*Felis concolor*) was common from coast to coast of North America as recently as A.D. 1600. Now it is virtually confined to a few remote parts of North America because its food supply has been decimated and because it has been persistently hunted by man.

In summary, the aspect of a whole fauna and of.the related flora as well should be considered before the ecosystem (including the climate) to which it belonged can be confidently inferred. The climatic significance of plant or animal groups is best understood when existing and former distributions are plotted on maps and compared.

PLANTS

Although evolutionary trends can be discerned in it, the Quaternary history of plants is characterized more obviously by repeated displacement of populations in the manner already described, accompanied by partitioning, especially around ice sheets and in regions of high and abrupt relief. These and other aspects of plant history are emphasized in a thoughtful review by E. B. Leopold (p. 203 in P. S. Martin and Wright, eds., 1967).

In chap 16 and in subsequent chapters we have emphasized the ecologic and stratigraphic significance of fossil pollen. We need to repeat here that the Quaternary stratigraphy of Europe has reached its present state in large part through studies of pollen. In some fossil localities plant megafossils are associated with pollen; in others they constitute the sole basis of information in the absence of pollen. In most localities where invertebrate or vertebrate fossils occur associated with plants or pollen, ecologic agreement among them is satisfactory.

INVERTEBRATE ANIMALS

Marine Invertebrates

As we noted in chap 27, planktonic Foraminifera probably have given more reliable indications of former temperature than other marine kinds. Littoral mollusks are generally more difficult to work with, although by employing temperature-sensitive taxa Feyling-Hanssen (1955) read climatic variations through very late Pleistocene and Holocene time, with correlation among the coasts of Scandinavia, Spitsbergen, and East and West Greenland. Invertebrates that reflect change in water temperature at least qualitatively occur in early Pleistocene sediments in the Mediterranean region and in sediments of Eem and Holstein age along the North Sea coast. Population shifts, among marine mollusks, that imply climatic change have been claimed for coastal areas of North America, southern South America, and Europe, but there is little broad integration of results. Evolution has occurred, but is not conspicuous and is marked primarily by extinctions.

A representative but extremely incomplete list of publications (chiefly summaries) on Quaternary marine mollusks, would include these titles: Pacific North America: Durham, 1950; Valentine, 1961; Wagner, 1959; Hopkins, ed., 1967, p. 47–90, 92, 326–349; Atlantic North America: Richards, 1962; Wagner, 1967; Wagner, p. 7 in Hood, ed., 1968; Argentina: Feruglio, 1949–1950; Mediterranean France: Mars, 1963.

Nonmarine Mollusks [4]

Fossil nonmarine mollusks are locally abundant in alluvium, colluvium, loess, and sediments of lakes and marshes. Much of the literature on them consists of faunal lists and descriptions, but the most fruitful studies have been those pursued quantitatively and in an ecologic context. Not until the 1950's did such studies begin to appear in numbers. Groups of mollusks of steppe, forest and woodland, and aquatic habitat are recognized, and fossil mollusks fortunately are present in numbers in those areas in which fossil-pollen sequences are rare. To some extent such mollusks reflect evolution through Quaternary time, but probably their greatest usefulness lies in their indications of change of environment with time and hence of change of climate.[5] Thus in central and eastern Eu-

---

[4] General references: Ložek, p. 201 in H. E. Wright and Frey, eds. 1965b; LaRocque, 1966–1970, pt. 1; Frye and Leonard, p. 429 in Teichert and Yochelson, eds., 1967; Sparks, 1961.

[5] Opinions differ as to the degree of environmental resolution that can be obtained in this way, because of relatively great adaptability of many terrestrial mollusks.

rope glacial ages are commonly marked by steppe assemblages and interglaciations by forest assemblages. A Holocene stratigraphic sequence of mollusk groups not unlike the Holocene pollen sequence has been recognized in Europe. However, although repeated changes of climate are reflected in mollusk groups, apparently evolution has not been sufficiently rapid to permit identification of the individual glacial and interglacial strata. For such identification fossil mammals are more useful than fossil mollusks. Nevertheless, the ecological implications of mollusks can throw light on climatic history. It has been claimed that mollusks in the Great Plains region of North America indicate a semiarid late-Pliocene climate followed by a dramatic change to a cooler and moister climate early in the Pleistocene (Frye and Leonard, p. 434 in Teichert and Yochelson, eds., 1967). As this is apparently the earliest record in that region of the comparatively cool, moist climates of glacial-age type, the occurrence might repay minute stratigraphic re-examination with the object of dating the change either directly or indirectly.

Modern quantitative/ecologic or stratigraphic studies of nonmarine mollusks in North America, apart from those mentioned at the beginning of the present section, include LaRocque, 1966–1970, pts. 2,3; Taylor, 1960; 1966; Wayne, 1959. A study in western Morocco is contained in Biberson and Jodot, 1965.

Insects [6]

Like planktonic animals in the ocean, insects are the most numerous animals on the lands. They live in all latitudes. As Quaternary fossils they are likewise abundant (largely, of course, in interglacial rather than glacial layers), the Coleoptera (beetles) apparently most so. Insect fossils occur most commonly in the form of detached wings and other parts, which must be concentrated by washing and screening fine sediments. As many as 140 species of fossil beetles have been identified from a single locality.

Although insects have been less widely studied than have other kinds of terrestrial invertebrates, a good deal has been learned about them. One basic fact is that so far little or no evolution in beetles through Quaternary time has been established. The same taxa as those living today extend at least as far back as the Tegelen Age. Hence evolutionary change is not a basis for the stratigraphic zoning of these insects.

On the other hand insects are very useful for determining former environments. This follows from the sensitivity of insects to climatic parameters, particularly temperature. They respond promptly to changes in cli-

[6]References: Coope, 1968; 1969; p. 359 in Cushing and Wright, eds., 1967; Shotton, p. 17 in H. E. Wright and Frey, eds., 1965b; West, 1968, p. 328.

mate because their rates of dispersal are generally rapid, and unlike plants they are not closely dependent on the characteristics of soils, which mature slowly. Indeed there is reason to believe that in a favorable locality the effect of climatic change is registered by beetles before it is recorded in the local flora.

Despite their dependence on temperature, beetles cannot easily be related to specific isotherms for the reconstruction of paleoclimate as is possible with certain kinds of plants, because their present-day distribution is related also to several environmental factors other than temperature. Hence in many cases groups of taxa occurring together as a faunule are assessed for their environmental significance, which is probably more accurate than one derived from a single kind. The results in most cases are wholly compatible with those derived from the fossil floras at the same places, and occasionally they enrich the environmental interpretation with details not obtainable from plants alone.

With this discussion of general principles, plants, and lower animals, we turn to chap 29 for a brief inquiry into Quaternary mammals, including man, on which much information is available.

CHAPTER 29

Quaternary Fossils II: Vertebrates

VERTEBRATES, ESPECIALY MAMMALS

Among the vertebrates we must deal almost exclusively with mammals because the Quaternary vertebrate fossil record consists mainly of mammals. Fossil fishes, birds, and a few reptiles and amphibians exist, but except perhaps for birds (cf. Mayr, 1946) the material is hardly adequate to provide a connected account. Accordingly we turn to the stratigraphy and ecology of Quaternary mammals, by continents, bearing in mind the comments we have made on them in several foregoing chapters.

EURASIA[1]

Europe. The Eurasiatic mammal fauna in Quaternary strata is indigenous; that is, it evolved out of pre-Quaternary stocks already resident in Eurasia. As mentioned before, the Pleistocene fauna is customarily subdivided into three successive groups, on the basis of which the Pleistocene is subdivided into Lower, Middle, and Upper units. Following these is a Holocene fauna. Table 29-A conveniently assembles data on the large mammals of Europe, but we must remember that any list of this kind distorts continuity because of unavoidable gaps between fossil-bearing strata. Further brief information on each of the taxa named in the table is given in Kurtén, 1967. To give some slight visual impression of the faunas we include outline sketches (figs 29-1, 29-2, 29-3, and 29-4) of typical taxa of the Early, Middle, and Late Pleistocene as well as Holocene strata.

As table 29-A shows, the European Pleistocene of most paleontologists begins with a long-enduring Villafranchian or Lower Pleistocene fauna

---

[1] General references: Kurtén, 1967; Kowalski in Turekian, ed., 1971; Kowalski, 1959; Thenius, 1962; Toepfer, 1963; Hescheler and Kuhn in Tschumi, ed., 1949; Zeuner, 1959, p. 308–338.

Table 29-A

European Upper Pliocene and Quaternary Mammal Faunas. Stratigraphically Arranged, with First and Last Appearances [a]

Stratigraphic Units	Fauna (Chief elements only)	Most significant localities of occurrence
HOLOCENE	First appearance: Bison bonasus, Apodemus agrarius and numerous synanthropic and introduced mammals, e.g. Rattus rattus, Rattus norvegicus, Mus musculus, Fiber zibethicus, Nyctereutes procyonoides, Dama dama. Last appearance: Megaloceros giganteus, Bos primigenius, Equus gmelini, Myotragus balearicus, Prolagus sardus. Arctic mammals (e.g. Rangifer tarandus) successively disappear from Middle Europe replaced by forest species. In the period of climatic optimum Bison bonasus appears in Middle Europe and many other species (e.g. Myotis bechsteini, Felis silvestris, Eliomys quercinus) extend their ranges beyond their present northern limits. With the replacement of forest by arable fields steppe mammals (e.g. Apodemus agrarius, Cricetus cricetus) spread to the West. Many synanthropic and introduced mammals appear in Europe.	
PLEISTOCENE — Upper — Weichsel	First appearance: Saiga tatarica, Felis manul, Allactaga jaculus. Last appearance: Palaeoloxodon antiquus, Mammuthus primigenius, Dicerorhinus kirchbergensis, Coelodonta antiquitatis, Equus germanicus, Equus hemionus, Hippopotamus amphibius, Dama dama, Ovibos moschatus, Bison priscus, Crocuta crocuta, Homotherium latidens, Felis leo, Felis pardus, Aonyx antiqua, Cuon alpinus, Ursus spelaeus, Microtus gregalis, Lagurus lagurus, Dicrostonyx torquatus, Ochotona pusilla. In Middle and Western Europe arctic and steppe mammals prevail: Mammuthus primigenius, Coelodonta antiquitatis, Ovibos moschatus, Bison priscus, Rangifer tarandus, Felis leo, Crocuta crocuta, Ursus spelaeus, Mustela nivalis, Dicrostonyx torquatus, Lemmus lemmus, Microtus gregalis, Microtus oeconomus, Ochotona pusilla, Lepus timidus. In the time of the interstadials forest mammals regain part of their areas in Middle Europe. For particular places the succession of many tundra–, steppe– and forest stages can be stated in the composition of the mammalian fauna.	Cave sediments, loess, peat bogs, river and glacial sediments throughout Europe, too numerous for a general list
PLEISTOCENE — Upper — Eem	First appearance: Dama dama, Rupicapra rupicapra, Felis lynx, Lutra lutra, Vulpes corsac, Ursus maritimus, Hystrix cristata, Lepus europaeus, Lepus timidus. Last appearance: Dicerorhinus hemitoechus, Hyaena hyaena In Middle and Western Europe, forest fauna with: Palaeoloxodon antiquus, Dicerorhinus kirchbergensis, Hippopotamus amphibius, Cervus elaphus, Capreolus capreolus, Felis chaus, Hystrix cristata etc.	England: Hyaena Stratum in Tornewton Cave Germany: travertines of Cannstatt and Taubach Czechoslovakia: travertines of Ganovce Deep layers of cave sediments throughout Europe

[a]Assembled for this book by Kazimierz Kowalski (Polish Academy of Sciences), 1970.

Stratigraphic Units			Fauna (Chief Elements Only)	Most significant localities of occurrence
P L E I S T O C E N E	Middle	Saale	First appearance: Mammuthus primigenius, Coelodonta antiquitatis, Equus hemionus, Alces alces, Capra ibex, Marmota marmota, Lagurus lagurus. Last appearance: Dama clactoniana, Ursus thibetanus In Middle and Western Europe, arctic fauna with: Mammuthus primigenius, Coelodonta antiquitatis, Rangifer tarandus, Alopex lagopus, Gulo gulo, Lemmus lemmus, Dicrostonyx torquatus, Microtus nivalis, Microtus gregalis, Lagurus lagurus, Ochotona pusilla.	Italy: San Agostino France: La Fage, Lazaret, Grimaldi England: Glutton Stratum in the Tornewton Cave Hungary: Uppony I Poland: lowest layers in Nieto-perzowa and other caves
		Holstein	First appearance: Dicerorhinus hemitoechus, Equus germanicus, Megaloceros giganteus, Dama clactoniana, Bos primigenius, Macaca sylvana, Hyaena hyaena, Felis silvestris, Martes martes, Aonyx antiqua, Vulpes vulpes, Ursus spelaeus, Ursus arctos, Oryctolagus cuniculus Last appearance: Mammuthus trogontherii, Trogontherium cuvieri, Mimomys cantianus. In Middle and Western Europe, forest fauna with: Palaeoloxodon antiquus, Dicerorhinus kirchbergensis, Dama clactoniana, Bubalus murrensis.	Italy: Spessa II France: Lunel-Viel England: Swanscombe, Gray's Thurrock, Clacton, Hoxne Germany: Heppenloch
		Elster	First appearance: Dicerorhinus kirchbergensis, Gulo gulo, Mustela nivalis, Dicrostonyx torquatus. Last appearance: Dicerorhinus etruscus, Equus mosbachensis, Megaloceros savini, Alces latifrons, Praeovibos priscus, Soergelia elisabethae, Macaca florentina, Felis toscana, Ursus deningeri. Arctic mammals appear in Middle and Western Europe (Rangifer, Ovibos, Gulo, Lemmus, Dicrostonyx).	Germany: Mosbach upper layers France: Cagny, Estève Janson Hungary: Tarkö, Vertesszollos Czechoslovakia: Koneprusy
		Cromer	First Appearance: Palaeoloxodon antiquus, Equus mosbachensis, Felis leo, Felis pardus, Pitymys gregaloides, Microtus arvaloides. Last appearance: Archidiskodon meridionalis, Dama nestii, Hyaena perrieri, Felis lunensis, Gulo schlosseri, Pannonictis pliocaenica, Lutra simplicidens, Mimomys savini, Hypolagus brachygnathus. In Middle and Western Europe, forest fauna with Hippopotamus amphibius, Macaca florentina, numerous Cervids, etc.	England: Cromer Forest Bed Germany: Voigstedt, Mauer Austria: Hundsheim Czechoslovakia: Stranska Skala

Table 29-A

European Upper Pliocene and Quaternary Mammal Faunas. Stratigraphically Arranged, with First and Last Appearances *(continued)*

Stratigraphic Units			Fauna (Chief Elements Only)	Most significant localities of occurrence
P L E I S T O C E N E	Middle	Menap	**First appearance of arctic mammals:** Gulo schlosseri, Rangifer tarandus, Praeovibos priscus, Ovibos moschatus, Lemmus lemmus and also of: Mammuthus trogontherii, Equus sussenbornensis, Sus scrofa, Megaloceros savini, Capreolus capreolus, Alces latifrons, Soergelia elisabethae, Bison priscus, Crocuta crocuta, Homotherium latidens, Felis pardina, Meles meles, Aonyx bravardi, Lutra simplicidens, Canis lupus, Lycaon lycaonoides, Ursus deningeri, Ursus thibetanus, Hystrix vinogradovi, Allophaiomys pliocaenicus, Arvicola terrestris. **Last appearance:** Anancus arvernensis, Equus süssenbornensis, Hyaena brevirostris, Homotherium sainzelli, Acinonyx pardineusis, Baranogale antiqua, Enhydrictis ardea, Cuon majori, Lycaon lycanoides, Vulpes alopecoides, Vulpes praecorsae, Citellus primigenius, Lepus terraerubrae.	Germany: Mosbach (lower layers) Süssenborn France: Valerots Czechoslovakia: Chlum 6 Hungary: Nagyharsanyhegy 2 Poland: Kamyk Roumania: Betfia Soviet Union: Nogaisk, Kair, Chortkov
	Villafranchian (Lower)	Upper	**First appearance:** Archidiskodon meridionalis, Equus stenonis, Equus hydruntinus, Hippopotamus amphibius, Dama nesti, Leptobos etruscus, Felis lunensis, Canis arnensis, Cuon majori, Citellus primigenius, Trogontherium cuvieri, Lagurus pannonicus, Lepus terraerubrae. **Last appearance:** Tapirus arvernensis, Equus stenonis, Equus bressanus, Sus strozzii, Cervus perrieri, Cervus etuerarium, Alces gallicus, Gazella borbonica, Deperetia ardea, Leptobos elatus, Leptobos etruscus, Dolichopithecus arvernensis, Euryboas lunensis, Megantereon megantereon, Felis issiodorensis, Canis etruscus, Nyctereutes megamastoides, Ursus etruscus, Hystrix refossa, Dolomys milleri, Oryctolagus lacosti.	Spain: Olivola Italy: Val d'Arno Superiore, Leffe France: Saint Vallier, Senèze Holland: Tegelen Hungary: Beremend 1, Villany 5 Poland: Kadzielnia Soviet Union: Chapry, Sinaia Balka
		Lower	**First appearance:** Dicerorhinus etruscus, Hipparion crusafonti, Equus bressanus, Cervus perrieri, Alces gallicus, Leptobos elatus, Gazella borbonica, Deperetia ardea, Macaca florentina, Dolichopithecus arvernensis, Hyaena perrieri, Euryboas lunensis, Ursus etruscus, Megantereon megantereon, Homotherium sainzelli, Felis issiodorensis, Felis toscana, Acinonyx pardinensis, Enhydrictis ardea, Aonyx bravardi, Canis etruscus, Nyctereutes megamastoides, Castor fiber, Hystrix refossa, Mimomys pliocaenicus, Oryctolagus lacosti. **Last appearance:** Zygolophodon borsoni, Dicerorhinus megarhinus, Sus arvernensis, Parailurus anglicus, Agriotherium insigne.	Spain: Villaroya Italy: Villafranca d'Asti France: Vialette, Mt. Perrier, Chagny England: Red Crag Czechoslovakia: Hajnačka Roumania: Mălusteni, Berešti, Čapeni Soviet Union: Stavropol
PLIOCENE		Astian	Anancus arvernensis, Zygolophodon borsoni, Dicerorhinus megarhinus, Hipparion crassum, Tapirus arvernensis, Pliohyrax occidentalis, Parailurus anglicus, Agriotherium insigne, Ursus ruscinensis, Nyctereutes sinensis, Trilophomys pyrenaicus, Stachomys trilobodon, Mimomys stehlini.	France: Montpellier, Roussillon, Sète, Nîmes Czechoslovakia: Ivanovce Poland: Weże, Rebielice Hungary: Gödöllö, Csarnota

References (listed in short form):

References

Astian

Fejfar O., 1960. Die plio-pleistozänen Wirbeltierfaunen von Hajnáčka und Iva-
novee /Slowakei/, ČSR. I. Die Fundumstände und Stratigraphie. Neues
Jahrb.f. Geol.u. Paläont., Abh.111:257–273.Stuttgart.
Kowalski K., 1960. Cricetidae and Microtidae /Rodentia/ from the Pliocene of
Węże/Poland/. Acta zool.cracov., Kraków, 5:447–505.
Kretzoi M., 1959. Fauna und Faunenhorizont von Csarnota. All.Földt. Intezet
Jelentesa, Budapest, p. 297–395.
Nikiforova K. V., 1964. On the stratigraphic position of the Astian. Rep. of the
VI INQUA Congress, Lódź, 2:547–557.
Thaler L., 1966. Les rongeurs fossiles du Bas-Languedoc dans leurs rapports
avec l'histoire de faune et la stratigraphie du tertiaire de'Europe.
Mém.Mus.Nat.d'Hist.Nat., Paris, n.s., C, 17:1–295.
Thenius E., 1959. Wirbeltierfaunen. In: A. Papp & E. Thenius, Tertiär, 2:I-
XI+1–328. Stuttgart.

Villafranchian

Fejfar O., 1964. The Lower Villafranchian vertebrates from Hajnáčka near Fila-
kovo in Southern Slovakia. Rozpravy Ustr.Ust. Geol., Praha, 30:1–115.
Kowalski K, 1958. An early pleistocene fauna of small mammals from the Kad-
zielnia Hill in Kielce. Acta paleomt.pol., Warszawa, 3:1–47.
Kurtén B., 1963. Villafranchian faunal evolution. Comment. biol., Helsinki,
26/3/:1–18.
Nikiforova K. W., 1965. Stratigraphische equivalente des Villafranchians
in the Sowjetunion. Proceedings Koninkl. Nederl. Akad., Amsterdam,
68/4/:237–248.
Samson P. M. & Radulesco C., 1963. Les faunes mammalogiques du Pléistocéne
inférieur et moyen de Roumanie. C. R. Acad., Paris, 257: 1122–1124.

Middle and Late Pleistocene

Kurtén B., 1968. Pleistocene Mammals of Europe. Weidenfeld and Nicholson,
London, 1–317.
Chaline J., 1969. Les rongeurs du pléistocène moyen et superieur de France.
Thèse, Université de Dijon/3 vols.//preprint/.
Fejfar O., 1961. Review of Quaternary Vertebrata in Czechoslovakia. Prace Inst.
Geol., Warszawa, 34/1/:109–118.
Kretzoi M., 1961. Stratigraphie und chronologie. Prace Inst. Geol., Warszawa,
34/1/:313–332.
Kahlke H.-D., 1961. Revision der Säugetierfaunen der klassischen deutschen
Pleistozän-Fundstellen von Süssenborn, Mosbach und Taubach. Geologie,
Berlin, 10/4–5/:493–532.
West R. G. & Wilson D. G., 1966. Cromer Forest Bed Series. Nature, London,
209/5022/:497–498.
Das Pleistozän von Voigtstedt. Paläont. Abh., Abt. A, Berlin, 2/2–3/, 1966,
227–692, Taf. I-XL.
Kowalski K., 1966. Stratigraphic importance of rodents in the studies on Euro-
pean Quarternary. Folia quatern., Kraków, 22:1–16.

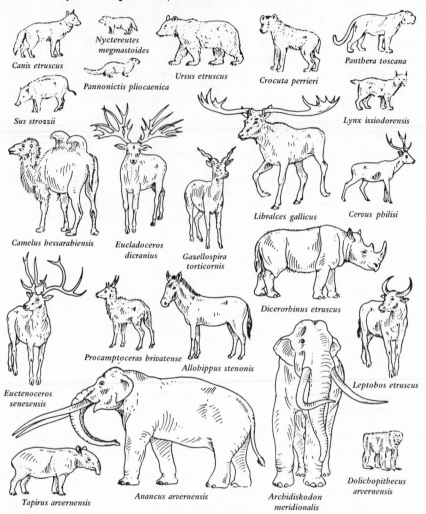

Nyctereutes megmastoides

Canis etruscus

Pannonictis pliocaenica

Ursus etruscus

Crocuta perrieri

Panthera toscana

Sus strozzii

Lynx issiodorensis

Camelus bessarabiensis

Eucladoceros dicranius

Gazellospira torticornis

Libralces gallicus

Cervus philisi

Euctenoceros senezensis

Procamptoceras brivatense

Allohippus stenonis

Dicerorhinus etruscus

Leptobos etruscus

Tapirus arvernensis

Anancus arvernensis

Archidiskodon meridionalis

Dolichopithecus arvernensis

Figure 29-1 Sketch restorations of representative members of a Villafranchian mammal fauna from Central Europe. Scale preserves relative size. (Thenius, 1962.)

(fig 29-1), itself divided, not very sharply, into upper and lower units.[2] Apart from kinds that suggest grassy steppe, the majority of the taxa are of the sort expectable in moist, warm-temperate forest environments.

[2] Some paleontologists think the base of the Pleistocene should be placed at the base of the Astian (now included in the Pliocene), which contains significant modern immigrants (Nikiforova, 1964).

These constituted an ecosystem that spread across Europe at least as far north as the North Sea, where it is represented in the Tegelen strata in the Netherlands, and in Britain as well. Nevertheless, although closely related to earlier forms, the Villafranchian mammals show regional variations that imply change. Kowalski (in Turekian, ed., 1971) sees the beginning of the change farther back in late-Cenozoic time and the Villafranchian changes as a continuation of the earlier evolution. Be that as it may, evidently there was evolution from earlier cosmopolitan conditions to a more varied world. This suggests that compartmentation of the continents, mainly through the rise of highlands, was splitting both flora and fauna into regional groups, a process that was carried further as Pleistocene time progressed. Furthermore, it by no means excludes the influence of secular climatic changes unrelated to tectonics.

This evolution is expressed in the appearance of new mammal groups such as true elephants, zebrine horses, and cattle, which diluted the otherwise Pliocene aspect of the fauna. Even with these changes, most kinds of Villafranchian animals were smaller in size than their counterparts of a later time. The principal European mammals listed in table 29-A show that the fauna changed with time, both through extinctions and through the introduction of new kinds. It is such changes that led to subdivision of the fauna into upper and lower Villafranchian units. Although the literature concerns itself primarily with large mammals, more climatic information could be extracted from fossil rodents, a group that evolved rapidly and that are sensitively related to vegetation.

Within the time between the Waal and "Cromer" Interglaciations (table 24-A) the fauna underwent so many extinctions and introductions that a new fauna can be said to have come into existence, a fauna noteworthy for new kinds of elephants and rhinoceroses. This was the Middle Pleistocene fauna (fig 29-2), well represented in sediments at two important localities in Germany, Mosbach and Süssenborn. It implies climates distinctly cooler than those of Villafranchian time, climates that apparently were related to the earliest extensive glaciations in Europe. The first "arctic" mammals appeared: first reindeer, musk-ox, and lemming, followed much later by woolly mammoth, woolly rhinoceros, and moose. This new fauna witnessed and was affected by the Elster and Saale glaciations, as well as by the Günz glaciation in the Alps and whatever correlative that may have had in northern Europe.

The Upper Pleistocene fauna (fig 29-3) continued the "arctic" forms and added to them the Siberian saiga antelope. Early in Upper Pleistocene time (that is, during the Eem Interglaciation) the straight-tusked elephant *Palaeoloxodon antiquus* disappeared, as well as the hippopotamus and other far-from-"cold" taxa. It was as though the stress of

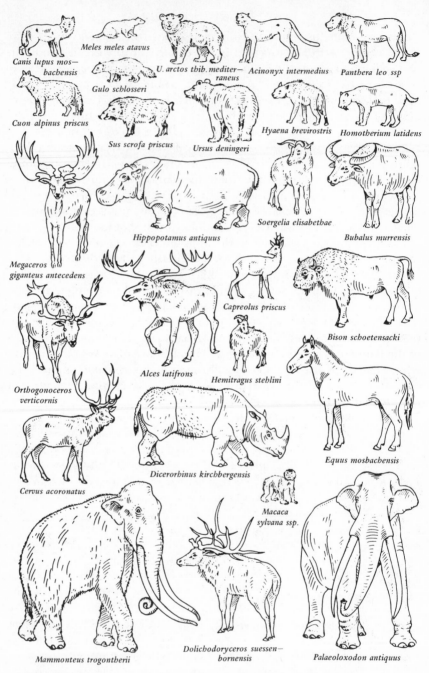

Figure 29-2 Sketch restorations of Middle Pleistocene mammals from Central Europe. Scale preserves relative size. (Thenius, 1962.)

Figure 29-3 Sketch restorations of Late Pleistocene mammals from Central Europe. Scale preserves relative size. (Thenius, 1962.)

repeated cold climates, with all the ecological changes such repetition engendered, was too much for less well-adapted mammals. At the end of the Weichsel Glaciation, and with it the end of the Upper Pleistocene fauna, many more large mammals became extinct, among them cave bear, cave lion, woolly mammoth, and woolly rhinoceros. An incomplete though long faunal list, covering the entire USSR, is given by Vereshchagin (p. 389 in P. S. Martin and Wright, eds., 1967).

The extensive movements of populations induced by the glacial-interglacial changes of climate were superposed on the evolutionary changes; both of course shared, to some extent, a single cause.

From what has been said about extinctions, it is apparent that the Holocene fauna (fig 29-4) was relatively impoverished. In Europe it included only one large mammal that was new—the bison. However, the human invasions of Europe in postglacial time brought with them many smaller mammals such as rats.

This brief review of the stratigraphy of European mammals leads us to recall a theory that prevailed in the early part of the present century. It was that European Quaternary fossil mammals could be sorted into a "warm" fauna and a "cold" fauna, and that the two groups of animals shifted northward or southward as climates changed. Developing knowl-

Figure 29-4 Sketch restorations of Holocene mammals from Central Europe. Scale preserves relative size. Except for *Bos primigenius,* all these taxa are still living. (Thenius, 1962.)

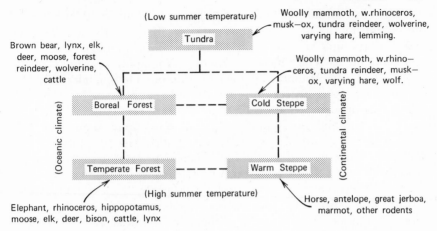

Figure 29-5   Normal ecologic connections (dashed lines) between late-Pleistocene biomes in Europe, with representative mammals. Faunal overlaps between adjacent biomes are common. Some taxa ranged from tundra to temperate forest. (After Zeuner, 1959.)

edge of both stratigraphy and ecology has made it clear that the difference exists only in the upper half of the Pleistocene Series and that the supposed "cold" fauna is mainly an evolutionary group, on which are superposed biologic facies related to different biomes. Hence the old theory, to the extent that it is true, is merely a part of a more complicated whole. In fig 29-5 the truly "cold" biologic facies is seen to be related to the tundra biome. It includes such animals as woolly mammoth, musk-ox, woolly rhinoceros, reindeer, arctic fox, snowshoe rabbit, and lemming. The ranges of each of these animals differed; indeed, the reindeer ranged so far beyond the tundra zone and into such different environments that, though a very common constituent of the facies, it is in no sense an indicator of tundra. In contrast the temperate-forest biome was inhabited at various times in the later Pleistocene by the straight-tusked elephant, Kirchberg's rhinoceros, moose, deer, bison, hippopotamus, lynx, and bear. Although some of these animals, too, ranged outside into subarctic forest or steppe, the group as a whole indicates a temperate forest environment. Three cave-dwelling examples of Pleistocene mammals of such wide range that they tell virtually nothing about habitat are: cave bear, cave lion, and hyena. A group known to have lived on the steppe, because of the geographic position and pollen spectra of their occurrence, includes horse, bison, Irish elk, and various specialized rodents. Of these some if not all ranged into the tundra biome.

Asia. When we turn to Asia, we find far less information than is available for the European part of the continent. Mammal faunas of Siberia and North China are shown in generalized form in table 25-A; these faunas are the basis on which the Chinese stratigraphic sequence is subdivided into Early, Middle, and Late Pleistocene. The boundaries of the subdivisions apparently occupy somewhat different positions from those of the European sequence. Probably the difference results from paucity of fossil localities and less complete physical stratigraphy. Some fossil lists for localities in South China are given by Kahlke (1961; see also Chang, 1968), and some taxa in stratigraphic context in Java are mentioned by Hooijer (1962). Takai, 1952, gives lists, subdivided into faunal zones, for the region Japan-Southeast Asia-India as a whole. Evidently a fauna of Villafranchian aspect, followed by one resembling the Middle-Pleistocene fauna of Europe, is recognizable throughout southeast Asia. In a brief summary of those faunas, Colbert (1942) gave the following characteristics:

> *Upper Pleistocene:* Red dog (dhole), stegodont, mammoth, rhinoceros, horse, hippopotamus, camel, cattle, man (*Homo sapiens*).
> *Middle Pleistocene:* Porcupine, bamboo rat, giant panda, stegodont, mammoth, rhinoceros, horse, hippopotamus, camel, cattle, orang, man (*Homo erectus*).
> *Lower Pleistocene* ("Villafranchian"): Stegodont, mammoth, rhinoceros, horse, hippopotamus, camel, cattle, orang.

Presumably the similarities to European faunas are more apparent than real, resulting in part from lack of identification to species level.

Comparing tables 25-A and 29-A we find many genera and some species that are identical. This fact illustrates the wide freedom of movement, especially among inhabitants of open country such as steppe and tundra, that is possible on a large continent. It shows also the basis of correlation of strata between Europe and eastern Asia.

Africa [3]

In Africa as in other continents Neogene time was marked by broad uplifts. In the southeastern region late Pliocene upwarping alone seems to have amounted to nearly 1800m. More intense local movements in the Rift Valley belt created altitudes as great as 5000m. With those movements climates became drier and more seasonal in some areas, and the differences between one region and another increased.

---

[3] General references include H. B. S. Cooke, 1958; Cooke, p. 65 in Howell and Bourlière, eds., 1963; H. B. S. Cooke, 1968; Arambourg, 1962; Howell, 1960.

Table 29-B indicates that the African Quaternary mammal faunas are far from being closely related to the European. The species and a great many of the genera are different. The explanation lies in the Cenozoic history of the African mammals (H. B. S. Cooke, 1968). Some lineages extend far back in the Cenozoic; others indicate immigration from Asia as late as late-Miocene or Pliocene time. Fewer lineages seem traceable back into Europe. These relationships suggest the existence, during the greater part of the Cenozoic Era, of effective barriers to intercontinental exchange. Such exchange seems never to have occurred in a wholesale manner and at best was infrequent; hence the peculiarly African aspect of the Quaternary mammals of Africa.

Knowledge of the Quaternary mammals is confined mainly to fossils from three regions: (1) the North African Mediterranean coastal belt, (2) East Africa (chiefly Kenya and Uganda), and (3) South Africa, mostly its eastern part. Each of these regions is ecologically distinct, and the first is separated from the other two by the formidable barrier created by the Sahara region, so effective that the fauna north of it now has Eurasian elements not present in the east and south. The great western region has yielded little information, partly at least because in the rain-forest belt exposures are few and moist ground conditions are unfavorable for the preservation of fossils.

As indicated in table 29-C, the fossil finds from the three regions mentioned have been classified into four stratigraphic units, the "faunal spans" of Cooke which are essentially mammal ages of North American usage. The lowest corresponds in a general way with the Villafranchian zone in the European sequence, but although the name Villafranchian has been applied repeatedly to assemblages of African mammals, it should, in the opinion of H. B. S. Cooke (1968, p. 251), be written "Villa-franchian" when used in an African context. However, radiometric dates of European and African occurrences are sufficient to indicate that both belong stratigraphically in the lower part of the Pleistocene. Representative African mammals from that position are depicted in fig 29-6.

Above the "Villafranchian" in Africa is a mammal unit of uncertain correlation. It may be equivalent to either an upper part of the Villafranchian or a lower part of the Middle Pleistocene in Europe. The uncertainty emphasizes the differences between the faunas of the two continents, but the doubt should be capable of resolution through radiometric dating.

Still higher up we find units resembling the Middle and Upper Pleistocene units in Europe, though with vague correlative boundaries. At this point we note that the faunas north and south of the Sahara involve some similarities at the "Villafranchian" level, fewer at the Middle

Table 29-B
African Plio-Pleistocene and Quaternary Mammal Faunas [a]

Division		North Africa	East Africa	South Africa
Holocene		Living fauna only	Living fauna only	Living fauna only
Upper Pleistocene		*Homo sapiens sapiens* *Homo sapiens* *neanderthalensis* *Loxodonta atlantica* *Elephas iolensis* *Equus mauritanicus*	*Homo sapiens sapiens* *Loxodonta africana* *Elephas iolensis* *Equus burchelli* *Equus grevyi*	*Homo sapiens sapiens* *Homo sapiens* *rhodesiensis* *Loxodonta africana* *Elephas iolensis* *Equus burchelli* *Equus quagga* *Equus zebra* *Equus plicatus* *Equus helmei*
		Hippopotamus amphibius *Sus scrofa* *Phacochoerus africanus*	*Hippopotamus amphibius* *Potamochoerus porcus* *Phacochoerus africanus*	*Hippopotamus amphibius* *Potamochoerus porcus* *Phacochoerus africanus* *Phacochoerus aethiopicus*
		Megaceroides algericus *Homoioceras antiquus* *Bos primigenius* *Gazella tingitana*	*Homoioceras nilssoni* *Gazella granti*	*Homoioceras bainii* *Gazella bondi*
Pleistocene		*Homo erectus* *Simopithecus sp.* *Megantereon sp.* *Panthera leo* *Ursus arctos lateti* *Loxodonta atlantica* *Elephas recki*	*Homo erectus* *Simopithecus jonathani* *Panthera leo* *Elephas recki*	*Homo erectus* *Simopithecus sp.* *Megantereon cf gracile* *Panthera leo* *Loxodonta atlantica* *Elephas iolensis*
		Ceratotherium simum simum *Equus mauritanicus* *Libytherium maurusium*	*Ceratotherium simum simum* *Diceros bicornis* *Equus olduvaiensis* *Equus burchelli* *Stylohipparion libycum* *Libytherium olduvaiensis* *Ancylotherium hennigi*	*Ceratotherium simum simum* *Diceros bicornis* *Equus helmei* *Equus plicatus* *Equus burchelli* *Stylohipparion steytleri*
Middle		*Afrochoerus sp.* *Camelus thomasi* *Homoioceras antiquus* *Gazella atlantica*	*Hippopotamus gorgops* *Hippopotamus amphibius* *Afrochoerus nicoli* *Mesochoerus olduvaiensis* *Orthostonyx brachyops* *Giraffa jumae* *Giraffa camelopardalis* *Libytherium olduvaiensis* *Bularchus arok* *Phenacotragus recki* *Damaliscus angusticornis* *Pultiphagonides africanus*	*Hippopotamus gorgops* *Hippopotamus amphibius* *Stylochoerus compactus* *Mesochoerus paiceae* *Orthostonyx sp.* *Giraffa camelopardalis* *Libytherium olduvaiensis* *Homoioceras bainii* *Gazella wellsi* *Lunatoceras mirum* *Megalotragus eucornutus*
Lower Pleistocene (Upper Villafranchian)		*Australopithecus sp.*	*Australopithecus spp.* *Simopithecus oswaldi* *Papio spp.* *Paracolobus chemeroni*	*Australopithecus spp.* *Simopithecus danieli* *Parapapio jonesi* *Parapapio whitei* *Cercopithecoides williamsi*

(continued)

[a]Assembled for this book by H. B. S. Cooke (Dalhousie University), 1970.

Division	North Africa	East Africa	South Africa
Lower Pleistocene (Upper Villafranchian)	*Meganthereon sp.*	*Machairodontinae*	*Meganthereon eurynodon* *Therailurus barlowi* *Lycyaena silberbergi*
	Anancus osiris *Elephas recki*	*Stegodon kaisensis* *Loxodonta atlantica* *Elephas recki* *Deinotherium bosazi*	 *Elephas recki*

	Ceratotherium simum
	   germanoafricanum
		Ceratotherium simum
		   germanoafricanum
		 Diceros bicornis
			Ceratotherium simum
			   germanoafricanum
			 Diceros bicornis

	Equus numidicus *Stylohipparion libycum*	*Equus olduvaiensis* *Stylohipparion libycum* *Hipparion albertense* *Ancylotherium hennigi*	*Equus sp.* *Stylohipparion steytleri* *Ancylotherium hennigi*
		Hippopotamus protamphibius *Hoppopotamus imagunculus* *Pronotochoerus jacksoni* *Notochoerus scotti* *Notochoerus euilus* *Mesochoerus olduvaiensis*	*Hippopotamus sp.* *Potamochoeroides shawi* *Notochoerus of capensis*
	Mesochoerus maroccanus	*Mesochoerus limnetes*	*Phacochoerus antiquus*
		Giraffa gracilis	*Giraffa sp.*
	Libytherium maurusium *Camelus sp.*	*Libytherium olduvaiensis* *Camelus sp.*	*Libytherium olduvaiensis*
	Numidocapra crassicornis	*Tragelaphus nakuae* *Beatragus antiquus* *Menelikia lyrocera*	*Redunca darti* *Makapania broomi*
	Gazella setifensis	*Gazella wellsi* *Redunca ancystrocera*	*Gazella wellsi*
Plio—Pleistocene (Including Lower Villafranchian)		*Australopithecus spp* *Parapapio jonesi*	*Australopithecus* *Parapapio jonesi* *Parapapio whitei*
	Meganthereon sp.	*Machairodontinae*	*Meganthereon gracile*

	Anancus osiris
	
	 Elephas recki (early
	   form)
		Stegodon kaisensis
		 Elephas recki (early
		   form)
			
			 Elephas recki (early
			   form)

	Mammuthus africanavus	*Loxodonta adaurora* *Mammuthus subplanifrons* *Deinotherium sp.*	*Loxodonta adaurora* *Mammuthus subplanifrons*

	Ceratotherium simum
	   germanoafricanum
		Ceratotherium simum
		   germanoafricanum
		 Diceros bicornis
		 Stylohipparion sp.
		 Hipparion albertense (early)	

	Hippopotamus hipponensis	*Hippopotamus spp.* (hexaprotodont) *Nyanzachoerus kanamensis* *Notochoerus euilus* *Notochoerus capensis*	 *Notochoerus capensis*
	Camelus sp.	*Camelus sp.* *Tragelaphus nakuae* (and many undescribed bovid species)	

(continued)

Table 29-B (*continued*)

Division	North Africa	East Africa	South Africa
Upper Pliocene		*Australopithecus spp.* *Anancus kenyensis* *Stegodon kaisensis* *Elephas ekorensis* *Mammuthus subplanifrons* *Primelephas gomphotheroides* *Ceratotherium praecox* *Stylohipparion sp.* *Hippopotamus spp.* (hexaprotodont) *Nyanzachoerus spp.* *Notochoerus capensis* (many undescribed Bovidae)	

level, and fewest at the Upper level. These relationships suggest that the Saharan region became a barrier to the interchange of population only by degrees through Pleistocene time, apparently by gradual increase of aridity, although the causes are still obscure. Toward the end of the Pleistocene, faunal interchange between Mediterranean Africa and Mediterranean Europe occurred. Superposed on the major faunal subdivisions in Europe are the glacial and interglacial ages, with their movements of populations clearly reflected in the places of occurrence and the numbers of various Pleistocene mammals. In Africa, in contrast, pluvial/nonpluvial fluctuation probably involved smaller changes of temperature and "served mainly to shift the boundaries of the . . . ecological zones, allowing the mammals to move with them" (H. B. S. Cooke, 1968), without causing either notable extinction or speciation. In other words, evolution among mammals was slower in Africa than it was in Europe.

A very important element among the Quaternary mammals of Africa consists of the hominids, which are discussed more appropriately elsewhere in this chapter.

North America

The Pleistocene mammal faunas of the Americas are linked with each other by the Isthmus of Panama land bridge and with those of Eurasia by a former land bridge across the existing Bering Strait. At various times movement of taxa occurred across both bridges in both directions, imparting cosmopolitan features to faunas that were otherwise provincial. Thus Quaternary faunas in North America include Asiatic elements, and to a lesser degree South American elements as well.

Table 29-C

Tentative Correlation of Provincial Quaternary Sequences in Africa on a Basis of Fossil Mammals. The *Spans* Named Are Essentially Provincial Mammal Ages. Compare Table 26-D.[a]

South Africa — Faunal Unit	South Africa — Localities	East Africa	North Africa	Possible Age
Recent	Cave and surface sites without extinct faunas	Cave and surface sites without extinct faunas	Cave and surface sites without extinct faunas	Recent
Florisbad–Vlakkraal faunal span	Vlakkraal and other deposits Florisbad and Chelmer Cave of Hearths	"Later Gamblian" beds and Eyasi beds	Sediments with Micoquian and Aterian cultures	Upper Pleistocene
		"Early Gamblian" beds		
Vaal–Cornelia faunal span	Hopefield Cornelia	Olduvai Bed IV + Olduvai Beds III to IV Semliki Beds	Lac Karar, etc.	Middle to Upper Pleistocene
	"Younger Gravels" of Vaal river	Olduvai Beds II to III	Ternifine	Middle Pleistocene
Swartkrans faunal span	Kromdraai Swartkrans Sterkfontein extension	Peninj Beds Olduvai Bed II (in part)	Yayo (Chad)	Cromerian or uppermost "Villafranchian"
Sterkfontein faunal span	Makapansgat, Taung and Sterkfontein	Olduvai Bed I, Omo, Later Kaiso	Ain Hanech, Koro–Toro, Bel Hacel etc.	
		Kanapoi and Earlier Kaiso	Lac Ichkeul, Ain Brimba Ain Boucherit, Fouarat	"Villafranchian"

[a]H. B. S. Cooke, 1968; modified 1970.

763

Dinopithecus 1/26
Papio 1/26
Machairodont 1/56
Notochoerus 1/70
Phacochoerus
1/70
Crocuta 1/26
Hipparion 1/48
Hippopotamus
Gazella 1/32
Libytherium 1/62
Deinotherium 1/120
Stegodon 1/120
Anancus 1/120
Mammuthus 1/12

Figure 29-6 Sketch restorations of representative mammals from Early Pleistocene of Africa. Scale is shown beneath each restoration. (H. B. S. Cooke.)

The stratigraphic study of Quaternary North American mammals has evolved on two different bases. The older basis, still widely in use, is the correlation of local faunas with glacial/interglacial stratigraphic units on whatever evidence may be available. It is represented in table 29-D. To a considerable extent it reflects broad movements of taxa under the influence of changing climate. Another basis, also indicated in that table and again in fig 21-2, consists of a sequence of four provincial ages (Savage, 1951). At present the two bases are reconcilable with each other only in the region of southwestern Kansas (Taylor, 1960, p. 21). The provincial ages better reflect the evolutionary characteristics of North American mammals, including exchange of taxa with Eurasia. In table 29-D one sees many non-European genera, and in addition genera such as *Equus, Bison, Bos, Mammuthus, Ursus,* and *Lepus.* These are common to both America and Europe through courtesy of the Bering land bridge described elsewhere in this chapter. Some of the leading genera are represented in fig 29-7.

Let us now look at the provincial ages individually. The oldest unit, the Blancan, is named for a locality in western Texas and is represented by fossils almost exclusively in western United States. Its fauna (McGrew, p. 549 in Colbert et al., 1948) distantly resembles the Villafranchian fauna of Europe in somewhat the same way as does the lower Quaternary fauna in Africa. It is marked, as in Europe, by the first appearance of zebrine horses as immigrants from Asia, as well as by the extinction of many Pliocene genera. However, many Pliocene relicts such as *Stegomastodon* and the three-toed horse *Nannippus* still persisted, as did similar relicts in Europe. Modern camelids (*Camelops*) made their first appearance, as did *Borophagus,* a "hyaenoid dog," and *Procastoroides,* a beaver. Most of the Asiatic immigrants were comparatively small. From South America came large ground sloths, porcupines, and glyptodonts. As a whole the fauna possessed a recognizably Pliocene character and lacked "cold" elements. Existing K/Ar dates indicate that the base of the Blancan is around 4 million years old, somewhat older than the base of the Villafranchian zone as dated hitherto.

The appearance of the Irvingtonian fauna, named for a locality near San Francisco, is currently thought to have coincided with a time late in the Kansan glaciation. This fauna is marked by an influx of new immigrants from Asia including mammoths, caballine horses, brown bear, saber-tooth cat, various nonarctic musk-oxen, antelopes, and an extensive variety of rodents. Bison followed shortly afterward. From South America came capybaras, giant armadillo, and *Glyptodon.* The mammoths immigrant in Irvingtonian time included *Mammuthus imperator* and led to the development of *M. columbi.* These elephants were common in south-

Table 29-D
Stratigraphic Ranges of Late Pliocene and Quaternary Mammal Genera in North America, Arranged in Order of Their Appearance in the Strata[a]

Blancan				Irving-tonian	Rancholabrean			Re-cent
Pliocene	Nebraskan	Aftonian	Kansan	Yarmouth	Illinoian	Sangamon	Wisconsin	Living

Sorex, longtailed shrew
Bassariscus, ringtail cat
Peromyscus, white-footed mouse
Mammut, American mastodon
Hypolagus, rabbit
Machairodus, sabre-tooth cat
Scapanus, western mole
Sciurus, tree squirrel
Citellus, ground squirrel
Canis, coyote and wolf
Vulpes, fox
Martes, marten and fisher
Mustela, weasel and mink
Nannippus, three-toed horse
Rhynchotherium, mastodon
Pliauchenia, camel
Felis, cat
Lynx, lynx
Taxidea, badger
Marmota, woodchuck and marmot
Castor, beaver
Perognathus, pocket mouse
Onychomys, grasshopper mouse
Ochotona, pika
Prodipodomys, kangaroo rat
Bensonomys, mouse
Dipoides, beaver
Buisnictis, mustelid
Ogmodontomys, vole
Notolagus, rabbit
Baiomys, pigmy mouse
Odocoileus, deer
Megalonyx, ground sloth
Tanupolama, llama
Paenemarmota, giant woodchuck
Blarina, short-tailed shrew
Cryptotis, least shrew
Notiosorex, desert shrew
Lasiurus, tree bat
Urocyon, gray fox
Procyon, raccoon
Spilogale, spotted skunk
Lutra, otter
Geomys, pocket gopher
Thomomys, pocket gopher
Sigmodon, cotton rat
Reithrodontomys, harvest mouse
Zapus, jumping mouse
Cuvieronius, mastodon
Platygonus, peccary
Stegomastodon, mastodon
Titanotylopus, giant camel
Pliophenacomys, vole
Plesippus, zebrine horse
Borophagus, bone-eating dog
Trigonictis, grison
Ischyrosmilus, sabre-tooth cat
Chasmaporthetes, hyaena
Procastoroides, beaver
Nebraskomys, vole

[a]Compiled by C. W. Hibbard et al., p. 513 in H. E. Wright and Frey, 1965a; Revised by J. E. Guilday and C. W. Hibbard in 1970. By permission, Princeton University Press.

	Blancan				Irving-tonian	Rancholabrean		Re-cent
Pliocene	Nebraskan	Aftonian	Kansan	Yarmouth	Illinoian	Sangamon	Wisconsin	Living

Hesperoscalops, mole
Neterogeomys, gopher
Pratilepus, rabbit
Paracryptotis, shrew
Brachyopsigale, mustelid
Symmetrodontomys, mouse
Nekrolagus, rabbit
Ceratomeryx, pronghorn
Cosomys, vole
Canimartes, mustelid
Pliopotamys, vole
Synaptomys, bog lemming
Cervus, wapiti
Camelops, camel
Equus (Asinus?)*, ass-like horse
Pliolemmus, vole
Glyptotherium, glyptodon
Ursus, bear
Panthera, jaguar
Cynomys, prairie dog
Paramylodon, ground sloth
Capromeryx, pronghorn
Castoroides, giant beaver
Simonycteris, bat
Hayoceros, pronghorn
Coendou*, porcupine
Arctodus, giant short-faced bear
Microsorex, shrew
Eptesicus, brown bat
Gulo, wolverine
Dipodomys, kangaroo rat
Ondatra, muskrat
Phenacomys, phenacomys
Microtus, vole
Pedomys, prairie vole
Pitymys, pine vole
Neofiber, water rat
Erethizon, porcupine
Sylvilagus, cotton-tail rabbit
Lepus, hare
Nothrotherium, small ground sloth
Chlamytherium, giant armadillo
Dinobastis, sabre-tooth cat
Smilodon, sabre-tooth cat
Hydrochoerus,* capybara
Mammuthus, mammoth
Mylohyus, woodland peccary
Euceratherium, shrub-ox
Preptoceras, shrub-ox
Tetrameryx, extinct pronghorn
Equus (Equus),* modern horse
Equus (Hemionus?),* hemionus-like horse
Tapirus,* tapir
Stockoceros, pronghorn
Brachyprotoma, skunk
Glyptodon, glyptodon
Osmotherium, skunk
Tremarctos,* spectacled bear
Neotoma, woodrat

Table 29-D (continued)

	Blancan					Irving-tonian	Rancholabrean		Re-cent	
	Pliocene	Nebraskan	Aftonian	Kansan	Yarmouth	Illinoian	Sangamon	Wisconsin	Living	
										Atopomys, vole
										Etadonomys, kangaroo rat
										Platycerabos, bovid
										Scalopus, eastern mole
										Parascalops, hairytail mole
										Antrozous, pallid bat
										Plecotus, big-eared bat
										Mephitis, striped skunk
										Tamias, eastern chipmunk
										Tamiasciurus, red squirrel
										Clethrionomys, redback vole
										Bison, bison
						?				Desmodus, vampire bat
						?				Dasypterus, yellow bat
						?				Dasypus, armadillo
						?	?			Didelphis, opossum
										Myotis, little brown bat
						?	?			Tadarida, freetail bat
						?	?			Conepatus, hognose skunk
										Glaucomys, flying squirrel
										Bootherium, bovid
										Cervalces, moose
						?				Brachyostracon glyptodon
						?	?			Boreostracon, glyptodon
						?	?			Eremotherium, giant ground sloth
										Paradipoides, beaver
										Oryzomys, rice rat
							?			Neochoerus, capybara
										Candylura, starnose mole
										Mormoops, leafchin bat
										Leptonycteris, longnose bat
										Pipistrellus, pipistrel bat
										Eumops, mastiff bat
										Aplodontia, sewellel
										Eutamias, western chipmunk
										Pappogeomys, Mexican pocket gopher
										Liomys, Mexican pocket mouse
										Dicrostonyx, collared lemming
										Napaeozapus, woodland jumping mouse
										Rangifer, caribou
										Antilocapra, pronghorn
										Oreamnos, mountain goat
										Ovibos, musk-ox
										Ovis, bighorn sheep
										Heterogeomys,* tropical gopher
										Saiga,* asiatic antelope
										Bos,* yak
										Sangamona, deer
										Symbos, woodland musk-ox
										Alces, moose
										Homo, man

* Still living, but extinct in the Nearctic region

ern North America and persisted until the close of the Pleistocene. The related woolly mammoth, *M. primigenius,* developed in Eurasia and apparently did not reach North America until about the time of the Wisconsin glaciations. Its range was primarily northerly, but mainly in open country.

The Rancholabrean fauna (Stock, 1956), named for a locality in Los

Figure 29-7 Sketch restorations of representative North American Quaternary animals. Scale preserves relative size. (Modified from Martin and Guilday, p. 1 in P. S. Martin and Wright, eds., 1967.)

Angeles, had its inception apparently in Illinoian time. Among the new arrivals at one time or another in the Rancholobrean provincial age were bison, another bovid, skunks, bats, and a good many rodents. *Mammut americanum,* the American mastodon, a common and wide-ranging browser, may have been derived from an earlier mastodont in North America. It persisted until 10,000 BP. At the end of the Rancholabrean, coinciding (within a few thousand years) with the end of the Pleistocene, the North American mammal fauna was decimated by extinctions, discussed in a following section. These left the fauna impoverished to a spectacular extent.

Most of the taxa mentioned represent localities lying between the Mississippi River and the Pacific. Less is known about the fauna of the forested country to the east. Despite its age, the most comprehensive source of information on the eastern region is Hay, 1923. A number of the stratigraphic attributions given there have been superseded by later knowledge, but the taxonomic information has much value. Fossil vertebrates of Michigan, all of them of Rancholabrean age, are described by R. L. Wilson (1967). For other regional discussions see Schultz and Frye, eds., 1968, p. 115 (Nebraska) and Hibbard and Taylor, 1960 (Kansas).

Changes in the ranges of mammals in North America with changing climates are not as well documented as those in Europe, partly because stratigraphy and vegetational ecology are less well known. Reviewing some of the lists and comments in chaps 19 to 22, we note that the belt of tundra that fringed the Laurentide Ice Sheet was populated by a group of characteristic mammals. Chief among them was the woolly mammoth, whose circumpolar range spanned both North America and Eurasia, though surprisingly there is no evidence in America of the Eurasian woolly rhinoceros. Caribou (reindeer) of the variety living on the North American tundra today occur as fossils in eastern and central North America. In the subarctic forest belt beyond the tundra typical large mammals included mastodon, moose, stag-moose, giant beaver, deer, and bears. Still farther away from the ice sheet were other elephants of the mammoth line, tapirs, peccaries from South America, and a variety of deer and carnivores. On the steppe, represented by the central Great Plains, were elephants, horses, bison, elk, antelope, ground sloths from South America, and a host of small mammals. The rich glacial-age fauna of southeastern United States is best represented by the late-Pleistocene faunas from Melbourne and Vero in Florida (Weigel, 1962; G. G. Simpson, 1929, p. 262). It includes a mastodon, elephants, camels, a huge bison, peccaries, deer, tapirs, horses, ground sloths, armadillos, several carnivores including a saber-tooth cat, wolves, and bears, to mention only the larger mammals.

The extensive silty alluvium, now frozen, in central Alaska contains a numerous mammal fauna. The stratigraphic position of the alluvium is not well known, although C^{14} dates show that the sediment antedates, at least in part, the Late Wisconsin drift. Freezing has preserved the skin and tissue of some of the mammals. The faunal list (Péwé and Hopkins in Hopkins, ed., 1967) includes dog, wolf, fox, badger, wolverine, a large cat, lynx, woolly mammoth, mastodon, horses, camel, saiga antelope (*Saiga*), bison, caribou, moose, stag-moose, elk, mountain sheep, musk-ox, musk-ox and yak types, ground sloth, beaver, and other rodents. The number of individuals is so great that the assemblage as a whole must represent a rather long time. The large cats and the ground sloth may seem surprising in a cold country, but their significance must remain unexplained until their stratigraphic position is better known. The general rarity of fossil mammals in glaciated as compared with nonglaciated North America suggests that the rich Alaskan faunas are probably interglacial.

A list (Maldonado-Koerdell, 1948) of the Pleistocene mammals known from localities in Mexico represents a vast territory and an unknown stratigraphic range, and thus lacks climatic and age significance, but it includes genera as old as Blancan. Its broad aspect is that of a modified North American fauna with conspicuous elements of South American origin.

South America

Information on the Pleistocene floras and faunas of South America is spotty. Probably the most important body of data pertains to Argentina, thanks to extensive pioneer studies by Florentino Ameghino late in the 19th century. The nonmarine stratigraphic units in Argentina, shown in table 26-E, contain mammals and were characterized briefly by Harrington (in Jenks, ed., 1956). The faunas are gathered into a list, by regions (Castellanos, 1956). Holocene forms are mentioned in Castellanos, 1962.

An impressive monograph by Hoffstetter (1952) deals systematically with the rich Pleistocene faunas of the interandine region of Ecuador and sets up faunal ages. The late-Pleistocene fauna of tar seeps in northern coastal Peru, much like the seep at Rancho La Brea in California, was reported by Lemon and Churcher (1961). Paragraphs on the Quaternary in several countries are given in Jenks, ed., 1956. A rich fauna in Bolivia is described briefly by Hoffstetter (1963).

The mammal faunas of North and South America differed in detail because each included numerous indigenous kinds. But in many respects the two were similar because of two-way traffic over the Panama land bridge. South American faunas included no "glacial" kinds; representatives that ranged into North America include several genera of ground sloths, armadillos, and perhaps opossums. Southbound traffic included, among many others, mastodonts, camels of llama type, tapirs, peccaries, dogs, and saber-tooth cats.

Australasia [4]

The fossil Quaternary vertebrates (and the living vertebrates as well) of Australia are unique in that they constitute a provincial fauna in which marsupial mammals are the most conspicuous element and even two genera of monotremes are represented. This situation is believed to have resulted from the separation of Australia from the Asiatic mainland in late-Cretaceous or slightly later time. Thereafter the dominantly marsupial fauna, protected against competiton from placentals, became endemic and evolved adaptively nearly as much as placentals did elsewhere. A variety of herbivores and specialized carnivores developed, filling the available niches. At some time during the Tertiary at least two groups of placentals, bats and rats, managed to reach Australia, possibly by island hopping, and in turn developed endemic forms. Finally, late in the Quaternary (but apparently before 25,000 BP) aboriginal man arrived, and with him the dog *Canis familiaris dingo,* originally an Asiatic wolf.

The Quaternary fauna, then, included well-adapted monotremes and marsupials, not only herbivores such as kangaroos and wallabies, wombats, koalas, bandicoots, and phalangers ("flying opossums"), but also carnivores such as the "Tasmanian devil," "Tasmanian wolf," and "marsupial lion." Some of these attained very large size; the largest, *Diprotodon,* a wombat, reached a length of more than 3m. In addition there were huge terrestrial lizards as much as 5 or 6m long, crocodiles, and an array of large flightless birds.

A great majority of the known fossil localities are in southeastern Australia, but a similar fauna must have existed far more widely. Environments, at least in the broad coastal belt, embraced both grasslands and xeric woodlands, with overwhelming predominance (nearly 600 species) of eucalypts and about 400 species of acacias. This assemblage seems to have replaced, in late-Cenozoic time, an earlier forest flora dominated by the live beech, *Nothofagus.* The change is thought to have been caused by increasing drought. The matter is more complicated, however, because evidence of relatively recent greater moisture exists in the form of isolated, apparently relict patches of *Nothofagus* forest, and fossil crocodiles and lungfishes (*Epiceratodus*) in areas now nearly waterless. It is easy to suggest that such occurrences indicate pluvial/interpluvial fluctuations superposed on a major drying trend that began far earlier. However, they are too few and too widely scattered to support anything other than the barest hypothesis.

[4] E. D. Gill, 1965; Woods, 1962.

In New Zealand there are and were no indigenous mammals at all other than bats. But there was a large and varied fauna of flightless birds, some of which were 3m or more in height. The fauna is now much reduced, and one of the largest genera, the moa *Euryapteryx,* became extinct only a few hundred years ago (C. A. Fleming, 1962b). As in Australia and elsewhere, extinction affected the larger kinds more than the smaller.

Land Bridges

Many areas now submerged beneath shallow seawater are known or believed to have been emerged eustatically during one or more glacial ages (chap 12). The emergence of some of them may have been aided by movement of the crust. Similarities of biota, living or fossil, on two separate lands are the chief reason for supposing the shallow sea floor between them was formerly emergent, constituting a land bridge. The greater the number of similar taxa, especially those seemingly incapable of traversing a strait, the more probable the land bridge. Besides biota, other features that suggest former land bridges include submerged land forms such as branching stream valleys.

Beringia. The most thoroughly researched former land bridge was Beringia, now the Bering Strait and its neighboring lands, fully discussed in Hopkins, ed., 1967. Not only was it the sole connection between the two continents of Eurasia and North America in late Cenozoic time, but also it was a broad land in its own right. An emergence of 100m today would re-create a Beringia nearly as wide as Alaska. Indeed an isthmus connected the two continents during much of Cenozoic time, and from the Miocene onward a strait opened and closed repeatedly. Before the continental ice sheets in the northern hemisphere came into being, these changes in the strait probably were caused mainly by crustal movement. thereafter the chief cause seems to have been eustatic fluctuation of sealevel. The existing strait opened ~12,000 BP.

Beringia, in late-Pleistocene time at least, was treeless country, tundra and grassland, with isolated areas of boreal forest, the whole apparently with low precipitation. Such vegetation made it possible for both tundra mammals and steppe mammals to traverse the bridge. Although boreal, the climate was probably not greatly different from that of the Aleutian Islands today because of proximity of the North Pacific Drift (current). The climate and vegetation acted as a filter that permitted the transit of mammals (listed elsewhere in this chapter) of primarily grazing habit, which could also withstand boreal temperatures. Successive waves of immigration into North America can be discerned in table 29-D.

English and Irish Channels. (Shotton, 1962.) Table 29-E lists some of

Table 29-E
Some Former Quaternary Land Bridges Inferred from Biogeographic
Data

Site of Former Bridge	Lands Formerly Connected	Minimum depth of existing strait (m)
English Channel	France : Britain	38
Irish Channel	Britain : Ireland	45
Mediterranean straits (several)	Europe : Africa	various
Bering Strait	Siberia : Alaska	38
Gulf of Tartary—La Pérouse (Soya) Strait	Siberia : Japan	60
Sunda Shelf	Malay Peninsula : Sumatra, Java, Borneo	40
Bass Strait	Australia : Tasmania	75
Torres Strait	Australia : New Guinea	75
Cook Strait	South Island : North Island (New Zealand)	90
Palk Strait	India : Ceylon	11

the larger and more obvious Quaternary land bridges. Among them are
the straits that separate the British Isles from the European Continent.
Their floors are related to that of the North Sea, which, as we noted in
chap 12, was emergent during later glaciations, when and where it was
not itself covered by the ice sheet. The English Channel, or at least its
narrow part, is thought to have been an isthmus throughout much or all
of the Pleistocene, even during interglacial times of higher sealevel. The
evidence consists mainly of fossils recording interchange of biota, particularly elephant, rhinoceros, deer, and hippopotamus. Breaching of the
isthmus seems to have been a late event, possibly ~8000 BP. The Irish
Channel, on the other hand, is thought to have remained a strait even
during glacial ages, until the Weichsel glaciation, mainly on the evidence
of fossil mammals: no land mammals certainly antedating the Weichsel
have yet been found in Ireland. A narrow land bridge between England
and Ireland apparently was cut by the Flandrian rise of sealevel around
8000 BP.

Southeast Asia. As is evident in fig 29-8, wide shelf areas could have
been converted to land bridges or near-bridges by moderate lowering of
sealevel or moderate crustal movement or both. Remarkable similarities
between living faunas of Borneo and Sumatra (Molengraaff and Weber,
1921), between fossil faunas of Southeast Asia, Malaya, and Java, and between those of Taiwan, the Philippines, and Celebes suggest two routes
of immigration southeastward from mainland to islands (Hooijer, 1951).

Similarities among the Pleistocene mammal faunas of Australia, Tasmania, and New Guinea suggest that at times free communication occurred there.

Three kinds of extinct dwarf elephants, variants of *Palaeoloxodon antiquus* (fig 28-1), on the Mediterranean islands Sardinia, Sicily, Malta, Crete, and Cyprus imply two or more land bridges postdating the Villafranchian fauna and antedating the last major glaciation. Probably crustal movement was the principal factor in the disjunctions.

Early in the Pleistocene at least, a land connection existed between Japan and the Asiatic mainland, because several species of mainland mammals, especially proboscideans, occur in lower Pleistocene sediments in Japan. A sealevel 60m lower than today's would create a land bridge from the lower Amur River region in coastal Siberia via Sakhalin Island to Japan. However, as a bridge existed also in pre-Pleistocene times, the emergence need not have resulted from glacial lowering of sealevel. A land bridge between Korea and Japan, likewise, can be suggested on the basis of a strait now <100m deep.

Panama. Among the Pleistocene fossil faunas in North America we find also South American animals such as ground sloths occurring as far

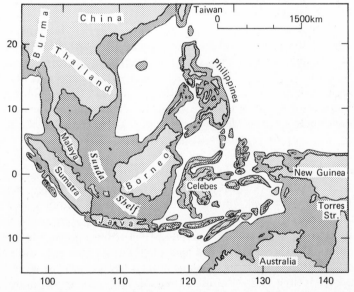

Figure 29-8 Sketch map showing sea-floor area less than about 180m (100 fathoms) deep (dark shading) off parts of southeast Asia and Australasia. (Based on Hooijer, 1951.)

north as central Alaska. Conversely we find in South America horse and deer, which can only have reached that continent from North America. Therefore, at times if not continuously during the Quaternary, the present Isthmus of Panama was in existence. There is no doubt that the isthmus was the route of faunal interchange. Today it is marked by a belt of rain forest, with swamps, more than 300km in length. Such terrain would constitute no barrier to an animal such as the peccary, one of the migrants, but it would have been a serious obstacle to ground sloths, horses, and deer. Likewise it would have hindered the movement of xerophylous grasses, shrubs, and herbs, which also succeeded in spreading from one continent to the other. Possibly the explanation lies in the displacement of climatic belts that took place during glacial ages and that brought pluvial climates to now-arid southwestern North America. A shift of the northern extratropical belt of high pressure southward through only 2° of latitude might have dried the isthmus sufficiently to replace the rain forest with savanna. This would have opened the isthmus to travel by plants and animals that could not easily, if at all, have crossed the tropical forests (Sauer, 1944, p. 557). Under this hypothesis the faunal movements could have occurred only during glacial ages, for only at such times could the climatic belts have moved toward the equator. The lowered sealevels of glacial times likewise should have aided the movements by slightly widening the isthmus.

The Pleistocene history of birds and fishes leads to inferences about bridges other than land bridges. The history of North American birds (Mayr, 1946) suggests that they traveled the Bering Strait route; but because birds can cross water barriers more easily than mammals can, inferences as to bridges must be drawn with caution. Disjunctions in the ranges of freshwater fishes are a great aid in reconstructing former drainage basins, such as those of which pluvial lakes in North America and Australia were a part (Hubbs and Miller, 1948).

Former land bridges have been hypothesized also in positions where the ocean today is several hundred to two thousand meters deep. A conspicuous example is the submarine Wyville-Thomson Ridge that connects Greenland, Iceland, and Scotland. The idea of a land bridge along its axis is based principally on the distribution of invertebrate land floras and faunas (e.g., Löve and Löve, eds., 1963), many of whose taxa are poorly adapted to long-distance dispersal. However, because the geology of the region affords no basis for the assumed bridge, at least within Miocene and later time (e.g., Trausti Einarsson, 1964), it seems more likely that any biotic interchanges within such time have occurred by means of water and overwater transport of various kinds.

Extinctions [5]

No review, however brief, of Quaternary fossils would be adequate without mention of the much-discussed problem of the numerous extinctions that occurred late in the Quaternary. Of course, extinctions of animal taxa occurred throughout the late Cenozoic, but their number increased sharply in Wisconsin/Weichsel time. According to P. S. Martin (1966), of the African genera of large mammals that existed 50,000 years ago, 40% have become extinct, and of the North American genera, 70%. Although believed by Krantz (1970) to be too high, these rates are greater than those for all earlier Quaternary time. On the other hand, the comparable rate for Europe and northern Asia is small. Besides large mammals (mainly herbivores), large flightless birds were affected; small mammals were affected principally, though not exclusively, on islands; the only marine mammals affected were littoral or seasonally terrestrial kinds. Finally, most extinctions in North America occurred between ~15,000 and ~9000 BP with a distinct peak around 11,000.

Hypotheses of the cause of these remarkable facts appeal to one or more of three factors: (1) disease or some other general catastrophe, (2) change of climate and therefore of ecology, and (3) the direct or indirect influence of man. The first almost totally lacks supporting evidence and for the present need not be considered further. This leaves (2) and (3) as the subject of current debate. The chief region under discussion is North America, where known extinctions were most numerous and were rather closely concentrated in time.

Essentially the hypothesis of climatic change rests on late-Wisconsin/Weichsel warming and drying of climates. This is held to have brought about changes in ecology, especially vegetation, in middle latitudes, and is thought to have involved forage changes with which many large herbivores, with their relatively great demands for space and nourishment, could not cope.

Against this view three arguments have been brought to bear. (1) Extinct genera include desert herbivores such as *Camelops* and *Nothrotherium*, which, far from being subjected to stress by enlargement of arid lands, should have benefited from it. Steppe grazers had only to move northward in order to maintain essentially a similar environment despite desiccation. An analogous opportunity was open to browsers such as mastodon or forest musk-ox. (2) The change of climate involved more time

[5] General references: P. S. Martin and Guilday, p. 1–62 in P. S. Martin and Wright. eds., 1967; C. A. Fleming, 1962b; Krantz, 1970.

than did the majority of extinctions. (3) Independent proof is lacking that Late Wisconsin climatic amelioration was radically different from the warming that introduced preceding interglacials, in which extinctions were comparatively few.

The hypothesis of the influence of man, first adumbrated by A. L. Wallace in 1911, includes both the direct slaughter of beasts by Paleolithic hunters and predatory pressure maintained by early man on large mammals as part of the overall ecology. Disturbance of the ecology by man could reduce animal populations through a variety of indirect means, such as selective killing of young individuals,[6] and selective killing of genera having exceptionally long periods of gestation (a characteristic not closely related to body weight). The strongest argument favoring the hypothesis is the chronology of extinction. This is based on an array of C^{14} dates, some of which can be questioned but which as a group are impressive (P. S. Martin and Wright, eds., 1967, p. 90–95, 103, 106, 108, 112), especially in that they pertain to a wide variety of climates and ecologies. Nowhere does extinction within a region antedate the arrival of man; indeed it follows closely the spread of man armed with effective weapons (idem, map, p. 114). Another argument concerns ecological niches. In the normal course of evolution no niche is left unfilled; when a genus dies out some other organism steps in to take its place. But in the Late Wisconsin extinctions most genera abandoned niches that were not thereafter filled naturally, although some grazing niches have since been filled artificially with cattle and sheep.

The argument most frequently advanced against the hypothesis of human agency is that in no territory was man sufficiently numerous to destroy the large number of mammals that became extinct. In reply it is claimed that Late Paleolithic man in many regions possessed stone-tipped projectiles launched with such velocity that they (demonstrably) penetrated skin, flesh, and bone, and that against such weapons the large mammals had evolved no system of defense. Killing methods, as evidenced in the USSR, are interestingly described by Vereshchagin (p. 372 in P. S. Martin and Wright, eds., 1967). They included pitfalls and snares, group hunts in which animals were driven over cliffs into ravines, into bogs, or onto river or lake ice. It is reported that at one Late-Paleolithic locality in the USSR nearly 1000 bison were killed with more than 300 spears during a single hunt. A somewhat similar description of

---

[6] In a thoughtful discussion Krantz (1970), expressed the opinion that human hunters could have practiced conservation by sparing females and young. He concluded that such practice would have increased the numbers of prey animals such as bison. The increase could have led to extermination of forms such as horse, antelope, and camel, that competed with bison for the same food.

mass killing in North America is given in J. D. Jennings, 1969, p. 104.

Although a number of North American fossil occurrences in which large mammals are associated with man's weapons are known, the number is not large. In New Zealand, on the other hand, 22 of 27 species of the extinct moa (a giant bird) occur in association with prehistoric man, who did not arrive in New Zealand until the first millenium A.D., long after the main postglacial warming had taken place (C. A. Fleming, 1962b).

It is very unlikely that a single cause can explain the entire pattern of extinction, and improbable that the same combination of causes was operative in every region in which extinction occurred. There is much more to be learned before the many aspects of the matter can be fully evaluated. But on a basis of the data marshaled thus far, it appears likely that in many regions the hand of man played an important if not decisive part.

EARLY MAN [7]

Hitherto we have omitted fossil man from our discussion of mammals because he is more conveniently treated in a special section. Special because fossil man, his development, and his industry constitute the wide field of *prehistory,* represented by an extensive literature with distinctive terminology, distinctive stratigraphic sequences based on cultural development, and a time scale (at least that part within the scope of radiocarbon dating) quoted in A.D./B.C. terms. Prehistory interlocks with the geologic stratigraphy of the late Cenozoic in the fields of human paleontology, of other fossils (mostly other mammals and mollusks) found with human remains, and of stone tools. As we have noted, not only human fossils but groups or classes of artifacts (although not single tools) are employed as means of stratigraphic correlation. Very commonly prehistorians and geologists work together to interpret the stratigraphy and paleoecology of a locality or a district. Many of the stratigraphic sequences of the prehistorian are very local, consisting of the microstratigraphy of man-made refuse in the floors of caverns, other dwelling places, and the sites of butchering of prey on a large scale. In such deposits two or three meters' thickness may span tens of thousands of years of history.

In this book we can hardly hope to do more than show how prehistory fits into late-Cenozoic stratigraphy, in terms of both human paleontology and cultural development as represented by artifacts. Although prehis-

[7] General references: Rouse, 1971 (broad survey); G. Clark, 1967; 1969; W. W. Howells, 1966; S. L. Washburn, 1960; Oakley, 1969; Haynes, 1969.

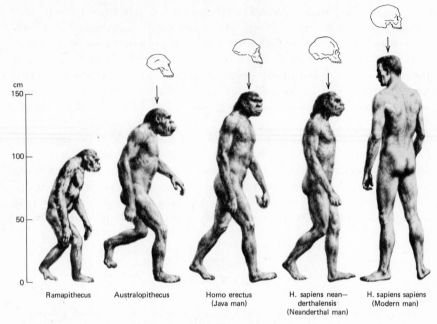

cm
150

100

50

0

Ramapithecus   Australopithecus   Homo erectus   H. sapiens nean—   H. sapiens sapiens
                                    (Java man)   derthalensis   (Modern man)
                                                    (Neanderthal man)

Figure 29-9 Key Late-Cenozoic hominids, restored, in stratigraphic and evolutionary sequence, representing a time lapse of nearly 10 million years. Outlines of corresponding skulls do not include that of *Ramapithecus*, which is still unknown. (Drawn by R. F. Zallinger under direction of E. L. Simons.)

tory, like geology, is worldwide, we shall have to content ourselves with a very generalized glance at fossils and artifacts, emphasizing a few regions of special interest.

Hominid Fossils

In order to grasp the phylogeny of hominids through Pliocene and Quaternary time as reconstructed from fossils, we must first examine fig 29-9, which depicts in restoration five taxa arranged in probable evolutionary sequence. In it we see the Pliocene early hominid *Ramapithecus* of northern India, succeeded by various forms of *Australopithecus* (Pliocene and Early Pleistocene), of *Homo erectus* (extending through much of the Pleistocene), and of *Homo sapiens* (later Middle Pleistocene and Upper Pleistocene). With these forms in mind we can turn to table 29-F, in which these taxa are represented in a less generalized way. There we see *Australopithecus* (presently known to extend back to between 3 and 4 million years BP) as a genus with several species. The evidence accumu-

Table 29-F
Some Generalized Phylogenetic, Stratigraphic, and Cultural Data on Early Man. Correlations are Only Approximate [a]

Homo erectus types	Phylogeny of hominids	Homo sapiens types	Periods and epochs	European mammal ages	European glacial/ interglacial units	Radiometric ages BP (10³ years)	Cultural ages (Europe)	Principal tool complexes (Europe)
		Asselar, Boskop Jericho, Choukou-tien (Upper C.), Ofnet'	Holocene	Holocene	Holocene	10	Bronze Iron Neolithic Mesolithic	Aurignacian Gravettian Solutrean Magdalenian
		Rhodesian Cro-Magnon, Skhul, Niah	Upper Pleistocene	Weichsel[d]	Late Middle	30 50		Mousterian
Solo		Tabun Neanderthal Ehringsdorf, Fontéchevade			Early Eem	70 100?		Acheulean
		Swanscombe Steinheim			Saale Holstein	200?		
Heidelberg		Vértesszöllös	Middle Pleistocene		Elster	400?	Paleolithic	Clactonian
					Cromer	600?		Chellean
Peking					Menap	800?		
Olduvai (Bed II), Ternifine Java (Trinil)	Homo				Waal	1,000?		
			Lower Pleistocene "Villafranchian"		Eburon			Oldowan
Swartkrans Java (Djetis) Lantian	Robust species of Australopithecus				Tegelen	2,000?		
	A. habilis				Brüggen			
	A. africanus					3,000?		
	Australopithecus		Tertiary Pliocene					
	Ramapithecus					9,000?		

[d] A more detailed table embracing the last 50,000 years appears in Leroi-Gourhan, 1960. Nonglacial units are in italic type.

[a] Sources include W. W. Howells, 1966; S. L. Washburn, 1960; Rouse, 1971; D. R. Pilbeam, unpublished.

lated thus far suggests that before the Pleistocene began, *Australopithecus* had started to differentiate, apparently into two lines. One is a robust line terminating in *A. robustus*. The other is a gracile line leading through *A. habilis* [8] to *Homo erectus,* a polytypic species represented by varieties such as Peking man and Heidelberg man. *Homo erectus* in turn led [9] to *Homo sapiens*, likewise polytypic, with varieties represented in

[8] Opinions differ as to the classification of this taxon. Some consider it to belong in the genus *Australopithecus* whereas others would place it in the genus *Homo.*

[9] Some believe that a *H. sapiens* branch existed simultaneously with a *H. erectus* line, whereas others view the relationship between these two taxa as a continuous polytypic development. The two possibilities are suggested by the vertical broken line in table 29-F.

many localities, one of the varieties being Neanderthal man, which became extinct ~35,000 years BP.

Because as far as we know *Australopithecus* is exclusively African, it is supposed that Africa was the scene of man's early development. Hominids spread out of Africa in the Early or Middle Pleistocene and occupied parts of Eurasia. Not until the later part of the late Pleistocene did they manage to reach the Americas and Australia, and apparently they did not penetrate to Oceania, New Zealand, or Madagascar until well on in Holocene time. Most of the early history of hominids after *Ramapithecus*, then, is African, and much of the known middle and later history is European and Asian. The Americas contain a comparatively very short record.

It has been suggested that shortening of the canine teeth and increase in the size of the molar teeth in *Ramapithecus*, as compared with its Miocene ancestors, indicates an ecological change from forests and swamps to drier forest and grassland. Such a change is implied independently by associated fossil mammals and by the contemporaneous rise of the Himalayan region. The change could have involved a change of diet, from the shredding of forest leaves and fruits to the chewing of grasses and seeds. This change was accompanied by a trend toward an erect posture (later manifested in the form of the pelvis) better adapted to grassland dwelling. Both changes led toward *Australopithecus*, whose environment consisted essentially of savanna and open woodland not greatly different from that of much of southern Africa today. By this time the hominid line had become omnivorous, because there is evidence in places that eggs, turtles, crabs, and birds were being eaten. The diet of *Australopithecus* also included mammals as large as baboons, which were killed by a blow on the skull. Thus there is evidence also that weapons (perhaps only long bones) were being employed to obtain food.[10]

As the place names at the far left of table 29-F indicate, *Homo erectus* had spread from Africa throughout Eurasia before Middle-Pleistocene time, reaching as far as coastal China and Java. The most obvious development in the skeletal evolution of *H. erectus* and *H. sapiens* is a gradual increase in roundness and volume (by a factor of 3 or 4) of the skull. This is portrayed in fig 29-9, and in much greater detail in W. W. Howells, 1966, p. 50. We can suppose that this change was related closely to increased skill in the making of tools and to increased sophistication in their use, mentioned in the following section.

[10] For a general statement on environment and diet see F. C. Howell, 1959.

Ancient Cultures

An *artifact* is any natural object to which a people have added at least one artificial feature in accordance with their customary manufacturing procedures (Rouse, 1971). Artifacts therefore record the culture of the people who made them. Although artifacts include carvings, paintings, and objects made of a wide variety of materials, most of those which are of chief interest in Quaternary stratigraphy are made from flint or other varieties of quartz. Such artifacts are of historic value first because they are much more abundant than human fossils and far more durable, and second because most of them were made to patterns that we can classify. In consequence artifacts can be grouped into styles and classified as *tool complexes,* somewhat as fossil mollusks can be grouped into taxonomic units. Hence it is evident that cultural evolution as read from tool complexes has proceeded contemporaneously with the anatomical evolution of peoples. This does not mean, however, that a particular cultural change is the result of a particular anatomical change or vice versa. As pointed out by Rouse (1971), changes in hominid morphology and changes in culture are independent variables; consequently an exact correlation between the two entities is not possible. This restriction is reflected in the slanting lines in the cultural-ages column in table 29-F. The setting up of cultural ages for various parts of the world has been made possible by the evolution of tool complexes such as those shown in the adjacent column. Cultural ages are crudely analogous to provincial mammal ages in that they are based on stratigraphic limits of occurrence of specific, recognizable assemblages. The boundaries of cultural ages are transitional, and are rather steeply time transgressive because peoples in some regions are slower to learn to make more advanced artifacts than are peoples elsewhere. The steep boundaries of cultural ages persist to the present time. The true date of inception of the oldest (Paleolithic) cultural age is unknown, but artifacts representative of it in northern Kenya are related to K/Ar dates as old as $\sim 2.5 \times 10^6$ years BP.

Individual tools are not named in table 29-F because they have less value for stratigraphy than tool complexes have. The Oldowan complex (named from the well-known Olduvai Gorge fossil locality in western Tanzania) is characterized by *pebble tools,* rounded or flattened cobbles or pebbles taken from stream or beach gravel and slightly worked by crude chipping along one edge. The complexes that were made during Middle Pleistocene and the earlier half of Late-Pleistocene time are varied but most of them are typified by hand axes. A *hand axe* is a tool with two faces, fashioned in whole or in part by flaking so as to create an edge around most of the perimeter, and made to be held in the hand. Hand axes of various kinds, accompanied by related tools, constituted tool complexes in Africa, southern Europe, and western Asia, and were made by techniques that gradually improved through the Paleolithic Age. Artifacts, tool complexes, and techniques of their manufacture and use are fully described in the literature of prehistory.

Although Paleolithic artifacts consist mainly of stone, Mesolithic objects include much wood, ivory, and other substances. They are transitional into the Neolithic tool kit and its related techniques, which included the pecking, grinding, and polishing of various kinds of stone and ivory, the making of ceramics and baskets, the domestication of animals (Ucko and Dimbleby, eds., 1969), and the introduction of agriculture. The inception of agriculture, as we noted earlier, is clearly evident in the pollen-stratigraphic record as well.

The succession of cultures occurring in sediments of many different kinds, often superposed upon each other at a single locality, constitutes a distinctive kind of stratigraphic sequence, called by prehistorians a chronologic sequence, that differs in meaning from the same term as used by geologists.

Early Man in the Americas [11]

The prehistory of man in the Americas seems to have been so short as to be hardly more than an addendum to the long prehistory of man in the Old World. Nevertheless, the peopling of two continents within a comparatively short time is a process of no small interest. We shall confine our very brief survey to three topics: (1) when and how man may have come to the Americas, (2) human fossils, and (3) prehistoric American cultures.

Arrival of Man in America. (General references: A. L. Bryan, 1969; Müller-Beck, no. 22, Hopkins, no. 24, and Laughlin, no. 23, in Hopkins, ed., 1967). Physical and cultural similarities establish the earliest Americans as immigrants from Asiatic rather than European sources, but neither their route(s) of travel nor the date(s) of their entry are established firmly. Although some think immigration was by boat or raft across some part of the Pacific Ocean, most authorities adhere to the opinion that man crossed from Asia via the Bering land bridge, the route followed by many other mammals large and small. Accepting the latter view, we can look at the matter in more detail. Once they had drifted into Alaska via the land bridge, what route did the immigrants follow toward the south

[11] J. D. Jennings, 1969; Haynes, 1969; Rouse, 1971.

for their radiative expansion across America? This question is intimately related to date of travel, because it involves consideration of the areas that were blocked by glaciers at the time (fig 18-1). Around 20,000 to 14,000 BP Cordilleran glaciers and the Laurentide Ice Sheet would have spread to their maximum extent, whereas some thousands of years earlier or somewhat later those glaciers would have been less extensive and travel would have been far less difficult.

The few C^{14} dates that bear on this matter tell us that ~13,000 BP, people with a distinctive hunting culture were already established in North America, at least in eastern Oregon and on the Great Plains. However, there is no comparable information as to when people were first present in Alaska, 2500 to nearly 4000km farther northwest, because the earliest radiometric dates there are more recent. Reasoning from indirect evidence, we note that a Late-Paleolithic hunting culture resembling that in North America had developed in Siberia before 15,000 BP, although how long before that date is not known. We note also that the Bering land bridge was free of lowland glaciers at all times, and that from ~12,000 back through a time at least as long ago as ~35,000 BP the bridge existed because of lowered sealevel. Hence people might perhaps have crossed dry shod from Siberia into Alaska between those dates, and conceivably even earlier. At later and earlier times they could have crossed by traversing sea ice. In this connection we note the existence of evidence of the presence of man in southeastern Australia at least as early as 25,000 C^{14} years BP; the evidence seems to demand over-water travel.

More difficulty is involved, however, in travel from Alaska to the country south of lat 40° than in travel across Beringia. Because of that difficulty some prehistorians have reasoned that a human population may have existed in Alaska for thousands of years before deglaciation could have permitted post-Late Wisconsin movement of people toward the south. Others have suggested that southward movement took place before Late-Wisconsin glaciation, despite a probably very harsh climate, a suggestion neither proved nor disproved by existing data.

As to routes from Alaska southward, it is unlikely that the coastal region could have been traversed, because it consists of continuous high mountains cut by deep valleys that were filled with glaciers or sea water during the entire time in question. No route through the Cordillera itself seems likely to have been free of ice between the time of spreading of Late Wisconsin glaciers and a time rather late in the deglaciation. The most likely route led southeastward out of the nonglaciated interior of Alaska along a corridor clothed with tundra, between the Cordilleran ice on the west and the Laurentide Ice Sheet on the east. In what is now Yukon Territory the corridor was open throughout Late-Wisconsin time, but in British Columbia, at least in some segments, and also west of Edmonton, Alberta, it may have been closed in places by ice until perhaps 12,000 BP, when the warming that characterized the Two Creeks interval had begun. Although as yet there are very few facts on which to base a firm opinion, it has been conjectured that in Two Creeks time the corridor might have been negotiable, that a glacier readvance of Valders age might have closed it again, and that the corridor was finally reopened between 11,000 and 10,000 BP.

According to this conjecture, the Alaskan region would have been somewhat like the segment of a canal between two locks. The lock at the western end consisted of the Bering Strait; that at the southea·tern end probably consisted of a corridor that could be closed by glaciers. Generally speaking, when one end of the segment was open the other was closed, but the actions of the two gates are unlikely to have been exactly synchronous. Hence it seems possible that people could have filtered into Alaska via Beringia well before the Late-Wisconsin glacial maximum and that some of them could have passed on through the southeastern gate before it closed (although by ~12,000 BP the use of boats is a possibility that can hardly be excluded). Other peoples might have remained in the Alaskan lock until the waning Late-Wisconsin glaciers reopened the path toward the south. We repeat, however, that all this is largely speculation.

Whatever the path and the calendar of events, it seems likely that the motive for the expansion of people from Siberia into America was the pursuit of game, by means of techniques that perhaps had been evolved only rather recently. *Migration* is not an appropriate word for the movement, which is thought to have been very gradual, involving small bands, family groups perhaps, that drifted persistently eastward. Regardless of the length of their stay in Alaska, any living places established along the sea shore would have been submerged by subsequent rise of the sealevel.

The spread of early man southward into South America seems to have been rapid. Artifacts indicate the presence of man in northern South America as early as ~14,000 BP and as far south as the Strait of Magellan earlier than ~10,000 BP. Knowledge of successive styles of tools in various parts of the continent is summarized by Lanning and Patterson (1967).

Human Fossils and Culture. The known fossils of early man in North America are extraordinarily few, possibly because the practice of cremation might have been widespread. Setting aside a number of occurrences with serious stratigraphic or chronologic uncertainties, we have left in the Early category only two occurrences that are widely accepted. These two, at Midland, Texas and at Tepexpan near Mexico City, are C^{14} dated at ~10,000 BP. Both are of a late *Homo sapiens* type, compatible with Late Wisconsin-age arrival from Siberia.

More can be learned from artifacts recovered from many parts of North America. Most of the early artifacts are stone projectile points, fluted for hafting to form a dart or arrow. Those found in the Great Plains region constitute the Llano culture, clearly an advanced hunting culture. Some prehistorians believe it implies a longer antecedent development, within North America, than is allowed by a Late-Wisconsin opening of an immigration route from Alaska. They suggest that the culture developed in Alaska or even south of lat 49° before Late-Wisconsin time. Whether or not this was the case, these hunters are believed by some to have been mainly responsible for the widespread extinctions of mammals discussed elsewhere in this chapter. The Llano culture was followed, through a few thousand years, by the Plano culture, characterized by spear points of different size and form, though still indicative of hunting as the principal means of livelihood. These cultures, together with others that are not yet fully integrated into a sequence, belong to a Paleo-Indian cultural age, which endured until around 5000 BP.

Thereafter new waves of immigration (perhaps now by boat across the Bering Strait) occurred, as indicated by complexes of artifacts like those in the Neolithic of Eurasia. These belong to a Neo-Indian cultural age, in which tools evolved through recognized sequences. The people who made the later Neo-Indian tools were among those encountered by the Norsemen, Columbus, John Smith, and the passengers on the *Mayflower*.[12]

[12] Details that fill in this bare outline are given in the three references footnoted at the beginning of the North American section. Rouse, 1971, also contains compact data on Meso-America and South America.

CHAPTER 30

The Problem of Causes [1]

THE PROBLEM

The solar heat received by the Earth arrives at a greater rate in equatorial than in polar regions. It is redistributed from lower to higher latitudes by the atmosphere and to a far smaller degree by the ocean. The atmosphere redistributes not only heat but also moisture. On the average it evaporates water from the subtropical ocean and carries it to both higher and lower latitudes. En route some of the moisture is precipitated as an effect of turbulence and through the blocking effect of highlands, in the lee of which relatively dry areas occur. Also cold, dry air flows outward from arctic regions toward middle latitudes. There warm air encounters and rises over it, and precipitation results. In the ocean warm currents flow toward higher latitudes while cold arctic and antarctic water flows along the bottom toward the tropics, where it wells up to replace warmer surface water lost through the movement of wind-driven currents. By these processes and others heat and moisture are continually redistributed.

This complex mechanism, however, does not operate at constant strength. There is firm factual ground for the opinion that temperature, precipitation, and winds have varied in the past and are varying now. There is indeed no good reason for doubting that climatic change has been continual at least throughout the whole length of Phanerozoic time. However, fluctuation of climate is not necessarily accompanied by glacial phenomena, for in some regions the range of temperature or precipita-

[1] The best basic background reference is a good textbook of climatology (e.g., Trewartha, 1968). An excellent, compact treatment of the aspects of climatology significant for understanding glacial-age climates is Lamb, p. 8 in Nairn, ed., 1961. See also J. M. Mitchell, ed., 1968; Flohn, ed., 1969b.

tion does not permit the existence of glaciers. The problem of the cause of climatic change therefore goes well beyond that of the cause of glacial ages, with which it is commonly identified in the older literature. The problem is indeed worldwide, and the fluctuations of glaciers constitute only one of its many aspects. We are therefore dealing primarily with the question: What makes climates change? and only secondarily with the question: What causes glacial ages in the broad sense?

Having posed these questions, let us quote C. E. Dutton (1884, p. 1), the first investigator of the High Plateaus of Utah:

"While it cannot be doubted that the climate of the Glacial period was in some important respects different from the present climates the moment any attempt is made to ascertain what would be the effect, if any one of the determinants of climate were to undergo a marked variation, it is found to be so intricately interwoven with many other conditions and determinants that the ingenuity of the investigator is generally baffled in his endeavors to assign a just and proper weight to them all."

Dutton's statement is still substantially true. Let us admit at once that we do not know what are the basic causes of climatic change. Although nearly 150 years have elapsed since the Glacial Theory was proposed, the causes remain elusive because insufficient attention has been concentrated on relevant geophysical factors, particularly as regards the atmosphere. Hence we lack firm data against which the great array of theory that permeates the literature can be evaluated. Even to categorize all the accumulated theory would exceed the scope of the present book. Because of the plethora of published material on climatic change, we shall have to cut through it in what may be thought an arbitrary fashion, and what certainly *is* an inadequate fashion in view of the importance of the subject. An interesting and most useful book could and should be written on climatic change alone. But it would not be a definitive book because of the obvious lack of basic geophysical data with which we must continue to live, probably for some time.

So let us proceed with a short review, starting with such basic data as we possess: the firm instrumental record and the geologic record, less firm because it consists mainly of inference, and then attempting to summarize briefly some of the more important theoretical material. If the reader should be dissatisfied, as he well may be, with the tentative conclusions expressed, he will find in this chapter references that will lead him more deeply into what is clearly a fundamental problem.

THE PRINCIPAL DATA

Climate Since A.D. 1000

The long temperature record from England (fig 30-1), with which shorter records from other regions agree in main features, shows that during the last 1000 years average annual temperature has fluctuated through a range of 1.3° in a rather systematic way, with a high in the 13th century and in the 17th a low representing the latest part of the Neoglaciation (chap 16). Fluctuation of precipitation in the same region rises and falls in a similar manner (Lamb, 1965). The establishment of many weather stations after the middle of the 19th century made possible more detailed curves representing very recent time. The curves in fig 30-2 are based on a network of stations throughout the world. In them we see an extreme range of fluctuation of temperature amounting to ~0.6°C. A long warming trend reached a climax in the early 1940's and was followed in most latitudes by significant cooling (J. M. Mitchell in Fairbridge, ed., 1961). The warming was greater in high latitudes than in low, greater in the northern hemisphere than in the southern, and greater at continental than at maritime stations. This pattern of distribution of temperature rise is thought to have resulted, in the northern hemisphere, from changes in the positions and intensities of the anticyclones that operate over northern Eurasia and North America in winter.

During the last millenium precipitation in England rose and fell with the temperature (Lamb, 1965). However, worldwide data for the time since the mid-19th century have made it possible to relate precipitation

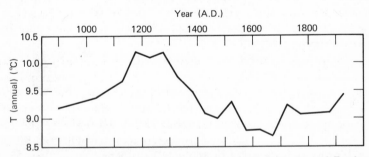

Figure 30-1   Fluctuation of mean annual temperature in central England, by 50-year averages from A.D. 900 to about 1930. The data for the time before 1680, when instrumental measurement began, are derived in various indirect ways explained in the source. (Lamb, 1965.) Compare fig. 16-13.

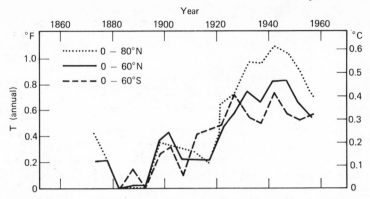

Figure 30-2 Fluctuation of annual temperature 1870–1959, by worldwide latitude belts (5-year means). (After J. M. Mitchell, p. 235 in Fairbridge, ed., 1961.)

changes to geographic position. From those data we see (Lamb, 1964, p. 496; p. 334 in Nairn, ed., 1964) that in England, precipitation increased with increase of temperature as in previous centuries, and fluctuated in a similar manner at stations on the Greenland and Antarctic Ice Sheets and generally at stations having oceanic climates. In contrast, at middle-latitude stations in the eastern parts of temperate continents and in arid regions the warmest time during the last century was marked by low precipitation values.

During the long warming, likewise, the major wind belts in both polar hemispheres increased in force while the strengths of the subtropical anti-cyclones increased as well. The Iceland low-pressure area broadened and moved toward the north. This enabled the storms that characterize the belt of westerlies to follow more northerly paths on the average, bringing warm air into the far-northern Atlantic region and causing increased melting of sea ice in the Atlantic; in the 1930's the melting became extreme. At the same time the Gulf Stream drift shifted northward. In general, since the early 1940's changes of opposite sign have occurred.

Responses to the long warming trend between the mid-19th century and the mid-20th are seen not only in reduced extent of North Atlantic/Arctic sea ice (which, however, has increased since the 1940's) but also in the regimens of glaciers. Most of the regimens measured have been negative overall during the warming trend (e.g., Thorarinsson, 1940; Drygalski and Machatschek, 1942, p. 200–248), with a few showing conspicuous positive regimens since about 1950 (e.g., Hoinkes, 1968). The warming trend is reflected also in the fluctuations of the Great Salt Lake

Table 30-A

Relation of Climatic Factors and Fluctuation of Glaciers and Lakes to Sunspot Number, A.D. 1750–1950 [a]

Date A.D.	Years with maximum sunspots	Sunspot number — For each year of maximum sunspots	Sunspot number — Average of maxima for bracketed periods	Sunspot number — Average annual by 40-yr. periods	Temperature [a] (+Above, ++Much above, −Below, −−Much below, = normal)	Precipitation [a] (+Above, ++Much above, −Below, −−Much below, = normal)	Glacier regimens (wide areas are glacial maxima)	Lakes without outlet (fluctuations) ← lower / higher →
1750	1750	83			≡ + +			Caspian Sea
1760	1761	86			+ +			
1770	1769	106	120	60.5	+ + ++ ++			
1780	1778	154			++ ++			
1790	1787	132			++ ++			
1800					++ +			
	1804	8	55	24.3	−− −−			Great Salt Lake, Utah
1810					−− −			Devils Lake, N. Dak.
1820	1816	46			− + +	≡		
1830	1830	71			+ + −	+ +		
1840	1837	138	124	61.3	−− −− −	+ +		
1850	1848	124			−− − + +	= −−		
1860	1860	96			+ +	−− −		
1870	1870	139			− −−	+ ++		+37.0
1880	1883-4	64	71	31.8	−− −−	++ ++		
1890	1893	85			−− −	+ +		
1900	1905	64			− −	−		
1910					+ +	−		
1920	1917	104	112	58.8	+ + ++	− −−		
1930	1928	78			++ ++	−− −−		
1940	1937	115			++ ++	−− −−		0.0
1950	1947	152			+ +	−		
1960					+		?	?

[a] For stations in middle and high-middle latitudes, northern hemisphere.

[a] Based on Willett, 1951, p. 2; Caspian Sea data compiled by G. E. Hutchinson, Yale University. For further data see Yamamoto, 1961; for projections through earlier time see Bray, 1968.

in Utah (table 30-A), which have been observed continuously since 1850. From 1875 to 1940 the surface of the lake became lower, despite intermediate fluctuation, possibly reflecting increased evaporation with regional rise of temperature.

Responses of organisms are shown by comparative observations, which range back from 1950 through various periods to about 1800, on various kinds of marine fishes, land animals, birds, forest trees, and smaller plants in Eurasia and North America. Many organisms extended their ranges into higher latitudes and altitudes (e.g., Ahlmann, 1953). Steppe invaded deciduous forest; subarctic forest invaded tundra; tree limits rose. The changes observed are those expectable under a warming climate.

All these changes are matters of observation. We suppose, by analogy, that similar changes accompanied temperature fluctuations not only in the 19th and 20th centuries but also throughout the last millenium. Indeed the record furnishes confirmation at a few points (Lamb, 1964, p. 501). When the high temperatures of the period A.D. ~1000 to 1300 gave way to low temperatures through the period 1300 to 1700, the characteristics of the belt of westerly winds changed. Likewise the storm tracks within that belt in Europe lay, on the average, farther north by 5° to 10° of latitude during the warm time than they did during the subsequent colder period. The shift had the affect of replacing a warm, dry climate in northwestern Europe with a colder, wetter climate; the change is confirmed by a number of indirect historical data. Changes of still earlier date, extending back through some 10,000 years, are examined by Lamb et al. (p. 174 in Sawyer, ed., 1966).

To recapitulate, the record of the last millenium tells us with various degrees of probability that when temperatures in middle north latitudes undergo a sustained decline, various related things happen:

1. The general atmospheric circulation is intensified.
2. In the northern-hemisphere belt of westerlies, storm tracks shift southward through substantial distances, and with the shift precipitation increases in oceanic areas and decreases in continental areas.
3. The Gulf Stream is displaced southward.
4. The area of Arctic sea ice enlarges southward into the Atlantic.
5. Glaciers expand generally.
6. Lakes in closed basins under dry climates deepen.
7. The belt of tundra enlarges at the expense of boreal forest.

Although these changes are far from equally well established, they appear to be consistent with each other. Also they represent as a group what might reasonably be expected to happen during the inception of a glacial age, based on the geologic record of the last 20,000 to 30,000 years.

In fact they could be accepted as basic points in a crude model of glacial-age transformations, one that requires much testing and much ampli-
fication.

Such a model would relate only to the northern hemisphere and
mainly to the middle and polar parts of that hemisphere, because information on the southern hemisphere and on low latitudes generally is too
scanty to permit much generalization. The most we can say at present is
that accumulated data on recent climatic patterns and on regimens of
glaciers around the world suggest (although they do not prove) that
within the two hemispheres temperatures have been fluctuating in a
broadly synchronous manner. That belief is supported by evidence, from
cores raised from beneath the eastern Pacific, that the Equatorial Counter
Current in that region has undergone no displacement through Quaternary time (Arrhenius, 1950, p. 87). It is supported also by C^{14}-dated pollen sequences that show that climatic changes in southern South America
were contemporaneous with those in the northern hemisphere (e.g.,
Heusser, p. 124 in Sawyer, ed., 1966; Salmi, 1955).

Relation of Climatic Fluctuation to Solar Activity [2]

The instrumental record through recent centuries yields another piece
of significant information: a correlation exists between variation of climate and variation of solar activity. Radiant energy from the Sun is received at the outer surface of the Earth's atmosphere at a mean rate of
~1.95 cal / cm^2 / min, the "solar constant." The disposition of this energy is
shown in fig 30-3. Of 100 units of it, 35 are lost to space as the direct and
indirect reflection of incoming radiation; the remaining 65 are lost as
outgoing radiation; hence the energy exchange at the top of the atmosphere is in equilibrium. However, at the Earth's solid surface there is a
net gain of 28 units. This is consumed mainly in evaporating water;
sooner or later the water vapor condenses in the atmosphere, releases heat
of condensation, and so balances the energy deficiency of 28 units in the
atmosphere.[3]

The temperature at which equilibrium between incoming and outgoing radiation at the top of the atmosphere is maintained is closely related
to temperatures at the Earth's solid surface. Hence factors that affect this
equilibrium temperature also affect climate. Among possible factors are:
(1) variation in amount and wavelength of solar radiation, (2) change in
amount of atmospheric gases that absorb radiation, and (3) change in reflectivity of the atmosphere as a result of cloudiness or turbidity.

Looking more closely at these factors, we note that solar emission of ra-

[2] Various authors in Fairbridge, ed., 1961; H. C. Willett, 1965.
[3] Discussion in Öpik, 1953.

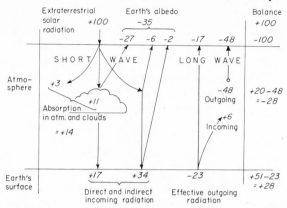

Figure 30-3 Mean radiation balance of the Earth's atmosphere. Values are expressed in arbitrary units. (From Wexler in Shapley, ed., 1953, p. 74, after Möller.) Other authorities give somewhat different values.

diant energy might vary as a whole, or parts of the solar spectrum from infra-red to ultraviolet might vary independently. Indeed, significant contemporary changes in climate seem to be related in time to variations of radiation that are concentrated in the ultra-violet part of the solar spectrum (papers by H. C. Willett; H. C. Willett and Prohaska, 1965). Atmospheric gases whose abundances might vary and thus affect climate include carbon dioxide (CO_2) and ozone (O_3). The latter gas is generated under the influence of ultraviolet radiation. CO_2 varies through interaction of the atmosphere with the ocean, through volcanic eruption, through weathering, through rates of fixation in carbonate by marine organisms, and through burning of fossil fuels.[4] Apart from the influence of man's activities, atmospheric turbidity varies principally with explosive volcanic activity, but because even fine-grained tephra does not remain in suspension for long, fluctuation of turbidity is commonly of rather short duration. Of these substances in the atmosphere ozone is considered to have the greatest potential for affecting climate.

Fluctuation of the intensity of the stream of charged particles emitted by the Sun has long been recognized. The fluctuation is in phase with changes in the number of sunspots (vortical effects of solar flares) that are counted at the solar surface. Sunspot number, then, measures solar activity in crude approximation. It varies continually, and the variation is

[4] An unknown fraction of the increase in temperature shown in the recent part of the curve, fig 30-2, results from CO_2 added to the atmosphere by artificial burning of fossil fuels.

known with fair accuracy through the period since A.D. 1700. During the period 1750–1950 the maximum number was 155 (in 1947), the minimum 46 (in 1816). The variations appear to be cyclic; the time between occurrences of average maxima varies between 7 and 17 years and averages about 11 years. When disturbance in the solar surface is maximum, flares and therefore sunspots are maximum, whereas during periods of "quiet Sun" sunspots are minimum.

If we compare this variation of solar activity with climatic and related events at the Earth's surface we find some correlation. When the Sun is quiet, anticyclonic polar wind systems are intensified, middle-latitude storm tracks tend to shift toward the equator, and temperature gradients from tropics toward poles are thereby shortened and steepened, thus intensifying circulation in middle latitudes, increasing evaporation and storminess, and redistributing precipitation. Hence middle latitudes, broadly speaking, become colder and wetter. At the same time the rate of production of C^{14} in the atmosphere increases (e.g., Suess, 1965; Olsson, ed., 1970), as is indicated by comparison of C^{14} dates with tree-ring dates of samples of ancient wood. The variation in rate of production implies variation in the cosmic-ray flux from the Sun.

Conversely, during periods of greater solar activity expressed in maximum sunspot number, the anticyclonic polar wind systems weaken and contract. Climatic belts are displaced poleward, the temperature gradient becomes gentler, circulation becomes slower, and middle-latitude storminess diminishes. The result is warm, dry climates in middle latitudes. As the circumpolar vortex contracts, masses of warm moist air increasingly invade the high latitudes. There, despite increased precipitation, glaciers shrink because of increased heat brought to them by invasions of tropical air. These conditions are observed as incipient tendencies in the short-term fluctuations of the present day and are present also in the historic record. Accordingly it is argued (Huntington, 1914; H. C. Willett in Malone, ed., 1951, p. 382) that by analogy glacial and interglacial ages represent tendencies of the same kind, similarly energized, but sustained over longer periods.

As indicated by the geologic record, glacial/interglacial fluctuation of temperature was much greater than fluctuation during the last millenium as shown in fig 30-1. If we assume that those larger-amplitude changes of temperature at the Earth's surface result likewise from variation in the flux of solar energy, we can explain them in three possible ways. (1) Increased amplitude of variation of the flux. (2) Variation of the flux through small amplitude (possibly no greater than that observed through the last few centuries) but sustained through longer times. (3) Some combination of (1) and (2). The only present basis for choosing among these possibilities is that (2) involves the smallest change in rate of transfer of energy and therefore might be thought the most probable.

Although a connection between solar radiation and the Earth's atmospheric-pressure pattern seems to exist, the "drive"—the linkage between the incidence of any changed input of radiant energy at the top of the atmosphere and the distribution of the heat by circulation of the lower atmosphere—is not known. The mechanism of the drive awaits elucidation, and indeed our knowledge of solar behavior is extremely slight. Meanwhile, however, variation in the atmospheric abundance of ozone appeared to Kraus (1960) to be capable at least qualitatively of acting as the link, and H. C. Willett (1965) expressed a similar opinion.

Actual measurement of the sunspot/atmospheric-circulation complex is the result of studies mostly postdating 1945. But that a general relationship exists among sunspots, temperatures, and degree of storminess, supported by very meager observational data, was stated with remarkable insight by Ellsworth Huntington (Huntington and Visher, 1922) as early as 1922. Huntington emphasized storminess rather than low temperature; possibly this emphasis, in view of the geologic evidence of lowered snowline in very low latitudes, deprived his concept of the attention it deserved.

Pattern of Temperature Fluctuation in Time and Area

Although we know that temperature has fluctuated, at least at times, through at least the last 10 million years or so (chap 16), we possess a good picture of the fluctuation only through the last few hundred years and an approximate picture only through the period calibrated by radiocarbon. The rest consists mainly of widely spaced K/Ar dates, extrapolation, and conjecture. Bearing in mind that amplitude of fluctuation at sealevel seems to have been greater in middle and high latitudes in the northern hemisphere than in the tropics, we can appreciate the complexity of extending and filling in the record. At present we cannot assert that fluctuation has been or has not been periodic.

Effect of Geometric Variation in Elements of the Earth's Orbit

The idea that climatic change might arise from geometric variation of the distribution of insolational heating of the Earth's solid surface was suggested by J. Adhémar in 1842, but it received detailed attention when it was elaborated by Croll (1875). Croll's argument ran as follows. Disturbances of the Earth as a planet by the Moon and the Sun cause a periodic shift in the position of perihelion. The shift has a period, the precessional period, of a little more than 21,000 years. The shift affects the *distribution* of solar heat received by the Earth, though it does not affect the *total amount* received. The result will be that any given point at a high latitude in the northern hemisphere will be affected accordingly. During a fraction of the precessional period it will have relatively long cold winters followed by relatively short hot summers. During the succeeding equal length of time this same point will have slightly shorter winters than before, each with a little more heat per hour, followed by slightly longer summers than before, each with a little less heat per hour. Croll believed that snow accumulation, resulting in the growth of glaciers, would be favored by the period of longer winters, and that snow wastage would be reduced by the accompanying shorter summers despite the greater summer heat. The theory requires that the postulated glacial ages alternate between the northern and southern hemispheres and that the duration of each glacial age be limited to no more than half the precessional period.

Croll's concept can not be accepted because the geologic evidence shows that the glacial and interglacial ages have been longer than the theory permits, and because the evidence, though incomplete, does not favor alternation of climatic changes between the hemispheres. Furthermore it was shown that the theory is not valid quantitatively in that the calculated temperature variations are far too small to bring about the results envisaged by Croll.

The basic idea, however, was extended and elaborated by others. Later versions recognized the effects of the precessional period of about 21,000 years but appealed also to two other planetary movements which likewise influence the distribution of heat on the Earth's surface. One is the regular variation in the eccentricity of the Earth's orbit, with a period of about 91,800 years. The other is the regular variation in the angle between the Earth's axis and the plane of the Earth's orbit, the period of this motion being about 40,000 years. The results of each of these changes, in terms of the heat received during the summer season at any given latitude on the Earth's surface at 1000-year intervals, are calculated mathematically. The results are combined and are plotted as an insolation curve, which is carried backward to any desired date. Because the periods of the three component movements differ, the resulting curve is non-periodic. It shows irregularly spaced maxima and minima of heat which, it is claimed, represent interglacial and glacial ages. However, the fluctuations in heat reception are regional; the total amount of heat received by the Earth as a whole remains constant.

Various calculations have been made, notably by Spitaler, Soergel, Köppen and Wegener, and Milankovitch (1941; references therein). In the last-mentioned work, published in 1920, improved in 1938, and widely quoted, four pairs or groups of temperature minima, calculated for the summer season only, are held to represent four principal glacial ages, together having occurred within the past 600,000 years. Amplitudes of the temperature fluctuations as calculated (for lat 65° and for the summer season only) are about equal to differences in mean summer temperature between points separated by 10° to 20° of latitude. The curve was later wholly recalculated from improved data (fig 30-4).

This scheme, widely known as the Astronomic Theory or Milankovitch Theory, avoids two serious objections to the Croll idea: the heat fluctuations are not periodic, and they do not alternate rigidly between the hemispheres. On the other hand, neither do they coincide. Two of the basic factors produce exact alternations of heat minima between the hemispheres. The third, the variation in the eccentricity of the orbit, results in minima in both hemispheres at the same time. The combined effect of all three factors is not to produce heat minima in the southern hemisphere coincidentally with heat minima in the northern, but to offset the heat minima at irregular time intervals between the hemispheres. Temperature fluctuation in the equatorial region would occur, but would be minimal.

The scheme has been criticized adversely by several astronomers, meterologists, and geologists.[5] The chief counts against it, some of which do not apply to the version shown in fig 30-4, are:

1. The scheme treats the atmosphere as a gray body, considers absorption of solar radiation to be proportional to the thickness of atmosphere traversed, treats the radiation reflected by the Earth as though all of it originated at the Earth's solid surface, assumes the atmosphere is at rest, assumes the Earth's surface includes no oceans, and assumes the effective factor in the accumulation of ice and snow is cool summer climate. The cumulative effect of all these simplifications on the accuracy of the calculations is probably serious.

2. The calculated temperature changes even at lat 65° appear too small to meet the quantitative evidence of Pleistocene climatic variations. At lower latitudes they would be even smaller. According to Simpson the actual changes probably are smaller than those computed, because the calculations neglect the smoothing effects of atmospheric circulation.

3. It requires slightly increased temperatures in the equatorial region during the glacial ages. Yet in the equatorial Andes and in equatorial East Africa, between glacial and nonglacial times, the snowline fluctuated

[5] Penck, 1938b; G. C. Simpson, 1940; Flint, 1947; C. E. P. Brooks in Malone, 1951, p. 1011; Öpik, 1966; H. C. Willett in Shapley, ed., 1953, p. 60–61; van Woerkom in Shapley, ed., 1953, p. 157; Shaw and Donn, 1968.

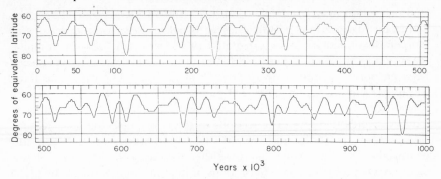

Figure 30-4   Insolation curve as derived by Brouwer and van Woerkom (van Woerkom in Shapley, ed., 1953, p. 156) and redrawn in Emiliani, 1955, fig 14. The curve shows variation of the calculated solar heat received at 65°N lat, expressed in degrees of equivalent latitude, during the last 1,000,000 years.

through a substantial amplitude, being low at times of cold climate in the northern hemisphere.

4. It requires that the changes be in part in opposite phase in the two polar hemispheres. The accumulated geologic evidence does not support the requirement.

On the basis of these objections the geometric scheme of distribution of insolational heating appears inadequate *in itself* to explain the Pleistocene climatic changes. Opinions differ as to how closely the insolation curve matches the Quaternary record. Emiliani (1966b, p. 856) and Broecker and van Donk (1970) saw correlations between the insolation curve, fig 30-4, and curves representing actual former fluctuation of temperature. Of course the geometric factors are real; they may well have influenced climate to some degree. But their quantitative influence remains to be assessed.

Relation of Glaciers to Altitude, Oceans, and Wind Belts

The close relation of existing glaciers to highlands is obvious, as is emphasized in chap 3. The Greenland and Antarctic Ice Sheets are only apparent exceptions, for at least parts of both cover very high lands. Although the central area of the Greenland body and much of the ice of East Antarctica overlie lower land, the relationship is reasonably explained as the result of coalescence of glaciers spreading downward from coastal mountains, like the glaciers of coastal Alaska today. The relation between former glaciers and highlands appears to be similar. The Laurentide and Scandinavian Ice sheets were apparent exceptions in that both spread widely over lowland areas. However, the highland origin of the latter body is demonstrated by provenance of erratics. That of the Laurentide body is deduced by analogy, and the spreading of both bodies from highland sources has been shown to be climatically reasonable.

The reasonableness of the growth of such bodies lies in the relationship among land areas, oceans, and wind belts. Today this relationship is such that winds bring moisture from oceanic sources to nourish the Greenland and Antarctic Ice Sheets. If summer temperatures were only slightly lower, apparently the Scandinavian and Laurentide Ice sheets could be re-created. The existing relation of lands to oceans and to wind belts and the positions of highlands on the continents are such as to maintain existing glaciers and to lead to the creation of additional glaciers whenever temperatures become low enough. So far as we know, these relationships have existed throughout late-Cenozoic time.[6] During that time they have been intensified to the extent that highlands have become more numerous and that many highlands have increased greatly in altitude.

The stage for the late-Cenozoic glacial ages, then, may well have been set by the late-Cenozoic development of high, mountainous continents (Flint, 1957, p. 500). This idea was clearly in the mind of J. D. Dana as early as 1856. The roster of highlands that have been created or heightened since the earlier part of Miocene time is impressive. A partial list was compiled by Flint (1957, table 27-E) but accumulated knowledge as of 1970 would increase the roster by a factor of two or more.

These movements, including broad upwarping, localized upfaulting, and intense orogeny, brought the lands to their highest average Cenozoic altitude, increasing their mean height from less than ~300m to ~800m. Coupled with an assumed eustatic lowering of sealevel of about 100m during a glacial age, this would have increased the mean altitude of the lands by ~600m. In terms of the lapse rate this is equivalent to a temperature reduction of more than $3°$ averaged over the land areas and to considerably more than that at the higher altitudes, although averaged over the world as a whole it would be only about $1°$. In addition the cloud formation induced by mountains reflects solar radiation and thus aids terrestrial

[6] On evidence of fossils it has been claimed that the general positions of climatic belts relative to poles have been measurably constant as far back as Eocene time. Isoflors reconstructed from the many known occurrences of Eocene plants in high northern latitudes form circumpolar girdles that parallel existing isotherms in the same region (Chaney, 1940). Similarly the Eocene marine invertebrate faunas of the northern hemisphere reveal that the northern limit of the tropics at that time generally paralleled the existing limit, but lay $10°$ to $20°$ north of it (Durham, 1952, p. 338).

cooling. Hence in both direct and indirect ways the rise of highlands increases the likelihood of glaciation.[7]

Of course the creation of new highlands could not have been the sole cause of the lowered temperatures of glacial ages. If it had been, the geologic record would contain evidence of only one glacial age extending, perhaps with gradual increase of intensity, from the later Miocene through the present. The conspicuous fluctuations of the snowline upon the flanks of stable or nearly stable highlands demand a different cause, one capable of producing substantial changes of temperature through intervals as short as a few hundred years. Clearly the balance between temperature change and response factors is metastable; a small change of temperature can produce conspicuous responses if other, independent factors are favorable.

Ideas stimulated by evidence of sea-floor spreading and the drifting of continents have extended the concept of late-Cenozoic glacial ages. Dott (1969) suggested that tectonic activity on a broad scale may be episodic and that the latest major pulse may have begun in Miocene time. If true this idea, by emphasizing strong concentration of uplifts and orogeny through the later Cenozoic, could provide an improved rationale for the concentrated building of glaciers. Crowell and Frakes (1970) pointed more specifically to mid-Cenozoic crustal movements that closed the Tethys Sea and created the Isthmus of Panama. This would have resulted in meridional in place of latitudinal circulation in ocean or wind system, thus aiding the transport of oceanic moisture toward the sites of high-latitude glaciers in the northern hemisphere. Menard (in Turekian, ed., 1971) suggested that Cenozoic motions of crustal plates increased the aggregate area of land in high latitudes and also widened the North Atlantic Ocean, with resulting influence on winds and ocean currents.

Pre-Late-Cenozoic Glaciation

These considerations make it appear likely that in later Cenozoic time a number of physical conditions were more favorable to the inception of glaciers than they had been for a long time previously. Nevertheless the stratigraphic record indicates that glaciation occurred at earlier times.[8] Where certain features are present in the rocks at a locality, we confidently accept them as evidence of glaciation (Flint, p. 140 in Nairn, ed., 1961). The most diagnostic of these consists of tillite overlying a striated and polished floor. More common is tillite alone. Again, there are diamictites that are not certainly tillites, followed by diamictites whose origin is probably nonglacial. High frequency of similar occurrences within a re-

---

[7] Quantitative estimates are given by C. E. P. Brooks (p. 1013 in Malone, ed., 1951).

[8] General reference: Schwarzbach, 1963.

gion favors a glacial interpretation. In very ancient rocks, even where glacial origin is not doubted, stratigraphic position is very difficult to fix unless radiometric dates are available; hence correlation is not very close.

Despite these difficulties, the evidence points to strong glaciation within the southern hemisphere in late-Paleozoic time, nearly worldwide glaciation in "Eocambrian" (latest Precambrian) time, and glaciation at various earlier times within the Precambrian. In addition, glacial features are reported from Ordovician, Silurian, and other strata in one region or another, without wide regional confirmation.

It seems unlikely that in low- and middle-latitude regions lacking highlands the snowline could have been low enough to intersect the land surface. Maritime climates would have prevailed, as they seem to have done to an extreme degree in Cretaceous time. On high mountains, however, glaciers may have formed at many times in the geologic past. But in the absence of evidence of lowland glaciation we suppose such glaciers were restricted to high altitudes, where rapid erosion is likely to have destroyed the evidence of their former existence.

THEORIES OF CLIMATIC CHANGE[9]

Grouping of Theories

Having examined, albeit very lightly, the data pertinent to the cause of climatic variation, we now consider theories of the cause. Although formerly it was customary to speak of "the cause of ice ages," today we speak, more realistically, of the causes of fluctuation of climate, because we realize that growth or decay of glaciers is only one of many effects of change of climate, although of course it contributes to the intensity of the change. Also we realize the probability that there is no one single cause, but rather a combination of two or more (perhaps several) causes. Research in the impressively wide field of possibilities has not yet progressed sufficiently far to enable us to choose among the various published theories, many of them conflicting. The most we can hope to do is to list some of those most frequently advanced and to appraise them in very general terms.

The theories are so varied that they can be grouped in a number of different ways and also can be subdivided differently. For our present purpose they are loosely arranged in six broad groups of changes, begin-

[9] These references include lists and (or) critiques of possible causes: Beckinsale in Whittow and Wood, eds., 1965; Chappell, 1968; Lamb and Johnson, 1959–1961; Mellor, 1964, p. 126; J. M. Mitchell, ed., 1968; Öpik, 1965; Schwarzbach, 1968, Sellers, 1967, p. 207; H. E. Wright, 1961a, p. 974.

ning with the Sun and interstellar space and continuing with the Earth itself:

1. Variations in solar emissivity.
2. Veils of cosmic dust.
3. Geometric variations in the Earth's motions.
4. Variations in transmissivity and absorptivity of the Earth's atmosphere.
5. Lateral and vertical movements in the Earth's crust.
6. Changes in the system of atmosphere/ocean circulation.

Some theories fit neatly within a single group whereas others bridge two or more groups. Let us first comment briefly, by number, on the groups of changes rather than on individual theories.

(1) The observational basis consists of the occurrence of solar flares and variation of sunspot number with time and correlative changes in factors in the Earth's climate. This basis is strong enough to accord an important place to (1) as a likely influence, at least qualitative, on climatic change.

(2) The occurrence from time to time of veils of interplanetary "dust" has been postulated, albeit on no evidence other than that of climatic variation. It is supposed that such particulate matter would intercept solar radiation and thus reduce the portion that reaches the Earth. Because a basis for evaluation has not been found, the postulate cannot be usefully discussed at present.

(3) This furnishes the base for the Astronomic Theory discussed in a foregoing section. We do not doubt that it affects temperatures at the Earth's surface but are uncertain whether it is adequate quantitatively to cause glacial ages. Because it embodies a specific timetable, a long series of reliable radiometric dates tied to stratigraphy should do much toward its further evaluation.

(4) Observational testing of theories in this field is possible, and should advance critical appraisal significantly. More information on variations of the amounts and positions of CO_2 and O_3 in the atmosphere and their relation to solar variations would be helpful. The fullest treatment of atmospheric CO_2 in relation to climatic change is that of Plass (1956). Data on variations of O_3 and their relation to sunspots are given by H. C. Willett and Prohaska (1965); see also Kraus, 1960.

The hypothesis that fine-grained tephra suspended in the stratosphere could intercept radiation and thereby affect temperature is of long standing. It was advocated during the first half of the present century in several publications by W. J. Humphreys, and was reviewed by Lamb

(1969), who found a correlation between temperature and major volcanic eruptions. The temperature effect, however, faded out after 15 years at most. In addition it is unlikely that even a major eruption could affect wide regions in both hemispheres; yet both the instrumental record and the geologic record indicate that temperature changes in both short and long time scales were of secular extent.

(5) The fact that the great ice sheets of today and of the glacial ages are and were as close to the poles as permitted by the disposition of high lands is mentioned in various parts of this book. Unquestionably the localization of glaciers could be influenced profoundly by drifting of continents with respect to polar positions. In pre-Cenozoic times this could have happened, but as an explanation of the late-Cenozoic glaciations lateral movement of continents must be ruled out. However, the close relation of vertical movements, particularly localized uplift and mountain making, to lowered temperature has been argued elsewhere in this chapter. Although such movements could easily result in glaciation they cannot explain the fluctuations of glaciers between glacial and interglacial times, and so are not in themselves the prime cause of the sequence of late-Cenozoic climatic changes.

(6) Major movements within atmosphere and ocean are importantly related to climate because they are the agencies by which heat is exchanged between low and high latitudes. Changes in storm tracks, wind belts, and ocean currents such as the Gulf Stream have occurred recently, but because the changes seem to be correlated with solar phenomena they may be among the effects of such phenomena rather than prime causes of change of climate. Likewise, during glacial ages vertical circulation in the oceans (in the Atlantic at least) was altered. The circulation cell was shortened to occupy a narrower range of latitude. Apparently, however, this was brought about mainly by the presence of more ice and colder water in high latitudes and was not in itself a prime factor in change of climate.

The oceanic circulation pattern could have been affected also by crustal movements that altered the configurations of ocean basins. The Isthmus of Panama and the Wyville-Thomson Ridge between Greenland and Scotland are examples. But like mountain ranges, such features cannot explain short-term climatic fluctuation.

An important aspect of the atmosphere/ocean system consists of variations in the volume and areal distribution of snow and ice. Such variations are themselves effects of changes of temperature. The covering of an area of land or sea with ice and snow increases its albedo by a large factor. During glacial ages, when as much as 30% of aggregate land area and substantial sea areas were thus covered, chilling of surface air was enhanced and further extension of the ice cover was promoted. But this too is a secondary effect rather than a primary cause.

Some Individual Theories

The literature is studded with glacial-age theories. Many are now of historical interest only, because improvements in data, mostly geologic, have made obvious some of the flaws in earlier ideas. As the process of improvement proceeds, most—perhaps even all—the ideas that survive today will be eliminated. The field is narrowing with the progress of research. It is difficult to categorize the individual proposals because most of them overlap more than one of the groups listed in the preceding section. It is almost equally difficult to maintain, at least for the present, that any single proposal is wholly untenable because most of them are complex rather than simple, and most contain elements that may be meaningful even though others may be improbable. We can be fairly certain that the cause of secular climatic change involves more than one basic factor. Indeed the record of the last century or so of observation indicates that the parameters of climate are varying continually at present. In the following paragraphs we characterize and briefly comment on eight specific ideas.

(1) The Astronomic or Milankovitch concept, already discussed together with the chief arguments against it, remains active mainly because temperature fluctuations of compatible duration seem to be visible in some oceanic-core sequences (e.g., Broecker, p. 140 in J. M. Mitchell, ed., 1968) and in some coral-reef terraces (e.g., Veeh and Chappell, 1970). There the matter rests, pending improvement in our stock of data, but if the concept proves to have played a part in climatic change, it is likely to have done so as a triggering mechanism rather than as a sole cause.

(2) The concept that solar variations, acting through some linkage in the atmosphere, provoke changes in the factors that govern climate has been mentioned previously. Because knowledge of solar behavior, apart from measurement of sunspot number, is meager in the extreme, this concept is based on little more than an assumption, albeit a reasonable one.

(3) The concept (Lorenz in J. M. Mitchell, ed., 1968) that the atmosphere itself might possess internal instability and so might furnish a built-in mechanism of change is alternative to (2). This idea views the atmosphere as a nearly intransitive dynamic system in which, in the words of Mitchell (op. cit., p. 158), "minor environmental disturbances may have sufficed to 'flip' the atmospheric circulation and climate from one state to another, and to 'flop' it back again."

(4) The equation of glacial ages with increased temperature, a concept

proposed well before the end of the 19th century, states that expanded glaciers require increased snowfall, hence greater atmospheric moisture, hence more rapid evaporation, and hence higher atmospheric temperatures. The concept therefore equates the expansion of glaciers with climates that are growing warmer. The basic fallacy in this idea was lucidly stated by Dutton (1884). Many variants of the equation of glaciation with relative warmth have appeared. A modification of it is embodied in a hypothesis (G. C. Simpson, 1934; 1940) that placed the glacial ages part way between the peaks and troughs of an assumed curve of fluctuation of atmospheric temperature. Although valuable in discussion of the important relations between cloudiness and climate, the scheme has inherent rigidities and does not fit the geologic data on past climates. For example, it does not explain lowered isotopic temperatures of tropical seas during glacial ages, nor the expansion of glaciers on high equatorial mountains, and on peaks such as Mauna Kea, Hawaii, in the middle of an ocean that was, by the hypothesis, becoming warmer at the time.

Simpson's concept was criticized adversely by several investigators (e.g., Öpik, 1953). It fails to recognize the probability of glacier expansion during climatic cooling, despite decreased overall precipitation on the glaciers, by virtue of increased proportional snowfall and reduced ablation. The concept attributes climatic variation to fluctuation of solar radiation.

(5) An imaginative scheme of a very different kind because it involves no extraterrestrial activity was proposed by A. T. Wilson (1964), is thoughtfully discussed by Hollin (1965), and was followed, with little critical comment, by Flohn (1969a). It assumes cyclic activity in the Antarctic Ice Sheet. The ice sheet grows until its base starts to melt; this permits surging on an enormous scale, with creation of a great ice shelf along the coast. As a result, and with the aid of the increased albedo effect, the world ocean becomes cooled sufficiently to cause the building of the great ice sheets in the northern hemisphere. The ice shelf would displace enough water to raise sealevel by a large amount within a short time; the rise should immediately precede each secular glacial age. However, the sediment cores raised from the Southern Ocean do not reveal the expectable volume of ice-rafted sediment, and the pre-glacial-age rises of sealevel are still to be confirmed. Also there are physical problems involved in glacier surging on a continental scale. Further evaluation of this idea is sure to materialize.

(6) A proposal that has received wide publicity (Ewing and Donn, 1956; 1958; Donn and Ewing, 1966) consists of several distinct facets and has undergone a number of changes; for these reasons it is difficult to summarize. It contains two leading ideas. (a) The first is the inception of

glaciation in Antarctica and the probable persistence of the Antarctic Ice Sheet through interglacial ages. (b) The second is the inception of northern glaciers by moisture derived from an ice-free Arctic Ocean and the control of glacial/interglacial oscillation of climates by alternate cooling and warming of the northern oceans as glaciers grow to a maximum. As the Arctic Ocean freezes over, glaciers become starved, and shrink, causing oceanic warming. This oscillatory scheme rests on variation of precipitation and appears to consider temperature changes as secondary.

Of these ideas, (a) is generally in accord with existing opinion but (b) has been debated. Skepticism has been expressed that the Arctic Ocean is an adequate reservoir for glacier-building moisture, and that it was ice-free during any part of a glacial age. According to Hunkins (in Turekian, ed., 1971), the evidence (see also Herman, 1970) favors an ice cover throughout the last million years or so. Again, the ocean is a far less effective transporter of heat than the atmosphere is, and the efficacy of the proposed oscillation mechanism has been questioned (Emiliani, 1967, p. 508). Indeed the concept seems incapable of explaining the short-term temperature variations of different amplitudes that are reflected in the geologic record.

(7) Like the two preceding ideas, a proposal (Adam, 1969) too recent to have stimulated published comment is based in part on exchange of heat stored in the ocean, with oscillations that induce and then end glaciations. The Astronomic Theory is relied on for initial variations of temperature in high latitudes. Basically the proposal represents a combination of earlier ideas rather than the introduction of a radically new concept, and the choice of the ocean rather than the atmosphere as the effective agency of heat exchange raises a quantitative question. The idea of forced oscillations inherent in this and the two preceding models is interestingly examined by Chappell (1968).

(8) The final model to be outlined is the solar-topographic concept. It is twofold, appealing to (a) continental uplift and mountain making to lift parts of the Earth's solid surface above the snowline, and (b) solar fluctuation to bring about variations of climate and therefore of glaciers. It was put forward (Flint, 1947; 1957) as a reasonable and comparatively simple explanation of the facts of the geologic record, and still appears to be an acceptable general model. The recognition of glacier building in Miocene and Pliocene time coincidentally with widespread uplift strengthens the validity of element (a) in the concept. The modest connections recently established between solar activity and climatic change through the last 100 years strengthen the validity of element (b). Thus the concept seems even more plausible than it did 25 years ago. It does not conflict with theories involving variation in the transport of

heat from low latitudes to high by atmosphere or ocean, differing from them only in that it ascribes changes of phase to solar activity rather than to self-reversing mechanisms. In this regard it seems the model best capable of explaining the widespread synchrony of climatic change with a high degree of resolution, reflected in curves such as those in fig 16-11. Also it seems compatible with the rather spotty occurrence of pre-Cenozoic glaciation through the geologic record. We could reasonably suggest that solar fluctuation acting on climates was able to cause extensive glaciation only when and where the upheaval of highlands reached up, as it were, to meet falling temperature part way. According to the concept, only when uplifts were unusually high and widely distributed, especially in the regions traversed by the belts of westerly winds, could small solar fluctuations reduce temperatures enough to bring about the building of glaciers. The record of terrestrial floras and marine invertebrate faunas is one of diminishing temperature throughout the later Cenozoic, possibly because growing glaciers were increasing the Earth's albedo. It is also one of diminishing rainfall on lowlands. Presumably the distribution of precipitation was changing, becoming greater on the windward flanks of growing highlands as it diminished on the intervening lower lands. Thus conditions for extensive glaciation were improving in terms of precipitation as well as in terms of temperature. Concentrations of rainfall on the flanks of highlands were gradually converted, by incremental reduction of temperature, into snowfall and thence into glaciers.

The solar-topographic concept does not pretend to be more than a very general model. Within it, the idea (item 3 in our list) of built-in fluctuations within the atmospheric system itself could be substituted for the solar concept (item 2), but this hardly seems worth while without more knowledge of both solar and atmospheric activity. The true explanation of late-Cenozoic climatic change, whatever it proves to be, will probably be complex and may include the Milankovitch mechanism, but the probability that both crustal uplift and solar variation will form part of it seems at present to be good.

Bibliography

Publications and Maps Frequently Consulted

(References are to the bibliography in this book)

1. Serials and Similar Publications

Association française pour l'Étude du Quaternaire. (Paris.) Bull. no. 1 (1964)–.
Biuletyn Peryglacjalny. (Łódź.) v. 1 (1954)–.
Eiszeitalter und Gegenwart. (Öhringen.) v. 1 (1951)–.
International Association for Quaternary Research. Congress. Trans., v. 1 (1928)–. Also special vols. with various titles issued in years following the 7th Congress (Boulder, Colorado), 1965.
Journal of glaciology. (Cambridge.) v. 1 (1947)–.
Quaternaria. (Roma.) v. 1 (1954)–.
Quaternary research. (New York.) v. 1 (1970)–.
Radiocarbon. (New Haven.) v. 1 (1959)–.
Zeitschrift für Geomorphologie. (Berlin.) v. 1–11 (1925–1943); n.s., v. 1 (1957)–.

2. Broad Textbooks and Regional Monographs

Charlesworth, J. K., 1957.
Flint, R. F., 1947.
———, 1957. (See also Russian section.)
Markov, K. K., and Lazukov, G. I., 1965. (See Russian section.)
———, and Velitchko, A. A., 1967. (See Russian section.)
The Quaternary, v. 1 (1965)–: Interscience Publishers, New York. At head of title: The geologic systems.
West, R. E., 1968.
Woldstedt, Paul, 1954–1965.
———, 1969.
Wright, H. E., and Frey, D. G., eds., 1965a.
International Geological Congress. Commission de Stratigraphie. Lexique stratigraphique international. Issued as parts, by countries and by geologic systems. (Incomplete.)

3. Regional Maps of Surficial or Glacial Geology

Among the maps currently in print are:
Glacial map of the United States east of the Rocky Mountains. 1:1,750,000. (Flint and others, 1959.)
Glacial map of Canada. 1:5,000,000. (Prest and others, 1968.)
Glacial map of North America. 1:4,555,000. 1945. (Geol. Soc. America Spec. Paper 60.)
Glacial- or surficial-geology maps, mostly on scales near 1:500,000, have been published by the U. S. Geol. Survey or by local authorities, of the States of Alaska (1960), Indiana (1958), Michigan (1955), Montana east of the Rockies (1961), New Hampshire (1950), North Dakota (1963), Ohio (1961), and Vermont (1970).
Glacial map of USSR. 1:5,000,000. 1959 (See Russian section.)
Glacial maps of Finland, Norway, and Sweden have been published by Government authorities within those countries.

References

This list is highly selective because it is impossible to cite all the references on the subject. I have favored references with extensive, well-chosen bibliographies that lead the reader further into the literature. Hence not all the references cited are the "most important" ones as far as their textual content is concerned. For this reason, or for others, many excellent titles, including modern ones, have been omitted. The majority of the references antedating 1950 that appear here are included because of their historical importance rather than their current scientific value.

An asterisk (*) denotes a title that has a particularly useful (although not necessarily large) bibliography. "Useful" is a relative term here, depending on the topic covered and on other factors. The superscript ^{R} indicates that the title cited is listed in the Russian language in a special section following the English-language references.

English-Language Titles

Adam, D. P., 1969, Ice ages and the thermal equilibrium of the Earth: Univ. of Arizona Dept. Geochronol., Interim Res. Rept. 15, 26 p.*

Agassiz, Louis, 1840, Études sur les glaciers: Neuchâtel: privately publ., 346 p., plates, atlas.

——, 1847, Système glaciaire ou recherches sur les glaciers. . . . Pt. I, Nouvelles études et expériences sur les glaciers actuels. . . . : v. 1 (8vo), p. 1–598; v. 2 (folio), 3 maps, 9 plates.

Agterberg, F. P., and Banerjee, I., 1969, Stochastic model for the deposition of varves in glacial Lake Barlow-Ojibway, Ontario, Canada: Canadian Jour. Earth Sci., v. 6, p. 625–652.*

Aguirre, Emiliano, 1969, Evolutionary history of the elephant: Science, v. 164, p. 1366–1376.*

Ahlmann, H. W., 1919, Geomorphological studies in Norway: Geograf. Annaler, v. 1, p. 1–210.*

——, 1948, Glaciological research on the North Atlantic coasts: Roy. Geogr. Soc., Res. Ser. no. 1, 83 p.*

——, 1953, Glacier variations and climatic fluctuations: American Geogr. Soc., New York, 53 p.

Ahlmann, H. W., and Thorarinsson, S., 1937, Vatnajökull. Scientific results of the Swedish-Icelandic investigations 1936–1937. Chapter 3, Previous investigations of Vatnajökull, marginal oscillations of its outlet-glaciers, and general description of its morphology: Geograf. Annaler, v. 19, p. 176–211.*

Albritton, C. C., 1958, Quaternary stratigraphy of the Guadiana Valley, Durango, Mexico: Geol. Soc. America Bull., v. 69, p. 1197–1216.

Alcock, F. J., 1944, Further information on glaciation in Gaspé: Roy. Soc. Canada, Trans., ser. 3, v. 38, sec. 4, p. 15–21.

Alden, W. C., 1918, The Quaternary geology of southeastern Wisconsin, with a chapter on the older rock formations: U.S. Geol. Survey Prof. Paper 106, 356 p.

———, 1924, The physical features of central Massachusetts: U.S. Geol. Survey Bull. 760, p. 13–105.

———, 1932, Physiography and glacial geology of eastern Montana and adjacent areas: U.S. Geol. Survey Prof. Paper 174, 133 p.

———, 1953, Physiography and glacial geology of western Montana and adjacent areas: U.S. Geol. Survey Prof. Paper 231, 200 p.*

Alešinskaja, Z. V., and Bondarev, L. G., 1969, Le Pléistocène de la depression d'Issyk-Koul et l'Histoire du climat du Tien-Chan: Assoc. française pour l'Étude du Quat., Bull., v. 6, p. 43–62.

Alexander, H. S., 1932, Pothole erosion: Jour. Geol., v. 40, p. 305–337.

Alimen, Henriette, 1961, Les témoins des glaciers disparus: Scientia, 6th ser., March, 1961, p. 1–9.

———, 1964, Le quaternaire des Pyrénées de la Bigorre: Mémoires Carte Géologique Détaillée de la France, Imprim. Nationale, Paris, 394 p.*

Altehenger, A., 1958, Klimaschwankungen im Pliozän von Wallensen (Hils): Eiszeitalter und Geg., v. 9, p. 104–109.

Altschuler, Z. S., and Young, E. J., 1960, Residual origin of the "Pleistocene" sand mantle in central Florida uplands and its bearing on marine terraces and Cenozoic uplift: U.S. Geol. Survey Prof. Paper 400, p. B202–B207.

Ambrose, J. W., 1964, Exhumed paleoplains of the Precambrian Shield of North America: American Jour. Sci., v. 262, p. 817–857.*

American Commission on Stratigraphic Nomenclature, 1961, Code of stratigraphic nomenclature: American Assoc. Petroleum Geologists Bull., v. 45, p. 645–665.

Andersen, B. G., 1960, Sørlandet i sen- og Postglacial tid: Norges Geol. Undersök. no. 210, 142 p.*

Andersen, S. A., 1931, Om Aase og Terrasser inden for Susaa's Vandomraade og deres Vidnesbyrd om Isafsmeltningens Forløb: Danmarks Geol. Undersøg., ser. 2, no. 54, 201 p. [Engl. summ. p. 169–201].

Andersen, S. T., 1954, A late-glacial pollen diagram from southern Michigan, U.S.A.: Danmarks Geol. Undersøg., v. 2, no. 80, p. 140–155.*

———, 1961, Vegetation and its environment in Denmark in the Early Weichselian Glacial (last glacial): Danmarks Geol. Undersøg., ser. 2, no. 75, 175 p.*

———, 1963, Pollen analysis of the Quaternary marine deposits at Tornskov in South Jutland: Danmarks Geol. Undersøg., ser. 4, no. 8, 23 p.

Anderson, R. C., 1957, Pebble and sand lithology of the major Wisconsin glacial lobes of the Central Lowland: Geol. Soc. America Bull., v. 68, p. 1415–1450.*

Andersson, J. G., 1906, Solifluction, a component of subaërial denudation: Jour. Geol., v. 14, p. 91–112.

———, 1939, Topographical and archaeological studies in the Far East: Östasiatiska Samlingarna (Stockholm), Bull. 11, 110 p.

Andrews, J. T., 1968a, Pattern and cause of variability of postglacial uplift and rate of uplift in arctic Canada: Jour. Geol., v. 76, p. 404–425.

———, 1968b, Postglacial rebound in Arctic Canada: similarity and prediction of uplift curves: Canadian Jour. Earth Sci., v. 5, p. 39–47.

Andrews, J. T., and King, C.A.M., 1968, Comparative till fabrics and till fabric variability in a till sheet and a drumlin; a small-scale study: Yorkshire Geol. Soc., Proc., v. 36, p. 435–461.

Antevs, Ernst, 1925, Retreat of the last ice sheet in eastern Canada: Canada Geol. Survey Mem. 146, 142p.*

———, 1929, Maps of the Pleistocene glaciations: Geol. Soc. America Bull., v. 40, p. 631–720.*

———, 1948, The Great Basin, with emphasis on glacial and postglacial times. III, Climatic changes and pre-white man: Univ. Utah, Bull., v. 38, p. 168–191.*

Arambourg, Camille, 1962, Les faunes mammalogiques du Pléistocène d'Afrique: p. 369–376 in Problèmes actuels de paléontologie (évolution des vertébrés), France. Centre Nat. de la Recherche Sci., Colloques Internat., no. 104, Paris, 474 p.

Armstrong, J. E., and Brown, W. L., 1954, Late Wisconsin marine drift and associated sediments of the Lower Fraser Valley, British Columbia, Canada: Geol. Soc. America Bull., v. 65, p. 349–364.

———, and Tipper, H. W., 1948, Glaciation in north central British Columbia: American Jour. Sci., v. 246, p. 283–310.

———, and others, 1965, Late Pleistocene stratigraphy and chronology in southwestern British Columbia and northwestern Washington: Geol. Soc. America Bull., v. 76, p. 321–330.*

Arneman, H. F., and Wright, H. E., 1959, Petrography of some Minnesota tills: Jour. Sed. Petrology, v. 29, p. 540–554.

Aronow, Saul, 1959, Drumlins and related streamline features in the Warwick-Tokio area, North Dakota: American Jour. Sci., v. 257, p. 191–203.

Arrhenius, Gustaf, 1950, Late Cenozoic climatic changes as recorded by the equatorial current system: Tellus, v. 2, p. 83–88.

———, 1952, Sediment cores from the east Pacific: Swedish Deep-Sea Expedition, repts, v. 5: Elander, Göteborg, 227 p.*

———, 1967, Deep-sea sedimentation. A critical review of U.S. work 1963–1967: Internat. Un. Geod. and Geophys., Rept. 14th Gen. Assem., Univ. of California/Scripps Inst. of Oceanogr. [publ.] 67–14, 25 p.

Atwood, W. W., 1909, Glaciation of the Uinta and Wasatch Mountains: U. S. Geol. Survey Prof. Paper 61, 96 p.

———, and Mather, K. F., 1932, Physiography and Quaternary geology of the San Juan Mountains, Colorado: U. S. Geol. Survey Prof. Paper 166, 171 p.

Atwood, W. W., Jr., 1937, Records of Pleistocene glaciers in the Medicine Bow and Park Ranges: Jour. Geol., v. 45, p. 113–140.

Aubert, Daniel, 1965, Calotte glaciaire et morphologie jurassiennes: Eclogae Geol. Helvet., v. 58, p. 555–578.

Auer, Väinö, 1958, The Pleistocene of Fuego-Patagonia. Part II. The history of the flora and vegetation: Finland Acad. Sci., Annals, ser. A, sec. III, no. 50, 239 p.*

Axelrod, D. I., and Bailey, H. P., 1969, Paleotemperature analysis of Tertiary floras: Palaeogeog., Palaeoclimatol., Palaeoecol., v. 6, p. 163–195.

Bader, Henri, 1961, The Greenland Ice Sheet: United States Army Corps of Engineers, Cold Regions Res. and Eng. Lab., Cold Regions Sci. and Eng. 1–B2, 18 p.

──────, and others, 1939, Der Schnee und seine Metamorphose: Beiträge zur Geologie der Schweiz, Geotechnische Serie, Hydrologie, pt. 3, 340 p. Bern: Kommissionsverlag Kümmerly and Frey. [Engl. transl.: Snow and its metamorphism. U.S. Army, Corps of Engineers, Snow, Ice and Permafrost Research Establishment, Transl. 14, Jan. 1954, 313 p.]

Baker, F. C., 1936, Quantitative examination of molluscan fossils in two sections of Pleistocene loess in Illinois: Jour. Paleont., v. 10, p. 72–76.

Bandy, O. L., 1968a, Cycles in Neogene paleoceanography and eustatic changes: Palaeogeography, Palaeoclimatology, Palaeoecology, v. 5, p. 63–75.

──────, 1968b, Paleoclimatology and Neogene planktonic foraminiferal zonation: Giornale di Geol., v. 35, p. 277–290.*

──────, and Wade, M. E., 1967, Miocene-Pliocene-Pleistocene boundaries in deep-water environments: Progr. In Oceanogr., v. 4, p. 51–66.

──────, and others, 1969, Alaskan Upper Miocene marine glacial deposits and the *Turborotalia pachyderma* datum plane: Science, v. 166, p. 607–609.

Barbour, G. B., 1935, Physiographic history of the Yangtze: China Geol. Survey, Mem., ser. A, no. 14, 112 p.

Bardin, V. I., and Suyetova, I. A., 1967, Basic morphometric characteristics for Antarctica and budget of the Antarctic ice cover: Tokyo. Nat. Science Mus., Jare Scientif. Repts., Spec. Issue 1, p. 92–100.

Barnett, D. M., 1966, A re-examination and re-interpretation of tide gauge data for Churchill, Manitoba: Canadian Jour. Earth Sci., v. 3, p. 77–88.

Barry, R. G., 1966, Meteorological aspects of the glacial history of Labrador-Ungava with special reference to atmospheric vapour transport: Geogr. Bull. (Ottawa), v. 8, p. 319–340. (Reprinted without change in v. 9, 1967, p. 25–46.)*

Bastin, E. S., and Hill, J. M., 1917, Economic geology of Gilpin County and adjacent parts of Clear Creek and Boulder counties, Colorado. U. S. Geol. Survey Prof. Paper 94, 379 p.

Bauer, Albert, 1954, Synthèse glaciologique: p. 27–56 in Contribution . . . à la connaissance de l'Inlandsis du Groenland, pt. II (no. N. II. 3), Paris, Expéditions Polaires Françaises.

──────, 1955, Über die in der heutigen Vergletscherung der Erde als Eis gebundene Wassermasse: Eiszeitalter und Geg. v. 6, p. 60–70.*

Bayrock, L. A., 1962, Heavy minerals in till of central Alberta: Alberta Soc. Petroleum Geologists, Jour., v. 10, p. 171–184.

──────, 1967, Catastrophic advance of the Steele Glacier, Yukon, Canada: Univ. of Alberta, Boreal Inst., Occasional Publ. 3. (Res. Council of Alberta Contrib. 377), 35 p.

Bellair, Pierre, 1965, Un exemple de glaciation aberrante. Les Iles Kerguelen: Com. Nat. Français des Recherches Antarctiques, Publ. 11, p. 1–27.

Bennema, I. J., 1954, Bodem- en Zeespiegelbewegingen in het Nederlandse Kustgebied (Holocene movements of land and sea-level in the coastal area of the Netherlands): Boor en Spade (Wageningen), v. 7, p. 1–96.*

Bentley, C. R., and Ostenso, N. A., 1961, Glacial and subglacial topography of West Antarctica: Jour. Glaciol., v. 3, p. 882–910.

————, and others, 1964, Physical characteristics of the Antarctic Ice Sheet: American Geogr. Soc. Antarctic Map Folio Ser., Folio 2, 10 p., 10 pl.*

Bequaert, J. C., 1943, The genus *Littorina* in the western Atlantic: Johnsonia, no. 7, 28 p.

Bergdahl, Arvid, 1953, Israndsbildningar i östra syd- och mellansverige med särskild hänsyn till åsarna: Lunds Universitets Geograf. Inst., Avh. 23, 208 p.

Bergsten, Folke, 1954, The land uplift in Sweden from the evidence of the old water marks: Geograf. Annaler, v. 36, p. 81–111.

Bergstrom, R. E., ed., 1968, The Quaternary of Illinois: Univ. of Illinois, Coll. of Agric., Spec. Publ. 14, 179 p.

Berkey, C. P., 1911, Geology of the New York City (Catskill) aqueduct. Studies in applied geology covering problems encountered in explorations along the line of the aqueduct from the Catskill Mountains to New York City: New York State Mus. Bull. 146, 283 p.

————, and Morris, F. K., 1927, Geology of Mongolia. A reconnaissance report based on the investigations of the years 1922–1923: Nat. Hist. of central Asia, v. 2. American Mus. Nat. Hist. (New York), 475 p.

Bernhardi, A., 1832, Wie kamen die aus dem Norden stammenden Felsbruchstücke und Geschiebe, welche man in Norddeutschland und den benachbarten Ländern findet, an ihre gegenwärtigen Fundorte?: Jahrb. für Mineralogie, Geognosie, und Petrefaktenkunde (Heidelberg), v. 3, p. 257–267.

Berry, E. G., and Miller, B. B., 1966, A new Pleistocene fauna and a new species of *Biomphalaria* (basommatophora: planorbidae) from southwestern Kansas, U.S.A.: Malacologia, v. 4, p. 261–267.

Beug, H. J., 1967a, On the forest history of the Dalmatian coast: Rev. Palaeobot. and Palynol., v. 2, p. 271–279.

————, 1967b, Probleme der Vegetationsgeschichte in Südeuropa: Deutsch. Botan. Ges. Bericht, v. 80, p. 682–689.

Biberson, Pierre, and Jodot, Paul, 1965, Faunes de mollusques continentaux du Pléistocène de Casablanca (Maroc): Service Géol. du Maroc, Notes, v. 25, p. 115–170.*

Birkeland, P. W., and others, 1970, Status of correlation of Quaternary stratigraphic units in the western conterminous United States: Quaternary Res., v. 1, no. 2.*

Birman, J. H., 1968, Glacial reconnaissance in Turkey: Geol. Soc. America Bull., v. 79, p. 1009–1026.*

Biscaye, P. E., 1965, Mineralogy and sedimentation of recent deep-sea clay in the Atlantic Ocean and adjacent seas and oceans: Geol. Soc. America Bull., v. 76, p. 803–832.*

Bishop, B. C., 1957, Shear moraines in the Thule area, northwest Greenland: United States Army, Corps of Engineers, Snow, Ice and Permafrost Research Establishment Research Rept. 17, 46 p.

Bishop, W. W., and Clark, J. D., eds., 1967, Background to evolution in Africa: Univ. of Chicago Press, 935 p.*

Blache, Jules, 1952, La sculpture glaciaire: Rev. de Géog. Alpine, v. 40, p. 31–123.

Black, R. F., 1964, Periglacial phenomena of Wisconsin, north-central United States: Internat. Quater. Assoc., 6th Congr., 1961, Rept., v. 4, p. 21–28.

——, 1965, Ice-wedge casts of Wisconsin: Wisconsin Acad. Sci., Arts and Letters, v. 54, p. 187–222.

Blackwelder, Eliot, 1931, Pleistocene glaciation in the Sierra Nevada and Basin Ranges: Geol. Soc. America Bull., v. 42, p. 865–922.

——, 1934, Supplementary notes on Pleistocene glaciation in the Great Basin: Nat. Acad. Sci., Jour., v. 24, p. 217–222.

Blagbrough, J. W., and Farkas, S. E., 1968, Rock glaciers in the San Mateo Mountains, south-central New Mexico: American Jour. Sci., v. 266, p. 812–823.

Blake, Weston, 1970, Studies of glacial history in Arctic Canada.I. Pumice, radiocarbon dates, and differential postglacial uplift in the eastern Queen Elizabeth Islands: Canadian Jour. Earth Sci., v. 7, p. 634–664.*

Blanc, A. C., 1937, Low levels of the Mediterranean Sea during the Pleistocene glaciation: Geol. Soc. London, Quart. Jour., v. 93, p. 621–651.

Blanchard, R. L., and others, 1967, Uranium and thorium series disequilibria in recent and fossil marine molluscan shells: Jour. Geophys. Res., v. 72, p. 4745–4757.

Bloom, A. L., 1967, Pleistocene shorelines: A new test of isostasy: Geol. Soc. America Bull., v. 78, p. 1477–1494.*

Blytt, A. G., 1876, Essays on the immigration of the Norwegian flora during alternating rainy and dry periods: Christiania [Oslo], A. Cammermeyer, 89 p.

Bobek, Hans, 1937, Die Rolle der Eiszeit in Nordwestiran: Zeitschr. für Gletscherk., v. 25, p. 130–183.

Boenigk, Wolfgang, 1970, Zur Kenntnis des Altquartärs bei Brüggen (westlicher Niederrhein): Köln Univ. Geol. Inst., Sonderveröffentl., no. 17, 141 p.*

Böhm, August, 1901, Geschichte der Moränenkunde: K. K. Geogr. Gesellsch. in Wien, Abh., v. 3, no. 4, 334 p.*

Boissonneau, A. N., 1966–1968, Glacial history of northeastern Ontario: Canadian Jour. Earth Sci., v. 3, p. 559–578; v. 5, p. 97–109.

Bojarskaja, T. D., 1969, Le Pléistocène de la Siberie orientale: Assoc. française pour l'Étude du Quat., Bull., v. 6, p. 63–73.

Bonatti, Enrico, 1966, North Mediterranean climate during the last Würm glaciation: Nature, v. 209, p. 984.

Bonifay, Eugène, 1964, Pliocène et Pléistocène méditerranéens. Vue d'ensemble et essai de corrélations avec la chronologie glaciaire: Ann. de Paléont. (Vertébrés), v. 50, 195–226.

Bonney, T. G., 1910, Some aspects of the glacial history of western Europe: Science, v. 32, p. 321–336, 353–363.

Borns, H. W., 1963, Preliminary report on the age and distribution of the late Pleistocene ice in north central Maine: American Jour. Sci., v. 261, p. 738–740.

Bostock, H. S., 1966, Notes on glaciation in central Yukon Territory: Canada Geol. Survey Pap. 65–36, 14 p.

Boswell, P.G.H., 1952, The Pliocene-Pleistocene boundary in the east of England: Geologists' Assoc., Proc., v. 63, p. 301–315.

Boulton, G. S., 1968, Flow tills and related deposits on some Vestspitsbergen glaciers: Jour. Glaciol., v. 7, p. 391–412.

Bourdier, Franck, 1961–1962, Le bassin du Rhône au Quaternaire. Géologie et préhistoire: Éditions Centre National de la Recherche Scientif. (Paris) v. 1, 364 p., 1961; v. 2, (ills., bibliog., index), 1962.*

Bowen, F. E., 1967, Early Pleistocene glacial and associated deposits of the west coast of South Island, New Zealand: New Zealand Jour. Geol. and Geophys., v. 10, p. 164–181.

Bowler, J. M., 1970, Late Quaternary environments; a study of lakes and associated sediments in southeastern Australia: Unpubl. Ph.D. thesis, Australian Nat. Univ., Canberra, 340 p.*

————, and Harford, L. B., 1966, Quaternary tectonics and the evolution of the Riverine Plain near Echuca, Victoria: Geol. Soc. Australia, Jour., v. 13, p. 339–354.

Boyd, L. A., 1948, The coast of northeast Greenland: American Geogr. Soc. Spec. publ. 30, 339 p.*

Bradley, W. C., and Addicott, W. O., 1968, Age of first marine terrace near Santa Cruz, California: Geol. Soc. America Bull., v. 79, p. 1203–1210.*

Bradley, W. H., 1936, Geomorphology of the north flank of the Uinta Mountains: U. S. Geol. Survey Prof. Paper 185, p. 163–204.

————, and others, 1942, Geology and biology of north Atlantic deep-sea cores between Newfoundland and Ireland: U.S. Geol. Survey Prof. Paper 196, 163 p.

Bramlette, M. N., 1946, The Monterey formation of California and the origin of its siliceous rocks: U. S. Geol. Survey Prof. Paper 212, 57 p.

Braun, E. L., 1955, Phytogeography of unglaciated eastern United States and its interpretation: Botan. Rev., v. 21, p. 297–375.

Bray, J. R., 1968, Glaciation and solar activity since the Fifth Century B.C. and the solar cycle: Nature, v. 220, p. 672–674.*

Brenner, Thord, 1944, Finlands åsars vittnesbörd om ytgestaltningen hos landisen: Fennia, v. 68, no. 4, 39 p.

Bretz, J. H., 1932, The Grand Coulee: American Geogr. Soc. Spec. Pub. 15, 89 p.

————, 1969, The Lake Missoula floods and the channeled scabland: Jour. Geol., v. 77, p. 505–543.*

Breuil, Henri, 1939, The Pleistocene succession in the Somme valley: Prehistoric Soc., Proc., p. 33–38.

Broecker, W. S., 1966, Glacial rebound and the deformation of the shorelines of proglacial lakes: Jour. Geophys. Res., v. 71, p. 4777–4783.

————, and Donk, Jan van, 1970, Insolation changes, ice volumes and the 0^{18} record in deep sea cores: Rev. Geophys. and Space Phys., v. 8, p. 169–198.*

————, and Farrand, W. R., 1963, Radiocarbon age of the Two Creeks forest bed, Wisconsin: Geol. Soc. America Bull., v. 74, p. 795–802.

————, and Kaufman, Aaron, 1965, Radiocarbon chronology of Lake Lahontan and Lake Bonneville II, Great Basin: Geol. Soc. America Bull., v. 76, p. 537–566.

————, and Orr, P. C., 1958, Radiocarbon chronology of Lake Lahontan and Lake Bonneville: Geol. Soc. America Bull., v. 69, p. 1009–1032.

————, and others, 1958, The relation of deep sea sedimentation rates to variations in climate: American Jour. Sci., v. 256, p. 503–517.

————, and others, 1960, Evidence for an abrupt change in climate close to 11,000 years ago: American Jour. Sci., v. 258, p. 429–448.

Brøgger, W. C., and Reusch, H. M., 1874, Giant's kettles at Christiania: Geol. Soc. London, Quart. Jour., v. 30, p. 750–771.

Brooks, C. E. P., 1949, Climate through the ages: 2d ed., Ernest Benn, London, 395 p.

Brothwell, Don, and Higgs, Eric, eds., 1963, Science in archaeology: Thames and Hudson, London, 595 p.*

Brouwer, Aart, 1956, Pleistocene transgressions in the Rhine delta: Quaternaria, v. 3, p. 83–88.

Brüggen, Juan, 1950, Fundamentos de la geología de Chile: Instituto Geográfico Militar, Santiago, 374 p.

Brun, Annik, 1966, Révision de la stratigraphie des dépots quaternaires dans la basse vallée de la Dranse (Haute-Savoie): Rev. de Géogr. Phys. et de Géol. Dynam., v. 8, p. 399–404.

Brunnacker, Karl, 1957, Bemerkungen zur Feinstgliederung und zum Kalkgehalt des Lösses: Eiszeitalter und Geg., v. 8, p. 107–115.

———, 1964a, Über Ablauf und Altersstellung altquartärer Verschüttungen im Maintal und nächst dem Donautal bei Regensburg: Eiszeitalter und Geg., v. 15, p. 72–80.

———, 1964b, Böden des älteren Pleistozäns bei Regensburg: Geologica Bavarica, v. 53, p. 148–160.*

Bryan, A. L., 1969, Early man in America and the Late Pleistocene chronology of western Canada and Alaska: Current Anthropol., v. 10, p. 339–367.*

Bryan, Kirk, 1941, Correlation of the deposits of Sandia Cave, New Mexico, with the glacial chronology: Smithsonian Misc. Coll. (Washington), v. 99, p. 45–64.

———, 1948, Los suelos complejos y fósiles de la altiplanicie de México, en relación a los cambios climáticos: Soc. Geol. México, Bull., v. 13. p. 1–20.

———, and Albritton, C. C., 1943, Soil phenomena as evidence of climatic changes: American Jour. Sci., v. 241, p. 469–490.*

Bryson, R. A., and others, 1967, Radiocarbon isochrones on the disintegration of the Laurentide Ice Sheet: Arctic and Alpine Res., v. 1, p. 1–14.

Büdel, Julius, 1944, Die morphologischen Wirkungen des Eiszeitklimas im gletscherfreien Gebiet: Geol. Rundschau, v. 34, p. 482–519.*

———, 1949, Die räumliche und zeitliche Gliederung des Eiszeitklimas: Die Naturwissenschaften, v. 36, p. 105–112, 133–139.*

———, 1953, Die "periglazial"-morphologische Wirkungen des Eiszeitklimas auf der ganzen Erde: Erdkunde, v. 7, p. 249–266. (Engl. transl.: Internat. Geology Rev., v. 1, 1959, no. 3, p. 1–16.)

———, 1959, Periodische und episodische Solifluktion im Rahmen der klimatischen Solifluktionstypen: Erdkunde, v. 13, p. 297–314. (Engl. summ.) *

Bull, A. J., 1940, Cold conditions and land forms in the South Downs: Geologists' Assoc., Proc., v. 51, p. 63–71.

Burri, Marcel, 1963, Le Quaternaire des Dranses: Univ. de Lausanne. Bull. des Laboratoires de Géol., etc., no. 142, 34 p.

Butler, B. E., 1958, Depositional systems of the Riverine Plain of southeastern Australia in relation to soils: Commonwealth Sci. and Ind. Res. Org., Soil Publ. 10, 35 p.

Butzer, K. W., 1958, Quaternary stratigraphy and climate in the Near East: Bonner Geogr. Abh., no. 24, 157 p.*

———, 1964, Environment and archeology. An introduction to Pleistocene geography: Aldine Publishing Co., Chicago, 524 p.*

———, and Cuerda, Juan, 1962, Coastal stratigraphy of southern Mallorca and its implications for the Pleistocene chronology of the Mediterranean Sea: Jour. Geol. v. 70, p. 398–416.

————, and Hansen, C. L., 1968, Desert and river in Nubia: Univ. of Wisconsin Press, Madison, 562 p.*

Cabot, T. D., 1939, The Cabot expedition to the Sierra Nevada de Santa Marta of Colombia: Geogr. Rev., v. 29, p. 587–621.

Caine, Nelson, 1968, The blockfields of northeastern Tasmania: Canberra. Research School of Pacific Studies, Dept. Geogr. Publ. G/6, 127 p.*

Caldenius, C. C., 1932, Las glaciaciones cuaternarias en la Patagonia y Tierra del Fuego: Geogr. Annaler, v. 14, p. 1–164. Also publ. as: Argentina. Dirección Gen. de Minas y Geología Pub. 95, 1932, 150 p.*

————, 1940, The Tehuelche or Patagonian shingle-formation. A contribution to the study of its origin: Geogr. Annaler, v. 22, p. 160–181.

Calkin, P. E., 1970, Strand lines and chronology of the glacial Great Lakes in northwestern New York: Ohio Jour. Sci., v. 70, p. 78–96.

Cameron, Dorothy, 1965, Goethe—discoverer of the ice age: Jour. Glaciol., v. 5, p. 751–754.

Capps, S. R., 1916, The Chisana-White River district, Alaska: U. S. Geol. Survey Bull. 630, 130 p.

————, 1940, Gold placers of the Secesh Basin, Idaho County, Idaho: Idaho Geol. Survey Pamph. 52, 43 p.

————, and Leffingwell, E. de K., 1904, Pleistocene geology of the Sawatch Range, near Leadville, Colorado: Jour. Geol., v. 12, p. 698–706.

Carey, S. W., and Ahmad, Naseeruddin, 1961, Glacial marine sedimentation: 1st Internat. Sympos. on Arctic Geology, Proc., v. 2, p. 865–894. Reprinted 1961 as Univ. of Tasmania, Dept. of Geol. Publ. 87, p. 865–894.

Carney, Frank, 1909a, The development of the idea of glacial erosion in America: Denison Univ. Sci. Lab., Bull., v. 14, p. 199–208.*

————, 1909b, The raised beaches of the Berea, Cleveland, and Euclid sheets, Ohio: Denison Univ. Sci. Lab., Bull., v. 14, p. 262–287.

Carol, Hans, 1947, The formation of *roches moutonnées:* Jour. Glaciol., v. 1, p. 58–59.

Carozzi, A. V., 1966, Agassiz's amazing geological speculation: the Ice-age: Studies in Romanticism v. 5, p. 57–83.

Carruthers, Douglas, 1913, Unknown Mongolia: Hutchinson & Co., London, 659 p.

Castellanos, Alfredo, 1956, Carácteres del pleistoceno en la Argentina: Internat. Quater. Assoc., 4th Congr., Rome-Pisa, 1953, Actes, v. 2, p. 942–948.

————, 1962, El Holoceno en la Argentina: Univ. Nac. del Litoral (Rosario), Publs., no. 45, 78 p.*

Chadwick, G. H., 1928, Adirondack eskers: Geol. Soc. America, Bull., v. 39, p. 923–929.

Chamberlin, T. C., 1883, Preliminary paper on the terminal moraine of the second glacial epoch: U. S. Geol. Survey 3d Ann. Rp., 1881–82, p. 291–402.

————, 1888, The rock-scorings of the great ice invasions: U. S. Geol. Survey 7th Ann. Rp., 1885–86, p. 147–248.

————, 1894, Proposed genetic classification of Pleistocene glacial formations: Jour. Geol., v. 2, p. 517–538.

————, 1895, Glacial phenomena of North America: p. 724–775 in Geikie, James, *The Great Ice Age,* 2d ed., Appleton, New York.

Chambers, Robert, 1853, On glacial phenomena in Scotland and parts of England: Edinburgh New Philos. Jour., v. 54, p. 229–281.

Chaney, R. W., 1940, Tertiary forests and continental history: Geol. Soc. America Bull., v. 51, p. 469–488.

Chang, K. C., 1962, New evidence on fossil man in China: Science, v. 136, p. 749–760.*

———, 1968, The archaeology of ancient China: Yale Univ. Press, New Haven, rev. ed., 483 p.*

Chapman, D. H. 1937, Late-glacial and postglacial history of the Champlain Valley: American Jour. Sci., v. 34, p. 89–124.

Chappell, John, 1968, Changing duration of glacial cycles from lower to upper Pleistocene: Nature, v. 219, p. 36–40.*

Charlesworth, J. K., 1956, The late-glacial history of the Highlands and Islands of Scotland: Royal Soc. Edinburgh Trans., v. 62, pt. 3, p. 103–928.

———, 1957, The Quaternary Era, with special reference to its glaciation: Edward Arnold, London, 2 v., 1700 p.*

Charpentier, Jean de, 1841, Essai sur les glaciers et sur le terrain erratique du bassin du Rhône: Marc Ducloux, Lausanne, 363 p.

Chavaillon, Jean, 1964, Étude stratigraphique des formations quaternaires du Sahara nord-occidental (de Colomb-Béchar à Reggane): Centre de Recherches sur les Zones Arides (Paris), sér. géol. no. 5, 393 p.*

Chemekov, Y. F., 1961a, The snowline of the late Upper-Quaternary glaciation in the south of the Soviet Far East: USSR, Akad. Nauk, Isvestia, Geogr. ser., no. 6, p. 73–88.[R]

———, 1961b, Quaternary transgressions of the far-eastern seas and of the northern Pacific: Estonian SSR. Akad. Nauk., Inst. Geol., Trudy, no. 8, p. 155–174.[R]

Chorley, R. J., 1959, The shape of drumlins: Jour. Glaciol., v. 3, p. 339–344.

Choubert, Georges, 1965, Quelques réajustements au tableau des corrélations des lignes de rivages de l'Atlantique (Maroc) et de la Méditérranée: Internat. Quater. Assoc. 6th, Congr., Warsaw 1961, Rept., v. 1. p. 269–278.

Clark, G. M., 1968, Sorted patterned ground: new Appalachian localities south of the glacial border: Science, v. 161, p. 355–356.

Clark, Grahame, 1967, The stone age hunters: McGraw-Hill, New York, 143 p.*

———, 1969, World prehistory, a new outline: 2d. ed., Cambridge Univ. Press, 331 p.*

Clark, J. D., 1965, The later Pleistocene cultures of Africa: Science, v. 150, p. 833–847.

———, and van Zinderen Bakker, E. M., 1964, Prehistoric culture and Pleistocene vegetation at the Kalambo Falls, Northern Rhodesia: Nature, v. 201, p. 971–975.

Clayton, Lee, and Freers, T. F., eds., 1967, Glacial geology of the Missouri Coteau and adjacent areas: North Dakota Geol. Survey, Misc. ser. 30, 170 p.

Clayton, Lee, and others, 1965, Intersecting minor lineations on Lake Agassiz Plain: Jour. Geol., v. 73, p. 652–656.

Clisby, K. H., and Sears, P. B., 1956, San Augustin Plains—Pleistocene climatic changes: Science, v. 124, p. 537–539.

Close, M. H., 1867, Notes on the general glaciation of Ireland: Roy. Geol. Soc. Ireland, Jour., v. 1, p. 207–242.

Coetzee, J. A., 1967, Pollen analytical studies in east and southern Africa: Palaeoecology of Africa, v. 3, A.A. Balkema, Cape Town, 146 p.*

Colbert, E. H., 1942, The Pleistocene faunas of Asia and their relationships to early man: New York Acad. Sci., Trans., ser. 2, v. 5, p. 1–10.

————, and others, 1948, Pleistocene of the Great Plains: Geol. Soc. America Bull., v. 59, p. 541–630.*

Coleman, A. P., 1921, Northeastern part of Labrador, and New Quebec: Canada Geol. Survey Mem. 124, 68 p.

————, 1941, The last million years. A history of the Pleistocene in North America: Univ. of Toronto Press, 216 p.*

Colinvaux, P. A., 1968, Reconnaissance and chemistry of the lakes and bogs of the Galápagos Islands: Nature, v. 219, p. 590–594.

Collomb, Edouard, 1847, Preuves de l'existence d'anciens glaciers dans les vallées des Vosges: Victor Masson, Paris, 246 p.

Colquhoun, D. J., and Johnson, H. S., 1968, Tertiary sea-level fluctuation in South Carolina: Palaeogeog., Palaeoclimatol., Palaeoecol., v. 5, p. 105–126.

Colquhoun, D. J., and others, 1968, A fossil assemblage from the Wicomico Formation in Berkeley County, South Carolina: Geol. Soc. America Bull., v. 79, p. 1211–1220.

Colton, R. B., and others, 1961, Glacial map of Montana east of the Rocky Mountains. Scale 1:500,000: U. S. Geol. Survey Misc. Geol. Inv., Map I–327.

————, 1963, Preliminary glacial map of North Dakota. Scale 1:500,000: North Dakota Geol. Survey and U. S. Geol. Survey.

Condra, G. E., and others, 1950, Correlation of the Pleistocene deposits of Nebraska: Nebraska Geol. Survey, Bull. 15A, 74 p.*

Conolly, J. R., and Ewing, Maurice, 1965, Pleistocene glacial-marine zones in North Atlantic deep-sea sediments: Nature, v. 208, p. 135–138.

Conrad, T. A., 1839, Notes on American geology: American Jour. Sci., v. 35, p. 237–251.

Cooke, H. B. S., 1958, Observations relating to Quaternary environments in east and southern Africa: South Africa Geol. Soc. Bull., v. 20, annexure, 73 p.*

————, 1968, Evolution of mammals on southern continents. II, The fossil mammal fauna of Africa: Quart. Rev. Biol., v. 43, p. 234–264.*

Cooke, H. C., 1930, Studies of the physiography of the Canadian Shield. II— Glacial depression and post-glacial uplift: Roy. Soc. Canada Trans., ser. 3, v. 24, sec. 4, p. 51–87.

Coope, G. R., 1968, Insect remains from silts below till at Garfield Heights, Ohio: Geol. Soc. America Bull., v. 79, p. 753–756.

————, 1969, The contribution that the Coleoptera of glacial Britain could have made to the subsequent colonisation of Scandinavia: Opuscula Entomol. (Lund), v. 34, p. 95–108.

————, and Sands, C. H. S., 1966, Insect faunas of the last glaciation from the Tame Valley, Warwickshire: Roy. Soc. London, Proc., sec. B, v. 165, p. 389–412.

Cooper, W. S., 1935, The history of the upper Mississippi River in late Wisconsin and postglacial time: Minnesota Geol. Survey Bull. 26, 116 p.

Costin, A. B., and others, 1964, Snow action on Mount Twynam, Snowy Mountains, Australia: Jour. Glaciol., v. 5, p. 219–228.*

————, 1967, Nonsorted steps in the Mt. Kosciusko area, Australia: Geol. Soc. America Bull., v. 78, p. 979–992.

Cotton, C. A., 1942, Climatic accidents in landscape-making: Whitcombe and Tombs, Christchurch, N.Z., 354 p. (Reprint, 1943, John Wiley and Sons, Inc., New York.*

———, and Te Punga, M.T., 1955, Solifluxion and periglacially modified landforms at Wellington, New Zealand: Roy. Soc. New Zealand, Trans., v. 82, p. 1001–1031.*

Couper, R. A., and McQueen, D. R., 1954, Pliocene and Pleistocene plant fossils of New Zealand and their climatic interpretation: New Zealand Jour. Sci. and Techn., sec. B, v. 35, p. 398–420.

Cox, Allan, 1969, Geomagnetic reversals: Science, v. 163, p. 237–245.

———, and others, 1968, Radiometric time-scale for geomagnetic reversals: Geol. Soc. London, Quart. Jour., v. 124, p. 53–66.

Craig, B. G., 1965a, Glacial Lake McConnell and the surficial geology of parts of Slave River and Redstone River Map-Areas, District of Mackenzie: Canada Geol. Survey Bull. 122, 33 p.

———, 1965b, Notes on moraines and radiocarbon dates in northwest Baffin Island, Melville Peninsula, and northeast District of Keewatin: Canada Geol. Survey Paper 65–20, 7 p.

———, and Fyles, J. G., 1960, Pleistocene geology of Arctic Canada: Canada Geol. Survey Paper 60–10, 21 p.*

Craig, Harmon, 1965, The measurement of oxygen isotope paleotemperatures: Part (24p) of Stable isotopes in oceanographic studies and paleotemperatures, Laboratorio di Geologia Nucleare (Pisa).

Crary, A. P., 1966, Mechanism for fiord formation indicated by studies of an ice-covered inlet: Geol. Soc. America Bull., v. 77, p. 911–930.

Crawford, C. B., and Eden, W. J., 1966, A comparison of laboratory results with *in-situ* properties of Leda clay: Canada. Nat. Res. Council, Div. Building Research, Res. Paper 274, 5 p.

Crittenden, M.D., 1963, New data on the isostatic deformation of Lake Bonneville: U.S. Geol. Survey Prof. Paper 454-E, 31 p.

Croll, James, 1875, Climate and time in their geological relations. A theory of secular changes of the Earth's climate: Edward Stanford, London, 577 p.

Crosby, W. O., 1928, Certain aspects of glacial erosion: Geol. Soc. America Bull., v. 39, p. 1171–1181.

Crowell, J. C., and Frakes, L. A., 1970, Phanerozoic glaciation and the causes of ice ages: American Jour. Sci., v. 268, p. 193–224.

Curray, J. R., 1964, Transgressions and Regressions: p. 175–203 in Papers in marine geology: Shepard commemorative volume, Macmillan Co., New York.*

———, and others, 1970, Late Quaternary sea-level studies in Micronesia: CARMARSEL Expedition: Geol. Soc. America Bull., v. 81, p. 1865–1880.

Cushing, E. J., and Wright, H. E., eds., 1967, Quaternary paleoecology: Yale University Press, New Haven, 433 p.*

Dahl, Eilif, 1955, Biogeographic and geologic indications of unglaciated areas in Scandinavia during the glacial ages: Geol. Soc. America Bull., v. 66, p. 1499–1520.*

Dahl, Ragnar, 1963, Shifting ice culmination, alternating ice covering and ambulant refuge organisms?: Geograf. Annaler, v. 45, p. 122–138.

Dainelli, Giotto, and Marinelli., Olinto, 1928, Le condizioni fisiche attuali:

Spedizione Italiana de Filippi nell' Himàlaia, Caracorum e Turchestàn Cinese (1913–1914), ser. 2, v. 4, 1928, 479 p. N. Zanichelli, Bologna.

Dalrymple, G. B. 1964, Potassium-Argon dates of three Pleistocene interglacial basalt flows from the Sierra Nevada, California: Geol. Soc. America Bull., v. 75, p. 753–757.

———, and others, 1965, Potassium-Argon age and paleomagnetism of the Bishop Tuff, California: Geol. Soc. America Bull., v. 76, p. 665–673.

Daly, R. A., 1912, Geology of the North American Cordillera at the forty-ninth parallel: Canada Geol. Survey Mem. 38, 857 p.

———, 1934, The changing world of the ice age: Yale Univ. Press, New Haven, 271 p.*

———, 1938, Architecture of the Earth: D. Appleton-Century Co., New York, 211 p.

Damuth, J. E., and Fairbridge, R. W., 1970, Equatorial Atlantic deep-sea arkosic sands and ice-age aridity in tropical South America: Geol. Soc. America Bull., v. 81, p. 189–206.

Dana, J. D., 1863, Manual of geology: Theodore Bliss and Co., Philadelphia, 798 p.

Dansgaard, Willi, and Tauber, Henrik, 1969, Glacier oxygen-18 content and Pleistocene ocean temperatures: Science, v. 166, p. 499–502.

Dansgaard, Willi, and others, 1969, One thousand centuries of climatic record from Camp Century on the Greenland Ice Sheet: Science, v. 166, p. 377–381.

Darlington, P. J., 1957, Zoogeography, The geographical distribution of animals: John Wiley and Sons, New York, 675 p.

Darton, N. H., and Salisbury, R. D., 1906, Description of Cloud Peak and Fort McKinney quadrangles, Wyoming: U. S. Geol. Survey Folio 142, 16 p.

Darton, N. H., and others, 1906, Description of the Bald Mountain and Dayton quadrangles, Wyoming: U. S. Geol. Survey Folio 141, 15 p.

Darwin, Charles, 1859, On the origin of species by means of natural selection . . . : 1st ed., J. Murray, London, 502 p. Another ed: 1963, New York, Doubleday (Dolphin Books C 172), 517 p.

David, T. W. E., 1914, Remarks on physiography and glacial geology of East Antarctica: Geogr. Jour., v. 44, p. 566–568.

———, 1950, The geology of the Commonwealth of Australia: Edward Arnold, London, v. 1, 747 p.; v. 2, 618 p.; v. 3, maps.*

Davis, J. R., and Nichols, R. L., 1968, The quantity of melt water in the Marble Point-Gneiss Point area, McMurdo Sound, Antarctica: Jour. Glaciol., v. 7, p. 313–320.

Davis, M. B., 1958, Three pollen diagrams from central Massachusetts: American Jour. Sci., v. 256, p. 540–570.

———, 1969, Palynology and environmental history during the Quaternary Period: American Scientist, v. 57, p. 317–332.

Davis, N.F.G., and Mathews, W. H., 1944, Four phases of glaciation with illustrations from south-western British Columbia: Jour. Geol., v. 52, p. 403–413.

Davis, W. M., 1906, The sculpture of mountains by glaciers: Scottish Geogr. Mag., v. 22, p. 76–89.

Dawson, G. M., 1890, On the glaciation of the northern part of the Cordillera . . . : American Geol., v. 6, p. 153–162.

Dawson, J. W., 1893, The Canadian ice age. Being notes on the Pleistocene geology of Canada, with especial reference to the life of the period and its climatal conditions: Wm. V. Dawson, Montreal, 301 p.

Deevey, E. S., 1949, Biogeography of the Pleistocene: Geol. Soc. America Bull., v. 60, p. 1315–1416.*

———, 1958, Radiocarbon-dated pollen sequences in eastern North America: Verh. 4th Internat. Tagung der Quartärbotaniker 1957: Veröff. Geobotanisches Institut Rübel in Zürich, Heft 34, p. 30–37.

———, and Flint, R. F., 1957, Postglacial Hypsithermal interval: Science, v. 125, p. 182–184.*

De Geer, Gerhard, 1897, Om rullstensåsarnas bildningssätt: Geolog. Fören. i Stockholm Förh., v. 19, p. 366–388.

———, 1912, A geochronology of the last 12,000 years: Int. Geol. Congr., 11th, Stockholm 1910, Compte Rendu, v. 1, p. 241–258.

———, 1940, Geochronologia Suecica Principles: K. Svenska Vetensk. Handl., 3d ser., v. 18, no. 6. Almqvist & Wiksells, Stockholm, Text [and] Atlas, 367 p., with pls. 1–53 and 65 text-figs.; Atlas, with pls. 54–90.

Demek, Jaromir, and Kukla, Jiri, eds., 1969, Periglzialzone, löss und Paläolithikum der Tschechoslowakei: Czechoslovakia. Acad. Sci., Geogr. Inst., Brno, 157 p.*

Denny, C. S., 1951, Pleistocene frost action near the border of the Wisconsin drift in Pennsylvania: [part of a symposium of Ecol. Soc. Am. on "The glacial border-climatic, soil, and biotic features"]: Ohio Jour. Sci., v. 51, p. 115–125.

———, 1956, Surficial geology and geomorphology of Potter County, Pennsylvania: U. S. Geol. Survey Prof. Paper 288, 72 p.

Denton, G. H., and Armstrong, R. L., 1969, Miocene-Pliocene glaciations in southern Alaska: American Jour. Sci., v. 267, p. 1121–1142.

Denton, G. H., and Porter, S. C., 1970, Neoglaciation: Scientif. American, v. 222, p. 101–110.

Denton, G. H., and Stuiver, Minze, 1967, Late Pleistocene glacial stratigraphy and chronology, northeastern St. Elias Mountains, Yukon Territory, Canada: Geol. Soc. America Bull., v. 78, p. 485–510.

Denton, G. H., and others, 1970, Late Cenozoic glaciation in Antarctica. The record in the McMurdo Sound region: Antarctic Journal of the United States, v. 5, p. 15–21.

Derbyshire, Edward, 1962, Fluvioglacial erosion near Knob Lake, central Quebec-Labrador, Canada: Geol. Soc. America Bull., v. 73, p. 1111–1126.

———, 1968, Glacial map of N.W.-Central Tasmania: Tasmania Dept. Mines, Geol, Survey Record 6, 47 p. +map, scale 1:125,000.

———, and others, 1965, Glacial map of Tasmania: Roy. Soc. Tasmania, Spec. Pub. 2, 10 p. + map, scale 1:250,000.*

Desnoyers, Jules, 1829, Observations sur un ensemble de depôts marins plus récens que les terrains tertiaires du bassin de la Seine, et constituant une formation géologique distincte; précédées d'un aperçu de la non-simultanéité des bassins tertiaires: Ann. des Sci. Nat. (Paris), v. 16, p. 171–214, 402–491.

De Terra, Helmut, and Movius, H.L., 1943, Research on early man in Burma. . . . : American Philos. Soc., Trans., v. 32, p. 265–464.

———, and Paterson, T. T., 1939, Studies on the ice age in India and associated human cultures: Carnegie Inst. (Washington) Publ. 493, 354 p.

Detterman, R. L., Bowsher, A. L., and Dutro, J. T., 1958, Glaciation on the

Arctic slope of the Brooks Range, northern Alaska: Arctic, v. 11, p. 43–61.

Diamond, Marvin, 1960, Air temperature and precipitation on the Greenland Ice Cap: Jour. Glaciol., v. 3, p. 558–567.

Dietrich, W. O., 1953, Neue Funde des etruskischen Nashorns in Deutschland und die Frage der Villafranchium-Faunen: Geologie, v. 2, p. 417–430.

Dijk, D. C. van, 1959, Soil features in relation to erosional history in the vicinity of Canberra: Melbourne, Commonw. Sci. and Ind. Res. Org., Soil Pub. 13, 41 p.

Diller, J. S., 1895, Mount Shasta, a typical volcano: Nat. Geog. Soc. Nat. Geog. Mon. 1, p. 237–268; also in The physiography of the United States (Nat. Geogr. Soc.), p. 237–268: American Book Co., New York, 1896.

Dimbleby, G. W., 1965, Post-glacial changes in soil profiles: Roy. Soc. London, Proc., ser. B, v. 161, p. 355–362.

Dionne, J.-C., 1968, Morphologie et sédimentologie glacielles, littoral sud du Saint-Laurent: Zeitschr. für Geomorph., Supplementbd. 7, p. 56–84.*

Dobrodeev, O. P., 1969, Paleogéographie du Pleistocène du sud de la Plaine Russe: Assoc. française pour l'Étude du Quat., Bull., v. 6, p. 75–86.

Dobrovolny, Ernest, 1962, Geología del Valle de La Paz: Bolivia. Dept. Nac. de Geol., Boletín 3, 153p.

Doering, J. A., 1958, Citronelle age problem: American Assoc. Petroleum Geologists Bull., v. 42, p. 764–786.

Dollfus-Ausset, Daniel, 1863–1870, Materiaux pour l'étude des glaciers: Paris, F. Savy, 1863–1872, 8 v; atlas.*

Donn, W. L., and Ewing, Maurice, 1966, A theory of ice ages III: Science, v. 152, p. 1706–1712.

Donn, W. L., and others, 1962, Pleistocene ice volumes and sea-level lowering: Jour. Geol., v. 70, p. 206–214.

Donner, J. J., 1957. The geology and vegetation of Late-glacial retreat stages in Scotland: Royal Soc. Edinburgh, v. 63, p. 221–264.*

———, 1962, On the post-glacial history of the Grampian Highlands of Scotland: Soc. Sci. Fennica, Comment. Biol., v. 24, no. 6, 29p.

———, 1963, The late- and post-glacial raised beaches in Scotland II: Acad. Sci. Fennicae, Ann., Ser. A, sec. III, no. 68, p. 1–12.

———, 1964a, Pleistocene geology of eastern Long Island, New York: American Jour. Sci., v. 262, p. 355–376.

———, 1964b, The late-glacial and post-glacial emergence of south-western Finland: Soc. Sci. Fennica, Commentationes Physico-Math., v. 15, no. 5, 47 p.

———, 1965, The Quaternary of Finland: p. 199–272 in Rankama, Kalervo (ed.), The Quaternary, v. 1. Interscience, New York.*

———, 1969, Land/sea level changes in southern Finland during the formation of the Salpausselkä endmoraines: Geol. Soc. Finland Bull., v. 41, p. 135–150.

Dort, Wakefield, 1957, Striated surfaces on the upper parts of cirque headwalls: Jour. Geol. v. 65, p. 536–542.

———, 1960, Glacial Lake Coeur d'Alene and berg-rafted boulders: Idaho Acad. Sci. Jour., v. 1, p. 81–92.

———, 1962, Glaciation of the Coeur d'Alene district, Idaho: Geol. Soc. America Bull., v. 73, p. 889–906.

———, 1965a, Glaciation in Idaho—a summary of present knowledge: Tebiwa (Jour. Idaho State Univ. Mus.), v. 8, p. 29–37.*

Dort, Wakefield, 1965b, Nearby and distant origins of glacier ice entering Kansas: American Jour. Sci., v. 263, p. 598–605.

———, 1966, Nebraskan and Kansan stades: complexity and importance: Science, v. 154, p. 771–772.

Dott, R. H., 1969, Circum-Pacific late Cenozoic structural rejuvenation. Implications for sea floor spreading: Science, v. 166, p. 874–876.

Douglas, R. J. W., ed., 1970, Geology and economic minerals of Canada, 5th. ed.: Canada Geol. Survey, Econ. Geol. Ser., no. 1.

Dozy, J. J., 1938, Eine Gletscherwelt in Niederländisch-Neuguinea: Zeitschr. für Gletscherk., v. 26, p. 45–51.

Dreimanis, Aleksis, 1953, Studies of friction cracks along shores of Cirrus Lake and Kasakokwog Lake, Ontario: American Jour. Sci., v. 251, p. 769–783.

———, 1959, Measurements of depth of carbonate leaching in service of Pleistocene stratigraphy: Geol. Fören. i. Stockholm Förh., v. 81, p. 478–484.

———, 1968, Extinction of mastodons in eastern North America: testing a new climatic- environmental hypothesis: Ohio Jour. Sci., v. 68, p. 257–272.

———, and Vagners, U. J., 1970, Bimodal distribution of rock and mineral fragments in basal tills: [To be publ. 1970 in Till Sympos. vol, Columbus].

———, and others, 1957, Heavy mineral studies in tills of Ontario and adjacent areas: Jour. Sed. Petrology, v. 27, p. 148–161.

———, and others, 1966, The Port Talbot interstade of the Wisconsin glaciation: Canadian Jour. Earth Sci., v. 3, p. 305–325.*

Drewes, Harald, and others, 1961, Geology of Unalaska Island and adjacent insular shelf, Aleutian Islands, Alaska: U.S. Geol. Survey Bull. 1028, p. 583–676.

Drygalski, Erich v., and Machatschek, Fritz, 1942, Gletscherkunde: Franz Deuticke, Vienna, 258 p.*

DuBar, J. R., 1958, Neogene stratigraphy of southwestern Florida: Gulf Coast Assoc. Geol. Societies, Trans., v. 8, p. 129–155.*

Dücker, Alfred, 1969, Der Ablauf der Holstein-Warmzeit in Westholstein: Eiszeitalter und Geg., v. 20, p. 46–57.

Duigan, Suzanne, 1963, Pollen analyses of the Cromer Forest Bed series in East Anglia: Roy. Soc. London, Philos. Trans., ser. B, v. 246, p. 149–202.

Dunbar, C. O., and Rodgers, John, 1957, Principles of stratigraphy: John Wiley and Sons, New York, 356 p.*

Duplessy, J. C., and others, 1970, Differential isotopic fractionation in benthic Foraminifera and paleotemperatures reassessed: Science, v. 168, p. 250–251.

Durham, J. W., 1950, Cenozoic marine climates of the Pacific coast: Geol. Soc. America, Bull., v. 61, p. 1243–1264.

———, 1952, Early Tertiary marine faunas and continental drift: American Jour. Sci., v. 250, p. 321–343.

Dury, G. H., 1964–1965. [Underfit streams]: U.S. Geol. Survey Prof. Paper 452, p. A1–A67, B1–B56, C1–C43.*

Dutton, C. E., 1884, The effect of a warmer climate upon glaciers: American Jour. Sci., v. 27, p. 1–18.

Dylik, Jan, 1966, Problems of ice-wedge structures and frost-fissure polygons: Biuletyn Peryglacjalny, no. 15, p. 241–291.

———, and Maarleveld, G. C., 1967, Frost cracks, frost fissures and related polygons: Netherlands. Geol. Stichting, Mededel., n.s. no. 18, p. 7–21.

Dyson, J. L., 1952, Ice-ridged moraines and their relation to glaciers: American Jour. Sci., v. 250, p. 204–211.

———, 1962, The world of ice: Alfred A. Knopf, New York, 292 p.

Eardley, A. J., 1968, Bonneville chronology: correlation between the exposed stratigraphic record and the subsurface sedimentary succession: Geol. Soc. America Bull., v. 78, p. 907–909.

———, and Gvosdetsky, Vasyl, 1960, Analysis of Pleistocene core from Great Salt Lake, Utah: Geol. Soc. America Bull., v. 71, p. 1323–1344.

Easterbrook, D. J., 1963, Late Pleistocene glacial events and relative sea-level changes in the Northern Puget Lowland, Washington: Geol. Soc. America Bull., v. 74, p. 1465–1483.

———, 1964, Void ratios and bulk densities as means of identifying Pleistocene till: Geol. Soc. America Bull., v. 75, no. 8, p. 745–750.

———, and others, 1967, Pre-Olympia Pleistocene stratigraphy and chronology in the central Puget lowland, Washington: Geol. Soc. America Bull., v. 78, p. 13–20.

Eberl, Barthel, 1930, Die Eiszeitenfolge im nördlichen Alpenvorlande: Dr. Benno Filser, Augsburg, 427 p.

Edelman, Nils, 1949, Some morphological details of the roches moutonnées in the archipelago of SW Finland: Comm. Géol. de Finlande, Bull. 144, p. 129–137.

Einarsson, Trausti, 1964, On the question of Late-Tertiary or Quaternary land connections across the North Atlantic, and the dispersal of biota in that area: Jour. Ecol., v. 52, p. 617–625.

Elson, J. A., 1969, Late Quaternary marine submergence of Quebec: Rév. de Géogr. de Montréal, v. 23, p. 247–258.*

Embleton, Clifford, and King, C.A.M., 1968, Glacial and periglacial geomorphology: St. Martin's Press, New York, 608 p.*

Emery, K. O., 1949, Topography and sediments of the Arctic basin: Jour. Geol., v. 57, p. 512–521.

———, 1960, The sea off southern California—a modern habitat of petroleum: John Wiley & Sons, New York, 366 p.

———, 1967, The Atlantic continental margin of the United States during the past 70 million years: Geol. Assoc. Canada, Spec. Paper 4, p. 53–70.

———, 1968, Relict sediments on continental shelves of world: American Assoc. Petroleum Geologists Bull., v. 52, p. 445–464.*

———, and others, 1965, A submerged peat deposit off the Atlantic coast of the United States: Limnol. and Oceanogr., v. 10, suppl., p. R97–R102.

Emiliani, Cesare, 1955, Pleistocene temperatures: Jour. Geol., v. 63, p. 538–578.

———, 1966a, Paleotemperature analysis of Caribbean cores P6304–8 and P6304–9 and a generalized temperature curve for the past 425,000 years: Jour. Geol. v. 74, p. 109–126.

———, 1966b, Isotopic paleotemperatures: Science, v. 154, p. 851–857.

———, 1967, The generalized temperature curve for the past 425,000 years; a reply: Jour. Geol., v. 75, p. 504–510.

———, 1970, Pleistocene paleotemperatures: Science, v. 168, p. 822–825.

———, and others, 1961, Paleotemperature analysis of the Plio-Pleistocene section at Le Castella, Calabria, southern Italy: Geol. Soc. America Bull., v. 72, p. 679–688.

Enquist, Fredrik, 1918, Die glaziale Entwicklungs-geschichte Nordwest-skandina-viens: Sveriges Geol. Undersökn., ser. C, no. 285, Årsbok v. 12, no. 2, 142 p.

Epstein, Samuel, and others, 1970, Antarctic Ice Sheet: Stable isotope analyses of Byrd Station cores and interhemispheric climatic implications: Science, v. 168, p. 1570–1572.

Ergenzinger, Peter, 1967, Die eiszeitliche Vergletscherung des Bayerischen Waldes: Eiszeitalter und Geg., v. 18, p. 152–168.

Ericson, D. B., and Wollin, Goesta, 1956, Correlation of six cores from the equatorial Atlantic and the Caribbean: Deep-Sea Res., v. 3, p. 104–125.

——, 1968, Pleistocene climates and chronology in deep-sea sediments: Science, v. 162, p. 1227–1234.

——, 1970, Pleistocene climates in the Atlantic and Pacific Oceans, a comparison based on deep-sea sediments: Science, v. 167, p. 1483–1485.

Ericson, D. B., and others, 1961, Atlantic deep-sea sediment cores: Geol. Soc. America Bull., v. 72, p. 193–286.*

Eriksson, K. G., 1960, Studier över Stockholmsåsen vid Halmsjön: Geol. Fören. i Stockholm förh., v. 82, p. 43–125 (Engl. summ., p. 120–123).

Erinç Sirri, 1954, The Pleistocene history of the Black Sea and the adjacent countries with special reference to the climatic changes: Univ. of Istanbul, Geogr. Inst., Rev. (Internat. Ed.), no. 1, 51 p.*

Eskola, Pentti, 1933, Tausend Geschiebe aus Lettland: Ann. Acad. Sci. Fennicae, ser. A, v. 39, no. 5, 41 p.

Evernden, J. F., and others, 1964, Potassium-argon dates and the Cenozoic mammalian chronology of North America: American Jour. Sci., v. 262, p. 145–198.*

Ewing, John, and others, 1963, Upper stratification of Hudson Apron region: Jour. Geophys. Res., v. 68, p. 6303–6316.

Ewing, Maurice, and Donn, W. L., 1956, A theory of ice ages: Science, v. 123, p. 1061–1066.

——, 1958, A theory of ice ages II: Science v. 127, p. 1159–1162.

Faegri, Knut, and Iversen, Johannes, 1964, Textbook of modern pollen analysis: 2nd ed., Ejnar Munksgaard, København, Hafner Pub. Co., New York, 237 p.*

Fairbridge, R. W., 1961, Eustatic changes in sea level: 99–185 in Physics and Chemistry of the Earth, v. 4.

——, ed., 1961, Solar variations, climatic change, and related geophysical problems: New York Acad. Sci., Annals, v. 95, p. 1–740.*

——, and Newman, W. S., 1968, Postglacial crustal subsidence of the New York area: Zeitschr. für Geomorph., v. 12, p. 296–317.*

Fairchild, H. L., 1896, Kame areas in western New York: Jour. Geol., v. 4, p. 129–159.

Falconer, G., and others, 1965, Major end moraines in eastern and central Arctic Canada: Geog. Bull., v. 7, p. 137–153.

Farrand, W. R., 1962, Postglacial uplift in North America: American Jour. Sci., v. 260, p. 181–199.*

——, 1964, The deglacial hemicycle: Geol. Rundschau, v. 54, p. 385–398.*

——, and Gajda, R. T., 1962, Isobases on the Wisconsin marine limit in Canada: Geogr. Bull., v. 17, p. 5–22.

Faure, Hugues, 1966, Evolution des grands lacs sahariens à l'Holocène: Quaternaria, v. 8, p. 167–175.

————, 1969, Lacs quaternaries du Sahara: Internat. Verein. Limnol., Mitt., v. 17, p. 131–146.

Ferguson, C. W., 1968, Bristlecone pine: Science and esthetics: Science, v. 159, p. 839–846.

Fernald, M. L., 1925, The persistence of plants in unglaciated areas of boreal North America: American Acad. Arts & Sci., Mem., v. 15, p. 237–342.*

Feruglio, Egidio, 1949–1950, Descripción geológica de la Patagonia: Argentina. Dirección. Gen. Yacimientos Petrolíferos Fiscales. v. 1, 1949, 334 p.; v. 2, 1949, 349 p.; v. 3, 1950, 431 p.*

Feyling-Hanssen, R. W., 1955, Stratigraphy of the marine late-Pleistocene of Billefjorden, Vestspitsbergen: Norsk Polarinstitutt (Oslo) Skrifter, no. 107, 186 p.*

————, and Olsson, Ingrid, 1960, Five radiocarbon datings of Post Glacial shorelines in Central Spitsbergen: Norsk Geogr. Tidsskr., v. 17, p. 122–131.

Fidalgo, Francisco, and Riggi, J. C., 1965, Los rodados Patagónicos en la Meseta del Guenguel y alrededores (Santa Cruz): Asoc. Geol. Argentina, Rev., tomo 20, p. 273–325.*

Field, W. O., 1947, Glacier recession in Muir Inlet, Glacier Bay, Alaska: Geogr. Rev., v. 37, p. 369–399.

Firbas, Franz, 1949–1952, Spät- und nacheiszeitliche Waldgeschichte Mitteleuropas nördlich der Alpen: Gustav Fischer, Jena, v. 1, 480 p., v. 2, 256 p.* Engl. summ. in Firbas, New Phytol., v. 49, 1950, p. 163.*

Fisk, H. N., 1938, Geology of Grant and La Salle Parishes [Louisiana]: Louisiana Geol. Survey, Geol. Bull. 10, 246 p.

————, 1944, Geological investigation of the alluvial valley of the lower Mississippi River: United States War Dept., Corps of Engineers (Miss. R. Comm.), 78 p. + plates.

————, 1951, Loess and Quaternary geology of the lower Mississippi Valley: Jour. Geol., v. 59, p. 333–356.

Fleming, C. A., 1962a, New Zealand biogeography. A paleontologist's approach: Tuatara (Wellington), v. 10, p. 53–108.*

————, 1962b, The extinction of moas and other animals during the Holocene period: Notornis (Wellington) v. 10, p. 113–117.

Fleming, W. L. S., 1935, Glacial geology of central Long Island: American Jour. Sci., v. 30, p. 216–238.

Flint, R. F., 1928, Eskers and crevasse fillings: American Jour. Sci., v. 15, p. 410–416.

————, 1937, Pleistocene drift border in eastern Washington: Geol. Soc. America Bull., v. 48, p. 203–232.

————, 1940a, Late Quaternary changes of level in western and southern Newfoundland: Geol. Soc. America Bull., v. 51, p. 1757–1780.

————, 1940b, Pleistocene features of the Atlantic Coastal Plain: American Jour. Sci., v. 238, p. 757–787.*

————, 1943, Growth of the North American ice sheet during the Wisconsin Age: Geol. Soc. America Bull., v. 54, p. 325–362.*

————, 1947, Glacial geology and the Pleistocene Epoch: John Wiley and Sons, New York, 589 p.*

————, 1948, Glacial geology and geomorphology [of parts of East Greenland]: p. 90–210 in Boyd, L. A., The coast of Northeast Greenland. [The Louise A. Boyd Arctic Expeditions of 1937 and 1938]: Am. Geogr., Soc., Spec. Publ. 30.*

Flint, R. F., 1949, Leaching of carbonates in glacial drift and loess as a basis for age correlation: Jour. Geology, v. 57, p. 297–303.

――――, 1955a, Rates of advance and retreat of the margin of the late-Wisconsin ice sheet: American Jour. Sci., v. 253, p. 249–255.

――――, 1955b, Pleistocene geology of eastern South Dakota: U.S. Geol. Survey Prof. Paper 262, 173 p.*

――――, 1957, Glacial and Pleistocene geology: John Wiley and Sons, New York, 553 p.*

――――, 1959a, Pleistocene climates in eastern and southern Africa: Geol. Soc. America Bull., v. 70, p. 343–374.

――――, 1959b, On the basis of Pleistocene correlation in East Africa: Geol. Mag., v. 96, p. 265–284.

――――, 1961, Two tills in southern Connecticut: Geol. Soc. America Bull., v. 72, p. 1687–1691.

――――, 1965, The Pliocene-Pleistocene boundary: p. 497–533 in Wright, H. E., and Frey, D. G., eds., International Studies on the Quaternary. Geol. Soc. America, Spec. Paper 84.*

――――, 1966, Comparison of interglacial marine stratigraphy in Virginia, Alaska, and Mediterranean areas: American Jour. Sci., v. 264, p. 673–684.

――――, and Bond, Geoffrey, 1968, Pleistocene sand ridges and pans in western Rhodesia: Geol. Soc. America Bull., v. 79, p. 299–314.

――――, and Brandtner, Friedrich, 1961, Climatic changes since the last interglacial: American Jour. Sci., v. 259, p. 321–328.

――――, and Denny, C. S., 1958, Quaternary geology of Boulder Mountain, Aquarius Plateau, Utah: U.S. Geol. Survey Bull. 1061, p. 103–164.

――――, and Dorsey, H. G., 1945, Glaciation of Siberia: Geol. Soc. America Bull., v. 56, p. 89–106.

――――, and Fidalgo, Francisco, 1964, Glacial geology of the east flank of the Argentine Andes between Latitude 39°10′S. and Latitude 41°20′S.: Geol. Soc. America Bull., v. 75, p. 335–352.

――――, 1969, Glacial drift in the eastern Argentine Andes between latitude 41°10′S. and latitude 43°10′S: Geol. Soc. America Bull., v. 80, p. 1043–1052.*

Flint, R. F., and Gale, W. A., 1958, Stratigraphy and radiocarbon dates at Searles Lake, California: American Jour. Sci., v. 256, p. 689–714.

――――, and Rubin, Meyer, 1955, Radiocarbon dates of pre-Mankato events in eastern and central North America: Science, v. 121, p. 649–658.

――――, and others, 1942, Glaciation of Shickshock Mountains, Gaspé Peninsula: Geol. Soc. America Bull., v. 53, p. 1211–1230.*

――――, 1945, Glacial map of North America: Geol. Soc. America, Spec. Paper 60, pt. 1—Glacial map (79 × 52½ in.); pt. 2—Explanatory notes, 37 p.*

――――, 1959, Glacial map of the United States east of the Rocky Mountains: Scale 1:1,750,000: Geol. Soc. America, 2 sheets.

――――, 1960a, Symmictite: A name for nonsorted terrigenous sedimentary rocks that contain a wide range of particle sizes: Geol. Soc. America Bull., v. 71, p. 507–510.

――――, 1960b, Diamictite, a substitute term for symmictite: Geol. Soc. America Bull., v. 71, p. 1809–1810.

Flohn, Hermann, 1953, Studien zur atmosphärischen Zirkulation in der letzten Eiszeit: Erdkunde, v. 7, p. 266–275.

————, 1963, Zur meteorologischen Interpretation der pleistozänen Klimaschwankungen: Eiszeitalter und Geg., v. 14, p. 153–160.

Flohn, Hermann, 1969a, Ein geophysikalisches Eiszeit-Modell: Eiszeitalter und Gegenwart, v. 20, p. 204–231.*

Flohn, Hermann, ed., 1969b, General climatology, 2: Elsevier Pub. Co., Amsterdam; New York, 266 p.*

Forbes, Edward, 1846, On the connexion between the distribution of the existing fauna and flora of the British Isles, and the geological changes which have affected their area, especially during the epoch of the Northern Drift: Great Britain Geol. Survey, Mem., v. 1, p. 336–432.

Foster, Helen, and Holmes, G. W., 1965, A large transitional rock glacier in the Johnson River area, Alaska Range: U.S. Geol. Survey Prof. Paper 525, p. B112–B116.

Frechen, J., and Lippolt, H. J., 1965, Kalium-Argon-Daten zum Alter des Laacher Vulkanismus, der Rheinterrassen und der Eiszeiten: Eiszeitalter und Geg., v. 16, p. 5–30.

Fredén, Curt, 1967, A historical review of the Ancylus Lake and the Svea River: Geol. Fören. i Stockholm Förh., v. 89, p. 239–267.

Frenguelli, Joaquin, 1950, Rasgos generales de la morfología y la geología de la Provincia de Buenos Aires: La Plata. Laboratorio de Ensayo de Materiales e Investigaciones Tecnológicas, ser. 2, no. 33, 72 p.

Frenzel, Burkhard, 1967, Die Klimaschwankungen des Eiszeitalters: Vieweg, Braunschweig, 291 p.*

————, 1968, The Pleistocene vegetation of northern Eurasia: Science, v. 161, p. 637–649.*

Fristrup, Børge, 1966, The Greenland Ice Cap: Rhodos, Copenhagen, 312 p.*

Fritts, H. C., 1966, Growth-rings of trees: their correlation with climate: Science, v. 154, p. 973–979.

Fromm, Erik, 1938, Geochronologisch datierte Pollendiagramme und Diatoméenanalysen aus Ångermanland: Geol. Fören. i Stockholm Förh., v. 60, p. 365–381.

Frye, J. C., and Leonard, A. B., 1963, Pleistocene geology of Red River basin in Texas: Texas Bur. Econ. Geol., Rept. Inv. 49, 48 p.

Frye, J. C., and others, 1968, Mineral zonation of Woodfordian loesses of Illinois: Illinois Geol. Survey Circ. 427, 44 p.

Fuchs, V. E., 1950, Pleistocene events in the Baringo Basin, Kenya Colony: Geol. Mag., v. 87, p. 149–174.

Fuller, M. L., 1914, The geology of Long Island, New York: U. S. Geol. Survey Prof. Paper 82, 231 p.*

Fulton, R. J., 1967, Deglaciation studies in Kamloops Region, an area of moderate relief, British Columbia: Canada Geol. Survey Bull. 154, 36 p.

Furness, F. N., ed., 1961, Solar variations, climatic change, and related geophysical problems: New York Acad. Sci., Ann., v. 95, p. 1–740.*

Fyles, J. G., 1963, Surficial geology of Horne Lake and Parksville map-areas. Vancouver Island, British Columbia: Canada Geol. Survey Mem. 318, 142 p.*

Gadd, N. R., 1960, Surficial geology of the Bécancour map-area, Quebec: Canada Geol. Survey Paper 59-8, 34 p.

————, 1961, Surficial geology of the Ottawa area report of progress: Canada Geol. Survey Paper 61-19, 14 p.

Gage, Maxwell, 1961, On the definition, date, and character of the Ross glaciation, early Pleistocene, New Zealand: Roy. Soc. New Zealand, Trans., v. 88, p. 631–637.

Galloway, R. W., 1956, The structure of moraines in Lyngsdalen, north Norway: Jour. Glaciol., v. 2, p. 730–733.

――――, 1961, Periglacial phenomena in Scotland: Geograf. Annaler, v. 43, p. 348–353.

――――, 1963, Glaciation in the snowy mountains: a re-appraisal: Linnean Soc. New South Wales, Proc., v. 88, p. 180–198.

――――, 1965, Late Quaternary climates in Australia: Jour. Geol., v. 73, no. 4, p. 603–618.*

Galon, Rajmund, ed., 1968, Last Scandinavian glaciation in Poland: Poland. Acad. Sci., Geogr. Studies No. 74, 216 p.*

Garbell, M. A., 1947, Tropical and equatorial meteorology: Pitman Publ. Corp., New York, 237 p.

Garwood, E. J., 1932, Speculation and research in Alpine glaciology: an historical review: Geol. Soc. London, Quart. Jour., v. 88, p. xciii–cxviii.*

Geikie, Archibald, 1863, On the phenomena of the glacial drift of Scotland: Geol. Soc. Glasgow, Trans., v. 1, 190 p.

Geikie, James, 1874, The great ice age and its relation to the antiquity of man: 1st ed., W. Isbister, London, 575 p.

――――, 1894, The great ice age and its relation to the antiquity of man: 3rd ed., Stanford, London, 850 p.

Geitzenauer, K. R., 1969, The Pleistocene calcareous nannoplankton of the subantarctic Pacific Ocean: Florida State Univ., Tallahassee, Ph. D. dissertation.

Gellert, J. F., ed., 1965, Die Weichsel-Eiszeit im Gebiet der Deutschen Demokratischen Republik: Berlin, Akademie-Verlag, 261 p.*

George, T. N., and others, 1969, Recommendations on stratigraphical usage: Geol. Soc. London, Proc. for 1969, p. 139–166.*

Gerasimov, I. P., ed., 1965, Last European glaciation: U.S.S.R. Acad. Sci., Nauka, Moscow, 220 p.*

Getz, L. L., and Hibbard, C., 1965, A molluscan faunule from the Seymour Formation of Baylor and Knox counties, Texas: Michigan Acad. Sci., Arts, and Lett., v. 50, p. 275–297.

Giegengack, R. F., 1968, Late-Pleistocene history of the Nile valley in Egyptian Nubia: Yale Univ., New Haven, Ph.D. dissert., 185 p.*

Gignoux, Maurice, 1950, Géologie stratigraphique: 4th ed., Masson & Cie., Paris, 735 p.*

――――, 1952, Pliocène et Quaternaire marins de la Méditerranée occidentale: Internat. Geol. Congr., 19th, Alger 1952, Rept., pt. 15, p. 249–258.

Gilbert, G. K., 1890, Lake Bonneville: U.S. Geol. Survey, Mon. 1, 438 p.

――――, 1906a, Crescentic gouges on glaciated surfaces: Geol. Soc. America Bull., v. 17, p. 303–314.

――――, 1906b, Moulin work under glaciers: Geol. Soc. America Bull., v. 17, p. 317–320.

Giles, A. W., 1918, Eskers in vicinity of Rochester, New York: Rochester Acad. Sci., Proc., v. 5, p. 161–240.*

Gill, E. D., 1961, Cainozoic climates of Australia: New York Acad. Sci., Ann., v. 95, p. 461–464.*

————, 1965, The paleogeography of Australia in relation to the migrations of marsupials and men: New York Acad. Sci., Trans., ser. 2, v. 28, p. 5–14.

Gill, W. D., 1952, The stratigraphy of the Siwalik series in the northern Potwar, Punjab, Pakistan: Geol. Soc. London Quart. Jour., v. 107, p. 375–394.

Gillberg, Gunnar, 1955, Den glaciala utvecklingen inom Sydsvenska höglandets västra randzon. I. Glacialerosion och moränackumulation: Geol. Fören. i Stockholm Förh., v. 77, p. 481–524.

————, 1965, Till distribution and ice movements on the northern slopes of the south Swedish highlands: Geol. Fören. i Stockholm Förh., v. 86, p. 433–484.

————, 1967, Further discussion of the lithological homogeneity of till: Geol. Fören. i Stockholm Förh., v. 89, p. 29–49.

————, 1968, Lithological distribution and homogeneity of glaciofluvial material: Geol. Fören. i Stockholm Förh., v. 90, p. 189–204.

————, 1969, A great till section on Kinnekulle, W Sweden: Geol. Fören. i Stockholm, Förh., v. 91, p. 313–342.

Gillispie, C. C., 1951, Genesis and geology. A study in the relations of scientific thought, natural theology, and social opinion in Great Britain, 1790–1850: Harvard Univ. Press, Cambridge, 315 p.

Giterman, R. E., and others, 1968, The main development stages of the vegetation of north Asia in Anthropogen: USSR. Acad. Sci., Geol. Inst., Trans., v. 177, 270 p.[R]

Gleason, H. A., and Cronquist, Arthur, 1964, The natural geography of plants: Columbia Univ. Press, New York, 420 p.

Glen, J. W., and others, 1957, On the mechanism by which stones in till become oriented: American Jour. Sci., v. 255, p. 194–205.

Godwin-Austen, R.A.C., 1851, On the superficial accumulation of the coasts of the English Channel . . . : Geol. Soc. London, Quart. Jour., v. 7, p. 118–136.

Goldthwait, J. W., 1924, Physiography of Nova Scotia: Canada Geol. Survey Mem. 140, 179 p.

————, 1925, The geology of New Hampshire: New Hampshire Acad. Sci., Handbk. no. 1, 86 p.

————, and Kruger, F. C., 1938, Weathered rock in and under the drift in New Hampshire: Geol. Soc. America Bull., v. 49, p. 1183–1198.

Goldthwait, R. P., 1940, Geology of the Presidential Range: New Hampshire Acad. Sci., Bull. 1, p. 1–41.

————, 1951, Development of end moraines in east-central Baffin Island: Jour. Geol., v. 59, p. 567–577.

————, 1958, Wisconsin Age forests in western Ohio; I. Age and glacial events: Ohio Jour. Sci., v. 58, p. 209–230.

————, 1959, Scenes in Ohio during the last Ice Age: Ohio Jour. Sci., v. 59, p. 193–216.

————, 1963, Dating the Little Ice Age in Glacier Bay, Alaska: Internat. Geol. Congr., 21st, Norden 1960, Rept., pt. 27, p. 37–46.

————, 1970, Mountain glaciers of the Presidential Range in New Hampshire: Arctic and Alpine Res., v. 2, p. 85–102.

————, and others, 1961, Glacial map of Ohio: U. S. Geol. Survey Misc. Geol. Invs., Map I–316.

Gonzalez, Enrique, and others, 1965, Late Quaternary glacial and vegetational sequence in Valle de Lagunillas, Sierra Nevada del Cocuy, Colombia:

Leidse Geol. Mededel., deel 32, p. 157–182.

Goodchild, J. G., 1875, The glacial phenomena of the Eden valley, etc: Geol. Soc. London, Quart. Jour., v. 31, p. 55–99.

Goodell, H. G., and others, 1968, The Antarctic glacial history recorded in sediments of the Southern Ocean: Palaeogeogr., Palaeoclimatol., Palaeoecol., v. 5, p. 41–62.*

Gooding, A. M., 1966, The Kansan glaciation in southestern Indiana: Ohio Jour. Sci., v. 66, p. 426–433.

Goretskiy, G. I., and Kriger, N. I., eds., 1967, Lower Pleistocene of the glaciated regions of the Russian Plain: USSR Acad. Sci., Comm. on Study of Quater. Period, Nauka, Moscow, 183 p.*ᴿ

Grahmann, Rudolf, 1932, Der Löss in Europa: Gesell. für Erdkunde zu Leipzig, Mitt., 1930–1931, p. 5–24.*

Gravenor, C. P., 1951, Bedrock source of tills in southwestern Ontario: American Jour. Sci., v. 249, p. 66–71.

———, and Kupsch, W. O., 1959, Ice-disintegration features in western Canada: Jour. Geol., v. 67, p. 48–64.

Gregory, J. W., 1931, A deep trench on the floor of the North Sea: Geogr. Jour., v. 77, p. 548–554.

Grichuk, V. P., and others, 1965, Report of. the Subcommission on the Plio-Pleistocene boundary: Internat. Quater. Congr., 6th, Warsaw 1961, Rept., v. 1, p. 311–329.

Grindley, G. W., 1967, The geomorphology of the Miller Range, Transantarctic Mountains, with notes on the glacial history and neotectonics of East Antarctica: New Zealand Jour. Geol. and Geophys., v. 10, p. 557–598.*

Gromov, V. I., and Nikiforova, K. V., 1969, Riss-Würm auf dem Gebiet der UdSSR: Deutsch. Ges. geol. Wiss., Ber., sec. A, v. 14, p. 471–475.

———, eds., 1969, The main problems of Anthropogen geology in Eurasia: Nauka, Moscow, 130 p.ᴿ

Gromov, V. I., and others, 1969, Scheme of subdivision of the Anthropogene: Commission on subdivisions of the Quaternary Period, Bull., no. 36, p. 41–55.ᴿ

Grossman, Karl, and Lomas, Joseph, 1895, On the glaciation of the Faröe Islands: Glacialists' Mag. (Liverpool), v. 3, p. 1–15.

Grosval'd, M. G., and Kotlyakov, V. M., 1969, Present-day glaciers in the U.S.S.R. and some data on their mass balance: Jour. Glaciol., v. 8, p. 9–22.*

Grötzbach, Erwin, and Rathjens, Carl, 1969, Die heutige und die jungpleistozäne Vergletscherung des Afghanischen Hindukusch: Zeitschr. für Geomorphol. Supplbd. 8, p. 58–75.

Guilday, J. E., 1971, The Pleistocene history of the Appalachian mammal fauna:* in P.C. Holt, ed., Distributional history of the biota of the Southern Appalachians, Virginia Polytechnic Institute Res. Div. Monogr, pt 3. (In course of publication.)

———, and others, 1964, A Pleistocene cave deposit in Bedford County, Pennsylvania: Nat. Speleol. Soc., Bull., v. 26, p. 121–194.

Gunn, C. B., 1968, A descriptive catalog of the drift diamonds of the Great Lakes region, North America: Gems and Gemology, 1968, p. 297–303, 333–334.

Gutenberg, Beno, 1941, Changes in sea level, post-glacial uplift, and mobility of the Earth's interior: Geol. Soc. America Bull., v. 52, p. 721–772.*

Guthrie, R. D., 1968, Paleoecology of a Late Pleistocene small mammal community from interior Alaska: Arctic, v. 21, p. 221–244.

Hack, J. T., 1941, Dunes of the western Navajo Country: Geogr. Rev., v. 31, p. 240–263.

———, 1957, Submerged river system of Chesapeake Bay: Geol. Soc. America Bull., v. 68, p. 817–830.

Hafsten, Ulf. 1964, A standard pollen diagram for the southern High Plains, U.S.A., covering the period back to the Early Wisconsin Glaciation: Internat. Quaternary Congr., 6th, Warsaw 1961, Rept., v. 2, p. 407–420.

———, 1969, A proposal for a synchronous sub-division of the late Pleistocene Period having global and universal applicability: Nytt Mag. for Botanikk, v. 16, p. 1–13.

Häkli, T. A., and Kerola, P., 1966, A computer program for boulder train analysis: Soc. Géol. Finlande, Comptes Rendus, no. 38, p. 219–235.

Hamilton, R. A., and others, 1956, British North Greenland Expedition 1952–4: Scientific results: Geogr. Jour., v. 122, p. 203–239.

Hammen, Thomas van der, 1963, A palynological study on the Quaternary of British Guiana: Leidse Geol. Mededel., v. 29, p. 125–180.

———, and Gonzalez, Enrique, 1960, Upper Pleistocene and Holocene climate and vegetation of the "Sabana de Bogotá" (Colombia, South America): Leidse Geol. Mededel., v. 25, p. 261–315.

———, and Vogel, J. C., 1966, The Susacá-interstadial and the subdivision of the Late-glacial: Geologie en Mijnbouw, v. 45, p. 33–35.

———, and others, 1967, Stratigraphy, climatic succession and radiocarbon dating of the last glacial in the Netherlands: Geologie en Mijnbouw, v. 46, p. 79–95.*

Hanna, S. R., 1969, The formation of longitudinal sand dunes by large helical eddies in the atmosphere: Jour. Applied Meteorol., v. 8, p. 874–883.

Hansen, A. M., 1894, The glacial succession in Norway: Jour. Geol., v. 2, p. 123–144.

Hansen, Bert, 1970, The early history of glacial theory in British geology: Jour. Glaciol., v. 9, p. 135–141.

Hansen, Edward, and others, 1961, Décollement structures in glacial-lake sediments: Geol. Soc. America Bull., v. 72, p. 1415–1418.

Hansen, Sigurd, 1940, Varvity in Danish and Scanian late-glacial deposits; with special reference to the system of ice-lakes at Egernsund: Danmarks Geol. Undersøg., ser. 2, no. 63, 478 p.

———, 1965, The Quaternary of Denmark: p. 1–90 in Rankama, Kalervo, ed., The Quaternary, v. 1, Interscience, New York, 300 p.*

Hardy, C. T., and Muessig, Siegfried, 1952, Glaciation and drainage changes in the Fish Lake Plateau, Utah: Geol. Soc. America Bull., v. 63, p. 1109–1116.

Harland, W. B., and others, 1966. The definition and identification of tills and tillites: Earth-Sci. Rev., v. 2, p. 225–256.*

Harris, A. H., and Findley, J.S., 1964, Pleistocene-Recent fauna of the Isleta Caves, Bernalillo County, New Mexico: American Jour. Sci., v. 262, p. 114–120.

Harris, S. E., 1943, Friction cracks and the direction of glacial movement: Jour. Geol., v. 51, p. 244–258.*

Harrison, P. W., [Wyman], 1957, A clay-till fabric: its character and origin: Jour. Geol. v. 65, p. 275–308.*

Harrison, Wyman, 1958, Marginal zones of vanished glaciers reconstructed from the preconsolidation-pressure values of overridden silts: Jour. Geol. v. 66, p. 72–95.

———, 1960, Original bedrock composition of Wisconsin till in central Indiana: Jour. Sed. Petrology, v. 30, p. 432–446.

Hartshorn, J. H., 1958, Flowtill in southeastern Massachusetts: Geol. Soc. America Bull., v. 69, p. 477–482.

Haselton, G. M., 1966, Glacial geology of Muir Inlet, southeastern Alaska: Ohio State Univ., Inst. Polar Stud., Rept 18, 34 p.

Hatai, Kotora, ed., 1967, Tertiary correlations and climatic changes in the Pacific: Pacific Sci. Congr., 11th, Tokyo, 1966, 102 p.*

Hatherton, Trevor, ed., 1965, Antarctica: Methuen and Co., London, 511 p.*

Hattersley-Smith, Geoffrey, 1960, Some remarks on glaciers and climate in northern Ellesmere Island: Geograf. Annaler, v. 42, p. 44–46.

Haug Émile, 1907–1911, Traité de géologie: Armand Colin, Paris, 2024 p.; v. 1, 1907; v. 2, 1908–1911.

Havens, J. M., 1964, Climatological notes from Axel Heiberg Island, N.W.T., Canada: Arctic, v. 17, p. 261–263.

Hay, O. P., 1923, The Pleistocene of North America and its vertebrated animals from the States east of the Mississippi River and from the Canadian provinces east of longitude 95°: Carnegie Inst. Washington, Publ. no. 322, 499 p.*

Haynes, C. V., 1969, The earliest Americans: Science, v. 166, p. 709–715.

Hays, J. D. & Berggren W. A., Quaternary boundaries (unpublished).

Hays, J. D., and others, 1969, Pliocene-Pleistocene sediments of the equatorial Pacific. Their paleomagnetic, biostratigraphic, and climatic record: Geol. Soc. America Bull., v. 80, p. 1481–1514.

Heer, Oswald, 1858, Die Schieferkohlen von Uznach und Dürnten: Orell Füszli and Co., Zürich, 40 p.

———, 1865, Die Urwelt der Schweiz: F. Schulthess, Zürich, 622 p.

Heide, Simon van der, and Zagwijn, W. H., 1967, Stratigraphical nomenclature of the Quaternary deposits in the Netherlands: Netherlands. Geol. Stichting, Mededel., n.s. no. 18, p. 23–29.

Heim, Albert, 1919–1922, Geologie der Schweiz: C. H. Tauchnitz, Leipzig, v. 1, 1919, 704p.; v. 2, pt. 1, 1921, p. 1–476; v. 2, pt. 2, 1922, p. 477–1018.

Heim, G. E., and Howe, W. B., 1963, Pleistocene drainage and depositional history in northwestern Missouri: Kansas Acad. Sci., v. 66, p. 378–392.

Hellaakoski, Aaro, 1931, On the transportation of materials in the esker of Laitila: Fennia, v. 52, no. 7, 41 p.

Henderson, E. P., 1959a, A glacial study of central Quebec-Labrador: Canada Geol. Survey Bull. 50, 94 p.*

———, 1959b, Surficial geology of Sturgeon Lake Map-Area, Alberta: Canada Geol. Survey Mem. 303, 108 p.

Hendy, C. H., and Wilson, A. T., 1968, Palaeoclimatic data from speleothems: Nature, v. 219, p. 48–51.

Herm, Dietrich, and Paskoff, Roland, 1967, Vorschlag zur Gliederung des marinen Quartärs in Nord- und Mittel-Chile: Neues Jahrb. für Geol. und Paläont., Monatsh. 10, p. 577–588.

Herman, Yvonne, 1970, Arctic paleo-oceanography in late Cenozoic time: Science, v. 169, p. 474–477.

Hershey, O. H., 1896, Ozarkian epoch—a suggestion: Science, v. 3, p. 620–622.

————, 1897, Eskers indicating stages of glacial recession in the Kansan epoch in northern Illinois: American Geologist, v. 19, p. 197–209, 237–253.

Heusser, C. J., 1955, Pollen profiles from the Queen Charlotte Islands, British Columbia: Canadian Jour. Bot., v. 33, p. 429–449.*

————, 1960. Late-Pleistocene environments of North Pacific North America: American Geogr. Soc., Spec. Publ. 35, 308 p.*

————, 1963, Pollen diagrams from three former cedar bogs in the Hackensack tidal marsh, northeastern New Jersey: Torrey Bot. Club Bull., v. 90, p. 16–28.

————, 1964, Palynology of four bog sections from the western Olympic Peninsula, Washington: Ecology, v. 45, no. 1, p. 23–40.

————, 1966, Late-Pleistocene pollen diagrams from the Province of Llanquihue, southern Chile: American Philos. Soc., Proc., v. 110, p. 269–305.*

————, and others, 1954, Geobotanical studies on the Taku Glacier anomaly: Geogr. Rev., v. 44, p. 224–239.

Hey, R. W., 1963, Pleistocene screes in Cyrenaica (Libya): Eiszeitalter und Geg., v. 14, p. 77–84.

Hibbard, C. W., 1955, Pleistocene vertebrates from the upper Becerra (Becerra Superior) Formation, valley of Tequixquiac, Mexico, with notes on other Pleistocene forms: Univ. of Michigan Mus. Paleontol., Contribs., v. 12, p. 47–96.

————, 1956, Vertebrate fossils from the Meade Formation of southwestern Kansas: Michigan Acad. Sci., Arts, and Lett., Papers, v. 41, p. 145–200.

————, 1958, Summary of North American Pleistocene local faunas: Michigan Acad. Sci., Arts, and Lett., Papers, v. 43, p. 1–32.*

————, and Taylor, D. W., 1960, Two late Pleistocene faunas from southwestern Kansas: Univ. of Michigan Mus. Paleont., Contribs., v. 16, p. 1–223.

Hickox, C. F., 1962, Pleistocene geology of the Central Annapolis Valley, Nova Scotia: Nova Scotia Dept. Mines Mem. 5, 36 p.

Higgins, C. G., 1965, Causes of relative sea-level changes: American Scientist, v. 53, p. 464–476.

Hill, M. N., ed., 1962–1963, The sea: Interscience Publishers, New York and London, v. 1, 1962, 864p, v. 2, 1963, 554p., v. 3, 1963, 963p.*

Hitchcock, C. H., 1878, Glacial drift [in New Hampshire]: Concord: The geology of New Hampshire. . . , pt. 3 [v. 3], p. 177–338.

Hitchcock, Edward, 1841, First Anniversary Address before the Association of American Geologists: American Jour. Sci., v. 41, p. 232–275.

Hjulström, Filip, and others, 1955, Problems concerning the deposits of wind-blown silt in Sweden: Geograf. Annaler, v. 37, p. 86–117.

Ho, T. Y., and others, 1969, Radiocarbon dating of petroleum-impregnated bone from tar pits at Rancho La Brea, California: Science, v. 164, p. 1051–1052.

Hobbs, W. H., 1910, The cycle of mountain glaciation: Geogr. Jour., v. 36, p. 146–163, 268–284.

Hoffstetter, Robert, 1952, Les mammifères pléistocènes de la République de l'Équateur: Soc. Géol. de France, Mém., n.s., v. 31, no. 66, 391 p.*

————, 1963, La faune pleistocène de Tarija (Bolivie). Note preliminaire: Mus.

Nat. d'Histoire Naturelle (Paris), Bull, v. 35, p. 194–203.

Högbom, Bertil, 1914, Über die geologische Bedeutung des Frostes: Univ. Uppsala, Geol. Inst. Bull., v. 12, p. 257–389.*

Högbom, Ivar, 1923, Ancient inland dunes of northern and middle Europe: Geograf. Annaler, v. 5, p. 113–242.*

Hoinkes, H. C., 1968, Glacier variation and weather: Jour. Glaciol., v. 7, p. 3–19.

Hole, F. D., 1943, Correlation of the glacial border drift of north central Wisconsin: American Jour. Sci., v. 241, p. 498–516.

————, ed., 1965, Pedological record of the Quaternary: Soil Science, v. 99, p. 1–72.

Hollin, J. T., 1962, On the glacial history of Antarctica: Jour. Glaciol., v. 4, p. 173–195.*

————, 1965, Wilson's theory of ice ages: Nature, v. 208, p. 12–16.

Hollingworth, S. E., 1931, The glaciation of western Edenside and adjoining areas and the drumlins of Edenside and the Solway basin: Geol. Soc. London Quart. Jour., v. 87, p. 281–359.

Holmes, C. D., 1941, Till fabric: Geol. Soc. America, Bull., v. 52, p. 1299–1354.

————, 1947, Kames: American Jour. Sci., v. 245, p. 240–249.

————, 1952, Drift dispersion in west-central New York: Geol. Soc. America Bull., v. 63, p. 993–1010.

————, 1960, Evolution of till-stone shapes central New York: Geol. Soc. America Bull., v. 71, p. 1645–1660

Holmes, G. W., and Moss, J. H., 1955, Pleistocene geology of the southwestern Wind River Mountains, Wyoming: Geol. Soc. America Bull., v. 66, p. 629–654.*

Holst, N. O., 1876–1877, Om de glaciala rullstensåsarna: Geol. Fören. i Stockholm Förhandl., v. 3, p. 97–112.

Holtedahl, Hans, 1959, Geology and paleontology of Norwegian sea bottom cores: Jour. Sed. Petrology, v. 29, p. 16–29.

————, 1967, Notes on the formation of fjords and fjord-valleys: Geograf. Annaler, v. 49, Ser. A, p. 188–203.*

Hood, P. J., ed., 1968, Earth science symposium on Hudson Bay: Canada Geol. Survey Paper 68–53, 386 p.*

Hooijer, D. A., 1951, Pygmy elephant and giant tortoise: Sci. Monthly, v. 72, p. 3–8.

————, 1956, The lower boundary of the Pleistocene in Java and the age of Pithecanthropus: Quaternaria, v. 3, p. 5–10.*

————, 1962, Pleistocene dating and man: Advancement of Sci., v. 18, p. 485–489.

Hopkins, D. M. 1963, Geology of the Imuruk Lake area, Seward Peninsula, Alaska: U. S. Geol. Survey Bull. 1141-C, 101p.

————, ed., 1967, The Bering land bridge: Stanford Univ. Press, 495 p.*

————, and others, 1965, Quaternary correlations across Bering Strait: Science, v. 147, p. 1107–1114.*

Hoppe, Gunnar, 1948, Isrecessionen från Norrbottens kustland i belysning av de glaciala formelementen [The retreat of the ice from the lower-lying regions of Norrbotten, north Sweden, as illustrated by end moraines, other marginal glacial deposits and striae]: Geographica. Skrifter från Uppsala

Universitets Geografiska Institution, no. 20, 112 p.* (Engl. summ.)

———, 1951, Drumlins i Nordöstra Norrbotten: Geograf. Annaler, v. 33, p. 157–165.

———, 1952, Hummocky moraine regions, with special reference to the interior of Norrbotten: Geograf. Annaler, v. 34, 1952, p. 1–72.*

———, 1963, Some comments on the "ice-free refugia" of northwestern Scandinavia: p. 321–335 in Löve and Löve, eds., 1963.

———, 1968, Grímsey and the maximum extent of the last glaciation of Iceland: Geograf. Annaler, v. 50, p. 16–24.

———, and Liljequist, G. H., 1956, Det sista nedisningsförloppet i Nordeuropa och dess meteorologiska bakgrund: Ymer (Stockholm), 1956, p. 43–74 (Engl. summ.)

Horberg, Leland, 1938, The structural geology and physiography of the Teton Pass area, Wyoming: Augustana Library Pub. 16, 86 p.

———, 1940, Geomorphic problems and glacial geology of the Yellowstone Valley, Park County, Montana: Jour. Geol., v. 48, p. 275–303.

———, 1950, Bedrock topography of Illinois: Illinois Geol. Survey, Bull. 73, 111 p.

———, and Anderson, R. C., 1956, Bedrock topography and Pleistocene glacial lobes in central United States: Jour. Geol., v. 64, p. 101–116.

Horie, Shoji, 1965, Late Pleistocene glacial fluctuations and changes of sea level in the Japanese Islands and their tentative correlation with oscillations in North America and Europe: Int. Quaternary Congr., 6th Warsaw, 1961, Rept., v. 1, p. 175–184.

Hörnsten, Åke, 1964, Ångermanlands kustland under isavsmältningsskedet: Geol. Fören. i Stockholm Förh., v. 86, p. 181–205.

Hossfield, P. S., 1951, Calcareous tufa deposits in northern New Guinea: Royal Soc. Australia Trans., v. 74, p. 108–114.

Hough, J. L., 1963, The prehistoric Great Lakes of North America: American Scientist, v. 51, p. 84–109.*

———, 1966, Correlation of glacial lake stages in the Huron-Erie and Michigan basins: Jour. Geol., v. 74, p. 62–77.

Howard, A. D., 1965, Pseudo superglacial till: Science, v. 148, p. 1461–1462.

Howell, F. C., 1959, The Villafranchian and human origins: Science, v. 130, p. 831–844.*

———, 1960, European and northwest African Middle Pleistocene hominids: Current Anthropol., v. 1, p. 195–232.*

———, and Bourlière, François, (eds.), 1963, African ecology and human evolution: Viking Fund Publs in Anthrop. no. 36, Aldine Publ. Co., Chicago, 666p.*

Howell, J. V., and others, 1960, Glossary of geology and related sciences: American Geol. Inst., Washington, 2nd ed., 325+72 p.

Howells, W. W., 1966, Homo erectus: Scientif. American, v. 215, p. 46–53.

Hubbard, G. D., 1934, Unity of physiographic history in southwest Norway: Geol. Soc. America, Bull., v. 45, p. 637–654.

Hubbs, C. L., and Miller, R. R., 1948, The zoological evidence: Correlation between fish distribution and hydrographic history in the Desert basins of western United States. Part II, The Great Basin: Univ. of Utah Bull. v. 38, no. 20, 166 p.*

Hughes, O. L., and others, 1969, Glacial limits and flow patterns, Yukon Territory, south of 65 degrees North Latitude: Canada Geol. Survey Paper 68–34, 9p.

Hull, Edward, 1885, Mount Seir, Sinai and western Palestine: Richard Bentley and Son, London, 227 p.

Hultén, Eric, 1937, Outline of the history of arctic and boreal biota during the Quaternary period: Bokförlag Thule, Stockholm, 168 p.*

Hume, W. F., and Craig, J. I., 1911, The glacial period and climatic change in north-east Africa: British Assoc. Adv. Sci., Rept. 81st Mtg., p. 382–383.

Hummel, D., 1874, Om rullstenbildningar: K. Svenska Vetenskapsakad., Bihang till Handl., v. 2, no. 11, 36 p.

Hunt, C. B., 1953, Pleistocene-Recent boundary in the Rocky Mountain region: U.S. Geol. Survey Bull. 996, p. 1–25.

——, 1954, Pleistocene and Recent deposits in the Denver area, Colorado: U. S. Geol. Survey Bull. 996, p. 91–140.

——, 1967, Physiography of the United States: W. H. Freeman and Co., San Francisco, 480 p.

Huntington, Ellsworth, 1906, Pangong: a glacial lake in the Tibetan Plateau: Jour. Geol., v. 14, p. 599–617.

——, 1907, Some characteristics of the glacial period in non-glaciated regions. Geol. Soc. America Bull., v. 18, p. 351–388.

——, 1914, The solar hypothesis of climatic changes: Geol. Soc. America Bull., v. 25, p. 477–590.*

——, and Visher, S. S., 1922, Climatic changes: Yale Univ. Press, New Haven, 329 p.

Hutchinson, G. E., 1939, Ecological observations on the fishes of Kashmir and Indian Tibet: Ecol. Monogrs., 9, p. 145–182.

Hutton, James, 1795, Theory of the Earth: v. 2, William Creech, Edinburgh, 567 p. (Reprinted 1959 in facsimile, New York, Hafner Publishing Co.)

Hyyppä, Esa, 1955, On the Pleistocene geology of southeastern New England: Soc. Geogr. Fenniae, Acta Geographica 14, p. 155–225.

——, 1966, The Late-Quaternary land uplift in the Baltic sphere and the relation diagram of the raised and tilted shore levels: Ann. Acad. Scient. Fennicae, ser. A, sec. 3, no. 90, p. 153–168.*

Ikebe, Nobuo, 1969, A synoptical table on the Quaternary stratigraphy of Japan: Osaka City Univ., Jour. Geosci., v. 12, p. 45–51.*

INQUA, 1969a, Études françaises sur le Quaternaire: Assoc. française pour l'étude du Quaternaire, Bull., 1969, suppl., 274 p.*

——, 1969b, La stratigraphie des loess d'Europe: Assoc. française pour l'étude du Quaternaire, Bull., 1969, suppl., 176 p.*

Iversen, Johannes, 1952–1953, Origin of the flora of western Greenland in the light of pollen analysis: Oikos, (Copenhagen) v. 4, p. 85–103.

——, 1960, Problems of the early post-glacial forest development in Denmark: Danmarks Geol. Undersøg., ser. 4, no. 4, p. 1–32.

Ives, J. D., 1960a, Former ice-dammed lakes and the deglaciation of the middle reaches of the George River, Labrador-Ungava: Geogr. Bull. no. 14, p. 44–70.

——, 1960b, The deglaciation of Labrador-Ungava—an outline: Cahiers de Géogr. de Québec, v. 4, p. 323–344.

——, 1966, Block fields, associated weathering forms on mountain tops and the nunatak hypothesis: Geograf. Annaler, v. 48A, p. 220–223.

Ives, R. L., 1938, Glacial geology of the Monarch Valley, Grand County, Colorado: Geol. Soc. America Bull., v. 49, p. 1045–1066.

Jäckli, H. C. A., 1965, Pleistocene glaciation of the Swiss Alps and signs of postglacial differential uplift: Geol. Soc. America, Spec. Paper 84, p. 153–157.

Jaeger, Fritz, 1926, Forschungen über das diluviale Klima in Mexiko: Petermanns Geogr. Mitt. Ergänzungsh. 190, 64 p.

————, 1939, Die Trockensee der Erde. . . . : Petermanns Geogr. Mitt., Ergänzungsb. 236, 159 p.*

Jahn, Alfred, 1960, The oldest periglacial period in Poland: Biuletin Peryglacjalny, v. 9, p. 159–162.

Jahns, R. H., 1943, Sheet structure in granites. Its origin and use as a measure of glacial erosion in New England: Jour. Geol., v. 51, p. 71–98.

————, and Willard, M. E., 1942, Late Pleistocene and Recent deposits in the Connecticut Valley, Massachusetts: American Jour. Sci., v. 240, pp. 161–191, 265–287.

————, and others, 1955, The geology of southern California: California Div. Mines Bull. 170, 878 p.*

Jamieson, T. F., 1862, On the ice-worn rocks of Scotland: Geol. Soc. London Quart. Jour., v. 18, p. 164–184.

————, 1863, On the parallel roads of Glen Roy, and their place in the history of the glacial period: Geol. Soc. London Quart. Jour., v. 19, p. 235–259.

————, 1865, On the history of the last geological changes in Scotland: Geol. Soc. London Quart. Jour., v. 21, p. 161–203.

————, 1874, On the last stage of the Glacial Period in North Britain: Geol. Soc. London Quart.. Jour., v. 30, p. 317–338.

Järnefors, Björn, and Fromm, Erik, 1960, Chronology of the ice recession through middle Sweden: Internat. Geol. Congr., 21st., Norden 1960, Rept., pt. 4, p. 93–97.

Jelgersma, Saskia, 1961, Holocene sea level changes in the Netherlands: Maastricht, Ernest van Aelst. Uitgeversmij, 101 p.*

————, 1962, A late-glacial pollen diagram from Madelia, south-central Minnesota: American Jour. Sci., v. 260, p. 522–529.

Jenkins, D. G., 1967, Planktonic foraminiferal zones and new taxa from the Lower Miocene to the Pleistocene of New Zealand: New Zealand Jour. Geol. and Geophys., v. 10, p. 1064–1078.

Jenks, W. F., ed., 1956, Handbook of South American geology: Geol. Soc. America Mem. 65, 378 p.*

Jennings, J. D., 1969, Prehistory of North America: McGraw-Hill Book Co., New York, 402 p.*

Jennings, J. N., 1968, A revised map of the desert dunes of Australia: Australian Geogr., 10, p. 408–409.

————, and Banks, M. R., 1958, The Pleistocene glacial history of Tasmania: Jour. Glaciol., v. 3, p. 298–303.

Jessen, Knud, and Jonassen, H., 1935, The composition of the forests in northern Europe in epipalaeolithic time: Kgl. Danske Vidensk. Selsk., Biol. Medd., v. 12, no. 1, 64, p.

Jessen, Knud, and Milthers, Vilhelm, 1928, Stratigraphical and paleontological studies of interglacial fresh-water deposits in Jutland and northwest Germany: Danmarks Geol. Undersög., ser. 2, no. 48, 379 p.

Jessup, R. W., 1961, A Tertiary-Quaternary pedological chronology for the south-eastern portion of the Australian arid zone: Jour. Soil Sci. (Brit.), v. 12, p. 199–213.

Jewtuchowicz, Stefan, 1953, La structure du sandre: Łódź. Soc. des Sci. et des Lettres, classe 3, v. 4, no. 4, 23 p.

John, B. S., 1964, A description and explanation of glacial till in 1603: Jour. Glaciol. v. 5, p. 369–370.

Johnsson, Gunnar, 1959, True and false ice-wedges in southern Sweden: Geograf. Annaler., v. 41, p. 15–33.

———, 1962, Periglacial phenomena in southern Sweden: Geograf. Annaler, v. 44, p. 378–404.*

Johnston, W. A., 1916, Late Pleistocene oscillations of sea-level in the Ottawa valley: Canada Geol. Survey Mus. Bull. 24, 14 p.

———, 1926, The Pleistocene of Cariboo and Cassiar districts, British Columbia, Canada: Roy. Soc. Canada, Trans., 3d ser., v. 20, sec. 4, p. 137–147.

———, 1946, Glacial Lake Agassiz, with special reference to the mode of deformation of the beaches. Canada Geol. Survey Bull. 7, 20 p.

Jordan, R. R., 1964, Columbia (Pleistocene) sediments of Delaware: Delaware Geol. Survey Bull. 12, 69 p.

Judson, Sheldon, 1949, The Pleistocene stratigraphy of Boston, Massachusetts and its relation to the Boylston Street Fishweir: In Barghoorn, E. S., and others, The Boylston Street Fishweir II. A study of the geology, palaeobotany, and biology of a site on Stuart Street in the Back Bay district of Boston, Massachusetts, p. 7–48: Robert S. Peabody Found. for Archaeology, Papers, v. 4, no. 1.

———, 1950, Depressions of the northern portion of the Southern High Plains of eastern New Mexico: Geol. Soc. America Bull. 61, p. 253–274.

———, and Ritter, D. F., 1964, Rates of regional denudation in the United States: Jour. Geophys. Res., v. 69, p. 3395–3401.

Kahlke, H. D., 1961, On the complex of the *Stegodon-Ailuropoda-* fauna of southern China and the chronological position of *Gigantopithecus blacki* v. Koenigswald: Vertebrata Palasiatica (Peking), no. 2, p. 83–108.* [Chinese; Engl. summ.]

Kaiser, Karlheinz, 1960, Klimazeugen des periglazialen Dauerfrostbodens in Mittel- und Westeuropa: Eiszeitalter und Geg., v. 11, p. 121–141.

———, 1963, Die Ausdehnung der Vergletscherungen und "Periglazialen" Erscheinungen wären der Kaltzeiten des Quartären Eiszeitalters innerhalb der Syrisch-Libanesischen Gebirge und die Lage der Klimatischen Schneegrenze zur Würmeiszeit im östlichen Mittelmeergebiet: Int. Quater. Congr., 6th, Warsaw 1961, Rept., v, 3, p. 127–148.

Kampf, E. E., 1966, Das Holstein-Interglazial von Tönisberg im Rahmen des niederrheinischen Pleistozäns: Eiszeitalter und Geg., v. 17, p. 5–60.

Kano, Tadao, 1934–35, Contribution to the glacial topography of the Tugitaka Mountains, Formosa. I: Geogr. Rev. of Japan, v. 10 (1934), p. 606–623, 688–707, 816–835, 990–1017; v. 11 (1935), p. 244–263. (Engl. summ.)

Kapp, R. O., 1965, Illinoian and Sangamon vegetation in southwestern Kansas and adjacent Oklahoma: Univ. of Michigan Mus. Paleontol., Contribs., v. 19, p. 167–255.

———, and Gooding, A.M., 1964, Pleistocene vegetational studies in the

Whitewater Basin, Southeastern Indiana: Jour. Geol., v. 72, p. 307–326.

Karlstrom, T. N. V., and others, 1964, Surficial geology of Alaska. Map, scale 1:1, 584,000: U.S. Geol. Survey Misc. Geol. Inv., Map I–357.

Karrow, P. F., 1961, The Champlain Sea and its sediments: p. 97–108 in Soils in Canada: Royal Soc. Canada Spec. Pub. 3.

————, 1967, Pleistocene geology of the Scarborough area: Ontario Dept. Mines, Geol. Rept. 46, 108p.*

Kauranne, L. K., 1960, A statistical study of stone orientation in glacial till: Comm. Géol. de Finlande, Bull. 188, p. 87–97. Also in Soc. Géol. de Finlande, Comptes Rendus, v. 32, 1960, p. 87–97.

Kay, G. F., and Graham, J. B., 1943, The Illinoian and post-Illinoian Pleistocene geology of Iowa: Iowa Geol. Survey, v. 38, p. 1–262.*

Kay, G. F., and Leighton, M. M., 1933, Eldoran epoch of the Pleistocene period: Geol. Soc. America Bull., v. 44, p. 669–674.

Kay, G. F., and Pearce, J. N., 1920, The origin of gumbotil: Jour. Geol., v. 28, p. 89–125.

Kaye, C. A., 1959, Shoreline features and Quaternary shoreline changes, Puerto Rico: U. S. Geol. Survey Prof. Paper 317, p. 49–140.

————, 1960, Surficial geology of the Kingston Quadrangle, Rhode Island: U. S. Geol. Survey Bull. 1071–I, p. 341–396.

————, 1961, Pleistocene stratigraphy of Boston, Massachusetts: U.S. Geol. Survey Prof. Paper 424, p. B73–B76.

————, 1964a, Outline of Pleistocene geology of Martha's Vineyard, Massachusetts: U. S. Geol. Survey Prof. Paper 501, p. C134–C139.

————, 1964b, Illinoian and early Wisconsin moraines of Martha's Vineyard, Massachusetts: U. S. Geol. Survey Prof. Paper 501, p. C140–C143.

————, 1967, Kaolinization of bedrock of the Boston, Massachusetts, area: U. S. Geol. Survey Prof. Paper 575, p. C165–C172.

————, and Barghoorn, E. S., 1964, Late Quaternary sea-level change and crustal rise at Boston, Massachusetts, with notes on the autocompaction of peat. Geol. Soc. America Bull., v. 75, no. 2, p. 63–80.

Kempf, E. K., 1966, Das Holstein-Interglazial von Tönisberg im Rahmen des Niederrheinischen Pleistozäns: Eiszeitalter u. Geg., v. 17, p. 5–60.

Kendall, P. F., 1902, A system of glacier-lakes in the Cleveland Hills: Geol. Soc. London Quart. Jour., v. 58, p. 471–571.

Kendall, R. L., 1969, An ecological history of the Lake Victoria basin: Ecol. Monogrs., v. 39, p. 121–176.*

Kerr, F. A., 1934, Glaciation in northern British Columbia: Roy. Soc. Canada, Trans., ser. 3, v. 28, sec. IV, p. 17–31.

————, 1936, Quaternary glaciation in the Coast Range, northern British Columbia and Alaska: Jour. Geol., v. 44, p. 681–700.

King, L. H., 1969, Submarine end moraines and associated deposits on the Scotian Shelf: Geol. Soc. America Bull., v. 80, p. 83–96.

King, P. B., and Stupka, Arthur, 1950, The Great Smoky Mountains—their geology and natural history: Scientif. Monthly, v. 71, p. 31–43.

Kingery, W. D., ed., 1963, Ice and snow. Properties, processes, and applications: The M.I.T. Press, Cambridge, 684 p.

Kinsman, D. J., and Sheard, J. W., 1963, The glaciers of Jan Mayen: Jour. Glaciol., v. 4, p. 439–448.*

Kjartansson, Gudmundur, and others, 1964, C^{14} datings of Quaternary deposits in Iceland: Náttúrufraedingurinn (Reykjavik), v. 34, p. 97–145.

Klassen, R. W., 1969, Quaternary stratigraphy and radiocarbon chronology in southwestern Manitoba: Canada Geol. Survey Paper 69–27, 19p.

Klebelsberg, Raimund von, 1948–1949, Handbuch der Gletscherkunde und Glazialgeologie: Springer Verlag, Vienna, v. 1, Allgemeiner Teil, 1948; v. 2, Historischregionaler Teil, 1949, 1028 p.*

Klein, D. R., 1965, Postglacial distribution patterns of mammals in the southern coastal regions of Alaska: Arctic, v. 18, p. 7–20.

Klíma, Bohuslav, and others, 1961, Stratigraphie des Pleistozäns und Alter des Paläolithischen Rastplatzes in der Ziegelei von Dolní Věstonice (Unter-Wisternitz): Anthropozoikum (Prague), v. 11, p. 93–145.

Klute, Fritz, 1921, Uber die Ursachen der letzten Eiszeit: Geogr. Zeitschr., v. 27, p. 199–203.

———, 1951, Die Klimazonen des Eiszeitalters: Eiszeitalter und Geg., v. 1, p. 16–26.

Knechtel, M. M., 1942, Snake Butte boulder train and related glacial phenomena, north-central Montana: Geol. Soc. America Bull., v. 53, p. 917–936.

Knox, A. S., 1969, Glacial age marsh, Lafayette Park, Washington, D. C.: Science, v. 165, p. 795–797.

Kobayashi, Kunio, 1955, An introduction to periglacial or subnival morphology in Japan: Shinshu Univ., Jour. Fac. Liberal Arts and Sci., no. 5, p. 23–38.

———, 1958, Quaternary glaciation of the Japan Alps: Shinshu Univ., Jour. Fac. Liberal Arts and Sci., no. 8, pt. 2, p. 13–67.*

———, 1965 a, Problems of Late Pleistocene history of central Japan: Geol. Soc. America Spec. Paper 84, p. 367–391.

———, 1965b, The Quaternary of China and its recent studies: Chikyu Kagaku [Earth Science], no. 77, p. 40–45. (Japanese)

———, and Shimizu, Hideki, 1966, Significance of the Ikenotaira Interstadial indicated by moraines on Mt. Kumazawa of the Kiso Mountain Range, central Japan: Shinshu Univ., Jour. Fac. Sci., v. 1, no. 2, p. 97–113.

Koopmans, B. N., and Stauffer, P. H., 1968, Glacial phenomena on Mount Kinabalu, Sabah: Borneo Reg., Malaysia Geol. Survey Bull. 8, p. 25–35.

Korpela, Kauko, 1969, Die Weichsel-Eiszeit und ihr Interstadial in Peräpohjola (nördliches Nordfinnland) im Licht von Submoränen Sedimenten: Finland. Acad. Sci., Ann., ser. A III, no. 84, 109 p.*

Kotlyakov, V. M., 1966, Calculation of water volume accumulation in mountain glaciers of USSR: USSR, Akad. Nauk, Isvestia, Geogr ser., no. 3, p. 43–48.

Kowalski, Kazimierz, 1959, A catalogue of the Pleistocene mammals of Poland: Poland. Acad. Sci., Zool. Inst., Kraków Branch, Science Publ. House, Warszawa, 267 p.*

Kozarski, Stefan, 1961, Fossil congelifraction covers in the northern part of the Lushan (central China): Biuletyn Peryglacjalny, no. 10, p. 195–207.

———, 1963, Problem of Pleistocene glaciations in the mountains of east China: Zeitschr. für Geomorphol., v. 7, p. 48–70.

Krantz, G. S., 1970, Human activities and megafaunal extinctions: American Scientist, v. 58, p. 164–170.

Kraus, E. B., 1958, Recent climatic changes: Nature, v. 181, p. 666–668.

———, 1960, Synoptic and dynamic aspects of climatic change: Royal Meteorol. Soc., Quart. Jour., v. 86, p. 1–15.

Kriger, N.I., 1965, Loess, its characteristics and relation to geographical environment: USSR Acad. Sci., Comm. for Study of Quaternary Period. Moscow, "Nauka", 296 p.[*R]

Krinitzsky, E. L., and Turnbull, W. J., 1967, Loess deposits of Mississippi: Geol. Soc. America Spec. Paper 94, 64 p.

Krinsley, D. H., and Donahue, Jack, 1968, Environmental interpretation of sand grain surface textures by electron microscopy: Geol. Soc. America Bull., v. 79, p. 743–748.

Krumbein, W. C., 1933, Textural and lithological variations in glacial till: Jour. Geol., v. 41, p. 382–408.

———, 1939, Preferred orientation of pebbles in sedimentary deposits: Jour. Geol., v. 47, p. 673–706.

Kubiëna, W. L., 1953, The soils of Europe: Thomas Murby and Co., London, 318 p.[*]

Kuenen, P. H., 1955, Sea Level and crustal warping: Geol. Soc. America Spec. Paper 62, p. 193–204.

Kupsch, W. O., 1962, Ice-thrust ridges in western Canada: Jour. Geol. v. 70, p. 582–594.

———, 1964, Bedrock surface and preglacial topography of the Regina-Wynyard region, southern Saskatchewan: Regina, Third Int. Williston Basin Symposium, p. 274–283.

Kurtén, Björn, 1957, The bears and hyenas of the interglacials: Quaternaria, v. 4, p. 69–81.

———, 1960, Chronology and faunal evolution of the earlier European glaciations: Soc. Scient., Fennica, Commentationes Biologicae, v. 21, no. 5, 62 p.

———, 1968, Pleistocene mammals of Europe: Aldine Publishing Co., Chicago, 317 p.

———, and Vasari, Yrjö, 1960, On the date of Peking Man: Finland. Acad. Sci., Comment. Biol., v. 23, no. 7, 10 p.

La Fleur, R. G., 1965, Glacial geology of the Troy, N. Y., Quadrangle: New York State Mus., Map and Chart ser. no. 7., 22 p.

Låg, J., 1948, Undersøkelser over opphavsmaterialet for Østlandets morenedekker: Det norske Skogforsøksvesen, nr. 35, 223 p.

———, and Bergseth, H., 1957, Laboratory experiments on sedimentation of soil material in salt-water: Agric. College of Norway, Sci. Repts., v. 36, no. 7, 13 p.

Lagaaij, R., 1952, The Pliocene Bryozoa of the Low Countries and their bearing on the marine stratigraphy of the North Sea Sea region: Netherlands. Geol. Stichting, Medel., ser. C, V, no. 5, 233 p.

Lamb, H. H., 1964, Climatic changes and variations in the atmospheric and ocean circulations: Geol. Rundsch., v. 54, p. 486–504.

———, 1965, Climatic change with special reference to Wales and its agriculture: Univ. College of Wales, Aberystwyth, Dept. Geol. and Geogr., Memo. 8, 18 p.

———, 1969, Activité volcanique et climat: Rev. de Géogr. Phys. et de Géol. Dynam., v. 11, p. 363–380.[*]

———, and Johnson, A. I., 1959–1961, Climatic variation and observed changes in the general circulation: Geograf. Annaler, v. 41, p. 94–134; v. 43, p. 363–400.[*]

———, and others., 1962, A new advance of the Jan Mayen glaciers and a remarkable increase of precipitation: Jour. Glaciol., v. 4, p. 355–365.

Lamerson, P. R., and Dellwig, L. F., 1957, Deformation by ice push of lithified sediments in south-central Iowa: Jour. Geol., v. 65, p. 546–550.

Lamplugh, G. W., 1902, "Calcrete": Geol. Mag., dec. 4, v. 9, p. 575.

Land, L. S., and others, 1967, Pleistocene history of Bermuda: Geol. Soc. America Bull., v. 78, p. 993–1006.*

Lanning, E. P., and Patterson, T. C., 1967, Early man in South America: Scientif. American, v. 217, p. 44–50.

LaRocque, Aurèle, 1966–1970, Pleistocene mollusca of Ohio: Ohio Geol. Survey Bull. 62, pt. 1 (1966), pt. 2 (1967), pt. 3 (1968), pt. 4 (1970), 800 p.

Lartet, Louis, 1865, Sur la formation du bassin de la mer morte ou lac asphaltite, et sur les changements survenus dans le niveau de ce lac: (Paris), Acad. des Sci., Comptes Rendus, v. 60, p. 796–800.

Lasalle, Pierre, 1966, Late Quaternary vegetation and glacial history in the St. Lawrence Lowlands, Canada: Leidse Geol. Mededel., v. 38, p. 91–128.

Lawrence, D. B., 1950, Estimating dates of recent glacier advances and recession rates by studying tree growth layers: American Geophys. Un., Trans., v. 31, p. 243–248.

———, and Elson, J. A., 1953, Periodicity of deglaciation in North America since the Late Wisconsin maximum: Geograf. Ann., v. 35, p. 83–104.

Lawson, A. C., 1940, Isostatic control of fluctuations of sea level: Science, v. 92, p. 162–164.

Leavitt, H. W., and Perkins, E. H., 1935, Glacial geology of Maine. vol. II: Maine Technol. Expt. Sta., Bull. 30, 232 p.

Lee, H. A., 1959, Surficial geology of southern district of Keewatin and the Keewatin ice divide, Northwest Territories: Canada Geol. Survey Bull. 51, 42 p.

———, 1963, Glacial fans in till from the Kirkland Lake fault: a method of gold exploration: Canada Geol. Survey Paper 63–45, 36 p.

———, 1965, Investigation of eskers for mineral exploration: Canada Geol. Survey Paper 65–14, p. 1–17.

———, and Lawrence, D. E., 1968, A new occurrence of kimberlite in Gauthier Township, Ontario: Canada Geol. Survey, Paper 68–22, 16 p.

Lee, J. S., 1947, Quaternary glaciations in the Lushan area, central China: Acad. Sinica, Inst. Geol. (Nanking), Mon., ser. B, v. 2, 37 p. (Engl.), 60 p. (Chinese.)

Legget, R. F., 1962, Geology and engineering: 2d ed., McGraw-Hill Book Co., New York, 884 p.*

———, and Bartley, M. W., 1953, An engineering study of glacial deposits at Steep Rock Lake, Ontario, Canada: Econ. Geology, v. 48, p. 513–540.

Leighly, John, 1949, On continentality and glaciation: Geograf. Annaler, v. 31, p. 133–145.

Leighton, M. M., 1960, The classification of the Wisconsin glacial stage of north central United States: Jour. Geol. v. 68, p. 529–552.

———, and MacClintock, Paul, 1962, The weathered mantle of glacial tills beneath original surfaces in north-central United States: Jour. Geol., v. 70, p. 267–293.*

Lemke, R. W., 1958, Narrow linear drumlins near Velva, North Dakota: American Jour. Sci., v. 256, p. 270–283.

Lemon, R. R. H., and Churcher, C. S., 1961, Pleistocene geology and paleontology of the Talara region, northwest Peru: American Jour. Sci., v. 259, p. 410–429.

Leonard, A. B., and Frye, J. C., 1954, Ecological conditions accompanying loess deposition in the Great Plains region: Jour. Geol., v. 62, p. 399–404.

Leopold, E. B., 1956, Two late glacial deposits in southern Connecticut: Nat. Acad. Sci. (U.S.A.) Proc., v. 42, p. 863–867.

Leopold, L. B., 1951, Pleistocene climate in New Mexico: American Jour. Sci., v. 249, p. 152–167.

Leroi-Gourhan, Arlette, 1960, Flores et climats du Paléolithique récent: Congr. Préhist. de France, Monaco, 1959, Compte Rendu, p. 1–6.

Lessig, H. D., and Rice, W. A., 1962, Kansan drift of the Elkton, Ohio, rift: American Jour. Sci., v. 260, p. 439–454.

Leverett, Frank, 1899, The Illinois glacial lobe: U.S. Geol. Survey Mon. 38, 817 p.

———, 1908, Description of the Ann Arbor quadrangle, Mich: U.S. Geol. Survey, Geol. Atlas of United States, Folio 155, 15 p.+maps.

———, 1910, Comparison of North American and European glacial deposits: Zeitschr. für Gletscherk. v. 4, p. 241–316.

———, 1921, Outline of Pleistocene history of Mississippi Valley: Jour. Geol., v. 29, p. 615–626.

———, 1929, The Pleistocene of northern Kentucky: Kentucky Geol. Survey, ser. 6, v. 31, p. 1–80.

———, 1930, Problems of the glacialist: Science, v. 71, p. 47–57.

———, 1934, Glacial deposits outside the Wisconsin terminal moraine in Pennsylvania: Pennsylvania Geol. Survey Bull. G7, 123 p.

———, 1942, Shiftings of the Mississippi River in relation to glaciation: Geol. Soc. America Bull., v. 53, p. 1283–1298.

———, and Taylor, F. B., 1915, The Pleistocene of Indiana and Michigan and the history of the Great Lakes: U.S. Geol. Survey Mon. 53, 529 p.

Lewis, H. C., 1885, The direction of glaciation as ascertained by the form of the striae: Nature, v. 32, p. 557–558.

Lewis, R. G., 1935, The orography of the North Sea bed: Geogr. Jour., v. 86, p. 334–342.

Lewis, W. V., 1949a, An esker in process of formation: Böverbreen, Jotunheimen, 1947: Jour. Glaciol., v. 1, no. 6, p. 314–319.

———, 1949b, The function of meltwater in cirque formation: A reply: Geogr. Rev., v. 39, p. 110–128.

———, 1954, Pressure release and glacial erosion: Jour. Glaciol., v. 2, p. 417–422.

———, ed., 1960, Norwegian cirque glaciers: Royal Geogr. Soc. Res. Ser. 4, 104 p.

Libby, W. F., 1955, Radiocarbon dating: Univ. of Chicago Press, 2d ed., 175 p.

———, 1961, Radiocarbon dating: Science, v. 133, p. 621–627.

Lidz, Louis, 1966, Deep-sea Pleistocene biostratigraphy: Science, v. 154, p. 1448–1452.

Lind, A. O., 1969, Coastal landforms of Cat Island, Bahamas: Univ. of Chicago Dept. Geogr., Res. Paper 122, 156 p.

Lindgren, Waldemar, 1904, A geological reconnaissance across the Bitterroot Range and Clearwater Mountains in Montana and Idaho: U. S. Geol. Survey Prof. Paper 27, 123p.

Lindroth, C. H. 1963, The fauna history of Newfoundland illustrated by carabid beetles: Opuscula Entomol. (Lund), Suppl. 23, 112 p.*

Lipps, J. H., and others, 1968, Pleistocene paleoecology and biostratigraphy, Santa Barbara Island, California: Jour. Paleontol., v. 42, p. 291–307.

Liu, T.-S., and others, 1958, Chinese loess map of Shansi and Shensi regions: Science Record (Peiping), v. 2, p. 167–174.

Livingstone, D. A., 1964, The pollen flora of submarine sediments from Nantucket Shoals: American Jour. Sci., v. 262, p. 479–487.

————, 1967, Postglacial vegetation of the Ruwenzori Mountains in equatorial Africa: Ecol. Monogrs., v. 37, p. 25–52.

————, 1968, Some interstadial and postglacial pollen diagrams from eastern Canada: Ecol. Monogr., v. 38, p. 87–125.

Lizitzin, A. P., 1960, Bottom sediments of the eastern Antarctic and southern Indian Ocean: Deep-Sea Res., v. 7, p. 89–99.

Lliboutry, Louis, 1964–1965, Traité de glaciologie: Paris, Masson et Cie., 1040 p.*

Lockwood, W. N., and Meisler, Harold, 1960, Illinoian outwash in southeastern Pennsylvania: U.S. Geol. Survey Bull. 1121, p. B-1—B-9.

Løken, O. H., and Andrews, J. T. 1966, Glaciology and chronology of fluctuations of the ice margin at the south end of the Barnes Ice Cap, Baffin Island, N.W.T.: Geogr. Bull., v. 8, p. 341–359.

Lona, Fausto, 1950, Contributi alla storia della vegetazione e del clima nella Val Padana.—Analisi pollinica del giacimento Villafranchiano di Leffe (Bergamo): Società Italiana Sci. Nat., Atti, v. 89, p. 123–179.*

————, 1963, Floristic and glaciologic sequence (from Donau to Mindel) in a complete diagram of the Leffe deposit: Geobot. Inst. Rübel (Zürich), Berichte, v. 34, p. 64–66.

————, and Follieri, Maria, 1957, Successione pollinica della serie superiore (Günz-Mindel) di Leffe (Bergamo): Verh. 4th Internat. Tagung der Quartärbotaniker, Veröff. Geobot. Institut Rübel in Zürich, no. 34, p. 86–98.

Longwell, C. R., Flint, R. F., and Sanders, J. E., 1969, Physical geology: John Wiley and Sons, New York, 685 p.

Lorenzo, J. L., 1959, Los glaciares de México: Universidad Nac. Autónoma de México, Inst. de Geofísica, Mon. 1, 114 p. (Spanish, w Engl. transl.)

————, 1967, La etapa lítica en México: Inst. Nac. de Antropol. e Hist., Dept. de Prehist., publ. 20, 49 p.

Louis, Herbert, 1938, Eiszeitliche Seen in Anatolien: Zeitschr. Ges. für Erdkunde zu Berlin, p. 267–285.

————, 1944, Die Spuren eiszeitlicher Vergletscherung in Anatolien: Geol. Rundschau, v. 34, p. 447–481.

Löve, Áskell, and Löve, Doris, eds., 1963, North Atlantic biota and their history: The Macmillan Co., New York, Pergamon, Oxford; 430 p.*

Löve, Áskell, and Löve, Doris, 1967, The origin of the North Atlantic flora: Aquilo (Oulu), ser. botan., v. 6, p. 52–66.

Lovering, T. S., 1935, Geology and ore deposits of the Montezuma Quadrangle, Colorado: U.S. Geol. Survey Prof. Paper 178, 64 p.

Ložek, Vojen, 1961–1962, [Papers on Quaternary stratigraphy in Czechoslovakia]: Anthropozoikum (Prague), v. 9, p. 35–85; v. 11, p. 13–28.*

————, 1963, Růžový převis ve Vrátné dolině u Turčianské Blatnice: Československý Kras, v. 15, p. 105–117 (Engl. summ.)

————, 1964, Quartärmollusken der Tschechoslowakei. Prague: Geologischen Zentralanstalt, Rozpravy, v. 31, 374 p.

————, 1969, Über die malakozoologische Charakteristik der pleistozänen Warmzeiten mit besonderer Berücksichtigung des letzten Interglazials: Deutsch. Gesellsch. geol. Wiss., Bericht, sec. A, v. 14, p. 439–469.

Lüdi, Werner, 1953, Die Pflanzenwelt des Eiszeitalters im nördlichen Vorland der Schweizer Alpen: Geobot. Inst. Rübel (Zürich), Veröffentl., Heft 27, 208 p.

————, 1955, Die Vegetationsentwicklung seit dem Rückzug der Gletscher in den mittleren Alpen und ihrem nördlichen Vorland: Geobot. Inst. Rübel (Zürich) Bericht für 1954, p. 36–68.

————, 1957, Interglaziale Vegetation im schweizerischen Alpenvorland: Geobot. Inst. Rübel (Zürich), Veröff., Heft 34, p. 99–107.

Lum, Daniel, and Stearns, H. T., 1970, Pleistocene stratigraphy and eustatic history based on cores at Waimanalo, Oahu, Hawaii: Geol. Soc. America Bull., v. 81, p. 1–16.

Lumley, Henry de, 1963, Les niveaux quaternaires marins des Alpes-Maritimes. Corrélations avec les industries préhistoriques: Soc. Géol. de France, Bull., v. 5, p. 562–579.

Lundqvist, Gösta, 1935, Blockundersökningar, historik och metodic: Sveriges Geol. Undersök., ser. C, no. 390, 45 p.*

Lundqvist, Jan, 1962, Patterned ground and related frost phenomena in Sweden: Sveriges Geol. Undersök. Årsbok 55, no. 7, 101 p.*

————, 1965, Glacial geology in northeastern Newfoundland: Geol. Fören. i Stockholm Förh., v 87, p. 285–306.

————, 1967, Submoräna sediment i Jämtlands Län: Sveriges Geol. Undersök., ser. C, no. 618, 267 p.

Lustig, L. K., 1965, Clastic sedimentation in Deep Springs Valley, California: U.S. Geol. Survey Prof. Paper 352-F, p. 131–192.

Lüttig, Gerd, 1959, Stratigraphische Bemerkungen zum nichtmarinen Quartär Mittel-Italiens: Geol. Jahrb., v. 75, p. 651–662, pl. 40.

————, 1964a, Die Aufgaben des Geschiebeforschers und des Geschiebesammlers: Lauenburgische Heimat, v. 45, p. 6–26.

————, 1964b, Prinzipielles zur Quartär-Stratigraphie: Geol. Jahrb., v. 82, p. 177–202.

Lyell, Charles, 1830–1833, Principles of geology: J. Murray, London, v. 1, 1830, 511 p.; v. 2, 1832, 330 p.; v. 3, 1833, 398+109 p.

————, 1839, Nouveaux éléments de geologie: Pitois-Levrault & Cie., Paris, 648 p.

————, 1873, The geological evidences of the antiquity of man. 4th ed.: John Murray, London, 572 p.

Maarleveld, G. C., 1965, Frost mounds: Wageningen. Geologische Stichting, Mededelingen, n.s., no. 17, p. 3–16.*

MacClintock, Paul, 1922, The Pleistocene history of the lower Wisconsin River: Jour. Geol., v. 30, p. 673–689.

————, 1933, Correlation of the pre-Illinoian drifts of Illinois: Jour. Geol., v. 41, p. 710–722.

————, 1940, Weathering of the Jerseyan till: Geol. Soc. America Bull., v. 51, p. 103–116.

————, and Apfel. E. T., 1944, Correlation of the drifts of the Salamanca re-entrant, New York: Geol. Soc. America Bull., v. 55, p. 1143–1164.

———, and Richards, H. G., 1936, Correlation of late Pleistocene marine and glacial deposits of New Jersey and New York: Geol. Soc. America Bull., v. 47, p. 289—338.

———, and Stewart, D. P., 1965, Pleistocene geology of the St. Lawrence Lowland: New York State Mus. Bull. 394, 152 p.

———, and Twenhofel, W. H., 1940, Wisconsin glaciation of Newfoundland: Geol. Soc. America Bull., v. 51, p. 1729–1756.

Machatschek, Fritz, 1944, Diluviale Hebung und eiszeitliche Schneegren-zendepression: Geol. Rundschau, v. 34, pts. 7–8, p. 327–341.

Mackay, J. R., 1959, Glacier ice-thrust features of the Yukon coast: Geogr. Bull., no. 13, p. 5–21.

———, 1965, Glacier flow and analogue simulation: Geogr. Bull., v. 7, p. 1–6.

Maclaren, Charles, 1841, The glacial theory of Professor Agassiz of Neuchatel: The Scotsman Office, Edinburgh, 62 p. Reprinted, 1842, in Am. J. Sci., v. 42, p. 346–365.

MacNeil, F. S. 1957, Cenozoic megafossils of northern Alaska: U.S. Geol. Survey Prof. Paper 294-C, p. 99–126.

Macoun, Jaroslav, and others, 1965, Kvartér Ostravska a Moravské brány: Czechoslovakia. Geol. Survey and Acad. Sci. 419 p.* [German summ.]

Madsen, Victor, and others, 1928, Summary of the geology of Denmark: Danmarks Geol. Undersøg. ser. 5, no. 4, 219 p.*

Malde, H. E., 1964, Environment and man in arid America: Science, v. 145, p. 123–129.

———, 1968, The catastrophic late Pleistocene Bonneville Flood in the Snake River Plain, Idaho: U.S. Geol. Survey Prof. Paper 596, 52 p.

Maldonado-Koerdell, Manuel, 1948, Los vertebrados fósiles del Cuaternario en México: Soc. Méxicana de Hist. Nat., Revista, v. 9, p. 1–35.*

Malone, T. F., ed., 1951, Compendium of meteorology: American Meteorol. Soc., Boston, 1334 p.*

Manley, Gordon, 1951, The range of variation of the British climate: Geogr. Jour., v. 117, p. 43–68.

———, 1959, The late-glacial climate of north-west England: Liverpool and Manchester Geol. Jour., v. 2, pt. 2, p. 188–215.

Mannerfelt, C. M., 1945, Some glaciomorphological forms and their evidence as to the downwasting of the inland ice in Swedish and Norwegian mountain terrain: Geograf. Annaler, p. 1–239.

Mansfield, G. R., 1908, Glaciation in the Crazy Mountains of Montana: Geol. Soc. America Bull., v. 19, p. 558–567.

Mantell, G. A., 1822, The fossils of the South Downs, or illustrations of the geology of Sussex: L. Relfe, London, 327 p.

Markov, K. K., Lazukov, G. I., and Nikolaev, V.A., 1965, The Quaternary Period (Glacial Period—Antropogen Period), vols. 1 and 2: Moscow Univ., v. 1, 371 p.; v. 2, 435 p.*[R]

Markov, K. K., and Velitchko, A. A., 1967, The Quaternary Period (Glacial Period—Antropogen Period), v. 3: Nedra Publishing House, Moscow, v. 3, 438 p.*[R]

Mars, Paul, 1963, Les faunes et la stratigraphie du Quaternaire méditerranéen: Station Marine d'Endoume [Marseille], Recueil des Travaux, Bull. 28, fasc. 43. p. 61–97.*

Martin, A. R. H., 1959, The stratigraphy and history of Groenvlei, a South African coastal fen: Australian Jour. Bot., v. 7, p. 142–167.

Martin, P. S., 1958, Pleistocene ecology and biogeography of North America: American Assoc. Adv. Sci., Zoogeography, p. 375–420.

———, 1966, Africa and Pleistocene overkill: Nature, v. 212, p. 339–342.

———, and Wright, H. E., eds., 1967, Pleistocene extinctions. The search for a cause: Yale University Press, New Haven, 453 p.*

Mather, K. F., and others, 1942, Pleistocene geology of western Cape Cod, Massachusetts: Geol. Soc. America Bull., v. 53, p. 1127–1174.

Mathews, W. H., and Mackay, J. R., 1960, Deformation of soils by glacier ice and the influence of pore pressures and permafrost: Royal Soc. Canada, Trans., ser. 3, sec. 4, v. 54, p. 27–36.

Matthes, F. E., 1900, Glacial sculpture of the Bighorn Mountains, Wyoming: U.S. Geol. Survey, 21st An. Rp., 1899–1900 pt. 2, p. 167–190.

———, 1930, Geologic history of the Yosemite Valley: U.S. Geol. Survey Prof. Paper 160, 137 p.

———, 1931, How the Mississippi came to break through Crowleys ridge. (Abstr.): Assoc. Am. Geographers, Ann., v. 21, p. 131–132.

———, and Belmont, A. D., 1950, The glacial anticyclone theory examined in the light of recent meteorological data from Greenland—Part 2: American Geophys. Un. Trans., v. 31, p. 174–182.

Mawson, Douglas, 1943, Macquarie Island: Australian Antarctic Exped., Sci. Rept., ser. A, v. 5, 194p.

Mayer-Oakes, W. J., ed., 1967, Life, land and water: Univ. of Manitoba Press, Winnipeg, 414 p.*

Mayr, Ernst, 1946, History of the North American bird fauna: Wilson Bull., v. 58, p. 3–41.

McBurney, C.B.M., 1967, The Haua Fteah (Cyrenaica): University Press, Cambridge, 387 p.

McCallien, W. J., 1941, The birth of glacial geology: Nature, v. 147, p. 316–318.

McCoy, F. W., and others, 1967, Speculations on the origin of the Algodones Dunes, California: Geol. Soc. America Bull., v. 78, p. 1039–1044.

McDonald, B. C., 1967, Pleistocene events and chronology in the Appalachian region of southeastern Quebec, Canada: Yale Univ., New Haven, Ph.D. dissert., 161 p.

McGregor, V. R., 1967, Holocene moraines and rock glaciers in the central Ben Ohau Range, South Canterbury, New Zealand: Jour. Glaciol., v. 6, p. 737–748.

McGrew, P. O., 1944, An early Pleistocene (Blancan) fauna from Nebraska: Field Mus. Nat. Hist. (Chicago), Pub. 546, Geol. ser. v. 9, p. 33–66.

McKee, E. D., and Goldberg, Moshe, 1969, Experiments on formation of contorted structures in mud: Geol. Soc. America Bull., v. 80, p. 231–244.

McKenzie, G. D., 1969, Observations on a collapsing kame terrace in Glacier Bay National Monument, south-eastern Alaska: Jour. Glaciol., v. 8, p. 413–425.

Meade, R. H., 1969, Errors in using modern streamload data to estimate natural rates of denudation: Geol. Soc. America Bull., v. 80, p. 1265–1274.

Mehringer, P. J., 1967, Pollen analysis of the Tule Springs Area, Nevada: Nevada State Mus., Anthropol. Papers, no. 13, p. 129–200.

———, and others, 1968, Late-Pleistocene boreal forest in the western Ozark

Highlands?: Ecology, v. 49, p. 567–568.

Meier, M. F., 1951, Recent eskers in the Wind River Mountains of Wyoming: Iowa Acad. Sci., v. 58, p. 291–294.

———, 1958, The mechanics of crevasse formation: Internat. Un. Geod. and Geophys. Assemblée Générale, Toronto 1957, Comptes Rendus et Rapp., v. 4, p. 500–509.

———, 1960, Distribution and variations of glaciers in the United States exclusive of Alaska: Internat. Assoc. Scientif. Hydrol. Publ. 54, p. 420–429.

———, and Post, Austin, 1969, What are glacier surges?: Canadian Jour. Earth Sci., v. 6, p. 807–817.

Mei-Ngo, Jen, 1960, La glaciation du Yulungshan, Yunnan, Chine: Ann. de Géogr., v. 69, p. 50–56.

Meinzer, O. E., 1922, Map of Pleistocene lakes of the Basin-and-Range Province and its significance: Geol. Soc. America Bull., v. 33, p. 541–552.

Mellor, Malcolm, 1961, The Antarctic Ice Sheet: U.S. Army, Cold Regions Research and Eng. Lab., Cold Regions Science and Engineering, pt. 1, sec. B, no. 1. Hanover, New Hampshire, 50 p.*

———, 1964, Snow and ice on the Earth's surface: Cold Regions Science and Engineering, pt. 2, sec. C, no. 1: U.S. Army, Cold Regions Research and Eng. Lab., Hanover, New Hampshire, 163 p.

Melton, M. A., 1961, Multiple Pleistocene glaciation of the White Mountains, Apache County, Arizona: Geol. Soc. America Bull., v. 72, p. 1279–1282.

Menard, H. W., 1953, Pleistocene and Recent sediment from the floor of the northeastern Pacific Ocean: Geol. Soc. America Bull., v. 64, p. 1279–1294.

Menéndez Amor, Josefa, and Florschütz, Franz, 1964, Results of the preliminary palynological investigation of samples from a 50m boring in southern Spain: Réal Soc. Española Hist. Nat., Bol. (Geol.) v. 62, p. 251–255.

Menéndez Amor, Josefa, Florschütz, Franz, and Wijmstra, T. A. 1970 [in press]

Mercer, J. H., 1956, Geomorphology and glacial history of southernmost Baffin Island: Geol. Soc. America Bull., v. 67, p. 553–570.

———, 1962, Glacier variations of the Antarctic: IGY World Data Center A: Glaciology, Glaciological Notes no. 11, p. 5–29.

———, 1967, Glaciers of the Antarctic: American Geogr. Soc., Antarctic Map Folio 7, 10 p., 4 pl.

Mesolella, K. J., and others, 1969. The astronomical theory of climatic change. Barbados data: Jour. Geol., v. 77, p. 250–274.

Messerli, Bruno, 1967, Die eiszeitliche und gegenwärtige Vergletscherung im Mittelmeerraum: Geographica Helvetica, v. 22, p. 105–228. (Engl. summ., p. 219.)*

Meyer, H. H. J., 1904, Die Eiszeit in den Tropen: Geogr. Zeitschr., v. 10, p. 593–600.

Milankovitch, Milutin, 1941, Kanon der Erdbestrahlung und seine Anwendung auf des Eiszeitproblem: Acad. Roy. Serbe, Ed. spec. v. 133 (sec. des sci. math. et nat., v. 33), 633 p.

Miller, B. B., 1966, Five Illinoian molluscan faunas from the southern Great Plains: Malacologia, v. 4, p. 173–260.

Miller, D. J., 1953, Late Cenozoic marine glacial sediments and marine terraces of Middleton Island, Alaska: Jour. Geol., v. 61, p. 17–40.

Miller, Hugh, 1850, On peculiar scratched pebbles, etc., in the boulder clay in Caithness: British Assoc., Rept. for 1850, p. 93–96.

————, *fil.*, 1884, On boulder-glaciation: Royal Soc. Edinburgh, Proc., v. 8, p. 156–189.

Millette, J. F. G., and Higbee, H. W., 1958, Periglacial loess, I. morphological properties: American Jour. Sci., v. 256, p. 284–293.

Milliman, J. D., and Emery, K. O., 1968, Sea levels during the past 35,000 years: Science, v. 162, p. 1121–1123.

Milojević, B. Ž., 1950, Les plateaux de loess et les régions de sable de Yougoslavie: Belgrade, Soc. Serbe de Géographie, Mém., v. 6, 70 p.

Milthers, Keld, 1942, Ledeblokke og Landskabsformer i Danmark: Danmarks Geol. Undersøg., ser. 2, no. 69, 137 p.*

Milthers, Vilhelm, 1909, Scandinavian indicator boulders in the Quaternary deposits; extension and distribution: Danmarks geol. Undersøg., ser. 2, no. 23, 154 p.

Mitchell, G. F., 1960, The Pleistocene history of the Irish Sea: British Assoc. Adv. Sci. Rept., v. 17, p. 313–325.

Mitchell, J. M. ed., 1968, Causes of climatic change: Meteorol. Mon., v. 8, No. 30, 159 p.

Molengraaff, G. A. F., and Weber, Max, 1921, On the relation between the Pleistocene glacial period and the origin of the Sunda Sea . . . : Kon. Akad. van Wetenschappen, Amsterdam, Sec. Sci., Proc., v. 23, p. 395–439.

Montfrans, H. M. van, and Hospers, J., 1969, A preliminary report on the stratigraphical position of the Matuyama-Brunhes geomagnetic field reversal in the Quaternary sediments of the Netherlands: Geol. en Mijnbouw, v. 48, p. 565–572.

Moore, G. W., 1956, Aragonite speleothems as indicators of paleotemperature: American Jour. Sci., v. 254, p. 746–753.

Moore, R. C., 1949, The Pliocene-Pleistocene boundary: American Assoc. Petroleum Geols. Bull., v. 33, p. 1276–1280.

Moore, Sherman, 1948, Crustal movement in the Great Lakes area: Geol. Soc. America Bull., v. 59, p. 697–710.

Mooser, Federigo, and others, 1956, La Cuenca de México. Consideraciones geológicas y arqueológicas: Inst. Nac. de Antropol. e Hist., Dirección de Prehist., Publ. 2, 51 p.

Morawetz, Sieghard, 1955, Zur Frage der eiszeitlichen Temperaturniederung: Geogr. Gesellsch. Wien, Mitt., v. 97, p. 192–206.

Moreau, R. E., 1933, Pleistocene climatic changes and the distribution of life in East Africa: Jour. Ecology, v. 21, p. 415–435.

————, 1963, Vicissitudes of the African biomes in the late Pleistocene: Zool. Soc. London Proc., v. 141, pt. 2, p. 395–421.

Morlot, Adolphe de, 1856, Notice sur le Quaternaire en Suisse: Lausanne, Soc. vaudoise des Sci. Nat. Bull., v. 4, p. 41–45.

Mörner, Nils-Axel, 1969, The late Quaternary history of the Kattegatt Sea and the Swedish west coast: Sveriges Geol. Undersök., ser. C, no. 640, 487 p.

Morrison, R. B., 1965, Lake Bonneville: Quaternary stratigraphy of eastern Jordan Valley, south of Salt Lake City, Utah: U.S. Geol. Survey Prof. Paper 477, 80 p.*

————, and Wright, H. E., eds., 1967, Quaternary soils: Univ. of Nevada, Desert Res. Inst., 338 p.

————, 1968, Means of correlation of Quaternary succession: Univ. of Utah Press, Salt Lake City, 631 p.*

Moskvitin, A. I., 1970, Pleistocene stratigraphy of central and western Europe: Acad. Sci. USSR, Geol. Inst., Trans., v. 193, 287 p. (German summ.) *[R]

Movius, H. L., 1944, Early man and Pleistocene stratigraphy in southern and eastern Asia: Harvard Univ., Peabody Mus., Papers, v. 19, no. 3, 113 p.

———, 1949, Lower Paleolithic archaeology in southern Asia and the Far East: Wistar Inst., Studies in Phys. Anthropol., no. 1, p. 17–82.

———, 1960, Radiocarbon Dates and Upper Palaeolithic archaeology in central and western Europe: Current Anthropol., v. 1. p. 355–392.*

Müller, Paul, 1957, Zur Bildungsgeschichte der Mergel von Noranco bei Lugano: Geobot. Institut Rübel (Zürich), Bericht für 1956, p. 23–55.

Mulvaney, D. J., and Golson, Jack, eds., 1970, Aboriginal man and environment in Australia: Australian Nat. Univ. Press, Canberra. (In press.)

Murray, G. E., 1961, Geology of the Atlantic and Gulf Coastal province of North America: Harper & Bros., New York, 692 p.

Murray, R. C., 1953, The petrology of the Cary and Valders tills of northeastern Wisconsin: American Jour. Sci., v. 251, p. 140–155.

Musil, Rudolf, and Valoch, Karel, 1966, Beitrag zur Gliederung des Würms in Mitteleuropa: Eiszeitalter und Geg., v. 17, p. 131–138.

Nairn, A.E.M., ed., 1961, Descriptive palaeoclimatology: Interscience Publishers, New York, 380 p.*

———, ed., 1964, Problems in palaeoclimatology: John Wiley and Sons, New York, 705 p.*

Nangeroni, Giuseppe, 1950, Tre nuovi lembi di morenico Günz nelle Prealpi lombarde: Istituto Lombardo di Sci. e Lett., Rendiconti (Classe di Sci.), v. 83, p. 1–8.

Neev, David, and Emery, K. O., 1967, The Dead Sea: Israel Geol. Survey Bull. 41, 147 p.*

Nelson, R. L., 1954, Glacial geology of the Frying Pan River drainage, Colorado: Jour. Geol., v. 62, p. 325–343.

Newell, N. D., 1946, Geological investigations around Lake Titicaca: American Jour. Sci., v. 244, p. 357–366.

Newman, W. S., and others, 1969, Late Quaternary geology of the Hudson River estuary. A preliminary report: New York Acad. Sci., Trans., ser. 2, v. 31, p. 548–570.

Nichols, D.R., 1966, Permafrost in the Recent Epoch: Permafrost Internat. Conf. Proc., Nat. Acad. Sci.-Nat. Research Council, publ. 1287, p. 172–175.

Nichols, Harvey, 1967, Pollen diagrams from sub-arctic central Canada: Science, v. 155, p. 1665–1668.

Nikiforova, K. V., 1964, On the stratigraphic position of the Astian: Internat. Quaternary Congr., 6th, Warsaw 1961, p. 547–557.

———, ed., 1965, Correlation of Anthropogen deposits of northern Eurasia: USSR. Acad. Sci., Geol. Inst., Nauka, Moscow, 114 p.*[R]

Nilsson, Erik, 1953, Om södra Sveriges senkvartära historia: Geol. Fören. i Stockholm Förh., v. 75, p. 155–246; Engl. summ., p. 234–237.*

———, 1960, The recession of the land-ice in Sweden during the Alleröd and the Younger Dryas ages: Internat. Geol. Congr., XXI Sess., Norden, Rept., pt. 4, p. 98–107.

———, 1963, Pluvial lakes and glaciers in East Africa: Stockholm Univ., Acta, Stockholm Contribs. in Geol., v. 11, p. 21–57.

————, 1968, The late-Quaternary history of southern Sweden, geochronology, ice-lakes, land-uplift: K. Svenska Vetenskapsakad. Handl., ser. 4, v. 12, no. 1, 117 p.*

Nilsson, Tage, 1961, The Pleistocene-Holocene boundary and the subdivision of the late Quaternary in southern Sweden: Internat. Quater. Congr., 6th, Warsaw 1961, Rept., v. 1, p. 479–494.

Niskanen, Erkki, 1939, On the upheaval of land in Fennoscandia: Ann. Acad. Scient. Fennicae, ser. A, v. 53, no. 10, 30 p.

————, 1943, On the deformation of the Earth's crust under the weight of a glacial ice load and related phenomena: Ann. Acad. Scient. Fennicae, ser. A, v. 3, no. 7, 22 p.

Norin, Erik, 1958, The sediments of the central Tyrrhenian Sea: Uppsala Univ., Mineral.-Geol. Inst., Medd., 136 p. (Repts. Swed. Deep-Sea Exp., v. 8, no. 1.)

Norman, G. W. H., 1938, The last Pleistocene ice-front in Chibougamau district, Quebec: Roy. Soc. Canada, Trans., 3d ser., v. 32, sec. 4, p. 69–86.

Norris, S. E., and White, G. W., 1961, Hydrologic significance of buried valleys in glacial drift: U. S. Geol. Survey Prof. Paper 424-B, p. 34–35.

North, F. J., 1943, Centenary of the glacial theory: (Notes on manuscripts and publications relating to its origin, development and its introduction into Britain.) Geologists' Assoc., Proc., v. 54, p. 1–28.

Nugent, L. E., 1946, Coral reefs in the Gilbert, Marshall, and Caroline Islands: Geol. Soc. America Bull., v. 57, p. 735–780.

Nye, J. F., 1952, The mechanics of glacier flow: Jour. Glaciol., v. 2, p. 81–93.

————, 1959, The motion of ice sheets and glaciers: Jour. Glaciol., v. 3, p. 493–507.

Oakley, K. P., 1969, Frameworks for dating fossil man: 3d ed., Weidenfeld and Nicolson, London, 366 p.*

Oaks, R. Q., and Coch, N. K., 1963, Pleistocene sea levels, southeastern Virginia: Science, v. 140, p. 979–983.

Oba, Tadamichi, 1969, Biostratigraphy and isotopic paleotemperature of some deep-sea cores from the Indian Ocean: Tohoku Univ., Sci. Repts., 2d ser., v. 41, p. 129–195.

Odum, E. P., 1962, Fundamentals of ecology: 2d ed., W. B. Saunders Co., Philadelphia, 546 p.

Ogden, J. G., 1959, A late-glacial pollen sequence from Martha's Vineyard, Massachusetts: American Jour. Sci., v. 257, p. 366–381.

————, 1963, The Squibnocket Cliff peat: radiocarbon dates and pollen stratigraphy: American Jour. Sci., v. 261, p. 344–353.

————, 1965, Pleistocene pollen records from eastern North America: Botan. Rev., v. 31, p. 481–504.*

————, 1966, Forest history of Ohio. I. Radiocarbon dates and pollen stratigraphy of Silver Lake, Logan County, Ohio: Ohio Jour. Sci., v. 66, p. 387–400.

Okko, Marjatta, 1965, M. Sauramo's Baltic Ice Lake B IV-B V-B VI; a re-evaluation: Finland. Acad. Sci., Ann., ser. A III, no. 84, 63 p.

————, 1967, The relation between raised shores and present land uplift in Finland during the past 8000 years: Ann. Acad. Scient. Fennicae, ser. A, sec. III, no. 93, 59 p.

Okko, Veikko, 1945, Untersuchungen über den Mikkeli-Os: Fennia, no. 69, paper 1, 55 p.

Okko, Veikko, 1955, Glacial drift in Iceland. Its origin and morphology: Comm. Géol. de Finlande Bull., no. 170, 133 p.

——, and Peltola, Esko, 1958, On the Outokumpu boulder train: Soc. géol. de Finlande, Comptes Rendus, no. 30, p. 113–134.

Olausson, Eric, 1961, Remarks on some Cenozoic core sequences from the central Pacific, with a discussion of the role of coccolithophorids and foraminifera in carbonate deposition: Göteborgs Kungl. Vetenskaps- och Vitterhets-Samhälles Handl., Sjätte Följden, ser. B, v. 8, no. 10, 35 p.

——, 1965, Evidence of climatic changes in North Atlantic deep-sea cores, with remarks on isotopic paleotemperature analysis: Progress in Oceanogr., v. 3, p. 221–252.

Olsson, I. U., ed., 1970, Radiocarbon variations and absolute chronology: Almqvist & Wiksell, Stockholm; Wiley Interscience, New York, 660 p.*

Opdyke, N.D., and others, 1966, Paleomagnetic study of Antarctic deep-sea cores: Science, v. 154, p. 349–357.

Öpik, E. J., 1953, A climatological and astronomical interpretation of the ice ages and of the past variations of terrestrial climate: Armagh Observatory (N. Ireland), Contribs., no. 9, 79 p.

——, 1965, Climatic change in cosmic perspective: Icarus, v. 4, p. 289–307.

——, 1966, More on climatic change: Icarus, v. 5, p. 215–217.

Osmaston, H. A., 1965, The past and present climate and vegetation of Ruwenzori and its neighbourhood: Oxford Univ., Worcester College, D. Phil. thesis.

Østrem, Gunnar, 1964, Ice-cored moraines in Scandinavia: Geogr. Annaler, v. 46, p. 282–337.

——, 1965, Problems of dating ice-cored moraines: Geogr. Annaler, v. 47A, p. 1–38.

Ostry, R. C., and Deane, R. E., 1963, Microfabric analyses of till: Geol. Soc. America Bull., v. 74, p. 165–168.

Outcalt, S. I., and Benedict, J. B., 1965, Photo-interpretation of two types of rock glacier in the Colorado Front Range, U.S.A. Jour. Glaciol., v. 5, p. 849–856.

Page, B. M., 1939, Multiple Alpine glaciation in the Leavenworth area, Washington: Jour. Geol., v. 47, p. 785–815.

Parker, G. G., and others, 1955, Water resources of southeastern Florida: U. S. Geol. Survey Water-Supply Paper 1255, 965p.

Paschinger, Viktor, 1912, Die Schneegrenze in verschiedenen Klimaten: Petermanns Geogr. Mitt. Ergänzungsbd., v. 37, Heft 173, 93 p.

Paterson, W. S. B., 1969, The physics of glaciers: Pergamon Press Ltd., Oxford, 250 p.

Peabody, F. E., 1954, Travertines and cave deposits of the Kaap escarpment of South Africa, and the type locality of *Australopithecus africanus* Dart: Geol. Soc. America Bull., v. 65, p. 671–706.

Peach, B. N., and Horne, John, 1879, The glaciation of the Shetland Isles: Geol. Soc. London Quart. Jour., v. 35, p. 788–811.

Peacock, M. A., 1935, Fiord-land of British Columbia: Geol. Soc. America, Bull., v. 46, p. 633–696.

Pei, Wen-Chung, 1939, [Geochronological table no. 1,] An attempted correlation of Quaternary geology, palaeontology and prehistory in Europe and

China: London Univ., Inst. Archaeology, Occ. Paper 2, 16 p.

Pels, Simon, 1966, Late Quaternary chronology of the Riverine Plain of southeastern Australia: Geol. Soc. Australia, Jour., v. 13, p. 27–40.

Peltier, L. C., 1949, Pleistocene terraces of the Susquehanna River, Pennsylvania: Pennsylvania Geol. Survey, ser. 4, Bull. G 23, 158 p.

————, 1959, Late Pleistocene deposits [of Bucks County, Pennsylvania]: Pennsylvania Geol. Survey, ser. 4, Bull. C9, p. 163–184.

Penck, Albrecht, 1882, Die Vergletscherung der deutschen Alpen . . . : J. A. Barth, Leipzig, 483 p.

————, 1906, Süd-Afrika und Sambesifälle: Geogr. Zeitschr., v. 12, p. 600–611.

————, 1921, Die Höttinger Breccie und die Inntalterrasse nördlich Innsbruck: Preuss, Akad. Wiss. (Berlin), Abh. 1920, Phys.-Math. Kl., no. 2, 136 p.

————, 1931, Zentral-Asien: Gesellsch. für Erdkunde zu Berlin, Zeitschr., 1931, p. 1–13.

————, 1936, Europa zur letzten Eiszeit: p. 222–237 in Länderkundliche Forschung, J. Engelhorns Nachf., Stuttgart.

————, 1938a, Das Klima der Eiszeit: Internat. Quartärkonferenz, 3d, Wien, 1936, Verhandl., p. 83–97.

————, 1938b, Die Strahlungstheorie und die geologische Zeitrechnung: Gesellsch. für Erdkunde zu, Berlin, Zeitschr., 1938, p. 321–350.

————, and Brückner, Eduard, 1909, Die Alpen im Eiszeitalter: Tauchnitz, Leipzig, 1199 p.

Penny, L. F., 1964, A review of the last glaciation in Great Britain: Yorkshire Geol. Soc., Proc., v. 34, p. 387–411.

————, and Catt, J. A., 1967, Stone orientation and other structural features of tills in East Yorkshire: Geol. Mag., v. 104, p. 344–360.

————, and others, 1969, Age and insect fauna of the Dimlington silts, East Yorkshire: Nature, v. 224, p. 65–67.

Penttilä, Seppo, 1963, The deglaciation of the Laanila area, Finnish Lapland: Comm. Géol. de Finlande, Bull. 203, p. 7–71.

Permafrost International Conference (1966) Proceedings: Washington, Nat. Acad. Sci.-Nat. Research Council publ. 1287, 563 p.*

Peterson, J. A., 1965, Ice-push ramparts in the George River basin, Labrador-Ungava: Arctic Inst. North America, Jour., v. 18, p. 189–193.

Péwé, T. L., 1959, Sand-wedge polygons (tesselations) in the McMurdo Sound Region, Antarctica—a progress report: American Jour. Sci., v. 257, p. 545–552.

————, 1960, Multiple glaciation in the McMurdo Sound region, Antarctica—a progress report: Jour. Geol., v. 68, p. 498–514.

————, 1966, Paleoclimatic significance of fossil ice wedges: Biuletyn Periglacjalny, no. 15, p. 65–73.

————, ed., 1969, The periglacial environment, past and present: McGill-Queen's University Press, 487 p.

Pfannenstiel, Max, 1951, Quartäre Spiegelschwankungen des Mittelmeeres und des Schwarzen Meeres: Naturforsch. Gesell. in Zürich, Vierteljahrsschr., v. 96, p. 81–102.

Phleger, F. B., 1951, Ecology of Foraminifera, northwest Gulf of Mexico pt. 1, Foraminifera distribution: Geol. Soc. America Mem. 46, pt. 1, 88 p.

Picard, Leo, 1943, Structure and evolution of Palestine: Hebrew Univ., Jerusalem, Geol. Dept., Bull., v. 4, nos. 2–4 (Pub. 84), 187 p.

Piggot, C. S., and Urry, W. D., 1942, Time relations in ocean sediments: Geol. Soc. America, Bull., v. 53, p. 1187–1210.

Piskin, K., and Bergstrom, R. E., 1967, Glacial drift in Illinois: thickness and character: Illinois Geol. Survey Circ. 416.

Plass, G. N., 1956, Carbon dioxide and the climate: American Scient., v. 44, p. 302–316.*

Playfair, John, 1802, Illustrations of the Huttonian theory of the earth: W. Creech, Edinburgh, 528 p.

Polanski, Jorge, 1957, Sobre algunos metodos paleogeográficos de la investigación del cuartario pedemontano de Mendoza: Asoc. Geol. Argentina, Rev., v. 12, p. 211–232.

——, 1963, Estratigrafía, neotectónica y geomorfología del Pleistoceno pedemontano entre los rios Diamante y Mendoza (Provincia de Mendoza): Asoc. Geol. Argentina, Rev., v. 17, p. 127–349.

Porter, S. C., 1964, Composite Pleistocene snow line of Olympic Mountains and Cascade Range, Washington: Geol. Soc. America Bull., v. 75, p. 477–482.

——, 1966, Pleistocene geology of Anaktuvuk Pass, Central Brooks Range, Arctic Institute of North America Tech. Paper 18, 100 p.

——, 1970, The Quaternary glacial record in Swat Kohistan, northern West Pakistan: Geol. Soc. America Bull., v. 81, p. 1421–1446.

——, and Denton, G. H. (1967) Chronology of Neoglaciation in the North American Cordillera: American Jour. Sci., v. 265, p. 177–110.*

Poser, Hans, 1951, Die nördliche Lössgrenze in Mitteleuropa und das spätglaziale Klima: Eiszeitalter und Geg., v. 1, p. 27–55.

Post, Austin, 1960, The exceptional advances of the Muldrow, Black Rapids, and Susitna Glaciers: Jour. Geophys. Res., v. 65, p. 3703–3712.

——, 1966–1967, The recent surge of Walsh Glacier, Yukon and Alaska: Jour. Glaciol., v. 6, no. 45, 1966, p. 375–381; no. 47, 1967, p. 763–764.

——, 1969, Distribution of surging glaciers in North America: Jour. Glaciol., v. 8, p. 229–240.

Post, Hampus von, 1856, Om sandåsen vid Köping i Westmanland: Stockholm, K. Vetenskaps-Akad., Handl. för år 1854, p. 345.

Post, Lennart von, 1916, Forest tree pollen in south Swedish peat bog deposits: (Lecture to the 16th Convention of Scandinavian Naturalists, Kristiania.) English translation by M. B. Davis and K. Faegri, 1967, Pollen et Spores, v. 9, p. 375–401.

Potter, Noel, and Moss, J. H., 1968, Origin of the Blue Rocks block field and adjacent deposits, Berks County, Pennsylvania: Geol. Soc. America Bull., v. 79, p. 255–262.

Powers, H. A., and Wilcox, H. E., 1964, Volcanic ash from Mount Mazama (Crater Lake) and from Glacier Peak: Science, v. 144, p. 1334–1336.

Prest, V. K., and Grant, D. R., 1969, Retreat of the last ice sheet from the Maritime Provinces—Gulf of St. Lawrence region: Canada Geol. Survey Paper 69–33, 15 p.

Prest, V. K., and others, 1968, Glacial map of Canada. Scale 1:5,000,000: Geol. Survey of Canada, Map 1253A.

Prestwich, Joseph, 1886–1888, Geology, chemical, physical, and stratigraphical: Clarendon Press, Oxford, v. 1, 1886, 477 p.; v. 2, 1888, 606 p.

Priestley, Raymond, and others, eds., 1964, Antarctic research: Butterworths, London, 360 p.

Prinz, Gyula, 1909, Die Vergletscherung des nördlichen Teiles des zentralen Tien-schan-Gebirges: Wien, K. K. Geogr. Gesell., Mitt., v. 52, p. 10–75.

Prošek, František, and Ložek, Vojen, 1957, Stratigraphische Übersicht des tschechoslowakischen Quartärs: Eiszeitalter und Geg., v. 8, p. 37–90.

Pumpelly, Raphael, 1905, Explorations in Turkestan: Carnegie Inst., Washington, Pub. no. 26, 324 p.

Putnam, W. C., 1949, Quaternary geology of the June Lake district, California: Geol. Soc. America Bull., v. 60, p. 1281–1302.

———, 1950, Moraine and shoreline relationships at Mono Lake, California: Geol. Soc. America Bull., v. 61, p. 115–122.*

Raasch, G. O., ed., 1961, Geology of the Arctic: Univ. of Toronto Press, 1196 p.

Rampton, V. N., 1969, Pleistocene geology of the Snag-Klutlan area, southwestern Yukon Territory, Canada: Univ. of Minnesota Ph. D. dissert. 237 p.

Ramsay, A. C., 1852, On the superficial accumulations and surface-markings of North Wales: Geol. Soc. London Quart. Jour., v. 8, p. 371–376.

———, 1862, On the glacial origin of certain lakes in Switzerland, the Black Forest, Great Britain, Sweden, North America, and elsewhere: Geol. Soc. London Quart. Jour., v. 18, p. 185–204.

———, and Geikie, James, 1878, On the geology of Gibraltar: Geol. Soc. London Quart. Jour., v. 34, p. 505–541.

Rankama, Kalervo, ed., 1965, The Quaternary: Interscience Publishers, New York, v. 1, 300 p.*

———, 1967, The Quaternary: Interscience Publishers, New York, v. 2, 477 p.*

Rapp, Anders, 1960, Recent development of mountain slopes in Karkevagge and surroundings, northern Scandinavia: Geograf. Annaler, v. 42, p. 73–200.

Ray, C. E., and others, 1967, Fossil mammals and pollen in a late Pleistocene deposit at Saltville, Virginia: Jour. Paleontol., v. 41, p. 608–622.

Ray, L. L., 1940, Glacial chronology of the southern Rocky Mountains: Geol. Soc. America Bull., v. 51, p. 1851–1918.

———, 1965, Geomorphology and Quaternary geology of the Owensboro quadrangle, Indiana and Kentucky: U. S. Geol. Survey Prof. Paper 488, 72 p.

Reade, T. M., 1872, The post-glacial geology and physiography of west Lancashire and the Mersey estuary: Geol. Mag., v. 9, p. 111–119.

Reboul, Henri, 1833, Géologie de la période quaternaire, et introduction à l'histoire ancienne: F.-G. Levrault, Paris, 222 p.

Reed, Bruce, and others, 1962, Some aspects of drumlin geometry: American Jour. Sci., v. 260, p. 200–210.

Reeves, C. C., 1965a, Pleistocene climate of the Llano Estacado: Jour. Geol., v. 73, p. 181–189.

———, 1965b, Chronology of West Texas pluvial lake dunes: Jour. Geol. v. 73, p. 504–508.

———, 1968, Introduction to paleolimnology: Developments in sedimentology 11, Elsevier, Amsterdam, 228 p.

Reid, H. F. 1892, Studies of Muir Glacier, Alaska: Nat. Geogr. Mag., v. 4, p. 19–84.

Reiner, Ernst, 1960, The glaciation of Mount Wilhelm, Australian New Guinea: Geogr. Rev., v. 50, p. 491–503.

Repenning, C. A., and others, 1964, Tundra rodents in a Late Pleistocene fauna from the Tofty Placer District, central Alaska: Arctic, v. 17, p. 177–197.

Retzer, J. L., 1954, Glacial advances and soil development, Grand Mesa, Colorado: American Jour. Sci., v. 252, p. 26–37.

Richards, H. G., 1962, Studies on the marine Pleistocene: Pt. 1. The marine Pleistocene of the Americas and Europe. Pt. 2. The marine Pleistocene mollusks of eastern North America: American Philos. Soc., Trans., v. 52, pt. 3, 141 p.

————, and Fairbridge, R. W., 1965, Annotated bibliography of Quaternary shorelines (1945–1964): Acad. Nat. Sciences, Philadelphia, Spec. Publ. 6, 280 p.*

Richardson, H. L., 1943, The Ice Age in western China: West China Border Res. Soc., Jour., v. 14, ser. 3, p. 1–27. (Summary in Geogr. Rev., v. 34, 1944, p. 490.)

Richardson, J. L., 1966, Changes in level of Lake Naivasha, Kenya, during postglacial times: Nature, v. 209, p. 290–291.

Richmond, G. M., 1957, Three pre-Wisconsin glacial stages in the Rocky Mountain region: Geol. Soc. America Bull., v. 68, p. 239–262.

————, 1960, Correlation of alpine and continental glacial deposits of Glacier National Park and adjacent high plains, Montana: U.S. Geol. Survey Prof. Paper 400, p. B223–B225.

————, 1962, Quaternary stratigraphy of the La Sal Mountains, Utah: U. S. Geol. Survey Prof. Paper 324, 135 p.

————, 1963, Correlation of some glacial deposits in New Mexico: U. S. Geol. Survey Prof. Paper 450-E, p. 121–125.

————, 1964, Glaciation of Little Cottonwood and Bells Canyons, Wasatch Mountains, Utah. U. S. Geol. Survey Prof. Paper 454-D, 41 p.

————, ed., 1968, Geology of the Alps: Univ. of Colorado Studies, ser. in Earth Sciences, no. 7, 176 p.

————, and others, 1959, Application of stratigraphic classification and nomenclature to the Quaternary: American Assoc. Petroleum Geologists Bull., v. 43, p. 663–675.

Richter, Konrad, 1932, Die Bewegungsrichtung des Inlandeises rekonstruiert aus den Kritzen und Längsachsen der Geschiebe: Zeitschr. für Geschiebeforsch., v. 8, p. 62–66.

————, 1937, Die Eiszeit in Norddeutschland: Borntraeger, Berlin, 179 p.

————, 1953, Klimatische Vershiedenartigkeit glazialer Vorstossphasen in Norddeutschland: Internat. Quater. Congr., 4th, Roma-Pisa 1953, Actes., p. 809–818.

Richthofen, Ferdinand von, 1882, On the mode of origin of the loess: Geol. Mag., n.s., v. 9, p. 293–305.

Riedel, W. R., and others, 1963, "Pliocene-Pleistocene" boundary in deep-sea sediments: Science, v. 149, p. 1238–1240.

Robertson, Percival, 1938, Some problems of the middle Mississippi River region during Pleistocene time: St. Louis Acad. Sci., Trans., v. 29, p. 165–240.

Robin, G. de Q., 1964, Glaciology: Endeavour, v. 23, p. 102–107.

Rognon, Pierre, 1967, Le Massif de l'Atakor et ses bordures (Sahara central): Centre de Recherches sur les Zones Arides (Paris), ser. géol. no. 9, 559 p.

Rohmeder, Guillermo, 1942, La glaciación diluvial de los Nevados del Anconquija: Tucumán, Univ. Nac., Instit. Estudios Geográf., Mon. 2, 68 p.

Rominger, J. F., and Rutledge, P. C., 1952, Use of soil mechanics data in correlation and interpretation of Lake Agassiz sediments: Jour. Geol., v. 60, p. 160–180.

Rónai, Andreas, 1968, The Pliocene-Pleistocene boundary in the Hungarian basin: Hungary, Acad. Sci., Acta Geologica, v. 12, p. 219–230.

———, 1969, Eine vollständige Folge quartärer Sedimente in Ungarn: Eiszeitalter und Geg., v. 20, p. 5–34.

Ronca, L. B., and Zeller, E. J., 1965, Thermoluminescence as a function of climate and temperature: American Jour. Sci., v. 263, p. 416–428.

Roosma, Aino, 1958, A climatic record from Searles Lake, California: Science, v. 128, p. 716.

Ross, C. P., 1938, Geology and ore deposits of the Bayhorse region, Custer County, Idaho: U. S. Geol. Survey Bull. 877, 161 p.

Rouse, I. B., 1971, Prehistory. A systematic survey of man's unrecorded past: McGraw-Hill Book Co., New York (in course of publication).*

Rousseau, Jacques, 1950, Cheminements botaniques a travers Anticosti: Canadian Jour. Res., v. 28, sec. C, p. 225–272.

Różycki, S. Z., 1967a, Le sens des vents portant la poussière de loess a la lumière de l'analyse des formes d'accumulation du loess en Bulgarie et en Europe Centrale: Rev. Geomorph. Dynam., v. 17, no. 1, 9 p.

———, 1967b, Plejstocen Polski Środkowej: Państwowe Wydawnictwo Naukowe, Warszawa, 251 p. (Engl. summ.)

Rubey, W. W., 1952, Geology and mineral resources of the Hardin and Brussels quadrangles (Illinois): U. S. Geol. Survey Prof. Paper 218, 179 p.

Rudberg, Sten, 1958, Some observations concerning mass movement on slopes in Sweden: Geol. Fören. i Stockholm Förh., v. 1, p. 114–125.

Ruggieri, Giuliano, 1967, The Miocene and later evolution of the Mediterranean Sea: p. 283–290 in Aspects of Tethyan biogeography, Systematics Assoc. Pub. no. 7.

Ruhe, R. V., and others, 1968, Iowan drift problem, northeastern Iowa: Iowa Geol. Survey, Rept. Inv. 7, 40 p.*

Rühle, Edward, and Mojski, J. E., 1965, Stratigraphical atlas of Poland, stratigraphical and facial problems, pt. 12: Quaternary, 16 maps with text: Warsaw, Geological Institute.*

Ruske, Ralf, 1965, Mittelpleistozäne Lösse und Böden in Mitteleuropa und deren stratigraphische Einstufung: Geologie, v. 14, p. 554–563.

Russell, I. C., 1885, Geological history of Lake Lahontan, a Quaternary lake of northwestern Nevada: U. S. Geol. Survey Mon. 11, 288 p.

———, 1889, Quaternary history of Mono Valley, California: U. S. Geol. Survey, 8th An. Rp., 1886–1887, pt. 1, p. 261–394.

———, 1890, Notes on the surface geology of Alaska: Geol. Soc. America Bull., v. 1, p. 99–162.

———, 1893, Malaspina Glacier: Jour. Geol., v. 1, p. 219–245.

———, 1900, Geology of the Cascade Mountains in northern Washington: U. S. Geol. Survey 20th Ann. Rept., pt. 2, p. 83–210.

Sahni, M. R., and Khan, Ehsanullah, 1959, Stratigraphy, structure, and correlation of the Upper Shivaliks east of Chandigarh: Palaeontol. Soc. India, Jour., v. 4, p. 61–74.

Sakaguchi, Yutaka, 1961, Paleogeographical studies of peat bogs in northern Japan: Univ. of Tokyo, Jour. Fac. of Sci., sec. 2, v. 12, p. 421–513.*

Saks, V. N., ed., 1966, Quaternary period of Siberia: USSR Acad. Sci., Siberian Branch, Inst. of Geol. and Geophys. Nauka, Moscow, 514 p.[R]

Salisbury, R. D., and others, 1893, Surface geology—report of progress, 1892: New Jersey Geol. Survey, Ann. Rept. for 1892, pt. 1, p. 33–328.

Salmi, Martti, 1955, Additional information on the findings in the Mylodon Cave at Ultima Esperanza: Geogr. Soc. Finland, Acta Geogr., v. 14, p. 314–333.

Sasa, Yasuo, and Tanaka, Kaoru, 1938, Glaciated topography in the Kanbô Massif, Tyôsen (Korea): Hokkaido Imper. Univ., Jour. Fac. Sci., ser. 4, v. 4, p. 193–212.

Saucier, R. T., & Fleetwood, A. R., 1970, Origin and chronologic significance of late Quaternary terraces, Ouachita River, Arkansas and Louisiana: Geol. Soc. America Bull., v. 81, p. 869–890.*

Sauer, C. O., 1944, A geographic sketch of early man in America: Geogr. Rev., v. 34, p. 529–573.

Sauramo, Matti, 1923, Studies on the Quaternary varve sediments in southern Finland: Comm. Géol. de Finlande, Bull., no. 60, 164 p.

———, 1929, The Quaternary geology of Finland: Comm. Géol. de Finlande, Bull., no. 86, 110 p.

———, 1931, Zur Frage des inneren Baus des Salpausselkä in Finland: Zeitschr. für Gletscherk., v. 19, p. 300–315.

Saussure, Horace-Benédict de, 1786–1796, Voyages dans les Alpes, précédés d'un essai sur l'histoire naturelle des environs de Genève: Barde, Manget et cie., Genève, v. 1, 2 (1787), v. 3, 4 (1786), v. 5, 6, 7, 8 (1796).

Savage, D. E., 1951, Late Cenozoic vertebrates of the San Francisco Bay region: Univ. of California Dept. Geol. Sci., Bull., v. 28, p. 215–314.*

Savile, D. B. O., 1961, The botany of the northwestern Queen Elizabeth Islands: Canadian Jour. Bot., v. 39, p. 909–942.

Sawyer, J. S., ed., 1966, World climate from 8000 to 0 B.C.: Proc. Internat. Symposium on World Climate, Imperial College, London, 1966. Roy. Meteorol. Soc., London, 229 p.*

Scheidig, A., 1934, Der Löss und seine geotechnischen Eigenschaften: Th. Steinkopff, Dresden and Leipzig, 233 p.

Schimper, Karl, 1837, Ueber die Eiszeit: Société Helvétique des Sci. Nat., Actes, 22 Sess., Neuchâtel, p. 38–51.

Schmoll, H. R., 1961, Orientation of phenoclasts in laminated glaciolacustrine deposits, Copper River basin, Alaska: U. S. Geol. Survey Prof. Paper 424, p. C-192–C-195.

———, and Bennett, R. H., 1961, Axiometer—a mechanical device for locating and measuring pebble and cobble axes for macrofabric studies: Jour. Sed. Petrology, v. 31, p. 617–622.

Schofield, J. C., 1964, Postglacial sea levels and isostatic uplift: New Zealand Jour. Geol. and Geophys., v. 7, p. 359–370.

Scholl, D. W., and Stuiver, Minze, 1967, Recent submergence of southern Florida: A comparison with adjacent coasts and other eustatic data: Geol. Soc. America Bull., v. 78, p. 437–454.*

Scholl, D. W., and others, 1969, Florida submergence curve revised: its relation to coastal sedimentation rates: Science, v. 163, p. 562–564.

Scholtes, W. H., 1955, Properties and classification of the paha loess-derived soils in northeastern Iowa: Iowa State College Jour. Sci., v. 30, p. 163–209.

Schönhals, Ernst, 1953, Gesetzmässigkeiten im Feinaufbau von Talrandlössen mit Bemerkungen über die Entstehung des Lösses: Eiszeitalter und Geg., v. 3, p. 19–36.

Schott, W., 1935, Die Foraminiferen in dem äquatorialen Teil des atlantischen Ozeans: Wiss. Ergebn. d. Deutschen atlantischen Exped . . . "Meteor," 1925–27, v. 3, pt. 3, p. 42–134.

Schultz, C. B., and Frye, J. C., eds., 1968, Loess and related eolian deposits of the world: University of Nebraska Press, 355 p.*

Schultz, C. B., and Stout, T. M., 1948, Pleistocene mammals and terraces in the Great Plains: Geol. Soc. America Bull., v. 59, p. 553–588.

Schultz, J. R., 1937, A late Cenozoic vertebrate fauna from the Coso Mountains, Inyo County, California: Carnegie Inst., Washington, Pub. 487, p. 75–109.

Schulz, Werner, 1967, Über glazigene Schrammen auf dem Untergrund und sichelförmige Marken auf Geschieben in Norddeutschland: Geogr. Berichte, v. 2, p. 125–142.*

Schumm, S. A., 1968, River adjustment to altered hydrologic regimen— Murrumbidgee River and paleochannels, Australia: U. S. Geol. Survey Prof. Paper 598, 65p.*

————, and Bradley, W. C., eds., 1969, United States contributions to Quaternary research: Geol. Soc. America Spec. Paper 123, 305 p.

Schwarzbach, Martin, 1955, Allgemeiner Überblick der Klimageschichte Islands: Neues Jahrb. für Geol. und Paläontol., Monatsheft, p. 97–130.

————, 1963, Climates of the past: D. Van Nostrand Co., London, 328 p.*

————, 1968, Neuere Eiszeithypothesen: Eiszeitalter und Geg., v. 19, p. 250–261.

Schytt, Valter, 1959, The glaciers of the Kebnekajse-Massif: Geograf. Annaler, v. 41, 1959, p. 213–227.

————, and others, 1968, The extent of the Würm glaciation in the European Arctic: Stockholms Universitet. Naturgeografiska Instit., Medd., no. A 20, p. 207–216.

————, 1968, The extent of the Würm glaciation in the European Arctic: Int. Assoc. Sci. Hydrol., I.U.G.G., Gen. Assembly Bern, 1967, Comm. of Snow and Ice, publ. 79, p. 207–216.

Seddon, Brian, 1957, Late-glacial cwm glaciers in Wales: Jour. Glaciol., v. 3, p. 94–99.

Segerstråle, S. G., 1957, On the immigration of the glacial relicts of northern Europe, with remarks on their prehistory: Soc. Scient. Fennica, Comment. Biol., v. 16, no. 16, 117 p.*

Segerstrom, Kenneth, 1964, Quaternary geology of Chile: brief outline: Geol. Soc. America Bull., v. 75, p. 157–170.

Šegota, Tomislav, 1967, Paleotemperature changes in the upper and middle Pleistocene: Eiszeitalter und Geg., v. 18, p. 127–141.*

Sellers, W. D., 1967, Physical climatology: Univ. of Chicago Press, 2d impr., 272 p.

Selling, O. H., 1948, Studies in Hawaiian pollen statistics. Part III. On the late Quaternary history of the Hawaiian vegetation: Bishop Mus. (Honolulu), Spec. Pub. 39, 154 p.*

Sernander, Rutger, 1910, Die schwedischen Torfmoore als Zeugen postglazialer Klimaschwankungen: Internat. Geol. Congr., 11th, Stockholm, 1910, p. 203–211.

Servant, Michel, and others, 1969, Chronologie du Quaternaire récent des basses régions du Tchad: Paris. Acad. Sci., Comptes Rendus, v. 269, p. 1603–1606.

Shackleton, Nicholas, 1967, Oxygen isotope analyses and Pleistocene temperatures re-assessed: Nature, v. 215, p. 15–17.

Shaler, N. S., 1874, Preliminary report on the recent changes of level on the coast of Maine . . . : Boston Soc. Nat. History, Mem., v. 2, p. 320–340.

———, 1884, On the origin of kames: Boston Soc. Nat. History, Proc., v. 23, p. 36–44.

Shapley, Harlow, ed., 1953, Climatic change: Harvard Univ. Press, Cambridge, 318 p.

Sharp. R. P., 1938, Pleistocene glaciation in the Ruby-East Humboldt Range, northeastern Nevada: Jour. Geomorphol., v. 1, p. 296–323.

———, 1942, Multiple Pleistocene glaciation on San Francisco Mountain, Arizona: Jour. Geol., v. 50, p. 481–503.

———, 1949a, Pleistocene ventifacts east of the Big Horn Mountains, Wyoming: Jour. Geol., v. 57, p. 175–195.

———, 1949b, Studies of superglacial debris on valley glaciers: American Jour. Sci., v. 247, p. 289–315.*

———, 1951, Glacial history of Wolf Creek, St. Elias Range, Canada: Jour. Geol., v. 59, p. 97–117.

———, 1960a, Glaciers: Univ. of Oregon Press, 78 p.

———, 1960b, Pleistocene glaciation in the Trinity Alps of northern California: American Jour. Sci., v. 258, p. 305–340.

———, and others, 1959, Pleistocene glaciers on southern California mountains: American Jour. Sci., v. 257, p. 81–94.

Shaw, D. M., & Donn, W. L., 1968, Milankovitch radiation variations, a quantitative evaluation: Science, v. 162, p. 1270–1272.

Shaw, E. W., 1911a, High terraces and abandoned valleys in western Pennsylvania: Jour. Geol., v. 19, p. 140–156.

———, 1911b, Preliminary statement concerning a new system of Quaternary lakes in the Mississippi basin: Jour. Geol., v. 19, p. 481–491.

Shepard, F. P. 1937, Origin of the Great Lakes basins: Jour. Geol., v. 45, p. 76–88.

———, 1963a, Submarine geology: 2d ed., Harper, and Row, New York, 557 p.*

———, 1963b, Thirty-five thousand years of sea level: p. 1–10 in Essays in marine geology in honor of K. O. Emery, Univ. of Southern California Press, Los Angeles.

———, and Curray, J. R., 1967, Carbon-14 determination of sea level changes in stable areas: p. 283–291 in Progress in Oceanography, v. 4.

Shepps, V. C., 1953, Correlation of the tills of northeastern Ohio by size analysis: Jour. Sed. Petrology, v. 23, p. 34–48.

———, and others, 1959, Glacial geology of northwestern Pennsylvania: Pennsylvania Geol. Survey Bull. G 32, 59 p. + map.

Shotton, F. W., 1962, The physical background of Britain in the Pleistocene: Advancemt. of Sci., v. 19, p. 193–206.

———, 1967a, Age of the Irish Sea Glaciation of the Midlands: Nature, v. 215, p. 1366.

———, 1967b, The problems and contributions of methods of absolute dating within the Pleistocene period: Geol. Soc. London Quart. Jour., v. 122, p. 356–383.*

———, and others, 1970, [Discussion of principles of Quaternary stratigraphic subdivision in Britain]: Geol. Soc. London Proc., no. 1660, p. 377–392.

Simonson, R. W., 1954, Identification and interpretation of buried soils: American Jour. Sci., v. 252, p. 705–732.

——, and Hutton, C. E., 1954, Distribution curves for loess: American Jour. Sci., v. 252, p. 99–105.

Simpson, G. C., 1934, World climate during the Quaternary period: Roy. Meteorol. Soc. Quart. Jour., v. 60, p. 425–478.

——, 1940, Possible causes of change in climate and their limitations: Linn. Soc. London, Proc., v. 152, p. 190–219.

Simpson, G. G., 1929, Extinct land mammals of Florida: Florida Geol. Survey 20th Ann. Rept. 1927–1928, 294 p.*

Sindowski, K.-H., 1958, Das Eem in Wattgebeit zwischen Norderney und Spiekeroog, Ostfriesland: Geol. Jahrb., v. 76, p. 151–174.

Sirkin, L. A., 1967, Correlation of Late Glacial pollen stratigraphy and environments in the northeastern U.S.A.: Rev. Palaeobotan. and Palynol.. v. 2, p. 205–218.

Sissons, J. B., 1960, Subglacial, marginal, and other glacial drainage in the Syracuse-Oneida area, New York: Geol. Soc. America Bull., v. 71, p. 1575–1588.

——, 1962, A re-interpretation of the literature on late-glacial shorelines in Scotland with particular reference to the Forth area: Edinburgh Geol. Soc., Trans., v. 19, p. 83–99.

——, 1967, The evolution of Scotland's scenery: Oliver and Boyd, Edinburgh; Archon Books, Hamden, Connecticut, 259 p.

Slater, George, 1927, The structure of the disturbed deposits of Møens Klint, Denmark: Royal Soc. Edinburgh, Trans., v. 55, p. 289–302.

——, 1929, The structure of the drumlins exposed on the south shore of Lake Ontario: New York State Mus. Bull. 281, p. 3–19.

Smalley, I. J., 1966, Drumlin formation: a rheological model: Science, v. 151, p. 1379.

——, and Unwin, D. J., 1968, The formation and shape of drumlins and their distribution and orientation in drumlin fields: Jour. Glaciol., v. 7, p. 377–390.

Smith, G. D., 1942, Illinois loess—Variations in its properties and distribution: a pedologic interpretation: Univ. of Illinois, Agric. Expt. Sta., Bull. 490, p. 139–184.

Smith, G. I., and Haines, D. V., 1964, Character and distribution of nonclastic minerals in the Searles Lake evaporite deposit California: U. S. Geol. Survey Bull. 1181–P, 58 p.*

Smith, G. I., and others, 1967, Pleistocene geology and palynology, Searles Valley, California: Guidebook, Friends of the Pleistocene, Pacific Coast Section, September 1967, 66 p.

Smith, H. T. U., 1936, Periglacial landslide topography of Canjilon Divide, Rio Arriba County, New Mexico: Jour. Geol., v. 44, p. 836–860.

——, 1965, Dune morphology and chronology in central and western Nebraska: Jour. Geol., v. 73, p. 557–578

——, and Ray, L. L., 1941, Southernmost glaciated peak in the United States: Science, v. 93, p. 209.

Smith, W. D., 1927, Contributions to the geology of southeastern Oregon (Steens and Pueblo Mountains): Jour. Geol., v. 35, p. 421–440.

——, and others, 1941, Geology and physiography of the northern Wallowa

Mountains, Oregon Dept. Geol. and Mineral Ind., Bull. 12, 64 p.

Snyder, C. T., and Langbein, W. B., 1962, The Pleistocene lake in Spring Valley, Nevada, and its climatic implications: Jour. Geophys. Res., v. 67, p. 2385–2394.

Snyder, C. T., and others, 1964, Pleistocene lakes in the Great Basin: U. S. Geological Survey Misc., Geol. Inv. Map I-416.*

Soergel, Wolfgang, 1919, Lösse, Eiszeiten und paläolithische Kulturen. Eine Gliederung und Altersbestimmung der Lösse: Gustav Fischer, Jena, 177 p.

———, 1924, Die Diluvialen Terrassen der Ilm und ihre Bedeutung für die Gliederung des Eiszeitalters: Gustav Fischer, Jena, 79 p.

Soil Survey Staff, 1960, Soil classification. A comprehensive system. 7th approximation: U. S. Dept. Agric., Soil Cons. Service, 265 p.

Sparks, B. W., 1961, The ecological interpretation of Quaternary non-marine Mollusca: Linnean Soc. London, Proc., 1959–60, p. 71–80.

Spieker, E. M., and Billings, M. P., 1940, Glaciation in the Wasatch Plateau, Utah: Geol. Soc. America Bull., v. 51, p. 1173–1198.

Spreitzer, Hans, 1941, Die Eiszeitforschung in der Sowjetunion: Quartär (Berlin) v. 3, p. 1–43.

Sproule, J. C., 1939, The Pleistocene geology of the Cree Lake region, Saskatchewan: Royal Soc. Canada, Trans., 3d ser., v. 33, sec. 4, p. 101–109.

Stalker, A. M., 1956, The erratics train, foothills of Alberta: Canada Geol. Survey Bull. 37, 28 p.

———, 1960a, Ice-pressed drift forms and associated deposits in Alberta: Canada Geol. Survey Bull. 57, 38 p.

———, 1960b, Surficial geology of the Red Deer-Stettler map-area, Alberta: Canada Geol. Survey Mem. 306, 140 p.

Stearns, C. E., and, Thurber, D. L., 1967, Th^{230}/U^{234} dates of late Pleistocene marine fossils from the Mediterranean and Moroccan littorals: Progress in Oceanography, v. 4, p. 293–305.

Stearns, H. T., 1945, Glaciation of Mauna Kea, Hawaii: Geol. Soc. America Bull., v. 56, p. 267–274.

Stearns, S. R., 1966, Permafrost (perennially frozen ground): Cold Regions Science and Engineering, pt. 1, sec. A2. U. S. Army, Cold Regions Research and Eng. Lab., Hanover, New Hampshire, 77 p.*

Stewart, D. P., and MacClintock, Paul, 1969, The surficial geology and Pleistocene history of Vermont: Vermont Geol. Survey Bull. 31, 251 p.*

Stipp, J. J., and others, 1967, K/Ar age estimate of the Pliocene-Pleistocene boundary in New Zealand: American Jour. Sci., v. 265, p. 462–474.

Stock, Chester, 1956, Rancho La Brea, a record of Pleistocene life in California: Los Angeles Co. Mus., 6th ed., Sci. Ser., no. 20, Paleontology, no. 11, 83p.*

Stokes, M. A., and Smiley, T. L., 1968, An introduction to tree-ring dating: Univ. of Chicago Press, 73 p.*

Stone, G. H., 1880, The kames of Maine: Boston Soc. Nat. Hist., Proc., v. 20, p. 430–469.

Stone, K. H., 1963, Alaskan ice-dammed lakes: Assoc. American Geographers, Annals, v. 53, p. 332–349.

Stout, Wilber, and others, 1943, Geology of water in Ohio: Ohio Geol. Survey, 4th ser., Bull. 44, 694 p.

Streiff-Becker, Rudolf, 1951, Pot-holes and glacier mills: Jour. Glaciol, v. 1, p. 488–490.

Strelkov, S. A., 1965, Northern Siberia: Nauka, Moscow, 336 p.*B

Strøm, K. M., 1945, Geomorphology of the Rondane area: Norsk Geologisk Tidsskrift, v. 25, p. 360–378.

Strömberg, Bo, 1965, Mappings and geochronological investigations in some moraine areas of south-central Sweden: Geogr. Annaler, v. 47A, p. 73–82.

Stubbings, H. G., 1939, Stratification of biological remains in marine deposits: British Mus. (Nat. Hist.), Sci. Repts. John Murray Exped. 1933–34, v. 3, p. 159–192.

Stuiver, Minze, 1964, Carbon isotopic distribution and correlated chronology of Searles Lake sediments: American Jour. Sci., v. 262, p. 377–392.

Stuntz, S. C., and Free, E. E., 1911, The movement of soil material by the wind, with a bibliography of eolian geology: U.S. Bur. Soils, Bull. 68, 263 p.

Suess, H. E., 1965, Secular variations of the cosmic-ray-produced Carbon 14 in the atmosphere and their interpretations: Jour. Geophys. Res., v. 70, p. 5937–5952.*

Suggate, R. P., 1965, Late Pleistocene geology of the northern part of the South Island, New Zealand: New Zealand Geol. Survey Bull. 77, 91 p.*

Swineford, Ada, and Frye, J. C., 1951, Petrography of the Peoria loess in Kansas: Jour. Geol. v. 59, p. 306–322.

Synge, F. M., and Stephens, N., 1960, The Quaternary Period in Ireland—an assessment: Irish Geogr., v. 4, p. 121–130.

Szafer, Władysław, 1953, Pleistocene stratigraphy of Poland from the floristical point of view: Soc. Geol. de Pologne, Ann., v. 22, p. 1–60 [Polish]; (English summ.)

———, 1954, Pliocene flora from the vicinity of Czorszytyn (West Carpathians) and its relationship to the Pleistocene: Instytut Geologiczny, Prace, v. 11: Warszawa, Wydawnictwa Geologiczne, 238 p.*

Szupryczyński, Jan, 1963, Rzeźba strefy marginalnej i typy deglacjacji lodowców południowego spitsbergenu: Polskiej Akad. Nauk, Inst. Geogr., Prace Geogr. nr. 39, Warszawa, 163 p.

Taber, Stephen, 1953, Origin of Alaska silts: American Jour. Sci., v. 251, p. 321–336.

Takai, Fuyuji, 1952, A summary of the mammalian faunae of eastern Asia and the interrelationships of continents since the Mesozoic: Japanese Jour. Geol. and Geogr., v. 22, p. 169–205.

———, and Tsuchi, Ryuichi, 1963, The Quaternary [of Japan]: p. 173–196 in Takai et al. (eds.) Geology of Japan, Univ. of California Press, 179 p.*

Takaya, Yosikazu, 1967, Observations on some Pleistocene outcrops in Cambodia: Kyoto Univ., The Southeast Asian Studies, v. 5, p. 556–571.

———, 1968, Observations of some Pleistocene outcrops in Ceylon: Kyoto Univ., The Southeast Asian Studies, v. 6, p. 321–330.

Talent, J. A., 1965, Geomorphic forms and processes in the highlands of eastern Victoria: Roy. Soc. Victoria, n.s., v. 78, p. 119–135.

Tanner, Väinö, 1937, The problems of the eskers. V—The Tälisvuom'puoltska esker in Enontekis Lapmark: Fennia, v. 63, no. 1, 31 p.

———, 1944, Outlines of the geography, life and customs of Newfoundland-Labrador (the eastern part of the Labrador peninsula): Acta Geographica Fenniae, v. 8, no. 1, 909 p.

Tarr, R. S., 1897, The Arctic sea ice as a geological agent: American Jour. Sci., v. 153, p. 223–229.

———, 1908, Some phenomena of the glacier margins in the Yakutat Bay region, Alaska: Zeitschr. für Gletscherk., v. 3, p. 81–110.

———, 1909, The Yakutat Bay region, Alaska: U.S. Geol. Survey Prof. Paper 64, 183 p.

———, and Martin, Lawrence, 1914, Alaskan glacier studies: Nat. Geogr. Soc., Washington, 498 p.

———, and Woodworth, J. B., 1903, Post glacial and interglacial (?) changes of level at Cape Ann, Massachusetts: Harvard Coll., Mus. Comp Zoöl., Bull., v. 6, p. 181–191.

Taylor, D. W., 1960, Late Cenozoic molluscan faunas from the High Plains: U.S. Geol. Survey Prof. Paper 337, 94 p.*

———, 1966 Summary of North American Blancan nonmarine mollusks: Malacologia, v. 4, p. 1–172.

Teichert, Curt, and Yochelson, E. L., eds., 1967, Essays in paleontology and stratigraphy: Univ. of Kansas Press, Lawrence (Univ. of Kansas, Dept. Geol. Spec. Publ. 2), 626 p.

Temple, P. H., 1965, Some aspects of cirque distribution in the west-central Lake District, northern England: Geograf. Annaler, v. 47, p. 185–193.

Te Punga, M. T., 1964, Relict red-weathered regolith at Wellington: New Zealand Jour. Geol. and Geophys., v. 7, p. 314–339.

Terasmae, Jaan, 1960, Contributions to Canadian palynology, No. 2. Part 1-A palynological study of post-glacial deposits in the St. Lawrence Lowlands. Part II-A palynological study of Pleistocene interglacial beds at Toronto, Ontario: Canada Geol. Survey Bull. 56, 41 p.

Thayer, T. P., 1939, Geology of the Salem Hills and the North Santiam River basin, Oregon: Oregon Dept. Geol. & Mineral Ind., Bull. 15, 40 p.

Thenius, Erich, 1959, Handbuch der stratigraphischen Geologie, v. 3, Tertiär, pt. 2: Wirbeltierfaunen: Ferdinand Enke, Stuttgart, 328 p.

———, 1962, Die Groszsäugetiere des Pleistozäns von Mitteleuropa: Zeitschr. für Säugetierkunde, v. 27, p. 65–82.

Thorarinsson, Sigurdur, 1937, Vatnajökull. Chap. 2, The main geological and topographical features of Iceland: Geograf. Annaler, v. 19, p. 161–175.

———, 1940, Present glacier shrinkage, and eustatic changes of sea-level: Geograf. Annaler, v. 22, p. 131–159.

Thornbury, W. D., 1950, Glacial sluiceways and lacustrine plains of southern Indiana: Indiana Div. Geol., Bull. 4, 21 p.

———, and Deane, H. L., 1955, The geology of Miami County, Indiana: Indiana Geol. Survey Bull. 8, 49 p.

Thorp, James, and others, 1952, [Map of] Pleistocene eolian deposits of the United States, Alaska and parts of Canada. Scale 1:2,500,000. (In colors): Geol. Soc. America, New York.

Thwaites, F. T., 1947, Geomorphology of the basin of Lake Michigan: Michigan Acad. Sci., Arts, and Lett., Papers, v. 33, p. 243–251.

———, and Bertrand, Kenneth, 1957, Pleistocene geology of the Door Peninsula, Wisconsin: Geol. Soc. America Bull., v. 68, p. 831–880.

Tilley, P. D., 1964, The significance of loess in southeast England: Internat. Congr. on Quaternary, 6th., Warsaw 1961, Rept., v. 4, p. 591–596.

Toepfer, Volker, 1963, Tïerwelt des Eiszeitalters: Leipzig, Geest and Portig, 198 p.

Torell, Otto, 1877, On the glacial phenomena of North America: American Jour. Sci., v. 13, p. 76–79.

Totten, S. M., 1969, Overridden recessional moraines of north-central Ohio:

Geol. Soc. America Bull., v. 80, p. 1931–1946.

Trefethen, J. M., and Trefethen, H. B., 1945, Lithology of the Kennebec Valley Esker: American Jour. Sci., v. 242, p. 521–527.

Trewartha, G. T., 1968, An introduction to climate: 4th ed., McGraw-Hill, New York, 408 p.

Tricart, Jean, 1963, Géomorphologie des régions froides: Collection Orbis, Paris, Presses Universitaires de France, 289 p.

———, 1970, Geomorphology of cold environments: The Macmillan Co., London; St. Martin Press, New York, 320 p.*

———, and Cailleux, André, 1953, Le modelé glaciaire et nival: Paris, Centre de Documentation Universitaire, 408 p.

———, 1967, Traité de géomorphologie, v. 2, Le modelé périglaciare: Paris, Société d'Edition d'Enseignement Supérieur, 512 p.

Troitskiy, S. L., 1966, Quaternary deposits and relief of the plains and littoral of the Yenesey estuary and contiguous parts of the Byrranga Ridge: USSR. Acad. Sci., Siberian Branch, Inst. of Geol. and Geophys. Nauka, Moscow, 207 p.*[R]

———, 1969, General Review of the Siberian marine Pleistocene: p. 32–43 in Problems of glacial geology of Siberia, Acad. Sci, USSR., Siberian Br.*[R]

Troll, Carl, 1928, Die Zentralen Anden [Bolivia] . . . : Gesellsch. für Erdk. zu Berlin, Zeitschr., Jubiläums-Sonderbd, p. 92–118.

———, 1944, Strukturböden, Solifluktion und Frostklimate der Erde: Geol. Rundschau, v. 34, p. 545–694.

———, 1958, Structure soils, solifluction, and frost climates of the Earth: U.S. Army. Corps of Engineers, Snow, Ice, and Permafrost Research Establishment, Transl. 43, 121 p.*

Trotter, F. M., 1929, The glaciation of eastern Edenside, the Alston block, and the Carlisle plain: Geol. Soc. London Quart. Jour., v. 85, p. 549–612.

Tschumi, Otto, ed., 1949, Urgeschichte der Schweiz: Frauenfeld, Huber & Co., Leipzig, 751 p.

Tsukada, Matsuo, 1966, Late Pleistocene vegetation and climate in Taiwan (Formosa): Nat. Acad. Sci. (U.S.), Proc., v. 55, p. 543–548.

———, 1967, Vegetation and climate around 10,000 B.P. in central Japan: American Jour. Sci., v. 265, p. 562–585.

Turekian, K. K., ed., 1971, Late Cenozoic glacial ages: Yale Univ. Press, New Haven. (in press.)*

Turner, C., and West, R. G., 1968, The subdivision and zonation of interglacial periods: Eiszeitalter und Geg., v. 19, p. 93–101.

Tuthill, S. J., and others, 1964, Fossil molluscan fauna from the upper terrace of the Cannonball River, Grant County, North Dakota: North Dakota Acad. Sci., Proc., v. 18, p. 140–156.

Tylor, Alfred, 1868, On the Amiens gravel: Geol. Soc. London Quart. Jour., v. 24, p. 103–125.

Ucko, P. J., & Dimbleby, G. W., eds., 1969, The domestication & exploitation of plants and animals: Gerald Duckworth & Co., Ltd., London, 581 p.*

Udvardy, M. D. F., 1970, Dynamic zoogeography with special reference to land animals: Van Nostrand Reinhold Co., New York, 464 p.*

Upham, Warren, 1891, Criteria of englacial and subglacial drift: American Geologist, v. 8, p. 376–385.

Upson, J. E., 1951, Former marine shore lines of the Gaviota Quadrangle, Santa

Barbara County, California: Jour. Geology., v. 59, p. 415–446.

———, 1970, The Gardiners Clay of eastern Long Island, New York—a reexamination: U.S. Geol. Survey Prof. Paper 700, p. B157–B160.

———, and Spencer, C. W., 1964, Bedrock valleys of the New England coast as related to fluctuations of sea level: Geol. Survey Prof. Paper 454-M, 44 p.

U.S. Bur. Plant Industry, 1951, Soil survey manual: U.S. Dept. Agric. Handbk. 18, 503 p.

U.S. Department of Agriculture, 1938, Soils and Men. Yearbook of agriculture 1938: Washington, Govt. Printing Office, 1232 p.

Valentin, Hartmut, 1954, Gegenwärtige Niveauveränderungen im Nordseeraum: Petermanns Geogr. Mitt., v. 98, p. 103–108.

———, 1958, Die Grenze der letzten Vereisung im Nordseeraum: Deutscher Geographentag Hamburg 1955, Tagungsbericht und wiss. Abh. (Wiesbaden), p. 359–366.

Valentine, J. W., 1961, Paleoecologic marine geography of the California Pleistocene: Univ. of California Pubs. Geol. Sci. v. 34, p. 309–442.

———, 1965, Quaternary mollusca from Port Fairy, Victoria, Australia, and their palaeoecologic implications: Roy. Soc. Victoria, nos., v. 78, p. 15–73.

Van Andel, T. H., and others, 1967, Late Quaternary history, climate, and oceanography of the Timor Sea, northwestern Australia: American Jour. Sci., v. 265, p. 737–758.

van der Vlerk, I. M., and Florschütz, Franz, 1950, Nederland in het Ijstijdvak. De geschiedenis van flora, fauna en klimaat, toen aap en mammoet ons land bewoonden: W. De Haan, Utrecht, 287 p.

———, 1953, The palaeontological base of the subdivision of the Pleistocene in the Netherlands: K. Nederlandse Akad. van Wetenschappen, afd. Natuurkunde, Verh., ser. 1, v. 20, no. 2, 58 p.

Veeh, H. H., and Chappell, John, 1970, Astronomical theory of climatic change; support from New Guinea: Science, v. 167, p. 862–865.

Veenstra, H. J., 1963, Microscopic studies of boulder clays: N.V. Uitgeverij "Stabo", Groningen.

Vélez de Escalante, Silvestre, 1943, Journal and itinerary . . . of the route from Presidio de Santa Fé del Nuevo-Mexico to Monterey, in northern California [Engl. transl.]: Utah Historical Quart., v. 11, 1943, p. 27–113.

Venzo, Sergio, 1955, Le attuali conoscenze sul Pleistocene Lombardo con particolare riguardo al Bergamasco: Soc. Italiana Sci. Nat., Atti, v. 94, p. 155–200.

———, 1965, Rilevamento geologico dell'anfiteatro morenico frontale del Garda dal Chiese all'Adige: Soc. Italiana Sci. Nat., Mem., v. 14, pt. 1, 82 p.

Vernon, Peter, and Hughes, O. L., 1966, Surficial geology, Dawson, Larsen Creek, and Nash Creek map-areas, Yukon Territory: Canada Geol. Survey Bull. 136, 25 p.

Viete, Günter, 1957, Kritische Bemerkungen zur Bestimmung der pleistozänen Inlandeismächtigkeit mit Hilfe von Drucksetzungsmessungen: Eiszeitalter und Geg., v. 8, p. 97–106.

———, 1960, Zur Entstehung der glazigenen Lagerungsstörungen unter besonderer Berücksichtigung der Flözdeformationen im mitteldeutschen Raum: Freiberger Forschungsh. C 78, 257 p.*

Viret, Jean, 1954, Les loess a bancs durcis de Saint-Vallier (Drôme) et sa faune de mammifères villafranchiens: Mus. d'Hist. Nat., Lyon, Nouv. Archives, Fasc. 4, 200 p.*

Virkkala, Kalevi, 1952, On the bed structure of till in eastern Finland: Comm. Géol. de Finlande Bull. 157, p. 97–109. Also in Soc. Géol. de Finlande, Comptes Rendus, no. 25, p. 97–109.

——, 1958, Stone counts in the esker of Hämeenlinna, southern Finland: Soc. Géol. de Finlande, Comptes Rendus, no. 30, p. 87–103.

——, 1960, On the striations and glacier movements in the Tampere region, southern Finland: Soc. Géol. Finlande, Comptes Rendus, no. 32, p. 159–176.

——, 1963, On ice-marginal features in southwestern Finland: Comm. Geol. de Finlande Bull., no. 210, 76 p.

Visher, S. S., 1954, Climatic atlas of the United States: Harvard Univ. Press, Cambridge, 403 p.

Visser, P. C., 1938, Wissenschaftliche Ergebnisse der Niederländischen Expeditionen in den Karakorum und die Angrenzenden Gebiete in den Jahren 1922, 1925, 1929/30 und 1935. Band 2, Glaziologie: E. J. Brill, Leiden, 216 p.

Vita-Finzi, Claudio, 1959, A pluvial age in the Puna de Atacama: Geogr. Jour., v. 125, p. 401–403.

——, 1969, Late Quaternary alluvial chronology of Iran: Geol. Rundschau, v. 58, p. 951–973.*

Von Engeln, O. D., 1911, Phenomena associated with glacier drainage and wastage, with special reference to observations in the Yakutat Bay Region, Alaska: Zeitschr. für Gletscherk., v. 6, p. 104–150.

——, 1918, Transportation of débris by icebergs: Jour. Geology, v. 26, p. 74–81.

Voorthuysen, J. H. van, 1950, The Plio-Pleistocene boundary in the Netherlands based on the ecology of Foraminifera: Geologie en Mijnbouw, n.s. v. 12, p. 26–30.

Wagner, F. J. E., 1959, Palaeoecology of the marine Pleistocene faunas of southwestern British Columbia. Canada Geol. Survey Bull. 52, 67 p.

——, 1967, Published references to Champlain Sea faunas 1837–1966 and list of fossils: Canada Geol. Survey Paper 67–16, 82 p.

Wahrhaftig, Clyde, 1949, The frost-moved rubbles of Jumbo Dome and their significance in the Pleistocene chronology of Alaska: Jour. Geology, v. 57, p. 216–231.

——, and Cox, Allan, 1959, Rock glaciers in the Alaska Range: Geol. Soc. America Bull., v. 70, p. 383–436.*

Walker, Donald, and West, R. G., eds., 1970, Studies in the vegetational history of the British Isles: Cambridge Univ. Press, 258 p.

Walker, P. H., 1966, Postglacial environments in relation to landscape and soils on the Cary Drift, Iowa: Iowa State Univ., Agric. and Home Econ. Expt. Sta., Res. Bull. 549, p. 837–875.

Wallace, R. E., 1948, Cave-in lakes in the Nabesna, Chisana, and Tanana River valleys, eastern Alaska: Jour. Geol., v. 56, p. 171–181.

Ward, W. H., 1952, The physics of deglaciation in central Baffin Island: Jour. Glaciol., v. 2, p. 9–22.

Wardle, Peter, 1963, Evolution and distribution of the New Zealand flora, as affected by Quaternary climates: New Zealand Jour. Bot., v. 1, p. 3–17.

Wardle, Peter, 1964, Facets of the distribution of forest vegetation in New Zealand: New Zealand Jour. Bot., v. 2, p. 352–366.

Warnke, D. A., 1970, Glacial erosion, ice rafting, and glacial-marine sediments: Antarctica and the Southern Ocean: American Jour. Sci., v. 269, p. 276–294.*

Warren, C. R., 1952, Probable Illinoian age of part of the Missouri River, South Dakota: Geol. Soc. America, Bull., v. 63, p. 1143–1156.

Washbourn, Celia, 1967, Lake levels and Quaternary climates in the Eastern Rift Valley of Kenya: Nature, v. 216, p. 672–673.

Washburn, A. L., 1947, Reconnaissance geology of portions of Victoria Island and adjacent regions, Arctic Canada: Geol. Soc. America, Mem. 22, 142 p.

———, 1956, Classification of patterned ground and review of suggested origins: Geol. Soc. America Bull., v. 67, p. 823–866.

———, 1967, Instrumental observations of mass-wasting in the Mesters Vig District, Northeast Greenland: Meddelelser om Grønland, v. 166, no. 4, 296 p.

———, 1968, Weathering, frost action, and patterned ground in the Mesters Vig District, Northeast Greenland: Meddelelser om Grønland, v. 176, no. 4, 303 p.*

———, and Stuiver, Minze, 1962, Radiocarbon-dated postglacial delevelling in north-east Greenland and its implications: Arctic, v. 15, p. 66–73.

Washburn, S. L., 1960, Tools and human evolution: Scientif. American, v. 203, p. 66–77.

Watts, W. A., 1969, A pollen diagram from Mud Lake, Marion County, north-central Florida: Geol. Soc. America Bull., v. 80, p. 631–642.

———, and Bright, R. C., 1968, Pollen, seed, and mollusk analysis of a sediment core from Pickerel Lake, northeastern South Dakota: Geol. Soc. America Bull., v. 79, p. 855–876.

Wayne, W. J., 1952, Pleistocene evolution of the Ohio and Wabash valleys: Jour. Geol., v. 60, p. 575–585.

———, 1958, Early Pleistocene sediments in Indiana: Jour. Geol., v. 66, p. 8–15.

———, 1959, Stratigraphic distribution of Pleistocene land snails in Indiana: Sterkiana (Bloomington, Indiana), no. 1, p. 9–18.

———, 1963 Pleistocene formations in Indiana: Indiana Geol. Survey Bull. 25, 85 p.

Weaver, J. C., 1946, Ice atlas of the northern hemisphere: Washington: Hydrographic Office U.S. Navy, H. O. no. 550, 105 p.

Webb, S. D., and Tessman, Norman, 1967, Vertebrate evidence of a low sea level in the Middle Pliocene: Science, v. 156, p. 379.

Weed, W. H., 1902, Geology and ore deposits of the Elkhorn Mining District, Jefferson County, Montana: U. S. Geol. Survey, 22d Ann. Rept., pt. 2, p. 399–549.

———, and Pirsson, L. V., 1896, Geology of the Castle Mountain Mining District, Montana: U. S. Geol. Survey Bull. 139, 164 p.

Weidick, Anker, 1971, Quaternary map of Greenland: Greenland Geol. Survey, Copenhagen. Scale 1:2,500,000. (In course of publ.)

Weigel, R. D., 1962, Fossil vertebrates of Vero, Florida: Florida Geol. Survey Spec. Pub. 10, 59 p.*

Weischet, Wolfgang, 1954, Die gegenwartige Kenntnis von Klima in Mitteleuropa beim Maximum der letzten Vereisung: Geogr. Gesellsch. München, Mitt., v. 39, p. 95–116.

Wells, P. V., 1966, Late Pleistocene vegetation and degree of pluvial climatic change in the Chihuahuan Desert: Science, v. 153, p. 970–975.

————, 1970, Postglacial vegetational history of the Great Plains: Science, v. 167, p. 1574–1582.

————, and Berger, Rainer, 1967, Late Pleistocene history of coniferous woodland in the Mohave Desert: Science, v. 155, p. 1640–1647.

Welten, Max, 1944, Pollenanalytische, stratigraphische, und geochronologische Untersuchungen aus dem Faulenseemoos bei Spiez: Geobotan. Inst. Rübel, (Zürich), Veröffentlichungen, no. 21, 201 p.

Wendorf, Fred, and others, 1961, Paleoecology of Llano Estacado: Mus. of New Mexico Press, Fort Burgwin Res. Center, Publ. no. 1, 144 p.*

Wennberg, Gunnar, 1949, Differentialrörelser i Inlandsisen sista Istiden i Danmark, Skåne och Östersjön: Lunds Geolog-Mineralog. Instit., Medd., no. 114, 201 p. + Bilaga, 46 p.

Wenner, C.-G., 1968, Comparison of varve chronology, pollen analysis and radiocarbon dating: Acta Universitatis Stockholmiensis, Stockholm Contribs. in Geol., v. 18, p. 75–97.

Wentworth, C. K., 1928, Striated cobbles in southern States: Geol. Soc. America, Bull., v. 39, p. 941–953.

————, 1930, Sand and gravel resources of the coastal plain of Virginia: Virginia Geol. Survey Bull. 32, 146 p.

————, 1936, An analysis of the shapes of glacial cobbles: Jour. Sed. Petrology, v. 6, p. 85–96.

————, and Powers, W. E., 1941, Multiple glaciation of Mauna Kea, Hawaii: Geol. Soc. America Bull., v. 52, p. 1193–1217.

West, R. G., 1961, Late- and postglacial vegetational history in Wisconsin, particularly changes associated with the Valders Readvance: American Jour. Sci., v. 259, p. 766–783.

————, 1963, Problems of the British Quaternary: Geologists' Assoc., Proc., v. 74, p. 147–186.

————, 1968, Pleistocene geology and biology, with especial reference to the British Isles: Longmans Green and Co., Ltd., London; John Wiley and Sons, Inc., New York, 377 p.*

————, and Donner, J. J., 1956, The glaciations of East Anglia and the East Midlands: a differentiation based on stone-orientation measurements of the tills: Geol. Soc. London Quart. Jour., v. 112, p. 69–91.

Westgate, J. A., 1968a, Surficial geology of the Foremost-Cypress Hills area, Alberta: Res. Counc. of Alberta Bull. 22, 122 p.

Westgate, J. A., 1968b, Linear sole markings in Pleistocene till: Geol. Mag., v. 105, p. 501–505.

————, and Dreimanis, Aleksis, 1967, Volcanic ash layers of recent age at Banff National Park, Alberta, Canada: Canadian Jour. Earth Sci., v. 4, p. 155–161.

Wexler, Harry, 1961, Ice budgets for Antarctica and changes in sea-level: Jour. Glaciol., v. 3, p. 867–872.

Weyl, P. K., 1968, The role of the oceans in climatic change: a theory of the ice ages: Meteorol. Mon., v. 8, p. 37–62.

Weyl, Richard, 1956, Eiszeitliche Gletscherspuren in Costa Rica (Mittelamerika): Zeitschr. für Gletscherk., v. 3, p. 317–325.

White, G. W., 1962, Multiple tills of end moraines: U. S. Geol. Survey Prof. Paper 450-C, p. 96–98.

White, G. W., 1969, Pleistocene stratigraphy of northwestern Pennsylvania: Pennsylvania Geol. Survey Rept. G55, 88 p.

White, S. E., 1951, A geologic investigation of the late Pleistocene history of the volcano Popocatépetl, Mexico: Syracuse University, Abstr., Ph.D. dissert., 7 p.

————, 1962, Late Pleistocene glacial sequence for the west side of Iztaccihuatl, Mexico: Geol. Soc. America Bull., v. 73, p. 935–958.

Whitehead, D. R., and Barghoorn, E. S., 1962, Pollen analytical investigations of Pleistocene deposits from western North Carolina and South Carolina: Ecol. Monogr., v. 32, p. 347–369.

————, and Davis, J. T., 1969, Pollen analysis of an organic clay from the interglacial Flanner Beach Formation, Craven County, North Carolina: Southeastern Geol., v. 10, p. 149–164.

Whitehouse, F. W., 1940, Studies in the late geological history of Queensland: Queensland Univ., Dept. Geol. Papers, n.s., v. 2, p. 2–22, 62–74.

Whitney, J. D., 1865, Geological Survey of California. Geology, v. 1: Sherman and Co., Philadelphia, 498 p.

Whittlesey, Charles, 1868, Depression of the ocean during the ice period: American Assoc. Adv. Sci., Proc., v. 16, p. 92–97.

Whittow, J. B., & Wood, P. D., eds., 1965, Essays in geography for Austin Miller: Univ. of Reading [England], 337 p.

Wijmstra, T. A., 1970, Palynology of the first 30 metres of a 120m deep section in northern Greece: Acta Bot. Neerl., v. 18, p. 511–527.

Willett, H. C., 1951, Extrapolation of sunspot-climate relationships: Jour. Meteorol., v. 8, p. 1–6.

————, 1965, Solar-climatic relationships in the light of standardized climatic data: Jour. Atmosph. Sci., v. 22, p. 120–136.

————, and Prohaska, John, 1965, Further evidence of sunspot-ozone relationships: Jour. Atmosph. Sci., v. 22, p. 493–497.

Willett, R. W., 1950, The New Zealand Pleistocene snow line, climatic conditions, and suggested biological effects: New Zealand Jour. Sci. and Techn., sec. B, v. 32, p. 18–48.

Willman, H. B., and Frye, J. C., 1969, High-level glacial outwash in the Driftless Area of northwestern Illinois: Illinois Geol. Survey Circ. 440, 21 p.

Willman, H. B., and others, 1963, Mineralogy of glacial tills and their weathering profiles in Illinois. Part I. Glacial tills: Illinois Geol. Survey Circ. 347.

Wilmarth, M. G., 1925. The geologic time classification of the United States Geological Survey compared with other classifications. Accompanied by the original definitions of era, period and epoch terms: U. S. Geol. Survey Bull. 769, 138 p.

Wilson, A. T., 1964, Origin of ice ages: an ice shelf theory for Pleistocene glaciation: Nature, v. 201, p. 147–149.

Wilson, R. L., 1967, The Pleistocene vertebrates of Michigan: Michigan Acad. Sci., Arts, and Letters, Papers, v. 52, p. 197–234.

Wissmann, Hermann von, 1937, The Pleistocene glaciation in China: Geol. Soc. China Bull., v. 17, p. 145–168.

————, 1959, Die heutige Vergletscherung und Schneegrenze in Hochasien mit Hinweisen auf die Vergletscherung der letzten Eiszeit: Wiesbaden, Akad. der Wiss. u. der Lit., Abh. der Math.-Naturw. Kl., Jahrg. 1959, p. 1103–1407.*

Witting, Rolf, 1918, Die Meeresfläche, die Geoidfläche und die Landhebung dem Baltischen Meere entlang und an der Nordsee: Fennia, v. 39, no. 5, 346 p. [Swed., German summ.]

Woldstedt, Paul, 952, Die Entstehung der Seen in den ehemals vergletscherten Gebieten: Eiszeitalter und Geg., v. 2, p. 146–153.

⸻, 1954–1965, Das Eiszeitalter: 2d. ed., Stuttgart, Ferdinand Enke Verlag. v. 1, 1954, 374 p.; v. 2, 1958, 438 p.; v. 3, 1965, 328 p.*

⸻, 1956a, Die Geschichte des Flussnetzes in Norddeutschland und angrenzenden Gebieten: Eiszeitalter und Geg., v. 7, p. 5–12.

⸻, 1956b, Über die Gliederung der Würm-Eiszeit und die Stellung der Lösse in ihr: Eiszeitalter und Geg., v. 7, p. 78–86.

⸻, 1962, Über die Gliederung des Quartärs und Pleistozäns: Eiszeitalter und Geg., v. 13, p. 115–124.

⸻, 1969, Quartär (Handbuch der stratigraphischen Geologie, v. 2): Ferdinand Enke, Stuttgart, 263 p.*

Woodring, W. P., and others, 1946, Geology and paleontology of Palos Verdes Hills, California: U.S. Geol. Survey Prof. Paper 207, 145 p.

Woods, J. T., 1962, Fossil marsupials and Cainozoic continental stratigraphy in Australia. A review: Queensland Mus. (Brisbane), Mem., v. 14, p. 41–49.

Woodworth, J. B., 1899, The ice-contact in the classification of glacial deposits: American Geologist, v. 23, p. 80–86.

⸻, 1905, Ancient water levels of the Champlain and Hudson valleys [N.Y.]: New York State Mus., Bull., 84, 265 p.

⸻, and Wigglesworth, Edward, 1934, Geography and geology of the region including Cape Cod, the Elizabeth Islands, Nantucket, Marthas Vineyard, No Mans Land, and Block Island: Harvard Coll., Mus. Comp. Zoöl., Mem., v. 52, 338 p.

Worthen, A. H., 1873, Geology of Sangamon County: Illinois Geol. Survey, v. 5, p. 306–319.

Wright, C. S., and Priestley, R. F., 1922, British (Terra Nova) Antarctic Expedition, 1910–1913. Chapter 7—Glaciology: Harrison and Sons, London, 581 p.

Wright, H. E., 1957, Stone orientation in Wadena drumlin field, Minnesota: Geograf. Annaler, v. 39, p. 19–31.

⸻, 1961a, Late Pleistocene climate of Europe: a review: Geol. Soc. America Bull., v. 72, p. 933–984.*

⸻, 1961b, Pleistocene glaciation in Kurdistan: Eiszeitalter und Geg., v. 12, p. 131–164.

⸻, 1962, Role of the Wadena lobe in the Wisconsin glaciation of Minnesota: Geol. Soc. America Bull., v. 73, p. 73–100.

⸻, 1968a, The roles of pine and spruce in the forest history of Minnesota and adjacent areas: Ecology, v. 49, p. 937–955.

⸻, 1968b, Climatic change in the eastern Mediterranean region: Univ. of Minnesota. Contract Nonr-710 (33), Task no. 389–129, Final Rept., 83 p.

⸻, and Frey, D. G., eds., 1965a, The Quaternary of the United States: Princeton Univ. Press, 922 p.*

⸻, eds., 1965b, International studies on the Quaternary: Geol. Soc. America Spec. Paper 84, 565 p.*

Wright, H. E., and Osburn, W. H., eds., 1967, Arctic and alpine environments: Symposium 22 papers, Indiana Univ. Press, 308 p.*

Wright, W. B., 1914, The Quaternary ice age: Macmillan, London, 464 p.

878   References

Wright, W. B., 1937, The Quaternary ice age: 2nd ed., Macmillan, London, 478 p.*

Wu, Y. S., 1948, Quaternary glaciation in south-west China and its bearing on gold placers (*Abstr.*): Internat. Geol. Congr., 18th, London 1948, Rept., pt. 13, 1952, p. 298.

Wynne-Edwards, V. C., 1937, Isolated arctic-alpine floras in eastern North America: a discussion of their glacial and recent history: Roy. Soc. Canada, Trans., 3d ser., v. 31, sec. 5, p. 33–58.

Yamamoto, Takeo, 1961, Sunspot-climatic relationships in fluctuations of glaciers in the Alps and atmospheric precipitation in Korea: Meteorol. Soc. Japan, Jour., v. 39, no. 5.

Yardley, D. H., 1951, Frost-thrusting in the Northwest Territories: Jour. Geol., v. 59, p. 65–69.

Young, D. J., 1964, Stratigraphy and petrography of north-east Otago loess: New Zealand Jour. Geol. and Geophys., v. 7, p. 839–863.

Zagwijn, W. H., 1960, Aspects of the Pliocene and early Pleistocene vegetation in the Netherlands: Univ. of Leiden, PhD. dissert. Ernest von Aelst, Maastricht, 78 p.*

――――, 1963, Pleistocene stratigraphy in the Netherlands, based on changes in vegetation and climate: K. Nederlands Geol. Mijnbouwk. Genootschap, Verhandelingen, Geol. Ser., v. 21, pt. 2, p. 173–196.

――――, 1967, Ecologic interpretation of a pollen diagram from Neogene beds in the Netherlands: Rev. Palaeobot. and Palynol., v. 2, p. 173–181.

――――, and Paepe, Roland, 1968, Die Stratigraphie der Weichselzeitlichen Ablagerungen der Niederlande und Belgiens: Eiszeitalter und Geg., v. 19, p. 129–146.*

Zeist, W. van, 1967, Late Quaternary vegetation history of western Iran: Rev. Palaeobot. and Palynol., v. 2, p. 301–310.

Zeuner, F. E., 1952, Dating the past. An introduction to geochronology: Methuen & Co., London, 3d ed., 495 p.*

――――, 1958, Dating the past. An introduction to geochronology: Methuen and Co., London, 4th ed., 516 p.*

――――, 1959, The Pleistocene Period. Its climate, chronology and faunal successions: 2d ed., Hutchinson, London, 447 p.

Zhusé, A. P., 1961, Stratigraphy of bottom sediments in the north-western Pacific Ocean: Estonian SSR, Acad. Sci., Inst. of Geol. Trans., v. 8, p. 183–196.[R]

Zinderen Bakker, E. M. van, 1966, The pluvial theory—an evaluation in the light of new evidence, especially for Africa: The Palaeobotanist, v. 15, p. 128–134.

Zubakov. V. A., 1965, Correlation of glaciations and Pleistocene marine transgressions in the Arctic part of eastern Siberia and in northwest North America: Soviet Geology, v. 6, p. 54–75.

――――, 1969, La chronologie des variations climatiques au cours du Pleistocène en Sibérie occidentale: Rév. de Géog. Phys. et du Géol. Dynam., v. 11, p. 315–324.

Russian-Language Titles

Ганешин, Г. С., редактор, 1959, Карта Четвертичных Отложений СССР, масштаб 1 : 5.000.000: Всесоюзный Наука-Исследовательский Геологический Институт (всегеи).

Герасимов, И. П., редактор, 1965, Последний Европейский Ледниковый Покров: издательство «Наука», Москва, 216 стр.

Гитерман, Р. Е., и остальные, 1968, Основные Этапы развития растительности северной Азии в Антропогене: Акад. Наук. СССР, Инст. Геол., Труды, том 177, стр. 270.

Горецкий, Г. И., и Кригер, Н. И., редакторы, 1967, Нижний Плейстоцен ледниковых районов Русской Равнины: СССР, Комиссия по Изучению Четвертичного Период, издательство «Наука», Москва, 183 стр.

Громов, В. И., и Никифорова, К. В., редакторы, 1969, Основные проблемы геологии Антропогена Евразии: «Наука», Москва, 130 стр.

Громов, В. И., и другий, 1969, Схема Подразделений Антропогена: Бюллетень комиссии по изучению Четвертичного Периода, № 36, стр. 41–55.

Жузе, А. П., 1961, Стратиграфия донных отложений на северо-западе Тихого Океана: Эст. ССР, Акад. Наук., Институт Геологии, Труды, том 8, стр. 183–196.

Зубаков, В. А., 1965, Корреляция оледенений и Плейстоценовых морских трансгрессий Арктической части восточной Сибири и северо-западной части Северной Америки: Советская Геология, том 6, стр. 54–75.

Котляков, В. М., 1966, Опыт подсчета запасов воды, аккумулированой в горных ледниках Советского Союза: Акад. Наук. СССР, Известия, серия Географическая, № 3, стр. 43–48.

Кригер, Н. И., 1965, Лёсс, Его Своиства и Связь с географической Средой: издательство «Наука», Москва, 296 стр.

Марков, К. К., Лазуков, Г. И., и Николаев, В. А., 1965, Четвертичный Период Ледниковый Период, том 1, 2: Издательство Московского Университета, том 1, 371 стр.; том 2, 435 стр.

Марков, К. К., и Величко, А. А., 1967, Четвертичный Период (Ледниковый Период—Антропогеновый Период), том 3: издательство «Недра», Москва, 438 стр.

Москвитин, А. И., 1970, Стратиграфия Плейстоцена центральной и западной Европы: Акад. Наук. СССР, Инст. Геол., Труды, том 193, 287 стр.

Никифорова, К. В., редактор, 1965, Корреляция Антропогеновых отложений северной Евразии: СССР Акад. Наук., Геологический Институт, издательство «Наука», Москва, 114 стр.

Сакс, В. Н., редактор, 1966, Четвертичный период Сибири: СССР Акад. Наук., Сибирское Отделение, Инст. Геолог. и Геофиз., издательство «Наука», Москва, 514 стр.

Стрелов, С. А., 1965, Себер Сибири: издательство «Наука», Москва, 336 стр.

Троицкий, С. Л., 1966, Четвертичные отложения и рельеф равнинных побережий Енисейского залива и прилегающей части гор Бырранга: Акад. Наук. СССР, Сибирское Отделение, Инст. Геологии и Геофизики, издательство «Наука», Москва, 207 стр.

——, 1969, Общий Обзор морского Плейстоцена Сибири: стр. 32–43 по Проблемы четвертичной геологии Сибири, Акад. Наук. СССР, Сибирское Отделение, издательство «Наука», Москва.

Флинт, Р. Ф., 1963, Ледники и Палеогеография Плейстоцена: издательство Иностранной Литературы, Москва, 576 стр.

Чемеков, Ю. Ф., 1961, Снеговая линия последнего Верхнечетвертичного оледенения на юге дальнего Востока СССР: Известия Акад. Наук СССР, серия Географическая № 6, стр. 73–88.

——, 1961–5, Четвертичные трансгрессии дальневосточных морей и северной части тихого океана: Акад. Наук Эстон. СССР, Инст. Геол., Труды, № 8, стр. 155–174.

Index

References are to text material only. Contents of tables are not indexed because the tables are readily found through their geographic locations. An asterisk denotes an illustration.